Positron Emission Tomography

Peter E. Valk, Dominique Delbeke, Dale L. Bailey,
David W. Townsend, and Michael N. Maisey (*Eds*)

Positron Emission Tomography

Clinical Practice

With 256 Figures
including 131 in Color

 Springer

Peter E. Valk, MB, BS, FRACP
Northern California PET Imaging Center
Sacramento, CA, USA

Dominique Delbeke, MD, PhD
Professor and Director of Nuclear
  Medicine
Positron Emission Tomography
Department of Radiology and
  Radiological Sciences
Vanderbilt University Medical Center
Nashville, TN, USA

Dale L. Bailey, PhD
Associate Professor of Medicine
Department of Nuclear Medicine
Royal North Shore Hospital
Sydney, Australia

David W. Townsend, PhD
Director
Cancer Imaging and Tracer Development
  Program
The University of Tennessee Medical
  Center
Knoxville, TN, USA

Michael N. Maisey, BSc, MD, FRCR, FRCP
Professor of Emeritus
Department of Radiological Sciences
Guy's and St Thomas' Clinical PET
  Centre
Guy's and St Thomas' Hospital Trust
London, UK

Artwork and tables marked with 📖 symbol throughout the book are original to the 1st edition (Valk PE, Bailey DL, Townsend DW, Maisey MN. Positron Emission Tomography: Basic Science and Clinical Practice. Springer-Verlag London Ltd. 2003) and are being republished in this edition.

British Library Cataloguing in Publication Data
Positron emission tomography : clinical practice
  1. Tomography, Emission
  I. Valk, Peter E., 1940–
  616′.07575
ISBN-10: 1852339713

Library of Congress Control Number: 2005932896

ISBN-10: 1-85233-971-3   e-ISBN 1-84628-187-3
ISBN-13: 978-1-85233-971-5

Printed on acid-free paper.

Printed in China   (EXPO/EVB)

9 8 7 6 5 4 3 2 1

Springer Science+Business Media
springer.com

Foreword

The use of positron emission tomography (PET) in clinical practice is key to the successful management of many patients with a wide variety of diseases. Whereas in the 1980s and 1990s nuclear medicine physicians struggled to convince other doctors about the potential clinical value of PET, it is now a challenge to keep up with the requests of our clinical colleagues for the various applications of PET. In fact, in most cases the barriers are now related to reimbursement for a given PET procedure. At the time this foreword was written, many FDG/PET procedures are reimbursable, and the U.S. National PET Registry, which is about to be implemented, will allow FDG/PET imaging data to accumulate on rare tumors, while allowing for Center for Medicare and Medicaid Services (CMS) reimbursement. The challenges are now shifting toward solving some of the limitations of FDG/PET through the use of next generation instrumentation and newer tracers that hold the promise of improving on what is already a remarkable achievement with FDG/PET.

It is my hope that there will in fact be an enormous growth of nuclear medicine driven by new imaging tracers that will fuel a growing number of clinical applications. Many failures will occur in order to give rise to the next generation of PET imaging tracers, but these failures are bc necessary on the road to a better tomorrow. Perhaps a "Super FDG" will be born out of these efforts, along with very specific imaging tracers useful for very select groups of patients. Technology will continue to drive our ability to identify new cellular targets, new molecular imaging agents against those targets, and the routine high-throughput synthesis and use of those new imaging tracers. Just as PET/CT has markedly influenced the growth of FDG/PET, newer technologies will likely come into play. The debut of MR-PET, molecular imaging technologies such as molecular optical imaging, and many other technologies will likely change the landscape of nuclear medicine forever. To keep the correct perspective we must remember that it is not about the technologies per se, but really about optimal patient care. We must not slip into the future, but plan for the future so we maximize our ability to help patients through the power of nuclear medicine and molecular imaging. However technologies and new tracers evolve, the concepts of interrogation at the cellular and molecular level will continue to define the evolution of nuclear medicine and the interception of disease processes through functional imaging.

The use of PET and PET/CT in patient care is best learned from clinicians at the leading edge of imaging who also routinely interpret the images. This excellent book put together by leading clinicians, who have helped the field of PET to get to the current stage, is an enormous educational resource. It is very thorough with clear examples and covers all major aspects of PET/CT application in patient management. Someday, technologies will surely evolve so that authors will be seen and speak to us virtually and interactively through "digital books." For now we must be content to learn from the best teachers through their words and image examples on paper. This book is the next best thing to having all the contributing authors virtually teaching the student about all aspects of PET/CT. Till the day that digital interactive books arrive, I plan on keeping this book nearby.

Sanjiv Sam Gambhir MD, PhD
Stanford University

Preface

Peter E. Valk passed away on December 16, 2003 in Berkeley, California. David Townsend wrote in the "*In Memoriam*" that was published in the February 2004 issue of the *Journal of Nuclear Medicine*: "He will be deeply missed by his many friends and colleagues throughout the Nuclear Medicine and PET community world-wide for his insight, knowledge, integrity and humour."

Peter was a dear friend and we certainly miss him. In 2003, Peter coedited *Positron Emission Tomography: Basic Science and Clinical Practice,* a comprehensive contemporary reference textbook on positron emission tomography (PET). A few months before he died, Peter informed me that Springer intended to divide this nearly 900 pages textbook into two separate volumes for clinical and basic sciences. Peter was acutely aware of his prognosis and asked me if I would be willing to take over and edit the clinical volume. I willingly accepted. This book *Positron Emission Tomography: Clinical Practice* is a selected and updated version of the clinical chapters from the original book.

Positron Emission Tomography is an exceptional functional imaging tool. There has been a tremendous increase in interest in PET in the past decade, not only as a research tool but particularly in the clinical arena. The editors of the original book (Peter Valk, Dale Bailey, David Townsend and Michael Maisey) noted how they had collectively been involved in many aspects of PET development, including instrumentation, algorithms and protocol developments and applications, as well as the training of basic scientists and medical specialists, and efforts to convince health bureaucrats of the value of functional imaging in patient management. Through their extensive involvement in all aspects of PET, they progressively became aware of the lack of a comprehensive and contemporary reference work covering the science and clinical applications of PET. The original edition of their book arose from a desire to redress this situation.

The field of PET is progressing rapidly with the developments of multimodality imaging using integrated PET/CT systems. For this separate edition of clinical applications, the intent remains true to the aims of the first edition, namely, to provide a contemporary reference work covering the science and clinical applications of PET with extensive updating to include PET/CT technology. The book is designed to be used by residents and fellows training in medical imaging specialties as well as imaging experts in private or academic medicine who need to become familiar with this technology, and by those whose specialties carry over to PET and PET/CT such as oncologists, cardiologists, neurologists and surgeons.

Chapters 1 to 4 address the basic aspects of PET and PET/CT including physics and instrumentation, an overview of the clinical advantages of the PET/CT technology over PET or CT alone; the viewpoint of the technologist, radiation dosimetry and protection. Chapters 5 to 25 address oncologic applications and have been significantly updated with new data related to the PET/CT technology; many PET/CT illustrations are included. As in the first edition, a chapter is devoted to infectious diseases and another to PET imaging in pediatric disorders. Chapter 26 is an overview of the cardiac applications of PET, and Chapter 27 discusses cardiac PET/CT that some experts envision as the future one-stop-shop cardiac examination. Chapter 28 is an overview of PET imaging in clinical neurology and is probably the least influenced by recent development of PET/CT technology.

To conclude, I restate part of the preface from the first edition:

We are indebted to the many friends and colleagues who have contributed to this book, and who have willingly shared their knowledge and experience.

The functional nature of PET is based on its ability to target specific biochemical pathways through sophisticated radioactive probes and to record the time course of tracer uptake with highly sensitive instrumentation. PET is indeed a rich area in which to work, in part because of the multidisciplinary nature of the field. Developments in instrumentation, for example, are as much driven by radiochemistry and medical challenges as they are by progress in basic physics and instrumentation. Manufacturers of PET instrumentation have also played a major role in the development of the field by sharing many of their designs for critical appraisal at an early stage, and by being willing to listen to, support, and often fund novel prototype concepts. The development of the combined PET/CT scanner is a prime example of this collaboration.

PET is currently moving forward rapidly on a number of fronts: instrumentation is developing at a fast pace; synthetic radiochemistry is becoming more sophisticated and reliable; new reconstruction algorithms and processing methods are becoming more generally usable because of rapid advances in computer hardware and software; clinical applications are burgeoning as PET becomes affordable for more practitioners; and developments in molecular biology and functional genomics provide opportunities for monitoring gene expression and targets for gene therapy.

In this context, it is a challenge to produce a reference work which remains current even during the period from production of the original text to eventual publication, let alone for a significant number of years afterwards. We leave it up to the reader, and to future readers, to assess how successful we have been in this endeavour.

Dominique Delbeke, MD, PhD
December 2005

Contents

Contributors

Abass Alavi, MD
Department of Radiology
Division of Nuclear Medicine
Hospital of the University of
    Pennsylvania
Philadelphia, PA, USA

Elias J. Anaissie, MD
Myeloma Institute for Research and
    Therapy
University of Arkansas for Medical
    Sciences
Little Rock, AR, USA

Edgardo J.C. Angtuaco, MD
Section of Neuroradiology
Department of Radiology
University of Arkansas for Medical
    Sciences
Little Rock, AR, USA

Dale L. Bailey, PhD
Associate Professor of Medicine
Department of Nuclear Medicine
Royal North Shore Hospital
Sydney, Australia

Bart Barlogie, MD, PhD
Myeloma Institute for Research and
    Therapy
University of Arkansas for Medical
    Sciences
Little Rock, AR, USA

Frank M. Bengel, MD
Nuklearmedizinische Klinik
Technische Universität München
München, Germany

Trond V. Bogsrud, MD
Department of Medical Imaging
Division of Nuclear Medicine
University Clinic
The Norwegian Radium Hospital
Oslo, Norway

Thierry Bury, MD, PhD
Division of Pulmonary Medicine
Department of Medicine
University Hospital of Liège, Sart-Tilman
Liège, Belgium

R. Edward Coleman, MD
Department of Radiology
Division of Nuclear Medicine
Duke University Medical Center
Durham, NC, USA

Leonard P. Connolly, MD
Department of Radiology
Division of Nuclear Medicine
Children's Hospital
Boston, MA, USA

Gary J.R. Cook, MBBS, MSc, MD, FRCP,
    FRCR
Department of Nuclear Medicine and PET
Royal Marsden Hospital
Sutton, UK

Bernadette F. Cronin, DCR (R), DRI,
    FETC
Department of Nuclear Medicine and PET
The Royal Marsden Hospital
Sutton, UK

Farrokh Dehdashti, MD
Division of Nuclear Medicine
Edward Mallinckrodt Institute of
    Radiology
Washington University School of
    Medicine
St. Louis, MO, USA

Dominique Delbeke, MD, PhD
Professor and Director of Nuclear
    Medicine
Positron Emission Tomography
Department of Radiology and
    Radiological Sciences
Vanderbilt University Medical Center
Nashville, TN, USA

Marcelo F. Di Carli, MD
Department of Radiology
Division of Nuclear Medicine/PET
Brigham and Women's Hospital
Harvard Medical School
Boston, MA, USA

Stefan Eberl, BE, MSc, PhD
Department of PET & Nuclear Medicine
Royal Prince Alfred Hospital
Sydney, Australia

Joshua Epstein, DSc
Myeloma Institute for Research and
    Therapy
University of Arkansas for Medical
    Sciences
Little Rock, AR, USA

Frederic H. Fahey, DSc
Department of Radiology
Division of Nuclear Medicine
Children's Hospital
Boston, MA, USA

Ignac Fogelman, MD, FRCP
Department of Nuclear Medicine
Guy's and St Thomas' Hospital
Guy's and St Thomas' Hospital Trust
London, UK

Sharon F. Hain, BSc, MBBS, FRACP
Institute of Nuclear Medicine
Middlesex Hospital
UCH NHS Trust and Hammersmith
    Hospitals NHS Trust
London, UK

Roland Hustinx, MD
Division of Nuclear Medicine
Department of Medicine
University Hospital of Liège,
    Sart-Tilman
Liège, Belgium

Ora Israel, MD
Department of Nuclear Medicine
Rambam Medical Center
Haifa, Israel

Hossein Jadvar, MD, PhD, FACNM
Department of Radiology
Division of Nuclear Medicine
Keck School of Medicine
University of Southern California
Los Angeles, CA, USA

Guy H.M. Jerusalem, MD, PhD
Division of Medical Oncology
Department of Medicine
University Hospital of Liège,
    Sart-Tilman
Liège, Belgium

Laurie B. Jones-Jackson, MD
Department of Radiology
University of Arkansas for Medical
    Sciences
Little Rock, AR, USA

Lale Kostakoglu, MD
Department of Radiology
Division of Nuclear Medicine
The New York Presbyterian Hospital
Weill Medical College of Cornell
    University
Weill Cornell Medical Center
New York, NY, USA

Kenneth A. Krohn, PhD
Departments of Radiology and
    Radiation Oncology
University of Washington
Seattle, WA, USA

Val J. Lowe, MD
Department of Radiology
Division of Nuclear Medicine
Mayo Clinic
Rochester, MN, USA

Michael N. Maisey, BSc, MD, FRCR,
    FRCP
Professor of Emeritus
Department of Radiological Sciences
Guy's and St Thomas' Clinical PET
    Centre
Guy's and St Thomas' Hospital Trust
London, UK

William H. Martin, MD
Department of Radiology and
    Radiological Sciences
Vanderbilt University Medical Center
Nashville, TN, USA

I. Ross McDougall, MBChB, MD, PhD,
    FRCP
Division of Nuclear Medicine
Stanford University Medical Center
Stanford, CA, USA

Marisa Miceli, MD
Myeloma Institute for Research and
    Therapy
University of Arkansas for Medical
    Sciences
Little Rock, AR, USA

Michael J. O'Doherty, MA, MSc, MD,
    FRCP
Department of Nuclear Medicine
Clinical PET Centre
Guy's, King's and St Thomas' School of
    Medicine
Guys and St Thomas' NHS Trust
St Thomas' Hospital
London, UK

Paola Piccini, MD, PhD
Division of Neurosciences and Mental
    Health
Imperial College School of Medicine
Hammersmith Hospital
London, UK

Joseph G. Rajendran, MD
Division of Nuclear Medicine
University of Washington Medical
    Center
Seattle, WA, USA

Erik Rasmussen, MS
Cancer Research and Biostatistics
Seattle, WA, USA

Pierre Rigo, MD, PhD
Department of Nuclear Medicine
Centre Hospitalier Princesse Grace
Monaco

Christiaan Schiepers, MD, PhD
Molecular and Medical Pharmacology
David Geffen School of Medicine
    at UCLA
Los Angeles, CA, USA

Markus Schwaiger, MD
Nuklearmedizinische Klinik
Technische Universität München
München, Germany

George M. Segall, MD
Nuclear Medicine Services
Veterans Affairs Palo Alto Health Care
    System, Palo Alto CA, and
Department of Radiology
Stanford University School of Medicine
Stanford, CA, USA

Anthony F. Shields, MD, PhD
Department of Internal Medicine
Karmanos Cancer Institute
Wayne State University
Detroit, MI, USA

Paul D. Shreve, MD
Advanced Radiology Services PC
Grand Rapids, MI, USA

Barry L. Shulkin, MD, MBA
Division of Diagnostic Imaging
St. Jude's Children's Research Hospital
Memphis, TN, and
Division of Nuclear Medicine
Department of Radiology
University of Michigan Medical Center
Ann Arbor, MI, USA

Barry A. Siegel, MD
Division of Nuclear Medicine
Edward Mallinckrodt Institute of
    Radiology
Washington University School of
    Medicine
St. Louis, MO, USA

Michael A. Smith, MA, MB BChir, FRCS
Department of Orthopaedics
The St Thomas' Soft Tissue Tumor Unit
Guys and St Thomas' NHS Trust
Guy's Hospital
London, UK

Brendan C. Stack, Jr., MD, FACS
Division of Otolaryngology, Head and
    Neck Surgery
Pennsylvania State University College of
    Medicine and Milton S. Hershey
    Medical Center
Hershey, PA, USA

Susan M. Swetter, MD
Dermatology Services
Veterans Affairs Palo Alto Health Care
    System, Palo Alto CA, and
Department of Dermatology
Stanford University School of Medicine
Stanford, CA, USA

Yen F. Tai, MB BS, MRCP
Division of Neurosciences and Mental
    Health
Imperial College School of Medicine
Hammersmith Hospital
London, UK

David W. Townsend, PhD
Director
Cancer Imaging and Tracer
    Development Program
The University of Tennessee Medical
    Center
Knoxville, TN, USA

Jocelyn E.C. Towson, MA, MSc
Department of PET & Nuclear Medicine
Royal Prince Alfred Hospital
Sydney, Australia

Guido J. Tricot, MD, PhD
University of Arkansas for Medical
    Sciences
Myeloma Institute for Research and
    Therapy
Little Rock, AR, USA

†Peter E. Valk, MB, BS, FRACP
Northern California PET Imaging
    Center
Sacramento, CA, USA

Frits Van Rhee, MD, PhD
Myeloma Institute for Research and
    Therapy
University of Arkansas for Medical
    Sciences
Little Rock, AR, USA

Richard L. Wahl, MD
The Johns Hopkins PET Center
Russell H. Morgan Department of
    Radiology and Radiological Science
Johns Hopkins University (JHU)
    School of Medicine
Baltimore, MD, USA

Ronald C. Walker, MD
Department of Radiology
University of Arkansas for Medical
    Sciences
Little Rock, AR, USA

Terence Z. Wong, MD, PhD
Department of Radiology
Division of Nuclear Medicine
Duke University Medical Center
Durham, NC, USA

Hongming Zhuang, MD, PHD
Department of Radiology
Division of Nuclear Medicine
Hospital of the University of
    Pennsylvania
Philadelphia, PA, USA

†Deceased.

1

Basic Science of PET and PET/CT

David W. Townsend

Introduction

Historical Background

The past few years have seen the transition of positron emission tomography (PET) from the research domain into mainstream clinical applications for oncology (1). The emergence of PET as the functional imaging modality of choice for diagnosis, staging, therapy monitoring, and assessment of recurrence in cancer has led to increasing demand for this advanced imaging technology. The recognition that functional imaging modalities such as PET may provide an earlier diagnosis and more accurate staging than conventional anatomic imaging has promoted the technology, particularly as PET is now a reimbursed imaging procedure in the United States for many types of cancer. Although PET offers an extensive array of different radiopharmaceuticals, or molecular probes, to image different aspects of physiology and tumor biology, currently the most widely used PET tracer is the fluorinated analogue of glucose, ^{18}F-2-deoxy-D-glucose (FDG). The increased uptake of glucose in malignant cells has been well known for many years (2), and although FDG is not a specific probe for cancer, nonspecificity can be a useful property when identifying and staging disease by a survey of the whole body. The widespread use of FDG is facilitated by the half-life of ^{18}F (110 min), which is convenient for transportation from a remote cyclotron and compatible with typical whole-body PET imaging times of 20 min or more following a 60- to 90-min uptake period.

The recent development of combined PET/CT instrumentation is an important evolution in imaging technology. Since the introduction of the first prototype computed tomography (CT) scanner in the early 1970s, tomographic imaging has made significant contributions to the diagnosis and staging of disease. Rapid commercial development followed the introduction of the first CT scanner in 1972, and within 3 years of its appearance more than 12 companies were marketing, or intending to market, CT scanners; about half that number actually market CT scanners today. With the introduction of magnetic resonance imaging (MRI) in the early 1980s, CT was, at that time, predicted to last another 5 years at most before being replaced by MRI for anatomic imaging. Obviously this did not happen, and today, with multislice detectors, spiral acquisition, and subsecond rotation times, CT continues to develop and play a major role in clinical imaging, especially for anatomic regions outside the brain.

Functional imaging, as a complement to anatomic imaging, has been the domain of nuclear medicine ever since the early 1950s. Initially, planar imaging with the scintillation (gamma) camera invented by Anger in 1958 was the mainstay of nuclear medicine. Even today, the widely used scintillation camera still follows Anger's original design, comprising a large sodium iodide crystal and collimator, with photomultiplier tubes as the photodetectors. In modern nuclear medicine, planar scintigraphy has been extended to tomography by the development of Single Photon Emission Computed Tomography (SPECT), which can be helpful for certain clinical applications. Although early SPECT systems actually predated CT, the real growth in SPECT did not occur until after the appearance of CT when similar reconstruction algorithms to those used in CT were applied to the reconstruction of parallel projections from SPECT data acquired by a rotating gamma camera.

Functional imaging with positron-emitting isotopes was first proposed in the early 1950s as an imaging technique that could offer greater sensitivity than conventional nuclear medicine techniques with single photon-emitting isotopes. The SPECT collimator is eliminated and replaced by electronic collimation—the coincident detection of two photons from positron annihilation—greatly increasing the sensitivity of the imaging system. However, other than some early prototypes in the 1960s, instrumentation to image positron emitters did not emerge seriously until the 1970s, and the first commercial PET scanners date from around 1980, about the time MRI also became commercially available.

PET was initially perceived as a complex and expensive technology requiring both a cyclotron to produce the short-lived PET radioisotopes and a PET scanner to image the tracer distribution in the patient. Consequently, during the 1970s, PET did not experience the explosive growth of CT, nor, during the 1980s, the comparable growth of MRI. In fact, it was not until the 1990s that PET became recognized as an important technique for imaging cancer by mapping glucose utilization throughout the body with FDG. The elevated utilization of glucose by malignant cells (2) allows cancerous tissue to be identified anywhere in the body, even though it may have no anatomic correlate that would allow identification on a CT scan. The effectiveness of FDG-PET imaging for diagnosing and staging malignant disease was officially recognized when the Centers for Medicare and Medicaid Services (CMS) approved reimbursement for a number of cancers in 1998. Following that decision, the application of FDG-PET for imaging cancer expanded rapidly, although still not at a rate that has ever fully rivaled the growth of the dominant anatomic imaging modalities CT and MRI during the 1970s and 1980s, respectively.

Combining Anatomy and Function

The corporate environments that developed these different imaging technologies, the medical specialties of radiology and nuclear medicine that were responsible for acquiring and operating them, and the differing chronology of clinical acceptance described here each contributed to ensuring CT and PET followed separate and distinct developmental paths. Both modalities have their strengths. CT scanners image anatomy with high spatial resolution, although malignant disease can generally only be identified from the presence of abnormal masses or from size changes in lymph nodes. PET, on the other hand can identify a functional abnormality in, for example, a normal-sized lymph node, although accurate localization of the node may be difficult, or even impossible, from the PET scan alone. To initiate the evolution in imaging technology that was required to physically integrate CT and PET (3) in a single device, initial skepticism from both the corporate environment and the medical profession had to be overcome. The key was to design and develop a research prototype PET/CT scanner within the context of National Institutes of Health (NIH) grant-funded collaboration between academia and industry. The first combined PET/CT prototype scanner was completed in 1998 (4), and clinical evaluation began in June of that year. The initial studies with the prototype (5–8) demonstrated a number of significant advantages of PET/CT: that functional abnormalities could now be accurately localized, that normal benign uptake of a nonspecific tracer such as FDG could be distinguished from uptake resulting from disease, and that confidence in reading both the PET and the CT increases significantly by having the anatomic and functional images routinely available and accurately aligned for every patient.

Although it may seem that, in many cases, it would be equally effective to view separately acquired CT and PET images for a given patient on adjacent computer displays, with (9) or without software registration, experience in the past 4 years with commercial PET/CT scanners has highlighted numerous unique advantages of the new technology. A number of these advantages were anticipated in the original PET/CT proposal, but others were unexpected and have only emerged since the technology became more widely available for clinical imaging.

The purpose of this chapter is to present the physical basis of PET and review aspects of the instrumentation that have been developed to image positron-emitting distributions. The design objectives of emerging PET/CT technology are described, and the status of current instrumentation and CT-based attenuation correction methodology are reviewed. Operation of PET/CT technology within the clinical setting opens up new possibilities for disease diagnosis, staging, and monitoring response to therapy. Although specific clinical applications are discussed in depth elsewhere in this volume, some general protocol definitions and refinements are presented here.

Physical Principles of PET Imaging

The principles of imaging tissue function in vivo with PET are summarized in Figure 1.1. Multiple steps are involved in the PET process, beginning with the selection and production of a suitable molecular probe, a pharmaceutical labeled with a positron-emitting radionuclide, the administration of the probe to the patient, and finally the imaging of the distribution of the probe in the patient. Positron emitters are neutron-deficient isotopes that achieve stability through the nuclear transmutation of a proton into a neutron. As shown in Figure 1.1a, this process involves the emission of a positive electron, or positron (β^+) and an electron neutrino (ν_e). The energy spectrum of the emitted positron depends on the specific isotope, with typical endpoint energies varying from 0.6 MeV for ^{18}F up to 3.4 MeV for ^{82}Rb. After emission, the positron loses energy by interactions in the surrounding tissue until it annihilates with an electron, as shown schematically in Figure 1.1b. The range that the positron travels in tissue obviously depends on the energy with which it is emitted, and the (electron) density of the surrounding tissue. The two annihilation photons are emitted in approximately opposite directions and are detected in coincidence; in this example, a coincidence is defined by two photons that are registered within a time interval of 2τ ns, where τ is the electronic coincidence time window.

Positron emitters such as ^{18}F are used to label substrates such as deoxyglucose (DG) (Figure 1.1c) to create

Figure 1.1. The principles of PET imaging shown schematically: (**a**) the decay of a neutron-deficient, positron-emitting isotope; (**b**) the detection in co-incidence of the annihilation photons within a time window of 2τ ns; (**c**) the glucose analogue deoxyglucose labeled with the positron emitter ^{18}F to form the radiopharmaceutical FDG; (**d**) the injection of the labeled pharmaceutical and the detection of a pair of annihilation photons in coincidence by a multiring PET camera; (**e**) the collection of the positron annihilation events into sinograms wherein each element of the sinogram contains the number of annihilations in a specific projection direction; and (**f**) a coronal section of the final, reconstructed whole-body image mapping the utilization of glucose throughout the patient.

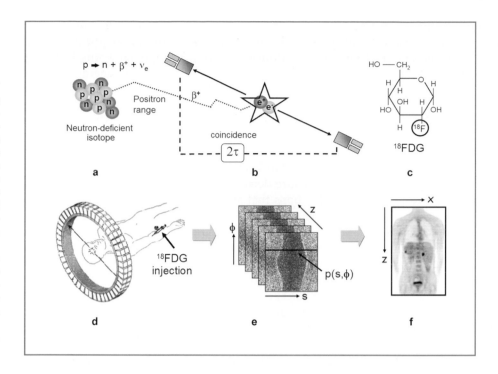

the radiopharmaceutical FDG. The radioactive tag is then transported by the circulation and incorporated into the organ of interest through metabolism of the pharmaceutical. For FDG, the relevant metabolic process is glucose utilization. The radiopharmaceutical is first injected into the patient, and the patient is then positioned in the PET scanner, a circular configuration of detectors (Figure 1.1d). The gamma ray pairs from positron annihilation are captured in coincidence by opposing detectors. The pairs of coincident photons (events) that are detected are stored in matrices (sinograms) where each row in the matrix represents a parallel projection, p(s, ϕ), of the activity distribution in the patient at a specific angle (ϕ) and axial position (z) (Figure 1.1e); s is the radial coordinate within a given parallel projection. An image reconstruction algorithm is applied to the sinogram data to recover the underlying radioactivity distribution, thus indirectly mapping the functional process that created the distribution of the positron emitter. For the radiopharmaceutical FDG, the images shown in Figure 1.1f are maps of FDG accumulation throughout the body reflecting glucose utilization by the different tissues.

The basis of PET is, therefore, that the pharmaceutical or substrate interacts with the body through a metabolic process; the radionuclide allows that interaction to be followed, mapped, and measured. For medical applications of PET, the most important radionuclides, with half-lives in parentheses, are ^{15}O (2.03 min), ^{13}N (9.96 min), ^{11}C (20.4 min), and ^{18}F (109.8 min). For clinical applications, ^{18}F is currently of greatest importance in oncology due to the widespread use of FDG. The maximum energy of the positron from decay of ^{18}F is 0.633 MeV, with a mean

range of 0.6 mm. It is, therefore, a nuclide with favorable properties for high-resolution PET imaging.

Spatial Resolution

The physics of positron emission imposes certain limitations on the spatial, temporal, and contrast resolution that can be attained in a particular imaging situation. Positron range (see Figure 1.1a) is obviously one limitation on the spatial resolution, because the goal is to map the distribution of positron-emitting nuclei and not the distribution of annihilation points. The energy carried by the positron may not be entirely dissipated during its journey in tissue, and the annihilating electron-positron system may have residual momentum. To conserve momentum, the annihilation photons are emitted slightly less than 180º apart, further contributing to a loss of spatial resolution because the two photons are assumed to be collinear and to form a straight line containing the point of emission of the positron. Neither of these assumptions is exact, and both positron range and acolinearity of the photons degrade spatial resolution. The contribution from acolinearity increases with increased separation of the coincident detectors, and the resolution degradation is a maximum at the center of the transverse field of view. For typical whole-body PET scanner designs, the contribution from photon acolinearity will be about 1.5 mm full-width at half-maximum (FWHM) for ^{18}F. Although these physical effects place a lower limit on the spatial resolution that can be achieved with PET, contributions from the size and design of the PET detectors further degrade the spatial resolution in clinical PET scanners. Physical effects contribute 2 mm or less, whereas the spatial resolution of a PET scanner in a clinical imaging situation is, at best, about 6 mm.

Scatter and Randoms

As with any imaging technique, not all acquired events contribute to the signal. Contributions to background noise include photons that scatter before detection and photons from two unrelated annihilations that are erroneously assigned to a single positron emission, as shown schematically in Figure 1.2. For scattered events (Figure 1.2a), one or both photons interact in the tissue before reaching the detectors and as a consequence the event is assigned incorrectly to the line joining the two detectors. The level of scatter, which at this energy is primarily caused by Compton scattering, is characterized by the scatter fraction (SF), the ratio of scattered to total events. Random coincidences (Figure 1.2b) are photon pairs from uncorrelated positron annihilations that arrive within the electronic time window (2τ) that defines a coincidence. The random coincidence rate increases linearly with the width of the time window and quadratically (except at very low rates) with the rate of single photons interacting in the detectors. Both these background processes introduce a bias into the reconstructed images, a bias that can be reduced or eliminated by measuring or modeling each process, although usually at the expense of increasing image noise.

Scattered photons can, in principle, be identified from the energy lost in the scattering process and rejected by applying a simple energy threshold. However, as seen later, the energy resolution of current PET detectors is unable to accurately distinguish scattered from nonscattered photons above a certain energy threshold that may be as low as 350 keV for BGO detectors. Thus, in addition to a lower energy threshold, sophisticated scatter correction models have been developed to remove the residual scatter bias (10–12). The scatter background cannot be measured directly and must, instead, be estimated from the data. In a typical clinical imaging situation, even after applying an energy threshold, the fraction of the total events in the image that are scattered (SF) is 8%–10% in two dimensions (2D) and up to 45% or greater in three dimensions (3D).

As mentioned, the random coincidence rate (R) is proportional to the square of the radiation (singles rate) incident on the detectors. This radiation arises not only from the radioactivity in the field of view of the scanner, but also from radioactivity outside the field of view when one of the two photons from a given positron annihilation enters the scanner field-of-view and reaches the detectors. For whole-body clinical imaging with FDG, the radiopharmaceutical distributes throughout the body and radioactivity that localizes in regions not within the field-of-view of the scanner increases the overall randoms rate to a level that may exceed 50% of the total acquisition rate. The randoms rate can be estimated from the singles rate and the coincidence time window (Figure 1.2b), or from a direct measurement of delayed events acquired in an out-

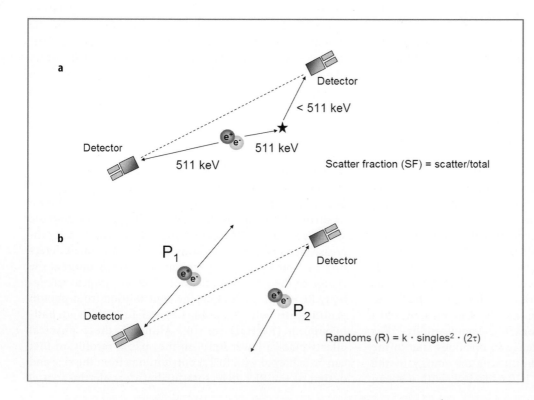

Figure 1.2. Contributions to PET data, acquired in addition to the true coincidences, where (**a**) one (or both) photons scatter and lose energy before they reach the PET detectors and (**b**) two annihilation photons from different positron decays, P_1 and P_2, are detected within the coincidence time window (2τ) appearing to form a true coincidence; such events are termed random coincidences.

of-time window. The advantage of a direct measurement of randoms is that it can account for any spatial variations in the distribution, whereas the disadvantage is that, as a measurement, it increases image noise when the randoms are subtracted.

Attenuation and Attenuation Correction

Annihilation photons that scatter are not only assigned to the incorrect line-of-response (LOR) (see Figure 1.2a), but they are also removed from the correct LOR. The removal of annihilation events from an LOR, either by Compton scattering or photoelectric absorption, is termed attenuation, as shown schematically in Figure 1.3a. Attenuation of 511-keV photons follows the usual exponential absorption law determined by the linear attenuation coefficient $\mu(x, E)$, where x is the pathlength in tissue and E is the photon energy; for PET, E = 511 keV, the rest mass of the positron or electron. A well-known advantage of PET is that, because *both* annihilation photons must traverse the tissue without interaction, attenuation is depth independent and is a function of the total thickness of tissue, greatly simplifying the attenuation correction procedure compared to SPECT. The attenuation correction factor (ACF) for a given LOR is obtained by integrating the linear attenuation coefficients along the path of the LOR (Figure 1.3a). The limits of integration, x_1 and x_2, are coordinates, measured along the LOR (k), at the entrance and exit of the patient. For PET, the correction is (1) independent of source depth, and (2) exact, since the ACFs can be measured directly (Figure 1.3b). The measurement is analogous to a CT scan acquired at 511 keV, rather than at 70 keV as in clinical CT. Up to three ^{68}Ge rod sources (R) covering the full axial extent of the scanner circulate around the patient to acquire the corresponding transmission data at 511 keV. The total transmission counts, I(k), acquired for a given LOR k is compared to the nonattenuated counts, $I_0(k)$, acquired in the absence of a patient (blank scan), and the ratio $I_0(k)/I(k)$ yields the ACF for LOR k. By measuring this ratio for all LORs, and applying the factors to the PET emission data, the effect of attenuation can be corrected.

For PET, the attenuation correction procedure is, in principle, exact. However, because the correction is based on a measurement involving photon counting statistics, additional noise is introduced into the PET data. The radioactivity in the rod sources is limited to avoid excessive dead time in adjacent PET detectors, and transmission scan times may represent 40% or more of the total scan duration.

For whole-body imaging of large patients, the ACFs are significant, exceeding a factor of 300 for LORs through the shoulders and abdomen, thereby amplifying the intrinsic noise in the PET emission data. Despite this noise amplification, the importance of correcting for attenuation in whole-body FDG scans is evident, as shown in Figure 1.4, comparing the uncorrected image (Figure 1.4a) and the corrected image (Figure 1.4b). Artifactually increased uptake in the skin and lungs, as well as nonuniform recovery of uptake in the liver and spleen, are well-known features of uncorrected FDG-PET images. The uncorrected images are not quantitative and, for interpretation purposes, estimates of standardized uptake values (SUVs) will be inaccurate.

Figure 1.3. The physics of PET attenuation and a procedure for correction of the attenuation effect. (**a**) An annihilation photon scatters in the patient, and the event is removed from that line-of-response (k). The attenuated activity I(k) is given by the unattenuated activity $I_0(k)$ multiplied by the integral along the LOR k of the linear attenuation coefficient $\mu(x, E_{PET})$ at the PET energy (511 keV). The integration variable x is integrated from x_1 to x_2, the limits of the intersection of the LOR k with the outline of the patient. (**b**) To correct for attenuation, the unattenuated activity $I_0(k)$ for LOR k is estimated from a blank scan B(k) acquired in the absence of a patient with up to three rotating rod sources (R) of ^{68}Ge. The attenuated value I(k) is obtained from the transmission scan T(k) acquired with the patient positioned in the scanner. The attenuation correction factor ACF for each line of response (LOR) k is given by the ratio B(k)/T(k).

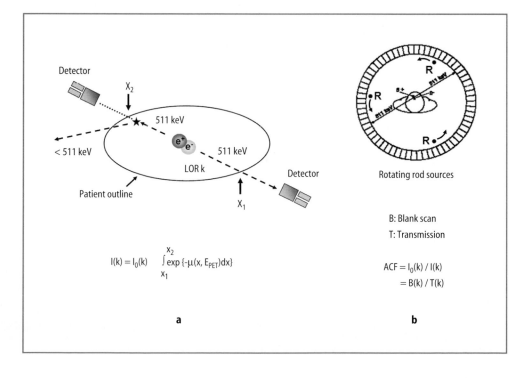

$$I(k) = I_0(k) \int_{x_1}^{x_2} \exp\{-\mu(x, E_{PET})dx\}$$

$$ACF = I_0(k) / I(k)$$
$$= B(k) / T(k)$$

a

b

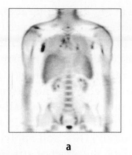 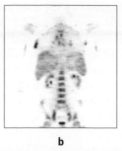

a   b

Figure 1.4. A coronal section through a whole-body FDG-PET image of a patient: the images have been reconstructed (**a**) without correction for attenuation and (**b**) with correction for attenuation, obtained as described in Figure 1.3. Note in the uncorrected image (**a**) the artifactual appearance of increased activity in the lungs, skin, and periphery of the liver as compared to the corrected image (**b**).

Imaging Technology for PET

The past decades has witnessed significant advances in the imaging technology available for PET, the pace increasing recently with the introduction of new fast scintillators and the combination of anatomic and functional imaging within the same scanner. This section briefly reviews some of these developments in PET imaging technology, including the basic block detector design, scintillator performance, two-dimensional (2D) and three-dimensional (3D) imaging, and reconstruction algorithms. Combined anatomic and functional imaging are discussed in the subsequent section.

PET Detectors

The PET block detector was first developed by Casey and Nutt in the mid-1980s (13). Previous efforts to improve PET spatial resolution through the use of smaller scintillation detectors, each coupled to a photomultiplier tube, became prohibitively expensive. In addition, the demand to increase the axial coverage of PET scanners by incorporating multiple detector rings into the design created complex and inconvenient coupling schemes to extract the scintillation signals. Multiplexing first 32, and then 64, detectors to four phototubes, Casey and Nutt decreased both complexity and cost in one design (13). A block of scintillator is cut into 8 x 8 detectors and bonded to four photomultipliers (Figure 1.5). Light sharing between the four phototubes (A to D) is used to localize the detector element in which the incident photon interacts. The block design shown in Figure 1.5 has been the basic detector component in all multiring PET scanners for more than 17 years.

Scintillators for PET

The first PET scanners, developed in the 1960s and early 1970s, were based on various geometric configurations of

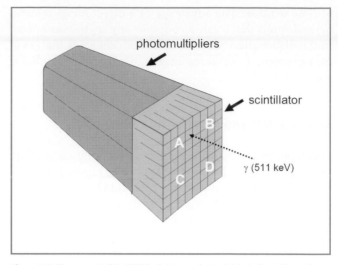

Figure 1.5. The concept of the PET block detector design. A block of scintillators is partitioned into 8 x 8 small elements with cuts of different depths. The scintillator is bonded to four photomultiplier tubes, A to D. An incident 511-keV annihilation photon is converted to light and detected by the phototubes. The distribution of light among the four phototubes localizes the position of the incident annihilation photon.

thallium-activated sodium iodide crystals [NaI(Tl)], a scintillator that is widely used in standard nuclear medicine gamma cameras to detect the 140-keV photons from the decay of technetium. To detect 511-keV annihilation photons, however, the low density of NaI(Tl) is a disadvantage unless thicker crystals are used to compensate for the reduced stopping power. In the late 1970s, bismuth germanate (BGO), a denser scintillator with greater stopping power than sodium iodide, was first considered as a prospective detector for PET (14). Compared to NaI(Tl), the higher density and increased photo fraction of BGO offer improved sensitivity by detecting a greater fraction of the incident photons. The light output of BGO is only about 15% of NaI(Tl) and the decay time is about 30% longer. Despite these suboptimal properties, BGO became established as the most widely used scintillator for PET for more than 25 years. Some of the physical properties of NaI(Tl) and BGO are compared in Table 1.1. Nevertheless, the use of NaI(Tl) for PET imaging continued and one company, UGM Medical (Philadelphia, PA, USA), successfully developed PET scanners based on NaI(Tl) crystals

Table 1.1. Physical properties of different scintillators for positron emission tomography (PET)

Property	NaI	BGO	LSO	GSO
Density (g/mL)	3.67	7.13	7.4	6.7
Effective Z	51	74	66	61
Attenuation length (cm)	2.88	1.05	1.16	1.43
Decay time (ns)	230	300	35–45	30–60
Photons/MeV	38,000	8,200	28,000	10,000
Light yield (%NaI)	100	15	75	25
Hygroscopic	Yes	No	No	No

NaI, sodium iodide; LSO, lutetium oxyorthosilicate; BGO, bismuth germanate; GSO, gadolinium oxyorthosilicate.

that were thicker than those used in the conventional gamma camera (15).

Thus, from the late 1980s, the BGO block detector became a standard for PET imaging. Blocks of scintillator 50 mm × 50 mm in size and 20 mm to 30 mm deep were cut into 8 × 8 crystals and bonded to four 1-inch photomultiplier tubes (Figure 1.5). Two contiguous rings of blocks covered an axial extent of 10 cm with 16 rings of crystals, each approximately 6 mm × 6 mm in size. A design with a third ring of blocks covering a total of 15 cm axially with 24 rings of detectors appeared in the early 1990s and became established as one of the most effective configurations for clinical PET scanning throughout the decade (16). The energy resolution and count rate performance of these multiring scanners were limited by the physical characteristics of BGO and thus a search for PET scintillators with improved characteristics was initiated, one that would have greater light output and shorter decay time than BGO.

The introduction of new, faster scintillators such as gadolinium oxyorthosilicate (GSO) (17) and lutetium oxyorthosilicate (LSO) (18), both doped with cerium, has recently improved the performance of PET scanners for clinical imaging. The physical properties of these newer scintillators are also compared in Table 1.1. Both GSO and LSO have shorter decay times than BGO by a factor of 6 to 7, reducing system deadtime and improving count rate performance, particularly at high activity levels in the field of view. Of even more importance for clinical imaging is the potential of faster scintillators to decrease the coincidence timing window, thereby reducing the randoms coincidence rate. The increased light output of the new scintillators improves the energy resolution because the increased number of light photons reduces the statistical uncertainty in the energy measurement. However, other physical effects contribute to the emission process and the improvement in energy resolution is not a simple func-tion of the number of light photons. The increased light output also improves the positioning accuracy of a block detector (see Figure 1.5), presenting the possibility to cut the blocks even finer into smaller crystals, thus further improving spatial resolution. Unlike NaI(Tl), BGO, LSO, and GSO are not hygroscopic, facilitating the manufacture and packaging of the detectors. GSO is somewhat less rugged and more difficult to machine than either BGO or LSO. LSO has an intrinsic radioactivity of about 280 Bq/mL with single photon emissions in the range 88 keV to 400 keV. Such a radioactive component is of little consequence for coincidence counting at 511 keV, except possibly at very low count rates.

Imaging in 3D

PET is intrinsically a 3D imaging methodology, replacing the physical collimation required for single photon imaging with the electronic collimation of coincidence detection. However, the first multiring PET scanners incorporated septa, lead or tungsten annular shields mounted between the detector rings (Figure 1.6a). The purpose of the septa was to shield the detectors from photons that scattered out of the transverse plane, thus restricting the use of electronic collimation to within the plane. This restriction also allows 2D reconstruction algorithms to be used on a plane-by-plane basis rather than requiring a full 3D reconstruction algorithm. However, by restricting annihilation events to a set of 2D planes, inefficient use is made of the emitted radiation. When the septa are extended (Figure 1.6b), only LORs with small angles of incidence are active; the remaining LORs intersect the septa and the photons never reach the detectors. When the septa are retracted (Figure 1.6c), many more LORs are active and the overall scanner sensitivity increases by a factor of 6 or greater, depending on the exact design (19).

Figure 1.6. PET acquisition in 2D and 3D modes. (**a**) Schematic of a multiring PET scanner with interring lead septa to shield the detector rings from out-of-plane scatter and randoms. (**b**) With the septa extended into the field-of-view, the number of active LORs is limited to those in-plane and small incident angles, whereas with the septa removed (**c**) the number of active LORs is greatly increased, thereby improving the sensitivity. (**d**) The noise equivalent count rate (NECR) as a function of activity concentration in the field-of-view shows significantly improved performance in 3D mode with the septa retracted, particularly at lower activity concentrations.

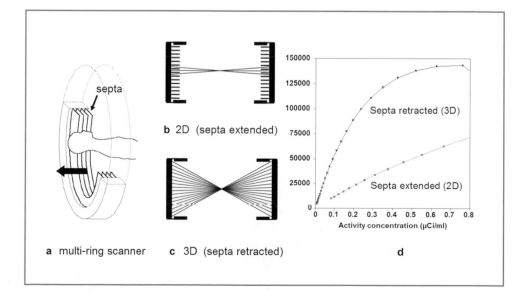

The first multiring PET scanners with retractable septa that included the capability to acquire data in either 2D or 3D mode appeared in the early 1990s (20).

Because the scatter and randoms rates also increase when the septa are retracted, any estimation of the net benefit of 3D imaging compared to 2D imaging must take into account these increases, and not just the increase in true coincidence events. The scatter fraction, for example, increases by a factor of at least 3 compared to 2D imaging and randoms rates increase by a similar or greater factor (19). A measure of actual improvement in signal-to-noise is the noise equivalent count rate (NECR) (21), defined by the expression $T^2/(T + S + \alpha R)$, where T, S, and R are the true, scattered, and random coincidence rates, respectively. The improvement in signal caused by the increase in true counts (from retracting the septa) is offset by the increase in statistical noise on T, the scattered events S, and the randoms R, as expressed in the denominator of the NECR. The factor α equals 2 for online randoms subtraction from the delayed coincidence window and 1 for noiseless randoms subtraction. Smoothing the randoms before subtraction therefore corresponds to $1 < \alpha < 2$. The NECR is shown in Figure 1.6d as a function of activity concentration in a 20-cm-diameter uniform cylinder for 2D and 3D acquisition mode. The significant improvement in the 3D NECR, and hence signal-to-noise, at lower activity concentrations is evident from the graphs.

While the curves in Figure 1.6d highlight the benefits of 3D imaging for the brain, 3D imaging for the rest of the body has been more problematic, mainly because of the difficulty of shielding the detectors from activity in the body outside the imaging field-of-view when the septa are retracted. Recently, however, a number of factors have significantly improved the image quality that can be achieved for whole-body 3D imaging. These factors include advances in reconstruction algorithms, more accurate scatter correction, and the introduction of the new, faster scintillators described previously.

The progress in reconstruction has primarily been the introduction of statistically-based algorithms into the clinical setting. Previously, one of the earliest and most widely used 3D reconstruction algorithms was the reprojection approach based on a 3D extension of the original 2D, filtered backprojection algorithm (22). Although this algorithm works well for the lower noise environment of brain imaging, the quality for whole-body imaging is less than optimal (Figure 1.7a). The development of Fourier rebinning (FORE) (23) enabled 3D data sets to be accurately rebinned into 2D data sets and reconstructed in 2D with a statistically-based algorithm such as ordered-subset expectation-maximization (OSEM) (24). The result of applying OSEM to the same data set as in Figure 1.7a is shown in Figure 1.7b. The improvement in image quality compared with Figure 1.7a is significant, although some nonuniformity remains in the liver and spleen. Further progress has been made by incorporating attenuation information directly into the reconstruction model in the form of weighting factors (Figure 1.7c). The activity in the liver and spleen shows improved uniformity, and the overall image quality is superior with the use of FORE and attenuation-weighted OSEM (AWOSEM) (25). The FORE + AWOSEM approach is an example of a hybrid 3D algorithm, where the data are acquired in 3D, rebinned to 2D, and reconstructed with a 2D algorithm. Other similar combinations are possible, such as the "2.5-dimensional" row action maximum-likelihood algorithm (2.5D RAMLA) developed by Daube-Witherspoon et al. (26) and used for clinical imaging. A second advance that has contributed to the improved image quality in 3D whole-body imaging has been progress in scatter correction algorithms. In particular, the development of faster, image-based algorithms (27) has improved accuracy. The elevated levels of scatter encountered in 3D imaging can be accurately estimated from the emission and transmission data and subtracted from the reconstructed images. Finally, one of the most significant factors contributing to the adoption of 3D acquisition for clinical whole-body imaging has been the introduction of the new, faster scintillators described earlier. For LSO, in particular, a shorter coincidence time window, reduced dead time, and improved energy resolution compared to the corresponding BGO scanner increases the maximum NECR and improves signal-to-noise.

Imaging Technology for PET/CT

The development of the combined PET/CT scanner (3, 4, 28) is an evolution in imaging technology whereby the

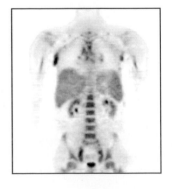

Figure 1.7. An FDG-PET whole-body scan acquired in 3D mode with septa retracted and reconstructed using (**a**) 3D filtered back-projection algorithm with reprojection (Hann window: cutoff, 0.5; 6-mm Gaussian smooth); (**b**) FORE plus OSEM (8 subsets, 2 iterations; 6-mm Gaussian smooth); and (**c**) FORE plus AWOSEM (8 subsets, 2 iterations; 6-mm Gaussian smooth). (These reconstructions were provided by Dr. David Brasse, now at the Institut de Recherches Subatomique, Strasbourg, France.)

a   b   c

Table 1.2. Some drawbacks of software fusion resolved by the hardware approach.

Software fusion	Hardware fusion
Access to image archives required	Images immediately available
Carefully controlled patient positioning	Single-patient positioning
Different scanner bed profiles	Same bed for both scans
Internal organ movement	Little internal organ movement
Disease progression in time	Scans acquired close in time
Limited registration accuracy	Improved registration accuracy
Inconvenience for patient (2 scans)	Single, integrated scan
Labor-intensive registration algorithms	No further alignment required

fusion of two established modalities becomes greater than the sum of the individual parts. It is well known, of course, that CT and PET scans of the same patient acquired on different scanners can be aligned using a number of available software methods (9, 29–31), even though the algorithms are often labor intensive and, outside the brain, may fail to converge to a satisfactory solution. An alternative approach, therefore, is to combine the imaging technology of CT and PET into one scanner such that both anatomy and function, accurately aligned, are imaged in a single scan session. This approach has been termed the hardware fusion approach. Some of the drawbacks of the software approach and the way in which they are addressed by the recently developed hardware approach are summarized in Table 1.2. Of particular importance is the positioning of the patient, the use of the same bed for both scans, and minimization of the effect of uncontrollable internal organ movement. For nonspecific tracers, such as FDG with normal, benign accumulation in tissue and organs, it is important, especially in the abdomen and pelvic region, to distinguish normal uptake from disease. Because accurate localization is routinely available with combined PET/CT scanning, such a distinction is generally straightforward.

Design Objectives

The development of the first PET/CT prototype was initiated in 1992 with the objectives to integrate CT and PET within the same device, to use the CT images for the attenuation and scatter correction of the PET emission data, and to explore the use of anatomic images to define tissue boundaries for PET reconstruction. Thus, the goal was to construct a device with both clinical CT and clinical PET capability so that a full anatomic and functional scan could be acquired in a single session, obviating the need for the patient to undergo an additional clinical CT scan. The original prototype (4) combined a single-slice spiral CT (Somatom AR.SP; Siemens Medical Solutions) with a rotating ECAT ART scanner (CPS Innovations, Knoxville, TN, USA). The components for both imaging modalities were mounted on the same mechanical support and

rotated together at 30 rpm. However, by the time the prototype became operational in 1998 (4), neither the CT nor the PET components were state-of-the-art. Nevertheless, the work convincingly demonstrated the feasibility of combining the two technologies into a single device that could acquire coregistered anatomic and functional images without the need for software realignment.

As mentioned, a number of important lessons emerged during the clinical evaluation program that followed the installation of the prototype and covered the years from 1998 until 2001 (5–8). More than 300 cancer patients were scanned, and the studies highlighted the advantages of being able to accurately localize functional abnormalities, to distinguish normal uptake from pathology, to minimize the effects of both external and internal patient movement, and to reduce scan time and increase patient throughput by using the CT images for attenuation correction of the PET data. Even during the initial evaluation it was evident that coregistered anatomy increases the confidence of physicians reading the study. Radiologists rapidly came to appreciate that coregistered functional images help to focus attention on regions of abnormal uptake, especially regions with no evident pathology on CT.

Despite concerns over the likely cost and operational complexity of combined PET/CT technology, the major vendors of medical imaging equipment nevertheless recognized a market for PET/CT. The first commercial design comprised a CT scanner and a PET scanner enclosed within a single gantry cover and operated from separate consoles. The design involved little integration at any level and was intended primarily to be the first commercial PET/CT scanner on the market, as indeed it was. The PET scanner included retractable septa, and standard PET transmission sources were offered as an alternative to CT-based attenuation correction. Retractable septa allowed the device to acquire PET data in either 2D or 3D mode. Within a few months, another PET/CT design (Figure 1.8) from a different vendor appeared that had no septa and acquired data fully in 3D (32). Because no mechanical storage was required for retractable septa and standard PET transmission sources were not offered, the design was compact; the patient port was a full 70-cm diameter throughout, and the overall tunnel length was only 110 cm. Integration of the control and display software allowed the scanner to be operated from a single console. As with these and most subsequent commercial designs, both the CT and the PET were clinical state-of-the-art systems, following the objectives of the original prototype. A more open concept PET/CT with spacing between the CT and PET scanners has since been offered by two other vendors, allowing greater access to the patient and reducing possible claustrophobic effects of the other designs.

The hardware integration of recent PET/CT designs has, therefore, remained rather minimal. The advantage is that vendors can then benefit more easily from separate advances in both CT and PET instrumentation. In the past

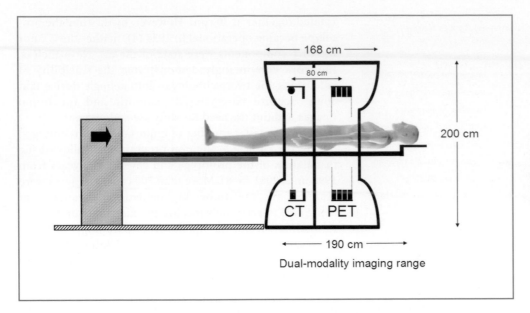

Figure 1.8. A schematic of a current PET-CT scanner design marketed by Siemens as the biograph (Siemens Medical Solutions, Chicago, IL, USA). The design incorporates a multidetector spiral CT scanner and a lutetium oxyorthosilicate (LSO) PET scanner. The dimensions of the gantry are 228 cm wide, 200 cm high, and 168 cm deep. The separation of the CT and PET imaging fields is about 80 cm. The co-scan range for acquiring both CT and PET is up to 190 cm. The patient port diameter is 70 cm.

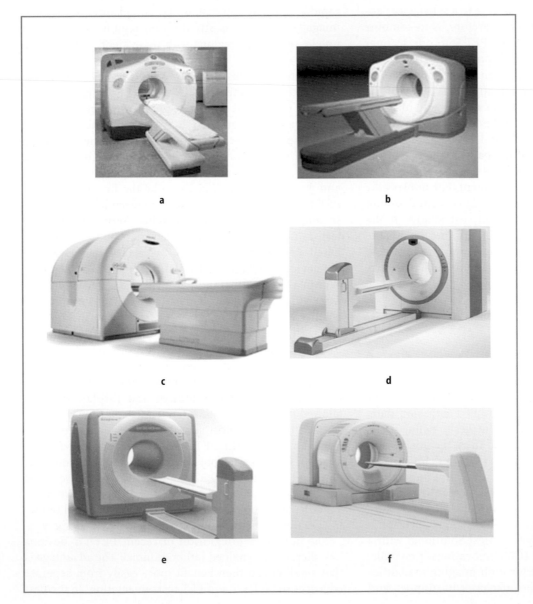

Figure 1.9. Current commercial PET/CT scanners from five major vendors of medical imaging equipment: (**a**) Discovery LS (GE Healthcare); (**b**) Discovery ST (GE Healthcare); (**c**) Gemini (Philips Medical Systems); (**d**) biograph (Siemens Medical Solutions); (**e**) SeptreP3 (Hitachi Medical Systems); (**f**) Aquiduo (Toshiba Medical Corporation).

few years, spiral CT technology has progressed from single to dual-slice, to 4, 8, 16, and, most recently, 64 slices; in parallel, CT rotation times have deceased to less than 0.4 s, resulting in very rapid scanning protocols. Advances in PET technology (as described earlier) have been equally dramatic with the introduction of the new faster scintillators GSO and LSO, faster acquisition electronics, and higher resolution detectors (smaller pixels). Currently, a top-of-the-line PET/CT configuration would comprise a 64-slice CT scanner and an LSO-based PET scanner with 4-mm pixels. However, although the 64-slice CT configuration is targeted primarily for cardiac applications, the greatest impact of PET/CT to date has been in the oncology field for which a 16-slice CT scanner is generally considered adequate.

Current Technology for PET/CT

Currently, five vendors offer PET/CT designs: GE Healthcare, Hitachi Medical, Philips Medical Systems, Toshiba Medical Corporation, and Siemens Medical Solutions. With the exception of the SceptreP3 (Hitachi Medical; Figure 1.9e), which is based on a 4-slice CT and rotating LSO detectors, all vendors offer a 16-slice CT option for higher performance, with some vendors also offering lower-priced systems with 2-, 4-, 6-, or 8-slice CT detectors. The specifications and performance of the PET components are vendor specific, with the biograph HI-REZ (Siemens Medical Solutions; Figure 1.9d) offering the best overall spatial resolution in 3D with 4 mm × 4 mm

LSO crystals; the original biograph design was based on 6 mm × 6 mm LSO detectors. The biograph is offered with 2-, 6-, 16-, and now 64-slice CT scanners. The same HI-REZ PET detectors are incorporated into the Aquiduo (Toshiba Medical; Figure 1.9f) in combination with the 16-slice Aquillion CT scanner (Toshiba Medical); a unique feature of this device is that the bed is fixed and the CT and PET gantries travel on floor-mounted rails to acquire the CT and PET data. The CT and PET scanners in the Aquiduo can be moved separately, and this is the only PET/CT design in which the CT tilt option has been preserved. The Discovery LS, the original PET/CT design from GE Healthcare, combined the Advance NXi PET scanner with a 4- or 8-slice CT (Figure 1.9a); note the size difference between the smaller CT scanner in front and the larger Advance PET scanner at the rear. The more recent Discovery ST from GE Healthcare has 6 mm × 6 mm BGO detectors in combination with a 16- or 64-slice CT scanner (Figure 1.9b); the gantry of the newly designed PET scanner now matches the dimensions of the CT scanner. The Gemini GXL (Philips Medical; Figure 1.9c) comprises 4 mm (in plane) and 6 mm (axial) GSO detector pixels, 30 mm in depth; the Gemini is also an open design with the capability to physically separate the CT and PET scanners for access to the patient, as in the Aquiduo. Each vendor has adopted a unique design for the patient couch to eliminate vertical deflection of the pallet (Figure 1.10) as it advances into the tunnel during scanning. All designs other than the SceptreP3 and the Discovery LS offer a 70-cm patient port for both CT and PET, thus facilitating the scanning of radiation therapy

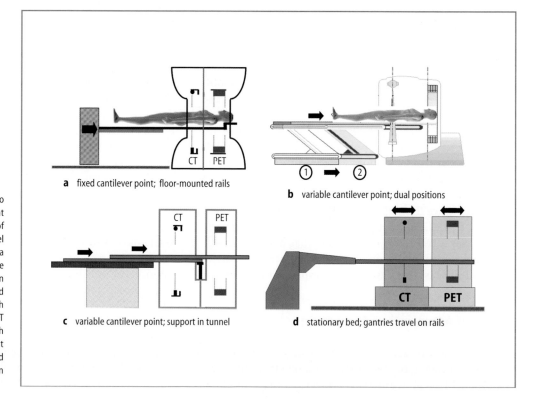

Figure 1.10. Four different solutions to the patient handling system (PHS) that eliminate variable vertical deflection of the pallet as it advances into the tunnel of the scanner. The designs include (**a**) a bed with a fixed cantilever point where the entire couch assembly moves on floor-mounted rails (biograph and SeptreP3), (**b**) a dual-position bed with one position for CT and one for PET (Discovery LS and ST), (**c**) a patient couch that incorporates a support throughout the tunnel (Gemini), and (**d**) a fixed couch with the scanner traveling on floor-mounted rails (Aquiduo).

a fixed cantilever point; floor-mounted rails

b variable cantilever point; dual positions

c variable cantilever point; support in tunnel

d stationary bed; gantries travel on rails

patients in treatment position. Although the Discovery and Gemini also offer standard PET transmission sources as an option, in practice most institutions use CT-based attenuation correction because of the advantage of low noise and short scan times that facilitate high patient throughput.

The Gemini, SceptreP3, Aquiduo, and biograph designs acquire PET data in 3D mode only, whereas the Discovery incorporates retractable septa and can acquire data in both 2D and 3D mode. While the debate continues as to whether 2D or 3D acquisition yields better image quality, particularly for large patients, significant improvement in 3D image quality has undoubtedly been achieved through the use of faster scintillators and statistically based reconstruction algorithms. The scintillators GSO (Gemini) and LSO (SceptreP3, Aquiduo, and biograph) result in lower rates of both scattered photons and random coincidences compared to BGO and offer superior performance for 3D whole-body imaging.

Although there has, to date, been little actual effort to increase the level of hardware integration, there has been significant effort to reduce the complexity and increase the reliability of system operation by adopting a more integrated software approach. In early designs, CT and PET data acquisition and image reconstruction were performed on separate systems accessing a common database. Increasingly, functionality has been combined so as to reduce cost and complexity and increase reliability. Similar considerations of cost and complexity for the hardware may lead, in the future, to greater levels of integration. The likelihood is that these designs will be application specific, incorporating an 8- or 16-slice CT for oncology and a 64-slice CT for cardiology. There will undoubtedly be a demand for more cost-effective, entry-level PET/CT designs for oncology such as the Hitachi SceptreP3, with the likelihood that PET/CT will eventually replace PET-only scanners entirely.

Even though all PET/CT designs offer clinical quality CT and PET, many centers elect to operate with low-dose, nondiagnostic CT for attenuation correction and localization only. This approach does not therefore use the PET/CT to its full, clinical potential, and it is hoped that, as PET/CT is introduced more widely into clinical routine, direct referrals for PET/CT will increase. With more than 1,000 PET/CT scanners installed worldwide, PET/CT now represents more than 95% of all PET sales and more than 10% of CT sales. In the 4 years since the first commercial PET/CT was introduced, the modality has had a far-reaching impact on medical imaging, particularly for staging malignant disease and monitoring response to therapy.

CT-Based Attenuation Correction

The acquisition of accurately coregistered anatomic and functional images is obviously a major strength of the combined PET/CT scanner. However, as mentioned, an additional advantage of this approach is the possibility to use the CT images for attenuation correction of the PET emission data, eliminating the need for a separate, lengthy PET transmission scan. The use of the CT scan for attenuation correction not only reduces whole-body scan times by at least 40%, but also provides essentially noiseless ACFs compared to those from standard PET transmission measurements. Because the attenuation values are energy dependent, the correction factors derived from a CT scan at mean photon energy of 70 keV must be scaled to the PET energy of 511 keV. The CT photon energy represents the mean energy of the polychromatic X-ray beam.

Scaling algorithms typically use a bilinear function to transform the attenuation values above and below a given threshold with different factors (33, 34). The composition of biologic tissues other than bone exhibit little variation in their effective atomic number and can be well represented by a mixture of air and water. Bone tissue does not follow the same trend as soft tissue because of its calcium and phosphorus content, and thus a different scaling factor is required that reflects instead a mixture of water and cortical bone. The breakpoint between the two mixture types has been variously set at 300 Hounsfield units (HU) (33) and at 0 HU (34). However, some tissue types, such as muscle (~60 HU) and blood (~40 HU), have Hounsfield units greater than zero and yet are clearly not a water–bone mix. A breakpoint around 100 HU would therefore appear to be optimal (Figure 1.11). Hounsfield units define the linear attenuation coefficients normalized to water and thus independent of the kVp of the X-ray tube. The scale factor for the air–water mix below about 100 HU will be independent of the kVp of the tube; this does not apply to the water–bone mixing and therefore the scale factor for bone is kVp dependent (35). The scaled CT images are then interpolated from CT to PET spatial resolution and the ACFs generated by reprojection of the interpolated images.

Intravenously injected iodinated contrast is used in CT to enhance attenuation values in the vasculature by increasing the photoelectric absorption compared with blood and resulting in a 40% change in attenuation. At the PET energy, where the photoelectric effect is negligible, the presence of contrast has only a 2% effect on attenuation. However, if contrast-enhanced pixels are misidentified as a water–bone mix, the scaling factor will be incorrect and the erroneously scaled pixels may generate artifacts in the PET image. Many thousands of PET/CT scans have now been performed in the presence of intravenous contrast, and experience has shown that contrast administration does not generally cause a problem that could potentially interfere with the diagnostic value of PET/CT (36). Oral contrast is administered to visualize the gastrointestinal tract and the distribution of the contrast material is rather variable, both in spatial distribution and in level of enhancement. Modifications to the basic scaling algorithm have been introduced to distinguish oral contrast enhancement from bone (37), and

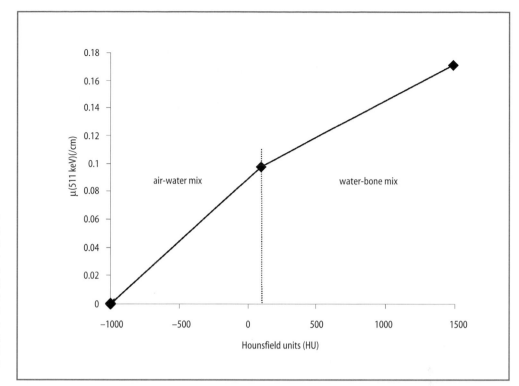

Figure 1.11. The bilinear scaling function used to convert CT numbers to linear attenuation values at 511 keV. The graph shows the linear attenuation coefficient at 511 keV as a function of the corresponding CT value (Hounsfield units, HU), based on measurements made with the Gammex 467 electron density CT phantom using tissue-equivalent materials. The separation between soft tissue (air–water mixing model) and bonelike tissue (water–bone mix) is about 100 HU.

strategies discussed elsewhere (38, 39) have been developed that minimize or eliminate problems from both intravenous and oral contrast. The modified algorithm (37) can, to some extent, also reduce artifacts caused by catheters and metallic objects in the patient. A more-detailed discussion of artifacts arising from metallic objects can be found elsewhere (40, 41).

Avoiding the administration of contrast would, of course, eliminate all such problems. However, standard-of-care in CT dictates the use of either intravenous or oral contrast, or both as in the case of the abdomen and pelvis. An obvious way to avoid such problems is to perform two CT scans: a clinical CT with appropriate contrast administration, and a low-dose, noncontrast CT for attenuation correction and coregistration. This two-scan approach, however, would further increase the radiation exposure to the patient.

PET/CT Protocols

Data acquisition protocols for PET/CT can, depending on the study, be relatively complex, particularly when they involve a clinical CT and a clinical PET scan. During the past 4 years since PET/CT technology first became commercially available, the initial rather simple and basic protocols have progressively become refined to correspond to accepted standards-of-care. Imaging protocols for oncology and cardiology are continuing to be refined and validated, and it will be some time before they become as well established as the corresponding protocols for CT only. Issues of respiration, contrast media, operating parame-

ters, scan time, optimal injected dose of FDG, and others must be carefully addressed before definitive PET/CT protocols for specific clinical applications emerge. Nevertheless, there are certain common features to the protocols, as shown schematically in Figure 1.12, which illustrates a typical PET/CT scan.

As for any FDG-PET scan, following an injection of 370–550 M of FDG and a 90-min uptake period, the patient is positioned in the scanner. A 90-min uptake period is preferred over a 60-min period because increased washout of background activity and improved tumor-to-background ratios are obtained even though a longer time is allowed for decay of the radionuclide. For all studies other than head and neck cancer, the patient is positioned in the scanner with the arms up to reduce attenuation that results from having the arms in the field-of-view; for head and neck cancer, the scan is acquired with arms down. The first step in the study (see Figure 1.12a) is the acquisition of a topogram, or scout scan, which takes 10 to 15 s and covers a range of up to 200 cm. The total range to be scanned by both PET and CT is then defined on the topogram, based on the specific indication for the study (that is, skull base to abdomen for head and neck malignancies, and neck through upper thigh for most other malignancies; for melanoma, the scan range covers head to toe, whenever possible). An appropriate respiration protocol must be defined and implemented to minimize the mismatch between CT and PET. In the absence of respiratory gating (42, 43), a good match is found if the CT is acquired with partial or full expiration and the PET with shallow breathing. This approach is feasible with the 16-slice CT scanner where a scan of the thorax and abdomen

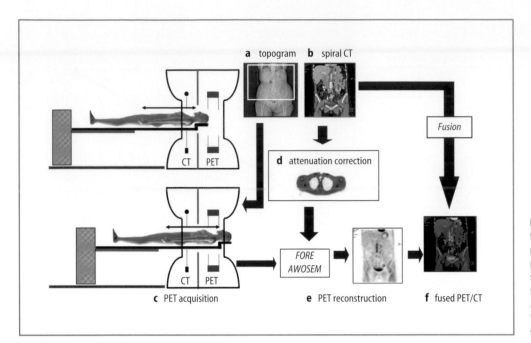

a topogram   **b** spiral CT

Fusion

d attenuation correction

FORE AWOSEM

c PET acquisition   **e** PET reconstruction   **f** fused PET/CT

Figure 1.12. A typical imaging protocol for a combined PET/CT study that comprises (**a**) a topogram, or scout scan, for positioning; (**b**) a spiral CT scan; (**c**) a PET scan over the same axial range as the CT scan; (**d**) the generation of CT-based ACFs; (**e**) reconstruction of the attenuation-corrected PET emission data; and (**f**) display of the final fused images.

can be as short as 12 seconds. Problems posed by respiration mismatch have been discussed (44). The whole-body CT scan is usually acquired with a slice width of about 5 mm, which may not always correspond to standard clinical protocols. In some studies such as head and neck, a slice width of 3 mm may be preferred.

Upon completion of the spiral CT scan (Figure 1.12b), the patient couch is advanced into the PET field-of-view and a multibed PET scan acquired over the same range as the CT (Figure 1.12c). When the emission data are acquired in 3D, consecutive imaging fields typically overlap by 25% to average out the variations in signal-to-noise that would otherwise result. The reconstruction of the CT images occurs in parallel with the acquisition of the PET data, allowing the calculation of scatter and attenuation correction factors to be performed during the PET acquisition. The CT-based ACFs are calculated as described earlier (Figure 1.12d), and once the acquisition for the first bed position is completed, PET reconstruction can begin (Figure 1.12e). the 3D reconstruction is performed using Fourier rebinning (FORE) (23) and the attenuation-weighted ordered-subset algorithm (AWOSEM) (25) mentioned previously. Thus, within a few minutes of the acquisition of the final PET bed position, attenuation-corrected PET images are reconstructed and available for viewing, automatically coregistered with the CT scan by simply accounting for the axial displacement between the CT and PET imaging fields-of-view (Figure 1.12f). The fused image is displayed as a combination of the individual CT and PET image pixel values v_{CT} and v_{PET}, respectively. Using an alpha blending approach, the fused image pixel value (v) is given by $\alpha v_{CT} + (1 - \alpha)v_{PET}$; for $\alpha = 0$, the fused image is PET, whereas for $\alpha = 1$ the fused image is CT. Obviously, for $0 < \alpha < 1$, the fused image represents the weighted combined pixel values of CT and PET.

Although it is not feasible with current designs to acquire the CT and PET data simultaneously, scan times have been reduced significantly by the replacement of the lengthy PET transmission scan with the CT scan. In addition, as mentioned previously, the introduction of new PET technology such as faster scintillators (LSO and GSO) has reduced the emission acquisition time so that whole-body PET/CT acquisition times of 5 to 10 min are feasible with state-of-the-art systems (45). Shorter imaging times lead to higher patient throughput, potentially creating logistical problems for imaging centers. To maintain high throughput, injection times must be carefully coordinated to ensure the constant availability of patients after a 90-min (±5 min) uptake period. It is essential to provide an adequate number of injection and uptake rooms.

In addition to the issues related to respiration (42, 43) and contrast (38, 39), a topic of ongoing debate is the clinical role of the CT scan. Obviously a low dose (40 mAs) CT scan would be adequate for attenuation correction and localization, whereas a full-dose CT scan (130 mAs) is required for clinical purposes. The decision to acquire a clinical CT scan depends on many factors, including whether such a scan was ordered by the referring physician, whether the patient has recently had a clinical CT, and whether the scan will be reviewed by a radiologist when the PET/CT study is read. Obviously the decision, which may be application specific, will dictate the protocol and the parameters of the CT scan. Increasingly, as PET/CT becomes established in clinical routine, the acquisition of both a clinical CT and a clinical PET scan should become standard practice. Referring physicians will be advised to consider a PET/CT immediately rather than ordering first a CT that might be equivocal and then requiring a PET scan. Such an approach will ensure the best use is made of costly imaging equipment. More details on these and other practical issues can be found in Coleman et al. (46).

Finally, although there are many technical reasons to prefer the combined PET/CT approach over software image fusion (see Table 2.2), the convenience to both patient and physician should not be underestimated. For the patient, one appointment and a single scan session is required to obtain complete anatomic and functional information related to his or her disease. For the physician, the potential to have accurately registered CT and PET images available at the same time and on the same viewing screen offers unique possibilities. The added confidence in reading and interpreting the study comes from the accurate localization of tracer accumulation, the distinction of normal uptake from pathology, and the verification that a suspicious finding can be confirmed by the other modality. In some cases, such a suspicious finding on one modality invites a closer examination of the other modality, a retrospective image review that can take place immediately after the PET/CT scan has concluded.

Conclusions

Even though combined PET/CT scanners have been in production for only 4 years, the technology is undergoing rapid evolution. For PET, the introduction of new scintillator materials, detector concepts, and electronics is resulting in performance improvements in count rate, spatial resolution, and signal-to-noise. At the same time, the increasing number of detector rows and reduction in rotation time are transforming whole-body CT performance. The combination of high-performance CT with high-performance PET is a powerful imaging platform for the diagnosis, staging, and therapy monitoring of malignant disease. Although the PET scanners incorporated into current PET/CT designs are still offered by some vendors for PET-only applications, more than 95% of PET sales are now PET/CT, and the likelihood is that PET-only scanners will be replaced entirely by PET/CT in the future. It is expected that there will then be a demand for a design that offers less performance at less cost. To meet this demand, an entry-level or midrange PET/CT is required, possibly in a form similar to the original prototype with PET detectors mounted on the same rotating assembly as the CT. Because the performance of the PET components is the limitation on the overall imaging time, institutions requiring high throughput and large patient volumes will always demand the highest PET performance. Nevertheless, a 6- or 8-slice CT scanner should be adequate for most oncology applications, with a 16- or 64-slice CT appropriate for PET/CT applications in cardiology. As the current PET/CT technology becomes more widespread, appropriate future designs of this concept will doubtless emerge.

Acknowledgments

The author is grateful to his colleagues at the University of Tennessee Medical Center, the University of Pittsburgh, the University of Essen, Germany, and CPS Innovations, Knoxville, Tennessee who have contributed significantly to the development of the combined PET/CT scanner. The PET/CT prototype was developed jointly with CPS Innovations, Knoxville, TN and funded in part by a grant from the National Cancer Institute (CA65856).

References

1. Valk PE, Bailey DL, Townsend DW, Maisey MN, editors. Positron Emission Tomography: Basic Science and Clinical Practice. Part IV: Oncology. London: Springer, 2003:481–688.
2. Warburg O. On the origin of cancer cells. Science 1956;123:306–314.
3. Townsend DW, Cherry SR. Combining anatomy with function: the path to true image fusion. Eur Radiol 2001;11:1968–1974.
4. Beyer T, Townsend DW, Brun T, Kinahan PE, Charron M, Roddy R, et al. A combined PET/CT scanner for clinical oncology. J Nucl Med 2000;41:1369–1379.
5. Charron M, Beyer T, Bohnen NN, Kinahan PE, Dachille M, Jerin J, Nutt R, et al. Image analysis in patients with cancer studied with a combined PET and CT scanner. Clin Nucl Med 2000; 25:905–910.
6. Meltzer CC, Martinelli MA, Beyer T, Kinahan PE, Charron M, McCook B, et al. Whole-body FDG PET imaging in the abdomen: value of combined PET/CT. J Nucl Med 2001;42:35P.
7. Meltzer CC, Snyderman CH, Fukui MB, Bascom DA, Chander S, Johnson JT, et al. Combined FDG PET/CT imaging in head and neck cancer: impact on patient management. J Nucl Med 2001;42:36P.
8. Kluetz PG, Meltzer CC, Villemagne VL, Kinahan PE, Chander S, Martinelli MA, et al. Combined PET/CT imaging in oncology: impact on patient management. Clin Positron Imaging 2000;3:223–230.
9. Hawkes DJ, Hill DL, Hallpike L, Bailey DL. Coregistration of structural and functional images. In: Valk P, Bailey DL, Townsend DW, Maisey MN, editors. Positron Emission Tomography: Basic Science and Clinical Practice. London: Springer, 2003:181–197.
10. Bailey DL, Meikle SR. A convolution-subtraction scatter-correction method for 3D PET. Phys Med Biol 1994;39:411–424.
11. Watson CC, Newport D, Casey ME. A single scatter simulation technique for scatter correction in 3D PET. In: Grangeat P, Amans J-L, editors. Three-Dimensional Image Reconstruction in Radiology and Nuclear Medicine. Dordrecht: Kluwer Academic, 1996:255–268.
12. Ollinger JM. Model-based scatter correction for fully 3D PET. Phys Med Biol 1996;41:153–176.
13. Casey ME, Nutt R. A multicrystal, two-dimensional BGO detector system for positron emission tomography. IEEE Trans Nucl Sci 1986;33:460–463.
14. Cho ZH, Farukhi MR. Bismuth germanate as a potential scintillation detector in positron cameras. J Nucl Med 1977;8:840–844.
15. Karp JS, Muehllehner G, Geagan MJ, Freifelder R. Whole-body PET scanner using curved-plate NaI(Tl) detectors. J Nucl Med 1998;39:50P.
16. Wienhard K, Eriksson L, Grootoonk S Casey M, Pietrzyk U, Heiss WD. Performance evaluation of the positron scanner ECAT EXACT. J Comput Assist Tomogr 1992;16:804–813.
17. Takagi K, Fukazawa T. Cerium-activated Gd_2SiO_5 single crystal scintillator. App Phys Lett 1983;42:43–45.
18. Melcher CL, Schweitzer JS. Cerium-doped lutetium oxyorthosilicate: a fast, efficient new scintillator. IEEE Trans Nucl Sci 1992;39:502–505.

19. Townsend DW, Isoardi RA, Bendriem B. Volume imaging tomographs. In: Bendriem B, Townsend DW, editors. The Theory and Practice of 3D PET. Dordrecht: Kluwer, 1998:111–132.

20. Spinks TJ, Jones T, Bailey DL, Townsend DW, Grootoonk S, Bloomfield PM. Physical performance of a positron tomograph for brain imaging with retractable septa. Phys Med Biol 1992;37:1637–1655.

21. Strother SC, Casey ME, Hoffman EJ. Measuring PET scanner sensitivity: relating count rates in the image to signal-to-noise ratios using noise equivalent counts. IEEE Trans Nucl Sci 1990;37:783–788.

22. Kinahan PE, Rogers JG. Analytic three-dimensional image reconstruction using all detected events. IEEE Trans Nucl Sci 1990;36:964–968.

23. Defrise M, Kinahan PE, Townsend DW, Michel C, Sibomana M, Newport DF. Exact and approximate rebinning algorithms for 3D PET data. IEEE Trans Med Imaging 1997;16:145–158.

24. Hudson H, Larkin R. Accelerated image reconstruction using ordered subsets of projection data. IEEE Trans Med Imaging 1994;13:601–-609.

25. Comtat C, Kinahan PE, Defrise M, Michel C, Townsend DW. Fast reconstruction of 3D PET data with accurate statistical modeling. IEEE Trans Nucl Sci 1998;45:1083–1089.

26. Daube-Witherspoon ME, Matej S, Karp JS, Lewitt RM. Application of the row action maximum likelihood algorithm with spherical basis functions to clinical PET imaging. IEEE Trans Nucl Sci 2001;48:24–30.

27. Watson CC. New, faster image-based scatter correction for 3D PET. In: IEEE Nuclear Science Symposium and Medical Imaging Conference, Toronto, Canada, November 1999, pp 1637–1641.

28. Townsend DW. A combined PET/CT scanner: the choices. J Nucl Med 2001;3:533–534.

29. Pelizzari CA, Chen GT, Spelbring DR, Weichselbaum RR, Chen CT. Accurate three-dimensional registration of CT, PET, and/or MR images of the brain. J Comput Assist Tomogr 1989;13:20–26.

30. Pietrzyk U, Herholz K, Heiss WD. Three-dimensional alignment of functional and morphological tomograms. J Comput Assist Tomogr 1990;14:51–59.

31. Woods RP, Mazziotta JC, Cherry SR. MRI-PET registration with automated algorithm. J Comput Assist Tomogr 1993;17:536–546.

32. Townsend DW, Beyer T, Blodgett TM. PET/CT scanners: a hardware approach to image fusion. Semin Nucl Med 2003;XXXIII(3):193–204.

33. Kinahan PE, Townsend DW, Beyer T, Sashin D. Attenuation correction for a combined 3D PET/CT scanner. Med Phys 1998;25:2046–2053.

34. Burger C, Goerres G, Schoenes S, Buck A, Lonn AH, von Schulthess GK. PET attenuation coefficients from CT images: experimental evaluation of the transformation of CT into PET 511-keV attenuation coefficients. Eur J Nucl Med Mol Imaging 2002;29:922–927.

35. Rappoport V, Carney J, Townsend DW. X-Ray tube voltage dependent attenuation correction scheme for PET/CT scanners. IEEE MIC Abstract Book 2004;M10-76:213.

36. Yau YY, Chan WS, Tam YM, Vernon P, Wong S, Coel M, et al. Application of intravenous contrast in PET/CT: does it really introduce significant attenuation correction error. J Nucl Med 2005;46:283–291.

37. Carney JP, Townsend DW. CT-based attenuation correction for PET/CT scanners. In: von Schultess G, editor. CLINICAL PET, PET/CT and SPECT/CT: Combined Anatomic-Molecular Imaging. Baltimore: Lippincott, Williams & Wilkins, 2003:46–58.

38. Antoch G, Freudenberg LS, Egelhof T, et al. Focal tracer uptake; a potential artifact in contrast-enhanced dual-modality PET/CT scans. J Nucl Med 2002;43:1339–1342.

39. Antoch G, Freudenberg LS, Beyer T, Bockisch A, Debutin JF. To enhance or not to enhance? ^{18}F-FDG and CT contrast agents in dual-modality ^{18}F-FDG PET/CT. J Nucl Med 2004;45(suppl):56S–65S.

40. Goerres GW, Hany TF, Kamel E, Von Schulthess GK, Buck A. Head and neck imaging with PET and PET/CT: artifacts from dental metallic implants. Eur J Nucl Med 2002;29:367–370.

41. Kamel EM, Burger C, Buck A, Von Schulthess G, Goerres G. Impact of metallic dental implants on CT based-AC in a combined PET/CT scanner. Eur Radiol 2003;13:724–728.

42. Goerres GW, Burger C, Kamel E, et al. Respiration-induced attenuation artifact at PET/CT; technical considerations. Radiology 2003;226:906–910.

43. Townsend DW, Yap JT, Carney JPJ, Long M, Hall NC, Bruckbauer T, et al. Respiratory gating with a 16-slice LSO PET/CT scanner. J Nucl Med 2004;45(5):165P (abstract).

44. Beyer T, Antoch G, Muller S, et al. Acquisition protocol considerations for combined PET/CT imaging. J Nucl Med 2004;45(suppl):25S–35S.

45. Halpern B, Dahlbom M, Quon A, et al. Impact of patient weight and emission scan duration on PET/CT image quality and lesion detectability. J Nucl Med 2004;45:797–801.

46. Coleman RE, Delbeke D, Guiberteau MJ, Conti PS, Royal HD, Weinreb JC, et al. Intersociety Dialogue on Concurrent PET/CT with an Integrated Imaging System. From the Joint ACR/SNM/SCBT-MR PET/CT Working Group. J Nucl Med 2005;46:1225–1239.

2

Incremental Value of Imaging Structure and Function

Dominique Delbeke

The rapid advances in imaging technologies are a challenge for both radiologists and clinicians who must integrate these technologies for optimal patient care and outcomes at minimal cost.

Since the early 1990s, numerous technologic improvements have occurred in the field of radiologic imaging, including (1) multislice spiral computed tomography (CT), which permits fast acquisition of CT angiographic images and multiphase enhancement techniques, and (2) positron emission tomography (PET), using ^{18}F-fluorodeoxyglucose (FDG) as a radiopharmaceutical that provides the capability for imaging glucose metabolism. Multiple indications for molecular imaging using FDG are now well accepted in the fields of neurology, cardiology, and oncology (1).

The widespread oncologic applications including differentiation of benign from malignant lesions, staging malignant lesions, detection of malignant recurrence, and monitoring therapy have contributed to the establishment of the PET technology in many medical centers in the United States, Europe, and progressively throughout the world. The goals of oncologic imaging are lesion detection, lesion characterization, evaluation of the extent of the neoplasm, staging for malignant lesions, and assessment of the therapeutic response. Staging includes lesion localization, evaluation of proximity to vessels, and detection of nodal and distant metastases. Some of these goals are better achieved with the high resolution of anatomic imaging techniques and others with molecular imaging using PET.

Molecular imaging using positron imaging is unique in that positron emitters allow labeling of radiopharmaceuticals that closely mimic endogenous molecules, and there are continuing efforts for development of new biologic tracers. FDG, allowing evaluation of glucose uptake and metabolism, is the radiopharmaceutical most widely used with the PET technology that has been approved by the Health Care Financing Administration (HCFA) for reimbursement by Medicare in the evaluation of patients with various body tumors, of myocardial viability, and in the presurgical evaluation of intractable epilepsy. Perfusion PET tracers [rubidium-28 (^{82}Rb) and ^{13}N-ammonia] have also been approved for reimbursement for evaluation of myocardial perfusion.

Instrumentation for Molecular Imaging with PET

The clinical utility of FDG imaging was first established using dedicated PET tomographs equipped with multiple rings of bismuth germanate oxide (BGO) detectors, but a spectrum of equipment is now available for positron imaging, including gamma camera-based PET capable of imaging conventional single photon emitters as well as positron emitters such as FDG and state-of-the-art PET systems with other crystal materials, for example, lutethium oxyorthosilicate (LSO) and gadolinium oxyorthosilicate (GSO).

Anatomic and Functional Imaging Are Complementary

Although numerous studies have shown that the sensitivity and specificity of FDG imaging is superior to that of CT in many clinical settings, the inability of FDG imaging to provide anatomic localization remains a significant impairment in maximizing its clinical utility. In addition, FDG is an analogue of glucose, and the distribution of FDG is not limited to malignant tissue (see Chapter 5). It has been well documented in the PET literature that, to avoid misinterpretations, the interpreter must be familiar with the normal pattern and physiologic variations of FDG distribution and with clinical data relevant to the patients (2, 3). Limitations of anatomic imaging with CT are well known and are related to size criteria to differentiate benign from malignant lymph nodes, difficulty with differentiating posttherapy changes from tumor recurrence, and difficulty in differentiating nonopacified loops of bowel from metastases in the abdomen and pelvis.

Close correlation of FDG studies with conventional CT scans helps to minimize these difficulties. In practice, for the past 10 years interpretation has been accomplished by visually comparing corresponding FDG and CT images. The interpreting physician visually integrates the two image sets to precisely locate a region of increased uptake on the CT scan. To aid in image interpretation, computer software has been developed to coregister the FDG-PET emission scans with the high-resolution anatomic maps provided by CT (4). These methods offer acceptable fusion images for the brain that is surrounded by a rigid structure, the skull. For the body, coregistration of two images, often obtained at different points in time, is technically much more difficult. Identical positioning of the patient on the imaging table is critical. Shifting internal organ movement and peristalsis compound the problem. However, software methods allow versatility in the choice of modalities and facilitate retrospective and selective application, but fully automatic registration algorithms are needed for routine clinical applications (5). Another approach that has gained wider acceptance recently is the hardware approach to image fusion using multimodality imaging with an integrated PET/CT imaging system. Townsend et al. (6, 7) have published reviews on the design of these systems.

Integrated PET/CT Systems

The recent technical development of integrated PET/CT systems provides CT and FDG-PET images obtained in a single imaging setting, allowing optimal coregistration of images. The fusion of anatomic and molecular images (PET and CT) obtained with integrated PET/CT systems sequentially in time but without moving the patient from the imaging table allows optimal coregistration of anatomic and molecular images, leading to accurate attenuation correction and precise anatomic localization of lesions with increased metabolism. The fusion images provided by these systems allow the most accurate interpretation of both PET and CT studies in oncology. Fusion FDG-PET/CT imaging is also a promising tool for optimizing radiation therapy. Because of the high photon flux of X-rays, CT attenuation maps from these integrated PET/CT systems also allow optimal attenuation correction of the PET images.

The first prototype of integrated PET/CT system was developed in collaboration with the group of investigators at the University of Pittsburgh (6, 8). Several manufacturers are now offering integrated PET/CT imaging systems combining different models of state-of-the-art dedicated PET tomographs and CT units with multiple rows of detectors side by side with a common imaging bed. The CT units in these integrated systems were first offered with 2 and 4 rows of detectors and are now available with 8 and 16 rows of detectors, opening the horizon for cardiac applications combining myocardial perfusion and viability with calcium scoring and CT coronary angiography (9). As this technology is rapidly progressing, CT units with 32, 40, and 64 rows of detectors will be incorporated in these integrated systems.

With these integrated systems, a CT scan and a PET scan can be acquired sequentially with the patient lying on the imaging table and simply being translated between the two systems. Accurate calibration of the position of the imaging table and the use of common parameters in data acquisition and image reconstruction permit the fusion of images of anatomy and metabolism from the same region of the body that are registered in space and only slightly offset in time. Because the CT scan is actually a high-resolution transmission map, these data can be used to perform a high-quality attenuation correction during image reconstruction of the emission data.

CT Acquisition Protocols for Integrated PET/CT Imaging

Beyer et al. (10) discussed the subject of CT acquisition protocols for integrated PET/CT imaging in the supplement of the *Journal of Nuclear Medicine* dedicated to PET/CT published in January 2004. Different CT transmission protocols can be applied if the transmission CT images are used for attenuation correction only, for anatomic localization, or optimized for diagnostic purposes.

CT transmission for Attenuation Correction

It has been demonstrated that attenuation effects are much more significant in coincidence imaging than in single photon emission tomography (SPECT) because both photons from an annihilation process must pass through the region without interaction. Attenuation correction has significant advantages for the clinical evaluation of FDG images from oncologic patients, the most important of which is improved anatomic delineation (mediastinum from lungs, lungs from liver). Therefore, localization of a lesion is easier on images with attenuation correction. The second advantage of attenuation correction is the ability to semiquantitatively measure the degree of uptake in a lesion using the standard uptake value (SUV), which may be helpful in some clinical settings.

For the body, various methods have been developed with measured attenuation using radioactive transmission sources. Measured attenuation correction is commonly performed by direct measurement of 511-keV photon attenuation through the body. The transmission scan adds to the length of the study, and motion of the patients during long scanning times is a problem because the quality of the image corrected for attenuation effects largely depends on accurate coregistration of the attenuation maps (transmission scan) and emission scan.

The use of an X-ray tube-based transmission scan provides attenuation-corrected emission images of high quality because of the high photon flux inherent with this technique. One advantage over the radioactive sources is the short duration for the transmission scan; with multidetector CT, CT acquisition from the base of the skull to midthigh typically lasts less than a minute, versus 20 to 25 min with a transmission scan performed using external radioactive sources. In addition, optimal coregistration between attenuation maps and emission images is possible using integrated PET/CT systems when the CT attenuation maps and FDG-PET emission images are obtained sequentially in time without moving the patient from the imaging table. Adequate CT transmission map can actually be obtained with very low current (10 mA) (11).

CT Transmission for Anatomic Localization

Higher CT currents are required to produce better-quality CT scans for anatomic localization. For optimized diagnostic CT, the optimal CT current depends on the size of the patient and the region of the body to be imaged but is typically greater than 140 mA for an adult. If the CT transmission scan performed concurrently to PET imaging is used for anatomic localization, at the present time most institutions use a lower CT current of 80 mA to reduce the radiation dose to the patients. The CT current can be adjusted according to the patient's weight (12).

CT Transmission for Diagnostic Purposes

Optimized diagnostic CT protocols involve administration of intravenous (IV) and oral contrast agents, which brings up the issue of the administration of contrast agents for the CT transmission scan (13). Intravenous contrast appears as regions of high contrast on CT images, especially during the arterial and arteriovenous phase of enhancement. If these CT images are used for attenuation correction, overcorrection may create artifacts of increased uptake on the FDG-PET images (14). High-density oral contrast agents (15) and metallic implants (16) can create similar artifacts. However, the administration of low-density oral contrast only result in minimal overcorrection and is not believed to interfere with accurate interpretation of the images (15, 17).

Because the FDG emission images must be acquired during normal breathing, there is still a debate as to whether optimal attenuation maps are provided by attenuation maps obtained with the breath-hold technique or during normal breathing (18, 19). Respiratory motion results in inaccurate localization of lesions at the base of the lungs or dome of the liver in 6 of 300 (2%) of patients (20, 21).

The complex issue of the optimal CT protocol to use in combination with PET imaging is still debated because of the implications related to the attenuation correction procedure to handle CT contrast agents and images artifacts resulting from mismatches in respiration and patient positioning between the CT and PET examinations (22, 23). In addition, there is an ethical issue in submitting all patients referred for a PET/CT study to the risk of intravenous contrast administration and radiation dose of an optimized diagnostic whole-body CT study.

Schaefer et al. (24) reviewed PET/CT and contrast-enhanced CT of 60 patients with lymphoma. For evaluation of lymph node involvement, sensitivity of PET/CT and contrast-enhanced CT was 94% and 88%, and specificity was 100% and 86%, respectively. For evaluation of organ involvement, sensitivity of PET/CT and contrast-enhanced CT was 88% and 50%, and specificity was 100% and 90%, respectively. Agreement of both methods was excellent (kappa = 0.84) for assignment of lymph node involvement but only fair (kappa = 0.50) for extranodal disease. They conclude that PET/CT performed with non-enhanced CT is more sensitive and specific than is contrast-enhanced CT for evaluation of lymph node and organ involvement, especially regarding exclusion of disease, in patients with Hodgkin disease and high-grade non-Hodgkin lymphoma.

Clinical Impact of integrated PET/CT Images in Oncology

For FDG-avid malignancies, it is well established that FDG-PET has a superior accuracy than CT for detection of malignant foci with a sensitivity in the 90% range. It is also well established that the addition of FDG-PET in the evaluation of these patients change the management in 10% to 30% of the cases (25). Because of the excellent performance of FDG-PET alone, it has been difficult to demonstrate the superiority of the PET/CT technology compared to PET alone or PET and CT acquired at different times and interpreted in conjunction with each other side by side. A definite advantage of the PET/CT technology is the availability of matching sets of images over the same field of view whether these images are interpreted side by side or with the availability of fusion images. With the PET/CT technology, all these sets of images are matching because they are obtained at the same time with the patient in the same position.

Published data regarding the incremental value of integrated PET/CT images compared to PET alone, or PET correlated with a CT obtained at a different time, are

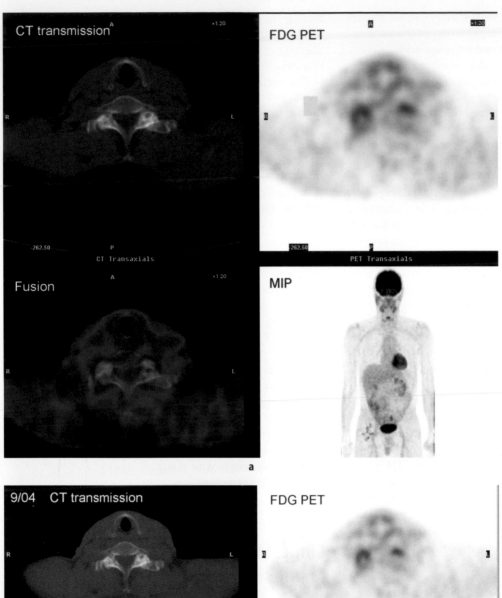

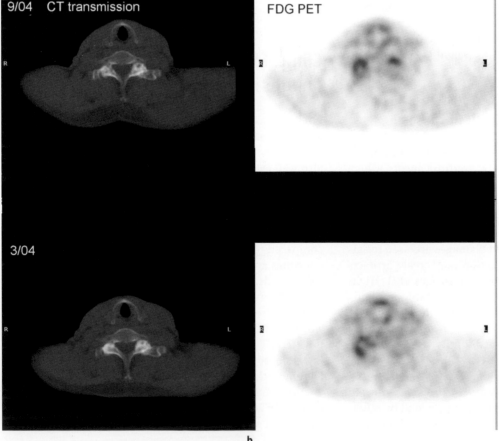

Figure 2.1. A 65-year-old man with a history of retroperitoneal lymphoma presented for follow-up 8 months after completion of chemotherapy. (**a**) The FDG-PET maximum intensity projection (MIP) image demonstrated physiologic FDG uptake in the head and neck, myocardium, and genitourinary (GU) and gastrointestinal (GI) tracts (*lower image*). There is some mild heterogeneous uptake around the right hip prosthesis and mild bilateral uptake at the basis of the neck projecting on each side of the vocal cords. Transaxial views through the low neck (*upper image*): the fusion image allowed precise localization of the foci of FDG uptake to facets joints of a low cervical vertebra, indicating probable inflammatory changes related to degenerative disease seen on the CT scan. (**b**) Comparison to the previous PET/CT scan performed 6 months earlier after completion of therapy showed no significant changes, confirming a benign etiology.

limited but conclude the following: (1) improvement of lesion detection on both CT and FDG-PET images, (2) improvement of the localization of foci of FDG uptake, resulting in better differentiation of physiologic from pathologic uptake, and (3) precise localization of the malignant foci, for example, in the skeleton versus soft tissue, or liver versus adjacent bowel or node. For example, Figure 2.1 illustrates mild abnormal FDG uptake at the base of the neck bilaterally, and the fusion image allowed precise localization to the facet joints of a lower cervical vertebra, indicating probable inflammatory changes related to degenerative disease seen on the CT scan. Figure 2.2a illustrates bilateral FDG uptake at the base of the neck corresponding to fatty tissue on the CT scan, indicating benign metabolically active adipose tissue. In Figure 2.2b, the fusion image allowed precise localization of the focus of FDG uptake in the right pelvis to the right adnexa. FDG uptake in a young woman without abnormalities on ultrasound is most likely physiologic (26).

PET/CT fusion images affect clinical management by guiding further procedures, excluding the need of further procedures, and changing both inter- and intramodality therapy. PET/CT fusion images have the potential to provide important information to guide the biopsy of a mass to active regions of the tumor and to provide better maps than CT alone to modulate field and dose of radiation therapy (27–29).

After performing 100 oncology studies using an integrated PET/CT system, the investigators at Pittsburgh University concluded that combined PET/CT images offer significant advantages, including more-accurate localization of foci of uptake, distinction of pathologic from physiologic uptake, and improvement in guiding and evaluating therapy (30, 31). A study of 204 patients at Rambam Medical Center (32) using integrated PET/CT systems concluded that the diagnostic accuracy of PET is improved in approximately 50% of patients. PET/CT fusion images improved characterization of equivocal lesions as definitely benign in 10% of sites and definitely malignant in 5% of sites. It precisely defined the anatomic location of malignant FDG uptake in 6% and led to retrospective lesion detection on PET or CT in 8%. The results of PET/CT images had an impact on the management of 14% of patients.

Antoch et al. (33) reviewed the accuracy of FDG-PET/CT imaging for tumor staging in 260 patients with solid tumors. The malignant diseases included lung ($n = 53$), head and neck ($n = 47$), gastrointestinal ($n = 44$), liver ($n = 23$), thyroid ($n = 22$), cancer of unknown primary ($n = 21$), genitourinary ($n = 18$), and other ($n = 32$). Tumor resection with T-stage verification was performed in 77 of 260 patients, operative assessment of N stage was performed in 72 of 260 patients, and M stage was pathologically verified in 57 of 260 patients. PET/CT was significantly more accurate for staging than CT alone, PET alone, and side-by-side PET plus CT. The stage was accurately determined with PET/CT in 84% of patients, with PET plus CT side by side in 76%, with CT alone in 63%, and with PET alone in 64% of patients. Integrated PET/CT had an impact on the treatment plan in 6%, 15%, and 17% of patients when compared with PET plus CT, CT alone, and PET alone, respectively.

Malignancies such as lymphoma and melanoma can arise anywhere in the body, and therefore advantages and limitations related to the regional location do not apply as much as for other malignancies. Integration of PET/CT into management of patients with lymphoma and melanoma has been recently reviewed (34). The performance of PET/CT imaging was evaluated in a group of 27 patients referred for restaging lymphoma using complete 12-month follow-up as standard of reference (35). Patient-based evaluation showed a superior sensitivity for combined FDG-PET/CT (93%) and FDG-PET and CT read side by side (93%) compared to FDG-PET alone (86%) and CT alone (78%).

It is also important to be aware of the potentially useful additional information provided by the independent interpretation by a radiologist experienced in body imaging of the non-contrasted CT portion of the study obtained with integrated PET/CT systems. An analysis of 250 patients demonstrated that these findings are uncommon (3% of patients) but could be important enough to warrant alterations in clinical management (36).

A prospective study of 98 patients compared whole-body magnetic resonance imaging (MRI) and whole-body FDG-PET/CT for staging various tumors (37). The study demonstrated that the overall TNM stage was correctly determined in 75 patients with PET/CT and 53 patients with MRI. PET/CT had a direct impact on the management on 12 patients compared to MRI, whereas MRI had an impact on the management on 2 patients compared to PET/CT. The authors suggested the use of FDG-PET/CT as a possible first-line modality for whole-body tumor staging.

Brain

Because of the symmetrical anatomy of the brain, PET cerebral emission images can be adequately be corrected for attenuation using calculated attenuation correction. In addition, the standard of care for morphologic examination of the brain is MRI with gadolinium enhancement, and cerebral PET images are typically interpreted in correlation with MRI. Therefore, many institutions do not perform a CT transmission scan in conjunction with a PET scan of the brain. At Vanderbilt University, the protocol for PET brain imaging includes a CT transmission scan as for whole-body PET/CT imaging because it is practical and time efficient. After performing approximately 200 cases in the past year, it became apparent that most anomalies on the PET emission images corresponded to an anomaly on the CT transmission image and/or helped localization on the MRI image or vice versa.

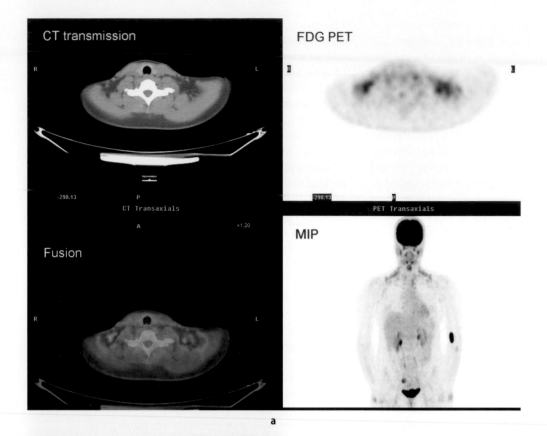

a

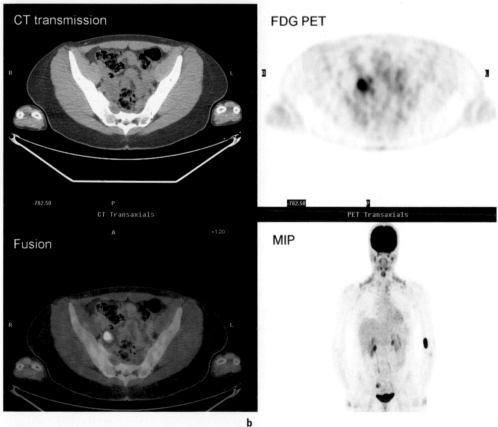

b

Figure 2.2. A 31-year-old woman with a history of breast cancer 4 years earlier treated by lumpectomy followed by chemotherapy presented with an episode of back pain. (**a**, **b**) FDG-PET maximum intensity projection image (MIP) demonstrated bilateral FDG uptake in the neck and supraclavicular and paravertebral regions in a distribution consistent with brown fat. There is also a focus of FDG uptake in the right lower pelvis. (**a**) Transaxial views through the supraclavicular regions confirm that the FDG uptake corresponds to fatty areas on the CT scan and therefore a benign etiology. (**b**) The transaxial fusion image through the pelvis demonstrated that the focus of FDG uptake corresponded to the right adenexa. A follow-up pelvic CT scan and ultrasound did not demonstrate any pathology. Adenexal FDG uptake in premenopausal females is commonly seen physiologically.

Head and Neck

Evaluation of the neck is extremely complex because of physiologic variations of uptake in muscular, lymphoid, glandular, and fatty tissue. The problem is compounded in postoperative patients because of the distorted anatomy. Therefore, interpretation of anatomic and molecular images in correlation with each other is critical. When the neck is the region of interest, the images should be acquired with the arms along the side, and it is critical that the neck be immobilized. Goerres et al. (38) have described FDG uptake in normal anatomy, benign lesions, and changes resulting from treatment.

The same group of investigators have recently reviewed the subject of PET and PET/CT in patients with head and neck cancer (39). The performance of PET/CT in head and neck cancer was compared to FDG-PET alone in a study of 68 patients with 157 foci of abnormal FDG uptake (40). The analysis was performed by consensus of two observers and accuracy was checked with histology and follow-up. With FDG-PET alone, 45 FDG-avid lesions were interpreted as benign, 71 as malignant, and 39 as equivocal. PET/CT images allowed exact localization of 100 lesions and decreased equivocal lesions by 53%. The accuracy of PET/CT was 96% compared to 90% with PET alone, and there was impact on management of 18% (12/68) patients.

Chest

In the chest, despite the limitations of relative inaccurate coregistration of PET and CT images resulting from motion of the diaphragm, the integrated PET/CT images are particularly helpful for localizing FDG-avid lymph nodes in the mediastinum or internal mammary chains and evaluating chest wall invasion. Figure 2.3 shows an example of a woman presenting with suspicion of recurrent breast cancer in the right axilla. The PET/CT images confirmed recurrence in the right axilla but also allow detection of a small metastatic lymph node in the right internal mammary chain. The transmission CT images should also be examined carefully for detection of lesions suspicious for malignancy that may not be FDG avid, such as bronchioalveolar carcinoma.

Goerres et al. (39) have recently reviewed the subject of PET and PET/CT in patients with lung cancer. Lardinois et al. (41), in a prospective study of 50 patients with suspected or proven non-small cell lung cancer compared CT and PET alone, PET and CT visually correlated and integrated PET/CT for staging. The standard of reference was histopathologic assessment of tumor stage and node stage. In this study, integrated PET/CT provided additional information in 41% of patients, and tumor staging was significantly more accurate with integrated PET/CT than with CT alone. The superior accuracy of FDG-PET/CT

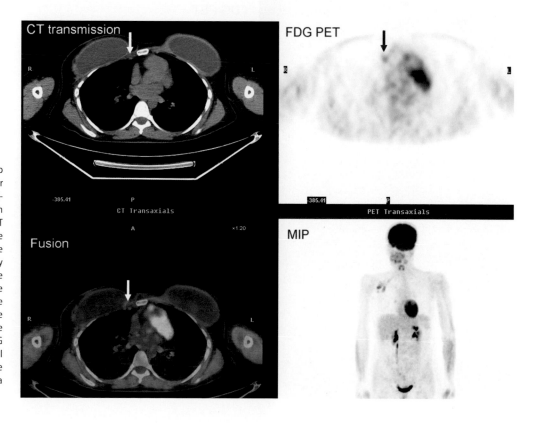

Figure 2.3. A 26-year-old woman who underwent bilateral mastectomies for breast carcinoma 4 years earlier, followed by chemotherapy, presented with right upper extremity pain. The FDG-PET maximum intensity projection image (MIP) demonstrated foci of FDG uptake in the right axilla, indicating axillary metastases. A small focus of FDG uptake is also noted projecting medially in the midchest at the level of the base of the heart. The transaxial fusion image through the focus of FDG uptake in the midchest demonstrated the focus of FDG uptake corresponded to a right internal mammary lymph node (arrows) on the CT transmission image, indicating a metastasis in another nodal basin.

imaging for TNM staging compared to CT alone and PET alone was documented in another study of 27 patients with non-small lung cancer using histopathology as the standard of reference (42). The overall TNM staging was correct in 19 patients with CT alone, in 20 patients with PET alone, and in 26 patients with PET/CT. In addition, PET/CT findings led to a treatment change for 19% of patients when compared to CT alone and 15% of patients when compared to PET alone.

Abdomen and Pelvis

PET/CT fusion images may be especially important in the abdomen and pelvis. PET images alone may be difficult to interpret owing to the absence of anatomic landmarks (other than the kidneys and bladder), the presence of nonspecific uptake in the stomach, small bowel, and colon, and urinary excretion of FDG. Hydration of the patient is an important consideration as well as emptying the bladder before the acquisition of the images. Optimal images are obtained with arms elevated above the head to avoid beam hardening artifact. Wahl (43) recently reviewed the subject of PET compared to PET/CT imaging for patient with abdominal and pelvic cancers.

A retrospective review of 45 patients with colorectal cancer referred for FDG-PET imaging using an integrated PET/CT system concluded that more definitely normal and definitely abnormal lesion characterizations (and fewer equivocal lesions) were made using the fused

PET/CT images than by using either modality alone. There was a decrease of equivocal lesions by 50% and increase in characterization by 30%. In addition, 25% more lesions could be definitively localized (44). There was no definite change in sensitivity and specificity, but correct staging increased from 78% to 89%. Figure 2.4 illustrates precise localization of a FDG-avid metastasis in the upper abdomen in a woman with a history of ovarian cancer and rising serum levels of tumor markers. A calcified left paraaortic lymph node was retrospectively identified on the corresponding CT scan.

Radiation Treatment Planning

Fusion PET/CT images from integrated systems have the potential to change significantly the field of radiation therapy. In a group of 39 patients with various extracranial body malignancies scheduled to be treated with radiation therapy, the gross target volumes (GTV) measured on CT images only were compared to the ones measured on fusion PET/CT images (28). Overall, GTV delineation was changed significantly (increased or decreased by 25% or more) in 56% (22/39) of patients based on PET/CT fusion images. In addition, the volume delineation variability between two independent oncologists decreased from a mean volume difference of 25.7 cm to 9.2 cm. Also, PET/CT changed the treatment from curative to palliative in 16% of patients because of distant metastases.

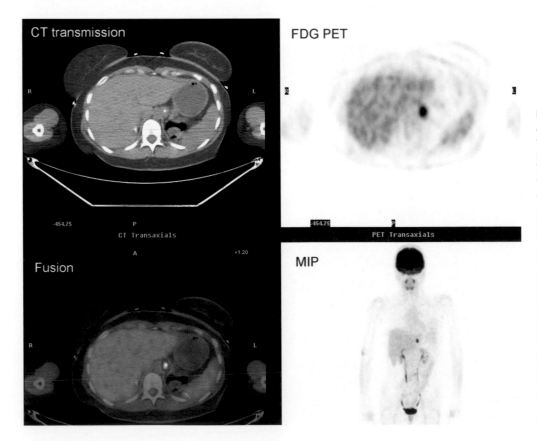

Figure 2.4. A 41-year-old woman with a history of stage IV ovarian cancer treated 2 years earlier presented with rising serum levels of tumor markers. The FDG-PET maximum intensity projection image (MIP) demonstrated a focus of FDG uptake in the upper abdomen near the gastroesophageal junction. The transaxial fusion image through the focus of FDG uptake in the upper abdomen demonstrated the focus of FDG uptake corresponded to a left paraaortic lymph node with calcifications located between the aorta and gastroesophageal junction on the CT transmission image, indicating a metastasis. The lesion was treated with radiation therapy. On a follow-up PET/CT scan, the lesion had decreased in size on the CT transmission images, and the degree of FDG uptake had decreased, but there was new pleural effusions with FDG uptake and cytology revealed adenocarcinoma.

Most of the work regarding the role of PET and PET/CT imaging to guide radiation therapy has involved patients with non-small cell lung cancer and has been reviewed by Bradley et al. (29). The potential benefit of incorporating PET data in the into the conventional radiation therapy treatment planning was documented in a study of 11 patients with non-small cell lung cancer (27). Patients were immobilized in the treatment position for acquisition of both CT and FDG-PET images. For all patients there was a change of the planned target volume outline based on CT images compared to PET/CT fused images. In 7 of 11 patients, the planned target volume was increased by an average of 19% to incorporate nodal disease. In the other 4 patients, the target planned volume was decreased by an average of 18% by excluding atelectasis and trimming the target volume to avoid delivering higher radiation doses to nearby spinal cord or heart.

Cardiac Imaging

Advanced cardiac applications are possible with integration of 16-slice CT and PET systems. Noninvasive visualization of the coronary tree has been investigated for a long time. MR angiography was promising in that regard, but in a multicenter trial only 84% of the vessel segments could be evaluated and the evaluation was limited to the proximal 3 to 5 cm of the major coronary arterics. In addition, the examination time was more than 1 h, which is a limitation in clinical practice (45). The multislice, CT technology allows measurement of calcium scoring and the performance of CT coronary angiography with visualization of the major branches of the coronary tree and definition of luminal narrowing (46, 47). The cardiac PET/CT technology can potentially allow evaluation in one imaging setting of (1) calcium scoring, (2) coronary artery anatomy with contrast-enhanced CT coronary angiography, (3) rest/stress myocardial perfusion with the possibility of quantitative measurement of coronary blood flow and blood flow reserve and localization of the hypo-perfused regions to specific coronary arteries, (4) evaluation of myocardial viability with FDG and precise anatomic localization, and (5) localization of atheromatous plaques with inflammation using FDG (48).

Preliminary data have demonstrated the feasibility of obtaining fusion images with such a PET/CT system superimposing perfusion ^{13}N-ammonia images to CT angiographic images displaying coronary anatomy (49).

With the introduction of the 16-slice integrated PET/CT systems (and CT units with higher performance in the future), PET/CT technology may become the long-awaited "one-stop shop" in cardiology. For example, patients could be just referred for myocardial perfusion with CT attenuation correction for the diagnosis or prognostic implications of coronary artery disease (CAD). Low-risk patients could be referred for calcium scoring and PET myocardial perfusion, and high-risk patients for PET myocardial perfusion and CT coronary angiography. The group of patients with poor left ventricular (LV) function evaluated for viability could be referred for PET perfusion/viability and CT coronary angiography. This subject is further addressed in a subsequent chapter.

Conclusion

The diagnostic implications of integrated PET/CT imaging include improved detection of lesions on both the CT and FDG-PET images, better differentiation of physiologic from pathologic foci of metabolism, and better localization of the pathologic foci. This new powerful technology provides more-accurate interpretation of both CT and FDG-PET images and therefore more-optimal patient care. PET/CT fusion images affect clinical management by guiding further procedures (biopsy, surgery, radiation therapy), excluding the need for additional procedures, and changing both inter- and intramodality therapy.

The combined approach of X-ray attenuation correction and image fusion with FDG imaging is a new and powerful diagnostic tool for nuclear medicine imaging, radiation therapy, and surgical planning. The applications for this technology, that have emerged with FDG as the radiotracer, will expand even further with the array of new promising PET tracers and in the field of cardiology.

References

1. Delbeke D, Martin WH, Patton JA, Sandler MP, editors. Practical FDG Imaging: A Teaching File. New York: Springer Verlag, 2002.
2. Cook GJR, Fogelman I, Maisey MN. Normal physiological and benign pathological variants of 18-fluoro-2-deoxyglucose positron emission tomography scanning: potential for error in interpretation. Semin Nucl Med 1996;26:308–314.
3. Engel H, Steinert H, Buck A, Berthold T, Boni RAH, von Schulthess GK. Whole body PET: physiological and artifactual fluorodeoxyglucose accumulations. J Nucl Med 1996;37:441–446.
4. Hutton BF, Braun M. Software for image registration: algorithms, accuracy, efficacy. Semin Nucl Med 2003;33(3):180–192.
5. Slomka PJ. Software approach to merging molecular with anatomic information. J Nucl Med 2004;45(suppl):36S–45S.
6. Townsend DW, Beyer T, Blodgett TM. PET/CT scanners: a hardware approach to image fusion. Semin Nucl Med 2003; 33(3):193–204.
7. Townsend DW, Carney JPJ, Yap JT, Hall NC. PET/CT today and tomorrow. J Nucl Med 2004;45(suppl):4S–14S.
8. Beyer T, Townsend DW, Brun T, et al. A combined PET/CT scanner for clinical oncology. J Nucl Med 2000;41:1369–1379.
9. Di Carli M. Advances in positron emission tomography. J Nucl Cardiol 2004;11(6):719–732.
10. Beyer T, Antoch G, Muller S, Egelhof T, Freudenberg LS, Debatin J, Bockisch A. Acquisition protocol considerations for combined PET/CT imaging. J Nucl Med 2004;45(suppl):25S–35S.
11. Kamel E, Hany TF, Burger, et al. CT vs. ^{68}Ge attenuation correction in a combined PET/CT system: evaluation of the effect of lowering the CT current. Eur J Nucl Med Mol Imaging 2002;29:346–350.

12. Cohade C, Wahl RL. Applications of positron emission tomography/computed tomography image fusion in clinical positron emission tomography: clinical use, interpretation methods, diagnostic improvements. Semin Nucl Med 2003;33(3):228–237.

13. Antoch G, Freudenberg LS, Beyer T, Bockisch A, Debatin JF. To enhance or not to enhance? ^{18}F-FDG andCT contrast agents in dual modality ^{18}F-FDG PET/CT. J Nucl Med 2004;45(suppl):56S–65S.

14. Antoch G, Freudenberg LS, Egeldorf T, et al. Focal tracer uptake: a potential artifact in contrasted-enhanced dual-modality PET/CT scans. J Nucl Med 2002;43:1339–1342.

15. Cohade C, Osman M, Nakamoto Y, et al. Initial experience with oral contrast in PET/CT : Phantom and clinical studies. J Nucl Med 2003;44:412–416.

16. Goerres GW, Hany TF, Kamel E, et al. Head and neck imaging with PET/CT: artifacts from dental metallic implants. Eur J Nucl Med Mol Imaging. 2002;29:367–370.

17. Dizendorf E, Hany TF, Buck A, et al. Cause and magnitude of the error induced by oral contrast agent in CT-based attenuation correction of PET emission studies. J Nucl Med 2003;44:732–738.

18. Goerres GW, Burger C, Schwitter MW, et al. PET/CT of the abdomen: optimizing the patient breathing pattern. Eur Radiol 2003;13:734–739.

19. Goerres GW, Burger C, Kamel E, et al. Respiration-induced attenuation artifact at PET/CT: technical considerations. Radiology 2003;226:906–910.

20. Cohade C, Osman M, Marshall LN, Wahl RN. PET/CT: accuracy of PET and CT spatial registration of lung lesions. Eur J Nucl Med Mol Imaging 2003;30:721–726.

21. Osman MM, Cohade C, Nakamoto Y, Wahl RH. Clinically significant inaccurate localization of lesions with PET/CT: frequency in 300 patients. J Nucl Med 2003;44:240–243.

22. Antoch G, Freudenberg LS, Beyer T, et al. To enhance or not to enhance? (18)F-FDG and CT contrast agents in dual-modality (18)F-FDG PET/CT. J Nucl Med 2004;45(I suppl):56S–65S.

23. Beyer T, Antoch G, Muller S, et al. Acquisition protocol considerations for combined PET/CT imaging. J Nucl Med 2004;45 (I suppl):25S–35S.

24. Schaefer NG, Hany TF, Taverna C, Seifert B, Stumpe KD, von Schulthess GK, Goerres GW. Non-Hodgkin lymphoma and Hodgkin disease: coregistered FDG PET and CT at staging and restaging. Do we need contrast-enhanced CT? Radiology 2004;232 (3):823–829.

25. Gambhir SS, Czernin J, Schimmer J, Silverman D, Coleman RE, Phelps ME. A tabulated summary of the FDG PET literature. J Nucl Med 2001;42(suppl):1S–93S.

26. Lerman H, Metser U, Grisaru D, Fishman A, Lievhitz G, Even-Sapir E. Normal and abnormal ^{18}F-FDG endometrial and ovarian uptake in pre- and postmenopausal patients: assessment by PET/CT. J Nucl Med 2004;45:266–271.

27. Erdi YE, Rosenzweig K, Erdi AK, et al. Radiotherapy treatment planning for patients with non-small cell lung cancer using positron emission tomography (PET). Radiother Oncol 2002;62:51–60.

28. Ciernik IF, Dizendorf E, Ciernik IF, Baumert B, et al. Radiation treatment planning with an integrated positron emission and computer tomography (PET/CT): a feasibility study. Int J Radiat Biol Phys 2003;57:853–863.

29. Bradley JD, Perez CA, Dehdashti F, Siegel BA. Implementing biological target volumes in radiation treatment planning for non-small cell lung cancer. J Nucl Med 2004;45(suppl I):96S–101S.

30. Charron M, Beyer T, Bohnen NN, et al. Image analysis in patients with cancer studied with a combined PET and CT scanner. Clin Nucl Med 2000;25:905–910.

31. Martinelli M, Townsend D, Meltzer C, Villemagne VV. Survey of results of whole body imaging using the PET/CT at the University of Pittsburgh Medical Center PET facility. Clin Posit Imaging 2000;3:161.

32. Bar-Shalom R, Yefremov N, Guralnik L, et al. Clinical performance of PET/CT in the evaluation of cancer: additional value for diagnostic imaging and patient management. J Nucl Med 2003; 44:1200–1209.

33. Antoch G, Saoudi N, Kuehl H, Dahmen G, Mueller SP, Beyer T, et al.. Accuracy of whole-body dual-modality fluorine-18-2-fluoro-2-deoxy-D-glucose positron emission tomography and computed tomography (FDG-PET/CT) for tumor staging in solid tumors: comparison with CT and PET. J Clin Oncol 2004;22(21):4357–4368.

34. Schoder H, Larson SM, Yeung HWD. PET/CT in oncology: integration into clinical management of lymphoma, melanoma, and gastrointestinal malignancies. J Nucl Med 2004;45(suppl):72S–81S.

35. Freudenberg LS, Antoch G, Schutt P, et al. FDG PET/CT in restaging of patients with lymphoma. Eur J Nucl Med Mol Imaging 2003;30:682–688.

36. Osman MM, Cohade C, Fishman EK, Wahl RL. Clinically significant incidental findings on the unenhanced CT portion of PET/CT studies: frequency in 250 patients. J Nucl Med 2005;46(8):1352–1355.

37. Barkhausen J, Dahmen G, Bockisch A, et al. Whole-body dual-modality PET/CT and whole body MRI for tumor staging in oncology. JAMA 2003;290(24):3199–3206.

38. Goerres GW, von Schultness GK, Hany TF. Positron emission tomography and PET CT of the head and neck: FDG uptake in normal anatomy, in benign lesions, and in changes resulting from treatment. Am J Roentgenol 2002;179:1337–1343.

39. Goerres GW, von Schultness GK, Steinert HC. Why most PET of lung and head and neck cancer will be PET/CT. J Nucl Med 2004;45(suppl 1):66S–71S.

40. Schoder H, Yeung HW, Gonen M, Kraus D, Larson SM. Head and neck cancer: clinical usefulness and accuracy of PET/CT image fusion. Radiology 2004;231(1):65–72.

41. Lardinois D, Weder W, Hany TF, et al. Staging of non-small cell lung cancer with integrated positron-emission tomography and compared tomography. N Engl J Med 2003;348:2500–2507.

42. Antoch G, Stattaus J, Nemat AT, et al. Non-small cell lung cancer: dual modality PET/CT in preoperative staging. Radiology 2003;229(2):526–533.

43. Wahl RL. Why nearly all PET of abdominal and pelvic cancers will be performed as PET/CT. J Nucl Med 2004;45(suppl I):82S–95S.

44. Cohade C, Osman M, Leal J, Wahl RL. Direct comparison of FDG PET and PET/CT in patients with colorectal cancer. J Nucl Med 2003;44:1797–1803.

45. Kim WY, Daniass PG, Stuber M, et al. Coronary magnetic resonance angiography for detection of coronary stenoses. N Engl J Med 2001;345:1863–1869.

46. Schoepf UJ, Becker CR, Ohnesorge BM, Yucel EK. CT of coronary artery disease. Radiology 2004;232(1):18–37.

47. Schoenhagen P, Halliburton SS, Stillman AE, Kuzmiak SA, Nissen SE, Tuzcu EM, White RD. Noninvasive imaging of coronary arteries: current and future role of multidetector row CT. Radiology 2004; 232(1):7–17.

48. Rudd JH, Warburton EA, Fryer TD, et al. Imaging atherosclerotic plaque inflammation with [^{18}F]-fluorodeoxyglucose positron emission tomography. Circulation 2002;105:2708–2711.

49. Namdar M, Hany TF, Koepfli P, Siegrist PT, Burger C, Wyss CA, Luscher TF, von Schulthess GK, Kaufmann PA. Integrated PET/CT for the assessment of coronary artery disease: a feasibility study. J Nucl Med 2005;46(6):930–935.

3

The Technologist's Perspective

Bernadette F. Cronin

Introduction

From the technologist's point of view, positron emission tomography (PET) combines the interest derived from the three-dimensional imaging of X-ray computed tomography (CT) and magnetic resonance imaging (MRI) with the functional and physiologic information of nuclear medicine. Until recently PET imaging was performed on "PET-only" scanners, and any direct comparison between the PET image and an anatomic one was either done by eye or required sophisticated hardware and software as well as extra manpower to create a registered image. Now all the key manufacturers have developed and are marketing combined PET/CT scanners that allow patients to have both a PET scan and a CT scan without leaving the scanning couch. This development creates an exciting imaging modality that offers the technologist the chance to develop a new range of skills.

Although clinical PET has been developing over the past 12 to 15 years, it is still relatively new, and the high cost of introducing such a service has limited access for staff wishing to enter the field. Thus, few staff entering the field now have previous experience. However, technologists with experience in other modalities bring with them useful knowledge and, although PET in most cases is developing as part of nuclear medicine, recruitment need not be confined to nuclear medicine technologists alone. In the United Kingdom, training for PET technologists is performed onsite with staff learning and gaining experience while working full time. Most recognized undergraduate and postgraduate courses cover PET in a limited capacity, providing theoretical information on a subject that is intensely practical. As with all new techniques there will be, for a while, a discrepancy between the number of centers opening around the country and the availability of trained staff. It is important for all concerned, and for the viability of PET itself, that training is formalized to create high national standards that can be easily monitored and maintained.

A common view is to assume that a PET scanner can be sited in an existing nuclear medicine department and that, once commissioned, will function with little extra input. Although there are obvious similarities between PET and nuclear medicine, there are also quite subtle differences, and it is not unusual for experienced nuclear medicine staff to find PET quite bewildering. The advent of PET/CT adds an extra dimension because many nuclear medicine technologists do not have any CT training or background. This factor also must be addressed in a structured way to ensure that the technique develops optimally. Currently, it is probably the least predictable imaging modality, and staff working in a PET facility need to possess certain characteristics if the unit is to be successful. PET is multidisciplinary, requiring input from various professional groups, and it is important that all these groups work together toward the same end. Teamwork is an essential component of a successful PET unit and, without this, departments can easily flounder. More people are gaining the required expertise to work in these units; however, there are still few people with the necessary expertise and they may not always be available to staff newly emerging departments. If departments are unable to recruit people with existing PET experience, it is advisable to send staff to comparable PET centers to obtain initial training.

Setting Up a PET Service

When setting up a PET service, several factors must be considered. First, and most important, what is the unit to be used for? If the only requirement is for clinical oncology work, a stand-alone scanner with the tracer being supplied by a remote site may well be sufficient. For the technologist, this option is in some ways the least attractive, as it always leaves the unit vulnerable to problems over which there is no control and provides the least diverse workload. At present, most of the work is done using fluorine-18-labeled tracers (mainly 2-[^{18}F]-fluoro-2-

deoxy-D-glucose) (FDG), and although the half-life is reasonably long (109.7 min) it does not provide much margin for error and thus patient scheduling can present added difficulties. If you are involved with a unit that has both scanner(s) and a cyclotron with chemistry facilities, the opportunities expand enormously. Scheduling, although not necessarily easier, can be more flexible, enabling better use of the equipment at your disposal. In addition, the variety of scans that can be performed increases, allowing a full clinical oncology, neurology, and cardiology service as well as the flexibility required for research work. Another factor to consider is the service you will offer if you are purchasing a PET/CT scanner. All commercially available PET/CT scanners combine state-of-the-art PET scanners with diagnostic CT scanners. The specification of the CT component varies between manufacturers, but all are capable of high-quality diagnostic work. Opinions vary as to whether the CT scanner should be used only for attenuation correction and image fusion or whether both the PET scan and the diagnostic CT scan can be performed on the one machine during a single hospital/clinic visit. The concept of "one-stop shops" has also been considered wherein patients receive their PET, CT, and radiotherapy planning in a single visit.

Staffing Requirements and Training

Staffing levels reflect the aims of the unit. If the intention is to have a stand-alone scanner with no onsite cyclotron and chemistry facilities, then, assuming a throughput of about 800 patients per annum, the department needs a minimum of two technologists. Two are needed to ensure that there will always be cover for annual leave and sickness and also to share the radiation dose. This number of staff assumes some additional scientific and clerical support staff. However, with a scanner plus cyclotron and chemistry facilities, the annual patient throughput is likely to increase because of the greater availability of radiopharmaceuticals for clinical studies and, possibly, research work. Assuming an annual workload of approximately 1,200 patients, at least three technologists are needed, with a corresponding increase in support staff. Provision of FDG for 8:00 A.M. injection, allowing scanning to start at 9:00 A.M., requires cyclotron and chemistry staff to begin work very early. Obvious revenue consequences are the need for out-of-hours or enhanced payment, which may make recruitment difficult. Thus, in many centers FDG may not be available for injection much before 9:30–10:00 A.M., with the first scan not starting until nearly 11:00 A.M. To maximize the resources available it will probably be necessary to extend the working day into the early evening by working a split-shift system.

Once the PET technologists have been selected, consideration must be given to how they will be trained. There is little point in sending a technologist destined for a clinical PET center to train in a center that undertakes research work only. The two units operate in very different ways with different priorities and likely are using a different range of tracers. To get an in-depth working knowledge of clinical PET, technologists should spend at least 4 and preferably 8 weeks at their "training" center. The amount of training required will depend on the existing knowledge of the technologist. It must be remembered that, to function successfully in a PET unit, the technologist should be experienced in patient care, handling of unsealed sources, intravenous cannulation, and administration of radiopharmaceuticals as well as the acquisition and subsequent reconstruction of the data.

A good basic understanding of physics and computing will also help, particularly when it comes to troubleshooting problems. CT training is also likely to be an advantage, and several short courses are available that provide a good grounding in basic CT. If the work is likely to include full diagnostic CT, there are staffing implications with regards to state registration and the operation of CT scanners that cannot be ignored.

Planning the PET Service

Clinical PET is divided into three main categories comprising oncology, neurology, and cardiology. Most clinical PET centers have found that the workload is split between these three areas in a similar way, with oncology taking about 80% to 90% and the balance being shared between the other two disciplines.

When setting up a clinical PET service it is important to consider the types of scans to be offered. When deciding which protocols will be used, the most important consideration is what question is being asked of PET. This requirement may not always be clear in the initial referral, and it is important that it is established before the study so that PET is not used inappropriately. Other things to consider are whether quantification [for example, a semi-quantitative standardized uptake value (SUV)] (1, 2) of the data will be required as this can only be performed on attenuation-corrected studies, which may not automatically be performed if working on a PET-only scanner. It is also important to know whether the patient is a new patient or is attending for a PET study as part of their follow-up. If patients are undergoing a follow-up scan, it is essential to know what treatment they have had and, critically, when this treatment finished or was last given. Following the completion of chemotherapy, approximately 4 to 6 weeks should elapse before the patient is scanned, and after radiotherapy it is ideal to wait longer, up to 3 to 6 months, although clinically this may not be practical. If patients have their PET scan midway through a chemotherapy regimen, the aim should always be to scan the patient as close to the next cycle of chemotherapy as possible, allowing at least 10 days to elapse since the previous cycle. Use should also be made of other diagnostics tests that the patient may have undergone, and so for the PET center it is valuable to have details about previous scans [CT, magnetic resonance imaging (MRI), etc.] and preferable to have the scan reports or images available before the appointment is made.

For PET, patient compliance must also be considered. Patients are accustomed to the idea that diagnostic tests on the whole are becoming quicker. However, with PET, the scanning times are significantly longer than other scanning techniques. Scans can take from 20 min up to 3 h for some of the more-complex studies, and it is important to establish whether the patient is able to tolerate this time period. The design of PET systems, although not so confined as MRI, can mean that patients who are claustrophobic may not be able to complete the scan without help and may need to be given some mild sedation to undergo this type of investigation. Recently developed PET/CT dual-modality systems will exacerbate this problem even more because of their increased axial length. Some patients, particularly younger subjects, are not be able to remain still for the length of time required for the scan and may well need heavier sedation or perhaps a general anesthetic.

The other key area that must be considered when setting up a PET service is which tracers are available. In centers with a cyclotron on site, a full range of clinical radiopharmaceuticals should be available. However, when operating a PET scanner at a site remote from a cyclotron, there will be much greater limitation as to available tracers (essentially, ^{18}F-labeled tracers or those labeled with longer-lived radioisotopes), and that in turn will limit the types of studies possible. The half-lives of these tracers can be very short, which, again, will influence the types of scan possible as well as the way the work is scheduled. Production times for the various tracers range from just a few minutes up to 2 or 3 h and scans must be booked accordingly. As with all imaging units, it is important to ensure that the best and most efficient use of the equipment and personnel available is made. PET scans are costly both financially and in terms of the time required, and a single scanner may only be able to accommodate 7 to 10 scans per day. The schedule should be arranged to ensure the scanner is in continuous use throughout the day to maximize throughput. In the United Kingdom, there are currently few centers producing FDG for distribution, and those centers that do so cannot produce it in large quantities. It is essential that tracer is used in a way that ensures that little, if any, is wasted, and this requires much thought on the part of the people booking the scans.

Information that should be given to the patient includes the time, date, and location where the examination will take place, with a clear explanation of what the scan involves and any special dietary requirements including special instructions for diabetics. An information booklet may be helpful but should be tailored to specific scan types and the working practice of individual departments. It is a good idea to ask the patient to confirm their inten-

tion to attend for their appointment. A patient not attending at the prearranged time can cause a huge disruption and waste of both tracer and scanner time, which is difficult to fill at short notice.

Patient Preparation and Scanning Protocols

Different centers have different methods for preparing and performing the tests, and these variations are determined by several factors, including the type of PET system available. However, regardless of the type of PET camera being used, there are similarities in technique with the greatest overlap in the area of patient preparation and management of the patient once they are in the department to obtain the best diagnostic data.

PET Scanning in Oncology

Today, FDG-PET scanning is regarded as useful in many oncologic conditions. The most frequent applications are in staging of disease, assessment and monitoring of treatment response, and evaluation of tumor recurrence, particularly when morphologic imaging techniques are equivocal or difficult to assess for technical reasons.

Patient preparation for FDG PET scans in oncology is fairly simple. Patient referrals are divided into two categories, those who are insulin-dependent diabetics (IDD) and all others. Insulin-dependent diabetics are asked to drink plenty of water in the 6 h before their scan appointment, but they are not asked to fast because it is both inappropriate and unnecessary to disrupt their blood glucose levels. All other patients are asked to fast for at least 6 h before their scan but are again encouraged to drink plenty of water. On arrival in the department, the patient's personal details should be checked and a clear explanation of what the scan involves should be given. It is important to give the patient the opportunity to ask questions at this stage to ensure they are completely relaxed. For this reason, the appointment time should include a preinjection period of about 15 min. For FDG-PET scanning it is advantageous to have the patient lying down and comfortably warm for both the injection and the uptake period as this should aid relaxation and reduce unwanted muscle and brown fat uptake. To obtain the best results from the FDG-PET scan, certain key areas are important. The site of the body chosen for injection of the FDG is important. Any extravasation at the injection site is unsatisfactory. A small amount of tissued radiotracer at the injection site will result in a local radiation dose caused primarily by the positrons themselves due to their short pathlength and high linear energy transfer. Also, the extravasated tracer will be cleared from the injection site via the lymphatic system and may result in

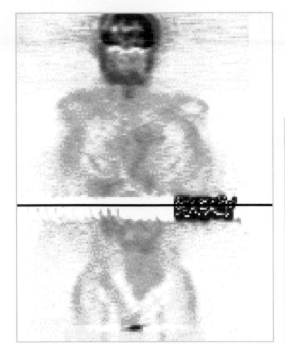

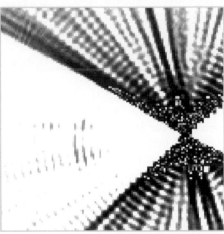

Figure 3.1. Reconstruction artifacts caused by extravasated tracer at the injection site can leave the final image difficult or even possible to interpret. 📖

uptake in more-proximal lymph nodes that could be confused with uptake from lymph node disease. Even small amounts of extravasated tracer can cause devastating artifacts on the three-dimensional (3D) reconstruction of the data and, in extremis, can render the scan nondiagnostic if it is adjacent to an area of interest (see example in Figure 3.1). Iterative reconstruction methods help by reducing these artefacts but may not remove them altogether.

Many of the patients attending for these scans have already undergone chemotherapy regimens so that their veins are extremely difficult to access. It is therefore essential that the technologist giving the injections is well trained and well practiced in venepuncture. A good, precise injection technique is essential with the emphasis on accuracy. Unfortunately, there will occasionally be mistakes, but with care it is possible to ensure these mistakes do not make the scan completely worthless. It is always advisable to keep the injection site as far away as possible from the area of interest. If patients are to be scanned with their arms down by their sides and the abdomen is under investigation, it is best to avoid the antecubital fossae and likewise the hands or wrists when the pelvic area is significant. However, with PET/CT scanning it is much more likely that patients will be scanned with their arms raised up above their head and so this is less likely to cause a problem. For many patients the arms may not be appropriate for the injection; for example, patients with a history of carcinoma of the breast, patients with bilateral axillary, neck, or supraclavicular disease, and patients whose primary lesion is on their arm. In any of these circumstances it is advisable to inject the patient in the foot so that if a problem occurs it will not affect the scan so much as to cause confusion. A note should always be made of the site of the injection for the reporting doctor's reference. A note should also be made if an unsuccessful attempt at venipuncture has been made before the successful one as leakage of injected tracer can occasionally be seen at this initial site. If extravasation of tracer is suspected, a quick test acquisition at that site should be performed to establish this as it is not helpful to remove the injection site from the field of view without establishing whether it is necessary. If it is, then remedial action can be taken; for example, lifting the arm above the patient's head and out of the field of view. With PET/CT it is likely that patients will be routinely imaged with their arms raised above their head to prevent beam hardening artifacts in the CT image. However, an image where the injection site (the arm, for example) is not seen but that does include axillary uptake will be difficult to assess unless the reporting clinician can be sure that the injection was without incident. For this reason, a quick acquisition of the injection site will avoid any subsequent confusion.

Normal physiologic muscle and brown fat uptake of FDG can also give rise to confusion when reporting PET scans (Figure 3.2) (3, 4). The causes of this increased uptake are not fully understood; however, it is clear that physical exertion and cold inevitably give rise to increased levels of FDG in these areas. Stress and nervous tension also play their part.

Unfortunately, it is not always clear which patients will produce high levels of muscle or brown fat uptake, and the calmest of patients can produce a very "tense"-looking scan. When uptake of this type occurs it is only the distribution that may give a clue as to the origins of the pattern of uptake, and it is almost impossible to distinguish this physiologic response from any underlying pathologic cause. To reduce this uptake, it is suggested that patients be given 5 to 10 mg diazepam orally 1 h before FDG injection. Clearly, this will increase the time

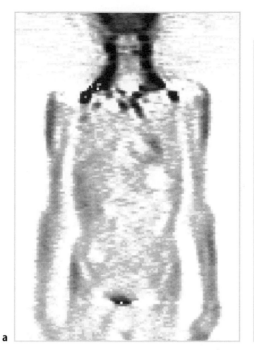

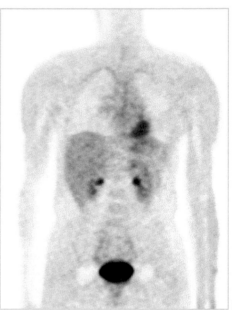

Figure 3.2. FDG uptake in tense muscle can mask or simulate underlying pathology (**a**). The use of diazepam before FDG injection can significantly reduce this effect (**b**). 📖

a

b

they remain in the department, as well as influencing their homeward journey, because they should not drive following this medication. Also, this step cannot be done at short notice as patients must arrive for their appointment an hour earlier than usual and will require an area where they can rest comfortably following administration of diazepam. Although the dosage suggested is relatively low, patients' reactions range from barely negligible to sleeping throughout most of their visit to the department. The unpredictability of this reaction means it is generally advisable to suggest that the patient bring someone with them to ensure that they are accompanied when they leave the department after the scan. A low ambient room temperature can also have an effect by increasing the incidence of physiologic uptake into brown fat, so it is also important that, on arrival, patients are injected and allowed to rest in a warm comfortable area. Although the likelihood of physiologically increased muscle/brown fat uptake is difficult to predict, for certain categories of patients diazepam should be given prophylactically. Obviously, any patient who has previously demonstrated physiologic uptake with FDG should be given diazepam on all subsequent visits. For patients attending their first PET scan, those with lymphoma for whom the axillae, neck, and supraclavicular area are of particular interest may well be given diazepam, and patients with carcinoma of the breast are also likely to be considered. Adolescents and young children also show significantly increased amounts of physiologic muscle/brown fat uptake and they too are routinely given diazepam, although in this group of patients this step is less successful than with adults. In addition to diazepam, every effort should be made to ensure that the total environment and experience are as relaxing as possible for the patient and that they feel comfortable and well informed about the proce-

dure. It is difficult to evaluate the true effect of oral diazepam because for each patient there is no control study. However, it does appear to reduce the incidence of this type of normal uptake and is therefore used frequently by many centers.

Uptake of FDG into normal myocardium is both unpredictable and difficult to manage. When investigating the thorax of a patient, particularly when looking for small-volume disease close to the myocardial wall, high levels of myocardial uptake can cause such severe artifacts that this can be almost impossible to assess, especially if the images are reconstructed with filtered back-projection (Figure 3.3).

Iterative reconstruction methods help decrease the magnitude of the artifact, but ideally myocardial uptake of the FDG should be minimized when necessary. The physiology of uptake of FDG into the myocardium is

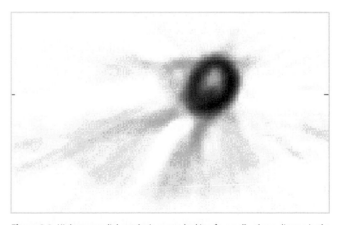

Figure 3.3. High myocardial uptake in a scan looking for small-volume disease in the chest can render the scan uninterpretable. 📖

complex and there are, as yet, no straightforward methods of reducing it. One suggestion is that by increasing the free fatty acid levels in the patient the heart may be persuaded to utilize this for its primary energy source and ignore the available glucose and FDG. One way of doing this is to ensure that the patients have a high caffeine intake before and immediately after the FDG injection, which can be achieved simply by encouraging the patient to drink unsweetened black coffee instead of water in the hours leading up to the scan and during the uptake period or, more palatably, a diet cola drink, as this contains no sugar but has high levels of caffeine.

In contrast to glucose, FDG is excreted through the kidneys into the urine, and artifacts can be produced if the radioactivity concentration in the urine is high. These artifacts can be seen around the renal pelvis and the bladder and can make it very difficult to assess these areas with confidence. In some centers, for patients in whom the upper abdomen or pelvis is in question, it is common practice to give 20 mg intravenous furosemide at the same time as the FDG injection. The effect of the furosemide is to dilute the radioactive concentration within the urine and thus make the images easier to interpret. The furosemide should be injected slowly, and patients should be encouraged to drink plenty of water to ensure they do not become dehydrated. Patients should be given an opportunity to void immediately before the start of the FDG scan and, if the pelvis is to be included in the scan, then the patient should be scanned from the bladder up. Obviously, it is not easy for patients to be given a diuretic and then have limited access to a toilet, but in the main, with a good explanation, patients tolerate this quite well. Again, iterative reconstruction techniques have improved image reconstruction and reduced the severity of artifacts caused by high levels of activity in the urine, and the advent of PET/CT scanners has allowed focal uptake in the ureters to be accurately assessed at the reporting stage, reducing the overall need for diuretics.

Finally, in patients in whom the laryngeal and neck area is being investigated, normal physiologic muscle uptake into the laryngeal muscles can cause confusion. Patients who speak before, during, and after injection are likely to demonstrate this type of uptake, whereas patients who are silent for this period do not (5). For this reason it is recommended that when this area is of clinical significance, the patient should be asked to remain silent for about 20 min before injection, during injection, and for the majority of the uptake period (the first half-hour being the most important time after injection). Although this may not be easy, most patients, if they understand why it is necessary, are sufficiently well motivated to comply.

PET Scanning in Neurology

For clinical PET in neurology, the most frequently used tracer is FDG. It features in the general workup for patients who have intractable epilepsy for whom the next option would be surgery, and PET is one of the many investigations these patients undergo to confirm the exact site of the epileptogenic focus. PET is also used in the investigation of primary brain tumors, mainly in patients who have had tumors previously treated by surgery, radiotherapy, or a combination of both. In these patients, standard anatomic techniques, for example, CT and MRI, are often difficult to interpret because of the anatomic disruption caused by the treatment. FDG-PET scanning may also be helpful when investigating patients with dementia-like symptoms because it can provide a differential diagnosis for the referring clinician.

When investigating partial epilepsy, [^{11}C]-flumazenil may also be used. This radiopharmaceutical shows the distribution of benzodiazepine receptors, and it can be useful in more accurately localizing the epileptogenic focus (6) than FDG alone. Similarly, in the investigation of primary brain tumors, the use of FDG alone may not always be sufficient to answer the clinical question. Normal cortical brain has a high uptake of FDG and so makes visualization of tumor tissue over and above this background level quite difficult. The grade of the brain tumor is correlated with the degree of uptake into the tumor (7), with low-grade tumors having a lower than normal uptake, intermediate-grade tumors having a similar uptake to normal brain, and high-grade tumors having a higher than normal uptake. For this reason, it is common for patients who are being assessed for recurrent tumor to have a [^{11}C]-L-methionine scan in addition to their FDG scan. Methionine is an amino acid that demonstrates protein synthesis rates, so, while the FDG gives information about the grade of the tumor, the methionine will provide information about the extent of any tumor present. To use any ^{11}C-labeled tracer obviously requires that the scanner be sited close to the cyclotron and radiochemistry unit where the tracer will be produced, as the 20-min half-life along with the low quantities currently produced precludes transport to a distant site.

The exact protocol for brain scanning will vary from center to center but again there are certain key points that can be taken from one site to another. For all these scans, patients may have to remain still for quite a significant period of time. When a patient is suffering from epilepsy, particularly if they are very young, their risk of having a seizure during the scan may be quite high. To ensure the minimum amount of disruption to the total study, it is advisable to acquire the data as a series of sequential dynamic time frames rather than as one long, single, static acquisition. Thus, if the patient does move during the scan it is possible to exclude that frame from the total scan data. If the patient is clearly marked at various points corresponding to the positioning laser lights, they can be repositioned so that only the affected frame is lost. However, any repositioning needs to be done with extreme accuracy to ensure coalignment of the transmission data for attenuation correction, which is often recorded before the start of the emission scan. Patient movement is a particularly important issue when using the shorter-lived ^{11}C tracers. If a patient moves during a single static acquisition, not only is the scan going to be corrupted, but the short half-life of the tracer makes it very difficult to repeat the scan as the resultant scan quality will be very poor. An alternative that is becoming increasingly popular on PET scanners that can operate in 3D mode is to acquire these as high-count, short-duration 3D scans only. Emission scan times in 3D for a brain may be of the order of only 5 to 10 min.

Preparation of patients for brain scans is again concerned with ensuring that the injected FDG does not have to compete with high levels of normal glucose in the patient. So, as for oncology scans, the patients are fasted and asked to drink water only in the few hours preceding the study, but because of the high avidity of uptake in the brain this fasting period is restricted to 3 h only. As with all other areas of clinical PET, insulin-dependent diabetics should be told to maintain their normal diet and insulin intake and should not be fasted. Following injection it is important that the patient remains quiet and undisturbed for the uptake period. External stimuli should be avoided, as these may cause "activation" within different areas of the brain, which, in extreme cases, might be evident in the final image produced. Ideally, it is best to ask the patient to remain in a darkened room with their eyes closed and no distractions. This restriction applies primarily to FDG scanning but should also be adhered to when using other tracers so that the preparation is standard and reproducible. Many patients being investigated for epilepsy may well be young children and may also have concurrent behavioral problems. As a result, it can be difficult to keep these patients still for the length of time required for the scan, and it is prudent to consider scanning them with some form of sedation. In general, it is preferable to scan under a full general anesthetic rather than a light sedation because the latter can be unpredictable and difficult to manage. However, a general anesthetic requires good cooperation with the anesthetics department and an anesthetist who has a reasonable understanding of PET and the time constraints associated with the technique. One option is to have a regular general anesthetic time booked, the frequency of which can be governed by the individual demands of the PET center. The anesthetic unfortunately results in a globally reduced FDG uptake, which will give a slightly substandard image, and the best results are achieved if the patient is awake for their FDG uptake period but under anesthetic for the scan. This protocol is not always possible, particularly when patients are having a dual-study protocol with FDG scans in conjunction with an ^{11}C-labeled tracer.

Sequential Dual-Tracer Neurologic Studies

In the usual technique for performing dual-tracer studies, the scan using the shorter half-life tracer is performed first, immediately followed by the FDG scan. When the patient is sedated or under a general anesthetic, the uptake period for the FDG inevitably is while the patient is anesthetized. However, the advantages of having the patient remaining still for the study usually outweigh the slightly reduced quality of the study caused by lower global uptake. In children under 16 years who are being investigated for epilepsy, it is advisable to perform an EEG recording for the whole duration of the uptake period.

FDG-PET brain scans for the investigation of epilepsy should always be performed interictally. The short half-life of the tracer makes ictal scanning technically very challenging, and the comparatively long uptake period of FDG into the brain could render ictal scanning unreliable. However, if a patient does enter status epilepticus during the uptake period, this may have an effect on the final image, and so it is worthwhile noting brainwave activity during this period. FDG scans in the interictal state in epilepsy are usually employed to locate regions of diminished uptake. Most adults can tolerate FDG brain scanning with or without an ^{11}C scan fairly well, and it is rare to require either general anesthetic or sedation for these patients. Once the acquisition is completed, the individual frames can be assessed for patient movement, and if there has been no movement they can be summed and reconstructed as a single frame study. Alternatively, it is possible, if there have been small amounts of movement between frames, to register sequential frames to each other before summing and reconstruction, although great care must be taken to ensure that the frames are coaligned with the transmission scan data used for attenuation correction. Where there has been significant movement on a frame, this frame will probably have to be excluded from the study and the remaining frames summed and reconstructed.

The production of ^{11}C radiopharmaceuticals tends to be less reliable than that of FDG and the yields tend to be lower. In addition to this, the half-life of ^{11}C is only 20.4 min, which means there is a limited period of time during which there is enough radioactivity to achieve a good diagnostic scan. Radiopharmaceuticals labeled with ^{11}C are produced on demand and can take approximately 1 h to make. Thus, it is common practice to book patients so that they arrive in the PET scanning department about 1 h before their scan appointment time. The production is not started until the patient presents for the study, thus removing the possibility of having tracer ready to inject but no patient. It is important that patients understand the reason for this timing so they are aware that, on arrival in the department, there will be a delay before their scan is started.

PET Scanning in Cardiology

Aside from some special research studies, PET cardiology studies are usually performed to assess hibernating myocardium or viability in patients with left ventricular (LV) dysfunction. Such studies can involve assessing myocardial perfusion using ammonia ([^{13}N]-NH$_3$) or [^{15}O]-H$_2$O and glucose metabolism with FDG. Hibernating myocardium may be thought of in terms of myocardium with reduced function caused by an adaptive response to chronic hypoperfusion and which, if revascularized, may recover function. Patients with poor LV function and ischemic heart disease represent only a small percentage of surgical candidates but are an important group. They have increased mortality and morbidity compared to patients with normal LV function but stand to gain most from revascularization in terms of increased survival. Separating those who may benefit from surgery (in whom the associated operative risks are worth taking) from those whom surgery cannot hope to benefit is vital.

Rest/stress flow studies can be used in the investigation of reversible ischemia in cases where [^{99m}Tc]-labeled perfusion agents or ^{201}Tl scanning is equivocal, difficult to interpret, or does not match the clinical findings. ^{13}N has a 9.97-min half-life, which means that this tracer can only be used if the scanner is situated close to the production facility. [^{13}N]-NH$_3$ is a highly diffusible tracer that, following intravenous injection, is efficiently extracted into perfused tissue where it remains for a considerable period (8, 9). Thus, it acts as a flow tracer with the uptake being proportional to perfusion. Dynamic [^{13}N]-NH$_3$ scanning allows quantification of blood flow to the myocardium in milliliters per minute per gram of tissue as well as giving qualitative information (10). Similarly, [^{15}O]-H$_2$O is a freely diffusible tracer that distributes in the myocardium according to perfusion. However, the

short half-life and rapid washout from myocardium makes it more difficult to use than [^{13}N]-NH$_3$.

FDG uptake gives information about the viability of the heart muscle. When the heart is normally perfused it relies on the oxidation of free fatty acids in the fasted state and aerobic glucose metabolism postprandially. When myocardial perfusion is reduced, the heart switches from free fatty acid metabolism almost entirely to glucose metabolism and, therefore, preferential uptake of FDG should occur in ischemic regions (Figure 3.4). As myocardial perfusion and oxygen consumption continue to reduce, myocardial contractility worsens and then stops while glucose consumption by the myocardium increases. However, in the fasted state, uptake of glucose can be very variable in both normal and ischemic myocardium, making some studies difficult to interpret. Following a glucose load (50 g orally), glucose becomes the primary energy source, with both normal and ischemic myocardium taking up FDG. To minimize any inhomogeneity of FDG uptake, it is best to glucose-load the patient before scanning. Uptake of FDG into the myocytes can be enhanced by insulin; however, the use of a feedback-controlled hyperinsulinemic clamp is technically time consuming. Using a sliding scale of insulin, a reduced number of uninterpretable studies has been observed, based on blood glucose before FDG injection (11). To obtain acceptable FDG uptake into the myocardium, it is vital that the patient is prepared correctly and managed in a controlled way throughout the PET scan. Uptake of FDG into the heart muscle is dependent on achieving the correct insulin–glucose balance in the patient before FDG

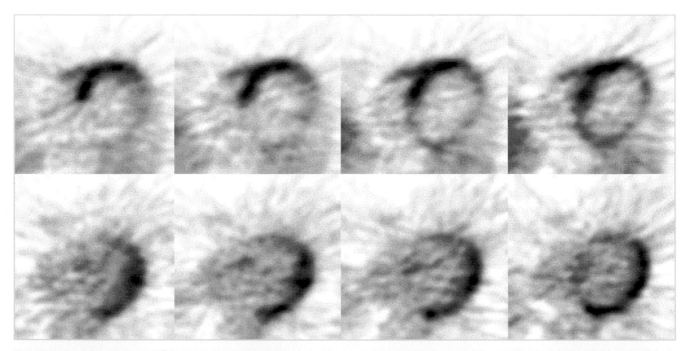

Figure 3.4. Myocardial perfusion with [^{13}N]-NH$_3$ (*top row*) and FDG (*bottom row*) demonstrates almost complete mismatch. The areas of low myocardial perfusion on the [^{13}N]-NH$_3$ scan demonstrate high FDG uptake, indicating hypoperfused, viable myocardium. (From Valk PE, Bailey DL, Townsend DW, Maisey MN. Positron Emission Tomography: Basic Science and Clinical Practice. Springer-Verlag London Ltd 2003, p. 785.) 📖

injection, and so the management of these patients is focused on this aspect. Practice guidelines for data acquisition and patient preparation have been published by the American Society of Nuclear Cardiology for PET myocardial glucose metabolism and perfusion imaging with FDG, [^{13}N]-NH$_3$, and ^{82}Rb. These guidelines include example of protocols used at different institutions in the United States for loading patients with glucose and insulin in preparation for a myocardial FDG study (12).

Cardiac PET scanning is perhaps the most complex and time consuming of the three main clinical groups. In a preparation protocol developed at the Guy's and St. Thomas' Hospital Clinical PET Center, London, UK, the patients are prepared slightly differently depending on whether they are nondiabetic, insulin-dependent diabetic (IDD), or non-insulin-dependent diabetic (NIDDM). All categories of patients are asked to refrain from caffeine-containing products for at least 24 h before their appointment time. This restriction both minimizes the inherent stress effect of caffeine and lowers the free fatty acid level that might otherwise compete with the FDG for provision of the myocardium's energy source. Nondiabetic patients should be asked to fast (water only) for 6 h before their appointment time. The non-insulin-dependent diabetics should fast for 6 h, but if their appointment is booked for the afternoon they are encouraged to eat breakfast and take their normal morning oral hypoglycemic as usual. The insulin-dependent diabetics are instructed to eat and take their insulin as normal with no modifications to their normal routine. Many patients attending for cardiac PET scanning are on medication, and it is not necessary for them to stop any of this before their scan.

After arrival in the department, the procedure is explained to the patient and a blood sample is taken to measure the baseline blood glucose level. If the blood glucose is below 7 mmol/L, the patient will be glucose loaded. The glucose can be given in the form of dextrose monohydrate powder dissolved in water, with the aim to give this drink approximately 1 h before the FDG is to be given. If the patient is not a diabetic, they are given 50 g dextrose monohydrate, and if they are diabetic (IDD or NIDDM) they receive 25 g. About 10 min before the FDG is planned to be injected, another blood glucose measurement is made, and depending on this result the patient may or may not be given insulin before FDG injection. The amount of insulin given will increase on a sliding scale, an example of which is shown in Table 3.1.

In the example shown in Table 3.1, the insulin used is human actrapid, which is diluted in about 1 mL normal saline and given intravenously to the patient about 5 min before the FDG injection. With this protocol, the key to achieving a good FDG scan is to push the glucose level up high enough so that either the patient's own regulatory system will produce insulin in response to this or the

Table 3.1. Sliding scale of insulin given dependent on blood glucose levels in cardiac PET FDG scanning.

Glucose concentration (mmol/L)	Action
<5.0	No insulin
5.0–8.0	3 units of insulin
8.0–12.0	4.5 units of insulin
>12.0	(At clinician's discretion)

Source: From Bailey DL, Townsend DW, Valk PE, Maisey MN. Positron Emission Tomography: Basic Sciences. Springer-Verlag London Ltd 2005, p. 304.

patient's glucose level will be high enough that exogenous insulin can be administered safely. Obviously great care must be taken when giving insulin to a patient. It is important that intravenous 50% glucose is on hand in case the patient becomes hypoglycemic, and it is also essential that the patient is not allowed to leave the department until they are safe and well enough to do so. If patients have been given insulin, they should be infused with about 50 mL 20% glucose to counteract any effects from the insulin during the last 10 to 15 min of the FDG scan. It is also sensible to make sure the patient has something to eat and drink and to check the patient's blood glucose level before he or she leaves.

If the blood glucose level is checked after the patient has been injected with the FDG, it is important to ensure that the blood sample is not taken from the same line through which the FDG was given. About 20 min after the FDG has been administered, a short test acquisition should be performed to assess whether FDG is getting into the heart muscle. Even though the uptake may still be quite low at this stage, and because of the short acquisition time the counting statistics will be poor, an experienced eye should be able to judge whether there is uptake. If uptake is not seen at this stage it may be worth waiting another 5 or 10 min and then repeating the test acquisition. If there is still no uptake after this time, a second administration of insulin (as prescribed by the doctor on site) may be enough to push the remaining circulating FDG into the myocardium. Although by this stage it may be more than 30 min after the FDG injection, it should not be assumed that the FDG scan is going to fail. This glucose loading protocol for cardiac imaging seems to work very well, with a very low (<5%) failure rate.

Further information for assessing the viability of the myocardium may be obtained by gating the FDG scans. Gating of FDG scans is done in the same way as gating of nuclear medicine studies. The patient preparation is identical to that already described as the tracer used is still FDG. The patient's electrocardiogram is monitored for a short time before the start of acquisition, and the average R–R interval is divided into a selected number of equal segments. During the acquisition, the same segment from each cardiac cycle acquired is stored in the

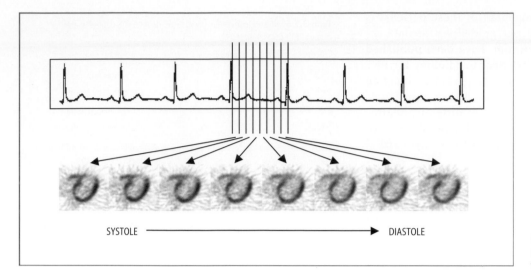

SYSTOLE ——————————————————→ DIASTOLE

Figure 3.5. Gating of the cardiac cycle produces a series of images of the heart that can be viewed as a ciné. 📖

same location of computer memory (Figure 3.5). The acquisition time for the FDG scan should be slightly longer than for static FDG cardiac scans to compensate for the poorer counting statistics in each individual gate. Thus, once the data have been reconstructed, a ciné image of the beating heart can be viewed tomographically.

Gated studies allow the assessment of myocardial viability as well as wall motion and contractility and are critical to evaluate regional and global function; they can help to identify dysfunctional regions of the myocardium. In the absence of perfusion images, localization of regional dysfunction is helpful to identify the region of the myocardium at risk where evaluation of viability is critical.

The most accurate interpretation of FDG images is obtained when correlated with perfusion images, as discussed in practice guidelines for interpretation and reporting that have been published by the American Society of Nuclear Cardiology (13). For correlation with FDG images, optimal perfusion images are obtained using a PET perfusion tracer, such as $[^{13}N]$-NH_3 or ^{82}Rb. However, if these tracers are not available, for example, at institutions that do not have a cyclotron or a ^{82}Rb generator, single photon emission computed tomography (SPECT) perfusion images using ^{99m}Tc tracers can be used for correlation. The combination of gated FDG-PET images and standard nuclear medicine SPECT imaging of myocardial perfusion makes it much more accessible because centers remote from a cyclotron can undertake clinical cardiac work more easily.

Radiation Protection for the PET Technologist

Over the past decade there has been a significant increase in interest in the clinical use of positron-emitting tracers. FDG, with its 109.8-min half-life, not only lends itself to use at sites close to a cyclotron and chemistry facility but also

can be transported to scanners at sites remote from the cyclotron. It is the most commonly used tracer in clinical PET at present, and the combination of these two factors means that hospitals are looking more and more at the possibility of introducing this radiopharmaceutical for use within existing departments. However, clinical units usually require high throughput (compared with research facilities), and this inevitably means increased numbers of scans and an increased radiation burden in the department.

With PET/CT, the radiation dose related to both the CT study and the PET tracers must be taken into consideration for both patients and technologists. The design and shielding of the PET/CT facilities must be modified appropriately as well, as further discussed in Chapter 4, Radiation Protection and Dosimetry in PET. Chapter 3 addresses only radiation protection for the technologist related to PET tracers.

Shielding

For most hospitals, the natural assumption is that PET scanning, either on a dedicated ring system or in a dual-headed coincidence gamma camera PET (GC-PET) system, is performed in the Nuclear Medicine Department. It should be remembered that although there are similarities between nuclear medicine and PET, the latter makes use of radiotracers with a much higher photon energy than those usually encountered in nuclear medicine (511 keV compared to 140 keV). Local legislation in the United Kingdom (14, 15) sets the annual whole-body dose limit for unclassified radiation workers at 6 mSv. This is a significant reduction from the previous limit of 15 mSv and consequently places a greater burden of responsibility on departments to ensure that staff members are provided with the means to keep their doses below this level. In addition to this, the newer PET/CT systems permit much faster acquisition times, which means that the number of scans per day can be increased.

As a result, the number of patients per technologist will inevitably increase and will bring an attendant increase in radiation dose. Doses to staff include not only the whole-body dose but also the dose to extremities. When handling unsealed sources, the key extremity dose is that to the hands. The photon energy of 511 keV means that conventional amounts of shielding with lead or tungsten will not be sufficient and more emphasis on the inverse square law, as a means of reducing dose, may need to be employed.

Currently, FDG is not produced in enormous quantities, and on arrival the activity in the multidose vial will range from 7 to 20 GBq. These multidose vials should be shielded within a lead pot with a minimum thickness of 18 mm, although thicker pots (34 mm) are being introduced by some departments, particularly in the light of increasing tracer yields. The design of the dose preparation area should be similar to that used in nuclear medicine. An L-shield with a lead glass insert and a lead base should be used. The thickness of the base and the shield should be of the order of 6 cm, and the lead glass window should have a 5.5-cm lead glass equivalent. Measurements of instantaneous dose rate taken from a shielded vial of FDG (7–8 GBq) placed behind the L-shield range from 20 μSv/h immediately in front of the L-shield to 2 μSv/h at 2.0 m from the shield. Clearly, the dose rate immediately in front of the shield is quite high and, as this is the place where the technologist will be standing when drawing up the tracer, it is essential that a good, reliable, and fast technique is developed so that the dose received is reduced to a minimum.

Dispensing

The suggested technique for drawing up the tracer differs from that commonly used in nuclear medicine, in that it is inadvisable to invert the lead pot and vial to withdraw the tracer. A better technique involves the use of a long flexible needle that will reach to the bottom of the vial, coupled with a filter needle for venting. The vial can be held firmly in place in the lead pot with a pair of long-handled forceps, and the syringe, once attached to the long needle, can be angled away from the top of the vial. Thus, both hands are kept out of the direct beam of the photons being emitted from the unshielded top of the vial. In this way, dose to the hands can be kept to a minimum and there is no risk of the vial sliding out from its shielding.

Various other methods have been used to try to reduce the hand dose further. For example, in some departments the needle within the FDG vial is connected to a long length of flexible tubing which itself is shielded and enables the technologist to put an even greater distance between their hands and the vial of activity. In addition, commercially available purpose-built units are available that provide a significantly higher degree of physical shielding. Each unit should be tested for their ease of use, especially in departments where more than one radio-pharmaceutical is likely to be used, as this will require the technologist swapping vials between the unit and their own lead pot.

Opinions vary slightly from department to department, but in the majority of PET centers the technologists draw up into unshielded syringes. Once the tracer has been drawn up the syringe can be placed into a syringe shield. Lead and tungsten syringe shields tend not to be used as the thickness of metal required to produce effective shielding would render the shield unmanageable. Instead, many centers make use of 1-cm-thick perspex (lucite) to manufacture the syringe shield. The effect of the perspex is to act as a barrier to any stray positrons and to increase the distance between the syringe and the technologist's hands, thus making use of the inverse square law to reduce the dose received. The shielded syringe can be transported to the patient for injection in an 18-mm (minimum) lead carry pot, which means the technologist need only be further exposed to the radiation during the injection itself.

Finger doses vary from center to center depending on the throughput. At the Guy's and St. Thomas' Hospital, a relatively busy clinical PET center in London, the monthly hand doses were compared to the activity handled by the technologists. Each technologist was injecting approximately 1 GBq of radioactivity per day, although this activity was being drawn up from vials containing much higher levels of activity. The average hand dose for these technologists was between 4 and 7 mSv per month, which, over a 12-month period, means that their extremity dose is comfortably within the current United Kingdom annual limit for unclassified workers (150 mSv). However, as FDG production and workload increases, technologists will find themselves handling more and more radioactivity, and so they must not become complacent but should continue to find ways to keep these doses to a minimum. The average hand doses given were with technologists handling FDG only; however, some of the shorter-lived tracers require higher injected activities to get a good diagnostic result. In centers where $[^{15}O]$-H_2O is used, a manual injection for a single scan (2.3 GBq) can result in a dose to each hand of nearly 1 mSv. It is important to ensure that these higher doses are shared between staff members and that centers do not end up with only one person performing this type of work.

Whole-Body Dose

The whole-body doses to staff in PET are slightly higher than those typically seen in nuclear medicine. Benatar et al. (16) studied the whole-body doses received by staff in a dedicated clinical PET center and then used measured data to estimate the doses that might be received by staff in departments using FDG but not originally designed to do so. The results showed that in a dedicated unit the average whole-body dose received by each technologist was around 5.5 μSv per patient study and approximately

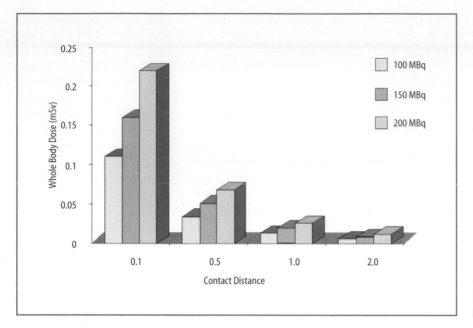

Figure 3.6. Estimated whole-body doses likely to be received by technologists during 1-h scan beginning 60 min after FDG injection. 📖

14 μSv per day. It should be noted that the new annual dose limit for unclassified workers of 6 mSv equates to an average daily dose of 24 μSv . With the exception of same-day rest/stress [^{99m}Tc]-Myoview studies (17), the average dose per patient study in nuclear medicine is only 1.5 μSv (18), which is much lower than for PET studies. However, the throughput per technologist in a PET unit is likely to be lower than that in nuclear medicine and, therefore, the daily whole-body dose received is comparable. This work was carried out in a dedicated purpose-built PET unit where the technologists are shielded from any radioactive sources for the majority of the working day and only had significant contact during injection and positioning of the patient. In this unit, the control room is separate to, and shielded from, the scanning area and the room where the patients wait after injection before the scan. This arrangement contrasts with the setup commonly seen in nuclear medicine where the control console is often sited adjacent to the gamma camera with no shielding between the two. To make estimates of the doses likely to be received by technologists using FDG in an existing department, Benatar et al. (16) measured the instantaneous dose rates at 0.1, 0.5, 1.0, and 2.0 m from the anterior chest wall in patients immediately following their FDG injection. The average dose rates per MBq injected at each of the four distances were used to estimate the whole-body dose that could be received by technologists. Figure 3.6 shows the whole-body dose received during a 1-h scan commencing 1 h after injection based on an injected activity between 100 and 200 MBq. These injected activities were used to try to reproduce the likely usage on a gamma camera PET system. At a distance of 2 m with an average injected activity of 200 MBq, the whole-body dose received is of the order of 12 μSv. Although in isolation this dose may not seem particularly high, it must be remembered that this is the dose from each individual PET patient, whereas each

technologist is likely to scan many more than this in a year, as well as a number of nuclear medicine patients. Five hundred PET patients in a year would take the technologist close to the 6 mSv annual limit. Therefore, if PET is going to be introduced into existing departments, some thought needs to be given to the positioning of the control console relative to the scanner itself. Figure 3.7 shows the instantaneous dose rates measured from a patient 1 h after being injected with FDG. It should also be remembered that this work related to the whole-body

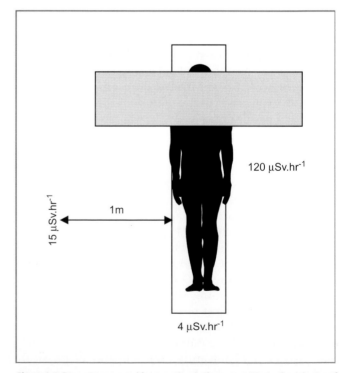

Figure 3.7. Dose rates measured from a patient in the scanner 55 min after injection of 334 MBq FDG. 📖

radiation doses that technologists would receive when coming into contact with patients who have been injected with approximately 350–400 MBq FDG. This level is currently the maximum injected activity in Europe, but in the United States, for example, it is not uncommon for patients to be injected with twice this amount, which will, of course, increase the radiation burden for the staff, patient, and carers.

The dose rate immediately next to the bed is quite high (120 μSv.h^{-1}). At 1 m from the bedside, it drops to 15 μSv.h^{-1} but, significantly, at the foot end of the bed the dose rate is only 4.0 μSv.h^{-1}. Obviously, the dose received by a technologist can be reduced by careful planning even within a preexisting unit. An alternative, if space is a constraint and the operator's console must be included in the scanning room, is to use a mobile lead shield. A practical shield, on wheels, has been reported as reducing the exposure rate by 90%, from approximately 20 μSv.h^{-1} at an operator's console roughly 1 m from a patient who had been injected with 370 MBq FDG to 2 μSv.h^{-1} (19). The comparable exposure rate from a typical [^{99m}Tc]-methylene diphosphate (MDP) bone scan patient at a similar distance is 6 μSv.h^{-1}. Consideration must also be given to the location where a patient waits during the uptake period after being injected with a positron-emitting tracer. With FDG, it is important that the patient is able to rest, preferably lying down, and should not be allowed to leave the department during this time. Ideally, a separate room should be provided, but in the event that this is not available and that they have to share a room, it should be remembered that the dose rates from patients immediately following injection can be quite high and may present a radiation hazard. Benatar et al. (16) used the dose rate data to estimate the doses that might be received by people coming into contact with injected patients during the 1-h uptake period. Figure 3.8 shows that, following administration of FDG, the dose received during the following hour can range from just over 0.3 mSv if sitting right next to the patient down to 0.016 mSv at a distance of 2 m.

Risk to the General Public

The United Kingdom annual dose limit for members of the general public is 1 mSv, and there is a recommendation that at any one exposure the dose should not exceed 30% of this (20). Clearly, in a general waiting area there will not be only one PET patient, and an escort or carer could receive in excess of 0.3 mSv when taking into account the contribution from all the patients present. Ideally it is best to have a "hot" waiting room separate to the "cold" waiting room, and the planning arrangements in the "hot" waiting room should ensure adequate separation of seating and preferably individually shielded cubicles for each patient.

The majority of clinical PET radiotracers have very short half-lives and therefore patients pose little radiation risk to members of the public once they have completed their scan and are ready to leave the department. FDG, with its longer half-life, will not have completely decayed away before the patient leaves and so consideration must be given to the severity of this risk. The critical group is likely to be health care workers who may have contact with several patients who have undergone PET investigations or with an individual patient who may be having other radiopharmaceutical studies in addition to their PET scan. United Kingdom guidelines on when patients can leave the hospital following administration of a radiopharmaceutical are currently based on the activity retained within the patient on exit rather than the total doses that would be received by others coming into contact with them. A study published in 1999 measured the instantaneous dose rates from patients just before leaving the department (21). The integrated doses were calculated for various contact times at differing distances. The results showed that for positron-emitting tracers with a half-life of 110 min or less the use of the activity-energy product is not a useful concept for restricting critical groups to an infrequent exposure. Instead, greater emphasis should be placed on the true estimate of actual radiation dose that could be received, or the predicted doses. With longer-lived PET tracers, for example, ^{124}I (half-life, 4.2 days), the recommendations would need to be reviewed. The recommendations given as a result of this work suggested that following an FDG-PET study using an injected activity of 350 MBq there was no need to prevent contact with a patient's partner following a scan. There was no need to restrict travel on public transport following the scan, despite the fact that the activity-energy product may exceed the local rules. Clearly, young children of patients should never accompany their parents to the PET or nuclear medicine department.

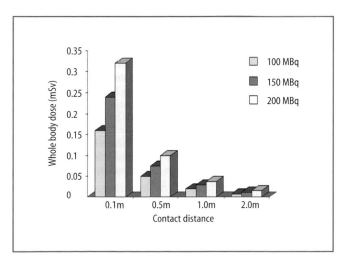

Figure 3.8. Estimated whole-body doses likely to be received by member of the public/accompanying person during the 1 h immediately following injection of FDG. 📖

Table 4.1. Dose limits recommended by the International Commission on Radiological Protection (ICRP).

	Occupational	Public
Effective dose	20 mSv year^{-1}, averaged over 5 years and not more than 50 mSv in any 1 year	1 mSv year^{-1}
Equivalent dose		
Lens of the eye	150 mSv year^{-1}	15 mSv year^{-1}
Skin	500 mSv year^{-1}	50 mSv year^{-1}
Hands and feet	500 mSv year^{-1}	—

Source: Data from ICRP Publication 60 (3).
From Valk PE, Bailey DL, Townsend DW, Maisey MN. Positron Emission Tomography: Basic Science and Clinical Practice. Springer-Verlag London Ltd 2003, p. 266.

a presumed risk of cancer similar to that of the postnatal infant (3, 8, 12–14), as summarized in Table 4.2. A review of all published studies on childhood cancer following fetal irradiation concluded that a fetal dose of 10 to 20 mGy can increase the risk by about 40%, the risk being greatest in the third trimester (15). As the baseline mortality rate for childhood cancer is so low, the effect of a fetal dose of 10 mGy is marginal, reducing the probability of *not* developing childhood cancer from 99.7% to 99.6% (12). A fetal dose of 25 mGy, quite possible with PET/CT, may double the natural risk (about 1 in 1,300 in the United Kingdom) of fatal childhood cancer, although it would have very little effect on lifetime risk (about 1 in 4) thereafter (8).

dose is 0 with an upper-bound 90% confidence limit between 60 and 100 mSv (6). An effective dose of 200 mSv to adults from their occupation adds about 1% in theory to the normal risk of dying from cancer of about 25%.

The data from the LSS are complemented by epidemiologic data from medical exposures, ranging from the trivial to high-dose radiotherapy, which also highlight the increased susceptibility of children (7). The age-dependence of risk is illustrated in Figure 4.1 (3, 8). There is understandable concern about the exposure to infants and children from CT (9–11). Concern about exposures in children, particularly for superficial radiosensitive organs including the lens of the eye, thyroid, and breast/anterior chest wall, can be extrapolated to exposures in utero, where the risk of developmental abnormalities is added to

Radiation from PET and CT Sources

PET Radionuclide Emissions

In terms of energy deposition in tissue, PET radionuclides have more in common with the radionuclides used for therapy than those used for diagnostic imaging. The amount of energy deposited locally or at a distance from disintegrating atoms in an infinite medium is indicated by the equilibrium absorbed dose constant, delta (Δ), as shown in Table 4.3 for a selection of radionuclides used for diagnosis and therapy (16, 17). Positrons, being non-penetrating charged particles, deposit their energy locally and account for most of the dose to the organs and tissues

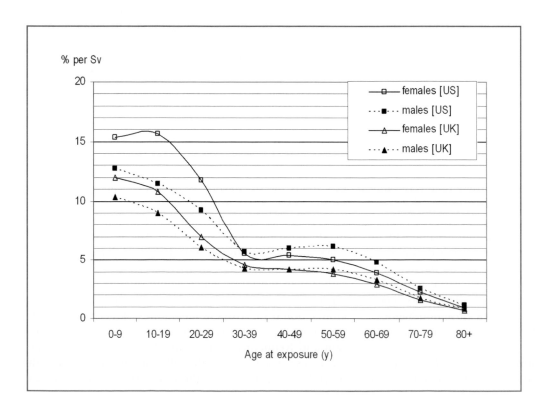

Figure 4.1. Age-dependent excess lifetime risk of fatal cancer following radiation exposure. [Data from ICRP 60 (International Commission on Radiological Protection) (U.S. data) and NRPB (National Radiological Protection Board) (U.K. data) (3, 8).]

Table 4.2. Risk of radiation exposure in utero and the normal risks of pregnancy.

Effect	Gestational age at exposure (weeks postconception)	Threshold (Gy)	Risk (Gy⁻¹)	Normal prevalence or incidence
Lethality	1–3 inclusive		Minimal	
Spontaneous abortion				>15% to term
Congenital abnormalities, growth retardation	4–7 inclusive	>0.1	50%	6%
Severe mental retardation	8–15 inclusive	>0.1	40%	0.5%
	16–25 inclusive	>0.2	10%	
Reduced IQ:	8–15 inclusive	>0.1	30 points	
Mental retardation				3%
Cancer	4 until term			
Incidence to 15 years		—	6%	0.077%
Mortality to 15 years		—	3%	0.154%
Mortality lifetime			~12%	~25%
Genetic disorders	4 until term		2.4%	8%

Source: Data from NRPB (8), ICRP (12).

of PET patients. Photons emitted in the annihilation of the positrons are penetrating and account for the exposure of persons nearby. The *activity* of a radioactive substance is measured by the number of radioactive atoms that disintegrate in a second, the becquerel (Bq) being an activity of one disintegration per second. The activity of ^{18}F-fluorodeoxyglucose (FDG) administered to adult patients is generally in the range from 200 to 750 MBq, or roughly 5 to 20 millicuries (mCi). (Activity in mCi is determined by dividing the value in MBq by 37.) The influence of half-life on the energy available from the total decay of a radioactive source is evident in Table 4.3. The short half-lives of clinical PET radionuclides limit the internal radiation dose to patients and the external radiation dose to persons who come in contact with the patient some time after the PET scan. However, they confer no particular benefit on PET staff, who must contend each day with high dose rates from patients and many patients to be scanned.

External exposure to gamma radiation is the most significant pathway for occupational exposure in PET facilities. The high dose rates from PET radionuclides relative to other radionuclides used for diagnostic imaging result from their high photon energy (511 keV) and photon yield (typically 197%–200%). Other potential pathways are these:

- A skin dose from surface contamination
- A deep dose from bremsstrahlung generated in lead or other material of high atomic number
- A superficial dose from positrons emitted from the surface of uncovered sources
- An immersion dose from a release of radioactive gas into the room air

The dose rate at a distance in air from a source of radiation can be expressed in various terms depending on the application. Dose rate constants at a distance of 1 m from a 1 GBq "point" source of various radionuclides are given

Table 4.3. Energy from decay of selected nuclear medicine radionuclides.

	Equilibrium absorbed dose constant, Δ^a (g Gy MBq⁻¹ h⁻¹)				
	Nonpenetrating Δn–p	Penetrating Δp	Total Δ	$T_{1/2}$	Energy from total decay of 1 MBq (μJ)
^{11}C	0.227	0.588	0.815	20.3 mins	397
^{13}N	0.281	0.589	0.870	10.0 mins	209
^{15}O	0.415	0.589	1.004	2.07 mins	50
^{18}F	0.139	0.570	0.709	1.83 h	1,868
^{90}Y	0.539	–	0.539	2.7 days	50,295
^{99m}Tc	0.010	0.072	0.082	6.0 h	708
^{131}I	0.109	0.219	0.328	8.05 days	91,250

[a]Data from Society of Nuclear Medicine MIRD Committee (16, 17).
Source: From Valk PE, Bailey DL, Townsend DW, Maisey MN. Positron Emission Tomography: Basic Science and Clinical Practice. Springer-Verlag London Ltd 2003, p. 267.

overall and along the z-axis, based on the preliminary scout view radiograph. ATCM also uses angular modulation, either sinusoidal or based on attenuation at projections from the previous rotation. ATCM increases dose in the region of metal implants but only to an extent similar to the dose reduction from using ATCM instead of fixed mA (24). Each manufacturer has a slightly different approach to ATCM, so it is important that the operator understands the connection between image quality, scan parameters, and dose when selecting the ATCM option (25). ATCM is a valuable means of reducing dose while maintaining image quality, presuming the operator selects an appropriate noise level or mA for a standard patient. Operators may err on the side of high image quality rather than low dose, as suggested by simulation studies in which noise was added to clinical CT images without impairing diagnostic quality (26). The challenge for PET/CT is to select a reference current or noise level for ATCM that is optimal for the purpose of the CT. If a diagnostic CT image is not required, tube current and time could be reduced by as much as 50% in some circumstances.

Independent evaluation reports on similar scanners, for example, 16-slice units (27), are available to assist anyone contemplating the purchase of a PET/CT scanner. As summarized by Nagel (28), the system characteristics that affect dose are the power waveform to the X-ray generator, the selectable tube current range and steps, the inherent beam filtration from the tube assembly, additional beam-shaping filters, the distance from the tube focus to the axis of rotation, the total slice or beam width (collimated at the tube, not selected at the detector array), the detector type and array, the scanner configuration, the scan field of view, and the selectable scan angle during a full rotation.

Medical Exposures

Estimation of Organ and Tissue Dose from PET

Despite the high energy of decay, the radiation dose from PET tracers, although limited by the short physical half-life, is similar to that from many imaging procedures using single photon emissions. The dose may also be limited by the maximum amount of activity that can be administered to the patient without taxing the response of the detector system, which is a particularly important consideration when the scanner is operated without septa in three-dimensional (3D) mode. For a whole-body study with a scanner with bismuth germanate (BGO) detectors operating in 3D mode, the administered activity should be less than 250 to 400 MBq, depending on the scanner and uptake period. Even with fast cerium-doped lutetium oxyorthosilicate (LSO) or gadolinium oxyorthosilicate (GSO) detectors, careful consideration needs to be given to the injected activity. Patient-based noise-equivalent count

rate (NECR) estimates can provide insight into maximum injected activities that should be used with the scanner to avoid operating beyond the peak NECR point (29, 30). Furthermore, near peak NECR the incremental improvement in signal to noise (S/N) with increasing activity is very small. Based on NECR analysis for an LSO scanner, the injected activity required to achieve peak S/N (or peak NECR) is more than 70% larger than that required to operate at 95% peak S/N (31). In other words, a very large increase (>70%) in injected activity yields only a small increase of 5% in S/N when operating close to the peak NECR of the scanner. Thus, there is potential for a considerable reduction in injected activity without noticeable degradation of image quality. Weight-based injected activities have also been suggested with very high activities (>600 MBq) for heavy patients (32, 33), but patient-based NECR analysis and image quality assessment of 3D whole-body studies have shown that little improvement in image quality can be gained from increasing the injected activity in heavy patients (29–31), and it is preferable to increase scan acquisition time. The larger injected activities could, however, result in increased radiation exposure to the patients and staff.

Radiopharmaceutical dose estimates are calculated using the methodology developed by the Medical Internal Radiation Dose (MIRD) Committee of the Society of Nuclear Medicine (34). Software for this purpose is available from the Radiation Dose Assessment Resource (RADAR) professional group at www.doseinfo-radar.com/OLINDA.html (35). The MIRD method requires an estimate of the spatial and temporal distribution of radioactivity in the body, which can be obtained from organ activity-time curves from images at various time points. Biokinetic models are used to model the movement of activity through anatomic and physiologic compartments. The cumulative activity in a "source" organ is multiplied by a dose factor to give the dose to a "target" organ. The total dose to each target organ is obtained by summing the contributions from all the identified source organs. Dose factors for PET radionuclides, and more than 800 others, can be accessed on the RADAR website www.doseinfo-radar.com, an extensive electronic resource which also includes a description of the various physiologic models and anthropomorphic phantoms for Monte Carlo modeling used in their derivation (36).

Image fusion allows a more accurate determination of organ uptake of PET radiopharmaceuticals than conventional gamma camera methods can do for non-PET radiopharmaceuticals. Tomography overcomes the need for geometric mean imaging with planar detectors and subtraction of "background" counts for under- and overlying tissues. PET/CT coregistration and edge contouring allow anatomically correct and consistent selection of regions of interest. Finer PET resolution reduces partial volume errors, and high sensitivity and attenuation correction of PET data gives better accuracy for radioactive tissue concentration. For example, organ delineation in combined

PET and MRI images has been used to measure the distribution of a carbon-11 ligand in maternal and fetal organs of macaque monkeys weighing less than 9 kg (37) and of FDG in adults (38).

Effective dose is calculated from the individual organ and tissue doses using the tissue weighting factors recommended by the ICRP in 1990 (3) (that replaced those used to calculate "effective dose equivalent" and which it may review again to reflect the latest estimates of organ/tissue susceptibility to radiation damage). Dose coefficients for radiopharmaceuticals have been published by the ICRP and others and are also available on the RADAR site (36, 39–43). Dose coefficients for various PET radiopharmaceuticals in the U.S. and British Pharmacopoeia and the ICRP compendia are shown in Table 4.6 (39, 41–46).

The ICRP noted the difficulties of applying conventional dosimetry methods to very short lived PET tracers and foreshadowed the development of novel ad hoc methods of dose estimation (39). For example, the radioactivity may not last long enough to allow true equilibration of the tracer in body compartments, or the highest dose may be received by organs or tissues such as the trachea or walls of major blood vessels that have not been assigned a specific tissue weighting factor. The dose estimates for injected ^{15}O-water and inhaled ^{15}O-gases are cases in point (47–49). The MIRD Committee has developed its own biokinetic model for the bladder and kidneys (50). Optional bladder parameters (static, dynamic) are relevant to FDG dosimetry because bladder dose can be reduced by hydration and frequent voiding (51). For the very short lived nuclides, biokinetics have little influence

Table 4.6. Dose coefficients for various positron emission tomography (PET) tracers listed in Pharmacopeia and ICRP.

	USP 2000	BP 2004	Effective dose (mSv GBq^{-1})	Organ of maximum dose	Maximum organ dose (mGy GBq^{-1})	Uterus dose (mGy GBq^{-1})	Source of data
^{11}C							
Acetate	✓	✓	5.0	Heart	100	2.0	RIDIC
			3.5	Liver	14	1.4	ICRP53/A4
Amino acids, generic			5.5	Pancreas	41	3.5	ICRP53/A5
Brain receptors, generic			4.5	Bladder	32	4.5	ICRP53/A6
Methionine	✓	✓	7.4	Bladder	91	5.7	ICRP53/A4
Methyl thymidine			3.5	Liver	32	1.5	ICRP80
Raclopride	✓	✓	5.3 (EDE)	—	—	—	Wrobel
Spiperone			5.3	Liver	22	2.2	ICRP53, ICRP80
Thymidine			2.7	Kidneys	11	2.4	ICRP80
All substances, realistic maximum			11	Bladder	170	9.2	ICRP53/A7
^{13}N							
Ammonia	✓	✓	2.0	Bladder	8.1	1.9	ICRP53, ICRP80
^{15}O							
CO gas		✓					ICRP53, ICRP80
20-min breath hold			0.81	Lungs	3.4	0.3	
1-h continuous			0.55	Lungs	2.3	0.2	
Water	✓	✓	1.1	Heart	1.9	0.35	ICRP53/A5 [a]
O_2 gas		✓					ICRP53, ICRP80
20-min breath hold			0.37	Lungs	2.4	0.057	
1-h continuous			0.4	Lungs	2.6	0.068	
^{18}F							
Fluoride	✓		24	Bone, RBM	40	19	ICRP53, ICRP80
FDG	✓	✓	19	Bladder	170	20	ICRP80
FDG			n/a	Bladder	73	n/a	MIRD19 [b]
FDG			29	Bladder	310	(19)	Deloar
Fluorodopa	✓		25	Bladder	300	28	ICRP53/A4 [c]
Amino acids, generic			23	Pancreas	140	17	ICRP53/A5

[a] ^{15}O-water: ICRP53 Addendum 5 is correction to ICRP80.
[b] ^{18}F-FDG MIRD19 is based on different biokinetics (45) to ICRP80.
[c] ^{18}F-Fluorodopa brain uptake is doubled by carbidopa pretreatment (46).
RBM, red bone marrow.
Source: Data from ICRP Publication 53 and Addenda; RIDIC; MIRD (44); Wrobel et al. (52); Deloar et al. (38).

Table 4.7. Selected diagnostic reference levels for single slice and multislice computed tomography (SSCT and MSCT) dose for adults protocols.

	Protocol	Reference	Level
NRPB (2005)	Head (cerebrum)	DLP	SSCT 760 / MSCT 930 mGy cm
		CTDI$_{vol}$	SSCT 55 / MSCT 65 mGy
	Lymphoma survey	DLP	SSCT 760 / MSCT 940 mGy cm
	Lung	CTDI$_{vol}$	SSCT 10 / MSCT 12 mGy
	Abdomen/pelvis	CTDI$_{vol}$	SSCT 12 / MSCT 14 mGy
MSCT (2004)	Head	CTDI$_{vol}$	MSCT 60 mGy
	Chest	CTDI$_{vol}$	MSCT 10 mGy
	Abdomen	CTDI$_{vol}$	MSCT 25 mGy
ACR (2004)	Head	CTDI$_w$	60 mGy
	Abdomen		35 mGy
EC (1999)	Head	CTDI$_w$ and DLP	60 mGy, 1,050 mGy cm
	Chest		30 mGy, 650 mGy cm
	Abdomen/pelvis		35 mGy, 780 mGy cm

Source: NRPB (National Radiological Protection Board) (UK) (61); MSCT (European Concerted Action on CT) (71); ACR (American College of Radiology) (72); EC (European Commission) (23).

Table 4.8. Effective dose from radionuclide imaging and diagnostic CT procedures.

		Protocol	mSv
Oncology[a]	^{18}F-FDG	370 MBq	7.0
	^{11}C-Methionine	400 MBq	2.1
	^{67}Ga-citrate	400 MBq	40.0
	^{201}Tl-chloride	120 MBq	19.2
	^{99m}Tc-mIBI	1 GBq	9.0
Brain[a]	^{15}O-water	1 GBq	0.93
	^{18}F-FDG	250 MBq	4.8
	^{99m}Tc-HMPAO	800 MBq	7.4
Myocardium[a]	^{13}N-Ammonia	550 MBq	1.1
	^{18}F-FDG	250 MBq	4.8
	^{99m}Tc-MIBI	1.3 GBq; 1-day rest/stress protocol	10.6
	^{201}Tl-chloride	140 MBq; stress/reinjection protocol	22.4
Bone[a]	^{18}F-NaF	250 MBq	6.0
	^{99m}Tc-MDP	800 MBq	4.6
CT[b]	Head	Acute stroke	1.7
	Chest	Lung cancer	6.9
	Abdomen	Liver metastases	7.1
	Abdomen/pelvis	Abscess	8.0
	Lung/abdomen/pelvis	Lymphoma survey, 636-mm length	12
CT in FDG PET/CT[c]	"Whole body" from neck to pelvis	Topogram	0.2–0.4
		Attenuation correction CT	1.3–4.5
		Diagnostic CT with contrast	14.1–18.6

[a]ICRP Publication 80 (41); data for ^{201}Tl from Thomas et al. (73).
[b]75th percentile doses from survey of CT practice in UK 2003; data from Shrimpton et al. (61).
[c]Data from survey of protocols at four centers by Brix et al. (75).

The Pediatric Patient

The ALARA principle is very important in pediatric applications. The activity of a radiopharmaceutical administered to a child is usually calculated by scaling down the adult dosage by the child's body weight or surface area (to maintain count density on planar imaging), subject to a minimum acceptable amount for very small children and infants (76, 77). Tumor-seeking radiopharmaceuticals used for imaging in pediatric oncology, radionuclide therapy, and for monitoring response to treatment have been summarized by Hoefnagel and de Kraker (78). Effective dose as a function of age for various radiopharmaceuticals is illustrated in Figure 4.3 (41, 73). Although the radiation dose is not the prime concern for these patients, FDG dose is clearly superior to ^{201}Tl-chloride or ^{67}Ga-citrate and is the radiopharmaceutical of choice for imaging, particularly when serial studies are indicated.

Pediatric CT factors are also required. As body size decreases and under the same exposure conditions, there is a marked increase in absorbed dose and the dose distribution from the periphery to the center becomes more uniform. It is important to reduce mA level to as few as 25 mAs for infants. Tube voltage should also be reduced for infants and small children to 100 kVp; at 80 kVp, an increase in beam hardening artifacts has been reported (79). For whole-body PET/CT scans, a protocol with a total beam width of 5 mm or more and a pitch greater than 1 is appropriate. Dose to children can be estimated using CT-Expo.xls phantoms for a 7-year-old and a baby, by applying scaling factors to the results calculated for adults using CTDosimetry.xls from v0.99r onward, or by using conversion factors at standard ages (0, 1, 5, and 10 years old) for DLP in various body regions (54, 55, 61).

The Lactating Patient

PET studies with FDG for oncology or epilepsy investigations are infrequently requested for a woman who is breast-feeding an infant. Avid uptake of FDG in lactating breast tissue has been reported in a small series of patients (80). The uptake of FDG appears to be mediated by the GLUT-1 transporter, which is activated by suckling, not by prolactin. By imaging the breast before and after the expression of milk and counting the activity in milk samples, it was confirmed that FDG, being metabolically blocked, is not secreted in milk to any significant amount but is retained in glandular tissue. The dose to glandular tissue will be higher than the value for the nonlactating breast of 0.0117 mGy per MBq injected (81). The ^{18}F concentration in milk, measured in samples from one patient, was about 10 Bq mL^{-1} at 1 h and 5 Bq mL^{-1} at 3 h postinjection. It was postulated that the ^{18}F activity is associated with the cellular elements in milk, mainly lymphocytes. Using the standard model of breast-feeding with the first feed at 3 h postinjection (82), it was estimated that the dose to the infant from ingested milk would not exceed 85 μSv following an injection of up to 160 MBq FDG (and by

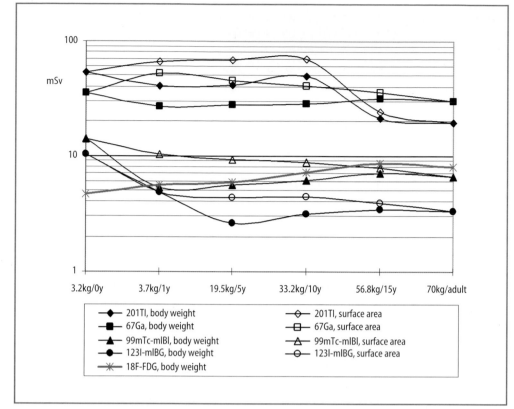

Figure 4.3. Dosimetry of radionuclide imaging in pediatric oncology. Administered activity scaled to adult dosage by body weight or surface area between limits of 20–120 MBq ^{201}Tl-chloride, 30–300 MBq ^{67}Ga-citrate, 100–720 MBq ^{99m}Tc-MIBI, and 70–250 MBq ^{123}I-MIBG, and by 6 MBq kg^{-1} body weight for ^{18}F-FDG (half this amount has been recommended for brain scans (107). Calculated from data in ICRP 80 (41) and Stabin (personal communication) for ^{18}F-FDG in newborn infants and for contaminant-free ^{201}Tl-chloride.

extrapolation up to 200 μSv after injection of 400 MBq). In addition, the infant would receive an external dose while being held while feeding from the breast or bottle. The dose rate against the chest could approach 200 μSv h^{-1} at 2 h after injection of 400 MBq FDG. Breast feeding and cuddling of the infant should be postponed for several hours after an FDG study if the infant's dose is to be kept below a dose constraint of 0.3 mSv for a single event (see following section).

The Pregnant Patient

Uterus dose from a diagnostic CT scan of the pelvis is typically about 30 to 40 mGy but could be considerably higher. One or two examinations of the abdomen/pelvis, for example, CT with and without contrast or followed by PET/CT, could cause the uterus dose to reach a level of 50 mGy, at which careful and individual fetal dosimetry assessment is recommended (83). In the zone between 100 and 500 mGy where adverse effects of radiation cannot be ruled out, the complex issues of risk and termination may be raised with the parents (12, 84). It is therefore understandable that the prospect of both CT and PET in a pregnant patient immediately raises concerns, especially if the pelvis will be in the field of view. The precautions used in radiology and nuclear medicine to discover before a scan if a female patient of childbearing capacity is or could be pregnant should be rigorously enforced in PET/CT to avoid the risk of inadvertent exposure of an embryo or

fetus. If there is doubt about a pregnancy, a urine pregnancy test should be obtained or the scan postponed until after the next menstrual period.

The exposure of an embryo or fetus from a radiopharmaceutical depends on the biodistribution. In the earliest stages of pregnancy, dose to the embryo is taken to be the same as dose to the uterus. After about 12 weeks, when trophoblastic nutrition has been replaced by placental nutrition, fetal dose depends on whether the radiopharmaceutical or any of its metabolites accumulates in or is transferred across the placenta, as well as on the distribution of activity in the mother. Where placental transfer of radioactivity occurs, the activity is generally assumed to be distributed uniformly in the fetus. Fetal dose at various stages of gestation has been calculated for a range of radiopharmaceuticals using the MIRD methodology and an anatomic model of a pregnant female at 3, 6, and 9 months gestation, but at that time there was no documented evidence of placental transfer of FDG (85). Fetal uptake of FDG has since been imaged in humans (86, 87) and of FDG and ^{11}C-cocaine in nonhuman primates (37, 88). FDG dose coefficients at early pregnancy and at 3, 6, and 9 months gestation are now available for placental transfer and bladder voiding at 2-h and 4-h intervals (89). The shorter interval is more appropriate, as patients are generally instructed to empty the bladder within an hour of injection and to drink fluids after the scan. Iodide is also known to cross the placenta. The fetal thyroid begins to concentrate iodine from about the 13th week of pregnancy and reach a maximum concentration at about the

Table 4.9. Radiation dose to embryo/fetus from PET radiopharmaceuticals.

	Absorbed dose per unit activity administered to mother (mGy MBq−1)			
	Early	3 months	6 months	9 months
18F-FDG, 2-h void[a]	0.018	0.018	0.016	0.015
18F-FDG, 4-h void[a]	0.022	0.022	0.017	0.017
18F-NaF[b]	0.022	0.017	0.0075	0.0068
124I-NaI[b]	0.14	0.1	0.059	0.046
124I-Na, fetal thyroid[c]	—	130	680	300

Source: Data from [a]Stabin (89), [b]Stabin (81), and [c]Watson (90).

5th to 6th month. Fetal thyroid dose from 124I-NaI is included in Table 4.9 (81, 89, 90).

CT dose to an embryo or fetus in early stages of pregnancy is assumed to be the same as the dose to the uterus, which can be estimated with CT dose calculation software. If the uterus is not in the scanned region, the dose falls off exponentially beyond 3 cm from the scan limit. In an early method of estimating dose to the fetus from CT, whether in the primary beam or outside the scanned region, correction factors at 1cm intervals from the fetal midpoint to the upper and lower scan limits were applied to the $CTDI_{100}$ (in the head phantom) for the same conditions of exposure including slice thickness and pitch (91). Monte Carlo simulation has been used to assess dose within and beyond the scanned region, from which look-up tables of dose normalized to 100 mAs at various kVp and beam shaping filters were prepared (92). In mid- to late pregnancy, when the body cross section more closely approximates a circle, the $CTDI_{vol}$ in the 32-cm body phantom at

the appropriate kVp should give a reasonable indication of fetal dose.

It is important to use low-dose CT protocols when PET/CT scanning is indicated in a pregnant patient whose condition allows for the possibility of fetal survival. The protocols should be established in advance. For whole-body oncology studies, the fetus is exposed during both PET and CT components. For brain and cardiac studies, fetal exposure is almost entirely caused by the PET radiopharmaceutical. Depending on the PET/CT scanner characteristics and the stage of pregnancy, it is possible to limit fetal dose from a whole-body scan to less than 15 mGy (350 MBq FDG injection, 120 kVp CT beam 10 mm wide, pitch of 1.5 and 80 mAs) and from a brain scan to less than 4 mGy (from 250 MBq FDG and negligible dose from CT) (Figure 4.4) (86). If the usual injected activity of FDG is reduced, scan time should be scaled up to maintain sensitivity for detection of small lesions. These fetal doses are below the threshold for radiation-induced abnormalities, would have minimal effect (about 1 in 1,000 or less) on the incidence of fatal cancer in childhood, and would not increase the normal risks of pregnancy.

The Volunteer Exposed for Medical Research

One area in which dose estimates are required is the recruitment of volunteers to participate in research studies. Regulatory authorities in many countries have adopted the recommendations of the ICRP (93): an exposure for research purposes is treated on the same basis as a medical exposure and therefore is not subject to a specific

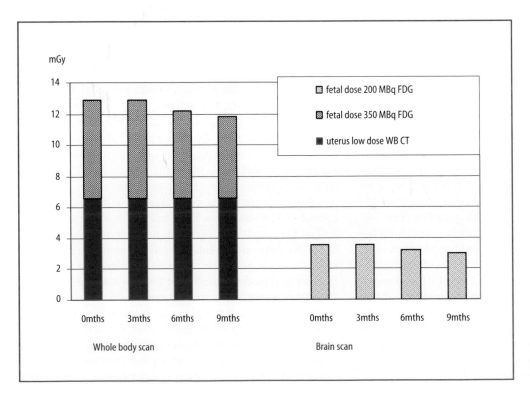

Figure 4.4. Fetal dose estimates for optimized low-dose FDG PET/CT whole-body and brain scan protocols in pregnancy. Calculated for FDG placental crossover and 2-h bladder voiding interval (89); CT whole-body scan Siemens Biograph, 110 kVp, 113 mA, 0.8-s rotation, 2 x 5 mm beam width, pitch 1.5 (55).

dose limit, the dose should be commensurate with the potential benefit of the research findings, and the study protocol should be approved by a properly constituted ethics committee. A constraint on effective dose may apply where the participant is not expected to benefit from the exposure, as in the exposure of "normal" subjects or patients enrolled in a clinical trial that involves additional or different exposures to what would otherwise be required for clinical management. The dose constraint would be considered, for example, in determining the maximum number of injections of ^{15}O-water for repeated tests of cognitive function in normal volunteers. It is unlikely that PET/CT research studies would fall below a dose constraint as low as 5 mSv, and these protocols may require further consideration by a regulatory authority or similar agency. The ICRP has published "realistic maximum" dose coefficients for ^{11}C-labeled substances and generic dose coefficients for ^{11}C- or ^{18}F-labeled amino acids and ^{11}C-brain receptor ligands, which are useful when preparing research submissions involving novel agents for which dosimetry data in humans are not yet available (43).

Occupational and Public Exposures

Health Care Workers Within and Outside the PET Facility

It has been known for many years that the radiation exposure to a technologist performing PET studies is generally higher than for conventional nuclear medicine imaging (94–96). Substantial shielding of syringes, vials, and transmission and quality control sources is standard practice in PET facilities. With inanimate sources effectively shielded, attention has turned to minimizing the exposure to staff from patients. Education of staff on the importance of distance and time is a key factor in dose control (97–100); see also Chapter 3. If operators need to be cross-trained to perform PET/CT procedures, it may be more of a challenge for a radiographer to adapt to handling unsealed sources and "hot" patients and receiving an occupational exposure of several mSv per year, than for a nuclear medicine technologist to prepare patients for a CT scan, follow preset CT protocols, and observe CT radiation safety precautions.

The dose to employees is likely to increase as PET/CT systems proliferate and it becomes feasible to scan 20 patients on one system in a day. In particular, employees who work in mobile PET/CT units may receive higher exposures because of space constraints and less opportunity to share duties requiring close contact with patients. Employee doses reported from different PET facilities are difficult to compare because of the variability in isotope supply, clinical workload, scan protocols, and physical accommodation. Surveys are most useful if quoted in terms of dose per GBq handled or per procedure stating the in-

jected activity. With due care and some rostering (say 50%) with other nuclear medicine duties, it should be possible to keep employee dose at or below 5 mSv year^{-1} while scanning up to 20 PET patients per day.

Task-specific monitoring can be used to identify actions that contribute most to staff exposure and to suggest areas for improvement (101–104). The most significant contribution to dose occurs during close contact with the patient, as is to be expected because vials and syringes can be shielded but dose rates within 0.5 m of the patient can be of the order of 4 to 8 μSv min^{-1} following an injection of FDG. Dose rates near the FDG patient have been measured in a number of studies, at different orientations to the patient, and at different times postinjection (95, 96, 99, 100, 105, 106). Representative values are shown in Table 4.10. From such data, the American Association of Physicists in Medicine (AAPM) proposes a mean normalized dose rate of 92 μSv GBq^{-1} h^{-1} at 1 m in any direction from the patient for radiation protection planning (see following). This value would also be applicable for other PET nuclides with no other gamma emissions after correcting for the branching ratio. An example of the pattern of exposure from individual tasks involving contact with patients is shown in Figure 4.5. Dose rates can be integrated to give an indication of dose per task as shown in Figure 4.6, which if normalized to the radionuclide activity can be used for comparison between tasks, individuals, and facilities.

Close attention should be given to strategies that eliminate, postpone, or shorten time in close contact with the patient. Important measures include explaining the procedure, pointing out where the patient will go, administering medications, setting up EEGs, and establishing intravenous access by cannula with a saline flush syringe on a three-way tap or with an infusion set before administering the dose; flushing the dose immediately from the line, postponing the removal of lines and catheters until the conclusion of scanning, and using a tourniquet or asking the patient to maintain pressure on the puncture site after removal of a line; keeping at a distance while

Table 4.10. External dose rates near FDG patients.

Patient position, measurement location	Distance (m)	Normalized dose rate (95th percentiles) (μSv h^{-1} per MBq injected)	
		Postinjection[a]	2 h post injection[b]
Standing, at anterior chest	0.5	0.60	0.20
	2.0	0.10	0.03
Supine, at side	0.5	0.85	—
	2.0	0.11	—
Supine, at head	0.5	0.36	—
	2.0	0.075	—
Supine, at feet	0.5	0.078	—
	2.0	0.023	—

Source: Data from [a]Benatar et al. (105) and [b]Cronin et al. (106).

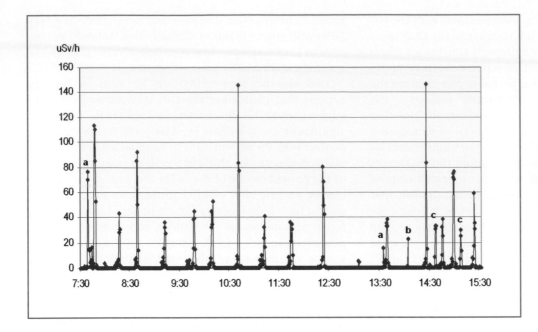

Figure 4.5. Exposure to a positron emission tomography (PET) technologist throughout the day. *Unlabeled peaks:* dispense–calibrate–inject 350 MBq ^{18}F-FDG for 15 patients. Dose rates are generally less than 60 μSv h^{-1} although can be as high as 150 μSv h^{-1} depending on proximity to the injection site. *Other peaks:* (**a**) unpacking two FDG deliveries; (**b**) directing a patient to toilet; (**c**) taking two blood samples. (Royal Prince Alfred Hospital, measurements with Eberline FH 41B-10 system; courtesy of R. Smart.)

directing patients to the toilet, using a wheelchair to move frail patients to and from the scanner as quickly as possible, and enlisting other persons to assist with patient handling. Nurses working within hospital PET facilities that scan many high-dependency patients may be the "critical group" so far as staff exposure is concerned. Specific practices are recommended for pediatric patients (107, 108). The quantitative measurement of cerebral glucose metabolism originally required a number of blood samples to be taken over a period of 30 to 40 min following the injection of FDG, resulting in significant operator exposure (109). A two-sample method has been developed with a fivefold reduction in operator dose per study (110).

Hand doses in PET may also be higher than in conventional nuclear medicine and are strongly influenced by technique because the radiation fields around partially shielded syringes and vials are highly directional. The report of a detailed study using thermoluminescent dosimeters (TLD) to measure dose to the hands during dispensing of FDG injections contains information that could be extrapolated to other facilities (111). The skin dose [H′(0.07)] measured at 18 locations on each hand ranged from about 100 to 300 μSv per GBq dispensed. The dose was higher on the nondominant hand. The dose at the base of the middle finger, a preferred position for wearing a ring dosimeter, was close to the average for the hand

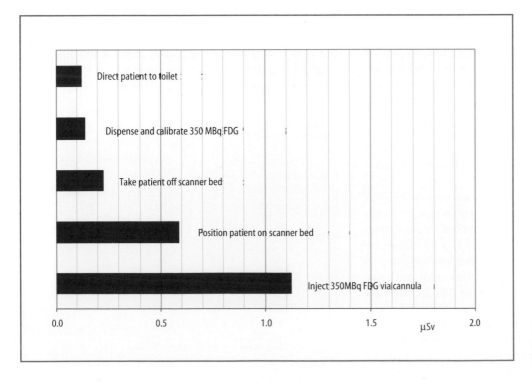

Figure 4.6. Dose to PET technologists performing specific tasks. (Royal Prince Alfred Hospital, measurements with Eberline FH 41B-10 dosemeter.)

and about half of the maximum dose location for the hand. The use of a tungsten syringe shield or an automated dispenser did not provide the expected improvement in protection because of the additional handling to remove the syringe shield to calibrate the filled syringe or to remove a filled syringe from the dispenser.

The duties and previous personal dosimetry history of an employee who declares that she is pregnant and wishes to continue at work during the pregnancy should be reviewed. The ICRP recommends that the fetus should be afforded the same level of protection as a member of the public (4) and has suggested that the use of shielding and/or isolation of sources in the workplace usually provides an adequate level of protection without the need for specific restrictions on the employment of pregnant women (3). However, this is not always possible in a PET workplace, and restrictions on the contact a pregnant employee has with patients are likely to be needed.

Following a PET scan, the patient may come into close contact with other health professionals. Dose rate measurements at various distances from the patient on leaving the PET facility, combined with modeling of potential patterns of close contact, indicated that nurses on a ward that regularly sends patients for PET scans are unlikely to receive more than 24 µSv per day (106). The exposure rate to a sonographer working at 0.5 m from a patient who received 400 MBq FDG 2 h prior would be about 40 µSv h^{-1} (112). In circumstances where a staff member may have frequent contact with PET patients, for example, nursing staff or porters of an oncology ward, personal dosimeters can be used to establish the level of exposure for informed guidance on policies and procedures. In most cases, contact is so infrequent that no special precautions are required.

Family Members, Carers, and Members of the Public

The dose limit of 1 mSv year^{-1} for members of the public recommended by the ICRP has been widely adopted. This limit is used as a criterion when discharging radionuclide therapy patients from hospital, although a dose constraint of 0.3 mSv from any single event has subsequently been proposed (by the European Union) to allow for several exposures occurring during the course of a year (113). The dose to persons near a PET patient depends on the activity in the patient and excretion (if any), and the pattern of close contact effectively within a distance of about 2 m or less. Only ^{18}F sources need be considered. The dose to persons near a patient has been modeled for a number of scenarios, showing there is no need to restrict the activities of patients for the remainder of the day of their PET scan to satisfy a dose constraint of 0.3 mSv per event (106).

Persons knowingly and willingly assisting with the support of the patient are regarded by the ICRP as carers, not subject to the dose limit for members of the public. A

dose constraint of 5 mSv per event has been proposed for carers (113). Other family members—especially children—should be subject to the same dose constraint as members of the public, as it is quite possible that the patient will undergo more than one radionuclide imaging procedure within a year. Not all the accompanying persons in a common waiting area of a PET facility may qualify as carers. For them, the 0.3-mSv dose constraint may be exceeded if they are seated among patients who have been injected with FDG and are waiting to be scanned. Patients should be advised at the time of booking that they should not be accompanied by pregnant women, infants, or children when attending for the scan. If this cannot be arranged, the accompanying persons should neither stay with the patient during the FDG uptake phase nor wait in an area used by other injected patients.

Facilities and Equipment

Facility Planning for Radiation Protection

The main impact of PET/CT on radiation safety is a significant increase in clinical workload that has implications for the layout and shielding of the whole facility. The scanner room is likely to be the busiest point of a facility. Injected patients are a mobile and significant source of exposure, so isolation and internal traffic are important aspects of facility planning. The layout should minimize the distance to escort patients between preparation rooms, toilets, scanner room, and postscan waiting areas and also avoid incidental contacts between patients in transit and other staff. Generously sized change cubicles, toilet areas, and preparation rooms, trolley bays with ready access to the scanner room, and handrails and support aids will reduce staff exposure while assisting frail patients moving to and from the scanner, toileting, or dressing. The scanner design and location should allow a patient to get on and off the scanner bed with minimal assistance, or to be transferred quickly to and from a trolley if necessary. A "cold" waiting area should be available for patients before injection and for accompanying persons. With a capacity of up to 20 FDG scans a day, a single PET/CT scanner could require three patient preparation areas for FDG injection and uptake. A single preparation area shared by all patients is not recommended as it would unnecessarily increase the exposure of the person giving the injections. Space may have to be found within the limits of an existing facility, remembering that other imaging and counting equipment should be separated or shielded to prevent interference from patients and syringes.

Shielding

The scanner room, preparation rooms, and possibly a postscan waiting area may need substantial shielding depending on the degree of isolation from occupied areas and other imaging and counting equipment. Professional organizations such as the AAPM may be consulted for authoritative information on design and shielding issues. PET radionuclides present more of a challenge than other radionuclides used for diagnostic imaging in nuclear medicine, or diagnostic X-rays for that matter, because of their higher photon energy and hence smaller cross section for photoelectric absorption. The objective of shielding design is to determine the transmission ratio (B) of dose rates, or of dose integrated over a specified interval, with and without the shield in place. The thickness of shielding material to achieve the desired value of B can then be calculated or obtained from published data.

Small sources in the workplace or during transport should be shielded to attenuate the maximum dose rate intensity to an acceptable value ($B = I/I_0$) at a nominal close distance, say less than 10 μSv h^{-1} at 0.3 m, or as specified in Transport codes. Dose rates at the distance of interest can be calculated from the rate constants in Table 4.4. The appropriate quantity to use for operational purposes is the ambient dose equivalent [H*(10)] or deep dose equivalent (DDE), which is the dose at 10-mm depth in tissue, rather than air kerma (114, 115). This criterion allows a small margin of safety above the effective dose E (or effective dose equivalent EDE or H_E) in which dose limits are defined using tissue weighting factors of various vintage, but which cannot be measured "in principle or in practice". If necessary, factors are available to convert from air kerma to H*(10) and E (116) or DDE and EDE (117).

Vials and syringes are treated as point sources, with the dose rate being inversely related to the square of the distance from the source. The dose rate at a distance from a line source can be calculated by the appropriate formula (118, 119). The values in Table 4.5 of mean half- and tenth-value layers (HVL and TVL) are suitable for estimating shield thickness in high-density materials such as lead and tungsten. Lead thickness in a clinical PET setting is typically 50 mm for bench shields, storage caves for waste, and PET camera quality control sources, 30 mm for vial containers located behind a bench shield, and 15 mm for syringe shields. The lead glass for a window in a bench shield has superior optical transmission if supplied as a single piece rather than stacked sheets. Vial and syringe shields are too heavy to manipulate with safety so mechanical supports are necessary when dispensing and injecting PET radiopharmaceuticals. Tungsten may be preferable to lead for small PET source containers. For example, a cylindrical pot for a vial 5 cm high and 2 cm in diameter designed for 1% transmission (two tenth-value layers) would be approximately 25% heavier and 30% wider if fabricated in lead rather than tungsten.

Plastic liners may be used within lead or tungsten vial and syringe shields to absorb all positrons, although most positrons from ^{18}F would be absorbed in the vial or syringe wall. ^{15}O positrons absorbed in lead could generate bremsstrahlung (X-ray) photons up to their maximum energy of 1.7 MeV. However the energy converted to bremsstrahlung radiation is a small fraction of the average positron energy incident on a lead or tungsten shield (from less than 2% in the case of ^{18}F to 5% in the case of ^{11}C). The practical value of plastic syringe shield inserts is to increase the distance of the fingers from the source and possibly screen the skin from longer-range positrons.

Rooms used by "hot" patients generally require shielding. The room used for storing and dispensing PET radiopharmaceuticals and storing transmission/QC sources may also need some shielding to supplement source containers. A conservative approach to shielding design may avoid the need for expensive retrofits as technology improves but must be balanced against the cost (expense and space) of the safety margin, which could be considerably more in PET than in radiology. A reasonable estimate of future workload is needed for both PET and CT in terms of number of patients, type of study, and protocol. With the advent of PET/CT, the number of patients scanned in a day is limited more by the time for patient handling, including changeover, positioning, and clinical purposes such as administration of contrast or markup for radiation therapy, than by the time for the transmission scan.

A similar design method can be used for X-ray and PET sources (120–122). Barriers between the source and an occupied area should attenuate the dose D_0 without shielding for the maximum anticipated workload during a specified interval, usually 1 week, to an acceptable design limit D ($B = D/D_0$). The value of D depends on who has access to the area in question, and for how long. Regulatory authorities should be consulted for local requirements. The limit may be adjusted for partial occupancy of the area by individuals while the source is present, normally taken to be a 40-h working week. By convention, full-time occupancy is assumed for "controlled" work areas, and the regulatory design limit for these areas could be as high as 100 or 120 μSv per week (96). However, because PET staff must also have close contact with radioactive patients, for example, about 30 min per day at centers scanning up to 10 patients per day (105, 123), their exposure at all other times should be kept as low as reasonably achievable by designing barriers to 20 μSv or less in a week, particularly for the control room where they spend a lot of time. A low level of ambient radiation in the workplace is reassuring when recruiting staff and reduces the need for rostering to meet dose constraints. For areas to which the public has access, the design limit is usually 20 μSv per week or a lesser dose constraint. Default values for occupancy factors (T= 1) recommended by the NCRP or other bodies for public areas (120) can be used if the anticipated use of the area does not allow a firm estimate of occupancy.

The unshielded dose (D_0) at a specified distance can be determined from the workload (W) expressed as GBq-h per week for radionuclides or mA-min per week for head and body scans at specified kVp for X-ray apparatus. PET workloads for each room are determined by the activity injected and excreted, the number of patients, and the time of entry and exit to the room. W, the activity-time integral in each room, should allow for radioactive decay and excretion losses before the patient's entry and radioactive decay while in the room. A reasonably conservative approach would be to assume a high throughput of patients and injected activity based on the scanner specifications and the experience of busy centers. Currently, about 90% of patients undergo oncology (whole-body) studies, being scanned from 45 to 90 min after injection of 150 to 800 MBq FDG. Immediately before the scan, the patient is asked to empty the bladder, removing about 10% to 20% of the injected activity that has been excreted in urine. Scan time can range from 15 to 45 minutes or more. The remaining 10% of patients generally undergo brain or cardiac studies of shorter duration. The daily pattern of unshielded dose rate in a PET/CT scanner room is illustrated in Figure 4.7.

D_0 is calculated from the product of W and the dose rate at 1 m from the patient, corrected for distance. The dose rate constants for a point source are not applicable for an extended source geometry and do not allow for self-attenuation in the source. A mean dose rate of 92 μSv h^{-1} per GBq at 1 m in any direction from an FDG patient, based on measurements reported in the literature, is proposed by the AAPM (D. Simpkin, personal communication). No allowance is made for attenuation of 511 keV in the scanner gantry or other hardware, which is a very conservative assumption. When calculating distance, the patient is regarded as a point source located 1 m above the floor, at the midpoint of the scan range or on the bed or

chair used during the uptake phase. The point of occupation is taken to be 0.5 m beyond a wall, 0.5 m above the floor level above, or 1.7 m above the floor level below. The inverse square law is generally used to adjust the dose for distance. Except for the scanner room itself, room dimensions and distances to the nearest occupied areas may be small, as in a mobile trailer facility. At close range (less than 3 m), an inverse 1.5 power of the distance is more appropriate. For example, to reduce the dose from a workload of 12 GBq-h per week to 20 μSv at 2 m or 4 m, the required B would be $20/(12 \times 92/2^{1.5}) = 0.05$ or $20/(12 \times 92/4^2) = 0.29$, respectively.

The upper limit on the estimate of CT workload would be to assume the system is used for diagnostic CT only (for example, if isotope supply were interrupted) with more patients and higher exposure factors, for example, 200 patients and 40,000 mA-min per week. It may be more realistic to assume the same number of patients as for the PET workload and diagnostic or non-diagnostic CT protocols as per local policy. The primary CT beam is absorbed in the patient or scanner, leaving only leakage and scattered radiation, which is of short duration and substantially lower energy than 511 keV, to consider. The unshielded dose can be estimated from the workload and isodose contours or a scatter distribution grid map (dose per unit workload) supplied by the manufacturer.

The thickness of a barrier for the required value of B can be determined from half- and tenth-value layers or published transmission curves. Under idealized "narrow beam" conditions with scatter excluded by collimation of the source and detector, the attenuation of a monochro-

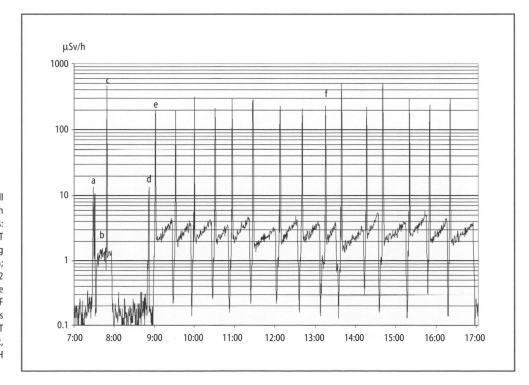

Figure 4.7. Pattern of exposure on wall at head end of Siemens Biograph scanner throughout the day. Dose rates: (**a**) CT warm-up and calibration; (**b**) PET phantom quality control; (**c**) CT during PET/CT acquisition with PET phantom; (**d**) CT calibration; (**e**) CT for 1 of 12 whole-body scans; (**f**) CT for one of three brain scans. During PET acquisitions, ^{18}F dose rate rises as scanner bed moves through the gantry toward the wall. PET and CT radiation detected at 1-m height, 2.1 m from isocenter with Eberline FH 41B-10.

matic beam of radiation through an absorbing medium is described by the following equation:

$$I = I_0 \, e^{-\mu x}$$

In this case, B is a simple exponential attenuation factor and µ is the linear attenuation coefficient at a given energy for the shielding material concerned. The thickness (x) of the shield for a given energy and material would be calculated from known values of µ, or of the total mass attenuation coefficient µ/ρ and the density ρ (118, 119). However a simple exponential function is not suitable for the "broad beam" geometry of extended radionuclide sources with energies of more than a few hundred keV and barrier materials of low atomic number in which Compton scattering is the predominant interaction, as for 511-keV PET photons incident on concrete barriers. Under broad beam conditions, scatter in the forward direction builds up in the barrier until an equilibrium depth is reached beyond which attenuation is more nearly exponential. Broad beam transmission can be estimated by point kernel or Monte Carlo modeling. Alternatively, the half- and tenth-value layers given in Table 4.5 can be used as they apply to broad beam conditions and moderate B values, not being derived from µ for narrow beams or from an average attenuation over many orders of magnitude as in NCRP Report No. 49 (19, 20, 124). The various estimates of transmission of 511-keV photons through lead and concrete are shown in Figure 4.8, indicating how narrow beam analysis significantly underestimates transmission through concrete (20, 118, 125).

PET barriers around patients generally require modest attenuation with B of the order of 0.1 to 0.4, but this can

translate to substantial thickness of lead or concrete. In marked contrast, typical values of B for a CT installation are in the range of 10^{-3} to 10^{-4}. Transmission curves in lead and concrete for CT secondary radiation (scatter and leakage) at various kVp calculated by Simpkin are available (120, 122, 126). Attenuation is not a single exponential function because the beam is not monoenergetic. The final slope represents the attenuation coefficient for the highest energies, as from leakage radiation. The thickness of lead required for CT installations is generally less than 2 mm, which has little effect on 511-keV photons, whereas the thickness of concrete may be similar for PET and CT requirements. Figure 4.9 illustrates the attenuation through barriers of lead or concrete for a notional weekly workload of 100 patients injected with 500 MBq FDG for whole-body scans using low-dose CT protocols. It is necessary to consider PET and CT workloads and transmission independently and estimate the contribution of each to the total dose for various barrier material options.

Cost factors apart, lead has the advantage over concrete for walls because it requires less space and reduces the floor loading by roughly half. A combination of the two may be an effective solution. It is preferable to position the doors to the scanner room where shielding is required for CT only. All the doors to the scanner room should have the usual radiation warning sign and light for CT and should not be interlocked to the operation of the scanner. Depending on the view of the regulatory authority, it may be possible to avoid the considerable expense of a large leaded glass viewing window to the control room, or at least minimize the size of the window, by using video camera surveillance to give the technologist a clear view of

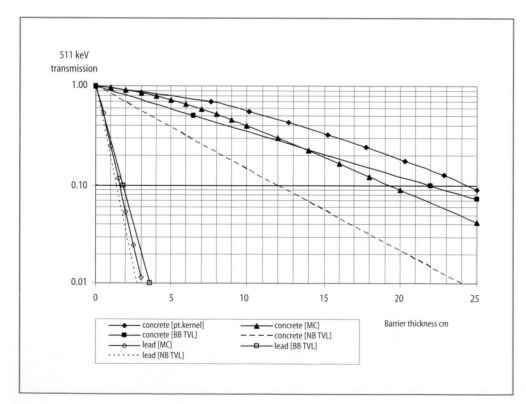

Figure 4.8. Attenuation of 511-keV photons in concrete (2.2 g cm^{-3}) and lead, calculated by Monte Carlo and point kernel modeling (125); Simpkin and Courtney, personal communications) and from broad and narrow beam linear attenuation coefficients (20, 118).

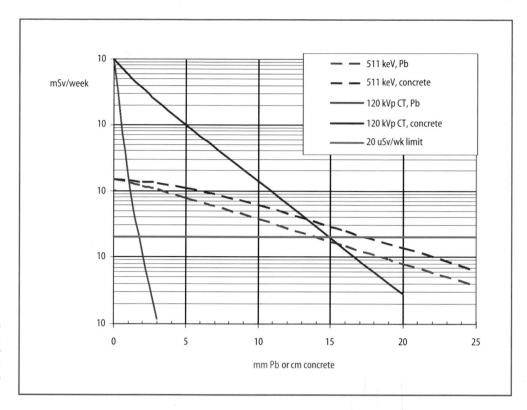

Figure 4.9. Shielding estimation for area adjacent to a PET/CT scanner that receives 0.15 mSv from ¹⁸F and 10 mSv from 120-kVp CT secondary radiation in a week from notional workload of 100 patients injected with 500 MBq ¹⁸F-FDG.

the patient and entry doors. If not, it is important that the attenuation properties of the window are specified at 511 keV as well as for the X-rays. The linear attenuation coefficient of leaded glass is the sum of the linear attenuation coefficients for lead and glass in the mixture, which can be calculated from their respective mass attenuation coefficients and their densities in the mixture if the percentages by weight are known (127). For example, a window of glass that is 55% lead by weight and 1.4 cm thick would have a transmission B of about 0.62, allowing for buildup (P. Brown, personal communication).

Containment of Sources

Spills are an uncommon event and usually result from mishaps with intravenous lines or urinary catheters. Strategically located dispensers for disposable gloves are convenient when dealing with radiopharmaceuticals, patients, and waste. The importance of gloves and monitoring can be seen from the dose rates for skin contamination in Table 4.5: a droplet from an FDG solution with a concentration of 100 MBq mL⁻¹ could deliver 500 mSv, the annual dose limit for the skin, in just 6 min.

In facilities located near a cyclotron, PET gas tracers may be used. The gas supply and return lines from the radiochemistry laboratory to the scanner room will require shielding; the scanner manufacturer may specify an allowable maximum dose rate adjacent to the gantry. In occupied areas and the PET scanner room itself, a thickness of 20 to 25 mm of lead around the lines may be sufficient. A mask over the patient's head should effectively contain

the administered gas and scavenge the exhaled gas for venting via a stack to the atmosphere. The air of the PET scanner room should be continuously monitored during a gas study. Because of the high background radiation level in the room, an air sampler is required to pass the air through a sensitive detector in a remote low-background area. The scanner room should be kept at negative pressure to the adjacent areas. The room air should not be recirculated but vented direct to atmosphere.

Radiation Instrumentation

The dose calibrator used in general nuclear medicine applications is adequate for PET in a clinical setting. A high ranging chamber may be required in a PET production laboratory if measuring very high activities. The chamber should be provided with additional shielding, up to 50 mm of lead, to protect the operator during PET nuclide measurements. With 511-keV photons, no corrections should be required for the geometry of the source container (e.g., syringe, vial) or volume; the manufacturer's settings possibly overestimate the activity of ¹⁸F by 3% to 6% depending on the geometry (128).

Radiation instrumentation should include a survey meter, preferably a dual-purpose instrument for measurement of dose rate and surface contamination. Geiger-Mueller (GM) detectors have good sensitivity to PET nuclide emissions, and their energy response is fairly uniform over the photon energy range of a few hundred keV. No energy response correction is necessary for a GM meter that has been calibrated at 660 keV with a ¹³⁷Cs source. Finally, an electronic personal dosimeter is a very useful investment for monitoring staff in training or performing tasks where dose rates are high or there is prolonged close contact with a source.

References

1. Beyer T, Antoch G, Muller S, et al. Acquisition protocol considerations for combined PET/CT imaging. J Nucl Med 2004;45:25S–35S.
2. Antoch G, Freudenberg L, Beyer T, et al. To enhance or not enhance? 18F-FDG and CT contrast agents in dual-modality 18F-FDG PET/CT. J Nucl Med 2004;45:56S–65S.
3. International Commission on Radiological Protection. 1990 Recommendations of the International Commission on Radiological Protection. ICRP Publication 60. Oxford: Pergamon, 1991.
4. International Commission on Radiological Protection. ICRP 2005 Draft Recommendations: www.icrp.org; 2005.
5. United Nations Scientific Committee on the Effects of Atomic Radiation. Sources and effects of ionizing radiation. Vol 1: Sources. UNSCEAR 2000 Report. UNSCEAR, 2000.
6. Preston D, Pierce D, Shimizu Y, et al. Dose response and temporal patterns of radiation-associated solid cancer risks. Health Phys 2003;85:43–46.
7. Ron E. Cancer risks from medical radiation. Health Phys 2003;85:47–59.
8. National Radiological Protection Board. Board Statement on Diagnostic Medical Exposures to Ionising Radiation During Pregnancy and Estimates of Late Radiation Risks to the UK Population. Chilton: NRPB, 1993.
9. Brenner DJ, Elliston CD, Hall EJ, et al. Estimated risks of radiation-induced fatal cancers from pediatric CT. AJR 2001;176:289–296.
10. Donnelly LF, Emery KH, Brody AS, et al. Minimizing radiation dose for pediatric body applications of the single-detector helical CT: strategies at a large children's hospital. AJR 2001;176:303–306.
11. Paterson A, Frush DP, Donnelly LF. Helical CT of the body: are settings adjusted for pediatric patients? AJR 2001;176:297–301.
12. International Commission on Radiological Protection. Pregnancy and Medical Radiation. ICRP Publication 84. Oxford: Pergamon, 2000.
13. Lowe SA. Diagnostic radiography in pregnancy: risks and reality. ANZJOG 2004;44:191–196.
14. Wakeford R, Little MP. Childhood cancer after low-level intrauterine exposure to radiation. J Radiol Prot 2002;22:A123–A127.
15. Doll R WR. Risk of childhood cancer from fetal irradiation. Br J Radiol 1997;70:130–139.
16. Dillman LT. Radionuclide decay schemes and nuclear parameters for use in radiation-dose estimation. MIRD Pamphlet No. 4. J Nucl Med 1969;10(suppl 2).
17. Dillman LT. Radionuclide decay schemes and nuclear parameters for use in radiation-dose estimation, Part 2. MIRD Pamphlet No. 6. J Nucl Med 1969;11(suppl 4).
18. Groenewald W, Wasserman H. Constants for calculating ambient and directional dose equivalents from radionuclide point sources. Health Phys 1990;58:655–658.
19. Delacroix D, Guerre JP, Leblanc P, et al. Radionuclide and Radiation Protection Handbook 1998. Radiat Prot Dosim 1998;76.
20. Wachsmann F, Drexler G. Graphs and Tables for Use in Radiology. Berlin: Springer-Verlag, 1976.
21. Seibert JA. X-ray imaging physics for nuclear medicine technologists. Part 1: Basic principles of X-ray production. J Nucl Med Technol 2004;32:139–147.
22. International Commission on Radiological Protection. Managing Patient Dose in Computed Tomography. ICRP Publication 87. Oxford: Pergamon, 2000.
23. European Commission. Report No. EUR 16262 EN. European Guidelines on Quality Criteria for Computed Tomography, 1999.
24. Rizzo SMR, Kalra MK, Maher MM, et al. Do metallic endoprostheses increase radiation dose associated with automatic tube-current modulation in abdominal-pelvic MDCT? A phantom and patient study. AJR 2005;184:491–496.
25. Keat N. CT scanner automatic exposure control systems. Report No. 05016. ImPACT CT Scanner Evaluation Centre, 2005.
26. Britten A, Crotty M, Kiremidjian H, et al. The addition of computer simulated noise to investigate radiation dose and image quality in images with spatial correlation of statistical noise: an example application to X-ray CT of the brain. Br J Radiol 2004;77:323–328.
27. Platten D, Keat N, Lewis MH, et al. Sixteen slice CT scanner comparison report. Report No. 05013. Version 12. ImPACT CT Scanner Evaluation Centre, 2005.
28. Nagel HD, editor. Radiation Exposure in Computed Tomography, 4th ed. Hamburg: CTB Publications, 2002.
29. Watson CC, Casey M, Beyer T, et al. Evaluation of clinical PET count rate performance. IEEE Trans Nucl Sci 2003;50:1379–1385.
30. Watson CC, Casey ME, Bendriem B, et al. Optimizing injected dose in clinical PET imaging by accurately modeling the count rate response functions specific to individual patient scans. J Nucl Med 2005;46(11):1825–1834.
31. Eberl S, Fulham MJ, Meikle SR et al. Optimization of injected dose and scanning protocol for 3D whole body PET/CT studies [abstract]. J Nucl Med 2004;45(suppl):426P.
32. Everaert H, Vanhove C, Lahoutte T, et al. Optimal dose of 18F-FDG required for whole-body PET using an LSO PET camera. Eur J Nucl Med Mol Imaging 2003;30:1615–1619.
33. Halpern BS, Dahlbom M, Quon A, et al. Impact of patient weight and emission scan duration on PET/CT image quality and lesion detectability. J Nucl Med 2004;45:797–801.
34. Loevinger R, Budinger TF, Watson EE. MIRD Primer for absorbed dose calculations. New York: Society of Nuclear Medicine, 1988.
35. Stabin M, Sparks R, Crowe E. OLINDA/EXM Personal Computer Code. Radiation Dose Assessment Resource (RADAR) www.dose-info-radar.com; 2004.
36. Stabin M, Siegel BA. Physical models and dose factors for use in internal dose assessment. Health Phys 2003;85:294–310.
37. Benveniste H, Fowler JS, Rooney WD, et al. Maternal-fetal in vivo imaging: a combined PET and MRI study. J Nucl Med 2003;44:1522–1530.
38. Deloar HM, Fujiwara T, Shidahara M, et al. Estimation of absorbed dose for 2-[F-18]fluoro-2-deoxy-D-glucose using whole-body positron emission tomography and magnetic resonance imaging. Eur J Nucl Med 1998;25:565–574.
39. International Commission on Radiological Protection. Radiation Dose to Patients from Radiopharmaceuticals. ICRP Publication 53. Oxford: Pergamon, 1987.
40. International Commission on Radiological Protection. Radiation Dose to Patients from Radiopharmaceuticals: Addendum 1 to ICRP53. ICRP Publication 62. Oxford: Pergamon, 1991.
41. International Commission on Radiological Protection. Radiation Dose to Patients from Radiopharmaceuticals Addendum 2 to ICRP53. ICRP Publication 80. Oxford: Pergamon, 1998.
42. International Commission on Radiological Protection. Radiation Dose to Patients from Radiopharmaceuticals Addendum 3 to ICRP53. www.icrp.org, 2000.
43. International Commission on Radiological Protection. Radiation Dose to Patients from Radiopharmaceuticals. Addenda 4-7 to ICRP Publication 53: www.icrp.org, 2001.
44. Hays MT, Watson EE, Thomas SR, et al. MIRD Dose Estimate Report No. 19: Radiation absorbed dose estimates from 18F-FDG. J Nucl Med 2002;43:210–214.
45. Hays MT, Segall GM. A mathematical model for the distribution of fluorodeoxyglucose in humans. J Nucl Med 1999;40:1358–1366.
46. Brown WD, Oakes TR, DeJesus OT, et al. Fluorine-18-fluoro-L-DOPA dosimetry with carbidopa pre-treatment. J Nucl Med 1998;39:1884–1891.

47. Brihaye C, Depresseux JC, Comar D. Radiation dosimetry for bolus administration of oxygen-15 water. J Nucl Med 1995;36:651–656.

48. Smith T, Carrison T, Lammertsma AA, et al. Dosimetry of intravenously administered oxygen-15 labelled water in man: a model based on experimental human data from 21 subjects. Eur J Nucl Med 1994;21:1126–1134.

49. Deloar HM, Watabe H, Nakamura T, et al. Internal dose estimation including the nasal cavity and major airway for continuous inhalation of C15O2, 15O2 and C15O using the thermoluminescent method. J Nucl Med 1997;38:1603–1613.

50. Thomas SR, Stabin M, Chen C-T, et al. MIRD Pamphlet No.14 revised: a dynamic urinary bladder model for radiation dose calculations. J Nucl Med 1999;40:102S–123S.

51. Dowd MT, Chen C-T, Wendel MJ, et al. Radiation dose to the bladder wall from 2-[18F] fluoro-2-deoxy-D-glucose in adult humans. J Nuc Med 1991;32:707–712.

52. Wrobel M, Carey JE, Sherman P, et al. Simplifying the dosimetry of carbon-11-labelled radiopharmaceuticals. J Nucl Med 1997;38:654–660.

53. Almeida P, Bendriem B, de Dreuille O, et al. Dosimetry of transmission measurements in nuclear medicine: a study using anthropomorphic phantoms and thermoluminescent dosimeters. Eur J Nucl Med 1998;25:1435–1441.

54. Stamm G, Nagel HD. CT-expo: a novel program for dose evaluation in CT. Rofo Fortschr Geb Rontgenstr Neuen Bildgeb Verfahr 2002;174:1570–1576.

55. ImPACT CT scanner evaluation centre. CTDosimetry.xls 0.99. Medicines and Healthcare Products Regulatory Agency (MHRA), UK, 2004.

56. Jones D, Shrimpton PC. Survey of CT Practice in the UK. Part 3: Normalised organ doses calculated using Monte Carlo techniques. NRPB-R250. Report No. NRPB-R250. National Radiological Protection Board, UK, 1991.

57. Jarry G, DeMarco JJ, Beifuss U, et al. A Monte Carlo-based method to estimate radiation dose from spiral CT: from phantom testing to patient-specific models. Phys Med Biol 2003;48:2645–2663.

58. Wu T-H, Huang Y-H, Lee JJS, et al. Radiation exposure during transmission measurement: comparison between CT- and germanium-based techniques with a current PET scanner. Eur J Nucl Med Mol Imaging 2004;31:38–43.

59. Huda W, Mergo PJ. How will the introduction of multi-slice CT affect patient doses? In: Proceedings, IAEA International Conference, Malaga 2001 Radiological Protection of Patients in Diagnostic and Interventional Radiology, Nuclear Medicine and Radiotherapy. Vienna: IAEA, 2001:202–205.

60. Hidajat N, Wolf M, Nunnemann A, et al. Survey of conventional and spiral CT doses. Radiology 2001;218:395–401.

61. Shrimpton P, Hillier MC, Lewis MH, et al. Doses from Computed Tomography (CT) Examinations in the UK: 2003 Review. NRPB-W67. Chilton, UK: National Radiological Protection Board, 2005.

62. Thomton FJ, Paulson EK, Yoshizumi TT, et al. Single versus multidetector row CT: comparison of radiation doses and dose profiles. Acad Radiol 2003;10:379–385.

63. Hamberg LM, Rhea JT, Hunter GJ, et al. Multi-detector row CT: Radiation dose characteristics. Radiology 2003;226:762–772.

64. Lewis MH. Radiation dose issues in multi-slice CT scanning. ImPACT technology update no. 3. ImPACT CT Scanner Evaluation Centre, 2004.

65. Platten D. CT issues in PET/CT scanning. ImPACT technology update no. 4. ImPACT CT Scanner Evaluation Centre, 2004.

66. Beyer T, Antoch G, Bockisch A, et al. Optimized intravenous contrast administration for diagnostic whole-body 18F-FDG PET/CT. J Nucl Med 2005;46:429–435.

67. Shrimpton PC, Jones DG, Hillier MC, et al. Survey of CT practice in the UK. Part 2: Dosimetric aspects. Report No. NRPB-R249. London: HMSO/NRPB, 1991.

68. Olerud H. Analysis of factors influencing patient dose from CT in Norway. Radiat Prot Dosim 1997;71:123–133.

69. McLean D, Malitz N, Lewis S. Survey of effective dose levels from typical paediatric CT protocols. Australas Radiol 2003;47:135–142.

70. International Commission on Radiological Protection. Radiological Protection and Safety in Medicine. ICRP Publication 73. Oxford: Pergamon, 1996.

71. MSCT. 2004 CT quality criteria: European Concerted Action on CT (FIGM-CT-2000-20078). Published at www.msct.info/CT_Quality_Criteria.htm, 2004.

72. McCollough CH, Bruesewitz MR, McNitt-Gray MF, et al. The phantom portion of the American College of Radiology (ACR) Computed Tomography (CT) accreditation program: Practical tips, artifact examples, and pitfalls to avoid. Med Phys 2004;31:2423–2442.

73. Thomas S, Stabin M, Castronovo F. Radiation-absorbed dose from 201Tl-thallous chloride. J Nucl Med 2005;46:502–508.

74. Thompson J, Tingey D. Radiation doses from computed tomography in Australia. Report No. ARL TR/123. Australian Radiation Laboratory (now Australian Radiation Protection and Nuclear Safety Agency), 1997.

75. Brix G, Lechel U, Glatting G, et al. Radiation exposure of patients undergoing whole-body dual-modality 18F-FDG PET/CT examinations. J Nucl Med 2005;46:608–613.

76. Paediatric Task Group of European Association of Nuclear Medicine. A radiopharmaceuticals schedule for imaging in paediatrics. Eur J Nucl Med 1990;17:127–129.

77. Towson JE, Smart RC, Rossleigh MA. Radiopharmaceutical activities administered for paediatric nuclear medicine procedures in Australia. Radiat Prot Australasia 2000;17:110–120.

78. Hoefnagel C, de Kraker J. Pediatric tumors. In: Ell P, Gambhir S, editors. Nuclear Medicine in Clinical Diagnosis and Treatment, 3rd ed. Edinburgh: Churchill Livingstone, 2004.

79. Cody DD, Moxley DM, Krugh KT, et al. Strategies for formulating appropriate MDCT techniques when imaging the chest, abdomen, and pelvis in pediatric patients. AJR 2004;182:849–859.

80. Hicks RJ, Binns D, Stabin MG. Pattern of uptake and excretion of 18F-FDG in the lactating breast. J Nucl Med 2001;42:1238–1242.

81. Stabin MG. Health concerns related to radiation exposure of the female nuclear medicine patient. Environ Health Perspect 1997;105(suppl 6):1403–409. Also at www.orau/ehsd/ridic.htm.

82. Stabin MG, Breitz HB. Breast milk excretion of radiopharmaceuticals: mechanisms, findings and radiation dosimetry. J Nucl Med 2000;41:863–873.

83. National Council on Radiation Protection and Measurements. Considerations regarding the unintended radiation exposure of the embryo, fetus or nursing child. NCRP Commentary No. 9. Bethesda, MD: NCRP, 1994.

84. Wagner LK, Lester RG, Saldana LR. Exposure of the Pregnant Patient to Diagnostic Radiations. A Guide to Medical Management, 2nd ed. Madison, WI: Medical Physics Publishing, 1997.

85. Russell J, Stabin M, Sparks R. Placental transfer of radiopharmaceuticals and dosimetry in pregnancy. Health Phys 1997;73:747–755.

86. Towson J, Fulham M, Eberl S, et al. Radiation dose from F18-FDG PET/CT scans in pregnancy [abstract]. Eur J Nucl Med 2004;31:S233.

87. Bohuslavizki KH, Kroger S, Klutman S, et al. Pregnancy testing before high-dose radioiodine treatment: a case report. J Nucl Med Technol 1999;27:220–221.

88. Benveniste H, Fowler JS, Rooney WD, et al. Maternal and fetal 11C-cocaine uptake and kinetics measured in vivo by combined PET and MRI in pregnant nonhuman primates. J Nucl Med 2005;46:312–320.

89. Stabin MG. Proposed addendum to previously published fetal dose estimate tables for 18F-FDG. J Nucl Med 2004;45:634–635.

90. Watson EE. Radiation absorbed dose to the human foetal thyroid. In: Proceedings, Fifth International Radiopharmaceutical Dosimetry Symposium, Oak Ridge TN, 1992. Oak Ridge: Oak Ridge Associated Universities, 1992:179–187.

91. Felmlee JP, Gray JE, Leetzow ML, et al. Estimated fetal radiation dose from multislice CT studies. AJR 1990;154:185–190.

92. Boone JM, Cooper VN, Nemzek WR, et al. Monte Carlo assessment of computed tomography dose to tissue adjacent to the scanned volume. Med Phys 2000;27:2393–2407.

93. International Commission on Radiological Protection. Radiological Protection in Biomedical Research. ICRP Publication 62. Oxford: Pergamon Press, 1991.

94. Bloe F, Williams A. Personnel monitoring observations. J Nucl Med Technol 1995;23:82–86.

95. Chiesa C, De Sanctis V, Crippa F, et al. Radiation dose to technicians per nuclear medicine procedure: comparison between technetium-99m, gallium-67 and iodine-131 radiotracers and fluorine-18 fluorodeoxyglucose. Eur J Nucl Med 1997;24:1380–1389.

96. Kearfott KJ, Carey JE, Clemenshaw MN, et al. Radiation protection design for a clinical positron emission tomography imaging suite. Health Phys 1992;63:581–589.

97. Benatar NA, Cronin BF, O'Doherty M. Radiation dose received by staff and carers/escorts following contact with 18F-FDG PET patients [abstract]. Nucl Med Commun 1999;20:462.

98. Bixler A, Springer G, Lovas R. Practical aspects of radiation safety for using fluorine-18. J Nucl Med Technol 1999;27:14–16.

99. Brown TF, Yasillo NJ. Radiation safety considerations for PET centers. J Nucl Med Technol 1997;25:98–102.

100. Dell MA. Radiation safety review for 511-keV emitters in nuclear medicine. J Nucl Med Technol 1997;25:12–17.

101. McElroy NL. Worker dose analysis based on real time dosimetry. Health Phys 1998;74:608–609.

102. Bird NJ, Barber RW, Turner KB, et al. Radiation doses to staff during gamma camera PET [abstract]. Nucl Med Commun 1999;20:471.

103. Towson J, Brackenreg J, Kenny P, et al. Analysis of external exposure to PET technologists [abstract]. Nucl Med Commun 2000;21:497.

104. Zeff B, Yester M. Patient self-attenuation and technologist dose in positron emission tomography. Med Phys 2005;32:861–865.

105. Benatar NA, Cronin BF, O'Doherty M. Radiation dose rates from patients undergoing PET: implications for technologists and waiting areas. Eur J Nucl Med 2000;27:583–589.

106. Cronin B, Marsden PK, O'Doherty MJ. Are restrictions to behaviour of patients required following fluorine-18 fluorodeoxyglucose positron emission tomographic studies? Eur J Nucl Med 1999;26:121–128.

107. Borgwardt L, Larsen HJ, Pedersen K, et al. Practical use and implementation of PET in children in a hospital PET centre. Eur J Nucl Med 2003;30:1389–1397.

108. Roberts EG, Shulkin BL. Technical issues in performing PET studies in pediatric patients. J Nucl Med Technol 2004;32:5–9.

109. McCormick VA, Miklos JA. Radiation dose to positron emission tomography technologists during quantitative versus qualitative studies. J Nucl Med 1993;34:769–772.

110. Eberl S, Anayat AA, Fulton RR, et al. Evaluation of two population-based input functions for quantitative neurological FDG PET studies. Eur J Nucl Med 1997;24:299–304.

111. Berus D, Covens P, Buls N, et al. Extremity doses of workers in nuclear medicine: mapping hand doses in function of manipula-

tion. In: IRPA 11 International Congress Proceedings, Madrid, 2004.

112. Griff M, Berthold T, Buck A. Radiation exposure to sonographers from fluorine-18-FDG PET patients. J Nucl Med Technol 2000;28:186–187.

113. Council of the European Union. Council Directive 96/29/Euratom on basic safety standards for the protection of the health of workers and the general public. Official J Eur Commun 1996;L159:1–114.

114. International Commission on Radiation Units and Measurements. Radiation quantities and units. ICRU Report No. 33. Bethesda, MD: ICRU, 1980.

115. International Commission on Radiation Units and Measurements. Determination of Dose Equivalents Resulting from External Radiation Sources. ICRU Report No. 39. Bethesda, MD: ICRU, 1985.

116. International Commission on Radiological Protection. Conversion Coefficients for Use in Radiological Protection Against External Radiation. ICRP Publication 74. Oxford: Pergamon, 1996.

117. American National Standards Institute/American Nuclear Society. Neutron and Gamma-Ray Fluence-To-Dose Factors. Report No. ANSI/ANS-6.1.1. ANSI/ANS, 1991.

118. Shleien B, editor. The Health Physics and Radiological Health Handbook, 2nd ed. Silver Spring, MD: Scinta, 1992.

119. Cember H. Introduction to Health Physics, 3rd ed. New York: McGraw-Hill, 1996.

120. National Council on Radiation Protection and Measurements. Structural Shielding Design for Medical X-ray Imaging Facilities. Report No. 147. Bethesda, MD: NCRP, 2005.

121. National Council on Radiation Protection and Measurements. Structural Shielding Design and Evaluation for Medical Use of X-rays and Gamma Rays of Energies up to 10 MeV. Report No. 49. Bethesda, MD: NCRP, 1976.

122. Madsen M, Anderson J, Halama J, et al. AAPM Task Group on PET and PET/CT Shielding Requirements. American Association of Physicists in Medicine, 2005 release.

123. Robinson CN, Young JG, Wallace AB, et al. A study of the personal radiation dose received by nuclear medicine technologists working in a dedicated PET center. Health Phys 2005;88(suppl):S17–S21.

124. Negin C, Worku G. Microshield version 4.0: a microcomputer code for shielding analysis and dose assessment. Rockville, MD: Grove Engineering, 1992.

125. Courtney J, Mendez P, Hidalgo-Salvatierra O, et al. Photon shielding for a positron emission tomography suite. Health Phys 2001;81(suppl):S24–S28.

126. Simpkin DJ. Transmission of scatter radiation from computed tomography (CT) scanners determined by a Monte Carlo Calculation. Health Phys 1990;58:363–367.

127. Shapiro J. Radiation Protection, 3rd ed. Cambridge: Harvard University Press, 1990.

128. Zimmerman BE, Kubicek GJ, Cessna JT, et al. Radioassays and experimental evaluation of dose calibrator settings for 18F. Applied Radiat Isot 2001;54:113–122.

5

Artifacts and Normal Variants in Whole-Body PET and PET/CT Imaging

Gary J.R. Cook

The number of clinical applications for positron emission tomography (PET) continues to increase, particularly in the field of oncology. In parallel with this is growth in the number of centers that are able to provide a clinical PET or PET/computed tomography (CT) service. As with any imaging technique, including radiography, ultrasound, CT, magnetic resonance imaging (MRI), and conventional single photon nuclear medicine imaging, there are a large number of normal variants, imaging artifacts, and causes of false-positive results that need to be recognized to avoid misinterpretation. It is particularly important to be aware of potential pitfalls while PET is establishing its place in medical imaging so that the confidence of clinical colleagues and patients is maintained. In addition, the advent of combined PET/CT scanners in clinical imaging practice has brought its own specific pitfalls and artifacts.

The most commonly used PET radiopharmaceutical in clinical practice is ^{18}F-fluorodeoxyglucose (FDG). As it has a half-life of nearly 2 h, it can be transported to sites without a cyclotron, and in view of this and the fact that there is a wealth of clinical data and experience with this compound, it is likely to remain the mainstay of clinical PET for the immediate future.

Mechanisms of Uptake of ^{18}F-Fluorodeoxyglucose

FDG, as an analogue of glucose, is a tracer of energy substrate metabolism, and although it has been known for many years that malignant tumors show increased glycolysis compared to normal tissues, its accumulation is not specific to malignant tissue. FDG is transported into tumor cells by a number of membrane transporter proteins that may be overexpressed in many tumors. FDG is converted to FDG-6-phosphate intracellularly by hexokinase, but unlike glucose does not undergo significant enzymatic reactions and, because of its negative charge, remains effectively trapped in tissue. Glucose-6-phos-phatase-mediated dephosphorylation of FDG occurs only slowly in most tumors, normal myocardium, and brain, and hence the uptake of this tracer is proportional to glycolytic rate. Rarely, tumors may have higher glucose-6-phosphatase activity resulting in relatively low uptake, a feature that has been described in hepatocellular carcinoma (1). Similarly, some tissues have relatively high glucose-6-phosphatase activity, including liver, kidney, intestine, and resting skeletal muscle, and show only low uptake. Conversely, hypoxia, a feature common in malignant tumors, is a factor that may increase FDG uptake, probably through activation of the glycolytic pathway (2).

Hyperglycemia may impair tumor uptake of FDG because of competition with glucose (3), although it appears that chronic hyperglycemia, as seen in diabetic patients, only minimally reduces tumor uptake (4). To optimize tumor uptake, patients are usually asked to fast for 4 to 6 h before injection to minimize insulin levels. This practice has also been shown to reduce uptake of FDG into background tissues including bowel, skeletal muscle, and myocardium (5). In contrast, insulin-induced hypoglycemia may actually impair tumor identification by reducing tumor uptake and increasing background muscle and fat activity (6).

In addition to malignant tissue, FDG uptake may be seen in activated inflammatory cells (7, 8), and its use has even been advocated in the detection of inflammation (9). An area where benign inflammatory uptake of FDG may limit specificity is in the assessment of response to radiotherapy (10). Here, uptake of FDG has been reported in rectal tumors and in the brain in relation to macrophage and inflammatory cell activity (11–13), which may make it difficult to differentiate persistent tumor from inflammatory activity for a number of months following radiotherapy in some tumors. Nonspecific, inflammatory, and reactive uptake has also been recorded following chemotherapy in some tumors (14, 15), and there is no clear consensus on the optimal time to study patients following this form of therapy.

Table 5.1. Normal distribution of ^{18}F-fluorodeoxyglucose (FDG).

Organ/system	Pattern
Central nervous system	High uptake in cortex, basal ganglia, thalami, cerebellum, brainstem. Low uptake into white matter and cerebrospinal fluid.
Cardiovascular system	Variable but homogeneous uptake into left ventricular myocardium. Usually no discernible activity in right ventricle and atria.
Gastrointestinal system	Variable uptake into stomach, small intestine, colon, and rectum.
Reticuloendothelial and lymphatic	Liver and spleen show low-grade diffuse activity. No uptake in normal lymph nodes but moderate activity seen in tonsillar tissue. Age-related uptake is seen in thymic and adenoidal tissue.
Genitourinary system	Urinary excretion can cause variable appearances of the urinary tract. Age-related testicular uptake is seen.
Skeletal muscle	Low activity at rest
Bone marrow	Normal marrow shows uptake that is usually less than liver.
Lung	Low activity (regional variation)

Source: From Valk PE, Bailey DL, Townsend DW, Maisey MN. Positron Emission Tomography: Basic Science and Clinical Practice. Springer-Verlag London Ltd 2003, p. 496.

Normal Distribution of FDG

The normal distribution of FDG is summarized in Table 5.1. The brain typically shows high uptake of FDG in the cortex, thalamus, and basal ganglia. Cortical activity may be reduced in patients who require sedation or a general anaesthetic, a feature that might limit the sensitivity of detection of areas of reduced uptake as in the investigation of epilepsy. It is not usually possible to differentiate low-grade uptake of FDG in white matter from the adjacent ventricular system (Figure 5.1).

In the neck, it is common to see moderate symmetrical activity in tonsillar tissue. This area may be more difficult to recognize as normal tissue if there has been previous surgery or radiotherapy that may distort the anatomy, resulting in asymmetrical activity or even unilateral uptake on the unaffected side. Adenoidal tissue is not usually noticeable in adults but may show marked uptake in children. Another area of lymphoid activity that is commonly seen in children is the thymus. This gland usually has a characteristic shape (an inverted V) and is therefore not usually mistaken for anterior mediastinal tumor (Figure 5.2). Clinical reports vary as to the incidence of diffuse uptake of FDG in the thyroid (16–18). This uptake may be a geographic phenomenon, since its presence is more likely in women and has been correlated with the presence of thyroid autoantibodies and chronic thyroiditis (18).

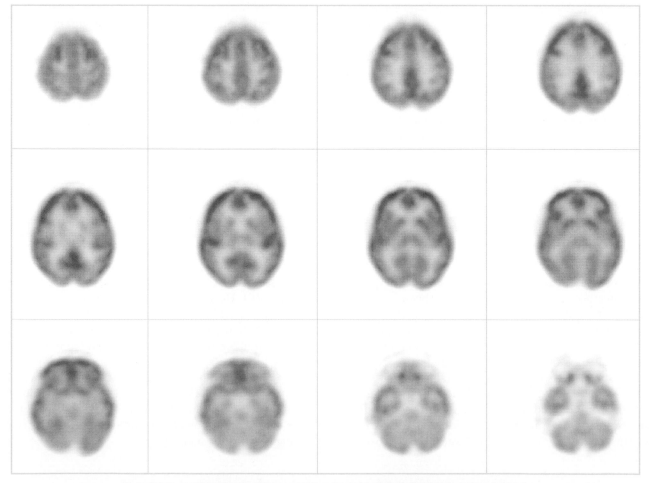

Figure 5.1. Normal FDG brain scan. The transaxial image is taken at the level of the basal ganglia and thalami.

In the chest, it has been recognized that there is variation in regional lung activity, this being greater in the inferior and posterior segments, and it has been suggested that this might reduce sensitivity in lesion detection in these regions (19). In the abdomen, homogeneous low-grade accumulation is seen in the liver and to a lesser extent in the spleen. Small and large bowel activity is quite variable, and, unlike glucose, FDG is excreted in the urine, leading to variable appearances of the urinary tract, both of which are discussed further here. Resting skeletal muscle is usually associated with low-grade activity, but active skeletal muscle may show marked uptake of FDG in a variety of patterns that are discussed further next.

Myocardial activity may also be quite variable. Normal myocardial metabolism depends on both glucose and free fatty acids (FFA). For oncologic scans, it is usual to try to reduce activity in the myocardium so as to obtain clear images of the mediastinum and adjacent lung. Although most centers fast patients for at least 4 to 6 h before FDG injection, reducing insulin levels and encouraging FFA acid metabolism in preference to glucose, myocardial activity may still be quite marked and varies among patients. Another possible intervention that has not been quantified or validated as yet is to administer caffeine to the patient to encourage FFA metabolism.

For cardiac viability studies it is necessary to achieve high uptake of FDG into the myocardium. Patients may receive a glucose load to encourage glucose (and hence FDG) rather than FFA metabolism, and it may also be necessary to administer insulin to enhance myocardial uptake, particularly in diabetic patients (20–22). The hy-

perinsulinemic euglycemic clamping method may further improve myocardial uptake but is technically more difficult (23–26); this allows maximum insulin administration without rendering the patient hypoglycaemic. An alternative method is to encourage myocardial glucose

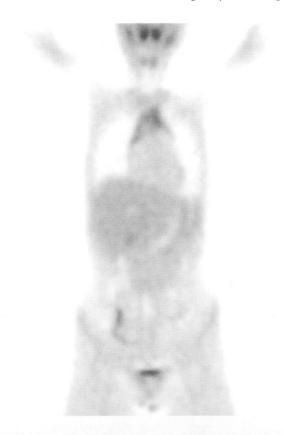

a

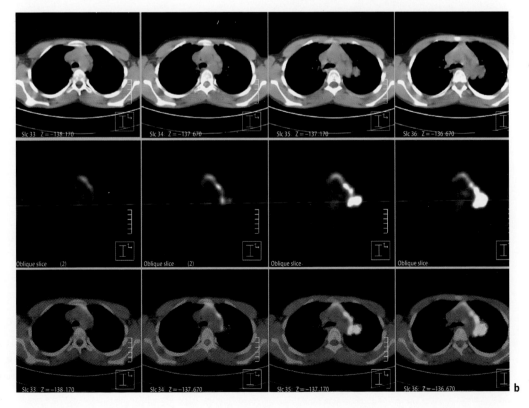

Figure 5.2. Transaxial (**a**) and coronal (**b**) FDG images in a child showing normal thymic activity.

b

metabolism by reducing FFA levels pharmacologically. Improved cardiac uptake of FDG has been described following oral nicotinic acid derivatives such as acipimox, a simple and safe measure that may also be effective in diabetic patients (27).

Variants That May Mimic or Obscure Pathology

A number of physiologic variations in uptake of FDG have been recognized, some of which may mimic pathology (16, 28–30) (Table 5.2).

Skeletal muscle uptake is probably the commonest cause of interpretative difficulty. Increased aerobic glycolysis associated with muscle activation, either after exercise or as a result of involuntary tension, leads to increased accumulation of FDG that may mimic or obscure pathology. Exercise should be prohibited before injection of FDG and during the uptake period to minimize muscle uptake.

A pattern of symmetrical activity commonly encountered in the neck, supraclavicular, and paraspinal regions (Figure 5.3) was initially assumed to be caused by invol-

Table 5.2. Variants that may mimic or obscure pathology.

Organ/system	Variant
Skeletal muscle	High uptake after exercise or resulting from tension, including eye movement, vocalization, swallowing, chewing gum, hyperventilation.
Adipose tissue	Uptake in brown fat may be seen particularly in winter months in patients with low body mass index.
Myocardium	Variable (may depend on or be manipulated by diet and drugs).
Endocrine	Testes, breast (cyclic, lactation, hormone replacement therapy), follicular ovarian cysts, thyroid.
Gastrointestinal	Bowel activity is variable and may simulate tumor activity.
Genitourinary	Small areas of ureteric stasis may simulate paraaortic or pelvic lymphadenopathy.

Source: From Valk PE, Bailey DL, Townsend DW, Maisey MN. Positron Emission Tomography: Basic Science and Clinical Practice. Springer-Verlag London Ltd 2003, p. 498.

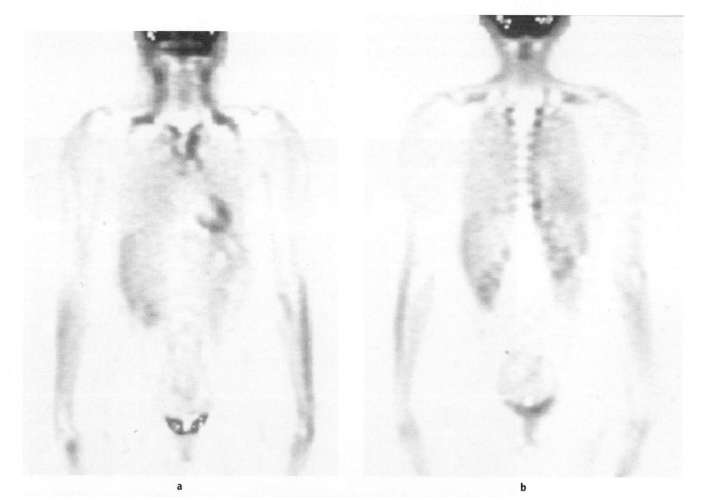

| a | b |

Figure 5.3. Coronal sections from a FDG study. Symmetrical brown fat activity is seen in the neck (**a**) and paraspinal (**b**) regions. Although this is a recognizable pattern, it can be appreciated that metastatic lymphadenopathy may be obscured, especially in the neck. PET/CT is helpful for the differentiation.

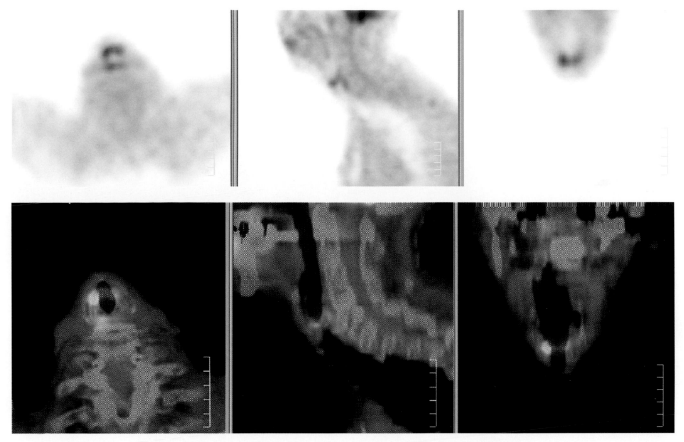

Figure 5.4. FDG uptake seen in laryngeal muscles in a patient who was talking during the uptake period.

untary muscle tension but with the advent of PET/CT it has become obvious that this activity originates in brown fat, a vestigial organ of thermogenesis that is sympathetically innervated and driven. To support this hypothesis, it has been noted that this pattern is commoner in winter months and in patients with lower body mass index (31). It appears that benzodiazepines are able to reduce the incidence of this potentially confusing appearance, possibly by a generalized reduction in sympathetic drive.

Even apparently innocent activities such as talking or chewing gum may lead to muscle uptake that simulates malignant tissue (Figures 5.4, 5.5). In patients being assessed for head and neck malignancies, it is therefore important that they maintain silence and refrain from chewing during the uptake period. In addition, anxious or breathless patients may hyperventilate, producing increased intercostal and diaphragmatic activity, and involuntary muscle spasm such as that seen with torticollis may lead to a pattern that is recognizable but may obscure diseased lymph nodes.

The symmetrical nature of most muscle uptake usually alerts the interpreter to the most likely cause, but occasionally unilateral muscle uptake may be seen when there is a nerve palsy on the contralateral side and may be mistaken for an abnormal tumor focus. This appearance has been described in recurrent laryngeal nerve palsy and in VIth cranial nerve palsy (30). Diffusely increased uptake

of FDG may also be seen in dermatomyositis complicating malignancy, a factor that may reduce image contrast and tumor detectability.

Uptake in the gastrointestinal system is quite variable and is most commonly seen in the stomach (Figure 5.6) and large bowel (Figure 5.7) and to a lesser extent in loops of small bowel. It is probable that activity in bowel is related to smooth muscle uptake as well as activity in intraluminal contents (32, 33). If it is important to reduce intestinal physiologic activity, pharmacologic methods to reduce peristalsis as well as bowel lavage could be useful. This procedure is too invasive and is unnecessary for routine patient preparation, and in most situations it is possible to differentiate physiologic uptake within bowel from abdominal tumor foci by the pattern of uptake, the former usually being curvilinear and the latter being focal (Figure 5.8). Some centers use a mild laxative as a routine in any patient requiring abdominal imaging, but improvement in interpretation has not been demonstrated.

Unlike glucose, FDG is not totally reabsorbed in the renal tubules, and urinary activity is seen in all patients and may be present in all parts of the urinary tract. This activity may interfere with a study of renal or pelvic tumors, either by obscuring local tumors or by causing reconstruction artifacts that reduce the visibility of abnormalities adjacent to areas of high urinary activity. Using iterative reconstruction algorithms rather than filtered

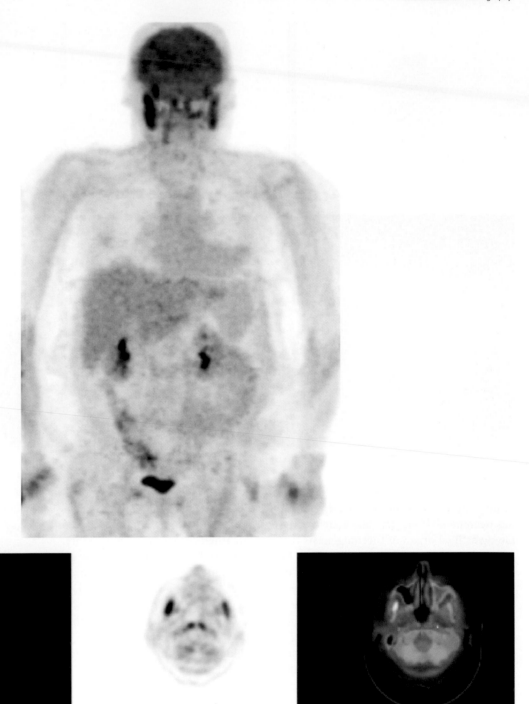

Figure 5.5. Symmetrical FDG uptake in the masseter muscles in a patient chewing gum, resembling bilateral lymphadenopathy.

back-projection can reduce this problem. Catheterization and drainage of urinary activity may reduce bladder activity that may obscure perivesical or intravesical tumors. However, this may still leave small pockets of concentrated activity that may resemble lymphadenopathy, causing even greater problems in interpretation. Bladder irrigation may help to some extent but is associated with increased radiation dose to staff and may introduce infection.

We have found it beneficial to hydrate the patient and administer a diuretic. This approach leads to a full bladder with *dilute* urine, making it easier to differentiate normal urinary activity from perivesical tumor activity and allowing the bladder to be used as an anatomic landmark. By diluting vesical FDG activity, reconstruction artifacts from filtered back-projection algorithms are also reduced. It is often helpful to perform image registration with either CT or MRI in the pelvis. Here it may be helpful

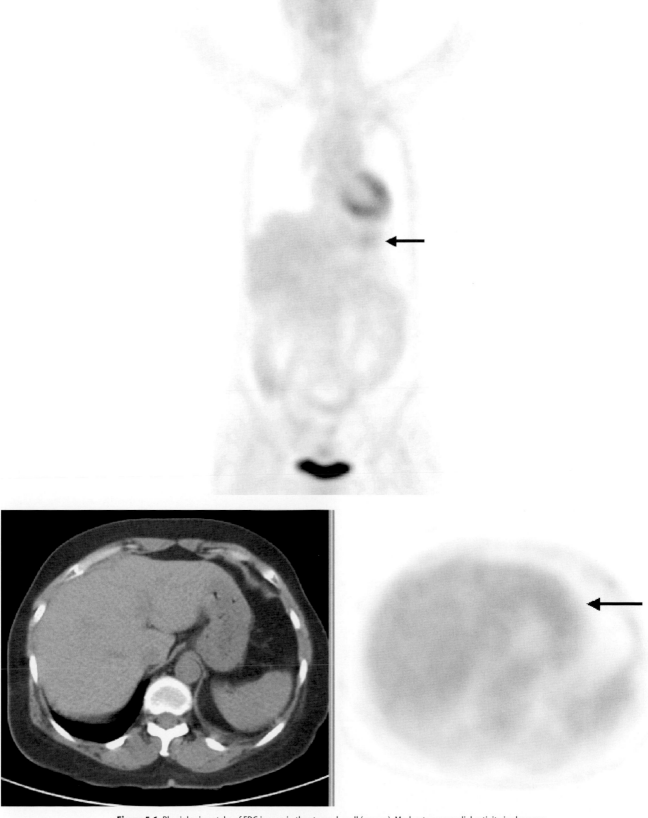

Figure 5.6. Physiologic uptake of FDG is seen in the stomach wall (*arrows*). Moderate myocardial activity is also seen.

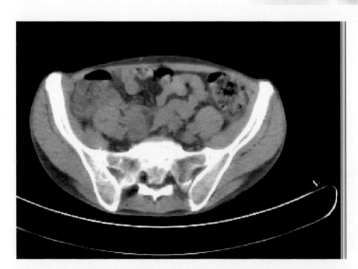

Figure 5.7. Moderate physiologic uptake that is seen in the region of the cecum and ascending colon in a patient with a primary lung cancer.

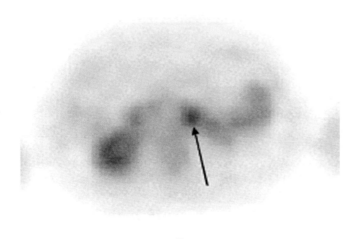

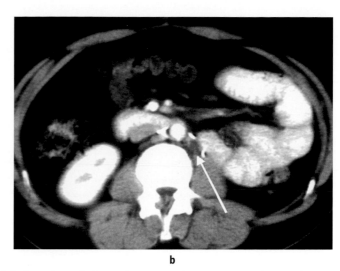

a   **b**

Figure 5.8. (**a**) Transaxial FDG slice through the upper abdomen and (**b**) corresponding CT slice in a patient with a history of seminoma and previous paraaortic lymph node dissection but rising tumor markers. The linear area of low-grade FDG activity can be seen to correspond to a barium-filled loop of bowel, but the more-focal area of high uptake (*arrow* on each part) corresponds to a small density located adjacent to the previous surgical clips, indicating recurrent disease at this site. The case demonstrates how normal bowel activity can be differentiated from tumor foci. 📖

to administer a small amount of ^{18}F-fluoride ion in addition to FDG, to allow easy identification of bony landmarks for registration purposes. Although excreted FDG may be seen in any part of the urinary tract, it is important to gain a history of any previous urinary diversion procedures, because these may cause areas of high activity outside the normal renal tract and may result in errors of interpretation unless this is appreciated.

Glandular breast tissue often demonstrates moderate FDG activity in premenopausal women and postmenopausal women taking estrogens for hormone replacement therapy. The pattern of uptake is usually symmetrical and easily identified as being physiologic, but there is the potential for lesions to be obscured by this normal activity. Breast-feeding mothers show intense uptake of FDG bilaterally (Figure 5.9). Similarly in males,

uptake of FDG may be seen in normal testes and appears to be greater in young men than in old (34).

Artifacts

Image reconstruction of PET images without attenuation correction may lead to higher apparent activity in superficial structures that may obscure lesions such as cutaneous melanoma metastases (28). A common artifact arising from this phenomenon is caused by the axillary skin fold, where lymphadenopathy may be mimicked in coronal image sections. However, the linear distribution of activity can be appreciated on transaxial or sagittal slices, which should prevent misinterpretation. Another

Figure 5.9. Transaxial FDG scan of a breast-feeding mother in whom intense symmetrical breast activity can be seen. 📖

Table 5.3. Artifacts.

Attenuation correction related	Apparent superficial increase in activity and lung activity if no correction applied.
Injection related	Lymph node uptake following tissued injection. Reconstruction artifacts due to tissued activity. Inaccuracies in standard uptake value calculation.
Attenuating material	Coins, medallions, prostheses.
Patient movement	Poor image quality. Artifacts on applying attenuation correction.

Source: From Valk PE, Bailey DL, Townsend DW, Maisey MN. Positron Emission Tomography: Basic Science and Clinical Practice. Springer-Verlag London Ltd 2003, p. 502.

major difference between attenuation corrected and non-corrected images is an apparent increase in lung activity in the latter caused by to relatively low attenuation by the air-containing lung (Table 5.3).

Filtered back-projection reconstruction leads to streak artifacts and may obscure lesions adjacent to areas of high activity. Many of these artifacts can be overcome by using iterative reconstruction techniques (Figure 5.10).

Patient movement may compromise image quality. In brain imaging it is possible to split the acquisition into a number of frames, so that if movement occurs in one frame then this can be discarded before summation of the data (35). When performing whole-body scans, unusual appearances may result if the patient moves between bed scan positions; this most commonly occurs when the upper part of the arm is visible in higher scanning positions but the lower part disappears when moved out of the field of view on lower subsequent scanning positions.

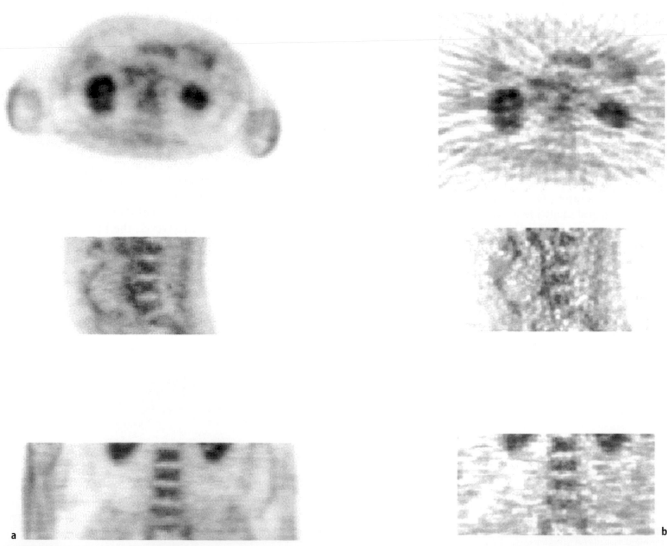

a   b

Figure 5.10. Transaxial, sagittal, and coronal abdominal FDG images from iterative reconstruction (**a**) and filtered back-projection (**b**) demonstrate the improved image quality and reduction in streak artifacts possible with the former. 📖

Special care is required in injecting FDG because soft tissue injection may cause reconstruction artifacts across the trunk and may even cause a low-count study or inaccuracies in standardized uptake value (SUV) measurements. Axillary lymph nodes, draining the region of tracer extravasation, may also accumulate activity following extravasated injections. The site of administration should be chosen carefully so as to minimize the risk of false-positive interpretation should extravasation occur.

Artifacts caused by prostheses are usually readily recognizable. Photon-deficient regions may result from metallic joint prostheses or other metallic objects carried by the patient. Ring artifacts may occur if there is misregistration between transmission and emission scans because of patient movement and are particularly apparent at borders where there are sudden changes in activity concentrations, for example, at a metal prosthesis. Misregistration artifacts between emission and transmission scans have become less frequent now that interleaved or even simultaneous emission/transmission scans are being performed.

Benign Causes of FDG Uptake

Uptake of FDG is not specific to malignant tissue, and it is well recognized that inflammation may lead to accumulation in macrophages and other activated inflammatory cells (7, 8). In oncologic imaging, this inflammatory uptake may lead to decrease in specificity. For example, it may be difficult to differentiate benign postradiotherapy changes from recurrent tumor in the brain unless the study is optimally timed or unless alternative tracers such as ^{11}C-methionine are used. Apical lung activity may be seen following radiotherapy for breast cancer, and moderate uptake may follow radiotherapy for lung cancer (36). It may also be difficult to differentiate radiation changes from recurrent tumor in patients who have undergone radiotherapy for rectal cancer within 6 months of the study (12).

Pancreatic imaging with FDG may be problematic. In some cases, uptake into mass-forming pancreatitis may be comparable in degree to uptake in pancreatic cancer. Conversely, false-negative results have been described in diabetic patients with pancreatic cancer. However, if diabetic patients and those with raised inflammatory markers are excluded, then FDG-PET may still be an accurate test to differentiate benign from malignant pancreatic masses (37).

A number of granulomatous disorders have been described as leading to increased uptake of FDG, including tuberculosis (38) and sarcoidosis (39) (Figure 5.11). It is often necessary to be cautious in ascribing FDG lesions to cancer in patients who are known to be immunocompromised. It is these patients who often have the unusual infections that may lead to uptake that cannot be differentiated from malignancy. PET remains useful in these patients, despite a lower specificity, as it is often able to locate areas of disease that have not been identified by other means and that may be more amenable to biopsy (40).

A more-comprehensive list of benign causes of abnormal FDG uptake is displayed in Table 5.4.

Specific Problems Related to PET/CT

One of the most exciting technologic advances in recent years is the clinical application of combined PET/CT scanners. However, this new technology has come with its own particular set of artifacts and pitfalls.

One of the biggest problems with PET/CT imaging in a dedicated combined scanner is related to differences in breathing patterns between the CT and the PET acquisitions. CT scans can be acquired during a breath-hold but PET acquisitions are taken during tidal breathing and represent an average position of the thoracic cage over 30 min or more. This difference may result in misregistration of pulmonary nodules between the two modalities, particularly in the peripheries and at the bases of the lungs where differences in position may approach 15 mm (41). Misregistration may be reduced by performing the CT scan while the breath is held in normal expiration (42, 43). It has been noted that deep inspiration during the CT acquisition can lead to deterioration of the CT attenuation-corrected PET image with the appearance of cold artifacts (Figure 5.12) and can even lead to the mispositioning of abdominal activity into the thorax (44). CT acquisition during normal expiration minimizes the incidence of such artifacts and also optimizes coregistration of abdominal organs.

High-density contrast agents, for example, oral contrast, or metallic objects (Figure 5.13) can lead to an artifactual overestimation of activity if CT data are used for attenuation correction (45–51). Such artifacts may be recognized by studying the uncorrected image data. Low-density oral contrast agents can be used without significant artifacts (52, 53), or the problem may be avoided by using water as a negative bowel contrast agent. Algorithms have been developed to account for the overestimation of activity when using CT-based attenuation correction that may minimize these effects in the future (53).

The use of intravenous contrast during CT acquisition may be a more-difficult problem. Similarly, the concentrated bolus of contrast in the large vessels may lead to overcorrection for attenuation, particularly because the concentrated column of contrast has largely dissipated by the time the PET emission scan is acquired. Artifactual hot spots in the attenuation-corrected image (48) or quantitative overestimation of FDG activity may result. When intravenous contrast is considered essential for a study, the diagnostic aspect of the CT scan is best performed as a third study with the patient in the same position, after, first, a low-current CT scan for attenuation correction purposes and, second, the PET emission scan.

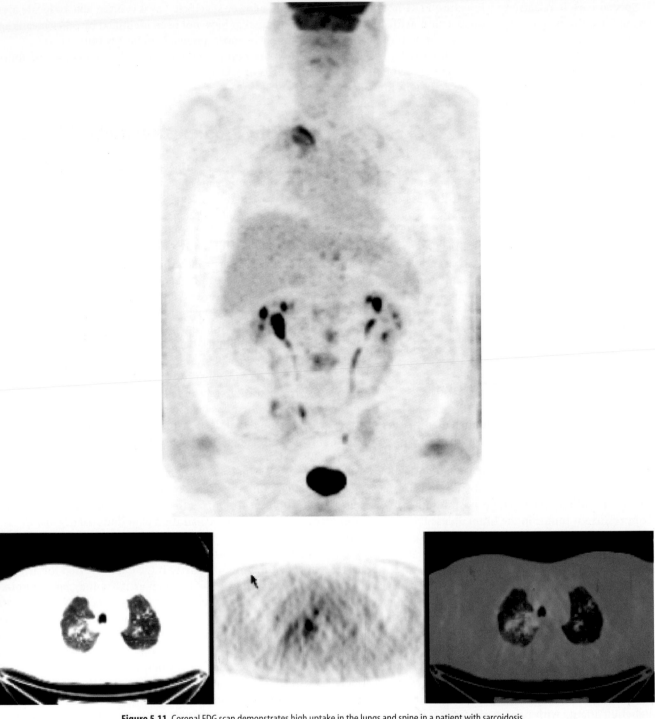

Figure 5.11. Coronal FDG scan demonstrates high uptake in the lungs and spine in a patient with sarcoidosis.

Although many centers have found low-current CT acquisitions to be adequate for attenuation correction and image fusion (54), it may be necessary to increase CT tube current in larger patients to minimize beam-hardening artifacts on the CT scan that may translate through to incorrect attenuation correction of the PET emission data (49). This effect can be caused by the patient's arms being in the field of view and may be minimized by placing the arms above the head for imaging. Differences in the field-of-view diameter between the larger PET and smaller CT parts of combined scanners can lead to truncation artifacts at the edge of the CT image, but these are generally small and can be minimized by the use of iterative image reconstruction methods (53).

Although some new artifacts are introduced by combined PET/CT imaging, it is likely that many pitfalls caused by normal variant uptake may be avoided by the ability to correctly attribute FDG activity to a structurally normal organ on the CT scan; this may be particularly evident in the abdomen when physiologic bowel activity

Table 5.4. Benign causes of FDG uptake.

Organ/type	Disease
Brain	Postradiotherapy uptake.
Pulmonary	Tuberculosis, sarcoidosis, histoplasmosis, atypical mycobacteria, pneumoconiosis, radiotherapy.
Myocardium	Heterogeneous left ventricular activity possible after myocardial infarction, increased right ventricular activity in right heart failure
Bone/bone marrow	Paget's disease, osteomyelitis, hyperplastic bone marrow.
Inflammation	Wound healing, pyogenic infection, organizing hematoma, esophagitis, inflammatory bowel disease, lymphadenopathy associated with granulomatous disorders, viral and atypical infections, chronic pancreatitis, retroperitoneal fibrosis, radiation fibrosis (early), bursitis.
Endocrine	Graves' disease and chronic thyroiditis, adrenal hyperplasia.

Source: From Valk PE, Bailey DL, Townsend DW, Maisey MN. Positron Emission Tomography: Basic Science and Clinical Practice. Springer-Verlag London Ltd 2003, p. 504.

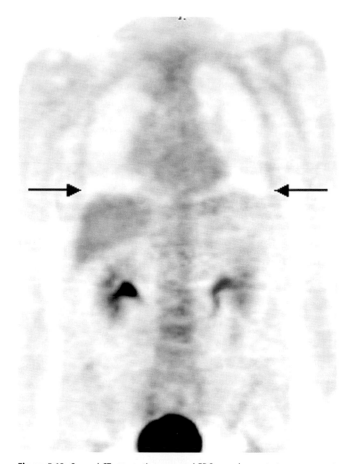

Figure 5.12. Coronal CT attenuation corrected FDG scan demonstrates an apparent loss of activity at the level of the diaphragm (*arrows*) resulting from differences in breathing patterns between the CT and PET scans.

Figure 5.13. Coronal FDG scan with (**a**) CT attenuation correction, (**b**) CT alone, and (**c**) uncorrected FDG of a patient with a metallic pacemaker placed over the right upper chest demonstrates artifactual increased uptake on the corrected images.

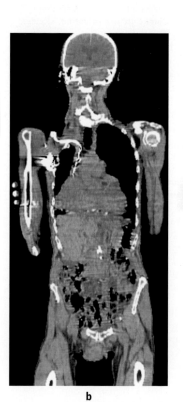

a

b

c

or ureteric activity can otherwise cause interpretative difficulties. PET/CT also has the potential to limit false-negative interpretations in tumors that are not very FDG avid by recognizing uptake as being related to structurally abnormal tissue and increasing the diagnostic confidence in tumor recognition by the use of the combined structural and functional data. Similarly, it may be possible to detect small lung metastases of a few millimeters on CT lung windows that are beyond the resolution of FDG-PET. The full use of the combined data, including the corrected and noncorrected PET emission data, and the inspection of soft tissue, lung, and bone windows on the CT data, may also allow the description and correct diagnosis of pertinent FDG-negative lesions, such as liver cysts, and incidental FDG-negative CT abnormalities, such as abdominal aortic aneurysm, to provide an integrated interpretation of all the available data resulting from this technology.

References

1. Torizuka T, Tamaki N, Inokuma T, et al. In vivo assessment of glucose metabolism in hepatocellular carcinoma with FDG-PET. J Nucl Med 1995;36(10):1811–1817.

2. Minn H, Clavo AC, Wahl RL. Influence of hypoxia on tracer accumulation in squamous cell carcinoma: in vitro evaluation for PET imaging. Nucl Med Biol 1996;23:941–946.

3. Wahl RL, Henry CA, Ethier SP. Serum glucose: effects on tumour and normal tissue accumulation of 2-[F-18]-fluoro-2-deoxy-D-glucose in rodents with mammary carcinoma. Radiology 1992;183:643–647.

4. Torizuka T, Clavo AC, Wahl RL. Effect of hyperglycaemia on in vitro tumour uptake of tritiated FDG, thymidine, L-methionine and L-leucine. J Nucl Med 1997;38:382–386.

5. Yasuda S, Kajihara M, Fujii H, Takahashi W, Ide M, Shohtsu A. Factors influencing high FDG uptake in the intestine, skeletal muscle and myocardium. J Nucl Med 1999;40:140P

6. Torizuka T, Fisher SJ, Wahl RL. Insulin induced hypoglycaemia decreases uptake of 2-[F-18]fluoro-2-deoxy-D-glucose into experimental mammary carcinoma. Radiology 1997;203:169–172.

7. Yamada S, Kubota K, Kubota R, et al. High accumulation of fluorine-18-fluorodeoxyglucose in turpentine-induced inflammatory tissue. J Nucl Med 1995;36:1301–1306.

8. Kubota R, Kubota K, Yamada S, et al. Methionine uptake by tumor tissue: a microautoradiographic comparison with FDG. J Nucl Med 1995;36:484–492.

9. Sugawara Y, Braun DK, Kison PV, et al. Rapid detection of human infections with fluorine-18 fluorodeoxyglucose and positron emission tomography: preliminary results. Eur J Nucl Med 1998;25:1238–1243.

10. Reinhardt MJ, Kubota K, Yamada S, Iwata R, Yaegashi H. Assessment of cancer recurrence in residual tumors after fractionated radiotherapy: a comparison of fluorodeoxyglucose, L-methionine and thymidine. J Nucl Med 1997;38:280–287.

11. Strauss LG. Fluorine-18 deoxyglucose and false-positive results: a major problem in the diagnostics of oncological patients. Eur J Nucl Med 1996;23:1409–1415.

12. Haberkorn U, Strauss LG, Dimitrakopoulou A, et al. PET studies of fluorodeoxyglucose metabolism in patients with recurrent colorectal tumors receiving radiotherapy. J Nucl Med 1991;32:1485–1490.

13. Kubota R, Kubota K, Yamada S et al. Methionine uptake by tumour tissue: a microautoradiographic comparison with FDG. J Nucl Med 1995;36:484–492.

14. Nuutinen JM, Leskinen S, Elomaa I, et al. Detection of residual tumours in postchemotherapy testicular cancer by FDG-PET. Eur J Cancer 1997;33:1234–1241.

15. Jones DN, McCowage GB, Sostman HD, et al. Monitoring of neoadjuvant therapy response of soft-tissue and musculoskeletal sarcoma using fluorine-18-FDG PET. J Nucl Med 1996;37: 1438–1444.

16. Shreve PD, Anzai Y, Wahl RL. Pitfalls in oncologic diagnosis with FDG PET imaging: physiologic and benign variants. Radiographics 1999;19:61–77.

17. Kato T, Tsukamoto E, Suginami Y, et al. Visualization of normal organs in whole-body FDG-PET imaging. Jpn J Nucl Med 1999;36:971–977.

18. Yasuda S, Shohtsu A, Ide M, et al. Chronic thyroiditis: diffuse uptake of FDG at PET. Radiology 1998;207:775–778.

19. Miyauchi T, Wahl RL. Regional 2-[18F]fluoro-2-deoxy-D-glucose uptake varies in normal lung. Eur J Nucl Med 1996; 23:517–523.

20. Kubota K, Kubota R, Yamada S, Tada M, Takahashi T, Iwata R. Re-evaluation of myocardial FDG uptake in hyperglycaemia. J Nucl Med 1996;37:1713–1717.

21. Knuuti MJ, Maki M, Yki-Jarvinen H, et al. The effect of insulin and FFA on myocardial glucose uptake. J Mol Cell Cardiol 1995;27:1359–1367.

22. Choi Y, Brunken RC, Hawkins RA, et al. Factors affecting myocardial 2-[F-18]fluoro-2-deoxy-D-glucose uptake in positron emission tomography studies of normal humans. Eur J Nucl Med 1993; 20:308–318.

23. Bax JJ, Visser FC, Raymakers PG, et al. Cardiac 18F-FDG-SPET studies in patients with non-insulin dependent diabetes mellitus during hyperinsulinaemic euglycaemic clamping. Nucl Med Commun 1997;18:200–206.

24. Huitink JM, Visser FC, van Leeuwen GR, et al. Influence of high and low plasma insulin levels on the uptake of fluorine-18 fluorodeoxyglucose in myocardium and femoral muscle assessed by planar imaging. Eur J Nucl Med 1995;22:1141–1148.

25. Locher JT, Frey LD, Seybold K, Jenzer H. Myocardial 18F-FDG-PET. Experiences with the euglycaemic hyperinsulinaemic clamp technique. Angiology 1995;46:313–320.

26. Ohtake T, Yokoyama I, Watanabe T, et al. Myocardial glucose metabolism in noninsulin dependent diabetes mellitus patients evaluated by FDG-PET. J Nucl Med 1995;36:456–463.

27. Bax JJ, Veening MA, Visser FC, et al. Optimal metabolic conditions during fluorine-18 fluorodeoxyglucose imaging: a comparative study using different protocols. Eur J Nucl Med 1997;24:35–41.

28. Engel H, Steinert H, Buck A, et al. Whole body PET: physiological and artifactual fluorodeoxyglucose accumulations. J Nucl Med 1996;37:441–446.

29. Cook GJR, Fogelman I, Maisey M. Normal physiological and benign pathological variants of 18-fluoro-2-deoxyglucose positron emission tomography scanning: potential for error in interpretation. Semin Nucl Med 1996;24:308–314.

30. Cook GJR, Maisey MN, Fogelman I. Normal variants, artefacts and interpretative pitfalls in PET imaging with 18-fluoro-2-deoxyglucose and carbon-11 methionine. Eur J Nucl Med 1999;26:1363–1378.

31. Hany TF, Gharehpapagh E, Kamel EM, et al. Brown adipose tissue: a factor to consider in symmetrical tracer uptake in the neck and upper chest region. Eur J Nucl Med 2002;29:1393–1398.

32. Bischof Delalove A, Wahl RL. How high a level of FDG abdominal activity is considered normal? J Nucl Med 1995;36:106P.

33. Nakada K, Fisher SJ, Brown RS, Wahl RL. FDG uptake in the gastrointestinal tract: can it be reduced? J Nucl Med 1999;40:22P–23P.

34. Kosuda S, Fisher S, Kison PV, Wahl RL, Grossman HB. Uptake of 2-deoxy-2-[18F]fluoro-D-glucose in the normal testis: retrospective PET study and animal experiment. Ann Nucl Med 1997;11:195–199.

35. Picard Y, Thompson CJ. Motion correction of PET images using multiple acquisition frames. IEEE Trans Med Imaging 1997; 16:137–144.

36. Nunez RF, Yeung HW, Macapinlac HA, Larson SM. Does post-radiation therapy changes in the lung affect the accuracy of FDG PET in the evaluation of tumour recurrence in lung cancer. J Nucl Med 1999;40:234P.

37. Diederichs CG, Staib L, Vogel J et al. Values and limitations of ^{18}F-fluorodeoxyglucose-positron-emission tomography with preoperative evaluation of patients with pancreatic masses. Pancreas 2000;20:109–116.

38. Knopp MV, Bischoff HG. Evaluation of pulmonary lesions with positron emission tomography. Radiologe 1994;34:588–591.

39. Lewis PJ, Salama A. Uptake of fluorine-18-fluorodeoxyglucose in sarcoidosis. J Nucl Med 1994;35:1–3.

40. O'Doherty MJ, Barrington SF, Campbell M, et al. PET scanning and the human immunodeficiency virus-positive patient. J Nucl Med 1997;38:1575–1583.

41. Goerres GW, Kamel E, Seifert B, et al. Accuracy of image coregistration of pulmonary lesions in patients with non-small cell lung cancer using an integrated PET/CT system. J Nucl Med 2002;43:1469–1475.

42. Goerres GW, Kamel E, Heidelberg TN, et al. PET/CT image co-registration in the thorax: influence of respiration. Eur J Nucl Med 2002;29:351–360.

43. Goerres GW, Burger C, Schwitter MR, et al. PET/CT of the abdomen: optimizing the patient breathing pattern. Eur Radiol 2003;13:734–739.

44. Osman MM, Cohade C, Nakamoto Y, et al. Clinically significant inaccurate localization of lesions with PET/CT: frequency in 300 patients. J Nucl Med 2003;44:240–243.

45. Dizendorf E, Hany TF, Buck A, et al. Cause and magnitude of the error induced by oral CT contrast agent in CT-based attenuation correction of PET emission studies. J Nucl Med 2003;44:732–738.

46. Goerres GW, Hany TF, Kamel E, et al. Head and neck imaging with PET and PET/CT: artefacts from dental metallic implants. Eur J Nucl Med 2003;29:367–370.

47. Kamel EM, Burger C, Buck A, von Schulthess GK, Goerres GW. Impact of metallic dental implants on CT-based attenuation correction in a combined PET/CT scanner. Eur Radiol 2003;13:724–728.

48. Antoch G, Freudenberg LS, Egelhof T, et al. Focal tracer uptake: a potential artifact in contrast-enhanced dual-modality PET/CT scans. J Nucl Med 2002;43:1339–1342.

49. Cohade C, Wahl RL. Applications of PET/CT image fusion in clinical PET: clinical use, interpretation methods, diagnostic improvements. Semin Nucl Med 2003;33:228–237.

50. Goerres GW, Ziegler SI, Burger C et al. Artifacts at PET and PET/CT caused by metallic hip prosthetic material. Radiology 2003;226:577–584.

51. Kinahan PE, Hasegawa BH, Beyer T. X-ray based attenuation correction for PET/CT scanners. Semin Nucl Med 2003;33:166–179.

52. Cohade C, Osman M, Nakamoto Y, et al. Initial experience with oral contrast in PET/CT: phantom and clinical studies. J Nucl Med 2003;44:412–416.

53. Dizendorf EV, Treyer V, Von Schulthess GK, et al. Application of oral contrast media in coregistered positron emission tomography-CT. AJR 2002;179:477–481.

54. Hany TF, Steinert HC, Goerres GW, et al. PET diagnostic accuracy: improvement with in-line PET/CT system: initial results. Radiology 2002;225:575–581.

6

PET Imaging in Brain Tumors

Terence Z. Wong and R. Edward Coleman

The epidemiology of intracranial tumors is complex, and includes a variety of benign and malignant histologies (1). For the year 2004, an estimated 18,400 primary malignant central nervous system (CNS) tumors were diagnosed in the United States, and these malignancies resulted in 12,690 deaths (2). Overall prognosis for these tumors remains poor, with an age-adjusted 5-year relative survival of 30.8%. Prognosis varies significantly with age; patients 19 years old or younger have a 5-year relative survival of 65%, and patients aged 44 or younger have a 5-year survival of 58.7%. Prognosis is very poor for patients aged 65 or older, with a 5-year survival of less than 6.5% (1).

Classification of Primary Intracranial Tumors

Data on nearly 60,000 patients with primary brain and CNS tumors diagnosed between 1997 and 2001 have been reported to the Central Brain Tumor Registry of the United State (CBTRUS) (3). In this population, gliomas accounted for 42% of all tumors and 77% of all malignant tumors. Histologically, glial tumors are graded on the basis of degree of cellularity, cellular pleomorphism, and number of mitotic figures, as well as other features such as the presence or absence of vascular proliferation or necrosis. The World Health Organization (WHO) has established a four-tier classification system (Grade I–Grade IV), with Grade I representing the most benign category and Grade II being semibenign histology. Grade III tumors, such as anaplastic astrocytoma, have increased cellularity and nuclear atypia and behave in a malignant fashion. Grade IV tumors, most notably glioblastoma multiforme, are the highest grade brain tumors and feature necrosis and vascular endothelial proliferation on histology. In the National Cancer Data Base (NCDB) series, overall 5-year survival was 30% for astrocytomas and only 2% for glioblastoma multiforme.

PET Radiopharmaceuticals for Imaging of Brain Tumors

The brain provides a unique environment for PET imaging. Because the brain is immobilized by the skull, motion can be minimized during imaging with a head-holder device. Accurate image registration with other imaging modalities such as computed tomography (CT) or magnetic resonance imaging (MRI) is possible. In addition, the tissues surrounding the brain have relatively uniform and predictable X-ray attenuation characteristics at 511 keV, allowing calculated attenuation-correction algorithms to be implemented.

Currently, 2-[^{18}F]-fluoro-2-deoxy-D-glucose (FDG) is the most commonly used radiopharmaceutical for PET imaging. FDG is currently the most widely available PET tracer and has proven efficacy in whole-body imaging for a variety of malignancies. FDG-PET is clinically relevant in brain tumors because of the relationship between glucose metabolism and malignant behavior.

Other radiopharmaceuticals have also been utilized to study various aspects of brain tumor physiology. PET studies with ^{15}O-labeled compounds can be used to quantify blood flow and tissue oxygen utilization. Inhaled ^{15}O-labeled carbon monoxide can be used to determine cerebral blood volume (CBV), intravenous ^{15}O-labeled water can be used to calculate cerebral blood flow (CBF), and ^{15}O-labeled oxygen can be then be used to estimate oxygen extraction (4–6). Radiolabeled amino acids such as ^{11}C-tyrosine and particularly ^{11}C-methionine have been used as markers for brain tumor metabolism (7). PET imaging with ^{11}C-methionine may be particularly useful for evaluating low-grade gliomas and oligodendrogliomas (8–10). 2-[^{11}C]-Thymidine has also been used as a PET imaging marker for cellular proliferation of brain tumors (11). Both ^{11}C-methionine and ^{11}C-thymidine may have higher specificity for tumor cell proliferation over inflammation and may therefore have an advantage over FDG for distinguishing recurrent tumor from necrosis following

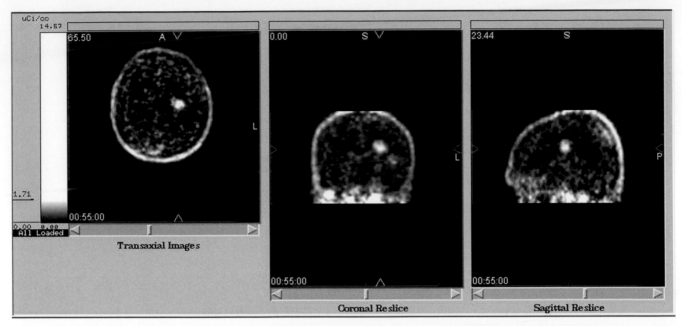

Figure 6.1. ¹⁸F-Fluorothymidine (FLT) positron emission tomography scan in a patient with glioblastoma. In contrast to 2-[¹⁸F]-fluoro-2-deoxy-D-glucose (FDG), radiolabeled thymidine, methionine, and choline PET tracers do not have high background cortical activity in the brain, making the hypermetabolic activity of the tumor clearly apparent.

high-dose radiation therapy (12). In general, these radiotracers are less practical to use in routine clinical practice because of the short half-lives of ¹⁵O (2 min) and ¹¹C (20 min) and require an onsite cyclotron for synthesis. More recently, the development of longer-lived ¹⁸F-labeled radiopharmaceuticals, such as ¹⁸F-thymidine (FLT) and ¹⁸F-choline (FCH), may enable commercial distribution and wider availability of these potentially important tumor markers. An example of an FLT-PET scan of a patient with a glioblastoma multiforme is shown in Figure 6.1. Because radiolabeled thymidine, methionine, and choline radiotracers do not have significant accumulation within normal brain tissues, image registration with MRI or other anatomic imaging is more challenging, and the use of combined PET/CT scanning will facilitate anatomic correlation and image registration with MRI.

FDG in Evaluation of Brain Tumors

Metabolic activity reflected by FDG-PET can provide important diagnostic information relating to tumor grade and prognosis from brain tumors (13), and FDG-PET can be useful for following patients for evidence of recurrent disease. However, even metabolically active brain tumors can be less conspicuous than malignancies elsewhere in the body because of the high background that results from glucose metabolism of the normal brain. Correlation of PET images with anatomic imaging (MRI or CT) is therefore essential for accurate interpretation. Although this correlation can be accomplished visually using side-by-side comparison of the MR or CT images and PET images, electronic coregistration of the anatomic images with the three-dimensional FDG-PET data provides much more accurate characterization of intracranial lesions.

Glucose Metabolism in the Brain

As in other tissues, accumulation of FDG in the brain is closely related to glucose metabolism. However, in contrast to other tissues, the brain utilizes glucose almost exclusively to meet its metabolic energy requirements. As a consequence, the normal brain has a very high background accumulation of FDG on FDG-PET scanning, particularly in gray matter structures such as the cerebral cortex and basal ganglia. Malignant tumors within the brain also have high glucose metabolism and avidly accumulate FDG. However, the abnormal hypermetabolism of high-grade brain tumors is often similar to that of normal gray matter structures, and this normal accumulation of FDG reduces the ability of FDG-PET to detect and evaluate small lesions. In addition, common pathologies within the brain such as infections or metastatic lesions frequently develop at the gray–white matter junction, and characterizing abnormalities in this area by FDG-PET can be particularly challenging. Therefore, precise anatomic localization of the intracranial lesion in question is an essential component for characterization of intracranial lesions using FDG-PET.

Image Acquisition

Patients are requested to have no caloric intake for at least 4 h before the FDG-PET study, but they are encouraged to drink water. Following intravenous injection of a standard dose of FDG (10 mCi for adults), the patient is allowed to rest quietly in a dimly lit room for at least 30 min during the uptake phase. The patients are instructed to keep their eyes open, and we do not occlude the ears. However, auditory and visual stimulation is minimized to avoid extraneous cortical activation in the associated areas. Scanning is performed on a dedicated PET scanner (GE Advance; GE Medical Systems, Milwaukee, WI, USA) using a single table position. An emission scan is obtained for 6 to 8 min using three-dimensional acquisition and subsequently reconstructed using three-dimensional (3D) filtered backprojection. Calculated attenuation correction is utilized, eliminating the need for a dedicated transmission scan. The resulting images have a transaxial slice thickness of 4.25 mm, with a 25.6-cm field-of-view represented in a 128 x 128 pixel matrix. Although these FDG-PET images can be reformatted in coronal and sagittal planes, our practice is to base our interpretation primarily on the coregistered axial PET/MR images. Patients who are unable to complete an MRI examination (e.g., because of a pacemaker or aneurysm clips) can undergo a combined contrast-enhanced CT and PET examination on a PET/CT scanner. In this case, the PET and CT images are intrinsically coregistered.

Image Registration with MRI

Evaluation of brain tumors with FDG-PET requires accurate correlation with anatomy for several reasons. Because the metabolic activity within brain tumors can resemble that of gray or white matter, high-grade tumors within gray matter structures or low-grade tumors within deep white matter may be difficult or impossible to identify on FDG-PET images alone. We have found that image registration is essential for evaluating small lesions in the brain. Because of the differences in head positioning for the MRI and PET scans, it is very difficult to visually correlate corresponding areas on the two modalities without performing formal registration. Anatomic correlation can also be very important for evaluating large tumors. Large tumors are frequently heterogeneous, containing smaller focal areas of contrast enhancement on MRI; each of these areas must be individually correlated with FDG-PET for evidence of high-grade features. In addition, brain tumors frequently lie in close proximity to the cerebral cortex, and anatomic correlation is necessary to distinguish high-grade tumor from normal adjacent cortical activity. For these reasons, FDG-PET images are routinely electronically coregistered with a recent MRI study for all brain tumor patients.

Axial T_1-weighted postgadolinium MR images are typically used for correlation with FDG-PET, although the PET images can also be coregistered with T_2-weighted images for evaluating nonenhancing tumors. Unenhanced MR images are not usually used for image registration, but it is prudent to examine the precontrast T_1-weighted images. The unenhanced T_1-weighted images are particularly important to review when examining patients who have had recent surgery as hemorrhagic components with high T_1 signal can otherwise be mistaken for contrast enhancement. The MR images are electronically transferred from any of the MRI scanners in our Radiology Department. Alternatively, for patients with MRI studies from outside institutions, we utilize a film scanner to digitize the appropriate images for coregistration. The semi-automated coregistration technique that we use is based on an iterative surface-fit algorithm (14) that has been implemented using software developed at Duke (15–17). The 3D FDG-PET data set is rotated and translated iteratively to maximize the correlation between the MR-defined and PET-defined surfaces. The image registration software then provides the FDG-PET image corresponding to each axial MR slice.

Image interpretation using the coregistered MR and PET images is accomplished using an interactive display that utilizes a single gray-scale window. The display software allows the user to scroll through the axial MR images and correlate these with the corresponding PET images. A "toggle" key allows rapid alternate display of the coregistered anatomic and functional images. Alternatively, the anatomic and functional correlation can be made by observing the region in question as the displayed image is adjusted gradually along a continuum between 100% MRI and 100% FDG-PET; an example of this technique is illustrated in Figure 6.2. These display techniques have enabled us to accurately characterize small MRI lesions and provide precise anatomic localization of PET abnormalities. Our general diagnostic strategy is to utilize the CT or MRI images to define the abnormality (e.g., contrast enhancement or T_2 signal abnormality); then, by applying the interactive display software, the coregistered PET image is used to characterize the metabolic activity *within the abnormal tissue*. Frequently, the anatomically defined signal abnormality is larger than the metabolic abnormality delineated by PET and will contain regions of heterogeneous FDG accumulation.

Diagnosis of Primary Brain Tumors

The primary value of PET in evaluating glial and other neuroepithelial tumors is the correlation of FDG metabolism with tumor grade. Early clinical studies demonstrated that the glycolytic rate of brain tumors as determined by FDG metabolism is a more-accurate reflection of tumor grade than contrast enhancement (18). The background FDG metabolism within normal gray and white matter structures can serve as convenient reference levels. Subjectively, the FDG uptake within the tumor is compared with the metabolism in the contralateral

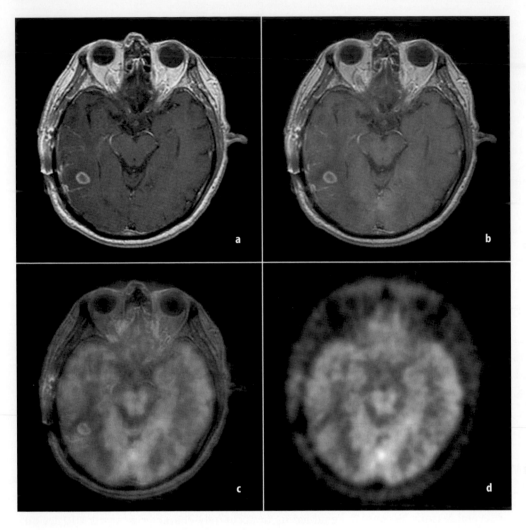

Figure 6.2. A 76-year-old man with history of glioblastoma multiforme and two prior surgical resections. Magnetic resonance imaging (MRI) demonstrates an enhancing nodule at the inferior aspect of the surgical resection cavity. Registered images demonstrate that this nodule has hypermetabolic activity similar to that of gray matter, consistent with high-grade tumor recurrence; this was confirmed at surgery. On the FDG-PET image alone (d), this nodule is indistinguishable from adjacent gray matter. (**a**) 100% MRI; (**b**) 70% MRI, 30% PET; (**c**) 30% MRI, 70% PET; (**d**) 100% PET. 📖

(presumably more normal) deep white matter and contralateral cortical gray matter. Qualitatively, low-grade tumors have FDG uptake similar to or below that of normal white matter, whereas the FDG accumulation in high-grade tumors either approaches or exceeds that of normal gray matter.

The FDG accumulation of low- and high-grade brain tumors relative to normal gray and white matter has been studied quantitatively by Delbeke et al. (19) in a study involving 58 consecutive patients with histologically proven high-grade (32 patients) and low-grade (26 patients) brain tumors. These authors measured the tumor to white matter (T/WM) and tumor to gray matter (T/GM) ratios to determine the optimum cutoff values for distinguishing low-grade (WHO Grades I and II) from high-grade (WHO Grades III and IV). They found that T/WM ratios greater than 1.5 and T/GM ratios greater than 0.6 were indicative of high-grade tumors with a sensitivity of 94% and specificity of 77%, respectively. This result supports the general subjective observation that low-grade tumors have metabolic activity similar to white matter and high-grade tumors have metabolic activity resembling gray matter.

Guidance for Stereotactic Biopsy

A characteristic feature of glial tumors is their heterogeneous nature. Their borders are generally poorly defined, and there is frequently geographic variation of tumor grade. In a study by Paulus and Peiffer (20), histologic features of 1,000 samples from 50 brain tumors (20 samples per tumor) were evaluated. They found that different grades were detected in 82% of tumors; moreover, 62% of the gliomas contained both high-grade (Grade III or IV) and benign (Grade II) features. This finding underscores the potential for sampling error and provides an explanation for the growth pattern of these tumors. Focal areas having higher grade may grow faster, resulting in an irregular contour to the tumor mass. High-grade tumors, notably glioblastoma multiforme, frequently originate from malignant degeneration of lower-grade tumors. These features cannot be distinguished on conventional anatomic imaging, such as MRI. Because of this underlying cellular heterogeneity, evaluation of these tumors by stereotactic biopsy, even with MR or CT guidance, is subject to significant sampling error and potential understaging. By mapping the metabolic pattern of these

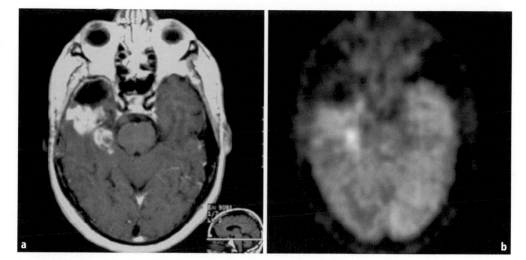

Figure 6.3. Coregistered MRI (**a**) and FDG-PET (**b**) images of a patient with recurrent high-grade tumor following surgical resection for anaplastic oligodendroglioma. Areas of contrast enhancement have high FDG accumulation compatible with high-grade tumor. Within this abnormality is an additional focus of very high metabolism posteromedially. 📖

heterogeneous tumors, FDG-PET can aid in targeting for stereotactic biopsy by selecting the subregions within the tumor that are most hypermetabolic and potentially have the highest grade (21–24).

This targeting may reduce the number of tissue samples required and may improve the accuracy of the biopsy for determining the actual tumor grade. At our institution, we can provide the neurosurgeon with coregistered MR/PET images, which enables the MRI coordinates for stereotactic biopsy to be selected based on the corresponding PET findings. Ultimately, we hope to fully integrate the registered MRI/PET data with the stereotactic guidance software to allow direct selection of the stereotactic biopsy coordinates based on PET-based targets.

An example of a patient with recurrent tumor is shown in Figure 6.3. This patient has had previous surgical resection of an anaplastic oligodendroglioma in the right temporal lobe and now has new regions of heterogeneous enhancement on MRI. The coregistered PET study reveals FDG accumulation that is similar to that of normal gray matter, consistent with recurrent high-grade tumor. In addition, there are focal subregions within the recurrent tumor that have varying degrees of hypermetabolic activity.

An example of a low-grade astrocytoma is illustrated in Figure 6.4. In this case, the tumor is poorly enhancing on MRI. The coregistered FDG-PET image shows corre-

sponding metabolism similar to that of the adjacent deep white matter, consistent with a low-grade tumor.

Prognostic Significance of FDG-PET

In addition to providing information relative to tumor grade, the degree of FDG metabolism carries prognostic significance. In a study of 29 patients with treated and untreated primary brain tumors, Alavi et al. (25) found that patients with hypermetabolic tumors had a significantly shorter survival than those with hypometabolic tumors. Moreover, within the subset of patients with high-grade gliomas, the patients with low tumor metabolism had a 1-year survival of 78%, whereas those with high tumor metabolism had a significantly poorer prognosis, with a 1-year survival of 29%. In another study involving 45 patients with high-grade tumors, Patronas et al. (26) showed similar results. In this study, patients having tumors with high metabolic activity had a mean survival of 5 months, whereas those patients with less glucose metabolism had a mean survival of 19 months.

The prognostic significance of FDG-PET findings in low-grade brain tumors has also been suggested. In patients with low-grade gliomas, the development of hypermetabolic features correlates with deleterious tumor

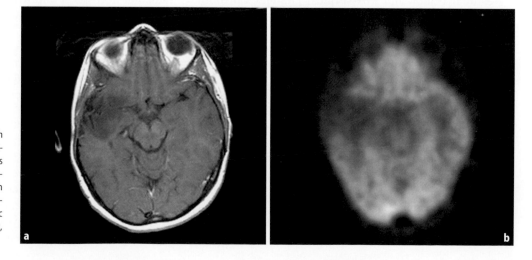

Figure 6.4. A 34-year-old woman with history of well-differentiated oligodendroglioma, status post two resections and chemotherapy. MRI (**a**) demonstrates residual nonenhancing tumor in right insular region. Coregistered FDG-PET (**b**) image demonstrates metabolic activity similar to that of white matter, compatible with low-grade tumor. 📖

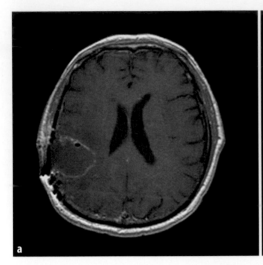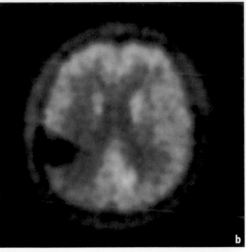

Figure 6.5. Same patient as in Figure 6.2. MRI (**a**) and coregistered FDG-PET image (**b**) of a The surgical resection cavity has a typical appearance with a small rim of enhancement on MRI. No corresponding hypermetabolic activity is noted on this axial slice to suggest residual high-grade tumor. Postsurgical changes alone do not result in hypermetabolic activity. 📖

evolution and poorer prognosis (27, 28). These studies suggest a relationship between the metabolic activity identified on FDG-PET scan and the biologic aggressiveness of both low- and high-grade primary brain tumors, independent of prior therapy. In addition, the prognostic information provided by FDG-PET imaging may be further improved by performing serial studies (29).

Evaluation of Brain Tumors Following Therapy

As with newly diagnosed brain tumors, image registration of FDG-PET images with MRI is important for evaluation of recurrent tumor. Suspicious areas of enhancement must be correlated with FDG metabolism. Images from a patient who had previous resections for glioblastoma multiforme are shown in Figures 6.2 and 6.5. In Figure 6.2, a nodular enhancing focus is present on MRI whereas no focal abnormality can be identified on the FDG-PET images alone. Coregistered images prove that this focus has FDG accumulation equal to that of gray matter, consistent with high-grade tumor. Without image coregistration, this activity could not be distinguished from adjacent normal cortex.

Residual tumor and postsurgical changes can both result in abnormal enhancement and can be indistinguishable on MRI following tumor resection. Postsurgical changes do not inherently result in increased metabolic activity, and FDG-PET is relatively unaffected for evaluating the postsurgical patient for residual tumor (30). Following surgery, a rim of contrast enhancement is often observed surrounding the resection cavity (see Figure 6.5); in this case, there is no associated hypermetabolic activity on the coregistered FDG-PET image. Abnormal hypermetabolic activity following surgery is compatible with residual high-grade tumor, and in the case of early recurrence, FDG-PET can help define the tumor target having the highest grade for stereotactic biopsy (31).

Following high-dose radiation therapy, radiation necrosis is usually reflected by diminished FDG metabolism within the treated field (32). Increased FDG activity following high-dose radiation therapy may be related to the metabolically active macrophages that accumulate at the therapy site. This activity is usually moderate (intermediate between white and gray matter) and relatively uniform in distribution. Occasionally, however, the metabolic activity may be equal to or greater than gray matter activity and can even have nodular characteristics; in these cases, radiation necrosis cannot be distinguished from recurrent high-grade tumor.

Barker et al. (33) studied 55 patients with high-grade tumors treated with surgery and radiation therapy who had enlarging areas of enhancement on MRI suggesting tumor recurrence or radiation necrosis. In this study, high FDG accumulation (equal or exceeding gray matter) correlated with significantly poorer prognosis compared to patients without hypermetabolic findings. In a study of 47 patients with primary and metastatic brain tumors who underwent stereotactic radiosurgery, Chao et al. (34) found FDG-PET to have a sensitivity of 75% and specificity of 81% for detecting recurrent tumor (versus radiation necrosis). For patient with brain metastases, coregistration of the FDG-PET images with MRI improved sensitivity from 65% to 86%.

Another study, by Valk et al. (35), found that the specificity of PET for distinguishing tumor recurrence from radiation necrosis is improved by using a combination of two PET studies. FDG-PET was used to evaluate metabolic activity, and a second PET study with ^{82}Rb was used to define regions of breakdown of the blood–brain barrier. In a series of 38 studies, the combined PET studies resulted in a sensitivity of 88% for detection of recurrent tumor and correctly identified radiation necrosis in 81% of cases.

Patients being treated for brain tumors are frequently receiving corticosteroids, which can influence the pattern of FDG metabolism. Patients on steroid therapy are reported to have decreased cerebral glucose metabolism (36). Steroid-induced hyperglycemia may be a contributing factor. A study by Roelcke et al. (37) found that patients with brain tumors in general have decreased glucose metabolism in the contralateral cortex and that the degree of this decrease correlates with tumor size.

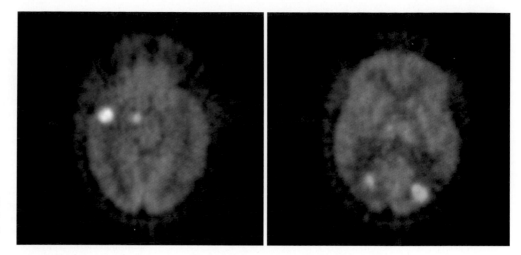

Figure 6.6. Patient with small cell lung cancer. Multiple intensely hypermetabolic lesions within the brain consistent with intracranial metastases. 📖

They suggest that tumor size may be a more important factor than corticosteroid dose in determining the degree of decreased metabolism in the contralateral cortex. Metabolism within brain tumors, on the other hand, was not affected by corticosteroid therapy.

Central Nervous System Lymphoma

FDG-PET is a sensitive imaging modality for evaluating Hodgkins and non-Hodgkins lymphoma in the body. Lymphoma in the central nervous system is also typically very metabolically active, whereas nonmalignant etiologies such as toxoplasmosis do not demonstrate high metabolic activity. In a study of 11 human immunodeficiency virus (HIV)-positive individuals at our institution with cerebral lesions on CT or MRI, FDG-PET accurately differentiated those patients with malignancy (lymphoma) from infectious etiologies (38). Other studies have supported these findings (39, 40).

Metastatic Brain Tumors

Soft tissue metastases in the brain and elsewhere in the body generally have high glucose metabolism, rendering them detectable by FDG-PET. Metastatic tumors in the brain, similar to high-grade primary brain tumors, generally have FDG accumulation comparable to that of normal cortical gray matter. When the metabolism exceeds that of normal gray matter, metastases can be readily identified on FDG-PET imaging (Figure 6.6). However, the high baseline glucose metabolism in the cortical brain provides a relatively low tissue-to-background environment for identifying intracranial metastases that are only mildly or moderately hypermetabolic. Furthermore, brain metastases are the result of hematogenous seeding, with a propensity to develop at the cortical gray–white junction. These early metastases are difficult if not impossible to distinguish from adjacent cerebral cortex on FDG-PET images alone. Finally, the cytotoxic edema that frequently surrounds metastatic deposits has relatively low FDG accumulation and may decrease conspicuity of the lesions through volume-averaging effects. The importance of coregistration with MRI or CT has been emphasized in the evaluation of primary brain tumors and would appear equally or more important for evaluating metastatic disease. An example of a patient with multiple intracranial metastases that are not evident on FDG-PET imaging is illustrated in Figure 6.7.

Larcos and Maisey (41) reviewed FDG-PET studies on 273 patients with various primary tumors to determine

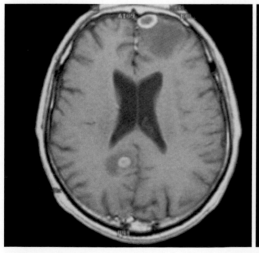

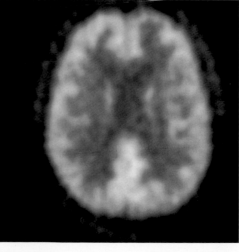

Figure 6.7. Intracranial metastases from non-small cell lung carcinoma. MR images demonstrate multiple enhancing lesions with surrounding edema. The metastatic lesions are not identified on the coregistered FDG-PET images. Metastases can be inconspicuous on PET images for reasons of small size, volume averaging with adjacent edema, variable metabolic grade, and tendency to occur at the gray–white junction. [Reprinted with permission from Wong et al. (42).]

the value of adding an abbreviated FDG brain scan to screen for intracranial metastases. They detected cerebral pathology in only 2% of the patients and unsuspected metastases in only 0.7%. They concluded that the addition of FDG brain scanning to routine whole-body imaging detected few clinically relevant lesions and should therefore not be performed routinely. This low detection rate has also been our experience at Duke University Medical Center, where we previously included brain imaging as part of our whole-body PET studies. We found that the additional FDG-brain study rarely contributed to clinical management and was less sensitive than MRI for detecting intracranial metastases. For these reasons, brain imaging is not currently included as part of our standard protocol for whole-body FDG-PET studies.

Conclusions

Compared to other organ systems, FDG-PET imaging of the brain presents unique challenges because of the high background glucose metabolism of normal gray matter structures. Coregistration of the MRI (or CT) and PET images is essential for accurate evaluation of brain tumors and is routinely performed at our institution.

FDG-PET is useful for characterizing primary brain tumors in terms of both tumor grade and prognosis. The heterogeneous nature of gliomas can result in significant sampling errors when patients are biopsied for primary tumor diagnosis or recurrent disease. FDG-PET can be used to define the most metabolically active targets for stereotactic biopsy; this in turn can improve diagnostic accuracy and reduce the number of biopsy samples required.

FDG-PET is also useful for evaluating residual or recurrent tumor following therapy and can be used to survey patients with low-grade brain tumors for evidence of degeneration into high-grade malignancy. In the case of suspected tumor recurrence or progression, PET can aid in defining appropriate targets for biopsy.

One limitation of FDG-PET is that radiation necrosis can occasionally be indistinguishable from recurrent high-grade tumor. A second limitation is that FDG-PET is not as sensitive as contrast-enhanced MRI for detecting intracranial metastases, and it is our experience that brain studies should not be included as part of routine whole-body PET studies.

Finally, this chapter has emphasized the general applications of FDG-PET for evaluation of brain tumors. Although FDG is currently the most readily available PET radiopharmaceutical for routine clinical applications, other tracers such as radiolabeled amino acids or nucleotides are potentially useful in certain situations.

References

1. Gurney JG, Kadan-Lottick N. Brain and other central nervous system tumors: rates, trends, and epidemiology. Curr Opin Oncol 2001;13(3):160–166.
2. Jemal A, Tiwari RC, Murray T, Ghafoor A, Samuels A, Ward E, et al. Cancer Statistics, 2004. CA Cancer J Clin 2004;54:8–29.
3. CBTRUS. Statistical Report: Primary Brain Tumors in the United States, 1997-2001. Central Brain Tumor Registry of the United States, 2004. *http://www.cbtrus.org/reports/reports.html*.
4. Herscovitch P, Markham J, Raichle ME. Brain blood flow measured with intravenous $H_2(^{15})O$. I. Theory and error analysis. J Nucl Med 1983;24(9):782–789.
5. Raichle ME, Martin WR, Herscovitch P, Mintun MA, Markham J. Brain blood flow measured with intravenous $H_2(^{15})O$. II. Implementation and validation. J Nucl Med 1983;24(9):790–798.
6. Mintun MA, Raichle ME, Martin WR, Herscovitch P. Brain oxygen utilization measured with O-15 radiotracers and positron emission tomography. J Nucl Med 1984;25(2):177–187.
7. Jager PL, Vaalburg W, Pruim J, de Vries EG, Langen KJ, Piers DA. Radiolabeled amino acids: basic aspects and clinical applications in oncology. J Nucl Med 2001;42(3):432–445.
8. Derlon JM, Chapon F, Noel MH, Khouri S, Benali K, Petit-Taboue MC, et al. Non-invasive grading of oligodendrogliomas: correlation between in vivo metabolic pattern and histopathology. Eur J Nucl Med 2000;27(7):778–787.
9. Ribom D, Eriksson A, Hartman M, Engler H, Nilsson A, Langstrom B, et al. Positron emission tomography (^{11}C-methionine and survival in patients with low-grade gliomas. Cancer (Phila) 2001;92(6):1541–1549.
10. Chung JK, Kim K, Kim Sk S, Lee J, Paek S, Yeo S, et al. Usefulness of (^{11}C-methionine PET in the evaluation of brain lesions that are hypo- or isometabolic on (^{18})F-FDG PET. Eur J Nucl Med 2002;29(2):176–182.
11. Eary JF, Mankoff DA, Spence AM, Berger MS, Olshen A, Link JM, et al. 2-[C-11]thymidine imaging of malignant brain tumors. Cancer Res 1999;59(3):615–621.
12. Reinhardt MJ, Kubota K, Yamada S, Iwata R, Yaegashi H. Assessment of cancer recurrence in residual tumors after fractionated radiotherapy: a comparison of fluorodeoxyglucose, L-methionine and thymidine. J Nucl Med 1997;38(2):280–287.
13. Di Chiro G. Positron emission tomography using [^{18}F] fluorodeoxyglucose in brain tumors. A powerful diagnostic and prognostic tool. Invest Radiol 1987;22(5):360–371.
14. Pelizzari CA, Chen GT, Spelbring DR, Weichselbaum RR, Chen CT. Accurate three-dimensional registration of CT, PET, and/or MR images of the brain. J Comput Assist Tomogr 1989;13(1):20–26.
15. Turkington TG, Hoffman JM, Jaszczak RJ, MacFall JR, Harris CC, Kilts CD, et al. Accuracy of surface fit registration for PET and MR brain images using full and incomplete brain surfaces. J Comput Assist Tomogr 1995;19(1):117–124.
16. Turkington TG, Jaszczak RJ, Pelizzari CA, Harris CC, MacFall JR, Hoffman JM, et al. Accuracy of registration of PET, SPECT and MR images of a brain phantom. J Nucl Med 1993;34(9):1587–1594.
17. Wong TZ, Turkington TG, Hawk TC, Coleman RE. PET and brain tumor image fusion. Cancer J 2004;10(4):234–242.
18. Patronas NJ, Brooks RA, DeLaPaz RL, Smith BH, Kornblith PL, Di Chiro G. Glycolytic rate (PET) and contrast enhancement (CT) in human cerebral gliomas. AJNR Am J Neuroradiol 1983;4(3):533–535.
19. Delbeke D, Meyerowitz C, Lapidus RL, Maciunas RJ, Jennings MT, Moots PL, et al. Optimal cutoff levels of F-18 fluorodeoxyglucose uptake in the differentiation of low-grade from high-grade brain tumors with PET. Radiology 1995;195(1):47–52.
20. Paulus W, Peiffer J. Intratumoral histologic heterogeneity of gliomas. A quantitative study. Cancer (Phila) 1989;64(2):442–447.

21. Hanson MW, Glantz MJ, Hoffman JM, Friedman AH, Burger PC, Schold SC, et al. FDG-PET in the selection of brain lesions for biopsy. J Comput Assist Tomogr 1991;15(5):796–801.

22. Pirotte B, Goldman S, Brucher JM, Zomosa G, Baleriaux D, Brotchi J, et al. PET in stereotactic conditions increases the diagnostic yield of brain biopsy. Stereotact Funct Neurosurg 1994;63(1-4):144–149.

23. Pirotte B, Goldman S, Bidaut LM, Luxen A, Stanus E, Brucher JM, et al. Use of positron emission tomography (PET) in stereotactic conditions for brain biopsy. Acta Neurochir 1995;134(1-2):79–82.

24. Pirotte B, Goldman S, David P, Wikler D, Damhaut P, Vandesteene A, et al. Stereotactic brain biopsy guided by positron emission tomography (PET) with [F-18]fluorodeoxyglucose and [C-11]methionine. Acta Neurochir Suppl 1997;68:133–138.

25. Alavi JB, Alavi A, Chawluk J, Kushner M, Powe J, Hickey W, et al. Positron emission tomography in patients with glioma. A predictor of prognosis. Cancer (Phila) 1988;62(6):1074–1078.

26. Patronas NJ, Di Chiro G, Kufta C, Bairamian D, Kornblith PL, Simon R, et al. Prediction of survival in glioma patients by means of positron emission tomography. J Neurosurg 1985;62(6):816–822.

27. De Witte O, Levivier M, Violon P, Salmon I, Damhaut P, Wikler D Jr, et al. Prognostic value positron emission tomography with [^{18}F]fluoro-2-deoxy-D-glucose in the low-grade glioma. Neurosurgery 1996;39(3):470–476; discussion 476–477.

28. Francavilla TL, Miletich RS, Di Chiro G, Patronas NJ, Rizzoli HV, Wright DC. Positron emission tomography in the detection of malignant degeneration of low-grade gliomas. Neurosurgery 1989;24(1):1–5.

29. Schifter T, Hoffman JM, Hanson MW, Boyko OB, Beam C, Paine S, et al. Serial FDG-PET studies in the prediction of survival in patients with primary brain tumors. J Comput Assist Tomogr 1993;17(4):509–561.

30. Hanson MW, Hoffman JM, Glantz MJ, et al. FDG-PET in the early postoperative evaluation of patients with brain tumor [abstract]. J Nucl Med 1990;31:799.

31. Glantz MJ, Hoffman JM, Coleman RE, Friedman AH, Hanson MW, Burger PC, et al. Identification of early recurrence of primary central nervous system tumors by [^{18}F]fluorodeoxyglucose positron emission tomography. Ann Neurol 1991;29(4):347–355.

32. Di Chiro G, Oldfield E, Wright DC, De Michele D, Katz DA, Patronas NJ, et al. Cerebral necrosis after radiotherapy and/or intraarterial chemotherapy for brain tumors: PET and neuropathologic studies. AJR Am J Roentgenol 1988;150(1):189–197.

33. Barker FG II, Chang SM, Valk PE, Pounds TR, Prados MD. ^{18}-Fluorodeoxyglucose uptake and survival of patients with suspected recurrent malignant glioma. Cancer (Phila) 1997;79(1):115–126.

34. Chao ST, Suh JH, Raja S, Lee SY, Barnett G. The sensitivity and specificity of FDG PET in distinguishing recurrent brain tumor from radionecrosis in patients treated with stereotactic radiosurgery. Int J Cancer 2001;96(3):191–197.

35. Valk PE, Budinger TF, Levin VA, Silver P, Gutin PH, Doyle WK. PET of malignant cerebral tumors after interstitial brachytherapy. J Neurosurg 1988;69(6):830–838.

36. Fulham MJ, Brunetti A, Aloj L, Raman R, Dwyer AJ, Di Chiro G. Decreased cerebral glucose metabolism in patients with brain tumors: an effect of corticosteroids. J Neurosurg 1995;83(4):657–664.

37. Roelcke U, Blasberg RG, von Ammon K, Hofer S, Vontobel P, Maguire RP, et al. Dexamethasone treatment and plasma glucose levels: relevance for fluorine-18-fluorodeoxyglucose uptake measurements in gliomas. J Nucl Med 1998;39(5):879–884.

38. Hoffman JM, Waskin HA, Schifter T, Hanson MW, Gray L, Rosenfeld S, et al. FDG-PET in differentiating lymphoma from nonmalignant central nervous system lesions in patients with AIDS. J Nucl Med 1993;34(4):567–575.

39. Heald AE, Hoffman JM, Bartlett JA, Waskin HA. Differentiation of central nervous system lesions in AIDS patients using positron emission tomography (PET). Int J STD AIDS 1996;7(5):337–346.

40. Roelcke U, Leenders KL. Positron emission tomography in patients with primary CNS lymphomas. J Neurooncol 1999;43(3):231–236.

41. Larcos G, Maisey MN. FDG-PET screening for cerebral metastases in patients with suspected malignancy. Nucl Med Commun 1996;17(3):197–198.

42. Wong TZ, Coleman RE. Brain Tumors. In: Bender H, Palmedo H, Biersack H-J, Valk PE, editors. Atlas of Clinical PET in Oncology. Heidelberg/Berlin: Springer-Verlag, 2000:153–170.

7

PET and PET/CT Imaging in Lung Cancer

Pierre Rigo, Roland Hustinx, and Thierry Bury

Every year nearly 1 million new cases of lung cancer are currently diagnosed, principally in developed countries. Most of these patients will die of their disease, the majority within 2 years of diagnosis. Lung cancer is the principal cause of cancer death, affecting primarily patients between 55 and 65 years of age (1). At the time of diagnosis, the disease has generally spread beyond the primary site. Twenty-five percent of patients have extension to adjacent hilar or mediastinal nodal areas and up to 35% to 45% of patients demonstrate the presence of distant metastases (2). Furthermore, the overall 5-year survival of patients with apparently limited disease and resectable tumor is only 35% to 40% (3, 4), making lung cancer one of the most important public health problems (5, 6).

Lung cancer has been a case study in the use of positron emission tomography (PET) using ^{18}F-fluorodeoxyglucose (FDG) for the clinical evaluation of cancer. Identification of the metabolic abnormalities of lung cancer for the differentiation of malignant from the nonmalignant nodule was the first non-central nervous system (CNS) oncologic application of FDG-PET to be established (7–11), leading the way for later advances in staging, therapy assessment, and follow-up, so that the use of PET is now generally accepted for the management of lung cancer.

Classification and Evaluation of Lung Cancer

Histologic Classification

The course and prognosis of lung cancer are not uniform and depend upon the histologic classification of the tumor. The World Health Organization (WHO) classification (Table 7.1) is most generally used (12). Malignant lung cancer, which excludes benign tumors as well as dysplasia and carcinoma in situ, consists primarily of small cell carcinoma and non-small cell lung carcinoma (NSCLC). These categories are further divided into several subtypes. Small cell carcinoma (approximately 25% of cases) is rarely limited at presentation and is usually disseminated at initial diagnosis. Surgery is seldom employed and only rarely results in cure. Chemotherapy is most often used but is only transiently effective. The incidence of small cell carcinoma is related to tobacco exposure, as is that of squamous cell carcinoma; the latter accounts for 30% of lung cancer cases. Squamous cell carcinoma originates primarily from proximal bronchial segments and develops slowly over time. It can be detected by cytologic examination.

Adenocarcinoma does not appear related to tobacco consumption, but its incidence is also rising, in particular in women. Most adenocarcinomas are peripheral in origin and disseminate faster than squamous cell carcinoma, both to hilar and mediastinal lymph nodes and to the brain. Bronchioloalveolar carcinoma is a form of adenocarcinoma, which may present in atypical fashion with multifocal tumor or a pneumonia-like presentation. The sensitivity of FDG-PET for detecting bronchioloalveolar or mucus-secreting carcinoma is lower than that for the other types, probably in part because of the low cellularity of these histologic forms (13, 14). Large cell carcinoma is the least common type of NSCLC and has a clinical course similar to that of adenocarcinoma.

TNM Classification

Lung cancer progresses by invading both lymphatic channels and vascular structures. Lymphatic spread is initially to lymph nodes that are situated around the segmental and lobar bronchi (N1); these in turn lead to the ipsilateral (N2) or contralateral (N3) mediastinal lymph nodes. Further extension can reach the supraclavicular lymph nodes (N3), the chest wall, the mediastinal organs, or the pleural surface (T3–T4) (15–18). Systemic metastases can be found in any organ system, but are seen predominantly in the brain, skeleton, liver, and adrenal glands. Pulmonary metastases are also commonly observed (5, 16).

Table 7.1. World Health Organization (WHO) histologic classification of epithelial bronchogenic carcinoma.

I. Benign		
II. Dysplasia and carcinoma in situ		
III. Malignant	A. Squamous cell carcinoma (epidermoid) and spindle (squamous carcinoma	
	B. Small cell carcinoma	1. Oat cell
		2. Intermediate cell
		3. Combined oat cell
	C. Adenocarcinoma	1. Acinar
		2. Papillary
		3. Bronchoalveolar
		4. Mucus secreting
	D. Large cell carcinoma	1. Giant cell
		2. Clear cell

Source: Adapted from Valk PE, Bailey DL, Townsend DW, Maisey MN. Positron Emission Tomography: Basic Science and Clinical Practice. Springer-Verlag London Ltd 2003, p. 518.

The TNM classification (Table 7.2) uses the anatomic extent of the primary tumor (T stage) and evidence of spread to regional lymph nodes (N stage) or distant sites (M stage) as a guide for prognosis and management. Prognosis varies according to the accuracy of the method used for staging. Typically, clinical staging understages the tumor as compared to final staging at surgery and pathologic examination (3), whereas FDG-PET and especially PET/CT staging approaches the final stage more closely (17).

Table 7.2. Non-small cell lung cancer (NSCLC): TNM staging.

Primary tumor	TX	Positive malignant cells; no lesion seen
	T0	No evidence of primary tumor
	TIS	Carcinoma in situ
	T1	<3 cm in greatest dimension
	T2	>3 cm in greatest dimension; distal atelectasis
	T3	Extension into the chest wall (including superior sulcus tumours); diaphragm, or mediastinal pleura or pericardium; <2 cm from carina or total atelectasis
	T4	Invasion of the mediastinal organs; malignant pleural effusion
Nodal involvement	N0	No involvement
	N1	Ipsilateral bronchopulmonary or hilar
	N2	Ipsilateral mediastinal or subcarinal; ipsilateral supraclavicular
	N3	Contralateral mediastinal, hilar or supraclavicular
Distant metastatic involvement	M0	None
	M1	Present

Source: From Valk PE, Bailey DL, Townsend DW, Maisey MN. Positron Emission Tomography: Basic Science and Clinical Practice. Springer-Verlag London Ltd 2003, p. 518.

Conventional Diagnostic and Staging Procedures

Diagnosis is the first clinical problem facing a clinician in charge of a patient suspected of having lung cancer. Accurate tumor staging is essential for selection of appropriate therapy. It is important both to avoid the morbidity and mortality of surgery in patients with nonresectable and rapidly progressing disease and to avoid denying patients with resectable disease the benefit of surgery based on incorrect diagnosis of mediastinal or systemic metastases. All evidence indicates that FDG-PET is the most accurate staging procedure that is available at present, but other staging procedures are commonly performed before PET as they are part of an established diagnostic process. It is important to understand the value and limitations of these other techniques to better define the place of PET (18–21).

Patient history and physical examination are important in the evaluation of patients with suspected lung cancer. Age , smoking history, and site, size, and growth of the lesion affect the cancer probability of a given lesion (22). Morphologic images can detect a nodule or a mass but frequently cannot determine its benign or malignant nature (18, 22). Chest X-ray is usually the first imaging study performed, but it is neither specific nor sensitive (22, 23). Standard and dynamic computed tomography (CT) imaging can detect smaller lesions and can also help to characterize some lesions. Smooth margins, small size (less than 1 cm), and central or popcorn calcifications favor a benign origin, whereas large lesions (>3 cm), spiculated margins, and the absence of "benign" calcifications favor a malignant tumor. Cytology (three sputum samples) is positive in 80% of central tumors but in only 20% of peripheral tumors (24). Diagnosis is enhanced by bronchoscopy with lavage or biopsy. Bronchoscopy can also be used to biopsy enlarged mediastinal lymph nodes (25). Any lymph node larger than 1 cm in diameter on CT is considered to be abnormal and requiring further investigations, but this criterion is insensitive and poorly specific. CT is the best anatomic imaging technique for determining tumor invasion of the peripheral organs, chest wall, and vertebrae (20), whereas magnetic resonance imaging (MRI) may be needed to evaluate the relationship of the tumor to the heart and large vessels (26, 27). The use of skeletal scintigraphy for detection of skeletal metastasis in asymptomatic patients is controversial because of the very low yield of the technique and its poor specificity (28–30).

Percutaneous fine-needle aspiration biopsy of pulmonary nodules can be used in cases where CT is not diagnostic. The procedure is very specific when positive, but sampling errors occur in 5% to 25% of cases, and negative or indeterminate biopsies must be further evaluated (31). Mediastinoscopy is the most accurate method of staging the superior mediastinal lymph nodes, but more-extensive nodal sampling and evaluation of contralateral lymph nodes are not possible in practice. Again, sampling errors occur, resulting in false-negative findings (32,33). Thoracotomy and mediastinotomy can also be used for diagnosis and staging as a last resort.

FDG-PET Imaging

FDG-PET Imaging Procedure

A whole-body imaging study, extending from the base of the brain to the inguinal regions, is indicated to detect skeletal and other distant metastases. Imaging of the brain is usually not performed because contrast-enhanced CT or MRI is more effective for diagnosis of cerebral metastases (34). Acquisition can be extended to the extremities in cases of more-peripheral suspected lesions. Acquisition should include transmission imaging to permit attenuation correction and quantitative evaluation. Attenuation-corrected imaging permits the use of the semiquantitative standardized uptake value (SUV). The SUV is determined by dividing the measured activity in a tissue region (MBq/mL) by injected dose and normalizing to body weight (MBq/g). SUV changes with time after injection, which determines tissue uptake of activity, and is influenced by the blood glucose concentration (35). In standardized conditions, the SUV is informative and reproducible (36). The use of the lean body weight instead of total body weight for normalization has been recommended because accumulation of FDG in fat is low (37). Normalization of activity to body surface area instead of weight has also been proposed (38).

Simultaneous evaluation of the PET and CT images is very helpful and even mandatory. This step can be performed visually, but the process is greatly facilitated by image registration or by concurrent imaging using a PET/CT system (39). In this case, as CT and PET are done sequentially rather than simultaneously, patient immobility is important and much attention must be given to patient comfort and breathing pattern to minimize position artifacts.

Limitations of FDG-PET Imaging

As do all diagnostic techniques, FDG-PET gives false-positive and false-negative results. False-positive results are related to nonspecific FDG uptake in normal tissues or in benign lesions. Macrophages, leukocytes, and activated lymphocytes are known to actively concentrate FDG in the setting of granulomatous inflammation (40). Acute and chronic bacterial infection, [bacterial pneumonia (41), pyogenic abscess], tuberculosis, fungal infection (histoplasmosis, aspergillosis, coccidioidomycosis), sarcoidosis (42), anthracosilicosis, and Wegener's disease

may lead to increased pulmonary or mediastinal uptake (43). Ischemic and prenecrotic lesions, for example, in pulmonary infarction, can also cause false-positive results (44). In some cases, these can be differentiated from tumor on the basis of the clinical context or the time course and intensity of FDG uptake (45). However, often there is overlap between the degree of FDG uptake of malignant and benign lesions. High prevalence of false-positive findings may be encountered in geographic areas where fungal infections are common and in patients who have a history of inflammatory or infectious disease. False-positive results can also be created by attenuation-correction artefacts, especially when incorrect attenuation values are calculated after CT transmission in the presence of a metallic prosthesis or supplementary contrast (46).

False-negative results may be caused by the small size of lesions relative to scanner resolution (partial-volume effect), low cellularity of lesions (mucinous tumors) (47), or low metabolic activity of tumors (bronchioloalveolar and some carcinoid tumors) (48). Partial-volume effect becomes significant as lesion size decreases below twice the full width at half-maximum scanner resolution, with resultant decrease in detected lesion activity (49, 50). With dedicated PET systems of current technology, detection sensitivity decreases as lesion volume decreases below approximately 1 cm^3. Improvement in scanner resolution will improve sensitivity for small lesions, but microscopic lesions will always be missed.

Partial-volume effect is also reflected in quantitative underestimation of tumor activity, with resulting decrease in SUV values. It should be noted that, when measuring the SUV of small lesions for therapy follow-up, decrease in SUV may occur as a result of decrease in lesion size rather than actual decrease in tumor uptake.

Indications for FDG-PET Imaging in Lung Cancer Management

Diagnosis of Solitary Pulmonary Nodules

Conventional Methods

Solitary pulmonary nodules or atypical focal abnormalities represent a common diagnostic indication for FDG-PET imaging. Approximately one-third of pulmonary nodules can be diagnosed as benign or malignant on the basis of standard CT criteria (51–53). Absence of growth over a 2-year period may suggests a benign lesion when comparative images are available (54), but its specificity is less than originally claimed. Two-thirds of solitary pulmonary nodules are indeterminate and require further diagnostic procedures. Lesion contrast enhancement using dynamic CT provides a highly sensitive but poorly specific

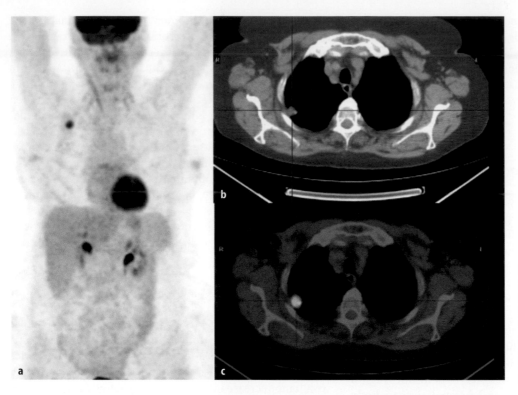

Figure 7.1. MIP image FDG-PET of an apical solitary pulmonary nodule (**a**). Lesion uptake [standard uptake value (SUV), 6.4] is typical of malignancy and corresponded to a non-small cell lung cancer. Transverse section of the computed tomography (CT) (**b**) and of the fused PET and CT images (**c**) are also shown.

approach to characterize malignant disease. In suitable cases (greater than 1 cm) and low-risk patients and pattern it can be used to exclude malignancy, but positive interpretations must be confirmed by PET (21).

Fiberoptic bronchoscopy and biopsy can diagnose central lung tumors, but diagnostic tissue samples are only obtained in 20% of patients with peripheral nodules (55). Conventionally, fine-needle biopsy under CT guidance is the next step (31). Needle biopsy detects malignancy with a sensitivity of more than 95% when an adequate sample is obtained, but sampling errors occur commonly, their frequency depending upon nodule location and size and on operator expertise. Malignancy can be excluded only when a specific benign diagnosis can be established on histologic examination. The most common and important complication of needle biopsy is pneumothorax, requiring hospitalization and insertion of a chest tube in 5% to 10% of patients (56). This complication occurs most commonly in patients who also have chronic obstructive airways disease and may cause acute deterioration in lung function in these patients. A significant number of benign nodules are eventually resected to obtain a definitive diagnosis (57).

Accuracy of FDG-PET Imaging in Solitary Pulmonary Nodules

Characterization of pulmonary nodules as benign or malignant was the first non-CNS oncologic use of FDG-PET to be established (Figure 7.1). The value of the technique has now been firmly established, first in small individual studies (7–11), then confirmed by an Institute for Clinical PET-sponsored multicenter study (58) and by numerous other studies from multiple institutions in prospective clinical studies (59–68). The results obtained in the principal studies are summarized in Table 7.3. In our own prospective study of 103 patients with indeterminate solitary pulmonary nodules (SPN) after CT imaging, 98 of 103 nodules were correctly identified with one false-negative and four false-positive findings (tuberculosis, anthracosilicosis, subacute abscess, and pneumonia scar) (69). Other published results confirmed these values of sensitivity, specificity, and negative predictive value. A recent study analysed the performance of PET in small pulmonary nodules (68). In 35 patients with nodules 10 mm in diame-

Table 7.3. Large prospective series of ¹⁸F-fluorodeoxyglucose-positron emission tomography (FDG-PET) in the evaluation of patients with lung nodules or masses.

Author (reference number)	Date	N	Sensitivity	Specificity	NPV	Prevalence
Patz (61)	1993	51	100	79	100	65
Duhaylongsod (62)	1995	87	97	82	92	68
Bury (63)	1996	50	100	88	100	66
Gupta (64)	1996	61	91	78	70	74
Sazon (65)	1996	107	100	52	100	77
Lowe (66)	1998	89	98	69	95	67
Prauer (67)	1998	50	90	83	86	57
Herder (68)	2004	36	93	77	94	39

Source: Adapted from Valk PE, Bailey DL, Townsend DW, Maisey MN. Positron Emission Tomography: Basic Science and Clinical Practice. Springer-Verlag London Ltd 2003, p. 519.

ter or less, sensitivity was 93% (13/14), specificity 77% (17/22), and negative predictive value 94% (see Table 7.3).

A meta-analysis by Gould et al. (70) that included 450 pulmonary nodules indicated mean sensitivity and specificity for detecting malignancy of 93.9% and 85.8%, and median sensitivity and specificity of 98% and 83.3%, respectively. Results in 1,474 focal pulmonary lesions were similar, with mean sensitivity and specificity of 96% and 73.5% and median sensitivity and specificity of 97% and 77.8%, respectively. Wide variations were observed between studies, especially in specificity. It has been suggested that some of these variations may be related to geographic variations in the prevalence of granulomatous diseases such as histoplasmosis or coccidioidomycosis, which are more common in North America than in Europe (70). Other potential causes include differences in interpretation of PET images, resulting in different diagnostic thresholds in different studies. A factor that may affect sensitivity is difference in study inclusion criteria, because a high prevalence of very small nodules can reduce measured sensitivity.

Gould et al. (71) showed that the maximum accuracy of FDG-PET was 91.2% for pulmonary lesions of any size and 90.0% for pulmonary nodules. Most FDG-PET interpreters operate at a threshold favoring sensitivity over specificity to minimize the number of false-negative results. Semiquantitative image interpretation by means of SUV determination has failed to improve the accuracy over subjective visual reading, but it could be useful for standardizing the threshold of positivity among readers and institutions.

The use of PET/CT helps confirm the exact localization of the nodule and allows for adjustment of subjective or SUV threshold criteria based on the size of the lesion and the probability of activity underestimation related to respiratory mismatch between the anatomic and functional images (inadequate attenuation correction) (72).

Role of FDG-PET in Solitary Pulmonary Nodule Management Strategy

FDG-PET characterization of solitary pulmonary nodules (SPN) may obviate the need for invasive biopsy in many patients (73, 74). Use of FDG-PET for this purpose requires a high negative predictive value (NPV), which depends on the pretest probability of cancer as much as it depends on the sensitivity and specificity of PET. The results obtained by Gambhir et al. (75) using decision analysis modeling have shown that only patients with 69% or lower pretest probability of cancer, as determined by factors such as lesion size, age, and smoking history, should undergo FDG-PET imaging for nodule diagnosis. In patients with a higher pretest probability, the posttest probability of cancer after a negative PET scan exceeds 10%, and histologic diagnosis will be needed regardless of the PET finding.

Also regardless of the NPV, patients with negative PET findings should always be managed by clinical and imaging follow-up because some risk of a false-negative result remains. In these patients the nodule should be evaluated at 6- to 12-month intervals for evidence of increase in size, which would be an indication for intervention to establish a histologic diagnosis. A combined clinicoradiologic strategy has been proposed by Rohren et al (21). Nodules are stratified on the basis of thin-section CT and clinical information in benign or indeterminate lesions. Low-risk patients with typical benign nodules need not be followed. Patients with indeterminate small nodules (less than 8 mm) require CT follow up. Stable nodules are followed up to 2 years at 3- to 6-month intervals. Progressive lesions require further evaluation or direct biopsy or excision. Larger (greater than 8 mm) indeterminate lesions in patients with low clinical and radiologic likelihood for malignancy may first be studied by dynamic CT. Absence of enhancement practically excludes malignancy, but the presence of enhancement is not specific and requires further testing. PET in that situation can confirm positivity and provide staging. Patients with high likelihood of malignancy should be studied by PET. A negative finding leads to CT follow up while a positive finding leads to biopsy and/or excision.

Cost-Effectiveness of FDG-PET in Solitary Pulmonary Nodules

The cost-effectiveness of FDG-PET for characterization of SPN has been assessed by several investigators. These studies are complex, as they need to take into account the prevalence of the disease, the prevailing management strategies in the country of the study, and the cost and accuracy of alternative diagnostic procedures, as well as the final surgical intervention. Decision analysis modeling has demonstrated the cost-effectiveness of PET in the United States in patients with low or intermediate prevalence of cancer (75). A watch-and-wait strategy is indicated when prevalence is less than 12%, whereas systematic surgery is more effective in high-prevalence situations (76). However, in these high-prevalence situations, FDG-PET is still required for pretreatment staging.

A Japanese modeling study did not find FDG-PET to be cost effective in comparison to CT plus mediastinoscopy using a lung cancer prevalence value of 71.4%, but PET yielded a life expectancy gain of 7.3 months (77). In Germany, PET appears to be cost effective if the cost of PET is Eu 1,620 or less; in the Unites States, because of higher costs of surgery and competing diagnostic modalities, cost-effectiveness is maintained up to a higher PET cost (76). The sensitivity of cost-effectiveness to the accuracy of PET has also been modeled, and the test appears to be robust, as it can sustain a decrease in both sensitivity and specificity up to 10% and remain cost effective.

Pretreatment Staging

A systematic approach to staging and treatment of NSC lung cancer based on knowledge of clinical and pathologic data should enable the oncologist to define the best therapeutic strategy adapted to the individual cases, aiming for cure or for the best possible palliative result. Definition of the extent of the disease both locoregionally and at distant sites is key to choosing the appropriate therapeutic option. Indeed, several studies have demonstrated a direct relationship between tumor staging, in particular, mediastinal staging, and the median-term survival (3). Patients without mediastinal lymph node and distant metastatic involvement should undergo surgical resection, whereas patients with mediastinal lymph node involvement require a more-complex strategy, which may include induction chemotherapy and/or radiotherapy and surgery. Anatomic imaging techniques have been inadequate in this regard, which is demonstrated by the large difference in tumor stage as determined preoperatively with conventional techniques and at surgery (3). There is a need for more-sensitive detection of tumor spread to reduce futile surgery, but this must be achieved without sacrificing specificity of diagnosis. The greatest detrimental effect of incorrect staging occurs when a patient with potentially curable tumor is denied resection because of a false-positive diagnosis of metastatic disease (77).

T Staging

Anatomic imaging by CT is used to define the extent of the primary tumor and to determine involvement of the pleural surfaces and the thoracic wall (20); this may be supplemented by MRI to evaluate the relationship of the tumor to the heart and large vessels (26, 27). FDG-PET alone cannot be used to assess the extent of primary tumor because of its lower spatial resolution and poor visualization of normal structures and anatomic details, but PET/CT now provides that information (Figure 7.2) PET/CT has proven useful to define involvement of the thoracic wall as well as to differentiate patients with reactive or metastatic pleural effusion (78). The interpretation must remain cautious, however, given the possibility of patient or respiratory motion between the two images. In the study by Lardinois et al. (79), 19 patients were evaluated for chest wall infiltration or mediastinal invasion. PET/CT staging was correct in 16 patients and equivocal in 3 whereas visual correlation was equivocal in 5 and incorrect in 4; specifically, PET/CT enabled correct definition of chest wall infiltration in 3 patients compared to PET alone or visual PET and CT correlation. Also PET/CT allowed differentiation of tumor and peritumoral inflammation or atelectasis in 7 patients. Cerfolio et al. (80) also reported improved evaluation of T status using PET/CT. As described by Schaffler et al. (81), a negative

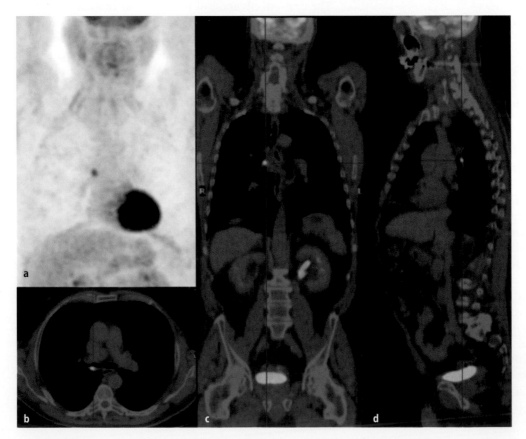

Figure 7.2. FDG-PET image (**a**) and fused PET/CT images [transverse (**b**); coronal (**c**); sagittal (**d**)] of a patient with a localized tumor in the right main bronchus (squamous cell carcinoma).

PET/CT indicates a benign lesion in patients with indeterminate pleural effusion by CT, whereas positive FDG-PET/CT fusion is sensitive for malignancy.

Local and Regional Nodal Staging

The limitations of anatomic imaging for hilar and mediastinal staging have been well documented: the size criterion of 1 cm that is used as the threshold for diagnosis of metastatic involvement leads to misdiagnosis of normal-sized nodes that are infiltrated by tumor and of nodes which are enlarged as a result of benign pathology. In a meta-analysis of mediastinal staging by CT and PET, Dwamena et al. (82) reported CT sensitivity and specificity of 60% and 77%, respectively, with a diagnostic accuracy of approximately 65%. Prospective data from the Radiology Diagnostic Oncology Group showed sensitivity of 52% and specificity of 69% (27). MRI has not been more accurate. Because of these limitations in imaging accuracy, mediastinoscopy has remained the reference technique for mediastinal lymph node staging despite its invasiveness and limited sampling capability (33).

Accuracy of FDG-PET in Nodal Staging

Numerous groups have reported comparisons of FDG-PET and CT for the assessment of mediastinal metastasis in NSCLC (Table 7.4a) (83–92). Most results have been reported on a positive/negative basis for the ipsilateral or contralateral mediastinum without reference to specific nodal stations (Figures 7.3, 7.4). In the study of 50 patients from Bury et al. (83), sensitivity and specificity of PET were 90% and 86%, respectively. CT and PET were discordant in 36 cases and PET was correct in 28 of these. In a similar evaluation, Pieterman et al. (89) reported sensitivity of 91% for overall mediastinal involvement and 75% on an individual lymph node basis; specificity was 86%.

Table 7.4a. FDG-PET for the evaluation of nodal staging: large series with independent PET and computed tomography (CT) readings.

Analysis per patients:

Author (reference number)	Date	N	Sensitivity	Specificity	Prevalence
Bury (83)	1996	50	90	86	32
Steinert (84)	1997	47	92	97	28
Vansteenkiste (85)	1997	50	67	97	30
Saunders (86)	1999	84	71	97	18
Kerstine (87)	1999	64	70	86	25
Weng (88)	2000	50	73	94	38
Pieterman (89)	2000	102	91	86	31

Table 7.4a. FDG-PET for the evaluation of nodal staging: large series with independent PET and CT readings.

Analysis per nodal station:

Author (reference number)	Date	N	Sensitivity	Specificity	Prevalence
Patz (90)	1995	62	83	82	37
Sasaki (91)	1996	71	76	98	24
Steinert (84)	1997	191	89	99	25
Berlangieri (92)	1999	201	80	97	10
Pieterman (89)	2000	516	76	–	07

Source: From Valk PE, Bailey DL, Townsend DW, Maisey MN. Positron Emission Tomography: Basic Science and Clinical Practice. Springer-Verlag London Ltd 2003, p. 522.

In their meta-analysis of staging by PET and CT, Dwamena et al. (82) reviewed 14 PET studies involving 514 patients and 29 CT studies with 2,226 patients and calculated summary receiver operating characteristic (ROC) curves. Pooled point estimates of diagnostic performance and the summary ROC curves indicated that PET was significantly more accurate than CT. Mean sensitivity and specificity were 79% ± 3% and 91% ± 2% for PET and 60% ± 2% and 77% ± 2%, respectively, for CT.

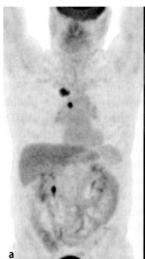
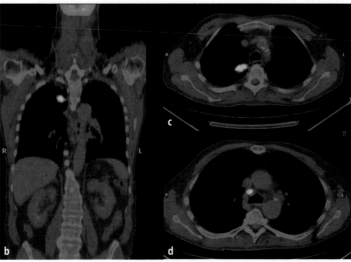

Figure 7.3. MIP image (FDG-PET) (**a**) and fused PET/CT images [coronal (**b**) and transverse (**c**, **d**)] of a patient presenting with a right superior lobe lesion and an ipsilateral lymph node focus (N2; nodal station 4R). Mediastinoscopy and surgery did not confirm the N2 status.

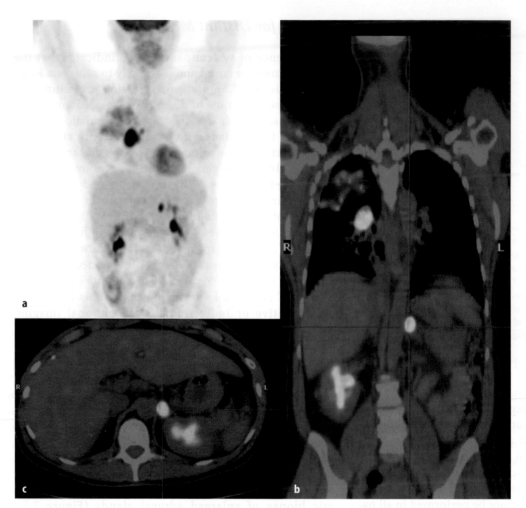

Figure 7.5. Parahilar right non-small cell lung cancer (NSCLC) with mediastinal lymph node and superior lobe inflammatory involvement. (**a**) MIP image of the FDG-PET. PET demonstrates the presence of a left adrenal metastasis [fused PET/CT images: (**b**) coronal; (**c**) transverse].

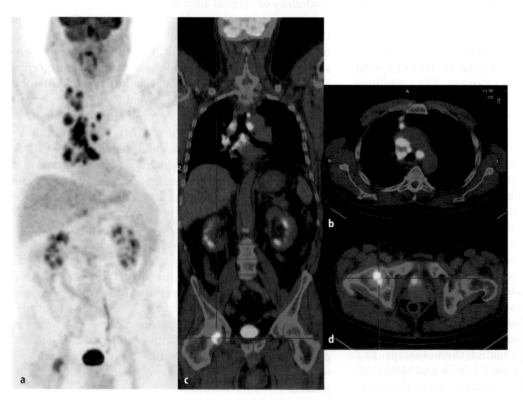

Figure 7.6. Right superior lobe NSCLC with multiple hilar, bilateral mediastinal, and supraclavicular lymph nodes (N3). (**a**) MIP image of the FDG-PET; (**b**) transverse section of the fused PET/CT images. Also note the presence of metastatic costal and femoral bone lesions in coronal (**c**) and transverse (**d**) sections of the fused PET/CT images.

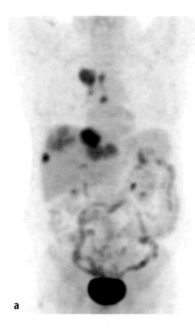

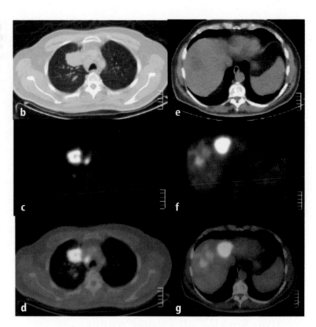

Figure 7.7. Newly diagnosed right upper lobe NSCLC, with ipsilateral mediastinal involvement: (**a**) MIP image; (**b**) CT transverse section; (**c**) PET transverse section; (**d**) fused PET/CT transverse section. There are also multiple liver metastases (**e–d**).

FDG-PET is also effective for detection of hepatic metastases (see Figure 7.7). The PET finding of hepatic metastasis is usually accompanied by abnormality on CT and ultrasound imaging as well. The advantages of PET over the other modalities in detection of hepatic metastasis are related to its high specificity and its ability to resolve abnormalities that remain indeterminate on conventional imaging. Hustinx et al. (116) showed in patients with hepatic metastases from a variety of primary tumors that PET was able to confirm or exclude metastasis accurately in 20% (13/64) of patients with equivocal conventional imaging findings.

FDG-PET is also effective for detecting pulmonary metastases, but CT has higher resolution and is less affected by respiratory blurring. In the lung, there is high attenuation contrast between a solid lesion and adjacent air-filled lung and the high resolution of CT can be used to full advantage. Elsewhere in the body, CT contrast between lesion and normal tissue is not so high as in the lungs, so that metabolic PET images provide higher contrast and better lesion detection than anatomic images. In the lung, CT can detect lesions of a few millimeters, which are not detectable by PET alone unless the lesion shows very high uptake of FDG. PET/CT makes effective use of these properties to identify lung lesions with minimal or even with no lung uptake.

Cerebral metastases are not optimally visualized on a whole-body PET survey and require a specific cerebral acquisition. This additional acquisition is probably not warranted in view of the low incidence of cerebral metastases in the asymptomatic patient and of the superior performance of contrast-enhanced CT or MRI for their detection (34). However, a review by Marom et al. (117) indicated that whole-body FDG-PET alone gave better results in staging than CT of the thorax, CT of the cerebrum, and skeletal scintigraphy combined.

Cost-Effectiveness of FDG-PET in Staging Lung Cancer

The effectiveness of FDG-PET in lung cancer staging reflects its ability to detect metastases that are not apparent on CT alone and to exclude metastases in benign lesions that cannot be differentiated from tumor by CT; this translates into changes in patient management in 19% to 41% of patients (92, 108, 117, 118). Of particular importance is the exclusion of surgery by demonstration of unsuspected distant metastases in 5% to 19% of patients (89, 93, 107, 108). Demonstration of mediastinal metastasis may lead to the addition of neoadjuvant chemotherapy to surgery if the patient is medically fit for these procedures, but distant disease is considered a contraindication to surgery in all cases, with the probable exception of solitary cerebral metastasis.

The cost-effectiveness of FDG-PET for lung cancer staging was analyzed using modeling by Scott et al. (119) and Dietlein et al. (120). Both studies showed PET to be cost effective when used for detection of mediastinal node involvement after a negative CT scan. They found that PET detects undiagnosed mediastinal disease in approximately 20% of cases, leading to a reduction in the number of surgical patients. In cases where the CT is positive for mediastinal node involvement, which represent about a third of the patients (see Table 7.5), a negative PET scan may be an indication for surgery without prior mediastinoscopy, whereas in the PET-positive patients mediastinoscopy is still needed. As a result, the savings in the CT-positive group will be insignificant compared to the added cost of PET. Scott et al. (119), however, did not include detection of distant metastasis in their model, whereas Dietlein et al. (120) used a metastatic prevalence of 5% only. In fact, a prevalence of 5% to 19% has been reported in all clinical studies that have addressed this question (89, 93, 107, 108), which is more in keeping with the 20% or greater rate of distant recurrence after resection of tumor in patients with stage I–II disease by conventional staging (16, 106).

If a realistically higher value was used in the model for the prevalence of unsuspected distant metastasis detected by PET only, whole-body PET for staging would probably be found cost effective in all patients who did not already have a diagnosis of distant metastasis. A randomized study has evaluated the incidence of "futile" surgery in 188 patients who were deemed operable after conventional evaluation. The rate of futile surgery in the group that underwent further staging by PET was 19 of 92 (21%), compared to 39 of 96 (41%) in the conventional group, which represents a reduction of 20%. There was no extra cost associated with the addition of PET imaging because of the improved accuracy in staging (121).

Staging Small Cell Lung Cancer

PET has also been used in the evaluation of patients with small cell lung cancer (122–131). Staging in these patients is usually limited to a binary distinction between patients with limited disease, with tumor confined to one hemithorax, and patients with extensive disease. Accurate staging is important because this determines subsequent therapy and prognosis (132). Several studies have demonstrated the potential usefulness of FDG-PET as a simplified staging tool in small cell lung cancer. Shumacher et al. (122) showed that FDG-PET detected additional metastasis resulting in upstaging 29% (7/24) of patients. Kamel et al. [125] showed that PET induced changes in management of 28% (12/42) of patients, including changes in radiation therapy (8 patients), cancellation of adjuvant therapy (3 patients), and indication for surgery (1 patient). In treated patients, the overall survival was lower in patients with PET-positive than in patients with PET-negative results. In addition, the SUV max shows a significant negative correlation with survival (130).

Evaluation of Pleural Tumor Involvement

FDG-PET imaging has also been proposed for differentiation of malignant from benign pleural disease and for staging of patients with mesothelioma of the pleura (133–136). PET appears to achieve acceptable separation of benign and malignant lesions despite the occurrence of some false-positive cases in patients with inflammatory pleural lesions. In particular, the study is unreliable after talc pleurodesis (137). In patients with malignant pleural disease, PET can define the anatomic pattern and extent of tumor but cannot distinguish between malignant mesothelioma, lymphoma, or metastatic carcinoma. It can, however, identify metastatic mediastinal lymph nodes and metastatic lesions at other sites. Flores et al.

(138) studied 60 patients preoperatively. FDG-PET correctly identified supraclavicular N3 or M1 disease in 6 patients but was unreliable to define either the extent of the primary tumor or the mediastinal nodal status. Another report by Wang et al. (139) suggests reliable use of PET for definition of mediastinal nodal status when PET was correlated with anatomical imaging modalities.

PET can, therefore, be used to exclude malignancy in pleural disease, to evaluate the extent of disease when present, and to identify the best site for biopsy when necessary (140).

FDG-PET for Determination of Prognosis

The prognosis of patients with NSCLC is primarily determined by the stage of the disease. However, other factors potentially related to the invasiveness and aggressivity of the tumor or its metabolic activity may also play a role in defining prognosis. Accordingly, several studies have examined the relationship between the SUV of the tumor on FDG-PET and prognosis after therapy. Three studies have reported an independent prognostic value of FDG uptake in addition to its impact on tumor staging. Patients with SUV values of greater than 7 to 10 had shorter survival compared to patients with SUV values below that threshold (141).

The prognostic value of FDG- PET was also evident when patients were re-examined after initial treatment, as demonstrated by Hicks et al. (142) and by Patz et al. (143). Patients with persistent or recurrent abnormalities had a poor prognosis (median survival, 12 months), whereas patients with a negative PET result had longer survival. Also patients with false-negative PET findings and pathologically proven primary lung cancer at presentation remained stage I and had a favorable prognosis, with the exception of a few patients with bronchioalveolar cell carcinoma or carcinoïds (144).

Use of FDG-PET in Therapy Planning and Monitoring

Assessment of Treatment Effect

After treatment of the primary tumor, it becomes necessary to detect residual or recurrent tumor to assess the results of the therapeutic regimen. In this situation, the structural changes produced by radiotherapy and surgery further reduce the accuracy of anatomic imaging. Assessment of persistent or recurrent disease leading to appropriate management changes is needed if the course of the disease is to be altered (142, 145).

In addition to its use in the selection of patients for surgery with curative intent, PET is also used for planning of radiotherapy with both therapeutic and palliative intent. Several studies have shown the ability of PET to better define the functional tumor volume and to provide an improved outline of the radiotherapy volume, both for inclusion of all active tumor and for sparing of adjacent structures. Standard radiotherapy planning relies primarily on CT to define the target volume and to separate active tumor from pulmonary tissue. However, CT notably lacks criteria for distinguishing cancer from abnormal lung tissue such as postobstructive atelectasis or consolidation or from normal or inflammatory lymph nodes. This limitation can result in inappropriate radiation dose administration with insufficient dose to the tumor or excessive irradiation of normal tissues. PET allows modification of the radiation field by increasing effective targeting and reducing toxicity. For instance, Hicks et al. (17) demonstrated changes in radiation treatment volumes in 25% of 153 patients with newly diagnosed NSCLC. The accuracy of PET in staging was confirmed in 88% of those patients where this could be evaluated, and PET resulted in significant changes in patient management and outcome. In patients with poorly demarcated

tumors, PET changed the irradiated volume in 41% (5/12) of the patients; decreased volume in 3 and increased volume in 2 (146). PET is especially useful in patients with postobstructive atelectasis. Nestle et al. (147) reported potential field reduction in 29% (10/34) of cases.

The treatment plan is also changed in regard to the involvement of lymph nodes (148). The impact of PET/CT coregistration was evaluated by Mah et al. (149). Radiotherapy intent and target volume were changed in 23% (7/30) and 22% (5/30) of patients, respectively. PET also was able to reduce the radiation dose to the spinal cord.

After surgery or radiotherapy, PET provides a sensitive means of differentiating scar from residual or recurrent tumor (Figures 7.8, 7.9). In a study by Bury et al. (145), PET was more sensitive and equally specific compared to other modalities for detection of residual disease or recurrence both after surgery and after radiotherapy. Hicks et al. (142) evaluated 63 patients with suspected relapse more than 6 months after definitive treatment of NSCLC and found that PET and conventional evaluation differed in 68% (43/63) of cases. PET was correct in 91% of these and induced major changes in diagnosis in 59% of cases. A review by Erasmus and Patz (150) indicated a high diagnostic accuracy for detection of recurrence (78%–98%), provided there was a delay between treatment and PET imaging of 2 months after surgery and of 4 to 6 months after radiotherapy.

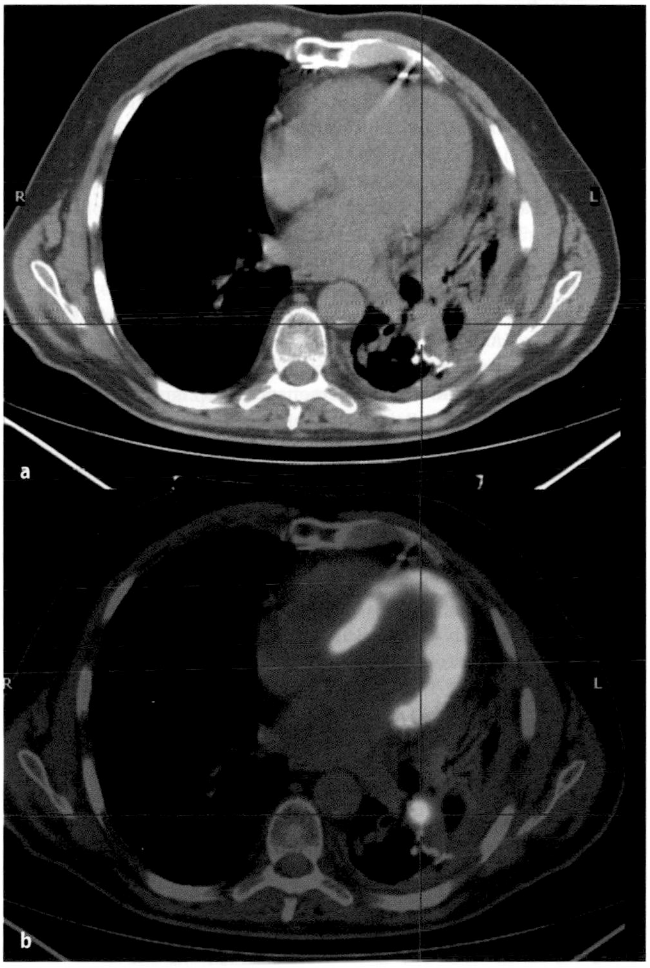

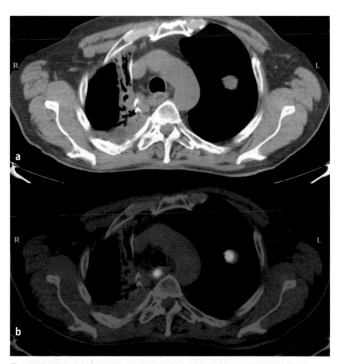

Figure 7.8. Non-small cell lung carcinoma of the left inferior lobe treated by surgery in 2000. Mediastinal recurrence treated by chemotherapy and radiotherapy in May 2002. PET in March 2003 demonstrates a local recurrence in an area of posttherapeutic scar. (**a**) Transverse CT section; (**b**) fused PET/CT image.

Figure 7.9. Patient treated by surgery and radiotherapy for a right apical non-small cell lung carcinoma. (**a**) CT demonstrates a recurrent nodule of the left apex. PET performed to assess the nature and extent of the nodule, and the status of the previous lesion, demonstrates a synchronous tumor in the upper oesophagus (**b**: fused PET/CT transverse section). No significant disease is present in the radiotherapy field.

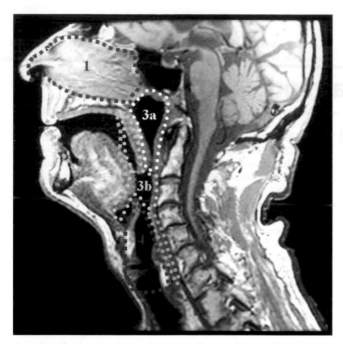

Figure 8.1. Anatomic sites and subsites of the head and neck. (1) Nasal cavity. (2) Oral cavity. (3a) Nasopharynx. (3b) Oropharynx. (3c) Hypopharynx (laryngopharynx). (4) Larynx.

Table 8.1. Classification of cervical lymph nodes into anatomic subsites and levels as recommended by American Joint Committee on Cancer (AJCC), the Union International Contre le Cancer (UICC), and the American Academy of Otolaryngology-Head and Neck Surgery (AAO-HNS).

Level	Lymph nodes	Radiologic landmarks
I	Submental Submandibular	Above hyoid bone, anterior to posterior edge of submandibular gland, superficial to mylohyoid muscle, between the anterior and posterior belly of the digastric muscle.
II	Upper jugular	Extending from the base of the skull to the inferior aspect of the hyoid bone (level II as long as you can see hyoid bone), between the posterior edge of the submandibular gland and the posterior border of sternocleidomastoid muscle.
III	Midjugular	Extending from just below the hyoid bone to the inferior aspect of the cricoid cartilage, between the posterior border of sternocleidomastoid muscle and the sternohyoid muscles.
IV	Lower jugular	Extending from below cricoid cartilage to the clavicle, between posterior border of sternocleidomastoid muscle and posterior to sternohyoid muscle.[a]
V	Posterior triangle	Extending from base of skull to clavicle, between anterior border of trapezius and posterior border of sternocleidomastoid muscle.
VI	Central compartment (prelaryngeal or Delphian, pretracheal and paratracheal nodes)	Extending from the hyoid bone to the suprasternal notch, between the medial border of the carotid sheet/anterior border of sternohyoid muscle.
Supraclavicular	Supraclavicular	At or caudal to the clavicle, above and medial to the ribs lateral to the lower jugular chain.

[a]Sternohyoid muscle origin from manubrium of sternum and medial end of clavicle and inserts at the body of the hyoid bone.
Source: Used with the permission of the American Joint Committee on Cancer (AJCC), Chicago, Illinois. The original source for this material is the *AJCC Cancer Staging Manual, Sixth Edition* (2002), published by Springer-Verlag New York, www.springeronline.com.

familiar to the head and neck surgeons should be used in the reports. Head and neck primary tumors are anatomically divided into in four major groups or levels based on anatomic sites (5–7). Each anatomic major site has a number of subsites. The four major sites and their subsites are shown in Figure 8.1):

1. *Nasal cavity and paranasal sinuses.*
2. *Lip and oral cavity including the tongue anterior to the circumvalate papillae:* mucosal lip (from the vermilion boarder of the lip), buccal mucosa, upper and lower alveolar ridges, retromolar gingiva, floor of the mouth, hard palate, oral tongue (tongue anterior to the circumvalate papillae).
3. *Pharynx:* Nasopharynx, parapharyngeal sites, oropharynx (soft palate including uvula, tonsils, base of tongue, pharyngeal walls), hypopharynx or laryngopharynx (pyriform sinus, postcricoid area, posterior hypopharyngeal wall).
4. *Larynx:* Supraglottic (suprahyoid epiglottis with both lingual and laryngeal epiglottic surfaces, aryepiglottic folds, arytenoid cartilages, the false vocal cords and the ventricles), glottis (the true vocal cords and the anterior and posterior commissures), and subglottis (from below the true vocal cords to the inferior margin of the cricoid).

The status of the regional lymph nodes in head and neck cancer has great prognostic importance and must be assessed in detail in each patient (5–7). According to the American Joint Committee on Cancer (AJCC), the Union International Contre le Cancer (UICC), and the American Academy of Otolaryngology-Head and Neck Surgery

(AAO-HNS), the lymph nodes are classified into specific groups based on anatomic location (Table 8.1, Figure 8.2) (6–8).

PET Imaging Technique in Head and Neck Cancer

PET imaging of head and neck cancer depends on the increased glucose metabolism of neoplasms, which was first described in the 1930s (9). Most PET imaging in head and neck cancer has been performed with ^{18}F-2-deoxy-D-glucose (FDG), which takes advantage of the increased anaerobic glucose metabolism of these malignancies. Squamous cell carcinoma (SCC) primary tumors as well as metastases of the head and neck, lung, and esophagus demonstrate especially avid FDG uptake. Other PET

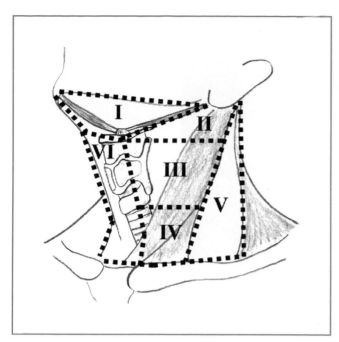

Figure 8.2. Schematic diagram of the anatomic *levels* of cervical lymph nodes (see Table 9.2). [Adapted. Used with the permission of the American Joint Committee on Cancer (AJCC), Chicago, Illinois. The original source for this material is the *AJCC Cancer Staging Manual, Sixth Edition* (2002) published by Springer-Verlag New York, www.springeronline.com.]

tracers such as ^{11}C-methionine, ^{11}C-thymidine, and ^{18}F-L-thymidine (FLT) have also been used to evaluate head and neck neoplasms but to a lesser extent. These tracers depend on abnormalities in protein or nucleic acid synthesis in tumor cells (10–15).

FDG imaging is performed in the fasting state (4 h or more since last meal) to reduce blood sugar and insulin level to minimize FDG uptake in skeletal and cardiac muscle. The patients can be encouraged to have a high-protein/restricted-carbohydrate meal the evening before the study. Tube feeding is stopped 6 h and nutritional supplementation stopped 8 h before injection of FDG. In poorly controlled diabetic patients, high muscle uptake can present diagnostic problems and FDG accumulation in tumors can be decreased. Torizuka et al. (16) showed that FDG uptake of untreated primary lung tumors was significantly reduced in diabetic patients with hyperglycemia, as compared to nondiabetic patients with normoglycemia. In addition, tumor/blood and tumor/muscle uptake ratios were significantly decreased. Blood glucose level should be checked before FDG injection and patients with whole blood values (plasma = whole blood × 1.12) greater than 150 to 200 dL/mg (8.3–11.1 mmol/L) should be rescheduled for imaging when their blood glucose value is under control. Diabetic patients should take their diabetic control medication and have a morning meal at least 4 h before their appointment time.

Tumor uptake of FDG continues to increase even up to 2.5 h after injection, so that standardizing the delay between injection and imaging is important. FDG emission scans should be performed at least 60 min after the intravenous administration of FDG. Improved target-to-background ratios can be obtained by waiting longer, and if patients can wait longer than 60 min without discomfort, there may be an advantage in waiting up to 90 min or more to perform the emission imaging (17–22).

Imaging should include at least the regions extending from the maxillary sinuses to the upper abdomen. Emission images should be corrected for attenuation using measured attenuation correction when performing head and neck imaging. The acute curvatures around the mouth, nose, mandible, and neck result in severe edge artifacts in non-attenuation-corrected PET images and therefore should not be used in diagnosis. Neck lymph nodes can lie near the skin surface, and edge artifacts resulting from noncorrected imaging can hamper their identification. The anatomic relationships of the airways and bony structures to the soft tissue can also be more reliably assessed when an attenuation map is available for comparison to the corrected emission image. However, combined PET/CT or PET fused with diagnostic CT is optimal, and this combination results in improved specificity and is more accurate than PET alone (23). Utilization of semiquantitative data analysis with standardized uptake value (SUV) calculation may also be helpful in identifying abnormal uptake, and this can only be performed when attenuation correction is used. Imaging of distant body regions should also be included to exclude distant metastatic disease.

Normal FDG Uptake, Variants, and Pitfalls

Interpretation of PET images of the head and neck region presents particular challenges. The region shows substantial variations in normal uptake that can present difficulties in the identification of pathology (Figure 8.1). Uptake in the abundant lymphoid tissue in the pharynx, forming the Waldeyer's ring (pharyngeal tonsil, palatine tonsils, lingual tonsillar tissue and lymphoid tissue along salpingopharyngeal folds), is easily recognized. Some asymmetry of the pharyngeal and palatine tonsils is not uncommon; however, especially when looking for a SCC with an unknown primary, an asymmetry should be noted as a possible site for the primary malignancy. A "V"-shaped high-uptake area in the floor of the mouth along the medial borders of the mandible is most likely the sublingual glands; however, uptake in the mylohyoid muscle is another possibility. The sublingual gland is a predominant mucous gland, which may explain the consistently much higher uptake compared to the parotid (serous gland) and submandibular glands (mixed serous and mucous) (24).

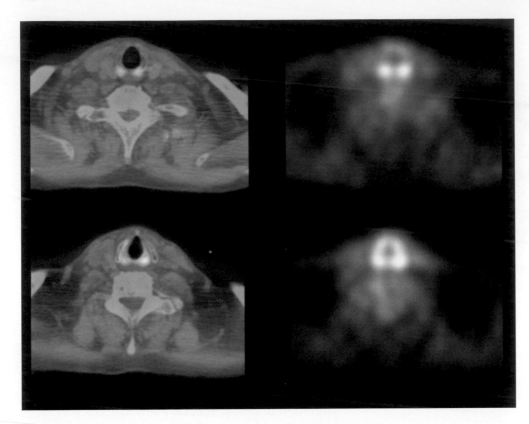

Figure 8.3. Transaxial fused PET/CT (*right column*) and PET (*left column*) of the neck at the level of glottis showing physiologic high uptake of the internal laryngeal muscles.

Intense uptake of the orbital muscles is common. With a high-resolution system, the individual eye muscles may be identified. On occasion, other muscle uptake can also be seen, most commonly including the sternocleidomastoid muscles, longus capitis, and longus colli. Scalene muscle uptake is often seen after neck dissection where the sternocleidomastoid muscle has been removed. Uptake in the intrinsic laryngeal musculature is a common finding and may be rather intense if the patient was talking during uptake of FDG. The typical finding is a symmetrical high uptake at the muscle origin and insertion of the arytenoid cartilage and some less-intense uptake along the course of the thyroarytenoid and vocalis muscle (Figure 8.3). Especially after unilateral surgery and radiation therapy, the cervical muscle uptake may be remarkably asymmetrical. The lower neck images sometimes demonstrate uptake only at the origin of the neck muscles, which can be confused with bilateral supraclavicular lymph node disease. The muscle uptake is, however, usually symmetrical and linear, and palpation demonstrates no nodes of sufficient size to account the amount of uptake seen on the scan. More rarely, temporalis, pterygoid, masseter, or other head and neck muscles accumulate tracer, sometimes depending on specific use by the patient, so that obtaining a history of physical activity or intense chewing may be helpful. Taking steps to reduce or eliminate muscle uptake can be crucial to correct image interpretation. Ensuring that the patient is not chewing gum, chewing tobacco, reading, or talking during the uptake phase is important. Also, some have advocated the use of a benzodiazepine drug to suppress muscle uptake. This step appears to be effective but deserves a note of caution because head and neck cancer patients may have existing compromise of the airway. Attention to the position of the head during the uptake phase is of prime importance. Experimenting with chair design and pillow placement to ensure slight head flexion and complete head relaxation is necessary to avoid unnecessary muscle uptake interference. Also, with three-dimensional (3D) surface projections of PET images and careful examination of all three orthogonal views, most muscle uptake can be differentiated from disease by its anatomic pattern of distribution. PET/CT or PET fused with a diagnostic CT is of great help to differentiate between physiologic muscle uptake and pathologic uptake in tumor or metastatic lymph nodes; this should not, however, replace appropriate patient preparation. If no combined PET/CT or a program for fusion of diagnostic CT and PET is available, CT and/or magnetic resonance imaging (MRI) images should be present for side-by-side comparison.

The cervical and upper thoracic spinal cord frequently shows mild uptake, gradually decreasing in caudal direction. Brown fat uptake is typically patchy, symmetrical, and may be very intense (Figure 8.4). Brown fat uptake is commonly located in the neck, supraclavicular, suboccipital, and paraspinal regions in the neck and chest and may also be located on scattered areas perivascular in the mediastinum. Brown fat, hypermetabolic tumor, and metastatic lymph nodes may coexist, and a differentiation may be difficult without PET/CT. Repeat imaging, making sure that the patient is warm and comfortable during injection and uptake, may be needed to resolve the issue. We typically use warm blankets and keep uptake rooms warm to prevent muscle uptake as much as possible (Figure 8.4). Benzodiazepine can be used as an extra measure to reduce fat uptake for repeated scans when necessary.

Degenerative joint disease in the cervical spine is very common and must not be mistaken for hypermetabolic

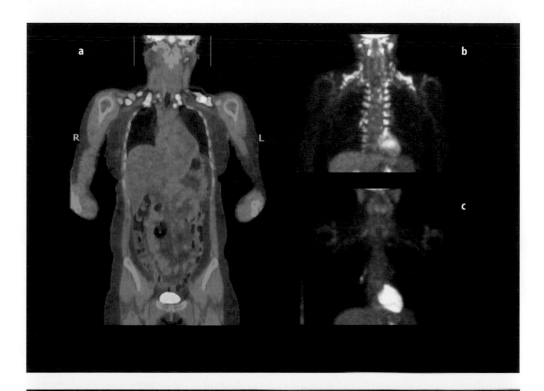

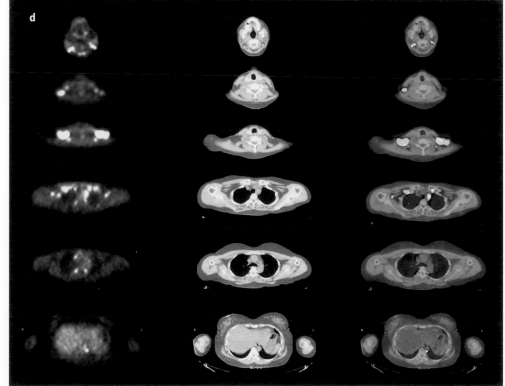

Figure 8.4. Brown fat uptake of a patient. (**a**) Coronal fused PET/CT showing extensive distribution of hypermetabolic brown fat. (**b**) Coronal PET with hypermetabolic brown fat compared with repeat study (**c**) when the patient was warmed 12 h before and comfortable during injection and uptake. (**d**) *Left row*, PET (cranial to caudal); *middle column*, CT; *right column*, fused PET/CT.

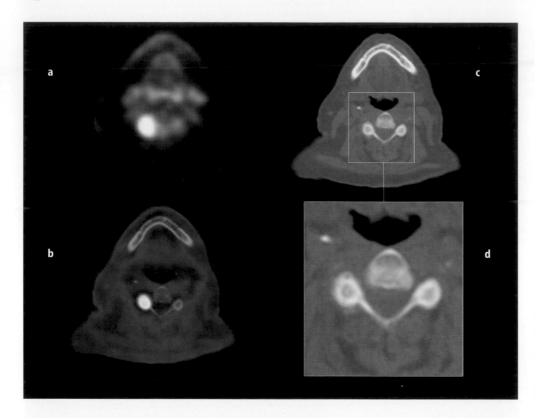

Figure 8.5. (**a**) Transaxial CT (*top*) and fused PET (*bottom*) show intense hypermetabolic foci in right cervical intervertebral joint. A slight misalignment between PET and CT is common. (**b**) The bony changes corresponding to the hypermetabolic foci is seen on CT a slice lower than on the fused images (**c**, **d**).

lymph nodes (Figure 8.5). Again, PET/CT or fusion of PET and a diagnostic CT is very useful. Degenerative joint disease uptake can be bilateral and asymmetrical, and FDG uptake may be very intense,

Maxillary sinusitis is common. Increased uptake along the bony lining of the sinus is a typical finding consistent with inflammatory sinusitis (Figure 8.6). Diffuse thyroid uptake is consistent with chronic thyroiditis and may be very intense in Hashimoto's thyroiditis. A hypermetabolic nodule in the thyroid gland as an incidental finding occurs in about 1% of oncologic patients and is associated with malignant primary or secondary thyroid cancer in as many as 47% of the cases (25–27).

Staging Head and Neck Cancer

Accurate anatomic staging is the single most important factor in patient assessment, treatment planning, and determination of prognosis (5, 28, 29). Standardized disease classification facilitates the exchange of clinical information and research results among different treatment

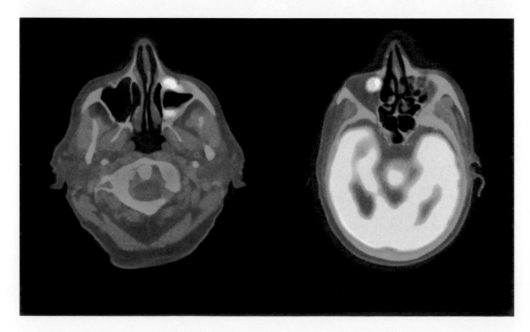

Figure 8.6. Typical finding on FDG-PET of maxillary sinusitis (*left*) and recurrent maxillary squamous cell carcinoma (SCC) (on CT changes from surgery and radiation).

centers and scientists. In general, staging groups from 0 to IV are used. Carcinoma in situ is categorized as 0, whereas a patient with advanced local disease and/or with distant metastases is categorized as stage IV. Staging is based on the anatomic TNM classification. The TNM classification of a tumor describes the anatomic extent of the disease based on clinical information including imaging and biopsies if available. The TNM classification and group staging for head and neck cancers are listed in Table 8.2. When the classification is based on additional information from pathologic examination (e.g., after surgery), the prefix "p" is used (e.g., pTNM) (5, 6). The current classification and staging guidelines in use for HNSCC are defined in the 2002 manual of the American

Table 8.2. TNM classification and group staging for head and neck cancers.

Table 8.2.1. T classification of lip and oral carcinomas.

T1	Tumor 2 cm or less in greatest dimension
T2	Tumor more than 2 cm but not more than 4 cm in greatest dimension
T3	Tumor more than 4 cm in greatest dimension
T4 (lip)	Tumor invades through cortical bone, inferior alveolar nerve, floor of mouth, or skin of face, i.e., chin or nose
T4a (oral cavity)	Tumor invades adjacent structures (e.g., through cortical bone, into deep [extrinsic] muscle of tongue [genioglossus, hyoglossus, hyoglossus, palatoglossus, and styloglossus], maxillary sinus, skin of face)
T4b (oral cavity)	Tumor invades masticator space, pterygoid plates, or skull base and/or encases carotid artery

Table 8.2.2. T classification of nasopharynx cancer.

T1		Tumor confined to the nasopharynx
T2		Tumor extends to soft tissues
	T2a	Tumor extends to the oropharynx and/or nasal cavity without parapharyngeal extension[a]
	T2b	Any tumor with parapharyngeal extension[a]
T3		Tumor involves bony structures and/or paranasal sinuses
T4		Tumor with intracranial extension and/or involvement of cranial nerves, infratemporal fossa, hypopharynx, orbit, or masticator space

[a]Parapharyngeal extension denotes posterolateral infiltration of tumor beyond the pharyngobasilar fascia.

Table 8.2.3. T classification of oropharynx cancer.

T1	Tumor 2 cm or less in greatest dimension
T2	Tumor more than 2 cm but no more than 4 cm in greatest dimension
T3	Tumor more than 4 cm in greatest dimension
T4a	Tumor invades the larynx, deep/extrinsic muscle of tongue, medial pterygoid, hard palate, or mandible
T4b	Tumor invades lateral pterygoid muscle, pterygoid plates, lateral nasopharynx, or skull base or encases carotid artery

Table 8.2.4. T classification of hypopharynx cancer.

T1	Tumor limited to one subsite of hypopharynx and 2 cm or less in greatest dimension
T2	Tumor invades more than one subsite of hypopharynx or an adjacent site, or measures more than 2 cm but not more than 4 cm in greatest diameter without fixation of hemilarynx
T3	Tumor more than 4 cm in greatest dimension or with fixation of hemilarynx
T4a	Tumor invades thyroid/cricoid cartilage, hyoid bone, thyroid gland, esophagus, or central compartment soft tissue[a]
T4b	Tumor invades prevertebral fascia, encases carotid artery, or involves mediastinal structures

[a]Central compartment soft tissue includes prelaryngeal strap muscles and subcutaneous fat.

Table 8.2.5. T classification of laryngeal supraglottic cancer.

T1	Tumor limited to one subsite of supraglottic with normal vocal cord mobility
T2	Tumor invades mucosa of more than one adjacent subsite of supraglottic or glottis or region outside the supraglottic (e.g., mucus of base of tongue, vallecula, medial wall of pyriform sinus) without fixation of the larynx
T3	Tumor limited to larynx with vocal cord fixation and/or invades any of the following: postcricoid area, pre-epiglottic tissues, paraglottic space, and/or minor thyroid cartilage erosions (e.g., inner cortex)
T4a	Tumor invades through the thyroid cartilage and/or invades tissues beyond the larynx (e.g., trachea, soft tissues of neck including deep extrinsic muscle of the tongue, strap muscles, thyroid, or esophagus)
T4b	Tumor invades prevertebral space, encases carotid artery, or invades mediastinal structures

Table 8.2.6. T classification of laryngeal glottic cancer.

T1		Tumor limited to the vocal cord(s) (may involve anterior or posterior commissures) with normal mobility
	T1a	Tumor limited to one vocal cord
	T1b	Tumor involves both vocal cords
T2		Tumor extends to supraglottic and/or subglottis, and/or with impaired vocal cord mobility
T3		Tumor limited to larynx with vocal cord fixation and/or invades paraglottic space, and or minor thyroid cartilage erosions (e.g., inner cortex)
T4a		Tumor invades through the thyroid cartilage and/or invades tissues beyond the larynx (e.g., trachea, soft tissues of neck including deep extrinsic muscle of the tongue, strap muscles, thyroid, or esophagus)
T4b		Tumor invades prevertebral space, encases carotid artery, or invades mediastinal structures

Joint Committee on Cancer (5, 6). Staging involves an accurate assessment of the anatomic extent of the primary tumor (T classification), regional lymphatic metastases (N classification), and distant metastases (M classification)

Table 8.2.7. T classification of laryngeal subglottic cancer.

T1	Tumor limited to the subglottis
T2	Tumor extends to vocal cord(s) with normal or impaired vocal cord mobility
T3	Tumor limited to larynx with vocal cord fixation
T4a	Tumor invades cricoid or thyroid cartilage and/or invades tissues beyond the larynx (e.g., trachea, soft tissues of neck including deep extrinsic muscle of the tongue, strap muscles, thyroid, or esophagus)
T4b	Tumor invades prevertebral space, encases carotid artery, or invades mediastinal structures

Table 8.2.8. T classification of maxillary sinus.

T1	Tumor limited to maxillary sinus with no erosion or destruction of bone
T2	Tumor causing bone erosion or destruction including extension into the hard palate and/or middle nasal meatus, except extension to posterior wall of maxillary sinus and pterygoid plates
T3	Tumor invades any of the following: bone of the posterior wall of maxillary sinus, subcutaneous tissues, floor or medial wall of orbit, pterygoid fossa, ethmoid sinuses
T4a	Tumor invades anterior orbital contents, skin of cheek, pterygoid plates, infratemporal fossa, cribriform plate, sphenoid or frontal sinuses
T4b	Tumor invades any of the following: orbital apex, dura, brain, middle cranial fossa, cranial nerves other than V_2, nasopharynx, or clivus

Table 8.2.9. T classification of nasal cavity and ethmoid sinus.

T1	Tumor restricted to any one subsite, with or without bony invasion
T2	Tumor invading two subsites in a single region or extending to involve an adjacent region within the nasoethmoidal complex, with or without bony invasion
T3	Tumor extends to invade the medial wall or floor of the orbit, maxillary sinus, palate, or cribriform plate
T4a	Tumor invades any of the following: anterior orbital contents, skin of nose or cheek, minimal extension to anterior cranial fossa, pterygoid plates, sphenoid or frontal sinuses
T4b	Tumor invades any of the following: orbital apex, dura, brain, middle cranial fossa, cranial nerves other than V_2, nasopharynx, or clivus

Table 8.2.10. T classification of nasal cavity and ethmoid sinus.

T1	Tumor 2 cm or less in greatest dimension without extraparenchymal extension*
T2	Tumor more than 2 cm but no more than 4 cm in greatest dimensions without extraparenchymal extension*
T3	Tumor more than 4 cm and/or having extraparenchymal extension*
T4a	Tumor invades skin, mandible ear canal, and/or facial nerve
T4b	Tumor invades skull base and/or pterygoid plates and/or encase carotid artery

Note: Extraparenchymal extension is clinical or macroscopic evidence of invasion of soft tissues. Microscopic evidence alone does not constitute extraparenchymal extension for classification purposes.

Table 8.2.11. The nodal classification for all head and neck cancers, except nasopharynx.

NX	Regional lymph nodes cannot be assessed
N0	No regional lymph node metastasis
N1	Metastasis in a single ipsilateral lymph node, 3 cm or less in greatest dimension
N2	Metastasis in a single ipsilateral lymph node, more than 3 cm but not more than 6 cm in greatest dimension; or in multiple ipsilateral lymph nodes, none more than 6 cm in greatest dimension; or in bilateral or contralateral lymph nodes, none more than 6 cm in greatest dimension
N2a	Metastasis in a single ipsilateral lymph node, more than 3 cm but not more than 6 cm in greatest dimension
N2b	Metastasis in multiple ipsilateral lymph node, none more than 6 cm in greatest dimension
N2c	Metastasis in bilateral or contralateral lymph node, none more than 6 cm in greatest dimension
N3	Metastasis in a lymph node more than 6 cm in greatest dimension

Table 8.2.12. The nodal classification of nasopharynx cancer.

NX	Regional lymph nodes cannot be assessed
N0	No regional lymph node metastasis
N1	Unilateral metastasis in a lymph node(s), 6 cm or less in greatest dimension, above the supraclavicular fossa*
N2	Bilateral metastasis in lymph node(s), 6 cm or less in greatest dimension;
N3	Metastasis in a lymph node(s), greater than 6 cm and/or to supraclavicular fossa
N3a	Greater than 6 cm in dimension
N2b	Extension to the supraclavicular fossa

Note: Midline nodes are considered ipsilateral nodes.

Table 8.2.13. Distant metastasis (M) for all head and neck cancers.

MX	Distant metastasis cannot be assessed
M0	No distant metastasis
M1	Distant metastasis

Table 8.2.14. Stage grouping of all head and neck cancers, except nasopharynx.

Stage 0	T*is*	N0	M0
Stage I	T1	N0	M0
Stage II	T2	N0	M0
Stage III	T3	N0	M0
	T1	N1	M0
	T2	N1	M0
	T3	N1	M0
Stage IVA	T4a	N0	M0
	T4a	N1	M0
	T1	N2	M0
	T2	N2	M0
	T3	N2	M0
	T4a	N2	M0
Stage IVB	T4b	Any N	M0
	Any T	N3	M0
Stage IVC	Any T	Any N	M1

Table 8.2.15. Stage grouping of nasopharynx cancer.

Stage 0	T*is*	N0	M0
Stage I	T1	N0	M0
Stage IIA	T2a	N0	M0
Stage IIB	T1	N1	M0
	T2	N1	M0
	T2a	N1	M0
	T2b	N0	M0
	T2b	N1	M0
Stage III	T1	N2	M0
	T2a	N2	M0
	T2b	N2	M0
	T3	N0	M0
	T3	N1	M0
	T3	N2	M0
Stage IVA	T4	N0	M0
	T4	N1	M0
	T4	N2	M0
Stage IVB	Any T	N3	M0
Stage IVC	Any T	Any N	M1

Source: The data presented in Table 8.2 are used with permission of the American Joint Committee on Cancer (AJCC), Chicago, IL. From Greene FL, Page DL, Fleming ID, Fritz AG, Balch CM, Haller DG, Morrow M, editors. *AJCC Cancer Staging Handbook, Sixth Edition.* New York: Springer-Verlag 2002:27–87.

(Figure 8.7). For all head and neck cancers, TX is defined as primary tumor cannot be assessed; T0, no evidence of primary tumor; Tis, carcinoma in situ; MX, distant metastasis cannot be assessed; M0, no distant metastasis; and M1, distant metastasis. The most common sites for distant metastasis are in the lungs, whereas bone, liver, and brain metastases occur less often. Mediastinal lymph node metastasis is classified as distant metastasis. Nodal (N) classification for all head and neck cancers is uniform except for nasopharynx cancer. The distribution and the prognostic impact of regional lymph node spread from nasopharynx cancer, particularly of the undifferentiated type, are different from those of other head and neck mucosal cancers and justify use of a different N classification. The group staging for all head and neck cancers is also uniform, again except for nasopharynx cancer: stage IVA is advanced resectable cancer, stage IVB is advanced unresectable, and stage IVC is advanced with distant metastases. The initial staging is based on all information available before treatment, including physical examination, diagnostic imaging [CT, MRI, ultrasound (US), nuclear medicine procedures including PET], fine-needle aspiration (FNA) biopsy, and result of surgical exploration if performed. The staging is important for selection of appropriate treatment and follow-up strategies and to facilitate the exchange of information between clinicians and scientists. The likelihood of local and regional spread varies with the site of the primary tumor and this must be appreciated when evaluating HNSCC patients. In additional to staging, a tumor will have a histologic grade. SCC are graded by histologic into well-differentiated (Grade I), moderately well differentiated (Grade II), or poorly differentiated to undifferentiated (Grade III) based on the amount of keratinization (5, 6).

Evaluation of the patient begins at the first consultation with a physical examination, which includes direct visualization of the primary tumor, either through the mouth or nose or by using outpatient endoscopy. Next, the neck is

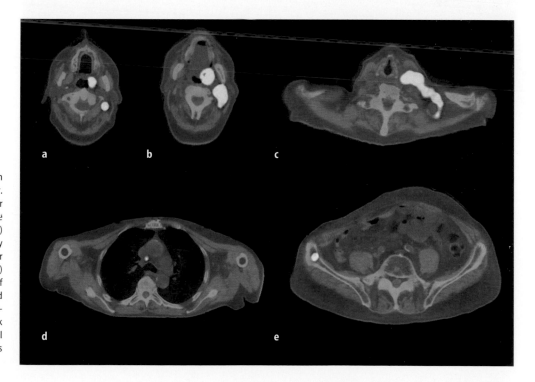

Figure 8.7. Eighty-five-year-old man with stage IVC SCC left tonsillar cancer. (**a**) Intense uptake in the primary tumor and in an upper posterior triangle metastatic lymph node (level V). (**b**) Intense uptake in the large primary tumor and in a conglomerate of upper jugular nodes metastasis (level II). (**c**) Intense uptake in a conglomerate of midjugular (between os hyoid and cricoid cartilage) (level III) and supraclavicular metastatic lymph nodes (neck flexed). (**d**) Hypermetabolic precarinal lymph node and (**e**) osseous metastasis in the right iliac bone.

palpated to determine the presence of cervical lymph node enlargement. Lymph nodes less than 1 cm in diameter cannot be reliably appreciated on physical examination. Body habitus, such as obesity or short neck, may make this assessment more difficult. Next, a histologic diagnosis must be obtained by direct biopsy of the tumor (often by use of endoscopy under general anesthesia), by needle biopsy of a neck mass, and/or the tumor must be assessed by anatomic imaging, particularly if it is not visible, palpable, or accessible to biopsy (5).

The most commonly used imaging modality for HNSCC is CT with intravenous injection of iodinated contrast material. MRI is being increasingly used (30). MRI is better in differentiating soft tissues, but longer imaging time results in imaging artifacts from patient motion (e.g., swallowing and breathing). Metal artifacts are a lesser problem with MRI than with CT. Lymph nodes more than 1 cm in diameter (1.5 cm for jugulodigastric nodes) or showing central necrosis are considered abnormal and suspicious for metastasis. The shortcoming to this anatomic approach to diagnosis is that it excludes the possibility of detecting early nodal metastases that have not caused enlargement of the lymph node. Ultrasound is shown to be superior to CT and MRI in evaluating cervical

lymph nodes, and US-guided fine-needle biopsy (FNAB) is shown to be superior to palpation-guided FNAB (30, 31). PET is a standard approach in many centers.

Distant metastases are unusual in a patient presenting with HNSCC, but hematogenous metastases are found in 20% to 40% during the course of the disease; this is an increasingly occurring phenomenon in the organ-sparing chemoradiation treatment era for HNSCC. The majority of the patients with distant metastases also have uncontrolled regional disease in the neck. The most common site for distant metastasis is the lungs, mediastinum, bone, and liver (Figure 8.7). Mediastinal lymph node metastases are considered distant metastases (5, 6). Standard assessment includes CT of the chest and abdomen.

Primary tumor at a second site, either synchronous or metachronous, is not an unusual finding in HNSCC (Figure 8.8). *Synchronous tumors* are defined as tumors that are diagnosed within 6 months of each other, and *metachronous tumors* are tumors that are diagnosed with an interval of more than 6 months. As many as 20% of patients with HNSCC develop a secondary SCC malignancy (5). These lesions are usually situated in the head and neck, lung, or esophagus. The standard approach to ruling out synchronous primary tumors in HNSSC has been

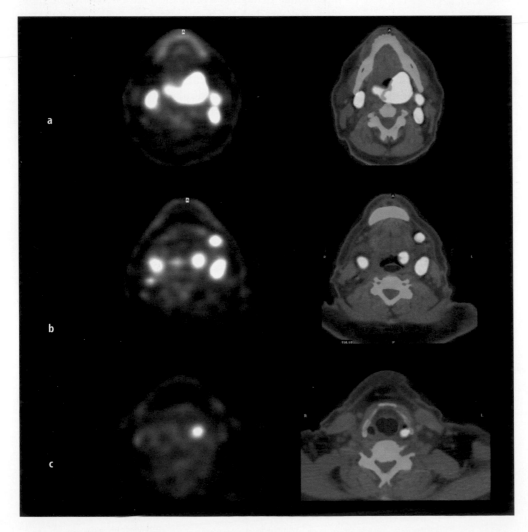

Figure 8.8. Transaxial PET/CT (PET, *left*; fused PET/CT, *right*) of a patient with a large pharyngeal SCC (**a**) shows a large, intensely hypermetabolic mass in the left soft palate with extension across midline into the right soft palate and two left and one right upper jugular (level II) metastatic lymph nodes. The tumor extends inferiorly (**b**) along the lateral aspect of the oropharynx, and at this level two right and one left upper jugular (level II) and one left submental (level I) metastatic lymph nodes are shown. A separate focus of hypermetabolic activity (**c**) is present in the left pyriform fossa with extension into the left aryepiglottic fold, representing a synchronous foci of SCC.

operative endoscopy (laryngoscopy, bronchoscopy, esophagoscopy), but improved techniques of outpatient endoscopy and CT imaging of the neck and chest have resulted in a decreased use of the operative approach. The greater accuracy of whole-body PET imaging has proved to be useful for detection of synchronous tumors and for surveillance for metachronous lesions.

Head and neck primary tumor location or subsite is a strong predictor for cervical metastases (Table 8.1). In many cases, failure to adequately treat the regional lymphatics will result in regional recurrence, which is the major cause of morbidity and mortality in these patients. Knowledge of the pattern of lymphatic metastasis for each tumor subsite is necessary to avoid undertreating HNSCC. For example, carcinoma of the supraglottic larynx has a high probability of metastasis to bilateral neck lymph nodes, even when only unilateral lymphadenopathy is detected by palpation or imaging. The true vocal cords are almost devoid of lymphatic drainage, and vocal cord tumors spread to regional lymph nodes late in the course of disease (4, 6, 7). Tumors in the oral cavity have a 30% prevalence of metastasis to regional lymph nodes at the time of presentation, but these often escape detection by palpation or CT. Nasopharyngeal tumors have a high probability of bilateral metastasis and metastasis to the posterior triangles of the neck. Laryngeal tumors, presumably because of a paucity of lymphatic drainage and other anatomic barriers to tumor spread, have a lower likelihood of nodal involvement, even when presenting at an advanced local stage.

Following adequate staging of the tumor, a treatment plan is formulated. Usually, the primary tumor is surgically removed, unless an organ-sparing approach has been selected. In conjunction with tumor removal, the primary site is reconstructed and the regional lymphatics are removed. If the neck is classified as N0 or N1, neck dissection and radiation are considered to be equally effective for treatment. The advantage of surgical neck dissection as treatment for the N0 neck is that a pathologic confirmation of the presence or absence of nodal metastases is determined. With primary irradiation of the neck, there is no pathologic diagnosis of metastatic disease. N2 disease is usually treated surgically either before or after chemoradiation. N3 disease almost always requires radical neck exploration.

The standard treatment for advanced head and neck cancer (stage III and IV) is surgery followed by postoperative radiation therapy (5, 32, 33), which may result in long-term morbidity and disfigurement. In an effort to diminish morbidity, organ preservation has been achieved by means of induction chemotherapy and radiation therapy, with surgical salvage if these are unsuccessful in eradicating the tumor. Despite this approach, patient survival has remained constant (3). Accordingly, disagreement remains over selection of advanced-stage head and neck tumors for treatment by standard surgical resection or by attempted curative chemotherapy. Chemotherapy or

radiation may be used alone to avoid the more-morbid surgical procedures such as laryngectomy, whereas partial glossectomy may be preferred to chemotherapy. A finding by PET of more-extensive disease than was suspected from the results of conventional procedures may provide a greater impetus to treat with chemotherapy rather than surgery because of its morbidity.

Primary Tumor Evaluation by PET

Evaluation of the anatomic extent of the primary tumor (T classification) using PET has been described in several published studies (34, 35). PET showed no advantage over conventional methods when the primary tumor was detectable by conventional techniques, so that classification of the primary tumor by PET will contribute little in most patients. Conventional tumor staging, using CT imaging and physical examination with endoscopy, provides important anatomic information that is not obtainable by PET. However, some studies have suggested that a high level of metabolic activity on FDG-PET imaging may be predictive of tumor response to treatment so that the PET result could also become a consideration in treatment selection. Tumors having high metabolic activity might be considered for more-aggressive multimodal therapy (35–38).

Diagnosis of Unknown Primary Tumor

In approximately 5% of cases, regional nodal metastasis is the presenting feature, and the primary tumor is not identified by standard techniques (8). Some of these primary tumors become apparent as the patient is followed over time, whereas others are thought to regress spontaneously and most are never diagnosed by conventional means. The most common sites for occult primary tumor are the pyriform sinuses, the palatine tonsils, base of the tongue, and nasopharynx. The site of a lymph node metastasis may give a clue to where the primary tumor might be. A metastatic node in the jugulodigastric area may indicate a primary tumor in the oropharynx, and a metastatic node in the posterior triangle may indicate a primary site in the nasopharynx. Following negative conventional evaluations, PET has identified the unknown primary tumor in 20% to 50% of cases, as reported by several groups (28, 39–47).

Regional Nodal Staging by PET

Many investigators have described the high sensitivity and specificity of FDG-PET imaging in regional nodal staging of head and neck cancer (34, 35, 48–56). All studies have shown PET to be equivalent or superior to all anatomic methods of imaging. In a study by Adams et al., 1,400 lymph nodes were sampled in 60 patients and PET had a 10% advantage in sensitivity and specificity over

CT, MRI, or US. The differences in accuracy between PET and anatomic imaging were found to be highly significant on statistical evaluation (48).

PET may be particularly valuable in deciding whether to perform neck dissection on patients with low-stage tumors where there is no clinical evidence of nodal spread (N0). Myers and Wax showed that PET was more than twice as sensitive as CT for identifying regional nodal metastasis on patients with clinical N0 findings in the neck (57). When PET is added to conventional imaging, patient management is changed in 8% to 15% of the patients (58–60). Hollenbeak et al. showed that adding a FDG-PET study in a patient classified as N0 by CT is cost effective (61).

Distant Staging by PET

Metastasis to distant regions is less common in HNSCC that in other malignant tumors, which could result from earlier detection of HNSCC due to clear-cut presenting symptoms. Because of the low prevalence of distant metastasis at initial diagnosis of HNSCC (less than 5%), PET is unlikely to identify distant disease in a clinically significant fraction of these patients.

Distant metastasis is seen more commonly in patients who have advanced-stage tumor than in patients with early-stage disease. Teknos et al. used PET to evaluate 12 patients with a new diagnosis of stage III or IV head and neck cancer (62). Two patients had true-positive mediastinal PET findings, with no evidence of mediastinal metastasis on CT. A third patient was found to have pulmonary metastasis on both CT and PET.

Whole-body PET imaging, orbits through the pelvis, is recommended as part of initial staging in all patients with head and neck cancer because such imaging may be important for diagnosis of synchronous or metachronous primary tumors and could reduce the need for other additional testing. Whole-body imaging may also be valuable to obtain a baseline evaluation for comparison with later scans. Subtle abnormalities resulting from chronic inflammatory lung lesions, for example, can be documented.

Evaluation of Therapy of Head and Neck Cancer

Conventional Evaluation

Findings obtained by conventional imaging are particularly difficult to interpret after surgery, chemotherapy, and radiation therapy, and PET has a major potential role in evaluating tumor response to treatment. Tissue changes produced by chemotherapy or radiation, such as fibrosis, edema, and hyperemia, are frequently difficult to

differentiate from residual viable tumor by physical examination or anatomic imaging. CT, MRI, and US imaging have limited ability to determine the benign or malignant nature of posttreatment changes, which include variable contrast enhancement and soft tissue distortion, that are seen following irradiation and chemotherapy. Also, changes in tumor size usually lag behind the metabolic effects of therapy, which determine tumor viability. Because FDG-PET is a metabolic imaging method, residual tumor may be diagnosed or excluded by persistence of high metabolic activity where there is no visible anatomic abnormality, or absence of high metabolic activity in the presence of persistent anatomic abnormality, respectively. PET evaluation of nonsurgical treatment results is optimized when a baseline scan is available for comparison.

PET Imaging in Therapy Evaluation

FDG-PET identifies changes in tumor metabolism that occur during therapy. The posttherapy findings that are obtained by FDG-PET depend on the treatment modality used. Radiation therapy may induce an early acute inflammatory reaction, associated with dense inflammatory infiltration and resulting increase in FDG uptake, that may be difficult to differentiate from tumor hypermetabolism (63). Highly cellular infiltrates containing large numbers of macrophages and fibroblasts, associated with high FDG activity, have been described following high-dose intracavitary therapy of brain tumors using radioactively labeled antitumor antibodies (64). Such uptake has also been seen following external irradiation of other tumors.

Increased FDG uptake may be seen in soft tissue regions that have been recently irradiated and appears to be more intense in tissues that receive the highest radiation dose. On the other hand, some normal deep structures may not show these radiation-related changes in metabolism (65). When such increase in FDG accumulation occurs in irradiated regions, it may remain apparent for as long as 12 to 16 months after treatment or more (66). The SUV of radiation-related uptake in normal tissues is generally less than the SUV values that are usually found in recurrent tumor, but overlap in FDG uptake may be seen and can lead to diagnostic confusion. Rege et al. found no change in uptake in adenoids, lingual and palatine tonsils, parotid, submandibular, and sublingual glands, nasal turbinates, soft palate, and gingiva 6 weeks or more after radiation therapy when these regions were included in the radiation field (65). Increased uptake following surgery including tissue inflammation and scarring may last for 4 months or longer.

Following treatment by chemotherapy, FDG hypermetabolism associated with inflammation does not present a comparable problem. Decrease in tumor uptake of FDG may be seen within days of commencing treatment, and uptake may diminish to normal tissue levels within 1 week

(67). Lowe et al. (68) compared the accuracy of FDG-PET to needle biopsy following chemotherapy/radiation as part of organ-sparing primary treatment of HNSCC in a prospective study. They found PET sensitivity for detection of residual tumor to be 90% (19/21), which was equal to the sensitivity of biopsy of the tumor region. When biopsy was repeated using FDG-PET guidance, biopsy sensitivity increased to 100%. The specificity of PET was 83% (5/6), compared to 100% for biopsy. Greven et al. (69) compared PET to histologic diagnosis in 31 patients with HNSCC who had been treated by radiation and found PET to have 80% sensitivity and 81% specificity for detection of residual or recurrent tumor.

A positive PET has a high predictive value for recurrent disease. If the PET scan is positive but the FNAB is negative, we recommend the biopsy be repeated with PET guidance. If the repeated FNAB is negative, decreased FDG uptake on a follow-up scan indicates unspecific inflammation. A negative PET scan is most likely predictive of a good treatment response. In a prospective study, Yao et al. studied 41 patients with advanced regional head and neck cancer treated with radiation therapy or chemoradiation (70). Treatment response was evaluated on FDG-PET and CT. Twelve of the patients had persistent lymphadenopathy on CT, and they all underwent either neck dissection or ultrasound-guided FNAB. Four of the 12 were found to have viable residual tumor. The finding of residual tumor did not correlate with the size of the involved lymph nodes on neither the preradiation nor the postradiation CT. The pathology correlated strongly, however, with the postradiation PET study. Patients with a SUV_{max} less than 3.0 on postradiation PET were all free from residual disease. They conclude that a negative postradiation therapy FDG-PET is very predictive of negative pathology in neck dissection or FNAB, even with large residual lymphadenopathy. In a nicely performed study, Goerres et al. (71) calculated disease probability based on data from their own institution and performed a systematic review of available data in the literature up to October 2001 on the accuracy of FDG-PET in primary staging and follow-up of head and neck cancer. They conclude that the main advantage of PET is the ability to reliably rule out the presence of disease in both primary staging and restaging. Rogers et al. (72) conducted a prospective study of 12 patients with stage III and IV cancer. FDG-PET was performed before and 1 month after radiation therapy. All patients underwent planned neck dissection. The presence of a positive PET 1 month after radiation therapy accurately indicated the presence of residual disease. However, in 6 of the 7 patients with negative PET, the pathology was positive. In 4 of the 6 patients with positive histology, there was significant necrosis and no comment could be made regarding the viability of the tissue. Further, the resolution of their camera was 8 mm and not quite optimal.

Assessment of Recurrent Head and Neck Cancer

Clinical Considerations and Conventional Evaluation

All HNSCC patients are at high risk for tumor recurrence and second primary tumor in the aerodigestive tract. All the cells of the mucosal lining of the upper aerodigestive tract have been exposed to the same carcinogenic agents, with consequently increased incidence of malignancy at multiple sites. The probability of tumor recurrence is related to specific tumor subsite and to the stage of the tumor at presentation, as already discussed. The pathologic features of the primary tumor that correlate with increased incidence of recurrence include nerve invasion, extracapsular spread in lymph nodes, and positive surgical margins. Persistence of smoking and alcohol consumption are ongoing risk factors for development of a second primary tumor.

Most tumor recurrences are diagnosed within the first 24 months following primary therapy for HNSCC. When local recurrences are detected early, they can often be treated by reexcision. Further surgical resection will further compromise any posttreatment dysfunction of speech, voice, swallowing, or airway and will have a negative impact on the patient's quality of life. Local and regional recurrences may occur following surgery, irradiation, or both. Direct carotid artery involvement by tumor may occur and can result in stroke or death from acute arterial hemorrhage. Treatment for local and regional tumor recurrence can include reoperation or reirradiation (external beam or implant) with curative intent, chemotherapy with palliative intent, or supportive measures only.

HNSCC patients require long-term surveillance for recurrence or second primary tumor and can by deemed "cured" only after 5 years of disease-free follow-up. Methods of surveillance can include primary physical examination or CT or MR imaging. Indirect or direct endoscopy is used when physical examination findings or the patient's symptoms arouse suspicion for recurrent or second primary tumor. FDG-PET has proved to be a powerful posttreatment surveillance tool and is shown to detect tumor at an earlier stage than by other methods.

Anatomic imaging by CT or MRI usually shows posttreatment abnormalities, and change from previous imaging findings is the most useful means of detecting recurrent tumor. Chong and Fan compared MRI and CT for identifying recurrent nasopharyngeal cancer and showed that both modalities had a sensitivity of approximately 50%, with substantial interobserver variation in image interpretation ($k = 0.563$) (33).

PET in Detection of Locoregional Recurrence

The superiority of PET to anatomic imaging modalities for identifying local and regional recurrence of head and neck cancer after therapy has been demonstrated in a number of published evaluations (33, 35, 52, 73–80). Perlow et al. (81) showed, in a small series of patients with advanced HNSCC evaluated for locoregional and/or distant metastases, that FDG-PET found unsuspected recurrences in half their patients. Shu-Hang et al. (82) showed that FDG-PET added significant information to MRI findings in 18 of 37 patients with MRI findings questionable for tumor recurrence. Kunkel el al. (83) studied 97 patients with previously resected HNSCC and found PET to be highly valuable for diagnosing recurrence in a postoperative setting.

Anzai et al. (73) used receiver operating characteristic (ROC) analysis to compare PET and CT/MRI in patients presenting with symptoms suggesting recurrence and no obvious tumor masses. Clearly higher sensitivity and specificity of PET were demonstrated, accompanied by a major difference in area under the ROC curve. When Fischbein et al. (78) used PET in routine surveillance of patients who had completed treatment for head and neck cancer, they found that nearly twice as many recurrences were detected by PET as by regular physical examination and routine CT imaging.

As expected, most additional recurrences that were detected by PET only were small, at times requiring repeated attempts at biopsy for confirmation. If an initial biopsy at the site of a PET abnormality is not positive, repeat biopsy or close follow-up should be undertaken. Surgical explo-ration should be considered with a clear-cut positive PET result because such results are not commonly incorrect. Currently instrumentation is lacking to help guide surgeons in the pursuit of these lesions. New device development may help with this scenario in the future (84).

PET in Detection of Distant Recurrence

Distant recurrences are more common in patients with locally recurrent disease than are distant metastases at initial staging. The lungs are the most common site of distant recurrence but the liver or skeleton may also be involved (5, 85). Before embarking on therapy of locally recurrent disease, whole-body FDG-PET imaging should be performed to exclude distant recurrence so as to avoid the morbidity of attempted curative surgery or radiation in the presence of noncurable distant tumor.

PET and Salivary Glands

About 70% of all salivary neoplasms arise in the parotid gland and about 75% of are benign (24). The most common parotid tumor is pleomorphic adenoma (benign mixed tumor). Other benign parotid tumors are monomorphic adenomas and Warthin's tumor. Cancer of the salivary glands is relatively rare, accounting for less than 1% of all cancers in the United States. Mucoepidermoid carcinoma and adenoid cystic carcinoma are the most common. About 50% of the submandibular tumors and almost all the sublingual tumors are malignant. Spread to regional lymph nodes is less

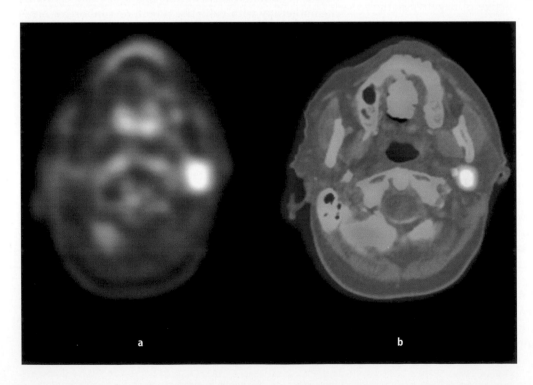

a   b

Figure 8.9. Grade 2 (of 4) adenocarcinoma possibly arising in a pleomorphic adenoma left parotid gland in 85-year-old woman with 10-year history of left fascial nerve palsy.

common than for head and neck SCC. Distant spread is most common to the lungs.

Only a few groups have studied the use of ^{18}F-FDG-PET in salivary gland cancers (85–91). Keyes et al. (86) studied 26 patients with parotid masses, finding false-positive scan for malignancy in 8 of 26 (31%), and concluded that FDG-PET was not useful in classifying parotid tumors as benign or malignant. Okamura et al. (88) studied 28 patients with parotid masses. High accumulation of FDG was found in all 5 carcinomas, in all Warthin's tumors, and in 8 of 12 pleomorphic adenomas, with mean SUV 5.92, 7.03, and 4.07, respectively. Five inflammatory lesions demonstrated only low accumulation of FDG. Because FDG concentrates in pleomorphic adenoma, the most common tumor in salivary glands, as well as in benign Warthin's tumor, it is not expected that PET will be useful in differentiation between benign and primary or secondary malignancy in the parotid gland. In fact, a hypermetabolic tumor in the parotid gland is more likely to be benign than malignant (Figure 8.9).

The incidence of lymph node metastases is less common than from HNSCC. Lymphatic spread tends to be orderly, first to intraglandular nodes, followed by periparotid and submandibular, then upper and midjugular nodes. Occasionally retropharyngeal nodes are involved, while bilateral spread is rare. Apparently no one has studied the use of PET for nodal staging, but we would expect FDG-PET to be useful for this purpose. We also expect PET to be useful for follow-up and restaging to differentiate between residual or recurrent tumor from posttreatment fibrosis.

Conclusion

PET imaging can improve the management of head and neck cancer. PET is highly accurate for nodal staging of the head and neck. TNM classification and staging of HNSCC by FDG-PET imaging may lead to more-appropriate patient treatment. Patients with more-extensive locoregional spread than is demonstrated by anatomic imaging may benefit from bilateral rather than unilateral neck dissection. Earlier detection of recurrence could provide a means of improving survival in recurrent HNSCC. Numerous comparisons of metabolic and anatomic imaging in diagnosis of recurrent tumor have shown higher sensitivity, specificity, and accuracy by PET compared to other imaging modalities. A negative PET scan excludes residual or recurrent disease with high certainty. FDG-PET is especially useful after initial treatment to differentiate residual or recurrent tumor from scar after surgery and postradiation and postchemotherapy changes. PET identifies unknown primary tumor in about 30% of cases. Diagnosis of unknown primary tumors of the head and neck may permit more-localized radiation therapy than irradiation with ports from the maxillary

sinuses to the lower neck, which is commonly undertaken without a primary diagnosis. Occasionally, unnecessary surgery of the head and neck or conjoint head and neck and thoracic procedures with curative intent might be avoided in patients who are found to have distant metastasis by PET.

References

1. Sankaranarayanan R, Masuyer E, Swaminathan R, Ferlay J, Whelan S. Head and neck cancer: A global perspective on epidemiology and prognosis. Anticancer Res 1998;18:4779–4786.
2. Kraus DH, Pfister DG, Harrison LB, et al. Larynx preservation with combined chemotherapy and radiation therapy in advanced hypopharynx cancer. Otolaryngol Head Neck Surg 1994;111:31–37.
3. Spaulding MB, Fischer SG, Wolf GT. Tumor response, toxicity, and survival after neoadjuvant organ-preserving chemotherapy for advanced laryngeal carcinoma. The Department of Veterans Affairs Cooperative Laryngeal Cancer Study Group. J Clin Oncol 1994;12:1592–1599.
4. Kacker A, Wolden S, Pfister DG, Kraus DH. Cancer of the larynx. In: Myers EN, Suen JY, Myers JN, Hanna EYN, editors. Cancer of the Head and Neck, 4th ed. Philadelphia: Saunders, 2003:333–377.
5. Myers EN, Suen JY, Myers JN, Hanna EYN, editors. Cancer of the Head and Neck, 4th ed. Philadelphia: Saunders, 2003.
6. Greene, FL, Page DL, Fleming ID, Fritz AG, Balch CM, Haller DG, Morrow M, editors. AJCC Cancer Staging Handbook, 6th ed. New York: Springer-Verlag, 2002:27–87.
7. Harnsberger HR. Handbook of Head and Neck Imaging, 2nd ed. St. Louis: Mosby, 1995.
8. Gavilán J, Herranz-Gonzales JJ, Lentsch EJ. Cancer of the neck. In: Myers EN, Suen JY, Myers JN, Hanna EYN, editors. Cancer of the Head and Neck, 4th ed. Philadelphia: Saunders, 2003:407–430.
9. Warburg O. The Metabolism of Tumors. London: Constable, 1930.
10. Geets X, Daisne JF, Gregoire V, Hamoir M, Lonneux M. Role of 11-C-methionine positron emission tomography for the delineation of the tumor volume in pharyngo-laryngeal squamous cell carcinoma: comparison with FDG-PET and CT. Radiother Oncol 2004;71:267–273.
11. Cobben DC, van der Laan BF, Maas B, Vaalburg W, Suurmeijer AJ, Hoekstra HJ, et al. ^{18}F-FLT PET for visualization of laryngeal cancer: comparison with ^{18}F-FDG PET. J Nucl Med 2004;45:226–231.
12. Chesnay E, Babin E, Constans JM, Agostini D, Bequignon A, Regeasse A, et al. Role of 11-C-methionine positron emission tomography for the delineation of the tumor volume in pharyngo-laryngeal squamous cell carcinoma: comparison with FDG-PET and CT. Radiother Oncol 2004;71:267–273.
13. De Boer JR, Pruim J, Burlage F, Krikke A, Tiebosch AT, Albers FW, et al. Therapy evaluation of laryngeal carcinomas by tyrosine-pet. Head Neck 2003;25:634–644.
14. Ninomiya H, Oriuchi N, Kahn N, Higuchi T, Endo K, Takahashi K, et al. Diagnosis of tumor in the nasal cavity and paranasal sinuses with ^{11}C choline PET: comparative study with 2-^{18}F] fluoro-2-deoxy-D-glucose (FDG) PET. Ann Nucl Med 2004;18:29–34.
15. Khan N, Oriuchi N, Ninomiya H, Higuchi T, Kamada H, Endo K. Positron emission tomographic imaging with ^{11}C-choline in differential diagnosis of head and neck tumors: comparison with ^{18}F-FDG PET. Ann Nucl Med 2004;18:409–417.
16. Torizuka T, Zasadny KR, Wahl RL. Diabetes decreases FDG accumulation in primary lung cancer. Clin Posit Imaging 1999;2:281–287.
17. Lowe VJ, Delong DM, Hoffman JM, Patz EF, Coleman RE. Dynamic FDG-PET imaging of focal pulmonary abnormalities to identify optimum time for imaging. J Nucl Med 1995;36:883–887.

18 Boerner AR, Weckesser M, Herzog H, Schmitz T, Audretsch W, Nitz U, et al. Optimal scan time for fluorine-[18] fluorodeoxyglucose positron emission tomography in breast cancer. Eur J Nucl Med 1999;26:226–230.

19 Thie JA, Hubner KF, Smith GT. Optimizing imaging time for improved performance in oncology PET studies. Mol Imaging Biol 2002;4:238–244.

20 Brink I, Klenzner T, Krause T, Mix M, Ross UH, Moser E, Nitzsche EU. Lymph node staging in extracranial head and neck cancer with FDG PET: appropriate uptake period and size-dependence of the results. Nuklearmedizin 2002;41:108–113.

21 Ma SY, See LC, Lai CH, Chou HH, Tsai CS, Ng KK, et al. Delayed ([18])F-FDG PET for detection of paraaortic lymph node metastases in cervical cancer patients. J Nucl Med 2003;44:1775–1783.

22 Kubota K, Yokoyama J, Yamaguchi K, Ono S, Qureshy A, Itoh M, Fukuda H. FDG-PET delayed imaging for the detection of head and neck cancer recurrence after radio-chemotherapy: comparison with MRI/CT. Eur J Nucl Med Mol Imaging 2004;31:590–595.

23 Schoder H, Yeung HW, Gonen M, Kraus D, Larson SM. Head and neck cancer: clinical usefulness and accuracy of PET/CT image fusion. Radiology 2004;231:65–72.

24 Hanna EY, Suen JY. Malignant tumors of the salivary glands. In: Myers EN, Suen JY, Myers JN, Hanna EYN, editors. Cancer of the Head and Neck, 4th ed. Philadelphia: Saunders, 2003:475–510.

25 Bogsrud TV, Karantanis D, Nahan MA, Mullan BP, Kasperbaur J, Strome S, et al. Focal high uptake in the thyroid gland as an incidental finding in ^{18}F-FDG PET. Thyroid 2004;14:760 (abstract).

26 Cohen MS, Arslan N, Dehdashti F, Doherty GM, Lairmore TC, Brunt LM, Moley JF. Risk of malignancy in thyroid incidentalomas identified by fluorodeoxyglucose-positron emission tomography. Surgery (St. Louis) 2001;130:941–946.

27 Kang KW, Kim SK, Kang HS, Lee ES, Sim JS, Lee IG, et al. Prevalence and risk of cancer of focal thyroid incidentaloma identified by ^{18}F-fluorodeoxyglucose positron emission tomography for metastasis evaluation and cancer screening in healthy subjects. J Clin Endocrinol Metab 2003;88:4100–4104.

28 Bocca E, Calearo C, Marullo T, DeVincentis I, Motta G, Ottaviani A. Occult metastases in cancer of the larynx and their relationship to clinical and histological aspects of the primary tumor: a four year multicentric research. Laryngoscope 1984;94:1086–1090.

29 Shuller DE, McGuirt WF, McCabe BF, Young D. The prognostic significance of metastatic cervical lymph nodes. Laryngoscope 1980;90:557–570.

30 Fukui MB, Curtin HD, Weissman JL. Radiological evaluation of cancer of the head and neck. In: Myers EN, Suen JY, Myers JN, Hanna EYN, editors. Cancer of the Head and Neck, 4th ed. Philadelphia: Saunders, 2003:67–100.

31 Senchenkov A, Staren ED. Ultrasound in head and neck surgery: thyroid, parathyroid, and cervical lymph nodes. Surg Clin N Am 2004;84:973–1000.

32 Al-Sarraf M, Hussein M. Head and neck cancer: present status and future prospects of adjuvant chemotherapy. Cancer Invest 1995;13:41–53.

33 Chong VFH, Fan Y-F. Detection of recurrent nasopharyngeal carcinoma: MR imaging versus CT. Radiology 1997;202:463–470.

34 Laubenbacher C, Saumweber D, Wagner MC, et al. Comparison of fluorine-18-fluorodeoxyglucose PET, MRI and endoscopy for staging head and neck squamous-cell carcinomas. J Nucl Med 1995;36:1747–1757.

35 Wong WL, Chevretton EB, McGurk M, et al. A prospective study of PET-FDG imaging for the assessment of head and neck squamous cell carcinoma. Clin Otolaryngol Appl Sci 1997;22:209–214.

36 Rege S, Safa AA, Chaiken L, Hoh C, Juillard G, Withers HR. Positron emission tomography: an independent indicator of radiocurability in head and neck carcinomas. Am J Clin Oncol 2000;23:164–169.

37 Kitagawa Y, Sano K, Nishizawa S, Nakamura M, Ogasawara T, Sadato N, Yonekura Y. FDG-PET for prediction of tumour aggressiveness and response to intra-arterial chemotherapy and radiotherapy in head and neck cancer. Eur J Nucl Med Mol Imaging 2003;30:63–71.

38 Allal AS, Slosman DO, Kebdani T, Allaoua M, Lehmann W, Dulguerov P. Prediction of outcome in head-and-neck cancer patients using the standarized uptake value of 2-[^{18}F]fluoro-2-deoxy-D-glucose. Int J Radiat Oncol Biol Phys 2004;59:1272–1273.

39 Kole AC, Nieweg OE, Pruim J, et al. Detection of unknown occult primary tumors using positron emission tomography. Cancer (Phila) 1998;82:1160–1166.

40 Braams JW, Pruim J, Kole AC, et al. Detection of unknown primary head and neck tumors by positron emission tomography. Int J Oral Maxillofacial Surg 1997;26:112–115.

41 Assar OS, Fischbein NJ, Caputo GR, Kaplan MJ, Price DC, Singer MI, et al. Metastatic head and neck cancer: role and usefulness of FDG PET in locating occult primary tumors. Radiology 1999;210:177–181.

42 Lassen U, Daugaard G, Eigtved A, Damgaard K, Friberg L. ^{18}F-FDG whole body positron emission tomography (PET) in patients with unknown primary tumors (UPT). Eur J Cancer 1999;35:1076–1082.

43 Safa AA, Tran LM, Rege S, et al. The role of positron emission tomography in occult primary head and neck cancers. Cancer J Sci Am 1999;5:214–218.

44 Jungehulsing M, Scheidhauer K, Damm M, et al. 2[F]-Fluoro-2-deoxy-D-glucose positron emission tomography is a sensitive tool for the detection of occult primary cancer (carcinoma of unknown primary syndrome) with head and neck lymph node manifestation. Otolaryngol Head Neck Surg 2000;123:294–301.

45 Greven KM, Keyes JJ, Williams Dr, McGuirt WF, Joyce Wr. Occult primary tumors of the head and neck: lack of benefit from positron emission tomography imaging with 2-[F-18]fluoro-2-deoxy-D-glucose. Cancer (Phila) 1999;86:114–118.

46 Johansen J, Eigtved A, Buchwald C, Theilgaard SA, Hansen HS. Implication of ^{18}F-fluoro-2-deoxy-D-glucose positron emission tomography on management of carcinoma of unknown primary in the head and neck: a Danish cohort study. Laryngoscope 2002;112:2009–2014.

47 Wong WL, Saunders M. The impact of FDG PET on the management of occult primary head and neck tumours. Clin Oncol (R Coll Radiol) 2003;15:461–466.

48 Adams S, Baum RP, Stuckensen T, Bitter K, Hor G. Prospective comparison of ^{18}F-FDG PET with conventional imaging modalities (CT, MRI, US) in lymph node staging of head and neck cancer. Eur J Nucl Med 1998;25:1255–1260.

49 Benchaou M, Lehmann W, Slosman DO, et al. The role of FDG-PET in the preoperative assessment of N-staging in head and neck cancer. Acta Otolaryngol 1996;116:332–335.

50 McGuirt WF, Williams DW 3rd, Keyes JW, et al. A comparative diagnostic study of head and neck nodal metastases using positron emission tomography. Laryngoscope 1995;105:373–375.

51 Myers LL, Wax MK, Nabi H, Simpson GT, Lamonica D. Positron emission tomography in the evaluation of the N0 neck. Laryngoscope 1998;108:232–236.

52 Rege S, Maass A, Chaiken L, et al. Use of positron emission tomography with fluorodeoxyglucose in patients with extracranial head and neck cancers. Cancer (Phila) 1994;73:3047–3058.

53 Kau RJ, Alexiou C, Laubenbacher C, Werner M, Schwaiger M, Arnold W. Lymph node detection of head and neck squamous cell carcinomas by positron emission tomography with fluorodeoxyglucose F-18 in a routine clinical setting. Arch Otolaryngol Head Neck Surg 1999;125:1322–1328.

54 Hannah A, Scott AM, Tochon-Danguy H, Chan JG, Akhurst T, Berlangieri S, et al. Evaluation of [18] F-fluorodeoxyglucose positron emission tomography and computed tomography with histopathologic correlation in the initial staging of head and neck cancer. Ann Surg 2002;236:208–217.

55 Bruschini P, Giorgetti A, Bruschini L, Nacci A, Volterrani D, Cosottini M, et al. Positron emission tomography (PET) in the

staging of head neck cancer: comparison between PET and CT. Acta Otorhinolaryngol Ital 2003;23:446–453.

56 Wax MK, Myers LL, Gona JM, Husain SS, Nabi HA. The role of positron emission tomography in the evaluation of the N-positive neck. Otolaryngol Head Neck Surg 2003;129:163–167.

57 Myers LL, Wax MK. Positron emission tomography in the evaluation of the negative neck in patients with oral cavity cancer. J Otolaryngol 1998;27:342–347.

58 Schmid DT, Stoeckli SJ, Bandhauer F, Huguenin P, Schmid S, von Schulthess GK, Goerres GW. Impact of positron emission tomography on the initial staging and therapy in locoregional advanced squamous cell carcinoma of the head and neck. Laryngoscope 2003;113:888–891.

59 Goerres GW, Schmid DT, Gratz KW, von Schulthess GK, Eyrich GK. Impact of whole body positron emission tomography on initial staging and therapy in patients with squamous cell carcinoma of the oral cavity. Oral Oncol 2003;39:547–551.

60 Sigg MB, Steinert H, Gratz K, Hugenin P, Stoeckli S, Eyrich GK. Staging of head and neck tumors: [^{18}F]fluorodeoxyglucose positron emission tomography compared with physical examination and conventional imaging modalities. J Oral Maxillofac Surg 2003;61:1022–1029.

61 Hollenbeak CS, Lowe VJ, Stack BC Jr. The cost-effectiveness of fluorodeoxyglucose 18-F positron emission tomography in the N0 neck. Cancer (Phila) 2001;92:2341–2348.

62 Teknos TN, Rosenthal EL, Lee D, Taylor R, Marn CS. Positron emission tomography in the evaluation of stage III and IV head and neck cancer. Head Neck 2001;23:1056–1060.

63 Hautzel H, Muller GH. Early changes in fluorine-18-FDG uptake during radiotherapy. J Nucl Med 1997;38:1384–1386.

64 Marriott CJC, Thorstad W, Akabani G, Brown MT, McLendon RE, Hanson MW, Coleman RE. Locally increased uptake of fluorine-18-fluorodeoxyglucose after intracavitary administration of iodine-131-labeled antibody for primary brain tumors. J Nucl Med 1998;39:1376–1380.

65 Rege SD, Chaiken L, Hoh CK, et al. Change induced by radiation therapy in FDG uptake in normal and malignant structures of the head and neck: quantitation with PET. Radiology 1993;189:807–812.

66 Lowe VJ, Heber ME, Anscher MS, Coleman RE. Chest Wall FDG Accumulation in serial FDG-PET images in patients being treated for bronchogenic carcinoma with radiation. Clin Posit Imaging 1998;1:185–191.

67 Rege SD, Chaiken L, Hoh CK, Choi Y, Lufkin R, Anzai Y, et al. Change induced by radiation therapy in FDG uptake in normal and malignant structures of the head and neck: quantitation with PET. Radiology 1993;189:807–812.

68 Lowe V, Dunphy F, Varvares M, et al. Evaluation of chemotherapy response in patients with advanced head and neck cancer using FDG-PET. Head Neck 1997;19:666–674.

69 Greven KM, Williams D3, Keyes JW, McGuirt WF, Watson NJ, Case LD. Can positron emission tomography distinguish tumor recurrence from irradiation sequelae in patients treated for larynx cancer? Cancer J Sci Am 1997;3:353–357.

70 Yao M, Graham MM, Hoffman HT, Smith RB, Funk GF, Graham SM, et al. The role of post-radiation therapy FDG PET in prediction of necessity for post-radiation therapy neck dissection in locally advanced head-and-neck squamous cell carcinoma. Int J Radiat Oncol Biol Phys 2004;59:1001–1010.

71 Goerres GW, Mosna-Firlejczyk K, Steurer J, von Schulthess GK, Bachmann LM. Assessment of clinical utility of ^{18}F-FDG PET in patients with head and neck cancer: a probability analysis. Eur J Nucl Med Mol Imaging 2003;30:562–571.

72 Rogers JW, Greven KM, McGuirt WF, Keyes JW Jr, Williams DW III, Watson NE, et al. Can post-RT neck dissection be omitted for patients with head-and-neck cancer who have a negative PET scan after definitive radiation therapy? Int J Radiat Oncol Biol Phys 2004;58:694–697.

73 Anzai Y, Carroll WR, Quint DJ, et al. Recurrence of head and neck cancer after surgery or irradiation: prospective comparison of 2-deoxy-2-[F-18]fluoro-D-glucose PET and MR imaging diagnoses. Radiology 1996;200:135–141.

74 Lapela M, Grenman R, Kurki T, et al. Head and neck cancer: detection of recurrence with PET and 2-[F-18]fluoro-2-deoxy-D-glucose. Radiology 1995;197:205–211.

75 Farber LA, Benard F, Machtay M, et al. Detection of recurrent head and neck squamous cell carcinomas after radiation therapy with 2-^{18}F-fluoro-2-deoxy-D-glucose positron emission tomography. Laryngoscope 1999;109 970–975.

76 Kao CH, ChangLai SP, Chieng PU, Yen RF, Yen TC. Detection of recurrent or persistent nasopharyngeal carcinomas after radiotherapy with 18-fluoro-2-deoxyglucose positron emission tomography and comparison with computed tomography. J Clin Oncol 1998;16:3550–3555.

77 Lowe VJ, Boyd JH, Dunphy FR, et al. Surveillance for recurrent head and neck cancer using positron emission tomography J Clin Oncol 2000;18: 651–658.

78 Fischbein NJ, Assar OS, Caputo GR, et al. Clinical utility of positron emission tomography with ^{18}F-fluorodeoxyglucose in detecting residual/recurrent squamous cell carcinoma of the head and neck. Am J Neuroradiol 1998;19:1189–1196.

79 Goerres GW, Schmid DT, Bandhauer F, Huguenin PU, von Schulthess GK, Schmid S, Stoeckli SJ. Positron emission tomography in the early follow-up of advanced head and neck cancer. Arch Otolaryngol Head Neck Surg 2004;130:105–109; discussion 120–121.

80 Cheon GJ, Chung JK, So Y, Choi JY, Kim BT, Jeong JM, et al. Diagnostic accuracy of F-18 FDG-PET in the assessment of post-therapeutic recurrence of head and neck cancer. Clin Posit Imaging 1999;2:197–204.

81 Perlow A, Bui C, Shreve P, Sundgren PC, Teknos TN, Mukherji SK. High incidence of chest malignancy detected by FDG PET in patients suspected of recurrent squamous cell carcinoma of the upper aerodigestive tract. J Comput Assist Tomogr 2004;28:704–709.

82 Shu-Hang N, Tung-Chieh JC, Sheng-Chieh C, Sheung-Fat K, Hung-Ming W, Chun-Ta L, et al. Clinical usefulness of ^{18}F-FDG PET in nasopharyngeal carcinoma patients with questionable MRI findings for recurrence. J Nucl Med 2004;45:1669–1677.

83 Kunkel M, Forster GJ, Reichert TE, Jeong JH, Benz P, Bartenstein P, Wagner W, Whiteside TL. Detection of recurrent oral squamous cell carcinoma by [^{18}F]-2-fluorodeoxyglucose-positron emission tomography: implications for prognosis and patient management. Cancer (Phila) 2003;98:2257–2265.

84 Tipinis SV, Nagarkar VV, Shestakova I, Gaysinskiy V, Entine G, Tornai MP, Stack BC. Feasibility of beta-gamma digital imaging probe for radioguided surgery. IEEE Trans Nuclear Sci 2004;51:110–116.

85 Manolidis S, Donald PJ, Valk PE, Pounds TR. The use of positron emission tomography in occult and recurrent head and neck cancer. Acta Otolaryngol 1998;Suppl:534.

86 Keyes JW Jr, Harkness BA, Greven KM, Williams DW III, Watson NE Jr, McGuirt WF. Salivary gland tumors: pretherapy evaluation with PET. Radiology 1994;192:99–102.

87 McGuirt WF, Keyes JW Jr, Greven KM, Williams DW III, Watson NE Jr, Cappellari JO. Preoperative identification of benign versus malignant parotid masses: a comparative study including positron emission tomography. Laryngoscope 1995;105:579–584.

88 Okamura T, Kawabe J, Koyama K, Ochi H, Yamada R, Sakamoto H, et al. Fluorine-18 fluorodeoxyglucose positron emission tomography imaging of parotid mass lesions. Acta Otolaryngol Suppl 1998;538:209–213.

89 Matsuda M, Sakamoto H, Okamura T, Nakai Y, Ohashi Y, Kawabe J, et al. Positron emission tomographic imaging of pleomorphic adenoma in the parotid gland. Acta Otolaryngol Suppl 1998;538:214–220.

90 Horiuchi M, Yasuda S, Shohtsu A, Ide M. Four cases of Warthin's tumor of the parotid gland detected with FDG PET. Ann Nucl Med 1998;12:47–50.

91 Mackie GC, Yarram SG. Pleomorphic adenoma simulating metastatic squamous cell carcinoma on F-18 fluorodeoxyglucose positron emission tomography. Clin Nucl Med 2004;29:743–744.

9

PET and PET/CT Imaging in Lymphoma

Guy H.M. Jerusalem, Roland Hustinx, and Pierre Rigo

The prognosis of patients with non-Hodgkin's lymphoma (NHL) and Hodgkin's disease (HD) has improved significantly, not only as a result of our increased knowledge of histopathologic patterns and new therapeutic concepts, but also because of advances in imaging techniques that have provided more-accurate staging information. The introduction of computed tomography (CT) was an important step forward in the staging of lymphoma. Magnetic resonance imaging (MRI) has provided no major advantage compared with CT and is not routinely used for this indication. These morphologic imaging techniques provide, in most instances, accurate definition of nodal involvement, but they do not identify lymphoma in normal-sized lymph nodes or differentiate nodes enlarged for other causes. They also have limitations for the detection of extranodal involvement such as spleen, liver, and bone marrow infiltration. In the posttherapy setting, the presence of a residual mass is an important clinical issue because further treatment is indicated only if viable tumor is still present. CT and MRI are unable to differentiate residual tumor from fibrosis. Gallium-67 (^{67}Ga) scintigraphy is useful in HD and high-grade NHL to assess response to treatment, but the method suffers from low spatial resolution of single photon scintigraphy (SPECT), lack of specificity, and difficulty in quantification of uptake. In addition, the sensitivity is low in infradiaphragmatic disease because of physiologic uptake in the abdomen. The limitations of ^{67}Ga scintigraphy in low-grade NHL are even greater. Moreover, ^{67}Ga scintigraphy should always be performed before treatment to determine if the individual patient has a gallium-avid lymphoma (1).

Recently, positron emission tomography (PET) using ^{18}F-fluorodeoxyglucose (FDG) has emerged as a new imaging procedure for staging and monitoring treatment of lymphoma. Despite the important role of ^{67}Ga scintigraphy in evaluation of response to treatment, FDG-PET imaging became the imaging modality of choice (2–8). FDG-PET imaging is more convenient than ^{67}Ga scintigraphy because the images can be acquired 1 h after the injection of FDG compared to several hours delay for ^{67}Ga. Furthermore, the radiation exposure from FDG is lower than with typical dosages of ^{67}Ga. This review of the current applications and perspectives of FDG-PET imaging for evaluation of lymphoma is an update of our previous work (9, 10).

Non-Hodgkin's Lymphoma

In 2003, approximately 53,400 new cases of NHL were diagnosed and 23,400 patients died of the disease in the United States (11). NHL is the leading cause of death from cancer in men between the ages of 20 and 39. The cause of NHL is unknown in most cases. Acquired immunodeficiency states, including acquired immunodeficiency syndrome (AIDS) or iatrogenic immunosuppression after solid organ transplantation, and some rare inherited disorders are associated with a increased risk of developing NHL. Epstein–Barr virus infection and environmental factors may increase the risk of genetic errors (12).

Various classifications have been used (for example, the Working Formulation, Kiel, and REAL) but the current World Health Organization (WHO) classification represents the first true international consensus on the classification of hematologic malignancies. The WHO classification is a list of distinct disease entities based on a combination of morphology, immunophenotype, genetic, and clinical features. It includes not only NHL, as did previous classifications, but all lymphoid malignancies, consisting of NHL, HD, lymphoid leukemias, and plasma cell neoplasms. Patient management depends on adequate biopsy for accurate diagnosis, patient history, careful clinical examination, laboratory tests, imaging studies, and further biopsies to determine accurate staging (for example, bone marrow biopsy, gastric biopsy).

The choice of treatment depends on patient characteristics, staging, laboratory tests, and the type of lymphoma.

The treatment choices for indolent NHL, such as small lymphocytic lymphoma, include no initial treatment ("wait and see"), radiotherapy for localized disease, systemic cytotoxic chemotherapy, and new biologic therapy. Except for rare localized disease, indolent NHL remains incurable, but long-term survival is frequently observed because the disease is only slowly progressive and responds well to various treatment options. Aggressive NHL, such as diffuse large B-cell lymphoma, is cured with anthracycline and cyclophosphamide-based polychemotherapy in 40% to 50% of all patients. More recently, the addition of rituximab, a chimeric human-murine monoclonal antibody (anti-CD20) improves the chance of survival. Salvage treatment with high-dose myeloablative chemotherapy followed by autologous stem cell transplantation can cure 10% to 20% of relapsing patients. Most patients with aggressive NHL who are not cured with polychemotherapy die some months or a few years after diagnosis because the disease is rapidly progressive after treatment failure. At present, the most valuable and widely used system to stratify patients with aggressive NHL is the International Prognostic Index (IPI) (13). Age, serum lactate dehydrogenase (LDH) levels, performance status, anatomic stage, and the number of extranodal sites determine the most appropriate treatment. As the stage and number of extranodal sites represent only two of five factors influencing the treatment, PET imaging is potentially more important in restaging at the end of induction chemotherapy or at the end of all treatment than at diagnosis.

Hodgkin's Disease

In 2003, approximately 7,600 new cases of HD were diagnosed, and 1,300 persons have died of the disease in the United States (11). An age-related bimodal incidence is observed, with a first peak occurring in the third decade of life and a second peak after the age of 50. A genetic risk for HD exists, and Epstein–Barr virus infection may be a predisposing factor. Patients with HD may have had more-severe initial Epstein–Barr virus infection or more-frequent viral replication associated with the development of HD. In contrast to NHL, the incidence is not increased in patients with immunosuppression. Most, if not all, cases of classic HD represent monoclonal B-cell disorders (14, 15). HD is defined as a lymphoma containing one of the characteristic types of Reed–Sternberg cells in a background of nonneoplastic cells. In contrast to NHL, HD spreads by contiguity from one lymph node chain to adjacent chains. However, the malignant cells may become more aggressive, invade blood vessels, and spread to organs in a manner similar to other malignancies.

The extent of disease is defined according to the "Cotswold Staging Classification" (16). The use of new imaging modalities and the administration of combined modality treatments have simplified staging procedures and made them less invasive. Laparotomy has disappeared as a routine staging procedure. Based on the stage and the presence or absence of B symptoms and various other prognostic factors (not uniform among large cooperative trial groups), patients can be divided into three prognostic groups. Patients with early-stage disease and very favorable prognostic factors are treated with extended-field radiotherapy alone. Patients with early- and intermediate-stage disease but unfavorable prognostic factors are treated with chemotherapy (usually four cycles) followed by involved-field radiotherapy. Patients with advanced-stage disease receive multiple cycles of polychemotherapy (usually eight) with or without consolidating radiotherapy. Alternative treatment options based on polychemotherapy alone in all stages including early-stage HD with favorable prognostic factors are being investigated. Sixty percent to 70% of patients with advanced disease are cured with standard polychemotherapy consisting of ABVD (adriamycin, bleomycin, vinblastine, and dacarbazine), and the prognosis is even more favorable for patients with early-stage disease. Relapsing patients can be cured by second-line salvage chemotherapy or by high-dose myeloablative chemotherapy followed by autologous stem cell transplantation, depending on various prognostic factors including previous treatment administered and disease-free interval before relapse. Current clinical trials are investigating new combinations of radiotherapy and chemotherapy to try to reduce late morbidity and mortality while maintaining or improving a high cure rate. Accurate staging is of particular importance in HD, especially if involved-field radiotherapy is used, because the stage of disease has major impact on the choice of treatment.

Staging in HD and NHL

All published studies comparing PET and CT imaging have similar methodologic problems because biopsy was performed in only a small number of the enlarged lymph nodes and histologic validation of results was generally not available. Additional lesions detected by FDG-PET imaging were biopsied only when the result has an impact on staging and treatment. In the absence of a histologic proof (gold standard), the calculation of sensitivity and specificity is not possible. Most studies examine concordance between routine staging procedures (standard of reference) and FDG-PET imaging and use response to treatment and follow-up data to assess the accuracy of the patient's original evaluation. However, this approach heavily biases results in favor of the least specific test, deceptively making it appear to be more accurate (17). Figures 9.1 and 9.2 illustrate FDG-PET/CT imaging in staging lymphoma.

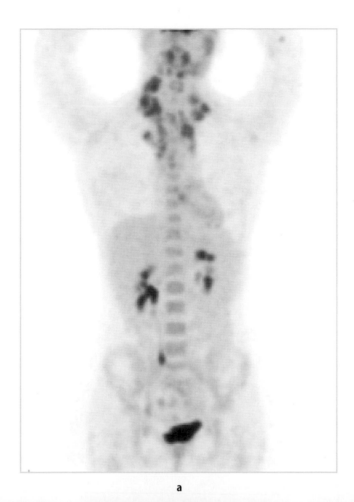

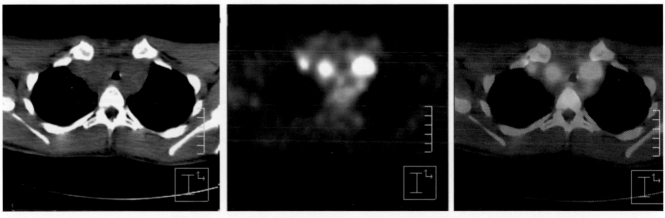

b

Figure 9.1. Sixteen-year-old patient with Hodgkin's disease (HD). FDG-PET imaging was performed at initial staging (**a**, projection image) and showed multiple foci of increased activity in the cervical and upper mediastinal areas. These foci correspond to nodal lesions on the positron emission tomography-computed tomography (PET/CT) (**b**, transaxial section), which also shows tracheal caliber reduced to 7 mm. There is also diffuse and mild uptake by the bone marrow, although bilateral bone marrow biopsies were negative.

Staging According to the Sites of Disease

Lymph Node Staging

Several studies showed that FDG-PET imaging is complementary to conventional imaging techniques (18–21). Newman et al. (18) compared FDG-PET and CT for imaging thoracoabdominal lymphoma. Fifty-four foci of abnormal uptake were detected by PET imaging in 13 patients. Forty-nine corresponding sites of lymphadenopathy and/or masses were detected by CT. All CT abnormalities suggestive of lymphoma corresponded to areas of focal increased uptake at FDG-PET imaging. The smallest lesion identified by PET and confirmed by CT imaging was 1 cm in maximal dimension. These preliminary data indicating good performances for PET imaging

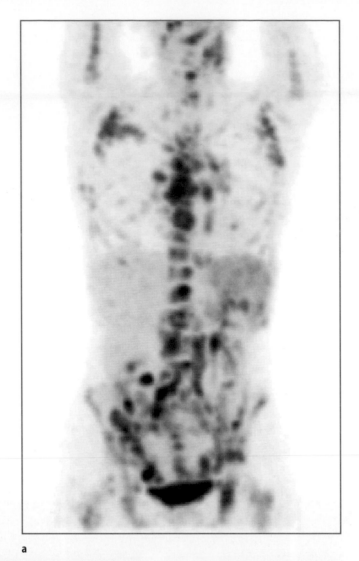

a

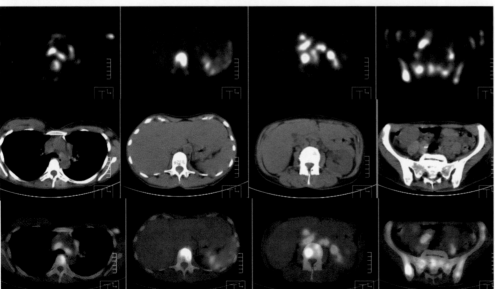

b

Figure 9.2. Recurrent HD. The 3D-projection FDG-PET image (**a**) shows massive skeletal involvement as well as multiple hypermetabolic nodal and splenic lesions. Selected transaxial sections of the PET images (**b**, *upper row*), corresponding CT sections (**b**, *middle row*), and fused images (**b**, *lower row*) are also shown.

for evaluation of lymphoma were confirmed by Hoh et al. (19). Whole-body FDG-PET imaging and conventional staging methods detected 33 of 37 disease sites in 18 patients, although not all sites were the same. Even more encouraging results were reported by Moog et al. (20). Both CT and FDG-PET imaging identified involvement of 160 nodal regions in 60 patients. Of the 25 additional regions seen by PET imaging, 7 were true positive, 2 were false positive, and 16 remained unresolved. CT showed six additional regions: three were false positive and three remained unresolved. The identification by FDG-PET imaging of additional lymph nodes involved by lymphoma was also observed by Jerusalem et al. (21). Peripheral lymph nodes (cervical, supraclavicular, axillary, and inguinal) were evaluated by physical examination and by PET imaging in 60 patients. Concordant results were observed in 41 patients. PET imaging detected more sites involved by lymphoma in 12 patients whereas clinically detected lymph nodes showed no FDG uptake in 7 patients. CT and FDG-PET imaging were also compared for evaluation of thoracoabdominal lymph nodes. CT detected enlarged lymph nodes that were not FDG avid in 8 patients, whereas lymph nodes of normal size were FDG avid in 5 patients. Based on these data, FDG-PET imaging and conventional staging procedures should be considered as complementary sources of diagnostic information in lymph node staging.

Extranodal Disease

There is clearly a need for better noninvasive methods for the detection of spleen, liver, and bone marrow involvement in patients with lymphoma (22). Twenty percent to 30% of patients with apparently localized supradiaphragmatic disease according to conventional staging procedures have infradiaphragmatic, mainly splenic, involvement discovered at staging laparotomy. Evaluation of bone marrow by CT is only feasible in exceptional cases, and findings must always be confirmed by bone marrow biopsy. The incidence of bone marrow involvement in newly diagnosed lymphoma is 10% in HD and 25% in NHL (22). In several studies, the sensitivity of CT ranged from 15% to 37% for the detection of splenic infiltration and from 19% to 33% for the detection of liver infiltration (23–25). Hepatic involvement is observed in 3.2% of patients with HD and in 15.1% of patients with NHL and splenic involvement in 23% of patients with HD and 22% of patients with NHL (22).

Moog et al. (22) showed that PET imaging may provide more information about extranodal lymphoma than CT. Forty-two extranodal manifestations of lymphoma were detected concordantly by both FDG-PET and CT imaging in 81 patients. PET imaging detected 24 additional sites of FDG uptake that were characteristic of lymphoma; positive confirmation by biopsy was obtained in 14 of these, including 9 in the bone marrow, 3 in the spleen, 1 in the liver, and 1 in the mesentery. The lesions that could not be verified were found in the lungs, in bone, and in spleen. Seven lesions were visualized only by CT; 5 were false positive and 1 remained unresolved.

Spleen

Rini et al. (26) compared FDG-PET with CT imaging for the evaluation of spleen involvement during initial staging. The findings were correlated with final diagnosis, which was obtained surgically for six patients and at autopsy for one patient. FDG-PET imaging was true positive for all five patients with splenic disease and true negative for both patients without splenic disease. CT was true positive for four of the five patients with splenic disease and false positive for both patients without splenic disease.

Bone Marrow Involvement

Detection of bone marrow involvement by FDG-PET imaging is a controversial issue (27). Prominent FDG-PET uptake in the bone marrow that is equal to or greater than the level of activity in the liver, uptake in the distant long bones, and heterogeneous and focal uptake represent abnormal bone marrow activity. Diffuse bone marrow uptake can be observed in patients with HD at diagnosis related to reactive myeloid hyperplasia characteristic of some HD patients. Diffusely increased uptake is usually observed in reactive bone marrow, particularly following chemotherapy and administration of growth factor, for example, granulocyte colony-stimulating factor (GCSF). However, diffuse bone marrow FDG uptake can also be caused by bone marrow involvement (28).

Carr et al. (29) reported that increased FDG uptake in the bone marrow can correctly assess marrow disease status in a high proportion of lymphoma patients. Unilateral iliac crest marrow aspirates and biopsies were performed in all 50 patients studied. Concordant results were observed in 78% (39 of 50) of patients; 13 were true positive and 26 true negative. In 8 patients with negative biopsy, PET imaging showed increased focal (*n* = 4) or diffuse (*n* = 4) FDG uptake, including 4 patients with focal FDG abnormalities that were distant from the site of biopsy. In one of the 4 patients, a subsequent biopsy the FDG-avid site confirmed marrow involvement. In the 4 patients with diffuse FDG uptake but normal biopsies, there was no other evidence indicating marrow involvement. The two patients with HD had reactive myeloid hyperplasia. PET imaging was false negative in 3 patients with positive biopsy. Two of these patients with low- and intermediate-grade NHL had no FDG uptake at the site of their primary lymph node disease. In only 1 patient, with mantle cell lymphoma, there was FDG uptake in at least one site (spleen).

Moog et al. (30) reported findings in 78 patients, 39 with NHL and 39 with HD. Seventy patients underwent bilateral iliac crest bone marrow biopsy. In addition to 7 concordant positive and 57 concordant negative findings, biopsy revealed 4 patients with bone marrow involvement (3 with low-grade NHL and 1 with intermediate-grade NHL) that was not detected by FDG-PET imaging. In 10 patients with negative biopsies, PET imaging showed focal bone marrow uptake of FDG, and the PET findings were confirmed in 8 of these patients, leading to an upgrade of the tumor stage in 10.3% (8 of 78) of patients. Unfortunately, not all studies found these promising results. Kostakoglu et al. (27) reported that FDG-PET imaging provided little information regarding bone marrow involvement in lymphoma with a positive predictive value of PET imaging of only 66.6%. False-negative results were observed in patients with limited involvement of bone marrow.

The study reported by Jerusalem et al. (31) was also unable to confirm previous encouraging results in smaller patient groups. They conclude that FDG-PET imaging cannot replace bone marrow biopsy, independently of histologic subtype. PET imaging missed bone marrow infiltration in 49 of 71 patients with a positive bone marrow biopsy. PET imaging alone was positive in 8 patients, but no further investigations were performed to confirm these findings. However, among these, bone marrow biopsy showed reactive hematopoietic changes without lymphoma infiltration in 4 of 5 patients with diffuse uptake and in 1 of 3 patients with focal uptake, suggesting a false-positive PET study in at least 5 of these patients.

Finally, Elstrom et al. (32) found that FDG-PET imaging was not reliable for the detection of bone marrow involvement in any lymphoma subtype. FDG-PET imaging rarely detected pathologically identifiable marrow involvement by follicular lymphoma and did not detect marrow involvement by mantle cell lymphoma or marginal zone lymphoma in any case. A potential explanation for these false-negative PET studies is a relatively low FDG uptake by these types of lymphoma or diffuse low-density marrow involvement by the tumor. Conversely, patients with HD or large B-cell lymphoma showed FDG uptake in bone marrow that was not confirmed by iliac crest biopsy in several cases. While these cases may represent false positives, patients may alternatively have had patchy bone marrow involvement that was not detected by blind iliac crest biopsy.

Bone Involvement

Moog et al. (33) showed that FDG-PET imaging could replace skeletal scintigraphy in the initial staging of HD and NHL. Of the 56 patients studied, 12 were found to have skeletal involvement by both methods. FDG-PET imaging detected 12 involved regions that were negative by scintigraphy in 5 patients; this was subsequently verified in 3 patients, and the other 2 cases remained un-

resolved. Conversely, skeletal scintigraphy revealed 5 abnormalities that were negative by PET imaging in 5 patients. Three of these abnormalities were found to be not related to lymphoma; final evaluation of the remaining 2 findings was not possible.

Gastric Involvement

In patients with gastric NHL patients, FDG-PET imaging can provide additional information complementary to endoscopy and CT (34). Rodriguez et al. (34) reported high FDG uptake in all six high-grade and one of two low-grade primary gastric lymphoma. No abnormal tracer uptake was seen in the patient with low-grade gastric NHL of the MALT type. In six of eight patients, FDG-PET imaging demonstrated larger tumor extension in the stomach compared to endoscopy. High FDG uptake was seen in two patients with a normal CT examination. Hoffmann et al. (35) confirmed false-negative results in gastric NHL of the MALT-type. All six patients (five low-grade and one high-grade) had false-negative FDG-PET studies, indicating that FDG-PET imaging is not useful for staging or follow-up of this histologic subtype of NHL.

Primary Cerebral Lymphoma

The clinical value of PET imaging in primary central nervous system (CNS) lymphoma remains to be established (36). Malignant lymphoma should be considered in the differential diagnosis of brain tumors showing high FDG uptake (37, 38). Unfortunately, PET imaging cannot differentiate between primary lymphoma of the CNS, malignant glioma or cerebral metastasis. It is important to take dexamethasone-induced changes of tumor cell metabolism into account (36, 38). Dexamethasone can induce inhibition of proliferative activity in CNS lymphoma producing decrease or even disappearance of abnormal FDG accumulation. In HIV-positive patients with contrast-enhancing cerebral lesions, FDG-PET imaging is useful to differentiate primary CNS lymphoma from nonneoplastic lesions such as toxoplasmosis (39–41). High FDG uptake most likely represents a malignant process that should be biopsied for confirmation rather than treated presumptively as infectious. All three studies (39–41) reported a good sensitivity (all 16 CNS lymphoma showed high FDG uptake), but some lesions in patients with progressive multifocal leucoencephalopathy had high FDG uptake, limiting the specificity (41).

Cost of Staging with PET Imaging and Conventional Methods

FDG-PET imaging is able to detect disease not seen by conventional imaging methods. Consequently, the stage of the disease is modified in some patients (21, 42, 43), and sometimes this will lead to change of treatment (42). FDG-PET imaging may have a major role for better defining the treatment of patients with HD, in particular when short-course chemotherapy is combined with involved-field radiotherapy. However, at this time, it remains unknown whether PET imaging at initial diagnosis can refine prognostic indices and stratification for treatment in HD or NHL. In the study of 60 patients by Jerusalem et al. (21), FDG-PET imaging provided comparable data to conventional staging procedures. Although PET imaging detected more lesions at initial presentation, this resulted in only few modifications of staging. Buchmann et al. (42) reported that PET imaging showed more lesions than CT except in infradiaphragmatic regions, in which the two methods produced equivalent results. In 8% (4 of 52) of patients, FDG-PET imaging led to upstaging and a change of therapy.

One of the major arguments against the widespread use of PET imaging is its high costs. However, PET imaging might reduce the costs of diagnostic workup in lymphoma patients by replacing other imaging procedures. Hoh et al. (19) found that FDG-PET imaging may be an accurate and cost-effective method for staging or restaging HD ($n = 7$) and NHL ($n = 11$). Accurate staging was achieved in 17 of 18 patients using a whole-body PET-based algorithm, compared to 15 of 18 patients using the conventional staging algorithm. The total costs of the whole-body FDG-PET-based algorithm was calculated to be almost half compared to the conventional staging algorithm. However, not all studies indicated cost saving by using FDG-PET imaging. Klose et al. (44) found that FDG-PET imaging led to upstaging in 4 of 22 patients, resulting in correct stage for 100% of patients with PET imaging and in only 81.8% of patients with CT. This finding resulted in an incremental cost-effectiveness ratio (additional costs per additionally correctly staged patient) of 3,133 _ for PET imaging.

Other important studies comparing staging based on PET with staging based on conventional imaging techniques are discussed in the following section, Results in Histologic Subtypes.

Results in Histologic Subtypes

Hodgkin's Disease

Several small, single-institution studies have shown that FDG-PET is useful for staging HD when it is performed in addition to conventional staging procedures (45–52) (Table 9.1). However, PET does not replace conventional staging techniques (53). False-negative (45–49, 51, 52) and false-positive (45) results have been reported. CT is also more accurate in defining bulk of disease. Integrated PET/CT imaging does overcome the inherent limitations

Table 9.1. Impact of FDG-PET imaging on staging and treatment in Hodgkin's disease (HD).

Authors	Modification of staging	Modification of treatment
Bangerter et al. (45)	6/44 (14%)	6/44 (14%)
Weidmann et al. (46)	3/20 (15%)	Not reported
Partridge et al. (47)	21/44 (48%)[a]	11/44 (25%)[a]
Jerusalem et al. (48)	7/33 (21%)[a]	1/33 (3%)[a]
Hueltenschmidt et al. (49)	10/25 (40%)	2/25 (4%)
Menzel et al. (50)	6/28 (21%)	Not reported
Weihrauch et al. (51)	4/22 (18%)	1/22 (5%)
Naumann et al. (52)	18/88 (20%)[b]	16/88 (18%)[b]
Overall	75/304 (25%)	37/256 (14%)

[a]Including 1 incorrect modification of staging and treatment based on false-negative PET.
[b]Including 6 incorrect modifications of staging and treatment based on false-negative PET.
Source: Adapted from Jerusalem et al. (121, 122).

of both imaging techniques. In the future, one can hope that it allows a better definition of radiation fields, minimizing radiation exposure to normal tissues.

Non Hodgkin's Lymphoma

The degree of FDG uptake in lymphoma was found to correlate with the grade of malignancy according to the Working Classification (54, 55), the KI-67 labeling index (56), and the number of cells in S phase (54). However, in some patients, the degree of FDG uptake correlated poorly with the S-phase fraction (57). Tumors are heterogeneous not only macroscopically (partially necrotic high-grade tumors), but also at a microscopic and a metabolic level, and measurements of FDG uptake reflects only the average metabolic activity. Koga et al. (58) showed different levels of FDG accumulation according to tumor location within the same patient suffering from untreated intermediate-grade NHL and the degree of glucose transporter 1 (GLUT-1) expression correlated with different degrees of FDG accumulation. In a study of 21 patients, Okada et al. (59) found that FDG uptake before treatment may be useful for predicting response to therapy and prognosis. However, in the absence of other studies that confirm these results, FDG-PET imaging is not routinely used as a prognostic factor before treatment.

Diffuse Large B-Cell Lymphoma (DLBCL)

Our group studied 53 patients (40 patients at initial diagnosis and 13 patients at relapse) with histologically proven high-grade NHL, most suffering from DLBCL (60). The results suggested that FDG-PET imaging was more sensitive than clinical examination and CT for the detection of involved lymph node regions. PET imaging was as effective as CT for the detection of extranodal lesions other than bone marrow involvement. It was concluded that FDG-PET imaging was an efficient, noninvasive method for staging and restaging high-grade NHL that should be used in conjunction with bone marrow biopsy. Elstrom et al. (32) found that FDG-PET imaging detected disease in at least one site in all 51 patients with DLBCL. Unfortunately, this study gave no information about the lesions that were missed or those only shown by FDG-PET imaging.

Follicular and Small Lymphocytic NHL

Jerusalem et al. (61) studied 36 patients with low-grade NHL. FDG-PET imaging detected more abnormal nodal regions than conventional staging in follicular lymphoma but was not accurate for the staging of small lymphocytic lymphoma. The detection rate of PET imaging for lymph node involvement was better than conventional staging for peripheral (34% more nodal regions detected) and thoracic (39% more nodal regions detected) but not for abdominal or pelvic nodal regions (26% fewer nodal regions detected). Attenuation correction was not performed in this study but the absence of attenuation correction may have been a limitation for detection of abdominal and pelvic lymph nodes. The sensitivity for detection of bone marrow involvement by lymphoma was unacceptably low for PET imaging, independently of histologic subtype. In contrast, PET imaging was as effective as conventional procedures for the detection of other extranodal localizations. There was a small number of cases in which the results by PET imaging and conventional procedures were discordant. FDG-PET imaging identified more cases of splenic or hepatic involvement but CT showed more cases of pleural or lung involvement.

Blum et al. (62) evaluated the impact of PET imaging on the management of patients with low-grade follicular NHL. Treatment was changed at initial presentation according to FDG-PET imaging in 6 of 12 patients, including modification of radiation fields and/or modification of the treatment modality (chemotherapy used instead or in combination with radiotherapy). Ten of the 37 patients who underwent restaging with FDG-PET imaging also had their management influenced. However, the conclusions drawn from this study are limited by the retrospective analysis and by the few staging modifications that were proven by a biopsy. Elstrom et al. (32) found that FDG-PET imaging detected disease in at least one site in 41 of 42 patients with follicular NHL. The single case of undetected follicular NHL consisted of an ileal tumor shown only by endoscopic biopsy. Although further studies are required before using FDG-PET imaging routinely, these data suggests that FDG-PET imaging may have a role in the initial staging of patients with follicular NHL.

Mantle Cell Lymphoma (MCL)

Elstrom et al. (32) detected disease in at least one site in all seven patients with MCL, but the routine use of PET imaging for this subtype of NHL is still debated. False-negative PET studies in MCL have been reported (63).

Other NHL Subtypes

FDG-PET imaging is not useful for staging and follow-up of extranodal B-cell lymphoma of the mucosa-associated lymphoid tissue (MALT) type (35). In contrast, nodal involvement in marginal zone lymphoma was detected by FDG-PET imaging in most patients and additional sites of disease were occasionally identified (64). Detection of follicular lymphoma of the duodenum was disappointing in a small group of patients (65). Tumor volume (thickness of the lesions in the gastrointestinal tract) rather than the subtype of lymphoma may be a critical factor explaining some of the false-negative studies. More encouraging results have been reported in enteropathy-type T-cell lymphoma (66) but not in peripheral T-cell lymphoma (32).

Evaluation of Response During or After Chemotherapy

One of the most challenging aspects for imaging lymphoma is the assessment of response to treatment. The rapidity of response during treatment appears to be an accurate predictor of response, with early tumor regression indicating a higher chance of cure (67). Patients in complete remission at the end of treatment present a longer disease-free and overall survival (68). Accurate restaging allows optimal selection of treatment options and could potentially improve outcome. Shorter treatment cycles are the goal of ongoing research in a subgroup of low-risk patients in an attempt to minimize side effects related to treatment. High-risk patients or those with persisting disease after first-line therapy may benefit from more-aggressive treatments, such as high-dose chemotherapy followed by stem cell transplantation. In fact, it seems reasonable to use salvage therapy at the time of minimal residual disease rather than at the time of a clinically overt relapse. Furthermore, earlier discontinuation of unsuccessful treatment would avoid the associated toxicity.

FDG-PET imaging is now the imaging technique of choice for determining chemosensitivity before high-dose chemotherapy followed by autologous stem cell transplantation and for evaluation after completion of therapy. Promising results have also been reported in assessment of early response to therapy. It is important to take into account the limitations of FDG-PET imaging for appropriate utilization of PET findings in the management of patients. Figure 9.3 illustrates FDG-PET/CT imaging in monitoring therapy of lymphoma.

Limitations

False-Negative Studies

The spatial resolution of the current PET systems is approximately 5 to 8 mm. Radiotracer uptake in structures less than twice the spatial resolution of the tomograph can be underestimated (=partial volume effect).

Immediately after chemotherapy, metabolic activity may be temporarily reduced. Therefore, for early assessment of response during chemotherapy, it is better to perform FDG imaging to wait as long as possible after the last day of chemotherapy (the last day before or even the same day of the start of a new cycle of chemotherapy). For evaluation after completion of therapy, the best timing is unclear and most investigators suggest waiting at least 1 month after the last day of chemotherapy.

Even if the spatial resolution of PET technology improves in the future, microscopic residual disease cannot be excluded with imaging modalities (69). Consequently, therapy can not be ended on the basis of a negative FDG-PET study, and periodic repeated imaging may be indicated to detect residual disease (70).

False-Positive Studies

Increased FDG uptake is not necessarily related to residual disease. Physiologic uptake can be misinterpreted in particular by an inexperienced observer. When PET imaging is first introduced at a center, a learning curve may have an impact on the false-positive rate. Dittmann et al. (71) observed four of eight false-positive PET findings in the region of cervical lymph nodes most likely related to muscular uptake.

High FDG uptake might be caused by immunologic hyperreactivity, especially by the thymus and other lymphatic tissues after chemotherapy or radiotherapy. Several authors have described rebound thymic hyperplasia as a potential cause of false-positive findings (49, 72–75). In most cases, the characteristic shape of an inverted V allows correct interpretation. Even when a recurrent pattern is shown by FDG-PET imaging in the mediastinum, it should be interpreted with caution. In the asymptomatic patient, multiple biopsies of suspected mediastinal lesions or a close clinical follow-up with frequent radiographic studies looking for changes in the size of residual tissue with enlarging masses suggestive of recurrent disease should be performed (75). Yoon et al. (76) reported a case where PET imaging strongly suggested recurrent disease. Excision of the mass revealed necrosis, fibrosis, and extensive infiltration by histiocytes, but there was no evidence of recurrent lymphoma or thymic hyper-

plasia. Weihrauch et al. (73) found a massive histiocytic reaction in one of three patients with a false-positive PET study in the mediastinum. Nonviable scar tissue was also detected by Naumann et al. (77) in a resected residual mediastinal mass that accumulated FDG.

Benign follicular hyperplasia of lymph nodes has been described by Mikosch et al. (70) as a cause of false-positive PET study. Benign tumors such as thyroid adenoma (49) can falsely suggest the presence of residual or recurrent lymphoma. Fracture at sites of bone marrow involvement before treatment is another cause of false-positive FDG-PET studies (78). Becherer et al. (79) described a false-positive PET study related to a fistula confirmed by surgery. Infectious or inflammatory lesions have to be excluded, in particular if abnormal FDG uptake is seen outside of the initially involved sites.

In most cases, the availability of precise anamnestic data such as the site of irradiation, clinical suspicion of infections, or inflammatory conditions aids in the differential diagnosis. Comparison to a pretreatment study can also be helpful in resolving equivocal results. If there is a discrepancy in the pattern of response on PET imaging, with areas of known disease responding to therapy while others deteriorate or new sites appear, then a cause other than residual disease should be excluded.

De Wit et al. (78, 80) found a low positive predictive value (PPV) due to a high number of false-positive PET findings outside residual masses such as pleural FDG uptake caused by inflammation. Naumann et al. (77) found two false-positive PET findings in patients suffering from HD related to FDG uptake outside of the residual mass (supraclavicular and thoracic paravertebral false-positive findings). de Wit et al. (78) described a case of false-positive PET study related to pneumonia caused by *Haemophilus influenzae*. Tuberculosis was found as source of pitfall in two patients reported by Mikosch et al. (70).

Sandherr et al. (81) reported a case of HD with hypodense splenic lesions and corresponding increased glucose metabolism in FDG-PET imaging 4 months after completion of treatment, suggesting early relapse. However, this patient suffered from toxoplasmosis. Cellular immunodeficiency associated with HD can result in severe infection. False-positive PET findings have been reported in epithelioid cell granulomas (82). The average frequency of epithelioid cell granulomas is 13.8% in HD. Granulomas without evidence of sarcoidosis may be caused by tuberculosis or other infectious diseases, especially fungal infections in immunosuppressed patients.

Naumann et al. (83) reported the simultaneous occurrence of eosinophilic granuloma and HD in two patients. Although Langerhans' cell histiocytosis (LCH) is not believed to be a neoplastic disorder, the association between LCH and acute leukemias, lymphomas, and solid tumors is well known. Unfortunately, FDG-PET imaging showed increased uptake both in HD and in granuloma.

Erythema nodosum may also be a source of a false-positive PET interpretation (84). Lorenzen et al. (85)

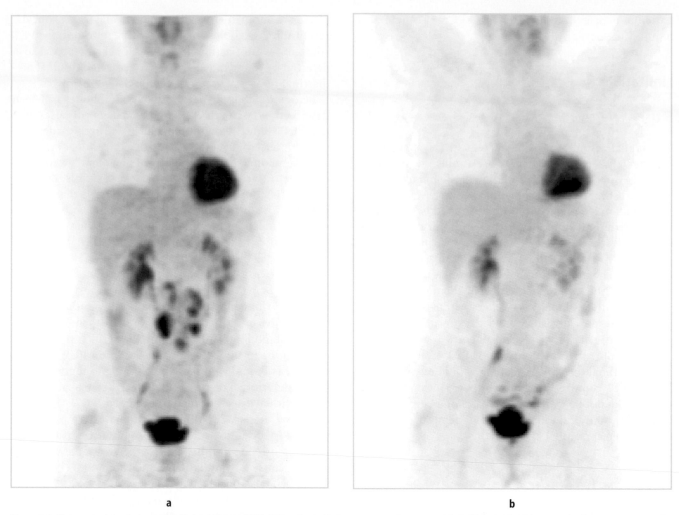

a   b

Figure 9.3. High-grade abdominal non-Hodgkin's lymphoma (NHL). PET performed before treatment shows several foci of increased activity corresponding to paraaortic nodes (**a**, projection image). PET/CT performed after two courses of chemotherapy shows a complete response (**b**, projection image), even though enlarged lymph nodes are still visible in the right paraaortic area

reported a case of false-positive FDG-PET study in the margin of an abdominal residual mass after chemotherapy in a patient suffering from NHL. Histology documented a necrotic centre surrounded by granulation tissue. Dittmann et al. (71) reported two cases of unspecific lymphadenitis in patients with false-positive PET study. One occurred in the iliac and one in the inguinal lymph node area. Hueltenschmidt et al. (49) reported two false-positive FDG-PET studies resulting from misinterpretation of an inflammatory lung process. Figure 9.4 illustrates a case of FDG-PET-avid cat scratch disease.

Transient nonspecific effects can occur following radiation therapy or chemotherapy. Cremerius et al. (86) and Weidmann et al. (49) described false-positive FDG-PET findings in patients with radiotherapy-induced pneumonitis. Most authors suggest that the time interval should be at least 3 months after radiotherapy to avoid false-positive findings caused by unspecific treatment effects. It is also important to exclude another malignant tumor that mimics lymphoma relapse. Jerusalem et al. (87) reported a case of rectal cancer; Mikosch et al. (70) observed a melanoma and a metastasis of melanoma in

two patients. Finally, in some cases, the reasons for false-positive PET findings remain unclarified (49, 70). All these false-positive cases emphasize the importance of obtaining a biopsy before starting salvage therapy.

Evaluation of Early Response to Therapy

Early response evaluation by FDG-PET imaging after one cycle of chemotherapy, a few cycles or at midtreatment can predict response, progression-free survival (PFS), and overall survival (OS) (8, 88–92). FDG-PET imaging has a higher predictive value of outcome when performed early during treatment than when performed after treatment. Early identification of nonresponders by FDG-PET imaging may lead to a change from an unsuccessful therapy to a more-effective one. The studies reported by Hoekstra et al. (93) and Dimitrakopoulou-Strauss et al. (94) showed that a decrease in FDG uptake was associated with successful response to therapy in NHL and HD, whereas increasing FDG uptake indicated progression. Römer et al. (95) observed a marked decrease in tumor

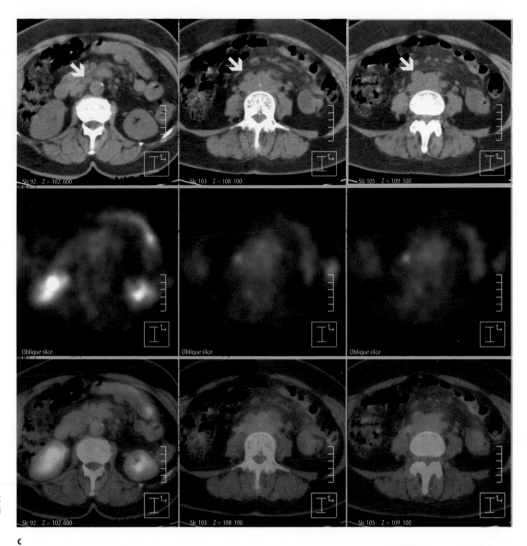

Figure 9.3 . (Contd.) (*arrows*, **c**: transaxial sections for CT, PET, and fused images).

c

uptake as early as 7 days after start of chemotherapy. Measurement of FDG uptake at 42 days, immediately before the third cycle of chemotherapy, was more predictive of long-term outcome than the seven-day uptake.

Jerusalem et al. (88) studied 28 patients with NHL after a median of three cycles of polychemotherapy (interim scan). Only 1 of 5 patients with persistent increased FDG uptake entered complete remission. All patients with a negative FDG-PET study entered complete remission, except for 2 patients who died of treatment toxicity during chemotherapy. By Kaplan–Meier analysis, PFS at 1 and 2 years was, respectively, 20% ± 18% and 0% for patients with positive FDG-PET findings and 81% ± 9% and 62% ± 12% for patients with negative findings. OS was also significantly better for FDG-PET-negative patients. Unfortunately, the sensitivity in identifying patients with a poor outcome was insufficient because only 5 of the 12 patients who relapsed had residual FDG uptake. The authors suggested that earlier evaluation after only one cycle of chemotherapy and quantitative analysis might increase the sensitivity of FDG-PET imaging to predict relapse.

Mikhaeel et al. (89) confirmed that an interim scan after two to four cycles of chemotherapy provided valuable information regarding early assessment of response and long-term prognosis in a study of 23 patients with aggressive NHL. Spaepen et al. (90) showed that the predictive value of FDG-PET imaging is independent of conventional prognostic factors (International Prognostic Index). Although the interim scan was only performed at midtreatment, the sensitivity to predict residual disease at the end of treatment was excellent. Only 6 of 37 patients with a negative FDG-PET study did not achieve a durable complete remission compared with 33 of 33 patients with a positive FDG-PET study. Although FDG uptake is not specific for tumoral tissue, the PPV for relapse was 100% (33 of 33). The authors confirmed the strong correlation between FDG-PET findings and PFS and OS.

Kostakoglu et al. (91) performed an earlier evaluation after only one cycle of chemotherapy in 30 patients with HD or NHL. Positive FDG-PET findings obtained both after the first cycle and at the completion of therapy were associated with a shorter PFS than were negative FDG-PET results. FDG-PET results obtained after the first cycle

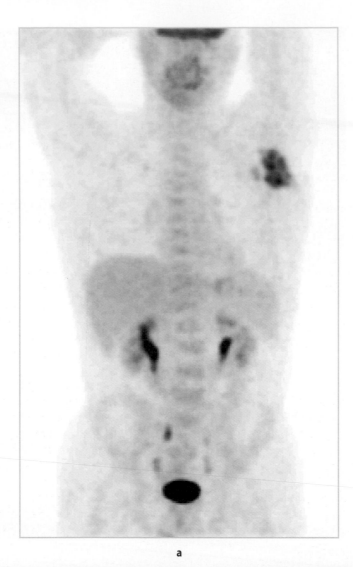

a

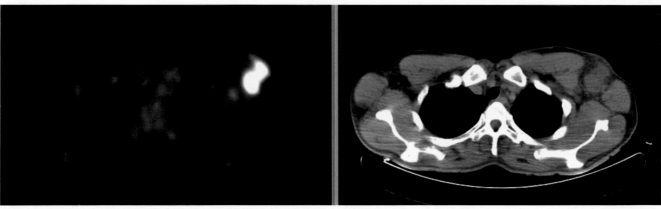

b

Figure 9.4. Patient with a previous history of lymphoma with clinical suspicion of recurrence (enlarged left axillary nodes). PET showed highly increased metabolism in the left axillary area (**a**, projection image), but the corresponding CT images showed a collected mass with a fistula to the skin, suggesting an infectious process rather than tumor recurrence (**b**, PET and CT transaxial sections). A biopsy revealed lesions consistent with cat's scratch disease.

correlated better with PFS than FDG-PET findings obtained after completion of chemotherapy.

In conclusion, FDG-PET imaging is the best imaging technique for assessment of early response during treatment. However, the published data are discordant about the best time (optimal sensitivity and prediction of PFS) to perform FDG-PET imaging, reflecting the fact that it depends on several factors and is not the same for all patients. Reduced numbers of viable cells or reduced metabolism of the damaged cells after treatment might be associated with decrease in tumor glucose utilization. Most studies are oversimplified by differentiating patients into only two categories (with or without residual FDG uptake). The probability that PET imaging remains positive depends on the sensitivity and resolution of the PET system (smallest lesion that can be detected), the biology of the tumor (more-rapid response in aggressive tumors), the tumor mass at diagnosis (tumor shrinkage below the detection level, later in large tumors), dose intensity of chemotherapy (more-rapid regression if higher doses), and interval between the last day of chemotherapy and FDG-PET imaging (risk of temporarily reduced metabolic activity early after chemotherapy).

Semiquantitative evaluation of the degree of FDG uptake is probably more accurate, but methodologic problems remain to be resolved (96). For example, decrease of FDG uptake may be influenced by a decrease of tumor size resulting in partial-volume effects. The use of standard uptake value (SUV) analysis must take into account the influence of different imaging times over multiple bed positions because FDG uptake in lymphoma does not reach a plateau in the 3-h postinjection time. Furthermore, monoclonal antibodies such as rituximab that have very long half-lives are now used routinely in association with chemotherapy, but the impact on metabolic activity evaluated by PET imaging remains unknown. On the other hand, a transient increase in inflammatory cells may result in overexpression of the fraction of viable cancer cells, as shown in a tumor mouse model (97). Further studies are clearly warranted to investigate the best timing for evaluation of early response to treatment before starting large prospective randomized trials in which treatment is changed in some patients to a more-aggressive or a more-experimental one based on FDG-PET findings.

Evaluation of Chemosensitivity Before High-Dose Chemotherapy

High-dose chemotherapy followed by autologous stem cell transplantation is the standard treatment for patients relapsing from lymphoma if the tumor remains chemosensitive. FDG-PET imaging is the imaging technique of choice for distinguishing responders from nonresponders to reinduction chemotherapy (79, 98–101).

Becherer et al. (79) observed that all five patients with a negative FDG-PET study and two of three patients with mild FDG uptake after reinduction chemotherapy remain in complete remission. On the other hand, of the eight patients with moderate or high FDG uptake, seven relapsed and one died early of treatment-related complications. In PET-negative patients, the overall and relapse-free 1-year survival were 100% compared to 55% and 18%, respectively, in PET-positive patients. A baseline study before reinduction chemotherapy provided no additional information with regard to a later relapse in comparison to a single assessment after reinduction chemotherapy. FDG-PET imaging

after reinduction chemotherapy provided additional information to established parameters for risk stratification.

Cremerius et al. (98) confirmed that chemoresistance as evidenced by metabolic response may be of major prognostic importance. FDG-PET imaging was performed at baseline, after three cycles of induction therapy, before high-dose chemotherapy (i.e., after five to seven cycles of induction therapy), and after high-dose chemotherapy. Partial metabolic response was defined as a greater than 25% decrease of SUV between successive FDG-PET studies. Twenty-four patients suffering from NHL were investigated, but the assessment of response from sequential PET scans using SUV was available in only 22 patients. Six of 7 patients who did not achieve at least a partial metabolic response after complete induction therapy developed lymphoma progression, whereas 10 of 15 patients with complete metabolic response or partial metabolic response remained in continuous remission. Three of 5 patients in whom residual disease was undetected by PET imaging suffered from follicular lymphoma. As reported after conventional first-line chemotherapy, the sensitivity of FDG-PET imaging to predict relapse was higher after induction chemotherapy than at the end of all treatment (after high-dose chemotherapy and autologous stem cell transplantation). The authors conclude that sequential PET scans after three cycles and before high-dose chemotherapy were indicated for response evaluation. A similar prognostic value was not obtained during the early phase of induction therapy, when all patients except 1 obtained a partial metabolic response with PET imaging. However, when the analysis was restricted to FDG-PET findings before high-dose chemotherapy, the results were similar: 5 of 15 patients with a negative PET study relapsed compared to 7 of 9 patients with a positive PET study (residual FDG uptake). Unfortunately, this easier and less-expensive approach (only one PET study without SUV calculation) was not discussed by the authors.

Filmont et al. (99) showed that FDG-PET imaging performed 2 to 5 weeks after the initiation of salvage chemotherapy can be used to predict outcome of patients suffering from lymphoma with high accuracy. Seven of 8 patients with a negative FDG-PET study remained in complete remission and 11 of 12 patients with a positive FDG-PET study relapsed after autologous stem cell transplantation. As in the previous studies, a PET study performed after autologous stem cell transplantation had no additional prognostic value.

Spaepen et al. (100) confirmed the important prognostic role of FDG-PET imaging in the pretransplant evaluation of patients with lymphoma. Twenty-five of 30 patients with a negative FDG-PET study before high-dose chemotherapy and autologous stem cell transplantation remained in complete remission compared to only 4 of 30 patients with a positive FDG-PET study.

The remaining question for the future is how to use the information provided by FDG-PET imaging. High-dose chemotherapy followed by autologous stem cell transplantation may be considered the only available therapeutic option. Even some patients with residual FDG uptake will have a good outcome after standard treatment with high-dose chemotherapy followed by autologous stem cell transplantation. On the other hand, it has not been shown that new approaches such as nonmyeloablative allogeneic transplants or other immunologic approaches are useful for the treatment of lymphomas that are refractory to chemotherapy.

Evaluation After Completion of Chemotherapy and/or Radiotherapy

Accurate documentation of residual disease after treatment is crucial for patient management. Even if complete disappearance of all clinical and laboratory evidence of disease is observed, some patients will present a residual mass after treatment. Morphologic imaging techniques such as MRI or CT cannot reliably differentiate between residual masses that essentially represent fibrosis, and thus do not require further treatment, and those containing viable tumor cells, which need salvage treatment. Residual masses are most frequently observed in patients with aggressive NHL who have a large tumor mass at diagnosis and in patients with nodular sclerosing HD. Two-thirds of patients with HD present with a residual mass

after completion of therapy, whereas only about 20% of patients with HD relapse. Residual masses are present in 25% to 40% of patients with aggressive NHL whereas only 25% of these patients will relapse.

FDG-PET imaging provides an accurate method for detection of residual tumor and can guide biopsies or help the planning of further treatment (69, 71–73, 78, 80, 89, 102–106, 123) (Table 9.2). A high accuracy and negative predictive value (NPV) of 80% to 100% was reported by most authors. This value depends not only on the sensitivity and resolution of the PET system and histologic subtype of lymphoma but also on the length of follow-up. We suggest reporting 6-month and 1-year PFS to allow comparison between studies. In contrast, there was a high variability in the PPV, ranging from 25% to 100%.

Mixed Population (Hodgkin's Disease plus Non-Hodgkin's Lymphoma)

de Wit et al. (78) studied 34 patients (17 HD, 17 NHL) after the end of chemotherapy. They observed 5 false-positive FDG-PET studies in 17 patients with abnormal FDG uptake. There were 3 false-positive results within and 2 false-positive results outside of residual masses after completed therapy. None of the 17 patients with a negative FDG-PET study relapsed, but the follow-up was short (median follow-up, 63 weeks).

Table 9.2. Predictive value of whole-body FDG-PET imaging for post-treatment evaluation.

Authors	Median follow-up (months)	Positive predictive value	Negative predictive value
Mixed population:			
de Wit et al. (78)	14	70% (12/17)	100% (17/17)
Jerusalem et al. (69)	23	100% (6/6)	83% (40/48)
Zinzani et al. (107)	20	100% (13/13)	97% (30/31)
Bangerter et al. (72)	31	56% (5/9)	93% (25/27)
Mikhaeel et al. (102)	38	89% (8/9)	91% (21/23)
Zinzani et al. (123)	12	87% (14/16)	100% (59/59)
Overall		83% (58/70)	94% (192/205)
HD:			
de Wit et al. (80)	26	67% (10/15)	100% (18/18)
Dittmann et al. (71)	6	87% (7/8)	94% (17/18)
Spaepen et al. (103)	32	100% (5/5)	91% (50/55)
Weihrauch et al. (73)	28	60% (6/10)	84% (16/19)
Guay et al. (104)	16	92% (11/12)	92% (33/36)
Friedberg et al. (105)	24	50% (4/8)	96% (23/24)
Overall		74% (43/58)	92% (157/170)
NHL:			
Spaepen et al. (106)	21	100% (26/26)	84% (56/67)
Mikhaeel et al. (89)	30	100% (9/9)	83% (30/36)
Overall		100% (35/35)	83% (86/103)
All studies:		83% (136/163)	91% (435/478)

Source: Adapted from Jerusalem et al. (121, 122).

Cremerius et al. (86) evaluated the use of FDG-PET imaging for assessment of residual disease in 27 patients (8 HD, 5 low-grade NHL, 12 high-grade NHL, 2 unclassified) at 7 days to 6 months (median, 54 days) after the last course of chemotherapy. They found that FDG-PET imaging had a significantly higher specificity (92% versus 17%), PPV (94% versus 60%), and accuracy (96% versus 63%) than CT. The sensitivity and the NPV were 100% with both imaging techniques.

Jerusalem et al. (69) confirmed the higher diagnostic and prognostic value of FDG-PET compared to CT imaging in a larger patient population (19 HD, 35 aggressive NHL). The 1-year PFS and OS were, respectively, 0% and 50% ± 20% in patients with a positive PET study compared with 86% ± 5% and 92% ± 4% in patients with a negative PET study. This study also showed that FDG-PET imaging can predict early progression but cannot exclude the presence of minimal residual disease, possibly leading to a later relapse.

Zinzani et al. (107) studied 44 patients with lymphoma (13 HD, 31 aggressive NHL) who had presented with abdominal involvement (41% with bulky mass) at the end of chemotherapy and/or radiotherapy. They found that FDG-PET imaging was the modality of choice for differentiating early recurrence or residual disease from fibrosis.

In a study of 72 patients (41 NHL, 29 HD, 2 unclassified), Cremerius et al. (108) found a better predictive value for FDG-PET than for CT imaging and serum LDH measurement. The sensitivity, specificity, and overall accuracy of FDG-PET imaging for the detection of residual disease were 88%, 83%, and 85%, respectively. The values for CT were 84%, 31%, and 54%, respectively and for serum LDH measurement 50%, 92%, and 73%.

Finally, in a study of 93 patients (44 HD, 49 NHL), Mikosch et al. (70) showed that FDG-PET imaging had a comparable sensitivity but a higher specificity and accuracy than CT and ultrasound. They found that the sensitivity, specificity, and accuracy of FDG-PET imaging were 91%, 81%, and 85%, respectively. However, FDG-PET imaging had a high accuracy when interpreted with appropriate clinical data, which can only be acquired through close cooperation with the referring clinicians. They suggested that, when FDG-PET imaging was equivocal regarding detection of residual viable tumor, repeated imaging at regular time intervals may be helpful.

Hodgkin's Disease

de Wit et al. (80) performed 50 concurrent CT and FDG-PET studies in 33 patients with HD. Seventeen studies were performed after completion of chemotherapy and 33 after completion of all treatments (chemotherapy and radiotherapy). Only 10 of 22 positive FDG-PET studies were true positive; these patients relapsed during follow-up after a median of 52 weeks. Among the 28 patients with a negative FDG-PET studies, 1 patient relapsed 3 years later.

The accuracy for predicting DFS for the whole group of patients was 74% for FDG-PET and 32% for CT imaging. If the analysis was restricted to studies performed after completion of all treatment, the accuracy was 85% for FDG-PET and 39% for CT imaging. The limited performance of FDG-PET imaging when the examination was performed before radiotherapy may be explained either by residual lymphoma that is eradicated by radiotherapy or by false-positive or false-negative findings related to a short time interval between the last chemotherapy and FDG-PET imaging. The authors pointed out that the rate of false-positive results after completed therapy is too high to justify intensive treatment such as following with autologous stem cell transplantation on the basis of FDG-PET results without histologic verification.

Dittmann et al. (71) compared the results obtained by CT and FDG-PET imaging in the assessment of residual masses (26 comparisons in 24 patients) and in case of suspected relapse (21 comparisons in 20 patients). In the evaluation of residual masses, FDG-PET imaging had a sensitivity of 87.5% (7 true-positive, 1 false-negative) and a specificity of 94.4% (17 true-negative, 1 false-positive) compared to 25% and 56%, respectively, for CT. FDG-PET imaging offered no additional information compared to CT in patients with suspected relapse (18 true-positive, 3 false-positive with both methods). Semiquantification of FDG uptake with SUV did not improve the discriminations of viable lymphoma from necrosis as compared to visual assessment alone. Measurement of SUV was also not useful to improve the accuracy of PET imaging for the diagnosis of relapse.

Hueltenschmidt et al. (49) found similar encouraging results for FDG-PET imaging in the restaging after treatment in 47 patients. The accuracy was 91% for FDG-PET imaging and 62% for conventional imaging methods. In contrast to the previous study, they also observed a better accuracy for FDG-PET compared with CT imaging in the confirmation of a suspected recurrence (83% and 56%, respectively).

In contrast to most other studies, Spaepen et al. (103) had no problems with false-positive FDG-PET findings. All 5 patients with a positive FDG-PET study relapsed, but they reported 5 relapses in 55 patients with a negative FDG-PET study, probably because of the long follow-up observation time. The 2-year actuarial PFS rate for FDG-PET-negative patients was 91% compared with 0% for FDG-PET-positive patients. The main weakness of this study was that 40 of 60 patients received additional radiotherapy after FDG-PET imaging. Furthermore, the 6 patients with a positive PFS after chemotherapy but a negative PET study after additional radiotherapy were classified as PET-negative patients.

Weihrauch et al. (73) evaluated 28 patients with a residual mediastinal mass of at least 2 cm after initial therapy or salvage chemotherapy. A negative FDG-PET study indicated a low likelihood of relapse before 1 year, if ever (NPV, 95%). On the other hand, a positive FDG-PET result indicated a significantly higher risk of relapse, justifying further diagnostic procedures and a closer follow-up (6 of 10 patients with a positive FDG-PET study relapsed).

Non Hodgkin's Lymphoma

Spaepen et al. (106) confirmed the high prognostic value of FDG-PET imaging after first-line chemotherapy in 93 patients with NHL. Among the 67 patients with a negative PET study, 56 remained in complete remission. All the 26 PET-positive patients relapsed. In 12 patients, confirmation of residual disease was not obtained because the FDG-PET uptake was localized in residual masses on conventional staging procedures and the patients received immediate second-line treatment. In the remaining 14 patients, relapse was proven either by biopsy ($n = 8$) or by progressive disease on CT or MRI. The 2-year PFS rate was 85% for FDG-PET-negative patients compared to 4% for FDG-PET-positive patients.

Jerusalem et al. (87) observed 11 true-positive, 2 false-positive, 44 true-negative, and 12 false-negative FDG-PET studies in 69 patients suffering from NHL. Seven of the 12 relapses in patients with a negative PET study occurred within 1 year after PET imaging and 5 of these 7 relapses were observed within 2 months after PET imaging. These results indicate that a negative FDG-PET study cannot exclude an early relapse in aggressive NHL.

Differences Between Hodgkin's Disease and Non-Hodgkin's Lymphoma

Some investigators suggested a different role for FDG-PET imaging in HD and NHL. However, Jerusalem et al. (87) found a comparable accuracy 90% (35 of 39) in HD and 80% (55 of 69) in NHL. This result is not surprising because FDG-PET imaging is able to identify correctly lymphomatous involvement before treatment in both HD and aggressive NHL. The differences observed according to histologic subtypes reflect the better prognosis for HD and the more-rapid tumor growth of aggressive NHL. False-positive PET findings have been reported in both HD and NHL. However, as relapse is a rare event in HD, the impact of a false-positive PET study on the PPV is much more important in HD than in NHL. On the other hand, because PET imaging cannot exclude minimal residual disease, the NPV of PET imaging is lower in NHL than in HD. As shown by Jerusalem et al. (87), these relapses can occur early after completing the treatment in NHL, even if the FDG-PET study is negative. In contrast, a relapse within 1 year after a negative FDG-PET study is uncommon in HD. The most important information from a clinical point of view is the sensitivity for detection of residual disease. Unfortunately, this sensitivity was low in both HD and NHL when patients with known residual disease were excluded from analysis. Jerusalem et al. (87) reported a sensitivity of 33% (1 in 3) in HD and 48% (11 in 23) in NHL.

Evaluation of Response After Radioimmunotherapy

Radioimmunotherapy (RIT) is a new form of treatment using both immunologic effects and radiation damage as methods to kill tumor cells. Torizuka et al. (109) showed in a small number of patients that FDG-PET findings at 1 to 2 months after ^{131}I-anti-B1 (CD20) RIT correlated well with the ultimate long-term response of NHL to RIT. In contrast, FDG-PET findings at 5 to 7 days after RIT may fail to reliably assess the long-term therapeutic effect. Further studies are clearly warranted before using FDG-PET imaging routinely for monitoring response to RIT.

Routine Follow-Up of Asymptomatic Patients

Relapse of HD or aggressive NHL is usually identified as a result of evaluation of symptoms rather than by routine screening of asymptomatic patients (110, 111). The early diagnosis of relapse allows more-rapid administration of salvage therapy, possibly improving the outcome. No published data are available regarding the value of PET imaging for the detection of preclinical relapse in the follow-up of patients suffering from aggressive NHL. In our experience, among the patients with a negative PET study at the end of treatment evaluation, early relapses were observed in several patients suffering from NHL (87). The sensitivity to detect preclinical relapses is probably low because the time interval between detectability by FDG-PET imaging and clinical symptoms is very short. Consequently, there is probably no role for PET (as for other conventional imaging techniques) in the routine follow-up of asymptomatic patients suffering from aggressive NHL. Our group reported results in 36 patients with HD who underwent PET imaging every 4 to 6 months for 2 to 3 years after the end of chemotherapy and/or radiotherapy (112). All 5 patients presenting residual tumor ($n = 1$) or relapse ($n = 4$) were correctly identified early (before 9 months) before confirmation by biopsy ($n = 4$) or by conventional imaging techniques ($n = 1$). However, FDG-PET imaging showed transient focal accumulations of tracer in 6 patients in whom subsequent examinations were negative. Thymic uptake was observed in 11 of 36 patients.

Further studies are required to evaluate the use of PET imaging for early detection of relapse in asymptomatic patients and the impact of such detection on outcome in HD.

Use of FDG-PET Imaging in Childhood Lymphoma

In young children, PET imaging presents a unique challenge because it requires not only fasting but also that the patient lies quietly for a prolonged period of time. Sedation of the patient (adding expense and risk) is sometimes necessary. PET imaging appears to have a good sensitivity for staging and evaluation of response to therapy in children and adolescents but there are some concerns about the PPV after treatment (113–115). Clinicians should be aware of FDG-PET findings in specific clinical conditions (116). FDG-avidity conforming to the normal distribution of the thymus gland in a child or teenager in the first year after completion of therapy may be monitored expectantly, if the clinical evaluation is otherwise negative for signs or symptoms of recurrence, because thymic hyperplasia accumulating FDG is a common finding appearing in the 6 to 12 months after treatment.

Reactive hyperplasia with progressive transformation of germinal centers is a pathologic entity characterized by chronic lymphadenopathy that accumulates FDG but requires only expectative monitoring. Progressive transformation of germinal centers predispose to the development of lymphocyte predominant HD and may also develop after treatment of lymphocyte predominant HD. Wickmann et al. (114) reported results of a retrospective multicenter study (27 centers) in 106 patients suffering from HD and participating to the GPOH-HD 95 study. Regarding the primary staging, there was a concordance of 92% between the findings on PET and on CT/MRI/ultrasound examinations per location, but in more than 50% of patients, a discrepancy occurred in at least 1 of the 9 investigated locations. In the follow-up studies, the results were less encouraging. In asymptomatic patients, the NPV of PET studies performed 2 to 26 weeks after treatment was 94% but the PPV was only 25%. In patients with suspected relapse at any time during follow-up, the PPV increased to 76% and the NPV decreased to 83%. In the GPOH-HD 95 trial, the indication for radiotherapy was limited to patients who did not show a complete remission after chemotherapy as determined radiographically. In the future protocol, the indication for radiotherapy in patients with early-stage HD should be further refined by using FDG-PET imaging for evaluating the response to chemotherapy (117). Furthermore, in patients at an advanced stage of the disease, it should be determined if sequential FDG-PET imaging during chemotherapy can separate patients into subgroups with an excellent or a poor prognosis.

In conclusion, FDG-PET imaging may play an important role in staging, assessment of response, planning radiation treatment fields, and routine follow-up after treatment in children and adolescents suffering from HD or NHL. However, further prospective studies are warranted to increase our knowledge about the role of PET imaging in staging and restaging pediatric patients, especially to indicate the value of PET imaging in the end of treatment assessment of patients presenting with a residual mass.

Advances in Technology: FDG-PET/CT Imaging and New Radiotracers

The quality of the images has rapidly improved by advances in technology over the past years, increasing the sensitivity of the method. Attenuation correction, most useful for the visualization of deep-seated lesions, in particular in the abdomen, is now routinely used. More recently, it became possible to perform sequential PET and CT studies on a single device in a same-day session.

Freudenberg et al. (118) performed FDG-PET/CT imaging for restaging in 27 patients with lymphoma. They observed some advantages with FDG-PET/CT imaging compared to side-by-side interpretation of FDG-PET and CT images. Integrated PET/CT imaging with fusion images not only improves the quality of the images but provides excellent anatomic maps to help with localization of the FDG-avid foci.

Schaefer et al. (119) also showed that FDG-PET/CT imaging performed with nonenhanced CT is more sensitive and specific than contrast-enhanced CT for the evaluation of lymph node and organ involvement in 60 patients with HD or high-grade NHL. It is highly probable that all PET devices will be integrated PET/CT devices in the near future.

New radiotracers such as ^{18}F-fluorothymidine (^{18}F-FLT) may improve the specificity of PET imaging. In a pilot study of 11 patients with both indolent and aggressive lymphoma, ^{18}F-FLT was suitable and comparable to FDG in the ability to detect malignant lesions by PET imaging (120).

Conclusions and Perspectives

Table 9.3 summarizes the current clinical indications of FDG-PET imaging in the evaluation of patients with lymphoma. FDG-PET imaging is complementary to but does not replace conventional staging techniques for the initial staging in HD and NHL. The stage of disease is modified based on PET imaging in some patients leading to a change of therapy. In the future, integrated FDG-PET/CT imaging may allow a better definition of the radiation

Table 9.3. Clinical indications.

Confirmed indications:

- Initial staging of patients with lymphoma
- Determination of chemosensitivity before high-dose chemotherapy followed by autologous stem cell transplantation
- Evaluation of treatment after completion of chemotherapy and/or radiotherapy

Perspectives:

- Evaluation early during the treatment after one or a few cycles of chemotherapy allowing modification of treatment strategy according to PET findings fields. FDG-PET is the best noninvasive imaging technique for evaluation of response during or after treatment. However, positive findings on PET imaging do not necessarily represent residual disease because infectious or inflammatory lesions can accumulate the radiotracer. On the other hand, a negative PET study cannot exclude minimal residual disease. PET imaging allows determination of the chemosensitivity of lymphoma before high-dose chemotherapy followed by autologous stem cell transplantation. In early treatment evaluation, persistent tumoral FDG uptake after a few cycles of chemotherapy seems to predict treatment failure. However, further studies are warranted before modifying routinely the treatment strategy according to PET findings.

References

1. Front D, Israel O. The role of Ga-67 scintigraphy in evaluating the results of therapy of lymphoma patients. Semin Nucl Med 1995;25:60–71.

2. Kostakoglu L, Leonard JP, Kuji I, et al. Comparison of fluorine-18 fluorodeoxyglucose positron emission tomography and Ga-67 scintigraphy in evaluation of lymphoma. Cancer (Phila) 2002;94:879–888.

3. Sasaki M, Kuwabara Y, Koga H, et al. Clinical impact of whole body FDG-PET on the staging and therapeutic decision making for malignant lymphoma. Ann Nucl Med. 2002;16:337–345.

4. Shen YY, Kao A, Yen RF. Comparison of ^{18}F-fluoro-2-deoxyglucose positron emission tomography and gallium-67 citrate scintigraphy for detecting malignant lymphoma. Oncol Rep 2002;9:321–325.

5. Wirth A, Seymour JF, Hicks RJ, et al. Fluorine-18 fluorodeoxyglucose positron emission tomography, gallium-67 scintigraphy, and conventional staging for Hodgkin's disease and non- Hodgkin's lymphoma. Am J Med 2002;112:262–268.

6. Van Den Bossche B, Lambert B, De Winter F, et al. [18]FDG PET versus high-dose 67Ga scintigraphy for restaging and treatment follow-up of lymphoma patients. Nucl Med Commun 2002; 23:1079–1083.

7. Bar-Shalom R, Yefremov N, Haim N, et al. Camera-based FDG PET and ^{67}Ga SPECT in evaluation of lymphoma: comparative study. Radiology 2003;227:353–360.

8. Zijlstra JM, Hoekstra OS, Raijmakers PG, et al. [18]FDG positron emission tomography versus ^{67}Ga scintigraphy as prognostic test during chemotherapy for non-Hodgkin's lymphoma. Br J Haematol 2003;123:454–462.

9. Jerusalem G, Rigo P. PET imaging in lymphoma. In: Valk PE, Bailey D, Townsend D, Maisey MN, editors. Positron Emission Tomography. Basic Science and Clinical Practice. London: Springer-Verlag, 2003:547–557.

10. Jerusalem G, Rigo P, Israel O. PET and PET/CT of lymphoma. In: von Schulthess GK, editor. Clinical Molecular Anatomic Imaging. Philadelphia: Lippincott Williams & Wilkins, 2003:350–361.

11. Cancer Facts and Figures 2003. Washington, DC: American Cancer Society, 2003. http://www.cancer.org/downloads/STT/CAFF 2003 Secured.pdf. 1-11-2004.

12. Magrath I. Molecular basis of lymphomagenesis. Cancer Res 1992;52:5529s–5540s.

13. Shipp M, Harrington D, Anderson J. A predictive model for aggressive non-Hodgkin's lymphoma. The International Non-Hodgkin's Lymphoma Prognostic Factors Project. N Engl J Med 1993; 329:987–994.

14. Kuppers R, Rajewsky K, Zhao M, et al. Hodgkin disease: Hodgkin and Reed-Sternberg cells picked from histological sections show clonal immunoglobulin gene rearrangements and appear to be derived from B cells at various stages of development. Proc Natl Acad Sci U S A 1994;91:10962–10966.

15. Vockerodt M, Soares M, Kanzler H, et al. Detection of clonal Hodgkin and Reed-Sternberg cells with identical somatically mutated and rearranged VH genes in different biopsies in relapsed Hodgkin's disease. Blood 1998;92:2899–2907.

16. Lister TA, Crowther D, Sutcliffe SB, et al. Report of a committee convened to discuss the evaluation and staging of patients with Hodgkin's disease: Cotswolds meeting. J Clin Oncol 1989;7:1630–1636.

17. Segall GM. FDG PET imaging in patients with lymphoma: a clinical perspective. J Nucl Med 2001;42:609–610.

18. Newman JS, Francis IR, Kaminski MS, et al. Imaging of lymphoma with PET with 2-[F-18]-fluoro-2-deoxy-D-glucose: correlation with CT. Radiology 1994;190:111–116.

19. Hoh CK, Glaspy J, Rosen P, et al. Whole-body FDG-PET imaging for staging of Hodgkin's disease and lymphoma. J Nucl Med 1997;38:343–348.

20. Moog F, Bangerter M, Diederichs CG, et al. Lymphoma: role of whole-body 2-deoxy-2-[F-18]fluoro-D-glucose FDG) PET in nodal staging. Radiology 1997;203:795–800.

21. Jerusalem G, Warland V, Najjar F, et al. Whole-body ^{18}F-FDG PET for the evaluation of patients with Hodgkin's disease and non-Hodgkin's lymphoma. Nucl Med Commun 1999;20:13–20.

22. Moog F, Bangerter M, Diederichs CG, et al. Extranodal malignant lymphoma: detection with FDG PET versus CT. Radiology 1998;206:475–481.

23. Munker R, Stengel A, Stabler A, et al. Diagnostic accuracy of ultrasound and computed tomography in the staging of Hodgkin's disease. Verification by laparotomy in 100 cases. Cancer (Phila) 1995;76:1460–1466.

24. Castellino RA, Hoppe RT, Blank N, et al. Computed tomography, lymphography, and staging laparotomy: correlations in initial staging of Hodgkin disease. Am J Roentgenol 1984;143:37–41.

25. Mansfield CM, Fabian C, Jones S, et al. Comparison of lymphangiography and computed tomography scanning in evaluating abdominal disease in stages III and IV Hodgkin's disease. A Southwest Oncology Group study. Cancer (Phila) 1990; 66:2295–2299.

26. Rini JN, Leonidas JC, Tomas MB, et al. ^{18}F-FDG PET versus CT for evaluating the spleen during initial staging of lymphoma. J Nucl Med 2003;44:1072–1074.

27. Kostakoglu L, Goldsmith SJ. Fluorine-18 fluorodeoxyglucose positron emission tomography in the staging and follow-up of lymphoma: is it time to shift gears? Eur J Nucl Med 2000;27:1564–1578.

28. Chiang SB, Rebenstock A, Guan L, et al. Diffuse bone marrow involvement of Hodgkin lymphoma mimics hematopoietic cytokine-mediated FDG uptake on FDG PET imaging. Clin Nucl Med 2003;28:674–676.

29. Carr R, Barrington SF, Madan B, et al. Detection of lymphoma in bone marrow by whole-body positron emission tomography. Blood 1998;91:3340–3346.

30. Moog F, Bangerter M, Kotzerke J, et al. 18-F-fluorodeoxyglucose-positron emission tomography as a new approach to detect lymphomatous bone marrow. J Clin Oncol 1998;16:603–609.

31. Jerusalem G, Silvestre RM, Beguin Y, et al. Does ^{18}F-FDG PET replace bone marrow biopsy (BMB) in patients with Hodgkin's disease (HD) or non-Hodgkin's lymphoma (NHL). Blood 2002;100:768a (abstract).

32. Elstrom R, Guan L, Baker G, et al. Utility of FDG-PET scanning in lymphoma by WHO classification. Blood 2003;101:3875–3876.

33. Moog F, Kotzerke J, Reske SN. FDG PET can replace bone scintigraphy in primary staging of malignant lymphoma. J Nucl Med 1999;40:1407–1413.

34. Rodriguez M, Ahlstrom H, Sundin A, et al. [^{18}F] FDG PET in gastric non-Hodgkin's lymphoma. Acta Oncol 1997;36:577–584.

35. Hoffmann M, Kletter K, Diemling M, et al. Positron emission tomography with fluorine-18-2-fluoro-2-deoxy-D-glucose (F-18-

FDG) does not visualize extranodal B-cell lymphoma of the mucosa-associated lymphoid tissue (MALT)-type. Ann Oncol 1999;10:1185–1189.

36. Roelcke U, Leenders KL. Positron emission tomography in patients with primary CNS lymphomas. J Neurooncol 1999;43:231–236.

37. Kuwabara Y, Ichiya Y, Otsuka M, et al. High [18F]FDG uptake in primary cerebral lymphoma: a PET study. J Comput Assist Tomogr 1988;12:47–48.

38. Rosenfeld SS, Hoffman JM, Coleman RE, et al. Studies of primary central nervous system lymphoma with fluorine-18-fluorodeoxyglucose positron emission tomography. J Nucl Med 1992;33:532–536.

39. Hoffman JM, Waskin HA, Schifter T, et al. FDG-PET in differentiating lymphoma from nonmalignant central nervous system lesions in patients with AIDS. J Nucl Med 1993;34:567–575.

40. Pierce MA, Johnson MD, Maciunas RJ, et al. Evaluating contrast-enhancing brain lesions in patients with AIDS by using positron emission tomography. Ann Intern Med 1995;123:594–598.

41. Heald AE, Hoffman JM, Bartlett JA, et al. Differentiation of central nervous system lesions in AIDS patients using positron emission tomography (PET). Int J STD AIDS 1996;7:337–346.

42. Buchmann I, Reinhardt M, Elsner K, et al. 2-(Fluorine-18)fluoro-2-deoxy-D-glucose positron emission tomography in the detection and staging of malignant lymphoma. A bicenter trial. Cancer (Phila) 2001;91:889–899.

43. Delbeke D, Martin WH, Morgan DS, et al. 2-Deoxy-2-[F-18]fluoro-D-glucose imaging with positron emission tomography for initial staging of Hodgkin's disease and lymphoma. Mol Imaging Biol 2002;4:105–114.

44. Klose T, Leidl R, Buchmann I, et al. Primary staging of lymphomas: cost-effectiveness of FDG-PET versus computed tomography. Eur J Nucl Med 2000;27:1457–1464.

45. Bangerter M, Moog F, Buchmann I, et al. Whole-body 2-[18F]-fluoro-2-deoxy-D-glucose positron emission tomography (FDG-PET) for accurate staging of Hodgkin's disease. Ann Oncol 1998;9:1117–1122.

46. Weidmann E, Baican B, Hertel A, et al. Positron emission tomography (PET) for staging and evaluation of response to treatment in patients with Hodgkin's disease. Leuk Lymphoma 1999;34:545–551.

47. Partridge S, Timothy AR, O'Doherty MJ. 2-Fluorine-18-fluoro-2-deoxy-D-glucose positron emission tomography in the pretreatment staging of Hodgkin's disease : influence on patient management in a single institution. Ann Oncol 2000;11:1273–1279.

48. Jerusalem G, Beguin Y, Fassotte MF, et al. Whole-body positron emission tomography using 18F-fluorodeoxyglucose compared to standard procedures for staging patients with Hodgkin's disease. Haematologica 2001;86:266–273.

49. Hueltenschmidt B, Sautter-Bihl ML, Lang O, et al. Whole body positron emission tomography in the treatment of Hodgkin disease. Cancer (Phila) 2001;91:302–310.

50. Menzel C, Dobert N, Mitrou P, et al. Positron emission tomography for the staging of Hodgkin's lymphoma: increasing the body of evidence in favor of the method. Acta Oncol 2002;41:430–436.

51. Weihrauch MR, Re D, Bischoff S, et al. Whole-body positron emission tomography using 18F-fluorodeoxyglucose for initial staging of patients with Hodgkin's disease. Ann Hematol 2002;81:20–25.

52. Naumann R, Beuthien-Baumann B, Reiss A, et al. Substantial impact of FDG PET imaging on the therapy decision in patients with early-stage Hodgkin's lymphoma. Br J Cancer 2004;90:620–625.

53. Jerusalem G, Beguin Y. Does positron emission tomography have a role in routine clinical practice in patients with Hodgkin's disease? Clin Lymphoma 2002;3:125–126.

54. Lapela M, Leskinen S, Minn HR, et al. Increased glucose metabolism in untreated non-Hodgkin's lymphoma: a study with positron emission tomography and fluorine-18-fluorodeoxyglucose. Blood 1995;86:3522–3527.

55. Rodriguez M, Rehn S, Ahlstrom H, et al. Predicting malignancy grade with PET in non-Hodgkin's lymphoma. J Nucl Med 1995;36:1790–1796.

56. Okada J, Yoshikawa K, Itami M, et al. Positron emission tomography using fluorine-18-fluorodeoxyglucose in malignant lymphoma: a comparison with proliferative activity. J Nucl Med 1992;33:325–329.

57. Leskinen-Kallio S, Ruotsalainen U, Nagren K, et al. Uptake of carbon-11-methionine and fluorodeoxyglucose in non-Hodgkin's lymphoma: a PET study. J Nucl Med 1991;32:1211–1218.

58. Koga H, Matsuo Y, Sasaki M, et al. Differential FDG accumulation associated with GLUT-1 expression in a patient with lymphoma. Ann Nucl Med 2003;17:327–331.

59. Okada J, Yoshikawa K, Imazeki K, et al. The use of FDG-PET in the detection and management of malignant lymphoma: correlation of uptake with prognosis. J Nucl Med 1991;32:686–691.

60. Najjar F, Jerusalem G, Paulus P, et al. Intérêt clinique de la tomographie à émission de positons dans la détection et le bilan d'extension des lymphomes non Hodgkiniens de malignité intermédiaire ou élevée. Médecine Nucléaire Imagerie Fonctionnelle et Métabolique 1999;23:281–290.

61. Jerusalem G, Beguin Y, Najjar F, et al. Positron emission tomography (PET) with 18F-fluorodeoxyglucose (18F-FDG) for the staging of low grade non-Hodgkin's lymphoma (NHL). Ann Oncol 2001; 12:825–830.

62. Blum RH, Seymour JF, Wirth A, et al. Frequent impact of [18F]fluorodeoxyglucose positron emission tomography on the staging and management of patients with indolent non-Hodgkin's lymphoma. Clin Lymphoma 2003;4:43–49.

63. Jerusalem G, Beguin Y. Positron emission tomography in non-Hodgkin's lymphoma (NHL): relationship between tracer uptake and pathological findings, including preliminary experience in the staging of low-grade NHL. Clin Lymphoma 2002;3:56–61.

64. Hoffmann M, Kletter K, Becerer A, et al. 18F-fluorodeoxyglucose positron emission tomography (18F-FDG-PET) for staging and follow-up of marginal zone B-cell lymphoma. Oncology 2003; 64:336–340.

65. Hoffmann M, Chott A, Puspok A, et al. 18F-Fluorodeoxyglucose positron emission tomography (18F-FDG-PET) does not visualize follicular lymphoma of the duodenum. Ann. Hematol 2004;83:276–278.

66. Hoffmann M, Vogelsang H, Kletter K, et al. 18F-Fluoro-deoxyglucose positron emission tomography (18F-FDG-PET) for assessment of enteropathy-type T cell lymphoma. Gut 2003;52:347–351.

67. Armitage JO, Weisenburger DD, Hutchins M, et al. Chemotherapy for diffuse large-cell lymphoma: rapidly responding patients have more durable remissions. J Clin Oncol 1986;4:160–164.

68. Coiffier B. How to interpret the radiological abnormalities that persist after treatment in non-Hodgkin's lymphoma patients? Ann Oncol 1999;10:1141–1143.

69. Jerusalem G, Beguin Y, Fassotte MF, et al. Whole-body positron emission tomography using 18F-fluorodeoxyglucose for posttreatment evaluation in Hodgkin's disease and non-Hodgkin's lymphoma has higher diagnostic and prognostic value than classical computed tomography scan imaging. Blood 1999;94:429–433.

70. Mikosch P, Gallowitsch HJ, Zinke-Cerwenka W, et al. Accuracy of whole-body 18F-FDG-PET for restaging malignant lymphoma. Acta Med Aust 2003;30:41–47.

71. Dittmann H, Sokler M, Kollmannsberger C, et al. Comparison of 18FDG-PET with CT scans in the evaluation of patients with residual and recurrent Hodgkin's lymphoma. Oncol Rep 2001; 8:1393–1399.

72. Bangerter M, Moog F, Griesshammer M, et al. Role of whole body FDG-PET imaging in predicting relapse of malignant lymphoma in patients with residual masses after treatment. Radiography 1999;5:155–163.

73. Weihrauch MR, Re D, Scheidhauer K, et al. Thoracic positron emission tomography using (18)F-fluorodeoxyglucose for the evaluation of residual mediastinal Hodgkin disease. Blood 2001;98:2930–2934.

74. Glatz S, Kotzerke J, Moog F, et al. Vortauschung eines mediastinalen Non-Hodgkin-Lymphomrezidivs durch diffuse Thymushyperplasie im 18F-FDG-PET. Fortschr Rontgenstr 1996;165:309–310.

75. Weinblatt ME, Zanzi I, Belakhlef A, et al. False-positive FDG-PET imaging of the thymus of a child with Hodgkin's disease. J Nucl Med 1997;38:888–890.

76. Yoon SN, Park CH, Kim MK, et al. False-positive F-18 FDG gamma camera positron emission tomographic imaging resulting from inflammation of an anterior mediastinal mass in a patient with non-Hodgkin's lymphoma. Clin Nucl Med 2001;26:461–462.

77. Naumann R, Vaic A, Beuthien-Baumann B, et al. Prognostic value of positron emission tomography in the evaluation of post-treatment residual mass in patients with Hodgkin's disease and non-Hodgkin's lymphoma. Br J Haematol 2001;115:793–800.

78. de Wit M, Bumann D, Beyer W, et al. Whole-body positron emission tomography (PET) for diagnosis of residual mass in patients with lymphoma. Ann Oncol 1997;8(suppl 1):S57–S60.

79. Becherer A, Mitterbauer M, Jaeger U, et al. Positron emission tomography with [18F]2-fluoro-D-2-deoxyglucose (FDG-PET) predicts relapse of malignant lymphoma after high-dose therapy with stem cell transplantation. Leukemia 2002;16:260–267.

80. de Wit M, Bohuslavizki KH, Buchert R, et al. 18FDG-PET following treatment as valid predictor for disease-free survival in Hodgkin's lymphoma. Ann Oncol 2001;12:29–37.

81. Sandherr M, von Schilling C, Link T, et al. Pitfalls in imaging Hodgkin's disease with computed tomography and positron emission tomography using fluorine-18-fluorodeoxyglucose. Ann Oncol 2001;12:719–722.

82. Bomanji JB, Syed R, Brock C, et al. Challenging cases and diagnostic dilemmas: case 2. Pitfalls of positron emission tomography for assessing residual mediastinal mass after chemotherapy for Hodgkin's disease. J Clin Oncol 2002;20:3347–3349.

(83. Naumann R, Beuthien-Baumann B, Fischer R, et al. Simultaneous occurrence of Hodgkin's lymphoma and eosinophilic granuloma: a potential pitfall in PET imaging. Clin Lymphoma 2002;3:121–124.

84. Cheong KA, Rodgers NG, Kirkwood ID. Erythema nodosum associated with diffuse, large B-cell non-Hodgkin lymphoma detected by FDG PET. Clin Nucl Med 2003;28:652–654.

85. Lorenzen J, de Wit M, Buchert R, et al. Granulation tissue: pitfall in therapy control with F-18-FDG PET after chemotherapy. Nuklearmedizin 1999;38:333–336.

86. Cremerius U, Fabry U, Neuerburg J, et al. Positron emission tomography with 18F-FDG to detect residual disease after therapy for malignant lymphoma. Nucl Med Commun 1998;19:1055–1063.

87. Jerusalem G, Warland V, Beguin Y, et al. Accuracy of end of treatment 18F-FDG PET for predicting relapse in patients with Hodgkin's disease (Hd) and non-Hodgkin's lymphoma (Nhl). Proc Am Soc Clin Oncol 2003;22:572 (abstract 2299).

88. Jerusalem G, Beguin Y, Fassotte MF, et al. Persistent tumor 18F-FDG uptake after a few cycles of polychemotherapy is predictive of treatment failure in non-Hodgkin's lymphoma. Haematologica 2000;85:613–618.

89. Mikhaeel NG, Timothy AR, O'Doherty MJ, et al. 18-FDG-PET as a prognostic indicator in the treatment of aggressive non-Hodgkin's lymphoma: comparison with CT. Leuk Lymphoma 2000;39:543–553.

90. Spaepen K, Stroobants S, Dupont P, et al. Early restaging positron emission tomography with (18)F- fluorodeoxyglucose predicts outcome in patients with aggressive non- Hodgkin's lymphoma. Ann Oncol 2002;13:1356–1363.

91. Kostakoglu L, Coleman M, Leonard JP, et al. PET predicts prognosis after 1 cycle of chemotherapy in aggressive lymphoma and Hodgkin's disease. J Nucl Med 2002;43:1018–1027.

92. Torizuka T, Nakamura F, Kanno T, et al. Early therapy monitoring with FDG-PET in aggressive non-Hodgkin's lymphoma and Hodgkin's lymphoma. Eur J Nucl Med Mol Imaging 2004;31:22–28.

93. Hoekstra OS, Ossenkoppele GJ, Golding R, et al. Early treatment response in malignant lymphoma, as determined by planar fluorine-18-fluorodeoxyglucose scintigraphy. J Nucl Med 1993;34:1706–1710.

94. Dimitrakopoulou-Strauss A, Strauss LS, Goldschmidt H, et al. Evaluation of tumour metabolism and multidrug resistance in pa-

tients with treated malignant lymphomas. Eur J Nucl Med 1995;22:434–442.

95. Römer W, Hanauske AR, Ziegler S, et al. Positron emission tomography in non-Hodgkin's lymphoma: assessment of chemotherapy with fluorodeoxyglucose. Blood 1998;91:4464–4471.

96. Young H, Baum R, Cremerius U, et al. Measurement of clinical and subclinical tumour response using [18F]-fluorodeoxyglucose and positron emission tomography: review and 1999 EORTC recommendations. European Organization for Research and Treatment of Cancer (EORTC) PET Study Group. Eur J Cancer 1999; 35:1773–1782.

97. Spaepen K, Stroobants S, Dupont P, et al. [(18)F]FDG PET monitoring of tumour response to chemotherapy: does [(18)F]FDG uptake correlate with the viable tumour cell fraction? Eur J Nucl Med Mol Imaging 2003;30:682–688.

98. Cremerius U, Fabry U, Wildberger JE, et al. Pre-transplant positron emission tomography (PET) using fluorine-18-fluorodeoxyglucose (FDG) predicts outcome in patients treated with high-dose chemotherapy and autologous stem cell transplantation for non-Hodgkin's lymphoma. Bone Marrow Transplant 2002;30:103–111.

99. Filmont JE, Czernin J, Yap C, et al. Value of F-18 fluorodeoxyglucose positron emission tomography for predicting the clinical outcome of patients with aggressive lymphoma prior to and after autologous stem-cell transplantation. Chest 2003;124:608–613.

100. Spaepen K, Stroobants S, Dupont P, et al. Prognostic value of pre-transplantation positron emission tomography using fluorine 18-fluorodeoxyglucose in patients with aggressive lymphoma treated with high-dose chemotherapy and stem cell transplantation. Blood 2003;102:53–59.

101. Schot B, van Imhoff G, Pruim J, et al. Predictive value of early 18F-fluoro-deoxyglucose positron emission tomography in chemosensitive relapsed lymphoma. Br J Haematol 2003;123:282–287.

102. Mikhaeel NG, Timothy AR, Hain SF, et al. 18-FDG-PET for the assessment of residual masses on CT following treatment of lymphomas. Ann Oncol 2000;11:S147–S150.

103. Spaepen K, Stroobants S, Dupont P, et al. Can positron emission tomography with [18F]-fluorodeoxyglucose after first-line treatment distinguish Hodgkin's disease patients who need additional therapy from others in whom additional therapy would mean avoidable toxicity? Br J Haematol 2001;115:272–278.

104. Guay C, Lepine M, Verreault J, et al. Prognostic value of PET using 18F-FDG in Hodgkin's disease for posttreatment evaluation. J Nucl Med 2003;44:1225–1231.

105. Friedberg JW, Fischman A, Neuberg D, et al. FDG-PET is superior to gallium scintigraphy in staging and more sensitive in the follow-up of patients with de novo Hodgkin lymphoma: a blinded comparison. Leuk Lymphoma 2004;45:85–92.

106. Spaepen K, Stroobants S, Dupont P, et al. Prognostic value of positron emission tomography (PET) with fluorine-18 fluorodeoxyglucose ([18F]FDG) after first-line chemotherapy in non- Hodgkin's lymphoma: is [18F]FDG-PET a valid alternative to conventional diagnostic methods? J Clin Oncol 2001;19:414–419.

107. Zinzani PL, Magagnoli M, Chierichetti F, et al. The role of positron emission tomography (PET) in the management of lymphoma patients. Ann Oncol 1999;10:1181–1184.

108. Cremerius U, Fabry U, Kroll U, et al. Clinical value of FDG PET for therapy monitoring of malignant lymphoma: results of a retrospective study in 72 patients. Nuklearmedizin 1999;38:24–30.

109. Torizuka T, Zasadny KR, Kison PV, et al. Metabolic response of non-Hodgkin's lymphoma to 131I-anti-B1 radioimmunotherapy: evaluation with FDG PET. J Nucl Med 2000;41:999–1005.

110. Radford JA, Eardley A, Woodman C, et al. Follow up policy after treatment for Hodgkin's disease: too many clinic visits and routine tests? A review of hospital records. BMJ 1997;314:343–346.

111. Weeks JC, Yeap BY, Canellos GP, et al. Value of follow-up procedures in patients with large-cell lymphoma who achieve a complete remission. J Clin Oncol 1991;9:1196–1203.

112. Jerusalem G, Beguin Y, Fassotte MF,, et al. Early detection of relapse by whole-body positron emission tomography (PET) in the follow-up of patients with Hodgkin's disease (HD). Ann Oncol 2003;14:123–130.

113. Montravers F, McNamara D, Landman-Parker J, et al. [(^{18})F]FDG in childhood lymphoma: clinical utility and impact on management. Eur J Nucl Med Mol Imaging 2002;29:1155–1165.

114. Wickmann L, Luders H, Dorffel W. 18-FDG-PET-findings in children and adolescents with Hodgkin's disease: retrospective evaluation of the correlation to other imaging procedures in initial staging and to the predictive value of follow up examinations. Klin Padiatr 2003;215:146–150.

115. Depas G, De Barsy C, Jerusalem G, et al. ^{18}F-FDG PET in children with lymphomas. Eur J Nucl Med Mol Imaging 2004;32:31–38.

116. Hudson MM, Krasin MJ, Kaste SC. PET imaging in pediatric Hodgkin's lymphoma. Pediatr Radiol 2004;34:190–198.

117. Korholz D, Kluge R, Wickmann L, et al. Importance of F^{18}-fluorodeoxy-D-2-glucose positron emission tomography (FDG-PET) for staging and therapy control of Hodgkin's lymphoma in childhood and adolescence: consequences for the GPOH-HD 2003 protocol. Onkologie 2003;26:489–493.

118. Freudenberg LS, Antoch G, Schutt P, et al. FDG-PET/CT in restaging of patients with lymphoma. Eur J Nucl Med Mol Imaging 2004;31:325–329.

119. Schaefer NG, Hany TF, Taverna C, et al. Non-Hodgkin lymphoma and Hodgkin disease: coregistered FDG PET and CT at staging and restaging—do we need contrast-enhanced CT? Radiology 2004;232:823–829.

120. Wagner M, Seitz U, Buck A, et al. 3'-[18F]Fluoro-3'-deoxythymidine ([^{18}F]-FLT) as positron emission tomography tracer for imaging proliferation in a murine B-cell lymphoma model and in the human disease. Cancer Res 2003;63:2681–2687.

121. Jerusalem G, Hustinx R, Beguin Y, et al. Whole-body positron emission tomography using ^{18}F-fluorodeoxyglucose for staging and response assessment in Hodgkin's disease. In: Columbus F, editor. Progress in Hodgkin's Disease Research. Nova Science 2005: in press.

122. Jerusalem G, Hustinx R. Nuclear Medicine. In: Canellos G, Lister TA, Young B, editors. The Lymphomas, 2nd ed. New York: Elsevier, 2005: in press.

123. Zinzani PL, Fanti S, Battista G, et al. Predictive role of positron emission tomography (PET) in the outcome of lymphoma patients. Br J Cancer 2004;91:850 Columbus F, editor. Progress in Hodgkin's Disease Research. Nova Science 2005:854.

10

PET and PET/CT Imaging in Colorectal Cancer

Christiaan Schiepers and Peter E. Valk

Colorectal cancer (CRC) is a common disease and belongs, with lung, breast, and prostate cancer, to the most frequently seen neoplasms in Western countries. In the United States, CRC frequency ranks third in men and second in women. CRC is the fourth leading cause of cancer mortality because it has a better prognosis than the other common cancers (1–3). During 1990–1995, the annual percent change in incidence was –2.3% and in mortality –1.1% compared to the previous half decade (2). In 2004, a total of 146,940 new CRCs are likely to be diagnosed, and 56,730 are expected to die of their disease (3). Incidence of CRC is similar in men and women (M/F ratio = 1.004), as is the mortality rate (M/F ratio = 0.997). About 70% of the patients have resectable disease, but only two-thirds can be cured by resection. In the remaining one-third of these patients, recurrence is diagnosed in the first 2 years after resection.

A 5-year survival rate of 62% is reported in the United States compared to 41% in the European Union (EU). The lifetime risk of developing colorectal cancer in the United States appears to be 5.78% (1 in 17) in men and 5.55% (1 in 18) in women.

Advances in diagnostic imaging technology have been directed to (1) help establish the diagnosis, (2) stage the extent of disease, and (3) enhance the development of accepted treatment protocols by monitoring response to therapy. Improved patient outcome may be expected if these goals are achieved. Various imaging modalities are available for this purpose, including the anatomic (e.g., radiography, computed tomography, sonography, magnetic resonance imaging) and functional modalities (e.g., molecular imaging, radioimmuno- and receptor scintigraphy, magnetic resonance spectroscopy). Positron emission tomography (PET) is based on imaging of biochemical processes in vivo, and creates tomographic images similar to imaging modalities such as computed tomography (CT) and magnetic resonance imaging (MRI). PET is unique because it supplies an image representing the metabolic activity of the underlying molecular processes (4, 5). The basics of PET are discussed elsewhere

in this volume. The place of PET in stratifying patients with colorectal cancer and in the workup of clinical problems is discussed in this chapter. In this respect, it is appropriate to speak of correlative imaging, in which all imaging modalities have their specific contribution and are not seen as competitive modalities (6).

The declining mortality rate of CRC is related to screening programs put in place, such as the hemoccult test and colonoscopy in asymptomatic individuals above age 50. Lifestyle changes contributed in part to early detection with subsequent surgery and a decrease in mortality. Sophisticated imaging such as PET is expensive and not a realistic option for screening purposes. An interesting study from Japan (7) reported an incidence of 2.1% neoplasms found within 1 year after screening an asymptomatic group of more than 3,100 people with ^{18}F-2-deoxy-D-glucose (FDG)-PET imaging. PET was true positive in 54% of tumors, and most of the false negatives were in the genitourinary (GU) tract. Obviously, PET cannot be cost effective under these circumstances.

The cost issue of PET for patient management in CRC has been addressed (8). The main advantage of PET in oncology is that with one injection and imaging session the whole body can be imaged in a tomographic mode. Generally, the more conventional approach with multiple CT or MR scans covering head, chest, abdomen, and pelvis are more expensive than a single whole-body PET.

The role of PET in colorectal cancer is a popular topic for review articles, in part because of the high incidence and mortality of this disease. Overviews emphasizing the clinical approach have been published by Arulampalam et al. (9) from a surgical point of view, by Akhurst and Larson (10) and by Anderson and Price (11) from the oncologist's perspective, and by Rohren et al. (12) from the imaging expert point of view.

The latest technologic development concerns dual-modality imaging, by combining a multislice CT and a PET scanner into one system. In addition, to providing high-resolution CT images in all planes and projections, CT-based attenuation correction can be performed (13).

Historical Perspective

Despite the advancement of conventional diagnostic methods, such as computed tomography (CT), both cross-sectional and helical, magnetic resonance (MR) imaging, and external and endoultrasonography (US), early detection of colorectal cancer remains problematic for primaries as well as recurrences. The addition of serum tumor markers or radioimmunoscintigraphy has not significantly improved detection at an early stage, hampering curative resection. Assessment of disease extent or tumor burden is necessary for proper patient selection of surgery with curative intent, or stratification to chemotherapy and/or radiation treatment for patients with advanced disease. Most distant metastases occur in the liver or lungs, and adequate staging is necessary to exclude patients with more extensive disease (14). Long-term survival after attempted curative resection of recurrent disease is only 35%. Therefore, appropriate noninvasive staging plays a pivotal role in selecting patients who would benefit from surgery and avoiding unnecessary surgery with major morbidity in those with unresectable disease.

Serum carcinoembryonic antigen (CEA) is used to monitor for possible recurrence. This technique has a reported sensitivity of 59% and specificity of 84% but cannot determine the location of recurrence (15). Lower gastrointestinal (GI) radiography with contrast has been used for detection of local recurrence with accuracy in the vicinity of 80%, but it is only 49% sensitive and 85% specific for overall recurrence. CT has been the conventional imaging modality used to localize recurrent disease with an accuracy of 25% to 73%. CT cannot reliably distinguish postsurgical changes from local recurrence and is often equivocal (16–18). CT of the abdomen misses hepatic metastases in about 7% of patients and underestimates the number of lobes involved in one-third of patients. In addition, CT commonly misses metastases to the peritoneum, mesentery, and lymph nodes. Among the patients with negative CT, half will have nonresectable lesions at the time of exploratory laparotomy. The results for MR imaging are comparable to CT (19).

FDG (2-^{18}F-fluoro-2-deoxy-D-glucose) has been used most frequently as the tracer for metabolic imaging (20). FDG is able to measure changes in glucose utilization, which is enhanced in cancer. After the introduction of PET for whole-body imaging (21), clinical studies have been aimed at staging CRC and evaluating the disease extent. In clinical practice, "PET whole-body imaging" usually extends from the base of the brain to the upper thighs. The yield in imaging the brain for metastases is low (1.5%), as was reported in the UK (22), mainly because of the high FDG uptake in gray matter. Metastases below the inguinal regions are uncommon in many cancers. Traditionally, three general regions of metastatic spread of CRC in the body are distinguished:

Table 10.1. Clinical performance of ^{18}F-2-deoxy-D-glucose (FDG)-positron emission tomography (PET) in recurrent colorectal cancer.

Body region	Patients	Sensitivity	Specificity
Local	366	94.5%	97.7%
Hepatic	393	96.3%	99.0%
Whole body	281	97.0%	75.6%

Source: Adapted from the meta-analysis of Huebner et al. (28).

local, hepatic, and extrahepatic. Hypermetabolic foci in these regions are suspicious for primary tumor, recurrence, or metastasis and form the basis of metabolic imaging in oncology.

The role of PET for primary CRC has not been studied systematically. The study by Abdel-Nabi et al. (23) showed excellent sensitivity for PET but poor specificity (43%). Their CT specificity of 37% is lower than reported in the literature, so that biased patient selection cannot be excluded. Falk et al. (24) showed in a mixed group of 16 patients that PET was superior to CT and had a positive predictive value (PPV) of 93% and a negative predictive value (NPV) of 50%. Ruhlmann et al. (25) directed their study to the clinical performance in a mixed group of patients and did not find differences between primary and recurrent disease. From the sparse evidence, there is so far no established role for FDG-PET in primary CRC.

The majority of studies deal with recurrent disease. The first series of systematically studied patients came from Heidelberg and demonstrated the impact of PET versus CT in the differential diagnosis of a pelvic mass (26). Ito et al. (27) reported the first study of PET versus MR imaging. Later, studies with more patients were published that confirmed these early findings. Table 10.1 provides the combined data for FDG-PET imaging in CRC as reported in the meta-analysis by Gambhir's group (28). They reviewed articles published during 1990–1999 that reported information on the use of new medical technology and selected 11 studies that fulfilled their strict guidelines for technology evaluation. Sensitivity and specificity in the meta-analysis were all 95% or higher, except for the specificity at extrahepatic sites, which revealed about 25% false-positive results. The studies included in the meta-analysis come from institutions in the United States, Europe, and Australia and showed remarkable consistency of results. From the reported literature in the previous decade, it is clear that PET has established a role in the evaluation and staging of patients with recurrent CRC.

Technique and Methodology

High rates of glycolysis are found in many malignant tumor cells (29) and high uptake of FDG is usually associated with a high number of viable tumor cells and high expression of glucose transporters (30). Increased FDG uptake is by no means specific for neoplasms (31). Inflammatory processes may also have increased uptake, and false-positive results have been reported in presacral abscess, tuberculosis, fungal infections, acute postoperative and radiation changes, pancreatitis, or diverticulitis. In the proper clinical context, many of these lesions will probably not be confused with local or regional disease, but all may contribute to false-positive distant findings.

The body imaging mode has become the standard for PET in oncology (21). After the uptake period, the patient is positioned in the scanner, and sequential acquisitions are performed along the length of the patient's body. As the emitted photons pass through the tissues, varying degrees of attenuation and scatter affect the final number that reach the detectors. A separate transmission scan is performed to correct for attenuation effects. The transmission scan can be acquired after tracer injection, enabling higher patient throughput. No difference in lesion detection was found between corrected and nonattenuation-corrected images for a variety of tumors (32, 33). However, these studies used filtered back-projection to reconstruct the images, instead of currently used iterative reconstruction methods. In the late 1990s, there was no consensus on the necessity of attenuation correction (34), but most facilities now use low noise attenuation correction with CT and scatter correction to provide more true-to-life depiction of metabolic activity in the body.

The spatial resolution of a modern CT or MR system is better than that of a PET system. However, this is not the only determining factor in detecting abnormalities. The "contrast resolution" or difference between the lesion and its surroundings helps determine the presence of disease. The accuracy of anatomic images is limited by a lower lesion to background contrast. The sole exception is the lung, where anatomic contrast between solid lesions and air-filled lung is high. The target-to-background ratio is usually much higher for FDG-PET. Except for the lung, the high metabolic contrast predominates over anatomic contrast. Sensitivity of PET is also affected by lesion size. Metabolically active lesions as small as 5 mm have been detected with FDG-PET.

True quantification of FDG metabolism is usually not performed for CRC because tumor kinetics are not known. The technique is more demanding of resources than imaging alone, and metabolic rate determination has been found to offer no diagnostic advantage. For therapy monitoring and prediction of prognosis, compartmental modeling may be able to discriminate responders from partial or nonresponders, but larger groups of patients need to be investigated (35).

Evaluation of static PET images can be performed visually, or semiquantitatively using the standardized uptake value (SUV) or a lesion-to-background ratio. The SUV is the measured activity in the lesion in mCi/mL divided by injected dose, expressed as mCi/kg of body weight. Strauss et al. (26) reported SUVs of 1.1 to 4.2 for pelvic recurrences, and Takeuchi et al. (36) showed that an SUV cutoff of 2.8

diagnosed local recurrences with 100% accuracy. Abdel-Nabi et al. (23) found SUVs of 2.8 to 14.5 for primary bowel cancers. Semiquantitative evaluation offers a more-objective way of reporting lesion uptake than visual image interpretation and is useful for comparing lesion activity in consecutive studies (treatment monitoring). However, visual interpretation appears sufficient for clinical needs and is equally effective for a one-time diagnosis.

Modifications that may improve the semiquantitative evaluation of FDG uptake include normalizing the dose to the body surface area (37) or the lean body weight (38) instead of the total weight of the patient. This SUV modification may be significant because the concentration of FDG is higher in muscle than in fat. For a concise comparison of the simplified quantitative analysis methods of FDG uptake, the article of Graham et al. (39) is recommended.

Patient Preparation and Diagnostic Protocol

Patients are studied in a prolonged fasting state to produce low insulin levels and induce low rates of glucose utilization by normal tissues, including voluntary muscles and myocardium. Malignant tissues are less dependent on hormone regulation, and thus will have higher uptake when compared to the surrounding normal tissues. The typical duration of a PET oncology protocol is about 2 h. A dose of 250 to 500 MBq (7–15 mCi) of FDG is the usual activity administered. After an uptake period of 45 to 75 min, the patient is asked to void and is positioned in the scanner. Interleaved emission (3–5 min per bed position) and transmission (2–3 min per bed) scans are acquired. A scan of the body has a total acquisition time of 35 to 50 min. Patients are usually scanned from feet to head, taking advantage of the low bladder uptake after voiding at the beginning of acquisition. The transmission scans can be acquired before, after, or interleaved with the emission scans. The typical duration of a dedicated "PET-only" oncology protocol is 1.5 to 2 h, depending on the duration of the postinjection uptake period.

Dual-modality imaging with PET/CT is used in many institutions. The imaging protocols are variable depending on the CT settings, that is, for attenuation correction only (low mAs sufficient) (40), diagnostic quality, and the use of contrast (intravenous and/or oral).

At University of California–Los Angeles (UCLA), oral contrast is given for the delineation of the bowel. A volume of 900 mL Ready-Cat (with 2% barium sulfate, but without glucose) is given orally in three portions during a period of 75 min before the acquisition starts. Our protocol comprises one CT acquisition of the torso, that is, from the base of the skull to the midthigh level, followed by a PET scan of the same area, with the arms up. Currently, we have a dual-slice CT that provides images of diagnostic quality with the following settings: 130 kVp, 80 mAs, pitch 1, and reconstruction slice thickness 5 mm. The helical CT takes about 80 sec to image the upper body. Intravenous contrast is given when ordered by the referring oncologist. Nonionic contrast (Omnipaque) in a volume of 100 to 130 mL at a rate of 1.5 mL/s is administered. The I.V. contrast may cause regions of high attenuation that lead to typical CT-based attenuation artifacts and pseudo-FDG uptake (41, 42). The FDG dose is 0.19 mCi/kg (7.0 MBq/kg) with a maximum of 15 mCi (555 MBq). The uptake interval between tracer administration and start of acquisition is 1 h. During the uptake period, the patient is comfortably resting in a chair with armrests in a dimly lit room without radio or television to minimize brain stimulation. The patient is covered with blankets to prevent shivering and activation of brown adipose tissue. Just before acquisition, the bladder is emptied. Our PET scanner has fast detectors (lutethium-ortho-silicate, LSO) allowing for imaging of 1 to 4 min per bed position to accumulate the necessary counts. Body weight determines the imaging time: less than 75 kg, 1 min/bed; less than 100 kg, 2 min/bed, etc. (43). The overall PET acquisition takes 6 to 26 min. With this setup, a standard whole-body PET/CT can be completed within 45 min. An additional CT of the chest is acquired during deep inspiration.

The cortex of the brain uses glucose as its substrate; therefore, FDG accumulation is high. Evaluation of metastatic disease to the brain with FDG is limited (22). Diffuse thyroid uptake can be a normal variant and is seen in patients with thyroiditis and Graves' disease. Metastatic cervical lymph nodes are occasionally seen in patients with colorectal carcinoma, and differentiation from thyroid uptake is important. In a typical fasting state, the myocardium primarily utilizes free fatty acids, but postprandially or after a glucose load, it favors glucose. When the chest is evaluated with FDG to assess the presence of metastases, a 12-h fasting state may be preferable to achieve low myocardial FDG uptake. Despite these measures, uptake in the myocardium is seen with high frequency in cancer patients.

FDG is filtered by the glomerulus and is only partly reabsorbed, unlike glucose. Thus, accumulation of FDG in the renal collecting system and urinary tract is normal. Hydration to promote diuresis, or administration of diuretics, can reduce the urinary activity seen on PET images. High bladder activity can result in positive and negative image artifacts if the images are not corrected for tissue attenuation. Correction for attenuation and iterative image reconstruction circumvent such artifacts, and bladder catheterization is to be avoided. Some centers use antiperistalsis and antimotility drugs to overcome prominent bowel uptake. The large bowel usually shows some FDG uptake, but this can be differentiated from abnormal activity by demonstrating a physiologic activity pattern along the bowel trajectory in the stack of 3D images. Many institutions in Europe utilize mild sedatives for patient comfort during the relatively lengthy acquisition.

To avoid misinterpretation of FDG images, it is critical to standardize the environment of the patient during the uptake period; to examine the patient for postoperative sites, tube placement, stoma, etc.; and to be aware of any invasive procedures or therapeutic interventions.

Accuracy of Metabolic and Anatomic Imaging in Recurrent Colorectal Cancer

Overall Accuracy of FDG-PET and Conventional Imaging

The aims and methods of studies evaluating FDG-PET in recurrent CRC have developed with time and experience. The objectives in recurrent disease are quite different from primary staging of CRC. The early studies of Strauss et al. (26) and Ito et al. (27) were aimed at feasibility of the technique and general comparison to CT and MR imaging. Subsequent studies addressed tumor staging for better stratification of patients before contemplated surgery (24, 44–47). The accuracy of PET in CRC was evaluated (47–49) as well as the impact of PET findings on management decisions in routine practice (25, 48–51). In presenting the findings of these studies, we included only reports of 50 or more patients that were published in peer-reviewed journals and that compared PET to CT. The referred patient populations vary from report to report, for instance, suspected versus diagnosed recurrence, single versus multiple sites of recurrence, first versus second recurrence, prospective versus retrospective study design, potentially resulting in referral bias. In addition, methods and PET systems used in the studies varied, for example, limited versus extended (i.e., upper body) field of view, acquisition duration, correction for attenuation, and scatter. Image reconstruction method, such as standard filtered back-projection versus iterative methods, was not always the same in the reported series. Moreover, lesion-based instead of patient-based analysis was frequently used. A number of studies were reported sequentially from the same institution and included some of the same patients; in which case we selected only one representative study. Despite the multiple variations, different objectives, and inconsistencies involved, the end results of the studies that met the criteria are collated in the tables, and are quite similar.

Table 10.1 shows the average sensitivity and specificity of PET as computed in the meta-analysis (28). The meta-analysis does not specify the results of conventional imaging, such as CT and ultrasound (US), and the comparison standard is not provided. For this reason, we pooled the data of individual reports, that is, table 3 from Schiepers and Hoh (52) and table 2 of Delbeke and Valk (53), to calculate the overall diagnostic accuracy of PET and CT for whole-body imaging. For the CRC data with direct comparison between modalities, 301 patients were accumulated between 1993 and 1997. The weighted average sensitivity was 95.5% for PET [95% confidence interval (CI), 93%–98%] and 71.4% for CT (95% CI, 66%–77%). The weighted average specificity was 88.3% for PET (95% CI, 84%–92%) and 85.4% for CT (95% CI, 81%–90%).

In the late 1990s, the single largest study of the accuracy of PET imaging with FDG was reported by Valk et al. (51), who compared the sensitivity and specificity of PET and CT for specific anatomic locations. They found that PET was more sensitive than CT in all locations except the lung, where the two modalities were equivalent. The largest differences between PET and CT were found in the abdomen, pelvis, and retroperitoneum, where more than one-third of PET-positive lesions were negative by CT. PET was also more specific than CT at all sites except the retroperitoneum, but the differences were smaller than the differences in sensitivity.

Local Recurrence

The results obtained by PET in detecting local recurrence that were reported in the early studies (26, 27) were corroborated in larger patient groups (Figure 10.1). Table 10.2 shows the results of the series evaluating local recurrence that met the criteria. PET was 17% more accurate than CT for this indication. Keogan et al. (54) studied recurrent rectal cancer, using pathology as comparison

Table 10.2. Detection of local recurrence in colorectal cancer.

First author	Reference number	Year	Patients	PET (sensitivity–specificity)	CT (sensitivity–specificity)
Schiepers	48	1995	76	93%–97%	60%–72%
Valk	51	1999	115	97%–96%	68%–90%
Whiteford	55	2000	70	90%–90%	71%–85%
Weighted average			261	94.0%–94.7%	66.5%–83.4%

Source: From Valk PE, Bailey DL, Townsend DW, Maisey MN. Positron Emission Tomography: Basic Science and Clinical Practice. Springer-Verlag London Ltd 2003, p. 562.

standard. They compared visual versus quantitative interpretation of PET studies and found the two methods of analysis to be equivalent. Whiteford et al. (55) studied 101 patients, 70 of which were evaluated for locoregional recurrence. They found that PET sensitivity was about 20% higher than for CT plus colonoscopy. They also found that PET sensitivity varied with histological tumor type, having a lower sensitivity for mucinous (58%) than nonmucinous CRC (92%).

Hepatic Metastasis

The aim of the presurgical workup is to distinguish isolated resectable disease, that is, local recurrence or soli-

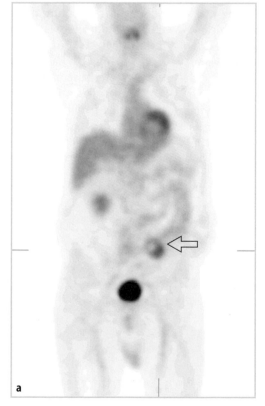
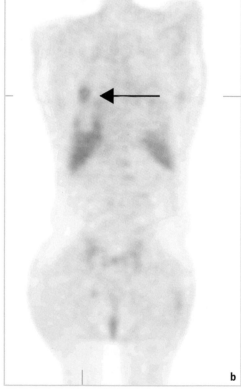

Figure 10.1. Coronal FDG-PET images in a 73-year-old man who underwent resection of colon cancer 4 years earlier. CT showed a mass in the lower lobe of the right lung. PET showed the pulmonary metastasis (**b**, ->) and also demonstrated local recurrence in the left lower quadrant (**a**, =>). Local recurrence was confirmed at surgery. Images are corrected for attenuation and scatter.

a

b

Table 10.3. Detection of hepatic metastases.

First author	Reference number	Year	Patients	PET (sensitivity–specificity)	CT (sensitivity–specificity)
Schiepers	48	1995	76	94%–100%	85%–98%
Delbeke	49	1997	61	91%–95%	81%–60%
Valk	51	1999	115	95%–100%	84%–95%
Whiteford	55	2000	101	89%–98%	71%–92%
Weighted average			353	92.4%–98.6	80.0%–88.7%

Source: From Valk PE, Bailey DL, Townsend DW, Maisey MN. Positron Emission Tomography: Basic Science and Clinical Practice. Springer-Verlag London Ltd 2003, p. 562.

tary liver metastasis, from advanced disease. By correct staging, patients with widespread metastasis may be identified in whom surgery is not an option, thus sparing the patient extensive surgery with its associated morbidity. The actual selection of patients with recurrent cancer results in 5-year survival rates of only 20% to 30% after secondary "curative" surgery (2).

Table 10.3 shows results obtained by PET and CT in diagnosis of hepatic metastasis, showing smaller differences than in diagnosing local recurrence. For detection of hepatic metastasis, PET was 12% more sensitive than CT (Figure 10.2). PET specificity is about equal to CT, except for Delbeke's study. Delbeke et al. (49) compared PET to conventional CT as well as CT portography, which is an invasive procedure. The sensitivity of PET (91%) was lower than CT portography (97%), but the specificity was 12% higher, particularly at postsurgical sites. Fong et al. (56) and Delbeke et al. (49) found that the sensitivity of PET in detecting hepatic metastases varied with lesion size, as would be expected. Fong found that only 25% of lesions smaller than 1 cm were detected by PET compared to 85% of lesions larger than 1 cm. By excluding the lesions less than 1 cm, Delbeke found that the sensitivity increased about 8% for both PET and CT. This 25% versus 8% detectability change, found between two series reported only 3 years apart, highlights some of the

difficulties in interpreting small lesions and may be attributed to the rapidly evolving technologic changes in multi-slice helical CT and high-resolution PET systems. Topal et al. (57) confirmed in a series of 99 patients that PET was a diagnostic tool complementary to CT and US in selecting patients for potentially curative liver resection. Similar findings were reported in a small series of 14 patients by Boykin et al. (58).

Fernandez et al. (59) investigated the 5-year survival after resection of metastasis from CRC. They established the 5-year survival of patients with conventional diagnostic imaging from the literature by pooling the data of 19 studies with a total of 6,090 patients. The 5-year survival rate was 30% and appeared not to have changed over time. These results were compared to their group of 100 patients with hepatic metastases, who were preoperatively staged for resection with curative intent. Addition of a preoperative FDG-PET study improved the 5-year survival rate to 58%, indicating that they were able to define a subgroup after conventional imaging that has a better prognosis (59). The main contribution was in detecting occult disease, leading to a reduction of futile surgeries.

Distant or Extrahepatic Metastasis

PET for staging of disease involvement may yield unexpected lesions that often appear to be metastases (Figures 10.3, 10.4); this is a feature of the PET imaging technique in which the whole body can be examined after one injection of an FDG dose. Table 10.4 provides results of detection of unknown metastases. In about 25% of patients referred for restaging during their preoperative evaluation, occult disease was found that was not suspected clinically or detected with conventional imaging.

PET using FDG has a proven record of accomplishment for characterizing indeterminate pulmonary nodules that can be metastases from CRC (60). Lai et al. (46), in their study of 34 patients, found that FDG PET was especially useful for detecting retroperitoneal and pulmonary metastases. Schiepers et al. (48) found that half the chest lesions on PET were false-positives.

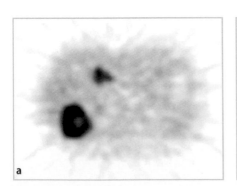
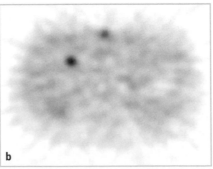
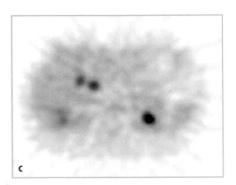

Figure 10.2. Transverse FDG-PET images in a 55-year-old man with a history of colon cancer 15 months earlier and status postresection. CT demonstrated a large metastasis in the right lobe of the liver posteriorly. PET shows this lesion (**a**) and also demonstrates multiple small metastases that were not seen on CT (**b**, **c**). 📖

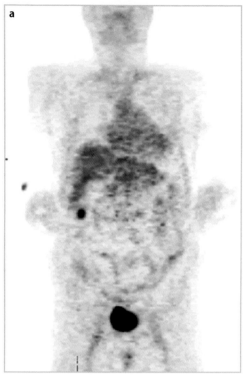

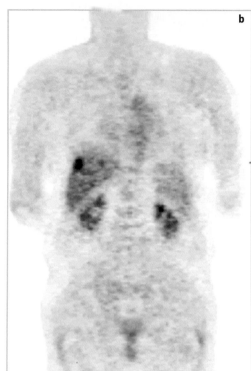

Figure 10.3. Coronal FDG-PET images in a 48-year-old man, with a history of resection of colon cancer 11 months earlier, who was found to have a lesion in the right lobe of the liver on CT examination. The remainder of the CT images of the abdomen and pelvis were normal. Needle biopsy confirmed the hepatic metastasis, and the PET study was performed for preoperative staging. PET images show the hepatic metastasis (**b**) with an extrahepatic focus of abdominal tumor, inferior to the right lobe of the liver (**a**). The PET findings were confirmed at surgery. 📖

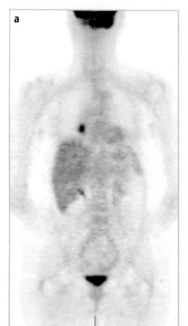

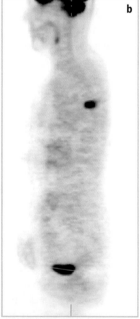

Figure 10.4. Coronal (**a**) and sagittal (**b**) FDG-PET image in a 51-year-old man with a history of resection of rectal cancer 3 years earlier. CT demonstrated a lesion in the lower zone of the right lung, and biopsy confirmed recurrent rectal cancer. CT imaging showed no other abnormality, and PET imaging was performed for preoperative staging. PET showed high uptake in the lung metastasis (**a**) and also showed metastasis in a thoracic vertebra, thereby excluding surgical resection of the lung lesion. The patient was treated with chemotherapy and irradiation. 📖

Clinical Indications for PET Imaging in Colorectal Cancer

Evaluation of Elevated Serum CEA Level

Flanagan et al. (61) reported the use of FDG PET in 22 patients with elevation of serum CEA level after resection of primary CRC and negative findings by conventional diagnostic procedures (Figure 10.5). Sensitivity and specificity of PET for tumor recurrence were 100% and 71%, respec-

Table 10.4. Detection of occult metastases with FDG-PET.

First author	Reference number	Year	Patients	Unsuspected metastases
Schiepers	48	1995	76	14 (18%)
Lai	46	1996	34	11 (32%)
Delbeke	49	1997	52	17 (33%)
Valk	51	1999	78	25 (32%)
Pooled average			240	67 (27.9%)

Source: Updated from Valk PE, Bailey DL, Townsend DW, Maisey MN. Positron Emission Tomography: Basic Science and Clinical Practice. Springer-Verlag London Ltd 2003, p. 563.

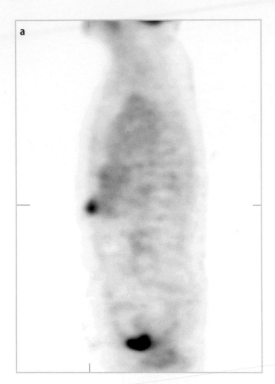
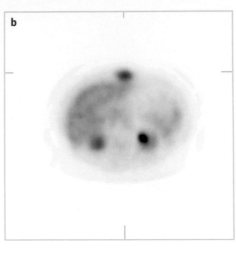

Figure 10.5. Sagittal (**a**) and transverse (**b**) FDG-PET image in a 73-year-old man with a history of colon cancer resection 22 months earlier, followed by 6 months of chemotherapy. The patient was found to have carcinoembryonic antigen (CEA) elevation, negative CT examination of the abdomen, and negative ultrasound (US) examination of the liver. PET demonstrates a hypermetabolic focus in the anterior epigastrium (coronal slice), anterior to the left lobe of the liver (transverse slice). Metastasis was confirmed by biopsy. 📖

tively. Valk et al. (51) reported a sensitivity of 93% and specificity of 92% in a similar group of 32 patients, 18 of whom proved to have recurrence on surgical evaluation or clinical follow-up. Maldonado et al. (62) reported for PET a sensitivity of 94% and specificity of 83% in a group of 72 patients. Flamen et al. reported, in a retrospective series of 50 patients, 79% sensitivity and 89% positive predictive value (63). From these studies, it is apparent that PET has a definite place in the workup of an unexplained CEA rise in CRC.

Preoperative Staging of Recurrent CRC

Table 10.4 shows results obtained by PET using FDG in detection of unsuspected distant metastases in patients undergoing preoperative evaluation, which is probably the most important function of PET in recurrent CRC. Table 10.4 shows that FDG-PET discovered disease at unsuspected sites in 28% of patients referred for preoperative staging. Lai et al. (46), in their study of 34 patients, found that FDG-PET was especially useful for detecting retroperitoneal and pulmonary metastases.

Surgical findings in patients who underwent surgery following staging by PET were reported by Valk et al. (51). Forty-two patients with PET evidence of localized recurrence were submitted to surgery. Seven of these patients (17%) had nonresectable tumor at surgery. Two patients with liver metastasis also had diffuse peritoneal tumor and 1 had multiple small liver lesions that had not been detected by PET. Three patients with focal pelvic recurrence and 1 with focal abdominal recurrence on FDG-PET had undetected diffuse tumors that could not be resected. Strassberg et al. (64) reported 43 patients with hepatic recurrence who were evaluated before surgery. PET findings contraindicated surgery in 6 of these patients (14%). Of the 37 patients who underwent laparotomy, liver resection was performed in all but 2 (5%). Median follow-up in the 35 patients who had resection was 24 months, and the Kaplan–Meyer estimate of overall 3-year survival was 77%, with a 3-year disease-free survival of 40%. These numbers were higher than had been reported previously (17). PET increased the rate of resectability at surgery and decreased the rate of rerecurrence following resection by detection of tumor preoperatively that had not been found by conventional imaging.

Three studies of preoperative staging by Beets et al. (44), Schiepers et al. (48), and Flamen et al. (50) reported results from the same institution over several years and contained overlapping groups of patients. In the series reported by Flamen et al. (50), the objective was to compare the number of lesions detected by PET and conventional imaging methods and study the discrepancies. All data were reprocessed with iterative reconstruction and reread by two nuclear medicine physicians. The accuracy of interpretation was similar to the earlier report of Schiepers et al. (48) indicating reproducibility of findings between different reconstruction techniques and different interpreting nuclear medicine specialists. In 10% of lesions, PET and conventional imaging findings were discordant, and in the other 90%, PET contributed additional diagnostic information. Almost half the discordant findings were locoregional, near the site of the original primary (50).

Impact of PET findings on Management Decisions

PET using FDG permits earlier evaluation for treatment by diagnosing recurrence when CT is still negative, and avoids unnecessary surgery by demonstrating nonresectable disease in some patients with known recurrence (65, 66). Whether PET will improve the resectability-rate by earlier tumor diagnosis or will reduce the rate of re-recurrence of undetected residual tumor at surgery remains to be established. Tabulating the studies that specifically addressed preoperative evaluation yields the results in Table 10.4. In about a quarter of patients, unsuspected metastases were detected. Although these findings may not always change the clinical stage of the disease, some cases will be reclassified to the nonsurgical category. Surgical decision making in CRC has been reported frequently from different countries (50, 51, 65, 66).

The question of change in patient management was addressed by several investigators, and the impact of PET varied from 23% in the series of Fong et al. (56) to 40% in the series of Beets et al. (44). Table 10.5 shows that, overall, a change in management was observed in about one-third of patients with recurrent CRC. A detailed analysis of the type of management changes can be found in the meta-analysis (25). Altogether, in 121 of 349 patients (34.7%), a management change was reported. Surgery was correctly avoided in more than half of the patients (53.7%), and unintended surgery was correctly initiated in 20.7% of patients. PET incorrectly suggested a change in management in 9.1% of patients.

Management change was addressed in a collaborative study from UCLA and the Northern California PET Imaging Center from the perspective of the referring physician. Questionnaires were sent to referring physicians after a PET study had been performed for diagnosis or staging of recurrent CRC determine whether the PET findings had had an impact on management decisions. Clinical stage was changed in 42% of patients. A treatment modality switch (e.g., from surgery to chemotherapy, from medical treatment to surgery) was reported in 37% of patients (67).

The great variability in study design of the published series evaluating management effect was addressed by Gambhir's group in their meta-analysis (28). Using rigorous selection guidelines based on U.S. medical payer source criteria, only 10 studies between 1990 and 1999 listed on MEDLINE qualified for inclusion. One additional study was selected after direct contact with the primary author for clarification of the text and tables (25). These studies were performed in four different countries: Australia, Belgium, Germany and the United States (four institutions), comprising seven different institutions in total. The reports from the same institution used overlapping patient groups. The results given in Table 10.1 were obtained from this meta-analysis.

In a prospective study, Kalff et al. (68) investigated the clinical impact in recurrent CRC. They found that FDG-PET directly influenced management in 59% of patients. Their findings confirmed earlier retrospective reports (46, 51, 56, 66). Management changes in recurrent and metastatic CRC was also studied by Desai et al. (69) in 114 patients. They found that FDG-PET altered therapy in 40% of patients. They concluded that PET should be used in the management of patients with recurrent CRC who are considered for potentially curative surgery.

The greater sensitivity of PET compared to CT in diagnosis and staging of recurrent tumor is the direct result of two factors: (1) early detection of abnormal tumor metabolism, before changes have become apparent on anatomic imaging, and (2) whole-body metabolic imaging, which permits diagnosis of tumor when it occurs in unusual and unexpected sites. PET imaging allows identification of areas with abnormal metabolism, which may guide subsequent CT and possible biopsy of these lesions. Thus, exact anatomic location and potential resectability can be assessed. In the region of the primary, PET is helpful in detecting nodal involvement and differentiating local recurrence from postsurgical changes, indications for which CT has known limitations. For hepatic lesions, PET is helpful to assess the number of lesions and liver lobes involved. For distant, extrahepatic abnormalities, PET may be used to characterize abnormal lesions diagnosed with conventional imaging procedures or may identify occult metastasis.

Three indications for metabolic imaging have been established in patients with known or suspected recurrent

Table 10.5. Change in patient management introduced by FDG-PET.

First author	Reference number	Year	Patients	Changed treatments
Beets	44	1994	35	14 (40%)
Lai	46	1996	34	10 (29%)
Valk	51	1999	78	24 (31%)
Fong	56	1999	40	9 (23%)
Whiteford	55	2000	101	26 (26%)
Pooled			288	83 (28.8%)

Source: Updated from Valk PE, Bailey DL, Townsend DW, Maisey MN. Positron Emission Tomography: Basic Science and Clinical Practice. Springer-Verlag London Ltd 2003, p. 566.

colorectal carcinoma: (1) rising serum CEA level with negative conventional diagnostic imaging; (2) characterization of equivocal lesion(s) on conventional imaging; and (3) staging of recurrent disease. FDG-PET scans for staging and restaging of CRC are reimbursed by national insurers in the United States, U.K., Germany, Italy, Belgium, and Switzerland.

Therapy Monitoring

The use of FDG-PET for monitoring chemotherapy has been reported (51). Accurate information about the effectiveness of radiotherapy and/or chemotherapy for colorectal cancer is important for continuation of the selected therapeutic regimen or switching to an alternative regimen. The number of studies in which PET has been used successfully for monitoring therapy is small, but monitoring chemotherapy in advanced colorectal cancer appears promising. Strauss and collaborators (70) studied the response to therapy with PET for various tracers. In 1991, they reported that tumor perfusion (studied with ^{15}O-water) is highly variable and does not correlate well with treatment response. Using labeled fluorouracil (FU) and FDG for measuring response to chemotherapy and/or radiotherapy, they found a correlation between uptake and outcome. High uptake of FU and low uptake of FDG correlated with good treatment response. The pharmacokinetics of FU suggest that FU may be superior to FDG in predicting the response to treatment on an individual patient basis (71). Strauss and collaborators (72–74) studied PET images of FU metabolite concentration and compared these with growth rate before and during chemotherapy, determined by CT volume measurements. Lesions with low FU uptake had a significant increase in volume and no response to treatment. Findlay et al. (75) evaluated the metabolism of liver metastases from CRC before and after treatment in 18 patients. The findings were compared to change in size on CT, and results were expressed as a tumor-to-liver uptake ratio. By this means they were able to differentiate eventual responders from nonresponders on both lesion-by-lesion and patient-based analysis. Regional therapy of hepatic metastases by chemoembolization can also be monitored with FDG (76). FDG uptake decreased in responding lesions and the presence of residual uptake was used to guide further regional therapy. The Sloan-Kettering group also found good correlation between the response of hepatic metastasis to chemotherapy and FDG uptake (10). Their most promising finding was the positive correlation between PET findings at 4 weeks and CT findings at 12 weeks, whereas the MR imaging at 4 weeks did not show a change in tumor volume.

Assessing response to radiation therapy (RT) was evaluated in 21 patients with recurrent disease by Haberkorn et al. (77). Reduction in FDG uptake was found in half the patients and correlated with the palliative benefit. PET was more accurate than serum CEA level in assessing response. These investigators found that FDG uptake immediately following radiation may be caused by inflammatory changes without residual tumor (77). They recommended a postradiation interval of 6 months before evaluating response. These findings mirror the results of an early report of Abe et al. (78). The time course of FDG uptake postradiation has not been studied systematically; but an interval of 6 months between RT and PET is recommended to assess presence of complete response.

Although the inflammatory response is real and may hamper image interpretation, early PET scanning may have a role to determine whether the tumor is sensitive to treatment. For assessing complete response, sufficient time needs to have elapsed, otherwise remaining FDG uptake may be incorrectly interpreted as residual or recurrent tumor. Schiepers et al. (79) investigated the effect of radiation after 10 fractionated doses of 3 Gy on primary rectal cancer (total dose, 30 Gy). In this pilot experiment, all cancers showed an effect of irradiation 2 weeks after finishing RT (Figure 10.6), and one tumor was no longer detectable in the resected, irradiated specimen. More research is needed to sort out the exact timing of cell damage, cell death, and cleanup of necrotic cells and its effect on metabolic activity of the cells mediating in the inflammatory response cascade. As a rule of thumb, we recommend assessing tumor response with FDG-PET about 3 months after completion of RT and 1 month after chemotherapy.

The European Organization for Research and Treatment of Cancer (EORTC) published a position paper on the use of FDG-PET for the measurement of tumor response (80). The EORTC-PET study group formulated standard criteria for reporting alterations in FDG uptake to assess clinical and subclinical response to therapy. Although there was wide variation in the methodology between PET centers surveyed, assessment of FDG uptake was thought to be a satisfactory method to monitor response to treatment. The group made initial recommendations on (1) patient preparation, (2) timing of PET scans,(3) methods to measure FDG uptake with SUV, and (4) definition of FDG tumor response.

Costs

PET using FDG has been shown to be cost effective for diagnosis and staging recurrent CRC in a study using clinical evaluation of effectiveness with modeling of costs (51) and a study using decision analysis (81). In both studies, all costs calculations were based on U.S. Medicare reimbursement rates for conventional diagnostic and therapeutic procedures and $1,800 cost for PET. The cost analysis by Valk et al. (51) assessed the use of PET in patients who had been diagnosed with recurrence and were

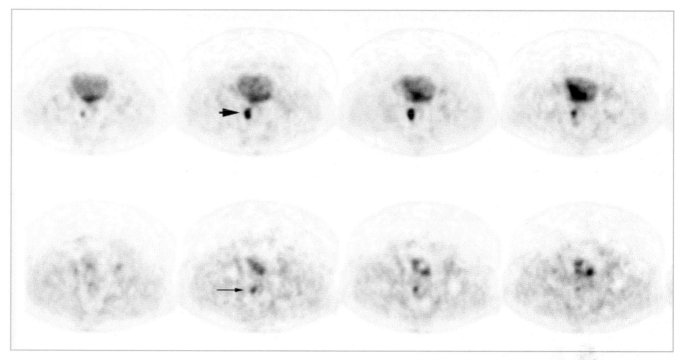

Figure 10.6. Transverse FDG-PET images in a 48-year-old man with primary rectum cancer, clinical stage cT3 N0 M0. Note the tumor focus posterior to the bladder (*upper row, arrowhead*). The *bottom row* shows the findings after 10 fractionated doses of 3-Gy external radiation (total, 30 Gy). The second PET scan was performed 2 weeks after completion of radiation. Note the marked decrease in FDG uptake after therapy (*arrow*). 📖

considered to have resectable tumor on the basis of conventional diagnostic imaging. In a management algorithm where recurrence at more than one site was treated as nonresectable, they found a cost saving of $3,003 per patient resulting from diagnosis of nonresectable tumor by PET.

The group of Gambhir used a decision analysis model to determine the cost-effectiveness of PET in management of recurrent CRC (82). In patients with an elevated serum CEA, who were evaluated for hepatic resection, the CT plus PET strategy was superior to the conventional CT-only strategy. The model showed an increased cost of $429 per patient, which resulted in an increased mean life expectancy of 9.5 days per patient. The assumptions that were used in this liver metastasis model (increased serum CEA with negative conventional imaging) included prevalence of recurrent CRC, 32%; CT liver sensitivity, 79%; CT liver specificity, 88%; PET-liver sensitivity, 96%; and PET-liver specificity, 99%. Frequency of liver metastasis was assumed to be 28.5%, and 81% of these were estimated to have extrahepatic involvement. This model showed that the CT plus PET strategy was cost effective for PET costs less than $1,200 and hepatic recurrence frequencies greater than 46%.

Based on the available data, all patients should undergo a PET scan with FDG for preoperative staging of recurrent CRC. In case of distant, extrahepatic hypermetabolic foci, a CT of the corresponding region should be performed for anatomic correlation. This approach allows more-accurate selection of patients who will benefit from surgery than conventional diagnostic methods and, more importantly, excludes patients who will not benefit from laparotomy and liver resection because of unrecognized distant disease.

Dual-Modality Imaging with PET/CT

PET using FDG has come of age for staging CRC and has emerged as a molecular imaging modality. Several limitations of dedicated PET, such as prolonged imaging times, lengthy transmission scans, and lack of anatomic landmarks, prompted for integration of imaging information. Traditionally, all imaging data was merged "inside the brain" of the imaging specialist. Subsequently, information from the different modalities was fused in computer memory. Patient position (e.g., arms up or down, tilt of image device relative to patient axis), breathing state (deep inspiration, breath-hold, or shallow breathing), resolution, etc., differ significantly among the various imaging modalities and imaging centers, and image fusion post hoc is, therefore, prone to errors. This limitation gave rise to the development of dual-modality systems such as PET/CT, in which the patient is positioned in a combined gantry and imaged during one *single* session. After the prototype PET/CT was introduced by the Pittsburgh group, dual-modality imaging has rapidly evolved as the imaging method of choice for oncology (13, 83). The combination of a PET and CT scanner

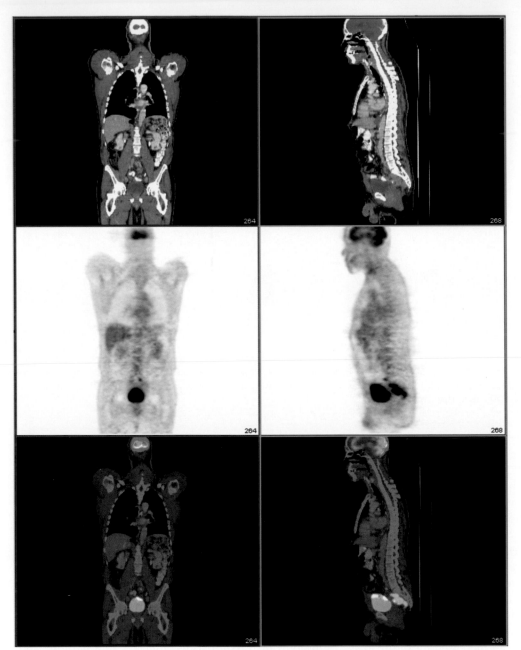

Figure 10.7. Body PET/CT scan in a 72-year-old man with recurrent colorectal cancer. Patient was 6 months past anterior resection of an adenocarcinoma of the rectum. He returned with abdominal pain and constipation. Proctoscopy did not reveal stenosis. Coronal planes (*left column*) and sagittal cuts (*right column*) are shown. The *top row* displays the CT images (soft tissue window), the *middle row* the PET images, and the *bottom row* the fusion images (CT in black and white with PET superimposed in color). Local recurrence was diagnosed (*green arrows*), and a regional lymph node metastasis was identified (*blue arrow*). Note the hot bladder, containing excreted FDG in the urine.

permits "mechanical" fusion of the images, greatly facilitating the localization of hypermetabolic foci as well as increasing specificity while characterizing lesions (Figures 10.7, 10.8, 10.9). In addition, the CT can be used for attenuation correction (see Chapter 1), eliminating the transmission scans and reducing acquisition time. In the dual imaging scenario, the purpose of the CT is to localize lesions in anatomic terms and of PET to categorize lesions in pathologic terms (benign versus malignant).

Lesions with abnormal metabolism can be assigned to normal or abnormal structures on CT and classified correctly as true or false positives. Thus far, a few well-controlled studies have been reported with PET/CT imaging and its clinical impact in diagnostic oncology (84, 85). The

reports show a 10% to 20% increase in both sensitivity and specificity.

Cohade et al. (85) reported their experience with an in-line PET/CT device in colorectal cancer. They focused on two aspects of the identified lesions: (1) type, that is, malignant or benign and (2) anatomic location. In their study, the "conventional" technique of attenuation correction was applied, that is, measured with a positron-emitting transmission source (^{68}Ge). The duration of the transmission scan was 3 min/bed, and a segmentation algorithm was used to calculate the attenuation map.

The purpose of the CT was to help the physician in interpreting the study, localize the lesion in anatomic

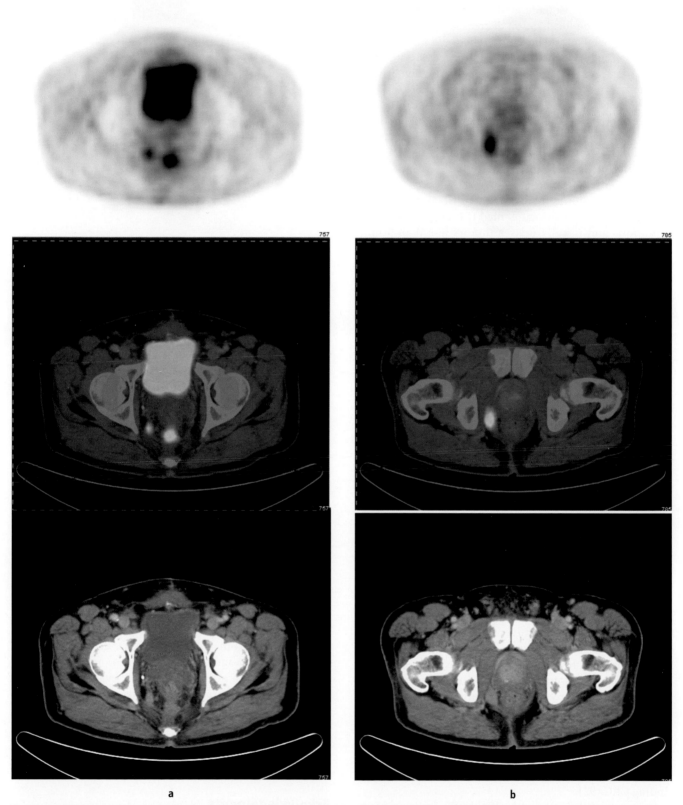

Figure 10.8. Axial images of the pelvis for PET (*top*), CT (*bottom*), and fusion (*middle*) of the same patient as in Figure 10.7. (**a**) Local recurrence near the anastomosis (*green arrow*) and rectal lymph node with increased FDG uptake (*blue arrow*). (**b**) Additional affected rectal lymph node at a lower level than (**a**).

terms, and categorize the lesion as benign or malignant. The results of PET/CT were compared to the interpretation of PET alone. Cohade et al. (85) found that the uncertainty in lesion localization decreased by 55% and that in lesion type by 50%. These differences did not translate into a different performance of PET alone versus PET/CT. There were no statistical differences in sensitivity, specificity, or accuracy. In evaluating the stage of CRC, an 11% increase in accuracy of staging was found. As Cohade et al. (85) pointed out, the gain by PET/CT is not "tremendously high" for CRC (85). However, one has to keep in mind that sensitivity and specificity of FDG-PET for staging of CRC is already quite high, as was established earlier (see Table 10.1).

Although a significant advantage of PET-alone over CT-alone has been amply documented (51, 86), PET has size limitations and usually misses tumors less than 5 mm. The Cohade paper demonstrates a benefit from registration and fusion of PET to CT. The certainty that a focus of FDG accumulation constitutes a lesion is clearly increased. In a group of 169 patients referred for staging of a mixture of neoplasms, we found a 12% improvement by PET/CT (87).

One of the main areas of contribution for PET/CT is the precise location of the bowel and lymph nodes, and association of FDG uptake with GI mucosa, which is quite variable. This consideration is even more important for restaging and for therapy monitoring after surgery when the anatomy has been changed. The effect of breathing on abdominal organs is less important than for the chest (88). Even artifacts in the liver, caused by breathing during the transport through the CT gantry, do not seem to pose clinical problems in staging of CRC. Non-attenuation-corrected tomograms and 2D projection images are always available to view images without artifacts and circumvent interpretation problems. To reduce bowel uptake, some centers use pharmacologic interventions to inhibit secretion and motility, but this step does not seem necessary on a routine basis.

Evaluation of the effects of intravenous and oral contrast used for CT on the PET images has just started (89, 90). PET/CT has already influenced the way we read standard FDG-PET scans. Accurate localization of muscle and brown fat uptake by PET/CT has been demonstrated (91). These patterns are found in young, tense, skinny, or shivering patients and are physiologic variants, which are now recognized while interpreting PET alone.

Therapy monitoring will become increasingly important and will have a major impact. In the near future, PET/CT will play an important role in planning of radiotherapy, which will be relevant for all types of cancer, including CRC (92).

As was recently outlined by Wahl (93), imaging of abdominal and pelvic cancer in the future will be almost exclusively done by PET/CT. This projection was based on the experience of the Johns Hopkins PET Center, where more than 2,700 studies were performed in a 2-year period.

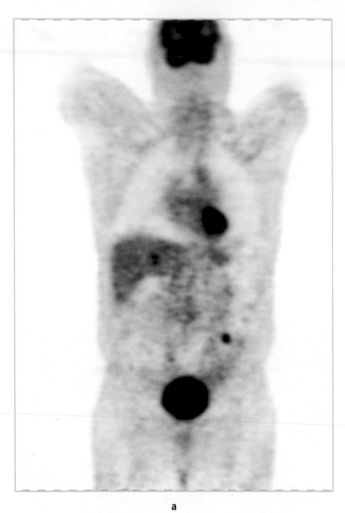

a

Figure 10.9. (a) Coronal plane with a liver metastasis (*blue arrow*) and left lower quadrant mesenteric metastasis (*green arrow*), in same patient as shown in Figure 10.7.

Conclusion

FDG-PET is indicated as the initial test for diagnosis and staging of recurrent CRC. In addition, it is indicated for preoperative staging of known recurrence that is considered to be surgically resectable.

PET imaging is valuable for distinguishing posttreatment changes from recurrent tumor, differentiation of benign from malignant lymph nodes, and monitoring therapy. Addition of FDG-PET to the evaluation of these patients reduces overall treatment cost by accurately distinguishing patients who will benefit from those who will not benefit from surgery. PET is evolving as a molecular imaging modality and will soon enter the realm of clinical gene imaging and gene therapy monitoring (5, 94).

Acknowledgments

The previous version of this chapter was co-authored by Peter Valk, who passed away in 2003. Tables were maintained and the old text incorporated in the current updated version. His advancement of the PET technology in the clinical scenario is well known. This text was prepared to honor his legacy.

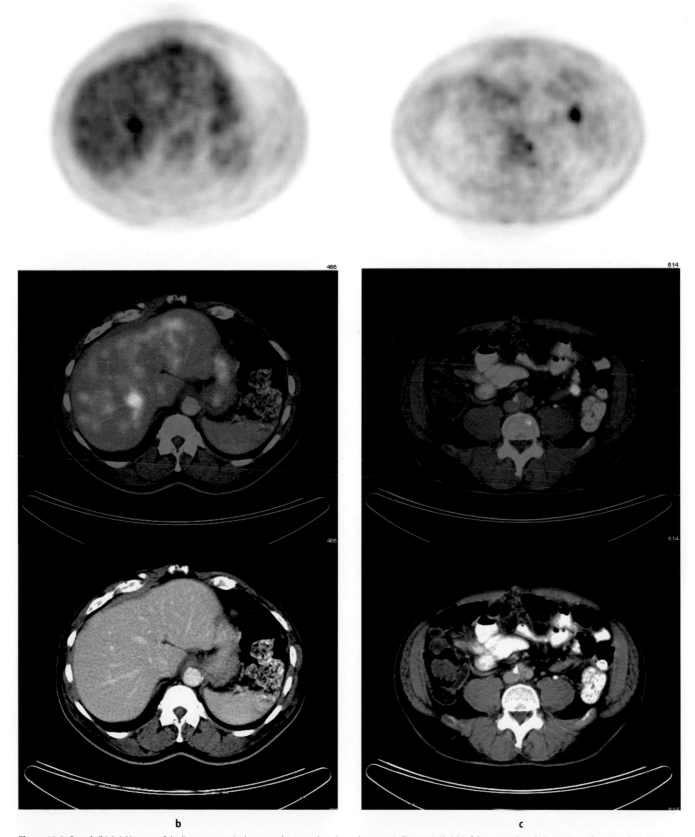

Figure 10.9. Contd. (b) Axial images of the liver metastasis shown on the coronal cut (same layout as in Figure 10.8). (**c**) Left lower quadrant lesion seen on the coronal cut (A) (same layout as in Figure 10.8).

References

1. Landis SH, Murray T, Bolden S, Wingo PA. Cancer statistics, 1999. CA Cancer J Clinicians 1999;49(1):8–31.
2. Wingo PA, Ries LAG, Rosenberg HM, Miller DS, Edwards BK. Cancer incidence and mortality, 1973–1995: a report card for the U.S. Cancer (Phila) 1998;82(6):1197–1207.
3. Jemal A, Tiwari RC, Murray T, Ghafoor A, Samuels A, Ward E, et al. Cancer statistics, 2004. CA Cancer J Clin 2004;54(1):8–29.
4. Hoh CK, Schiepers C, Seltzer MA, Gambhir SS, Silverman DH, Czernin J, et al. PET in oncology: will it replace the other modalities? Semin Nucl Med 1997;27(2):94–106.
5. Phelps ME. Inaugural article: positron emission tomography provides molecular imaging of biological processes. Proc Natl Acad Sci U S A 2000;97(16):9226–9233.
6. Schiepers C, Hoh CK. Positron emission tomography as a diagnostic tool in oncology. Eur Radiol 1998;8(8):1481–1494.
7. Yasuda S, Ide M, Fujii H, Nakahara T, Mochizuki Y, Takahashi W, et al. Application of positron emission tomography imaging to cancer screening. Br J Cancer 2000;83(12):1607–1611.
8. Valk PE, Pounds TR, Tesar RD, Hopkins DM, Haseman MK. Cost-effectiveness of PET imaging in clinical oncology. Nucl Med Biol 1996;23(6):737–743.
9. Arulampalam TH, Costa DC, Loizidou M, Visvikis D, Ell PJ, Taylor I. Positron emission tomography and colorectal cancer. Br J Surg 2001;88(2):176–189.
10. Akhurst T, Larson SM. Positron emission tomography imaging of colorectal cancer. Semin Oncol 1999;26(5):577–583.
11. Anderson H, Price P. What does positron emission tomography offer oncology? Eur J Cancer 2000;36(16):2028–2035.
12. Rohren EM, Turkington TG, Coleman RE. Clinical applications of PET in oncology. Radiology 2004;231(2):305–332.
13. Townsend DW, Cherry SR. Combining anatomy and function: the path to true image fusion. Eur Radiol 2001;11(10):1968–1974.
14. Hughes KS, Simon R, Songhorabodi S, Adson MA, Ilstrup DM, Fortner JG, et al. Resection of the liver for colorectal carcinoma metastases: a multi-institutional study of patterns of recurrence. Surgery (St. Louis) 1986;100(2):278–284.
15. Moertel CG, Fleming TR, MacDonald JS, Haller DG, Laurie JA, Tangen C. An evaluation of the carcinoembryonic antigen (CEA) test for monitoring patients with resected colon cancer. JAMA 1993;270(8):943–947.
16. Charnley RM. Imaging of colorectal carcinoma. Radiology 1990;174(1):283–283.
17. Steele G, Bleday R, Mayer RJ, Lindblad A, Petrelli N, Weaver D. A prospective evaluation of hepatic resection for colorectal carcinoma metastases to the liver. Gastrointestinal-Tumor-Study-Group Protocol-6584. J Clin Oncol 1991;9(7):1105–1112.
18. McDaniel KP, Charnsangavej C, Dubrow RA, Varma DGK, Granfield CAJ, Curley SA. Pathways of nodal metastasis in carcinomas of the cecum, ascending colon, and transverse colon: CT demonstration. Am J Roentgenol 1993;161(1):61–64.
19. Nelson RC, Chezmar JL, Sugarbaker PH, Bernardino ME. Hepatic tumors: comparison of CT during arterial portography, delayed CT, and MR imaging for preoperative evaluation. Radiology 1989;172(1):27–34.
20. Coleman RE. FDG imaging. Nucl Med Biol 2000;27(7):689–690.
21. Dahlbom M, Hoffman EJ, Hoh CK, Schiepers C, Rosenqvist G, Hawkins RA, et al. Whole-body positron emission tomography. 1. Methods and performance characteristics. J Nucl Med 1992;33(6):1191–1199.
22. Larcos G, Maisey MN. FDG-PET screening for cerebral metastases in patients with suspected malignancy. Nucl Med Commun 1996;17(3):197–198.
23. Abdel-Nabi H, Doerr RJ, Lamonica DM, Cronin VR, Galantowicz PJ, Carbone GM, et al. Staging of primary colorectal carcinomas with fluorine-18 fluorodeoxyglucose whole-body PET: correlation with histopathologic and CT findings. Radiology 1998;206(3):755–760.
24. Falk PM, Gupta NC, Thorson AG, Frick MP, Boman BM, Christensen MA, et al. Positron emission tomography for preoperative staging of colorectal carcinoma. Dis Colon Rectum 1994;37(2):153–156.
25. Ruhlmann J, Schomburg A, Bender H, Oehr P, Robertz-Vaupel GM, Vaupel H, et al. Fluorodeoxyglucose whole-body positron emission tomography in colorectal cancer patients studied in routine daily practice. Dis Colon Rectum 1997;40(10):1195–1204.
26. Strauss LG, Clorius JH, Schlag P, Lehner B, Kimmig B, Engenhart R, et al. Recurrence of colorectal tumors: PET evaluation. Radiology 1989;170(2):329–332.
27. Ito K, Kato T, Tadokoro M, Ishiguchi T, Oshima M, Ishigaki T, et al. Recurrent rectal cancer and scar: differentiation with PET and MR imaging. Radiology 1992;182(2):549–552.
28. Huebner RH, Park KC, Shepherd JE, Schwimmer J, Czernin J, Phelps ME, et al. A meta-analysis of the literature for whole-body FDG PET detection of recurrent colorectal cancer. J Nucl Med 2000;41(7):1177–1189.
29. Warburg O. On the origin of cancer cells. Science 1956;123:309–314.
30. Wahl RL. Targeting glucose transporters for tumor imaging: sweet idea, sour result. J Nucl Med 1996;37(6):1038–1041.
31. Strauss LG. Fluorine-18 deoxyglucose and false-positive results: a major problem in the diagnostics of oncological patients. Eur J Nucl Med 1996;23(10):1409–1415.
32. Bengel FM, Ziegler SI, Avril N, Weber W, Laubenbacher C, Schwaiger M. Whole-body positron emission tomography in clinical oncology: comparison between attenuation-corrected and uncorrected images [see comments]. Eur J Nucl Med 1997;24(9):1091–1098.
33. Kotzerke J, Guhlmann A, Moog F, Frickhofen N, Reske SN. Role of attenuation correction for fluorine-18 fluorodeoxyglucose positron emission tomography in the primary staging of malignant lymphoma. Eur J Nucl Med 1999;26(1):31–38.
34. Wahl RL. To AC or not to AC: that is the question. J Nucl Med 1999;40(12):2025–2028.
35. Metser U. ^{18}F-FDG PET in evaluating patients treated for metastatic colorectal cancer: can we predict prognosis? J Nucl Med 2004;45(9):1428–1430.
36. Takeuchi O, Saito N, Koda K, Sarashina H, Nakajima N. Clinical assessment of positron emission tomography for the diagnosis of local recurrence in colorectal cancer. Br J Surg 1999;86(7):932–937.
37. Kim CK, Gupta NC, Chandramouli B, Alavi A. Standardized uptake values of FDG: body surface area correction is preferable to body weight correction. J Nucl Med 1994;35(1):164–167.
38. Zasadny KR, Wahl RL. Standardized uptake values of normal tissues at PET with 2-[fluorine-L8]-fluoro-2-deoxy-D-glucose: variations with body weight and a method for correction. Radiology 1993;189(3):847–850.
39. Graham MM, Peterson LM, Hayward RM. Comparison of simplified quantitative analyses of FDG uptake. Nucl Med Biol 2000;27(7):647–655.
40. Burger C, Goerres G, Schoenes S, Buck A, Lonn AH, Von Schulthess GK. PET attenuation coefficients from CT images: experimental evaluation of the transformation of CT into PET 511-keV attenuation coefficients. Eur J Nucl Med Mol Imaging 2002;29(7):922–927.
41. Goerres GW, Hany TF, Kamel E, von Schulthess GK, Buck A. Head and neck imaging with PET and PET/CT: artefacts from dental metallic implants. Eur J Nucl Med Mol Imaging 2002; 29(3):367–370.
42. Halpern BS, Dahlbom M, Waldherr C, Yap CS, Schiepers C, Silverman DH, et al. Cardiac pacemakers and central venous lines can induce focal artifacts on CT-corrected PET images. J Nucl Med 2004;45(2):290–293.
43. Halpern BS, Dahlbom M, Quon A, Schiepers C, Waldherr C, Silverman DH, et al. Impact of patient weight and emission scan duration on PET/CT image quality and lesion detectability. J Nucl Med 2004;45(5):797–801.

44. Beets G, Penninckx F, Schiepers C, Filez L, Mortelmans L, Kerremans R, et al. Clinical value of whole-body positron emission tomography with [F-18]fluorodeoxyglucose in recurrent colorectal cancer. Br J Surg 1994;81(11):1666–1670.

45. Vitola JV, Delbeke D, Sandler MP, Campbell MG, Powers TA, Wright JK, et al. Positron emission tomography to stage suspected metastatic colorectal carcinoma to the liver. Am J Surg 1996;171(1):21–26.

46. Lai DT, Fulham M, Stephen MS, Chu KM, Solomon M, Thompson JF, et al. The role of whole-body positron emission tomography with [^{18}F]fluorodeoxyglucose in identifying operable colorectal cancer metastases to the liver. Arch Surg 1996;131(7):703–707.

47. Ogunbiyi OA, Flanagan FL, Dehdashti F, Siegel BA, Trask DD, Birnbaum EH, et al. Detection of recurrent and metastatic colorectal cancer: comparison of positron emission tomography and computed tomography [see comments]. Ann Surg Oncol 1997;4(8):613–620.

48. Schiepers C, Penninckx F, De Vadder N, Merckx E, Mortelmans L, Bormans G, et al. Contribution of PET in the diagnosis of recurrent colorectal cancer: comparison with conventional imaging. Eur J Surg Oncol 1995;21(5):517–522.

49. Delbeke D, Vitola JV, Sandler MP, Arildsen RC, Powers TA, Wright JK, Jr., et al. Staging recurrent metastatic colorectal carcinoma with PET. J Nucl Med 1997;38(8):1196–1201.

50. Flamen P, Stroobants S, Van Cutsem E, Dupont P, Bormans G, De Vadder N, et al. Additional value of whole-body positron emission tomography with fluorine-18-2-fluoro-2-deoxy-D-glucose in recurrent colorectal cancer. J Clin Oncol 1999;17(3):894–901.

51. Valk PE, Abella-Columna E, Haseman MK, Pounds TR, Tesar RD, Myers RW, et al. Whole-body PET imaging with [^{18}F]fluorodeoxyglucose in management of recurrent colorectal cancer. Arch Surg 1999;134(5):503–511; discussion 511–513.

52. Schiepers C, Hoh CK. Clinical positron imaging in oncology. In: Schiepers C, editor. Diagnostic Nuclear Medicine, 1st ed. Berlin: Springer-Verlag, 2000:159–176.

53. Delbeke D, Valk PE. Positron tomography imaging: diagnosis and management of colorectal carcinoma. In: Khalkhali IMJ, Goldsmith SJ, editors. Nuclear Oncology: Diagnosis and Therapy: Lippincott, Williams & Wilkins, 2001:351–360.

54. Keogan MT, Lowe VJ, Baker ME, McDermott VG, Lyerly HK, Coleman RE. Local recurrence of rectal cancer: evaluation with F-18 fluorodeoxyglucose PET imaging. Abdom Imaging 1997;22(3):332–337.

55. Whiteford MH, Whiteford HM, Yee LF, Ogunbiyi OA, Dehdashti F, Siegel BA, et al. Usefulness of FDG-PET scan in the assessment of suspected metastatic or recurrent adenocarcinoma of the colon and rectum. Dis Colon Rect 2000;43(6):759–767; discussion 767–770.

56. Fong YM, Saldinger PF, Akhurst T, Macapinlac H, Yeung H, Finn RD, et al. Utility of F-18-FDG positron emission tomography scanning on selection of patients for resection of hepatic colorectal metastases. Am J Surg 1999;178(4):282–287.

57. Topal B, Flamen P, Aerts R, D'Hoore A, Filez L, Van Cutsem E, et al. Clinical value of whole-body positron emission tomography in potentially curable colorectal liver metastases. Eur J Surg Oncol 2001;27(2):175–179.

58. Boykin KN, Zibari GB, Lilien DL, McMillan RW, Aultman DF, McDonald JC. The use of FDG-positron emission tomography for the evaluation of colorectal metastases of the liver. Am Surg 1999;65(12):1183–1185.

59. Fernandez FG, Drebin JA, Linehan DC, Dehdashti F, Siegel BA, Strasberg SM. Five-year survival after resection of hepatic metastases from colorectal cancer in patients screened by positron emission tomography with F-18 fluorodeoxyglucose (FDG-PET). Ann Surg 2004;240(3):438–447; discussion 447–450.

60. Lowe VJ, Fletcher JW, Gobar L, Lawson M, Kirchner P, Valk P, et al. Prospective investigation of positron emission tomography in lung nodules. J Clin Oncol 1998;16(3):1075–1084.

61. Flanagan FL, Dehdashti F, Ogunbiyi OA, Kodner IJ, Siegel BA. Utility of FDG-PET for investigating unexplained plasma CEA elevation in patients with colorectal cancer [see comments]. Ann Surg 1998;227(3):319–323.

62. Maldonado A, Sancho F, Cerdan J, Lozano A, Mohedano N, Jimenez J, et al. 16. FDG-PET in the detection of recurrence in colorectal cancer based on rising CEA level. Experience in 72 patients. Clin Positron Imaging 2000;3(4):170.

63. Flamen P, Hoekstra OS, Homans F, Van Cutsem E, Maes A, Stroobants S, et al. Unexplained rising carcinoembryonic antigen (CEA) in the postoperative surveillance of colorectal cancer: the utility of positron emission tomography (PET). Eur J Cancer 2001;37(7):862–869.

64. Strasberg SM, Dehdashti F, Siegel BA, Drebin JA, Linehan D. Survival of patients evaluated by FDG-PET before hepatic resection for metastatic colorectal carcinoma: a prospective database study. Ann Surg 2001;233(3):293–299.

65. Staib L, Schirrmeister H, Reske SN, Beger HG. Is F-18-fluorodeoxyglucose positron emission tomography in recurrent colorectal cancer a contribution to surgical decision making? Am J Surg 2000;180(1):1–5.

66. Imdahl A, Reinhardt MJ, Nitzsche EU, Mix M, Dingeldey A, Einert A, et al. Impact of ^{18}F-FDG-positron emission tomography for decision making in colorectal cancer recurrences. Langenbecks Arch Surg 2000;385(2):129–134.

67. Meta J, Seltzer M, Schiepers C, Silverman DH, Ariannejad M, Gambhir SS, et al. Impact of F-18-FDG PET on managing patients with colorectal cancer: the referring physician's perspective. J Nucl Med 2001;42(4):586–590.

68. Kalff V, Hicks RJ, Ware RE, Hogg A, Binns D, McKenzie AF. The clinical impact of (18)F-FDG PET in patients with suspected or confirmed recurrence of colorectal cancer: a prospective study. J Nucl Med 2002;43(4):492–499.

69. Desai DC, Zervos EE, Arnold MW, Burak WE Jr, Mantil J, Martin EW Jr. Positron emission tomography affects surgical management in recurrent colorectal cancer patients. Ann Surg Oncol 2003;10(1):59–64.

70. Strauss LG, Conti PS. The applications of PET in clinical oncology. J Nucl Med 1991;32(4):623–648.

71. Dimitrakopoulou-Strauss A, Strauss LG, Schlag P, Hohenberger P, Irngartinger G, Oberdorfer F, et al. Intravenous and intra-arterial oxygen-15-labeled water and fluorine-18-labeled fluorouracil in patients with liver metastases from colorectal carcinoma. J Nucl Med 1998;39(3):465–473.

72. Mantaka P, Strauss AD, Strauss LG, Moehler M, Goldschmidt H, Oberdorfer F, et al. Detection of treated liver metastases using fluorine-18-fluordeoxyglucose (FDG) and positron emission tomography (PET). Anticancer Res 1999;19(5C):4443–4450.

73. Moehler M, Dimitrakopoulou-Strauss A, Gutzler F, Raeth U, Strauss LG, Stremmel W. 18F-labeled fluorouracil positron emission tomography and the prognoses of colorectal carcinoma patients with metastases to the liver treated with 5-fluorouracil. Cancer (Phila) 1998;83(2):245–253.

74. Dimitrakopoulou-Strauss A, Strauss LG, Schlag P, Hohenberger P, Möhler M, Oberdorfer F, et al. Fluorine-18-fluorouracil to predict therapy response in liver metastases from colorectal carcinoma. J Nucl Med 1998;39(7):1197–1202.

75. Findlay M, Young H, Cunningham D, Iveson A, Cronin B, Hickish T, et al. Noninvasive monitoring of tumor metabolism using fluorodeoxyglucose and positron emission tomography in colorectal cancer liver metastases: correlation with tumor response to fluorouracil [see comments]. J Clin Oncol 1996;14(3):700–708.

76. Vitola JV, Delbeke D, Meranze SG, Mazer MJ, Pinson CW. Positron emission tomography with F-18-fluorodeoxyglucose to evaluate the results of hepatic chemoembolization. Cancer (Phila) 1996;78(10):2216–2222.

77. Haberkorn U, Strauss LG, Dimitrakopoulou A, Engenhart R, Oberdorfer F, Ostertag H, et al. PET studies of fluorodeoxyglucose

metabolism in patients with recurrent colorectal tumors receiving radiotherapy. J Nucl Med 1991;32(8):1485–1490.

78. Abe Y, Matsuzawa T, Fujiwara T, Fukuda H, Itoh M, Yamada K, et al. Assessment of radiotherapeutic effects on experimental tumors using fluorine-18 2-fluoro-2-deoxy-D-glucose. Eur J Nucl Med 1986;12(7):325–328.

79. Schiepers C, Haustermans K, Geboes K, Filez L, Bormans G, Penninckx F. The effect of preoperative radiation therapy on glucose utilization and cell kinetics in patients with primary rectal carcinoma. Cancer (Phila) 1999;85(4):803–811.

80. Young H, Baum R, Cremerius U, Herholz K, Hoekstra O, Lammertsma AA, et al. Measurement of clinical and subclinical tumour response using [F-18]-fluorodeoxyglucose and positron emission tomography: review and 1999 EORTC recommendations. Eur J Cancer 1999;35(13):1773–1782.

81. Gambhir SS, Valk P, Shepherd J, Hoh C, Allen M, Phelps ME. Cost-effective analysis modeling of the role of FDG PET in the management of patients with recurrent colorectal cancer. J Nucl Med 1997;38(suppl):90P–91P.

82. Park KC, Schwimmer J, Shepherd JE, Phelps ME, Czernin JR, Schiepers C, et al. Decision analysis for the cost-effective management of recurrent colorectal cancer. Ann Surg 2001; 233(3):310–319.

83. Beyer T, Townsend DW, Brun T, Kinahan PE, Charron M, Roddy R, et al. A combined PET/CT scanner for clinical oncology. J Nucl Med 2000;41(8):1369–1379.

84. Lardinois D, Weder W, Hany TF, Kamel EM, Korom S, Seifert B, et al. Staging of non-small-cell lung cancer with integrated positron-emission tomography and computed tomography. N Engl J Med 2003;348(25):2500–2507.

85. Cohade C, Osman M, Leal J, Wahl RL. Direct comparison of FDG-PET and PET/CT in patients with colorectal carcinoma. J Nucl Med 2003;44:1797–1803.

86. Delbeke D. Oncological applications of FDG PET imaging: brain tumors, colorectal cancer, lymphoma and melanoma. J Nucl Med 1999;40(4):591–603.

87. Schiepers C, Yap C S, Quon A, Giuliano P, Silverman DH, Dahlbom M, et al. Added value of PET/CT for cancer staging and lesion localization. Eur J Nucl Med Mol Imaging 2003;30(8):S227–S228.

88. Goerres GW, Burger C, Schwitter MW, Heidelberg TN, Seifert B, Von Schulthess GW. PET/CT of the abdomen: optimizing the patient breathing pattern. Eur Radiol 2003;13(4):734–739.

89. Dizendorf E, Hany TF, Buck A, Von Schulthess GK, Burger C. Cause and magnitude of the error induced by oral CT contrast agent in CT-based attenuation correction of PET emission studies. J Nucl Med 2003;44(5):732–738.

90. Cohade C, Osman M, Nakamoto Y, Marshall LT, Links JM, Fishman EK, et al. Initial experience with oral contrast in PET/CT: phantom and clinical studies. J Nucl Med 2003;44(3):412–416.

91. Cohade C, Osman M, Pannu HK, Wahl RL. Uptake in supraclavicular area fat ("USA-Fat"): description on ^{18}F-FDG PET/CT. J Nucl Med 2003;44(2):170–176.

92. Ciernik IF, Dizendorf E, Baumert BG, Reiner B, Burger C, Davis JB, et al. Radiation treatment planning with an integrated positron emission and computer tomography (PET/CT): a feasibility study. Int J Radiat Oncol Biol Phys 2003;57(3):853–863.

93. Wahl RL. Why nearly all PET of abdominal and pelvic cancers will be performed as PET/CT. J Nucl Med 2004;45(suppl 1):82S–95S.

94. Phelps ME. PET: the merging of biology and imaging into molecular imaging. J Nucl Med 2000;41(4):661–681.

11

PET and PET/CT Imaging in Esophageal and Gastric Cancers

Farrokh Dehdashti and Barry A. Siegel

Esophageal and gastric cancers are significant health problems. In the United States, it is estimated that 36,960 patients will be diagnosed in the year 2004 with esophageal or gastric cancer and that these cancers will be responsible for 4.5% of all cancer-related deaths (1). The incidence of gastric cancer is decreasing, whereas the incidence of esophageal cancer is increasing, mainly because of the increase in the frequency of adenocarcinoma of the distal esophagus.

Surgery is the mainstay of treatment of both esophageal and gastric cancer. Despite recent advances in surgical treatment, the overall prognosis of patients with esophageal or gastric cancer has not improved significantly because the neoplasm is often diagnosed at an advanced stage of the disease. Local and systemic recurrences are common, even after complete resection of the primary tumor and regional lymph nodes. Multimodality therapy, consisting of surgery with adjuvant or neoadjuvant radiotherapy, chemotherapy, or both, has been used recently as a means to improve survival of patients with esophageal or gastric cancer. Randomized clinical trials have shown that concurrent chemotherapy and radiotherapy are more effective than radiotherapy alone in treating advanced esophageal cancer (2). Current data suggest that these cancers are best managed with a tailored therapeutic regimen, based on thorough preoperative staging of the tumor and an understanding of established prognostic factors (3).

The TNM staging system is used for staging of both esophageal and gastric cancers: T stage refers to the depth of the invasion of the primary tumor, N stage refers to the extent of lymph node involvement, and M stage indicates the presence or absence of systemic metastases (Tables 11.1, 11.2). Current preoperative staging techniques, such as computed tomography (CT), are of limited accuracy, and invasive procedures often are used for better assessment of the stage of the disease. Beginning in the mid-1990s, positron emission tomography (PET) has been evaluated as a method for the staging of esophageal and gastric cancer. In the past few years, combined PET/CT

scanners have been rapidly replacing conventional PET for the evaluation of oncologic patients. Although there are still only limited data regarding the use of PET/CT in esophageal and gastric cancers, the combination of these two modalities is expected to improve the accuracy of image interpretation, and thus lead to better management of cancer patients (4, 5).

Esophageal Cancer

Over the past two decades in Western countries, the incidence of esophageal adenocarcinoma has increased and is currently higher than that of squamous cell carcinoma (2). Effective treatment and prediction of outcome in esophageal cancer are based on accurate tumor staging. Patients with early-stage disease may benefit from esophagectomy alone, whereas multimodality therapy with surgery and adjuvant or neoadjuvant chemotherapy and/or radiation may be indicated in patients with advanced locoregional disease. Nonsurgical palliative therapy is indicated in patients with distant metastatic disease because these patients have a poor prognosis regardless of the type of treatment. Cancer of the esophagus is most often diagnosed by endoscopic biopsy or brushing, and the size, location, and morphology of the tumor are evaluated by endoscopy and barium esophagography. Currently used staging methods include CT, endoscopic ultrasonography (EUS), and magnetic resonance imaging (MRI). However, these anatomic imaging techniques have significant limitations. EUS has been reported to have an accuracy of 85% for assessment of the depth of tumor invasion and 75% for detection of regional lymph node metastases (6). CT has also been found to have limited accuracy of 50% to 60% for staging esophageal cancer (7–9), and MRI has not significantly improved these staging results. The main shortcoming of these imaging modalities is their reliance on detection of structural changes for diagnosis of disease. The high rate of treatment failure after surgery with curative intent in patients with imaging

Table 11.1. TNM staging for esophageal carcinoma.

Primary Tumor (T)

Tis	=	carcinoma in situ
T1	=	tumor invades into but not beyond the submucosa
T2	=	tumor invades into but not beyond the muscularis propria
T3	=	tumor invades into the adventitia
T4	=	tumor invades adjacent structures

Regional Lymph Nodes (N)

N0	=	regional nodes not involved
N1	=	regional nodes involved

Distant Metastasis (M)

M0	=	no distant metastasis
M1	=	distant metastasis (including nodal involvement outside the mediastinum)

Stage Grouping

Stage 0	=	Tis, N0, M0
Stage 1	=	T1, N0, M0
Stage II	=	T1-2, N0-1, M0
Stage III	=	T3-4, N0-1, M0
Stage IV	=	Any T, Any N, M1

Source: From Valk PE, Bailey DL, Townsend DW, Maisey MN. Positron Emission Tomography: Basic Science and Clinical Practice. Springer-Verlag London Ltd 2003, p. 571.

Table 11.2. TNM sStaging for gastric carcinoma.

Primary Tumor (T)

Tis	=	carcinoma in situ
T1	=	tumor invades lamina propria or submucosa
T2	=	tumor invades muscularis propria
T3	=	tumor invades adventitia
T4	=	tumor invades adjacent structures

Regional Lymph Nodes (N)

N0	=	regional nodes not involved
N1	=	Metastasis in perigastric lymph nodes(s) within 3 cm of edge of primary tumor
N2	=	Metastasis in perigastric lymph nodes(s) more than 3 cm of edge of primary tumor, or in lymph nodes along left gastric, common hepatic, splenic, or celiac arteries

Distant Metastasis (M)

M0	=	no distant metastasis
M1	=	distant metastasis (including nodal involvement outside the mediastinum)

Stage Grouping

Stage 0	=	Tis, N0, M0
Stage 1	=	T1-2, N0-1, M0
Stage II	=	T1-3, N0-2, M0
Stage III	=	T2-4, N0-2, M0
Stage IV	=	Any T, Any N, M1

Source: From Valk PE, Bailey DL, Townsend DW, Maisey MN. Positron Emission Tomography: Basic Science and Clinical Practice. Springer-Verlag London Ltd 2003, p. 572.

evidence of only localized disease is likely related to current inaccurate staging procedures.

The addition of PET with 2-[^{18}F]fluoro-2-deoxy-D-glucose (FDG) to the current imaging techniques that are used for staging significantly improves accuracy. The use of FDG-PET in esophageal cancer was approved for reimbursement by the U.S. Medicare program in 2001. In many institutions, FDG-PET is now routinely employed for staging of esophageal cancer (10–17). There is very limited experience in the application of PET with other radiopharmaceuticals (e.g., ^{11}C-choline) for the evaluation of esophageal cancer (18). Accordingly, this chapter exclusively describes clinical results obtained with FDG.

Primary Tumor Staging (T Stage)

The T stage is determined by the depth of tumor infiltration into or through the esophageal wall, and this is one of the most important prognostic factors in esophageal cancer. Because of the ability of the esophagus to distend, dysphagia, which is a common presenting symptom, does not occur until the disease is advanced and the tumor bulk compromises the esophageal lumen. EUS is useful for accurate evaluation of the depth of primary tumor penetration within the wall and the invasion of periesophageal tissues. However, EUS is operator dependent and is unable to distinguish tumor from inflammation, so that tumor stage may be overestimated in the presence of peritumoral inflammation. The major limita-

tion of EUS is its inability to evaluate tumors that have caused stenosis of the esophageal lumen, thereby preventing passage of the endoscope. CT complements EUS in detecting macroscopic invasion of mediastinal fat and infiltration into the adjacent organs, particularly the trachea and bronchi (T3 and T4 stage). However, CT is limited for detection of early-stage (T1 and T2) tumors and for differentiating malignant from benign causes of esophageal wall thickening. The accuracy of CT is further limited by the diminished amount of mediastinal fat in many patients with esophageal cancer, who often have sustained significant weight loss by the time of presentation. In addition, accurate assessment of the local extent of the tumor may be hindered by partial-volume averaging consequent to the close proximity of the tumor to the pulsating aorta or heart (19).

FDG-PET can detect esophageal cancer before it becomes evident on CT, but PET is limited in its ability to determine the extent of tumor spread through the esophageal wall or tumor invasion of the adjacent structures. This limitation results chiefly from the poorer resolution of PET by comparison with anatomic imaging methods and its limited delineation of normal anatomic structures. In our experience, a heterogeneous pattern of FDG uptake at the primary site, especially when it has irregular margins, is suggestive of local extension of the tumor into the surrounding soft tissues. Several investigators have shown that FDG-PET has a higher sensitivity than CT for detection of primary esophageal cancer (83%–100% versus 67%–92%) (10–13, 15–17, 20, 21). The one exception was a study utilizing a partial-ring PET scanner without attenuation correction of images, where PET was found to have a lower sensitivity than CT (84% versus 97%) (22). In most studies, false-negative results of PET occurred in patients with small T1 lesions. Physiologic uptake of FDG in the normal esophagus also may be a limitation in detection of small or well-differentiated tumors.

FDG uptake in primary esophageal tumors has been assessed mainly by qualitative visual analysis. Yeung et al. (16) also assessed the FDG uptake in esophageal tumors by determination of the standardized uptake value (SUV) and found no difference in FDG uptake in adenocarcinomas and squamous cell carcinomas. Fukunaga et al. (23) found that 47 of 48 patients with esophageal cancer had a primary tumor SUV greater than 2.0 (sensitivity of 98%). The mean SUV in primary esophageal cancers (6.99 ± 3.05; $n = 48$) was greater than that of either normal esophagus (1.34 ± 0.37; $n = 10$) or a single benign esophageal tumor (0.86) (23). Flamen et al. (17) compared the primary tumor SUV with the T stage in 50 patients and found no correlation. In contrast, Kato et al. (24) reported a significant association between FDG uptake of the primary tumor, as measured by the SUV, and the depth of tumor invasion (P less than 0.05), tumor dimensions (P less than 0.01), the occurrence of lymph node involvement (P less than 0.01), and lymphatic invasion (P less than 0.01).

All currently used imaging techniques are limited in differentiating tumor from inflammatory disease and in detecting microscopic disease, so that histopathologic examination of the resected specimen remains the criterion standard for T-stage determination.

Regional Lymph Node Metastases (N Stage)

The status of regional lymph nodes is the most important prognostic factor in patients with esophageal cancer: patients with nodal metastases have a higher likelihood of systemic spread of the disease and a worse prognosis (25–27). Lymph node status has a major impact on treatment selection. Lymph node involvement, either regional or distant, commonly occurs before involvement of other distant organs. The high prevalence of lymph node involvement in esophageal cancer is the result of the rich network of lymphatics, which extends along the entire esophagus. The limitations of the current imaging techniques, CT and EUS, in accurate detection of lymph node involvement are related to their inability to detect tumor involvement in normal-sized lymph nodes and to differentiate whether lymph node enlargement is caused by metastatic or inflammatory disease. Although multidetector CT has better resolution and improved ability to

detect small lymph nodes, it has not been shown to improve the accuracy of esophageal cancer staging; this is no doubt related to the size criterion used by CT for detection of lymph node metastasis. To increase the accuracy of preoperative staging, the use of minimally invasive surgical staging, consisting of thoracoscopy with or without abdominal staging laparoscopy, has been recommended. However, because of their invasiveness, morbidity, and high cost, these procedures are not used routinely in clinical practice.

Clinical studies have shown that FDG-PET can significantly improve preoperative nodal staging (Table 11.3). Although most of these studies were retrospective and the studies employed different imaging protocols, the results demonstrate an important role for FDG-PET in staging esophageal cancer. In early work at Washington University, we studied 36 patients with esophageal cancer. In 29 patients who underwent esophagectomy with curative intent, we found that the accuracy for detection of nodal disease was 76% (22/29) for FDG-PET and 45% (13/29) for CT (10). Most subsequent studies including our own have demonstrated slightly lower sensitivity, with similar or higher specificity for detection of locoregional nodal disease by PET. The reported sensitivities have ranged from 22% to 71% (with one report of 92%) for PET, compared to 0% to 87% for CT (10–13, 15–17, 21, 22, 28–31) (Figure 11.1). Specificities ranged from 78% to 100% for PET and from 73% to 100% for CT (see Table 11.3) (10–13, 15–17, 21, 22, 28–32). A recent meta-analysis of 12 studies reported in the literature has demonstrated that the overall pooled sensitivity and specificity of FDG-PET for detection of locoregional disease were 51% [95% confidence interval (CI), 34%–69%] and 84% (95% CI, 76%–91%), respectively (33).

Although these reports suggest that FDG-PET may produce some small improvement in locoregional staging, the sensitivity of both PET and CT is too low for use in clinical decision making, and nodal sampling is used in all patients who are otherwise considered to be surgical candidates. False-negative results are chiefly found in nodes with small tumor burden (especially nodes less than 1 cm in diameter) and involved lymph nodes that lie in close proximity to the primary tumor. These adjacent lymph nodes are typically resected with the primary tumor, and their involvement usually does not alter management. False-positive results are chiefly caused by inflammatory disease or heterogeneous uptake in the primary tumor simulating periesophageal nodal metastasis. Inflammatory adenopathy should be suspected when there is other evidence of granulomatous disease on CT imaging (e.g., nodal calcification).

Distant Metastatic Disease (M Stage)

The prognosis for patients with metastatic esophageal cancer is very poor, and major surgery is not justified in

Table 11.3. Comparison of FDG-PET and CT for detection of regional lymph node involvement in esophageal cancer.

Study (year)	Number of patients Biopsy or surgery/total[a]	Sensitivity (%) PET	CT	Specificity (%) PET	CT
Flanagan et al. (1997)	29/36	72	28	82	73
Block et al. (1997)[b]	35/58	52	28	78	78
Luketich et al. (1997)	21/35	45	NA	100	NA
Kole et al. (1998)	22/26	92	38	88	100
Rankin et al. (1998)	18/25	37	50	90	80
Yeung et al. (1999)	NA/67	28	25	99	98
Flamen et al. (2000)	39/74	33	0	89	100
Lerut et al. (2000)[c]	42/74	22	83	91	45
Meltzer et al. (2000)	37/47	41	87	88	43
Kim et al. (2001)	50/53	52	15	94	97
Himeno et al. (2002)	31/36	37	31	96	88
Kato et al. (2002)	32/32	78	61	93	71
Wren et al. (2002)	21/24	71	57	86	71
Räsänen et al. (2003)	19/42	37	89[d]	100	54[d]
Yoon et al. (2003)	81/136	30	11	90	95
Liberale et al. (2004)	8/24	38	25	81	50

NA, information not available.
[a]Number of patients who underwent surgical resection of esophageal cancer.
[b]The PET and CT results of some of these patients are also reported in Flanagan et al. (1997).
[c]Reanalysis of 42 of 74 patients reported in Flamen et al. (2000).
[d]Endoscopic ultrasonography.
Source: Updated from Valk PE, Bailey DL, Townsend DW, Maisey MN. Positron Emission Tomography: Basic Science and Clinical Practice. Springer-Verlag London Ltd 2003, p. 574.

these patients. Therefore, it is essential to identify patients with advanced disease accurately to permit selection of the most effective and rational management approach and to avoid subjecting them to ineffective costly and debilitating therapeutic procedures. Esophageal cancer typically metastasizes to distant lymph nodes, liver, and lung, before metastasizing to other organs, such as bone and adrenal glands. Although evaluations of FDG-PET for detection of distant metastatic disease in patients with esophageal cancer have included only limited patient numbers, the results of these studies have demonstrated an important role for PET in pretreatment staging.

Clinical studies have shown that FDG-PET is more sensitive than conventional imaging such as CT, ultrasonography, and bone scintigraphy for demonstrating the true extent of metastatic disease (Table 11.4). In our initial study, we demonstrated that FDG-PET was superior to CT, detecting distant metastatic disease in 5 of 7 versus 0 of 7 patients (10). The positive PET findings were confirmed histologically in all 5 of these patients. The 2 patients with false-negative PET results were respectively found at laparotomy to have a small hepatic metastasis and a small pancreatic metastasis. Our subsequent evaluation of a larger patient group confirmed these findings, demonstrating a sensitivity of 100% (17/17) for PET versus 29% (5/17) for CT in detection of distant metastatic disease (11). Eleven of these 17 patients subsequently underwent

minimally invasive staging procedures, such as percutaneous biopsy or mediastinoscopy, with confirmation of the PET results in every case (Figure 11.2).

Luketich et al. (13) studied 35 patients with esophageal cancer and demonstrated that FDG-PET had a sensitivity of 88% (7/8) and specificity of 93% (25/27) for detection of distant metastatic disease. FDG-PET was falsely negative in 1 patient with a 2-mm hepatic lesion. CT demonstrated small (less than 1 cm) pulmonary lesions in 6 patients, all of whom had negative PET studies. Video-assisted thoracotomy confirmed the PET results in all 6 patients showing benign hamartoma in 2 patients and benign granuloma or fibrosis in 4. These investigators reported that FDG-PET results facilitated treatment planning by demonstrating unsuspected distant metastatic disease in up to 20% of patients with negative results by conventional imaging. In a later evaluation, Luketich et al. (14) prospectively compared PET and CT with minimally invasive staging in 91 patients (100 PET scans) with esophageal cancer. Seventy distant metastatic lesions in 39 patients were confirmed clinically or by biopsy. Sensitivity and specificity were 69% and 93%, respectively, for FDG-PET and 46% and 74%, respectively, for CT. Similar results are reported by others (12, 15).

In a prospective study of 74 patients, Flamen et al. (17) found that FDG-PET was superior to CT and EUS in detection of stage IV disease. The sensitivity and specificity

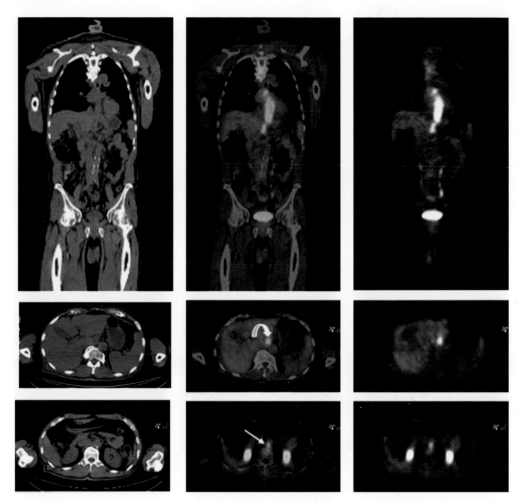

Figure 11.1. Staging esophageal cancer: 65-year-old man with an esophageal cancer. Coronal (*top*) computed tomography (CT), positron emission tomography (PET)-CT fusion, and PET images demonstrate intense ^{18}F-2-deoxy-D-glucose (FDG) uptake within the thickened distal esophagus, consistent with primary esophageal cancer. Transaxial (*middle* and *bottom*) CT, PET/CT fusion, and PET images demonstrate intense FDG uptake in an enlarged gastrohepatic ligament lymph node (2.0 × 2.5 cm) (*curved arrow;middle images*) and a small (7 mm) lymph node (*arrow, bottom images*), which showed increased FDG uptake, suspicious for metastatic disease.

Table 11.4. Comparison of FDG-PET and CT for detection of distant metastatic disease in esophageal cancer.

| Study (year) | Number of patients | Sensitivity (%) | | Specificity (%) | |
	Biopsy or surgery/total[a]	PET	CT	PET	CT
Flanagan et al. (1997)	7/36	71	0	NA	NA
Block et al. (1997)[a]	17/58	100	29	NA	NA
Luketich et al. (1997)	7/35	88	0	93	70
Kole et al. (1998)	8/26	100	62	92	92
Luketich et al. (1999)	39/91	69	46	93	74
Flamen et al. (2000)	34/74	74	47	90	78
Lerut et al. (2000)[b]	13/42	77	83	90	69
Meltzer et al. (2000)	10/47	71	57	93	93
Wren et al. (2002)	12/24	67	83	92	75
Räsänen et al. (2003)	15/42	47	33	89	96
Heeren et al. (2004)	24/74	78	37	98	87
Liberale et al. (2004)	7/58	88	44	88	95

CI, conventional imaging [CT and/or ultrasound (US)].
[a]The PET and CT results of some of these patients are also reported in Flanagan et al. (1997).
[b]Reanalysis of 42 of 74 patients reported in Flamen et al. (2000).

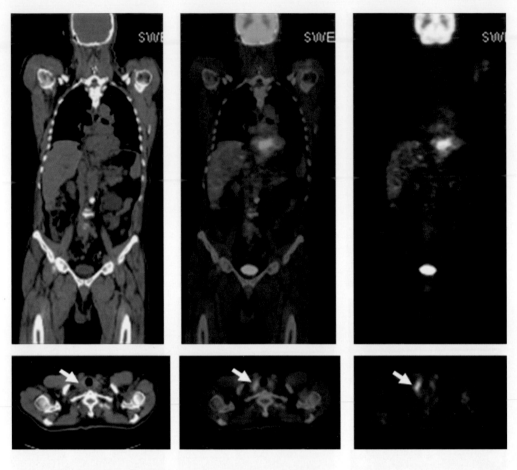

Figure 11.2. Staging esophageal cancer: 62-year-old man with adenocarcinoma of the distal esophagus and gastroesophageal junction. Coronal (*top*) CT, PET/CT fusion, and PET images demonstrate intense FDG uptake within the primary tumor mass Transaxial (*bottom*) CT, PET/CT fusion, and PET images show intense FDG uptake in a normal size (1 cm) right supraclavicular lymph node (*arrows*), which was easily accessible for biopsy to confirm inoperable disease.

for detection of stage IV disease were 74% and 90%, respectively, for FDG-PET, 41% and 83%, respectively, for CT, and 42% and 94%, respectively, for EUS. Combined use of CT and EUS had a sensitivity of 47% and a specificity of 78% in this group of patients with advanced disease. FDG-PET upstaged the tumor in 11 patients (15%) by detecting unsuspected metastatic disease and downstaged disease in 5 patients (7%). In a subsequent reanalysis of data in 42 of these 74 patients (28), these investigators also showed that FDG-PET had higher sensitivity (77% versus 83%) and specificity (90% versus 69%) than the combination of CT and EUS, specifically for detection of distant nodal disease. Similar results have been reported by others (22, 34, 35). For detection of metastatic disease, the reported sensitivities have ranged from 47% to 100% for PET, compared with 0% to 83% for CT (10–14, 17, 18, 22, 28, 31 ,34 ,35). Specificities ranged from 89% to 98% for PET and from 69% to 96% for CT [Table 11.4 (10–14, 17, 18, 22, 28, 31 ,34 ,35); Figure 11.3]. The recent meta-analysis by van Westreenen et al. (33) demonstrated that the overall pooled sensitivity and specificity of FDG-PET for detection of distant metastatic disease were 67% (95% CI, 58%–76%) and 97% (95% CI, 90%–100%), respectively.

Currently, there are no data available from multicenter trials that assess the role of PET in staging esophageal cancer, and the results of a recently completed multicenter trial (American College of Surgeons Oncology Group Study Z0060) are not expected to be available until 2005. The current literature demonstrates that FDG-PET at initial diagnosis assists in selection of the most appropriate mode of therapy for esophageal cancer. In particular, some patients with advanced disease, who are deemed to have resectable tumors on the basis of conventional imaging results, are excluded from attempted curative surgical procedures. However, because the positive predictive value of FDG-PET is less than 100%, histologic confirmation of PET findings indicating nonresectability is necessary before a patient is denied potentially curative surgery. FDG-PET diagnosis of nonresectable disease has the added advantage of also identifying the local or distant metastatic sites that are the most accessible to confirmation by minimally invasive surgical procedures (see Figure 11.3) (10, 11, 13, 14, 36). Therefore, the use of PET for staging esophageal cancer can reduce both the cost and morbidity of surgical management by reducing the number of ineffective surgical procedures. This practice may also be expected to increase the percentage cure rate of the resections that are undertaken.

It is expected that the interpretation accuracy of PET and CT will improve with the use of PET/CT scanners. These scanners provide accurately fused functional and morphologic data in a single examination (4, 5, 37). It also is expected that this improved accuracy will translate into improved patient management. There are only limited data yet available regarding the use of PET/CT in

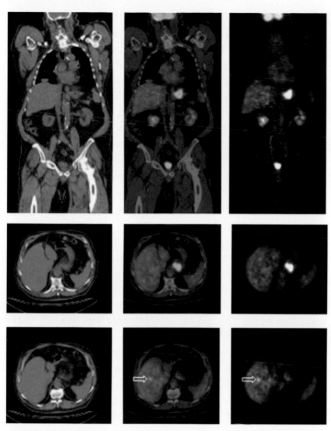

Figure 11.3. Staging esophageal cancer: 68-year-old man with adenocarcinoma of the gastroesophageal junction. Coronal (*top*) and transaxial (*middle*) CT, PET/CT fusion, and PET images demonstrate intense FDG uptake within the primary tumor mass. Transaxial (*bottom*) CT, PET/CT fusion, and FDG-PET images demonstrate a small focal area of increased FDG accumulation within a 1-cm metastasis in the right lobe of the liver (*open arrows*). The diagnosis of metastatic disease was confirmed by sonographically guided biopsy of the hepatic lesion.

esophageal cancer. Bar-Shalom et al. (38) studied 18 patients with esophageal cancer with PET/CT and demonstrated improved detection and characterization of 35% of suspicious lesions in 89% of patients. In addition, the PET/CT results affected management of 22% of patients.

In a recent study, Wallace et al. (39) compared the effectiveness of several different strategies for preoperative staging of patients with esophageal cancer. They compared the following six strategies: CT alone; CT + EUS with fine-needle aspiration biopsy; CT + thoracoscopy and laparoscopy (TL); CT + EUS with fine-needle aspiration biopsy + TL; CT + PET + EUS with fine-needle aspiration biopsy; and PET + EUS with fine-needle aspiration biopsy. The model was based on a third-party payer perspective and incorporated the following: the test characteristics for each staging technique; prevalence of local, regional, and distant disease; life expectancies and cost associated with the treatment for patients with local, regional, and distant disease; and probability of death for patients undergoing TL and those undergoing resection. The investigators found the combination of PET + EUS with fine-needle aspiration biopsy to be the most effective strategy.

Assessment of Prognosis with FDG-PET

Fukunaga et al. (23, 40) showed that semiquantitative and quantitative measures of FDG uptake in primary esophageal cancer can provide prognostic information. They found that patients with a primary tumor with SUV greater than 7.0 had a worse prognosis than those with SUV less than 7.0 (23). They also demonstrated a good correlation between hexokinase activity, assessed histochemically in the resected tumor specimens, and the preoperative tumor FDG uptake measured as SUV and as k3, the rate constant for phosphorylation of FDG. Kato et al. (24) also demonstrated that SUV for tumor FDG uptake can be used to predict prognosis. They reported that patients with a primary tumor with SUV greater than 3 had a worse prognosis than did those with SUV less than 3. The 2-year survival rate in the patients with high tumor FDG uptake was 48% versus 91% in those with low uptake.

Luketich et al. (14) reported that FDG-PET demonstration of local or distant metastatic disease at initial presentation was highly predictive of survival. The 30-month survival of patients with PET evidence of local disease only ($n = 64$) was 60% versus 20% for patients with PET evidence of distant disease ($n = 27$; $P = 0.01$). By comparison, when CT was used to stage the tumor, there was a lesser, but statistically insignificant, correlation between survival and CT findings: the 30-month survival of patients with CT evidence of local disease only ($n = 58$) was 52% versus 38% for patients with CT evidence of distant disease ($n = 33$). Choi et al. (41) recently demonstrated that several features of esophageal cancer on FDG-PET such as SUV of the primary tumor, the number of positive lymph nodes, the length of the tumor, and stage are independent prognostic predictors over other clinical features of patients with esophageal cancer who were undergoing curative surgery. The investigators studied 69 such patients, who were followed for disease recurrence and cancer-related death to assess survival. In univariant survival analysis, they found that the presence of adjuvant therapy, pathologic stage, number of CT-positive lymph nodes (0, 1, =2), tumor length on PET (cutoff: 3 cm, 5 cm), the number of PET-positive lymph nodes (0, 1, =3), and PET stage (N0M0, N1M0, M1) were significant prognostic predictor for disease-free survival. However, on multivariant survival analysis, only the number of PET-positive lymph nodes was an independent significant prognostic predictor for disease-free survival (hazard ratio, 1.87; P less than 0.001). These investigators suggested that a revised TNM staging system for esophageal cancer should be considered to include tumor length and the number of positive lymph nodes as important prognostic factors.

Assessment of Response to Therapy

The treatment for localized esophageal cancer is surgery; however, the long-term outcome for patients treated with surgery alone is very poor. Despite recent advances in surgical techniques and decreasing operative mortality, 5-year survival rates have remained low (5% to 23%) in patients treated by surgery alone (42–44). Recently, multimodality therapeutic approaches, combining surgery with neoadjuvant or adjuvant chemotherapy, radiation therapy, or both, have been used in patients with resectable and unresectable disease.

It has been shown that one of the strong predictors of long-term survival is the degree of response to chemotherapy and radiation therapy (45). Longer survival has been reported in patients showing complete response to chemotherapy and radiation than in patients with partial or no response. Conventional imaging techniques are limited for assessing the effectiveness of therapy, and a delay of several weeks after completion of therapy is necessary for evaluating response. Early assessment of tumor response during treatment would be valuable if nonresponders could be identified reliably, so that alternative treatment could be substituted in these patients and morbidity associated with ineffective treatment could be avoided. The role of PET in monitoring therapy of esophageal cancer has not been studied extensively, but FDG-PET has been used effectively to monitor therapy of several other cancers. Typically, a decrease in tumor FDG uptake is seen early during effective treatment whereas no significant decrease or even an increase is noted with ineffective therapy (46). The changes in FDG uptake generally occur earlier than corresponding anatomic changes on CT. It has been shown that FDG accumulates in sites of inflammation; thus, for assessment of response after therapy, a delay of several weeks or months after completion of therapy is believed to be needed to avoid false-positive results, although the optimal timing of such studies is still being investigated. There are two principal approaches for the use of FDG-PET in monitoring response to therapy in esophageal cancer. One approach involves the use of PET after completion of therapy to identify suitable candidates for surgical resection of esophageal cancer. The other approach involves the use of FDG-PET during the course of therapy to predict response to therapy and identify nonresponders so that an alternative therapy can be initiated.

Several studies have used FDG-PET at the completion of neoadjuvant therapy to assess response to therapy before surgical resection of esophageal cancer. Brücher et al. (47) correlated histopathologic tumor response to neoadjuvant radiation therapy and chemotherapy with changes in tumor FDG uptake, determined 3 weeks after completion of therapy in patients with esophageal carcinoma. The tumors of 13 patients showed a histopathologic response, defined as less than 10% residual viable tumor cells, whereas the tumors in 11 patients did not respond and had 10% or more residual viable tumor cells. The reduction in tumor FDG uptake was significantly more marked in responders than in nonresponders (mean ± standard deviation, –72% ± 11% versus –42% ± 22%; P

= 0.002). With a 52% reduction in tumor FDG uptake from baseline defined as the cutoff value, the sensitivity and specificity of FDG-PET for detection of response to neoadjuvant therapy were 100% and 55%, respectively. The positive predictive value and negative predictive value were 72% and 100%, respectively. After surgery, patients without evidence of response by PET had a significantly shorter survival (P less than 0.0001) (47). Similar results have been described by others (48). However, conflicting results were reported by Brink et al. (49), who demonstrated that the percent decrease in tumor FDG uptake 2.7 ± 0.6 weeks after neoadjuvant chemoradiotherapy did not correlate with tumor regression. Downey et al. (50) prospectively studied 24 patients with esophageal cancer before and after completion of neoadjuvant therapy. They demonstrated that patients with a median posttherapy decrease greater than 60% in primary tumor FDG uptake, as measured by the SUV, had better survival than did patients with a median decrease less than 60%. The 2-year disease-free survival in these two groups was 67% and 38%, respectively (P less than 0.05). In this relatively small study, the posttherapy FDG-PET study did not add to the assessment of locoregional resectability and did not detect new distant metastases. In our own experience, FDG-PET performed after induction therapy detected new metastatic disease, precluding resection, in about 10% of patients (Figure 11.4) (R. Battafarano et al., unpublished data). Swisher et al. (51) retrospectively studied 100 patients with PET, CT, and EUS before and 3 to 5 weeks after completion of neoadjuvant therapy. Fifty-eight patients had a pathologic response (=10% viable cells) to neoadjuvant chemoradiotherapy. After completion of therapy, the sensitivity, specificity, and accuracy were 51%, 69%, and 62%, respectively, for CT (response defined as esophageal thickness less than 14.5 mm); 56%, 75%, and 68%, respectively, for EUS (response defined as mucosal mass length less than 1 cm); 62%, 84%, and 76%, respectively, for PET of the primary tumor (response defined as SUV less than 4); and 69%, 78%, and 75%, respectively, for PET of primary tumor, regional, and distant metastatic disease (response defined as SUV less than 6). The investigators reported that only posttherapy SUV of the primary tumor predicted long-term survival. The 18-month survival of patients with posttherapy SUV of 4 or more was 34% compared with 77% for patients with an SUV less than 4.0 (P = 0.01).

Increased FDG uptake in inflamed tissue shortly after completion of therapy makes evaluation of response to cancer therapy by FDG-PET difficult. In a recent study, we found that the change in tumor FDG uptake 3 to 4 weeks after completion of neoadjuvant chemoradiotherapy was not reliable in distinguishing posttreatment inflammation from residual tumor (52). The time interval between chemoradiation therapy and follow-up PET should be carefully selected in future prospective studies to minimize false-positive results. It also is possible that differences in the false-positive rates between various studies

a Initial staging

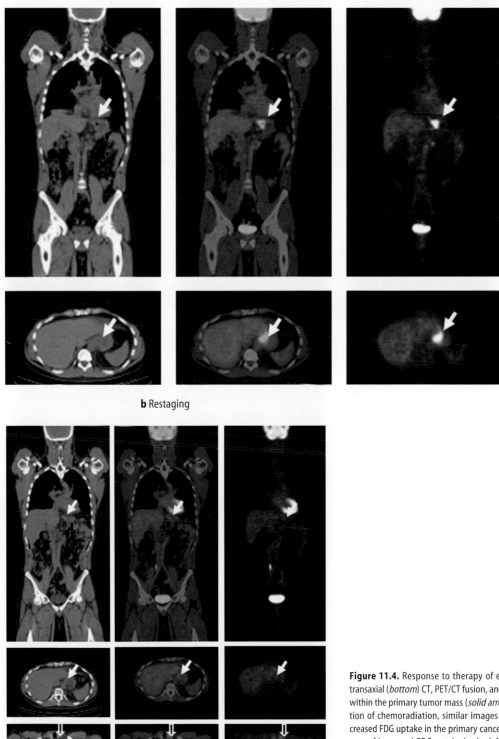

b Restaging

Figure 11.4. Response to therapy of esophageal carcinoma. (**a**) Coronal (*top*) and transaxial (*bottom*) CT, PET/CT fusion, and PET images demonstrate intense FDG uptake within the primary tumor mass (*solid arrows*). (**b**) Approximately 4 weeks after completion of chemoradiation, similar images show complete resolution of abnormally increased FDG uptake in the primary cancer (*solid arrows*). However, there is a new focal area of increased FDG uptake in the left supraclavicular region (*open arrows*), suspicious for interval development of distant metastatic disease. The diagnosis of metastatic disease was confirmed by biopsy of the left supraclavicular lymph node.

could be related to the dose of radiation therapy or the type of chemotherapeutic agents.

There is evidence that response to therapy can be predicted by PET as soon as 14 days after initiation of chemotherapy or chemoradiotherapy in esophageal cancer. Weber et al. (53), in a prospective study of 37 patients with locally advanced cancer of the esophagogastric junction, demonstrated that FDG-PET performed 14 days after the first cycle of cisplatin-based polychemotherapy was useful in predicting response to neoadjuvant chemotherapy. Clinical response was defined as a 50% reduction of tumor length and wall thickness as assessed by endoscopy and standard imaging techniques after 3 months of therapy. Histopathologic tumor response also was assessed in 32 patients who underwent surgery; complete response was defined as either no or a few scattered residual tumor cells. The decrease in tumor FDG uptake was significantly more marked in responders than in nonresponders (mean ± standard deviation, –54% ± 17% versus –15% ± 21%; P less than 0.001). When a 35% reduction in tumor FDG uptake relative to baseline was defined as the cutoff value between response and nonresponse, the early PET findings predicted clinical response with a sensitivity of 93% (14/15 patients) and a specificity of 95% (21/22 patient). Eight of 15 patients with a metabolic response, but only 1 of 22 patients without a metabolic response, had histologically complete or subtotal tumor regression. In addition, patients without evidence of response on PET had significantly shorter progression-free survival ($P = 0.01$) and overall survival ($P = 0.04$) (53). Wieder et al. (54), in a prospective study of 38 patients with intrathoracic esophageal cancer, demonstrated that the change in tumor FDG uptake at 14 days after initiation of neoadjuvant chemoradiotherapy was predictive of subsequent response to therapy and of survival. They observed that the decrease in tumor SUV was significantly higher in responders (defined as less than 10% viable cells in the resected specimen) than in nonresponders (decrease of 44% ± 15% in responders versus 21% ± 14% in nonresponders) ($P = 0.0055$). In addition, the change in tumor FDG uptake at this time point correlated with patient survival ($P = 0.011$). Using a 30% decrease of FDG uptake from baseline as the cutoff value, FDG-PET had a sensitivity of 93% and a specificity of 88% for distinguishing responders from nonresponders. Similarly, 3 to 4 weeks after completion of therapy, responders had greater reduction in tumor SUV than did nonresponders (70% ± 11% versus 51% ± 21%) (54). These investigators also reported that a radiation dose of 40 Gy and administration of fluorouracil as a continuous infusion causes a mild increase in FDG uptake in normal esophageal tissue during and early after completion of therapy (54).

These findings suggest that FDG-PET early during treatment or after completion of therapy has the potential to monitor response to therapy. It appears that the magnitude of the change in tumor FDG uptake after therapy in esophageal cancer is predictive of pathologic response and has long-term prognostic significance. With increasing use of multimodality therapy, such noninvasive assessment of tumor response has become more important. Further evaluations are required to define fully the role of PET in assessing response to therapy in esophageal cancer and to show that the use of PET to guide treatment decisions will result in improved patient outcomes.

Detection of Recurrent Disease

The long-term survival of patients with esophageal cancer remains poor despite aggressive therapy. Recurrence is common following presumed curative resection, mainly because of micrometastatic disease; thus, recurrence at distant sites is more common than local recurrence. Patients with recurrent disease have a poor prognosis, and the survival benefit of early detection of recurrent disease is uncertain. However, aggressive therapy of local recurrence may prolong disease-free survival or occasionally be curative. Although anatomic imaging modalities are limited in differentiating scar from recurrent disease, FDG-PET has the ability to detect and differentiate recurrent disease from posttherapy changes when disease has altered metabolism without any structural changes. Thus, PET is more suitable for early detection of recurrent disease.

Fukunaga et al. (40) studied 13 patients with suspected recurrent esophageal cancer; increased FDG uptake was noted in 6 of 7 patients with proven recurrent disease, whereas no significant FDG uptake was seen in the 6 patients who did not have recurrence. Flamen et al. (55) assessed the utility of FDG-PET in diagnosis of suspected recurrent esophageal cancer after initial curative resection in a study of 41 patients. Thirty-three patients proved to have recurrent disease. Forty lesions were identified: 9 local recurrences at the anastomotic site, 12 regional nodal metastases, and 19 distant metastatic lesions. For detection of local recurrence, the sensitivity and specificity were 100% and 43%, respectively, for PET and 100% and 93%, respectively, for conventional imaging (CT and EUS). For detection of metastatic disease, the sensitivity and specificity were 96% and 67%, respectively, for PET and 85% and 67%, respectively, for conventional imaging. FDG-PET provided additional information in 11 of 41 patients (27%); PET correctly detected recurrent disease in 5 patients with equivocal or negative conventional imaging results and demonstrated distant metastasis in 5 patients who had local recurrence. In another patient with equivocal conventional findings, PET correctly excluded recurrent disease. False-positive results attributable to inflammation were found in 4 patients with progressive anastomotic stenoses requiring repetitive dilatations. Thus, the patient's clinical history has to be considered to minimize false-positive results. Kato et al. (56) retrospectively compared PET and CT for detection of recurrent disease in 55 patients who had undergone esophagectomy for esophageal cancer. The sensitivity, specificity, and accuracy for detection of locoregional and distant metastatic disease were 96%, 68%, and 82%, respectively, for PET and 89%, 79%, and 84%, respectively, for CT. PET had higher sensitivity but lower specificity than did CT for detection of locoregional recurrence. In addition, PET had higher sensitivity for detection of bone metastasis than did CT, but had lower sensitivity than did CT for detection of lung metastasis. This study suggests that a combined PET/CT study will be the most sensitive tool for detection of recurrent disease.

Gastric Cancer

The incidence of gastric cancer has declined in the United States, but it is still the second most common cancer (57) and the second most common cause of cancer-related death in the world (58). Because of distensibility of the stomach, gastric cancers often become symptomatic only when the tumor is advanced and unresectable. Ninety-five percent of gastric cancers are adenocarcinomas, and the remaining 5% consist of leiomyosarcomas, lymphomas, carcinoids, squamous cell carcinomas, and other less-common tumors. Gastric cancers arising outside of the cardia have been classified by Lauren into intestinal and diffuse types. Intestinal-type cancers typically form gland-like structures and primarily involve the distal stomach (59). They typically occur in elderly individuals of low socioeconomic status and are believed to develop in a stepwise transition from normal mucosa to atrophic gastritis, intestinal metaplasia, dysplasia, and, finally, adenocarcinoma. The diffuse-type cancer is a poorly differentiated tumor that lacks glandular structures. This lesion is found more often in younger individuals and there appears to be a genetic predisposition (60, 61). Despite their differences, both types of gastric cancers are strongly associated with *Helicobacter pylori* infection (62). Over the past three decades, there has been a decrease in cancer of the antrum and an increase in cancer of the cardia and proximal stomach. Approximately 10% of gastric cancers involve the submucosa throughout the entire stomach, resulting in a rigid nondistendable organ (linitis plastica).

Surgical resection is the only curative therapy for gastric adenocarcinoma, and typically involves en bloc resection of the entire tumor and regional lymph nodes. Patients with early gastric cancer who are treated surgically have an excellent 5-year survival rate of approximately 90% (63). Limited surgery, such as endoscopic mucosal resection and laparoscopic wedge resection, are used in some patients with well-differentiated tumors of less than 2 cm that involve only the mucosal layer without nodal spread. Lymphatic involvement is a major determinant of prognosis in gastric cancer. Unfortunately, gastric cancer is all too often detected when the tumor is advanced and unresectable. Advanced disease is treated with neoadjuvant chemotherapy and has a very poor prognosis; the 5-year survival rate in the United State is reported to be 3% to 13% (63).

Gastric cancer is usually diagnosed by endoscopy or barium studies. These techniques provide excellent evalu-

ation of the mucosal surface of the stomach but are unable to determine the depth of mural invasion by tumor or the extent of metastatic disease. CT and EUS are the most common imaging methods used to determine the locoregional extent of disease. EUS is particularly useful for assessment of the depth of tumor invasion, but both imaging techniques are insufficiently accurate for nodal staging. Staging laparoscopy is performed routinely in patients who are thought to have resectable tumor on the basis of imaging findings to avoid surgery in patients with nonresectable tumor.

Several studies have evaluated the role of FDG-PET in gastric cancer (20, 64–67). Yeung et al. (65) studied 23 patients with gastric cancer. FDG-PET correctly identified all but 1 primary tumor (12/13) and identified recurrent disease at the anastomotic site in 2 patients. FDG-PET had a sensitivity of 93% and specificity of 100% for detection of tumor at the primary site (Figure 11.5). However, for detection of metastatic disease in intraabdominal lymph node stations, FDG-PET had a sensitivity of only 22% (2/9) with a specificity of 97% (32/33). FDG-PET correctly identified distant metastatic disease in 4 patients and was falsely negative in 4 additional patients who had peritoneal tumor spread. Stahl et al. (64) correlated tumor FDG uptake with histopathologic and endoscopic features of gastric cancer in 40 patients with locally advanced

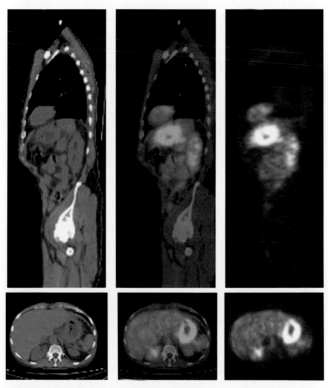

Figure 11.5. Staging of gastric cancer: 62-year-old woman with adenocarcinoma of the stomach. Coronal (*top*) and transaxial (*bottom*) CT, PET/CT fusion, and PET images of the torso demonstrate intense increased FDG uptake in the wall of the stomach. There is no evidence of locoregional or distant metastatic disease.

cancer and 10 controls. FDG-PET detected 24 of the 40 patients (60%) with gastric cancer, and the detection rate was significantly higher for intestinal-type tumors than for nonintestinal-type lesions (83% versus 41%; $P = 0.01$). The mean SUV was significantly different between the intestinal and nonintestinal types (6.7 ± 3.4 versus 4.8 ± 2.8; $P = 0.03$), between nonmucinous and mucinous cancers (2/18 intestinal type and 17/22 nonintestinal type contained extra-and intracellular mucin) (7.2 ± 3.2 versus 3.9 ± 2.1; P less than 0.01) and between grade 2 and grade 3 tumors (7.4 ± 2.3 versus 5.2 ± 3.3; $P = 0.02$). The survival rate was not significantly different between patients with PET-positive and PET-negative tumors ($P = 0.85$). The authors concluded that the low detection rate of gastric cancers by FDG-PET in their series was likely due to the greater number of nonintestinal-type tumors with high mucin content, which typically have lower FDG uptake than intestinal-type tumors. More recently, Mochiki et al. (66) studied 85 patients with gastric cancer with FDG-PET at initial diagnosis. FDG-PET detected the primary tumor in 75% (64/85) of the patients, and a significant correlation was noted between the primary tumor SUV and the depth of invasion (P less than 0.05), the size of the tumor (P less than 0.05), and the presence of lymph node metastasis (P less than 0.05). The primary tumor was not detected on FDG-PET in 21 patients; 15 patients of these patients had T1 primary lesions. The histology of the primary tumors (intestinal versus nonintestinal histology) did not significantly influence tumor detectability by PET ($P = 0.5$) in this study. CT was more sensitive (65% versus 23%) but less specific (77% versus 100%) than FDG-PET for detecting locoregional lymph node metastasis. This result was attributed to the difficulty of distinguishing FDG uptake in locoregional lymph nodes from that in the adjacent primary tumor on PET images. The intensity of FDG uptake was predictive of survival; patients with primary tumor SUVs greater than 4 had significantly poorer 2-year survival (51%) than did those with SUVs less than 4 (81%) (P less than 0.05). In addition, the 2-year survival of patients with PET-positive cancers was 66% and that for patients with PET-negative cancers was 94%. Yoshioka et al. (67) reported that FDG uptake in the primary tumor and in metastatic foci in the liver, lymph nodes, and lung is greater than that in bone metastases or peritoneal or pleural carcinomatosis. They also reported higher primary tumor FDG uptake in well-differentiated and moderately differentiated adenocarcinoma than in poorly differentiated adenocarcinoma and signet ring cell carcinoma (13.2 ± 6.3 versus 7.7 ± 2.6; P less than 0.05).

As in esophageal cancer, patients with gastric cancer who respond to neoadjuvant therapy have a more-favorable outcome. However, only 30% to 40% of patients respond to neoadjuvant therapy. Thus, the large fraction of patients with nonresponding tumors undergo several months of ineffective, toxic therapy without a survival benefit. Ott et al. prospectively evaluated 44 patients with locally advanced gastric cancer with FDG-PET before therapy and 14 days after initiation of cisplatin-based polychemotherapy (68); 35 of the 44 tumors had sufficient uptake on the PET images for quantitative analysis. Metabolic response was defined as a decrease of tumor FDG uptake by 30% or more. The PET findings after 14 days of therapy correctly predicted histopathologic response after 3 months of neoadjuvant therapy in 10 of 13 responders (77%) and in 19 of the 22 nonresponders (86%). Metabolic response also was predictive of survival. The median overall survival was 19 months for patients without a metabolic response (2-year survival of 25%) and had not been reached for patients with a metabolic response (2-year survival of 90%) ($P = 0.002$). If confirmed in larger studies, these results suggest that PET will be quite valuable for guiding the management of patients with locally advanced disease and for sparing nonresponders unnecessary morbidity and expenses of ineffective therapy.

The limitations of FDG-PET for evaluation of gastric cancer are similar to the limitations encountered in esophageal cancer and include poor sensitivity for detecting small tumor deposits. Sensitivity for detection of diffuse-type gastric cancer with high mucin content is lower than for detection of the intestinal type of gastric cancer (64, 68). Normal, moderately intense physiologic FDG uptake in the stomach may obscure tumors with low-level FDG uptake.

Gastrointestinal Stromal Tumors

Gastrointestinal stromal tumors (GISTs) are the most common mesenchymal tumors of the gastrointestinal tract and account for approximately 0.1% to 3% of all gastrointestinal (GI) tract tumors. GISTs are defined as immunohistochemically Kit-positive primary mesenchymal tumors of the GI tract. Approximately 95% of GISTs stain positively for Kit (CD117), a tyrosine kinase growth factor receptor (69). These tumors are thought to arise from Cajal cells in the gut wall, which are important for gut motor function. The majority of tumors previously diagnosed as GI smooth muscle tumors, such as leiomyoma, leioblastoma, or leiomyosarcoma, are now thought to have been GISTs. However, differentiation of leimyosarcoma from GIST is important to determine prognosis and the appropriateness of Kit-inhibitor therapy. Mature smooth muscle tumors have negligible mitotic activity and have benign behavior and do not express Kit. GISTs occur throughout the entire GI tract and also may arise from the omentum, mesentery, and retroperitoneum. GISTs arise predominantly in the stomach (60%) and small intestine (25%) and rarely in the rectum (5%), esophagus (2%), and a variety of other locations (5%) in the abdomen (69). They can be small benign tumors or sarcomas. Benign GISTs are more common in the stomach whereas malignant GISTs are more common in the intestines. Tumors that have metastasized at presentation have a very poor prognosis (70). Thirty percent of GISTs are malignant and 70% are benign. Mitotic rate, tumor size, and tumor site are three important prognostic factors. Tumors that are small (=2 cm) and show fewer than 5 mitotic figures per 50 high-power fields have an excellent prognosis, likely independent of site. Malignant GISTs tend to recur, and they metastasize commonly to the liver and peritoneum and less commonly to lung, pleura, peritoneum, bone, and subcutaneous tissues (71). Patients with metastatic or recurrent disease have a median overall survival in the range of 12 to 19 months (72, 73).

Radiologic features of GIST vary depending on their size and organ of origin. Most GIST that arise within the muscularis propria of the gut wall commonly manifest as dominant masses outside the organ of origin. Dominant intramural and intraluminal masses are less common. GISTs that occur in the GI tract and mesentery have areas of low attenuation on CT attributable to hemorrhage, necrosis, or cyst formation (74). After successful therapy, there is a significant decrease in CT attenuation values (75, 76).

Surgery is the mainstay of therapy for GISTs. These tumors are insensitive to conventional chemotherapy and radiation. GISTs are generally characterized by a gain-of-function mutation of the Kit receptor and, occasionally, of the platelet-derived growth factor (PDGF) receptor a. Mutant isoforms of the Kit or PDGF receptors uniformly expressed by GIST are considered the therapeutic targets for imatinib mesylate (STI571, Gleevec), a specific inhibitor of these tyrosine kinase receptors (77). This mutant receptor is thought to be the main reason for malignant transformation as well as for tumor growth in these patients (79). Imatinib inhibits tumor growth in GIST by competitive interaction at the adenosine triphos-

phate-binding site of the c-*kit* receptor (80). However, imatinib therapy is quite costly and is associated with several side effects, including anemia, periorbital edema, skin rash, fatigue, granulocytopenia, and diarrhea (81). The type of KIT or PDGFR-a mutation in advanced GIST is predictive of the response to imatinib therapy. Most GISTs express kinase oncoproteins that are intrinsically sensitive to imatinib and have an excellent overall clinical response to this drug. However, a minority of GISTs express kinase oncoproteins that are either intrinsically resistant to imatinib or are associated with a poor clinical response despite in vitro sensitivity to the drug (82). The response rates in GIST patients are 4% complete remission, 67% partial remission, 18% stable disease, and 11% progression, and 73% of GIST patients are free from progression at 1 year (81). It is important to assess response to therapy early to identify those patients with stable disease or disease progression, who can then be treated more aggressively with a higher dosage of imatinib or be considered for treatment with other investigational agents. FDG-PET has been shown to be superior to CT in detection of early metabolic changes indicative of tumor response induced by imatinib therapy. Gayed et al. (71) studied 54 patients with GIST and showed that FDG-PET detected 110 and CT detected 114 involved sites and/or organs, respectively, before initiation of therapy. The sensitivity and positive predictive values were 93% and 100%, respectively for CT and 86% and 98%, respectively, for FDG-PET. At 2 months after therapy, PET and CT findings agreed in 71% of patients (57% responders and 14% nonresponders). FDG-PET predicted response to therapy earlier than did CT in 22.5% of patients whereas CT predicted lack of response in 4% of the patients earlier than did PET (Figure 11.6). Stroobants et al. (83) studied 21 patients with GIST before and 8 days after initiation of

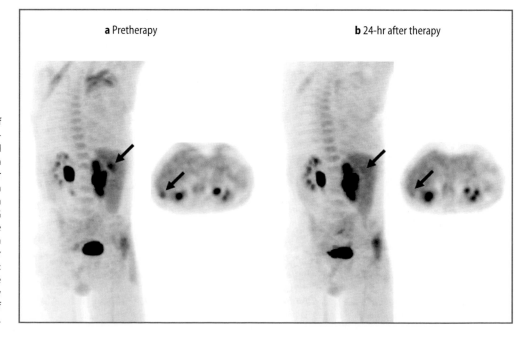

Figure 11.6. Response to therapy of gastrointestinal stromal tumor: 64-year-old man with metastatic gastrointestinal stromal tumor undergoing therapy with imatinib mesylate. (**a**) Left posterior oblique reprojection image (*left*) and a transaxial image (*right*) demonstrate a focal area (*arrows*) of increased FDG uptake in the posterior segment of the right lobe of the liver, consistent with hepatic metastasis. (**b**) Twenty-four hours after initiation of imatinib therapy: left posterior oblique reprojection image (*left*) and a transaxial image (*right*) show almost complete resolution (*arrows*) of the FDG uptake in the hepatic metastasis.

a Pretherapy   **b** 24-hr after therapy

Gleevec therapy. PET response based on the European Organization for Research and Treatment (EORTC) criteria (84) was compared with clinical response based on the RECIST criteria (85). PET response was associated with a longer progression-free survival (PFS) by comparison with PET nonresponse (92% versus 12% at 1 year; $P = 0.00107$).

In a recent study, Goerres et al. (86) evaluated 28 patients with PET/CT after treatment with imatinib. Patients without FDG uptake after the start of treatment had a better prognosis than did patients with residual activity. However, contrast-enhanced CT alone provided insufficient prognostic power. In another study, Antoch et al. (5) compared the value of PET alone, CT alone, and combined PET/CT for assessment of response to imatinib therapy in 20 patients with GIST. When patients were evaluated with FDG-PET/CT before and 1, 3, and 6 months after initiation of therapy, 135 lesions were detected by PET alone, 249 by CT alone, 279 by PET and CT side-by-side evaluation, and 282 lesions by fused PET/CT. PET/CT correctly determined tumor response in 95% of the patients after 1 month and in 100% after 3 months and 6 months of therapy. PET and CT images viewed side-by-side correctly determined response in 90% of the patients at 1 month and 100% at 3 and 6 months after therapy. PET alone correctly diagnosed response in 85% of the patients at 1 month and 100% at 3 and 6 months after therapy. CT was accurate in 44% of the patients at 1 month and 60% at 3 and 57% at 6 months after therapy. It appears that the combination of functional and anatomic imaging has the best performance for assessment of response to imatinib therapy in patients with GIST.

References

1. Jemal A, Tiwari RC, Murray T, et al. Cancer Statistics, 2004. CA Cancer J Clin 2004;54:8–29.
2. Koshy M, Esiashvilli N, Landry JC, Thomas CR Jr, Matthews RH. Multiple management modalities in esophageal cancer: epidemiology, presentation and progression, work-up, and surgical approaches. Oncologist 2004;9:137–146.
3. Stein HJ, Sendler A, Fink U, Siewert JR. Multidisciplinary approach to esophageal and gastric cancer. Surg Clin N Am 2000;80:659–682.
4. Bar-Shalom R, Yefremov N, Guralnik L, et al. Clinical performance of PET/CT in evaluation of cancer: additional value for diagnostic imaging and patient management. J Nucl Med 2003;44:1200–1209.
5. Antoch G, Saoudi N, Kuehl H, et al. Accuracy of whole-body dual-modality fluorine-18-2-fluoro-2-deoxy-D-glucose positron emission tomography and computed tomography (FDG-PET/CT) for tumor staging in solid tumors: comparison with CT and PET. J Clin Oncol 2004;22:4357–4368.
6. Rösch T. Endosonographic staging of esophageal cancer: a review of literature results. Gastrointest Endosc Clin N Am 1995;5:537–547.
7. Halvorsen RA Jr, Magruder-Habib K, Foster WL Jr, Roberts L Jr, Postlethwait RW, Thompson WM. Esophageal cancer staging by CT: long-term follow-up study. Radiology 1986;161:147–151.
8. Inculet RI, Keller SM, Dwyer A, Roth JA. Evaluation of noninvasive tests for the preoperative staging of carcinoma of the esophagus: a prospective study. Ann Thorac Surg 1985;40:561–565.
9. Urmacher C, Brennan MF. Preoperative staging of esophageal cancer: comparison of endoscopic US and dynamic CT. Radiology 1991;181:419–425.
10. Flanagan FL, Dehdashti F, Siegel BA, et al. Staging of esophageal cancer with FDG-PET. AJR 1997;168:417–424.
11. Block MI, Sundaresan SR, Patterson GA, et al. Improvement in staging of esophageal cancer with the addition of positron emission tomography. Ann Thorac Surg 1997;64:770–777.
12. Kole AC, Plukker JT, Nieweg OE, et al. Positron emission tomography for staging oesophageal and gastroesophageal malignancy. Br J Cancer 1998;74:521–527.
13. Luketich JD, Schauer P, Meltzer CC, et al. The role of positron emission tomography in staging esophageal cancer. Ann Thorac Surg 1997;64:765–769.
14. Luketich JD, Friedman DM, Weigel TL, et al. Evaluation of distant metastases in esophageal cancer: 100 consecutive positron emission tomography scans. Ann Thorac Surg 1999;68:1133–1137.
15. Rankin SC, Taylor H, Cook GJR, et al. Computed tomography and positron emission tomography in the pre-operative staging of oesophageal carcinoma. Clin Radiol 1998;53:659–665.
16. Yeung HWD, Macapinlac HA, Mazumdar M, Bains M, Finn RD, Larson SM. FDG-PET in esophageal cancer: incremental value over computed tomography. Clin Posit Imaging 1999;5:255–260.
17. Flamen P, Lerut A, Van Cutsem E, et al. Utility of positron emission tomography for the staging of patients with potentially operable esophageal carcinoma. J Clin Oncol 2000;18:3202–3210.
18. Kobori O, Kirihara Y, Kosaka N, Hara T. Positron emission tomography of esophageal carcinoma using ^{11}C-choline and ^{18}F-fluorodeoxyglucose. A novel method of preoperative lymph node staging. Cancer (Phila) 1999;86:1638–1648.
19. Duignan JP, McEntee GP, O'Connell DJ, Bouchier-Hayes DJ, O'Malley E. The role of CT in the management of carcinoma of the oesophagus and cardia. Ann R Coll Surg Engl 1987;69:286–288.
20. McAteer D, Wallis F, Couper G, et al. Evaluation of ^{18}F-FDG positron emission tomography in gastric and oesophageal carcinoma. Br J Radiol 1999;72:525–529.
21. Räsänen JV, Sihvo EIT, Knuuti J, et al. Prospective analysis of accuracy of positron emission tomography, computed tomography, and endoscopic ultrasonography in staging of adenocarcinoma of the esophagus and the esophagogastric junction. Ann Surg Oncol 2003;10:954–960.
22. Meltzer CC, Luketich JD, Friedman D, et al. Whole-body FDG positron emission tomographic imaging for staging esophageal cancer comparison with computed tomography. Clin Nucl Med 2000;25:882–887.
23. Fukunaga T, Okazumi S, Koide Y, Isono K, Imazeki K. Evaluation of esophageal cancers using fluorine-18-fluorodeoxyglucose PET. J Nucl Med 1998;39:1002–1007.
24. Kato H, Kuwano H, Nakajima M, et al. Comparison between positron emission tomography and computed tomography in the use of the assessment of esophageal carcinoma. Cancer (Phila) 2002;94:921–928.
25. Ellis FH, Eatkin E, Krasna MJ, Heatley GH, Balogh K. Staging of carcinoma of esophagus and gastric cardia: a comparison of different staging criteria. J Surg Oncol 1993;52:231–235.
26. Peters JH, Hoeft SF, Heimbucher J, et al. Selection of patients for curative or palliative resection of esophageal cancer based on pre-operative endoscopic ultrasonography. Arch Surg 1994;129:534–539.
27. Skinner DB, Little AG, Ferguson MK, Soriano A, Statszak VM. Selection of operation for esophageal cancer based on staging. Ann Surg 1986;204:391–401.
28. Lerut T, Flamen P, Ectors N, et al. Histopathologic validation of lymph node staging with FDG-PET scan in cancer of esophagus and gastroesophageal junction: a prospective study based on primary surgery with extensive lymphadenectomy. Ann Surg 2000;232:743–752.
29. Kim K, Park SJ, Kim B-T, Lee KS, Shim YM. Evaluation of lymph node metastases in squamous cell carcinoma of the esophagus with

positron emission tomography. Ann Thorac Surg 2001;71:1290–1294.

30. Himeno S, Yasuda S, Shimada H, Tajima T, Makuuchi H. Evaluation of esophageal cancer by positron emission tomography. Jpn J Clin Oncol 2002;32(9):340–346.

31. Wren SM, Stijns P, Srinivas S. Positron emission tomography in the initial staging of esophageal cancer. Arch Surg 2002;137:1001–1007.

32. Yoon YC, Lee KS, Shim YM, Kim B-T, Kim K, Kim TS. Metastasis to regional lymph nodes in patients with esophageal squamous cell carcinoma: CT versus FDG PET for presurgical detection—prospective study. Radiology 2003;227:764–770.

33. van Westreenen HL, Westerterp M, Bossuyt PMM, et al. Systematic review of the staging performance of ^{18}F-fluorodeoxyglucose positron emission tomography in esophageal cancer. J Clin Oncol 2004;22:3805–3812.

34. Heeren PAM, Jager PL, Bongaerts F, van Dullemen H, Sluiter W, Plukker JTM. Detection of distant metastases in esophageal cancer with ^{18}F-FDG PET. J Nucl Med 2004;45:980–987.

35. Liberale, Van Laethem JL, Gay F, et al. The role of PET scan in the preoperative management of esophageal cancer. Eur J Surg Oncol 2004;30:942–947.

36. Luketich JD, Schauer P, Urso K, et al. Future directions in esophageal cancer. Chest 1998;113:120S–122S.

37. Schöder H, Erdi YE, Larson SM, et al. PET/CT: a new imaging technology in nuclear medicine. Eur J Nucl Med 2003;30:1419–1437.

38. Leiderman M, Gaitini D, Keidar Z, et al. The value of PET/CT using FDG in patients with esophageal cancer. J Nucl Med 2003;44:21P.

39. Wallace MB, Nietert PJ, Earle C, et al. An analysis of multiple staging management strategies for carcinoma of the esophagus: computed tomography, endoscopic ultrasound, positron emission tomography, and thoracoscopy/laparoscopy. Ann Thorac Surg 2002;74:1026–1032.

40. Fukunaga T, Enomoto K, Okazumi S, et al. Analysis of glucose metabolism in patients with esophageal cancer by PET: estimation of hexokinase activity in the tumor and usefulness for clinical assessment using FDG. Nippon Geka Gakka Zasshi 1994;95:317–325.

41. Choi JY, Jang H-J, Shim YM, et al. ^{18}F-FDG PET in patients with esophageal squamous cell carcinoma undergoing curative surgery: prognostic implications. J Nucl Med 2004;45;1843–1850.

42. Orringer M. Transhiatal esophagectomy with thoracoscopy. In: Zuidema G (ed). Surgery of the Alimentary Tract, 3rd ed. Philadelphia: Saunders, 1991:408–433.

43. Katlic MR, Wilkins EW, Grillo HC. Three decades of treatment of esophageal squamous carcinoma at Massachusetts General Hospital. J Thorac Cardiovas Surg 1990;99:929–938.

44. Roder JT, Busch R, Stein HJ, et al. Ratio of invaded to removed lymph nodes as a predictor of survival in squamous cell carcinoma of esophagus. Br J Surg 1994;81:410–413.

45. Thomas CR. Biology of esophageal cancer and the role of combined modality therapy. Surg Clin N Am 1997;77:1139–1167

46. Smith TAD. FDG uptake, tumour characteristics and response to therapy: a review. Nucl Med Commun 1998;19:97–105.

47. Bücher BLDM, Weber W, Bauer M, et al. Neoajuvant therapy of esophageal squamous cell carcinoma: response evaluation by positron emission tomography. Ann Surg 2001;233:300–309.

48. Kato H, Kuwano H, Nakajima M, et al. Usefulness of positron emission tomography for assessing the response of neoadjuvant chemoradiotherapy in patients with esophageal cancer. Am J Surg 2002;184:279–283.

49. Brink I, Hentschel M, Bley TA, et al. Effects of neoadjuvant radio-chemotherapy on ^{18}F-FDG-PET in esophageal carcinoma. Eur J Surg Oncol 2004;30:544–550.

50. Downey RJ, Akhurst T, Ilson D, et al. Whole body ^{18}FDG-PET and the response of esophageal cancer to induction therapy: results of a prospective trial. J Clin Oncol 2003;21:428–432.

51. Swisher SG, Maish M, Erasmus JJ, et al. Utility of PET, CT, and EUS to identify pathologic responders in esophageal cancer. Ann Thorac Surg 2004;78:1152–1160.

52. Arslan N, Miller TR, Dehdashti F, Siegel BA, Battafarano RJ. Evaluation of response to neoadjuvant therapy by quantitative 2-deoxy-2-[^{18}F]fluoro-D-glucose with positron emission tomography in patients with esophageal cancer. Mol Imaging Biol 2002;4:320–329.

53. Weber WA, Ott K, Becker K, et al. Prediction of response to preoperative chemotherapy in adenocarcinomas of esophagogastric junction by metabolic imaging. J Clin Oncol 2001;19:3058–3065.

54. Wieder HA, Brücher BLDM, Zimmermann F, et al. Time course of tumor metabolic activity during chemoradiotherapy of esophageal squamous cell carcinoma and response to treatment. J Clin Oncol 2004;22:900–908.

55. Flamen P, Lerut A, Van Cutsem E, et al. The utility of positron emission tomography for the diasnosis and staging recurrent esophageal carcinoma. J Thorac Cardiovasc Surg 2000;120:1085–1092.

56. Kato H, Miyazaki T, Nakajima M, Fukuchi M, Manda R, Kuwano H. Value of positron emission tomography in the diagnosis of recurrent oesophageal carcinoma. Br J Surg 2004;91:1004–1009.

57. Landis S, Murray T, Bolden S, et al. Cancer statistics. CA Cancer J Clin 1999;49:8–31.

58. Neugut AI, Hayek M, Howe G. Epidemiology of gastric cancer. Semin Oncol 1996;23:281–291.

59. Lauren P. The two histological types of gastric carcinoma: diffuse and so-called intestinal-type carcinoma. Acta Pathol Microbiol Scand 1965;64:31–49.

60. Fuch CS, Meyer RJ. Gastric carcinoma. N Engl J Med 1995;333:32–41.

61. Hohenberger P, Gretschel S. Gastric cancer. Lancet 2003;362:305–315.

62. Zinvy J, Wang TC, Yantiss R, Kim KK, Houghton JM. Role of therapy or monitoring in preventing progression to gastric cancer. J Clin Gastroenterol 2003;36:S50–S60.

63. Roukos DH. Current advances and changes in treatment strategy may improve survival and quality of life in patients with potentially curable gastric cancer. Ann J Surg Oncol 1999;6:46–56.

64. Stahl A, Ott K, Weber WA, et al. FDG PET imaging of locally advanced gastric carcinomas: correlation with endoscopic and histopathologic findings. Eur J Nucl Med 2003;30:288–295.

65. Yeung HWD, Macapinlac HA, Karpeh M, Finn RD, Larson SM. Accuracy of FDG-PET in gastric cancer: preliminary experience. Clin Posit Imaging 1999;4:213–221.

66. Mochiki E, Kuwano H, Katoh H, Asao T, Oriuchi N, Endo K. Evaluation of ^{18}F-2-deoxy-2-fluoro-D-glucose positron emission tomography for gastric cancer. World J Surg 2004;28:247–253.

67. Yoshioka T, Yamaguchi K, Kubota K, et al. Evaluation of ^{18}F-FDG PET in patients with advanced, metastatic, or recurrent gastric cancer. J Nucl Med 2003;44:690–699.

68. Ott K, Fink U, Becker K, Stahl A, et al. Prediction of response to preoperative chemotherapy in gastric carcinoma by metabolic imaging: results of a prospective trial. J Clin Oncol 2003;21:4604–4610.

69. Corless CL, Fletcher JA, Heinrich MC. Biology of gastrointestinal stromal tumors. J Clin Oncol 2004;22:3813-3825.

70. Miettinen M, Lasota J. Gastrointestinal stromal tumors (GISTs): definition, occurrence, pathology, differential diagnosis and molecular genetics. Pol J Pathol 2003;54:3–24.

71. Gayed I, Vu T, Iyer R, et al. The role of ^{18}F-FDG PET in staging and early prediction of response to therapy of recurrent gastrointestinal stromal tumors. J Nucl Med 2004;45:17–21.

72. Casper ES. Gastrointestinal stromal tumors. Curr Treat Options Oncol 2000;1:267–273.

73. Dematteo RP, Maki RG, Antonescu C, Brennan MF. Targeted molecular therapy for cancer: the application of STI571 to gastrointestinal stromal tumor. Curr Probl Surg 2003;40:144–193.

74. Levy AD, Remotti HE, Thompson WM, Sobin LH, Miettinen M. Gastrointestinal stromal tumors: radiologic features with pathologic correlation. RadioGraphics 2003;23:283–304.

75. Choi H, Faria SC, Benjamin RS, Podoloff DA, Macapinlac HA, Charnsangavej C. Monitoring treatment effects of STI-571 on gastrointestinal stromal tumors (GIST) with CT and PET: a quantitative analysis [abstract]. Radiology 2002;225(suppl):583.

76. Chen MYM, Bechtold RE, Savage PD. Cystic changes in hepatic metastases from gastrointestinal stromal tumors (GISTs) treated with Gleevec (imatinib mesylate). Am J Radiol 2002;179:1059–1062.

77. Borg C, Terme M, Taïeb J, et al. Novel mode of action of c-*kit* tyrosine kinase inhibitors leading to NK cell-dependent antitumor effects. J Clin Invest 2004;114:379–388.

78. Lasota J, Jasinski M, Sarlomo-Rikala M, Miettinen M. Mutations in exon 11 of c-*kit* occur preferentially in malignant versus benign gastrointestinal stromal tumors and do not occur in leiomyomas or leiomyosarcomas. Am J Pathol 1999;154:53–60.

79. Taniguchi M, Nishida T, Hirota S, et al. Effect of c-*kit* mutation on prognosis of gastrointestinal stromal tumors. Cancer Res 1999;59:4297–4300.

80. Heinrich MC, Griffith DJ, Druker BJ, Wait CL, Ott KA, Zigler AJ. Inhibition of c-*kit* receptor tyrosine kinase activity by STI 571, a selective tyrosine kinase inhibitor. Blood 2000;96:925–932.

81. Verweij J, van Oosterom A, Blay J-Y, et al. Imatinib mesylate (STI-571 Glivec®, Gleevec™) is an active agent for gastrointestinal stromal tumours, but does not yield responses in other soft-tissue sarcomas that are unselected for a molecular target: results from an EORTC soft tissue and bone sarcoma group phase II study. Eur J Cancer 2003;39:2006–2011.

82. Heinrich MC, Corless CL, Demetri GD, et al. Kinase mutations and imatinib response in patients with metastatic gastrointestinal stromal tumor. J Clin Oncol 2003;21:4342–4349.

83. Stroobants S, Goeminne J, Seegers M, et al. [18]FDG-positron emission tomography for the early prediction of response in advanced soft tissue sarcoma treated with imatinib mesylate (Glivec). Eur J Cancer 2003;39:2012–2020.

84. Young H, Baum R, Cremerius U, et al. Measurement of clinical and subclinical tumour response using [^{18}F]-fluorodeoxyglucose and positron emission tomography: review and 1999 EORTC recommendations. European Organization for Research and Treatment of Cancer (EORTC) PET Study Group. Eur J Cancer 1999;35:1773–1782.

85. Therasse P, Arbuck SG, Eisenhauer EA, et al. New guidelines to evaluate the response to treatment in solid tumors. J Natl Cancer Inst 2000;92:205–216.

86. Goerres GW, Stupp R, Barghouth G, et al. The value of PET, CT and in-line PET/CT in patients with gastrointestinal stromal tumours: long-term outcome of treatment with imatinib mesylate. Eur J Nucl Med Mol Imaging 2004 Sept 4 (online).

12

PET and PET/CT Imaging in Tumors of the Pancreas and Liver

Dominique Delbeke and William H. Martin

Tumors of the Hepatobiliary System

A variety of benign and malignant tumors occur in the liver. The most common benign tumors found in the liver are cysts, followed by cavernous hemangiomas. Focal nodular hyperplasia and adenomas more often affect women on oral contraceptives, whereas fatty infiltration and regenerating nodules occur in patients with cirrhosis. Abscesses and angiomyolipoma are uncommon. Metastases to the liver from various primaries occur 20 times more often than primary hepatocellular carcinoma (HCC) and are often multifocal. Although many tumors may metastasize to the liver, the most common primaries producing liver metastases are colorectal, gastric, pancreatic, lung, and breast carcinoma. Ninety percent of malignant primary liver tumors are tumors from epithelial origin: HCC and cholangiocarcinoma.

Hepatocellular carcinoma (92% of epithelial tumors) arise from the malignant transformation of hepatocytes and are common in the setting of chronic liver disease such as viral hepatitis, cirrhosis, or patients exposed to carcinogens. Hepatocellular carcinoma most frequently metastasizes to regional lymph nodes, lung, and the skeleton.

Cholangiocarcinomas arise from biliary cells and represent only 10% of the epithelial tumors. About 20% of the patients who develop cholangiocarcinomas have predisposing conditions, including sclerosing cholangitis, ulcerative colitis, Caroli's disease, choledocal cyst, infestation by the fluke *Clonorchis sinensis*, cholelithiasis, or exposure to thorotrast among others. Approximately 50% of cholangiocarcinomas occur in the liver and the other 50% are extrahepatic. These tumors are often unresectable at the time of diagnosis and have a poor prognosis. Intrahepatic cholangiocarcinomas can further be subdivided in two categories: the peripheral type arises from the interlobular biliary duct and the hilar type (Klatskin's tumor) arises from the main hepatic duct or its bifurcation. In addition, they can develop in three different morphologic types: infiltrating sclerosing lesions (most common), exophytic lesions, and polypoid intraluminal masses. Malignant tumors can arise along the extrahepatic bile ducts; these are usually diagnosed early because they cause biliary obstruction. Tumors arising near the hilum of the liver have a worse prognosis because of their direct extension into the liver. Distant metastases occur late in the disease and most often affect the lungs.

Mesenchymal tumors such as angiosarcomas and epithelioid angioendothelioma (both of endothelial origin) and primary lymphoma are relatively rare malignant tumors that can affect the liver.

Gallbladder carcinoma is uncommon and is associated with cholelithiasis in 75% of the cases. These tumors are insidious, not suspected clinically, and often discovered at surgery or incidentally in the surgical specimen. They frequently spread to the liver and can perforate the wall of the gallbladder, metastasizing to the abdomen. Distant metastases affect the lungs, pleura, and diaphragm.

Methods of Diagnosis

The diagnostic issues include early detection of these tumors, differentiation of malignant tumors from benign tumors (lesion characterization), staging for potential resection that includes evaluation of proximity to vessels, invasion of adjacent structures, metastasis to regional lymph nodes and distant sites, and assessment of therapeutic response. Various imaging modalities available to achieve these goals include ultrasound (US), computed tomography (CT), magnetic resonance imaging (MRI), and functional imaging using radiopharmaceuticals (nuclear medicine). Tomographic imaging for functional radioisotopic studies can be performed using the single photon emission tomography technique (SPECT) if the radiopharmaceutical is a single photon emitter such as ^{99m}Tc and the positron emission tomography technique (PET) if the radiopharmaceutical is a positron emitter such as ^{18}F or ^{11}C.

Some of these tumors are associated with elevated serum levels of tumor markers that can be helpful for the diagnosis and surveillance of these patients, such as serum levels of alpha-fetoprotein for screening patients at risk for HCC and serum levels of Ca 19-9 for surveillance of patients with pancreatic carcinoma.

Transabdominal US is well established as a valuable screening technique that is inexpensive, portable, sensitive for evaluating bile duct dilatation, and able to detect hepatic lesions as small as 1 cm. It can also provide guidance for biopsy and drainage procedures. Its limitations include poor sensitivity (50%) for detection of small hepatic lesions and regional lymphadenopathy compared to CT and MRI.

Endoscopic ultrasound is a promising new technique for the evaluation of the extrahepatic bile ducts and pancreatic ducts. It is sensitive for the detection of choledocolithiasis and pancreatic masses. However, it is highly operator dependent, requires sedation, and is associated with significant morbidity. In addition, it has a limited field of view for the purposes of staging.

CT remains the first choice for a screening abdominal examination at many institutions. Recent developments in the field of CT and MRI for hepatic imaging are based on the dual perfusion of the liver: most of the blood flow to normal liver parenchyma is derived from the portal vein, whereas nearly all the blood flow to the hepatic neoplasms is derived from the hepatic artery. Therefore, some lesions are better seen at different times following administration of contrast material. Typically, hypervascular lesions (HCC, metastases of carcinoid carcinoma, islet cell tumor, malignant pheochromocytoma, renal cell carcinoma, sarcoma, melanoma, and breast carcinoma) may be best seen during the arterial phase of enhancement, or before contrast is administered, whereas hypovascular metastases (colorectal carcinoma and most metastases of other primaries) are best seen during the portal venous phase of enhancement. This triple-phase technique requires spiral (or helical) CT technology. Cavernous hemangiomas are best characterized using dynamic imaging with the appearance of puddling of contrast on delayed images. In summary, a triple-phase helical CT should be performed on all patients referred for evaluation of hepatic lesions to allow optimal detection and characterization (1).

After invasive arterial catheterization, contrast material can be injected into the superior mesenteric artery, which increases the sensitivity for detection of small hepatic lesions but decreases the specificity because of the frequent occurrence of nonspecific perfusion defects. CT portography is invasive and expensive (approximately eight times more than CT) but has the potential to detect HCCs and other hepatic lesions less than 1cm in diameter.

MR imaging is certainly as sensitive as CT for detection of focal hepatic lesions, but it too is inferior to CT portography. A multitude of pulse sequences have been developed to characterize lesions. Gadolinium chelate contrast agents are used similarly to the intravenous CT contrast agents, rapidly leaving the vascular space and reaching equilibrium throughout the extracellular fluid compartment after about 3 min (2). MR cholangiopancreatography (MRCP) permits visualization of the biliary tree noninvasively without the administration of contrast agents (3). Using a heavily T_2-weighted pulse sequence, solid organs and moving fluid have a low signal, whereas relatively stagnant fluid (such as bile) has a high signal intensity, resulting in the biliary tract appearing as a bright well-defined structure. Although MRCP does not provide the resolution of percutaneous transhepatic cholangiography (PTC) or endoscopic retrograde cholangiopancreatography (ERCP), it is able to clearly demonstrate intraluminal filling defects and luminal narrowing. MRCP provides invaluable information in both benign and malignant biliary tract disease.

Cholangiopancreatography, via PTC and ERCP, is an invasive technique but remains the procedure of choice for high-resolution assessment of the biliary tree anatomy. ERCP is performed by endoscopic cannulation of anatomic tracts and is therefore less invasive than PTC, which requires passage of a needle through the liver parenchyma. Contrast material is then injected directly into the biliary tree. Both techniques offer the advantage of allowing interventional procedures such as stent placement in the same setting as the imaging procedure. PTC demonstrates the intrahepatic ducts better than ERCP, which better depicts the extrahepatic ducts.

When a patient presents with a hepatic lesion seen with morphologic imaging techniques, functional imaging with conventional radiopharmaceuticals can help to characterize lesions. ^{99m}Tc-labeled red blood cells can be used to image the blood pool and is highly accurate in differentiating cavernous hemangiomas from other lesions. ^{99m}Tc-sulfur colloid accumulation in hepatic Kupffer cells allows characterization of focal nodular hyperplasia. [131]I- on ^{123}I-Metaiodobenzylguanidine (MIBG), which localizes through the norepinephrine reuptake mechanism into the catecholamine storage vesicles, can be used to image neuroendocrine tumors and their metastases. ^{111}In-Octreotide is a somatostatin analogue that accumulates in a variety of neuroendocrine tumors expressing somatostatin receptors but may also help characterize other pathologic processes such as lymphoma, sarcoidosis, and autoimmune diseases. [67]Gallium, [201]thallium, ^{99m}Tc-isonitriles (MIBI), and radiolabeled monoclonal antibodies are poor imaging agents for hepatic lesions because of their high physiologic liver background activity.

In summary, triple-phase CT functions as the standard imaging modality for the detection and characterization of hepatic lesions, whereas US, MRI, MRCP, and ERCP/PTC provide complementary techniques for further characterization of lesions in specific circumstances. Conventional radiopharmaceuticals with or without SPECT may contribute as well, but the development and proliferation of FDG-PET may yet provide unprecedented utility in the evaluation of these patients.

Role of Conventional Imaging to Evaluate Hepatobiliary Tumors

Metastases

Ultrasound can detect lesions as small as 1 cm in diameter, but it is operator dependent, inherently two dimensional, and suffers from poor specificity. Hepatic metastases can be hypoechoic, hyperechoic, cystic, or have mixed echogenicity. Isoechoic metastases go undetected. CT is the conventional method for screening the liver at many institutions. Metastases may be better seen during the arterial or portal venous phase after contrast injection, depending of the vascularity of the tumor. The development of the helical technique has resulted in a sensitivity comparable to that of MRI, although CT portography remains the most sensitive technique for detection of small lesions.

Hepatocellular Carcinoma

Differentiation of low-grade HCC from hepatic adenoma can be difficult even on core biopsy. Although a capsule is often present in small HCC, it is seldom seen in large ones. Invasion of the portal vein is often present, especially with large HCCs, although invasion of the inferior vena cava and hepatic vein occurs in only 5% of the cases. Hepatocellular carcinoma can undergo hemorrhage and necrosis or demonstrate fatty metamorphosis. The presence of these characteristics on imaging studies is suggestive of HCC but requires high-resolution techniques for successful detection. Dynamic multiphase gadolinium-enhanced MRI may be superior to dual-phase spiral CT to characterize HCC (4). Although large lesions (larger than 3 cm) may be visualized by CT and US with a sensitivity in the 80% to 90% range, smaller lesions may be difficult to distinguish from the surrounding hepatic parenchyma, especially in patients with cirrhosis and regenerating nodules, with a sensitivity in the range of 50%. CT portography is the most sensitive technique for detection of lesions less than 1 cm in size. [67]Gallium scintigraphy may help to identify HCC if the findings on other imaging modalities are equivocal. Seventy percent to 90% of HCC have gallium uptake greater than the liver, and gallium scintigraphy has been used in conjunction with a sulfur colloid scan to differentiate HCC from regenerating nodules in cirrhotic patients (5). It is not useful for lesions less than 2 to 3 cm in diameter.

Fibrolamellar carcinoma, a low-grade malignant tumor representing 6% to 25% of HCC, occurs in a younger population of patients without underlying cirrhosis. An avascular scar that may contain calcifications characterizes these tumors. In contrast focal nodular hyperplasia exhibit a vascular scar but without calcifications.

Cholangiocarcinoma

Cholangiocarcinomas are often not detected by CT because they are small and isodense. When this is the case (most hilar tumors and 25% of peripheral tumors), the level of biliary ductal dilatation infers the location of the tumor. When the tumor is visible on CT, it most often appears as a nonspecific hypodense mass. Delayed retention of contrast material is characteristic and must be differentiated from cavernous hemangioma. A central scar or calcification is seen in 25% to 30% of the cases. On MRI, these tumors are usually hypointense on T_1-weighted and hyperintense on T_2-weighted images. The central scar is best seen as a hypointense structure on T_2-weighted images. After gadolinium administration, there is early peripheral enhancement with progressive concentric enhancement, as with CT. MRI may demonstrate tumors not seen on CT and is being utilized as a problem-solving tool (4).

MRCP used as an adjunct to conventional MRI may provide additional information regarding the extension of hilar cholangiocarcinoma, for example. If MRCP can establish the resectability of the tumor, the patient may undergo immediate surgery and be spared a PTC and biliary drainage procedure (3).

PTC and/or ERCP are usually not indicated for peripheral tumors but can demonstrate hilar cholangiocarcinoma in most cases and are better than CT to evaluate the intraductal extent of the tumor. ERCP/PTC is the procedure of choice to demonstrate the infiltrating/sclerosing type of cholangiocarcinoma. Typically, a malignant stricture tapers irregularly and is associated with proximal ductal dilatation, although it is difficult to differentiate from sclerosing cholangitis, one of the preexisting conditions. Some tumors are seen as intraluminal defects, but mucin, blood clots, calculi, an air bubble, or biliary sludge may have a similar appearance. ERCP/PTC is often performed at the same time as a biliary drainage procedure.

PET and PET/CT Imaging in the Evaluation of Hepatic Lesions

The rapid advances in imaging technologies are a challenge for both radiologists and clinicians who must integrate these technologies for optimal patient care and outcomes at minimal cost. Since the early 1990s numerous technologic improvements have occurred in the field of radiologic imaging: these include (1) multislice spiral computed tomography (CT), which permits fast acquisition of CT angiographic images and multiphase enhancement techniques, and (2) positron emission tomography (PET) using ^{18}F-fluorodeoxyglucose (FDG) as a radiopharmaceutical that provides the capability for imaging glucose metabolism.

The clinical utility of FDG imaging was first established using dedicated PET tomographs equipped with multiple rings of bismuth germanate oxide (BGO) detectors, but a spectrum of equipment is now available for positron imaging, including multimodality imaging with an integrated PET/CT imaging system (6). Although no data are available at the time of this writing to document the incremental diagnostic of integrated PET/CT versus PET alone in hepatobiliary tumors, there are data available for many types of other tumors (7), as discussed in the overview chapter on PET/CT in oncology (chapter 2).

Role of FDG-PET in the Evaluation of Hepatic Metastases

A meta-analysis performed to compare noninvasive imaging methods (US, CT, MRI and FDG-PET) for the detection of hepatic metastases from colorectal, gastric, and esophageal cancers demonstrated that, at an equivalent specificity of 85%, FDG-PET had the highest sensitivity, 90% compared to 76% for MRI, 72% for CT, and 55% for US (8). Colorectal carcinoma commonly metastasize to the liver; this topic is addressed in chapter 10.

Role of FDG-PET in the Diagnosis and Staging of Hepatocellular Carcinoma

Differentiated hepatocytes normally have a relatively high glucose-6-phosphatase activity. Although experimental studies have shown that glycogenesis decreases and glycolysis increases during carcinogenesis, the accumulation of FDG in HCC is variable due to varying degrees of activity of the enzyme glucose-6-phosphatase in these tumors (9, 10). Therefore, it has been predicted that evaluation of liver tumors, especially HCCs, with FDG-PET would require dynamic imaging and blood sampling and kinetic analysis. Kinetic analysis is cumbersome to perform clinically and cannot be performed over the entire body, thus precluding its use for staging. Studies using kinetic analysis have shown that the phosphorylation kinetic constant ($k3$) is elevated in virtually all malignant tumors including HCC. The dephosphorylation kinetic constant ($k4$) is low in metastatic lesions and in cholangiocarcinomas, thus resulting in intralesional accumulation of FDG. However, $k4$ is similar to $k3$ for HCC that do not accumulate FDG (11–13). There are three patterns of uptake for HCC: FDG uptake higher, equal to, or lower than liver background (55%, 30%, and 15% respectively). FDG-PET detects only 50% to 70% of HCC but has a sensitivity greater than 90% for other malignant primary hepatic neoplasms and all metastatic tumors to the liver (14, 15). All benign tumors, including focal nodular hyperplasia, adenoma, and regenerating nodules, demonstrate FDG uptake at the same level as normal liver, except for rare abscesses with granulomatous inflammation. In ad-

dition, a correlation was found between the degree of FDG uptake, including both the standard uptake value (SUV) and $k3$, and the grade of malignancy (13, 14). Therefore, FDG imaging may have a prognostic significance in the evaluation of patients with HCC. Hepatocellular carcinomas that accumulate FDG tend to be moderately to poorly differentiated and are associated with markedly elevated alpha-fetoprotein levels (16, 17).

However, FDG-PET has limited value for the differential diagnosis of focal liver lesions in patients with chronic hepatitis C virus infection because of the low sensitivity for detection of HCC and the high prevalence of this tumor in that population of patients (18, 19).

Teefey et al. (20) did a prospective comparison of the diagnostic performance of CT, MRI imaging, US, and PET in the detection of HCC or cholangiocarcinoma in 25 liver transplant candidates and determined interobserver variability between the readers. Explanted liver specimens were examined histologically to determine presence and type of lesion. The sensitivity was as follows: US 89%, CT 60%, MRI 53%, and PET 0%. One or more imaging tests depicted 68 lesions. Histologic analysis revealed 18 HCC nodules; of these, 13 were correctly identified at CT, 14 at MR imaging, 13 at US, and none at PET. There were nine false-positive diagnoses of HCC with CT, five with MR imaging, and nine with US. They conclude that although US had the best diagnostic performance in depicting HCC on a patient-by-patient basis and was substantially better than were MR imaging and CT (which had nearly equivalent diagnostic performances), CT, US, and MR imaging performed similarly on a lesion-by-lesion basis. Small tumor nodules were the most common cause of missed HCCs with all tests. PET did not depict any HCCs.

In patients with HCC that accumulate FDG, PET imaging is able to accurately detect unsuspected regional and distant metastases, as with other tumors. In some cases, FDG-PET is the only imaging modality that can demonstrate the tumor and its metastases (14). In a series of 23 patients with HCC who underwent FDG-PET scanning in an attempt to identify extrahepatic metastases, 13 of the 23 (57%) patients had increased uptake in the primary tumor and four of the 13 had extrahepatic metastases demonstrated by FDG-PET images (21). In a larger study of 91 patients, despite the low sensitivity (64%) of FDG-PET imaging to detect HCC, there was a significant impact on management in 28% (26/91) of patients (22), including detection of unsuspected metastatic disease in high-risk patients, including liver transplant candidates, and monitoring response to hepatic-directed therapy.

^{11}C-Acetate for Hepatocellular Carcinoma

^{11}C-Acetate is a marker of membrane lipid synthesis and is a promising PET radiopharmaceutical for evaluation of some malignancies for which FDG has limitations such as HCC. Neoplastic cells incorporate acetate preferentially into lipids rather than into amino acids or CO_2 as a necessary condition for cell proliferation.

PET with ^{11}C-acetate has been shown to be useful in detection of HCC. Possible biochemical pathways that lead to accumulation of ^{11}C-acetate in tumors include (1) entry into the Krebs cycle from acetyl coenzyme A (acetyl CoA) or as an intermediate metabolite; (2) esterification to form acetyl CoA as a major precursor in ß-oxidation for fatty acid synthesis; (3) combining with glycine in heme synthesis; and (4) through citrate for cholesterol synthesis. Among these possible metabolic pathways, participation in free fatty acid (lipid) synthesis is believed to be the dominant method of incorporation into tumors. Ho et al. (23) used both FDG and ^{11}C-acetate PET imaging to study 57 patients with various hepatobiliary. For the 23 patients with HCC, the sensitivity of FDG and ^{11}C-acetate imaging were 47% and 87%, respectively, with a combined sensitivity of 100%. Well-differentiated tumors tended to be ^{11}C-acetate avid whereas poorly differentiated tumors tended to be FDG avid. Other malignant tumors were FDG avid but not ^{11}C-acetate avid. Benign tumors were not ^{11}C-acetate avid, except for mild uptake in focal nodular hyperplasia.

Role of FDG-PET in Monitoring Therapy of Hepatic Tumors

Because the majority of patients with HCC have advanced-stage tumors and/or underlying cirrhosis with impaired hepatic reserve, surgical resection is often not possible. Therefore, other treatment strategies have been developed, including hepatic arterial chemoembolization, systemic chemotherapy, surgical cryoablation, ethanol ablation, radiofrequency ablation, and, in selected cases, liver transplantation. In patients treated with hepatic arterial chemoembolization, FDG-PET is more accurate than lipiodol retention on CT in predicting the presence of residual viable tumor. The presence of residual uptake in some lesions can help in guiding further regional therapy (24–26). It is expected but not yet demonstrated that FDG-PET may surpass CT in determining the success of other ablative procedures in these patients. FDG-PET has been shown to be helpful monitoring regional therapy to liver tumors, primary and metastatic, with radiofrequency ablation, cryosurgery and ^{90}Y-labeled microspheres. Langenhoff et al. (27) prospectively monitored 23 patients with liver metastases following radiofrequency ablation and cryoablation. Three weeks after therapy, 51 of 56 metastases became FDG negative, and there was no recurrence during 16 months of follow-up, whereas among the 5 of 56 lesions with persistent FDG uptake, 4 of 5 recurred. Data in smaller series of patients support their findings (28, 29). Wong et al. (30) have compared FDG-PET imaging, CT or MRI, and serum levels of CEA to monitor the therapeutic response of hepatic metastases to ^{90}Y-glass microspheres. They found a significant difference between the FDG-PET changes and the changes on CT or MRI; the

changes in FDG uptake correlated better with the changes in serum levels of CEA.

In summary, preliminary data suggest that FDG-PET imaging may be able to effectively monitor the efficacy of regional therapy to hepatic tumors that are known to be FDG avid, but these data need to be confirmed in larger series of patients.

Role of FDG-PET in the Diagnosis and Staging of Cholangiocarcinomas

There is preliminary evidence that FDG-PET imaging may be useful in the diagnosis and management of small cholangiocarcinomas in patients with sclerosing cholangitis (31). Anderson et al. (32) reviewed 36 consecutive patients who underwent FDG-PET for suspected cholangiocarcinoma. Patients were divided into group 1, nodular type (mass larger than 1 cm) ($n = 22$) and group 2, infiltrating type ($n = 14$). The sensitivity for nodular morphology was 85% but only 18% for infiltrating morphology. Sensitivity for metastases was 65% with three false negative for carcinomatosis and one false-positive result in a patient with primary sclerosing cholangitis who had acute cholangitis. Seven of 12 (58%) patients had FDG uptake along the tract of a biliary stent. FDG-PET led to a change in surgical management in 30% (11/36) of patients because of detection of unsuspected metastases. Figure 12.1 illustrates detection of local recurrence and a metastasis with FDG-PET/CT imaging in a patient who underwent right hepatectomy for cholangiocarcinoma. The fusion images help to precisely localize the two foci of FDG uptake, one at the hepatic margin of resection and the other in a metastatic lesion posterior to the liver.

Kim et al. (33) reviewed FDG-PET images of 21 patients with cholangiocarcinoma and compared the PET findings to MRI ($n = 20$) and CT ($n = 12$). Intense FDG avidity was seen in all peripheral (11/11) but only in 2 of 10 hilar cholangiocarcinomas. For detection of lymph node metastases, FDG-PET and CT/MRI were concordant in 16 patients and discordant in 5 (FDG-PET was positive in 3, and CT and MRI in 2). FDG-PET identified unsuspected distant metastases in 4 of the 21 patients; all these patients had peripheral cholangiocarcinomas. They conclude that FDG-PET is useful in detecting the primary lesion in both hilar and peripheral cholangiocarcinomas and is of value in discovering unsuspected distant metastases in patients with peripheral cholangiocarcinomas. FDG-PET could be useful in cases of suspected hilar cholangiocarcinomas with nonconfirmatory biopsy and radiologic findings.

Role of FDG-PET in the Diagnosis and Staging of Gallbladder Carcinomas

Unsuspected gallbladder carcinoma is discovered incidentally in 1% of routine cholecystectomies. At present, the

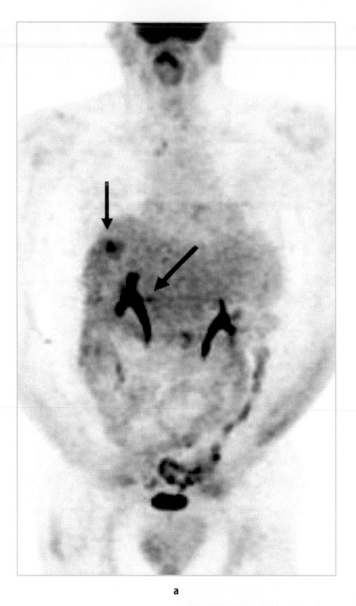

a

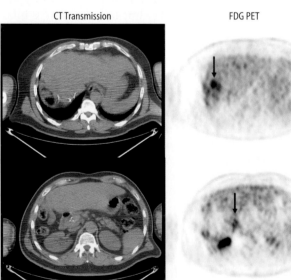

CT Transmission   FDG PET   Fusion

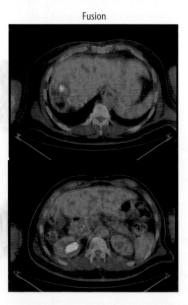

b

Figure 12.1. A 57-year-old man underwent a right hepatectomy for a cholangiocarcinoma and presented with suspicion of recurrence. ^{18}F-2-Deoxy-D-glucose (FDG)-positron emission tomography (PET)/computed tomography (CT) imaging was performed for restaging. (**a**) FDG-PET maximum intensity projection (MIP) image demonstrated two foci of FDG uptake in the region of the liver suspicious for recurrence (*arrows*). In addition, there is physiologic FDG uptake in the descending colon, kidneys, and bladder. (**b**) Transaxial views through the foci of FDG uptake in the region of the liver demonstrated that one focus of uptake is located at the margin of resection indicating local recurrence (*upper panel, arrow*). Images through the other focus of uptake demonstrated that this second focus of FDG uptake (*lower panel, arrow*) is located outside the liver posteriorly, indicating a metastasis.

majority of cholecystectomies are performed laparoscopically, and occult gallbladder carcinoma found after laparoscopic cholecystectomy has been associated with reports of gallbladder carcinoma seeding of laparoscopic trocar sites (34, 35). Increased FDG uptake has been demonstrated in gallbladder carcinoma (36) and has been helpful in identifying recurrence in the area of the inci-

sion when CT could not differentiate scar tissue from malignant recurrence (37). In a study reviewing 14 patients with gallbladder carcinoma, the sensitivity for detection of gallbladder carcinoma was 78%. Sensitivity for extrahepatic metastases was 50% in 8 patients; 6 of these had carcinomatosis (32). Figure 12.2 illustrates detection of a metastasis from gallbladder carcinoma with PET/CT

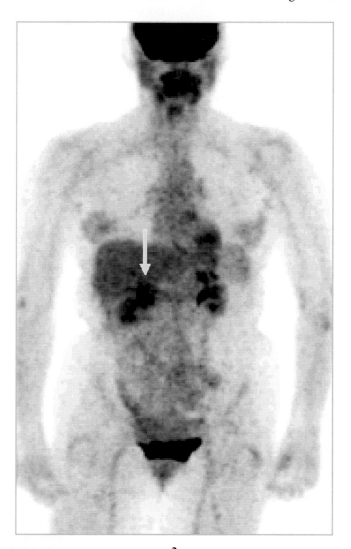

a

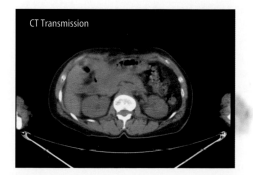
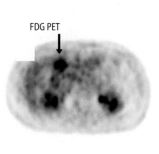
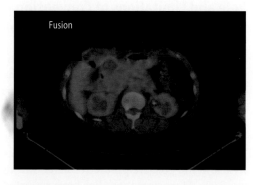

b

Figure 12.2. A 52-year-old woman underwent resection of a gallbladder carcinoma 2 months earlier and presented with abdominal pain. FDG-PET/CT imaging was performed for restaging. (**a**) FDG-PET maximum intensity projection (MIP) image demonstrated a focus of FDG uptake along the inferior border of the liver superimposed over the upper pole of the right kidney on the anterior projection of the MIP image (*arrow*). In addition, there is physiologic FDG uptake in the kidneys and bladder. (**b**) Transaxial views through the focus of FDG uptake in the abdomen demonstrated that the focus of uptake corresponded to a mass adjacent to the inferior border of the liver (*arrow*). Magnetic resonance imaging (MRI) confirmed an enhancing lesion in that location that was resected surgically and proven to be gallbladder carcinoma metastasis.

imaging in a patient who underwent resection of the primary earlier. Again, the fusion images facilitated precise localization of the metastasis that was further characterized by MRI and proven surgically.

Tumors of the Pancreas

Pancreatic carcinomas usually arise from the pancreatic ducts and are the third most common malignant tumor of the gastrointestinal tract and the fifth leading cause of cancer-related mortality. Most tumors arise in the head of the pancreas, and patients present with bile duct obstruction, pain, and jaundice. Carcinoma of the ampulla of Vater may be difficult to differentiate from those arising from the head of the pancreas. Tumors arising in the body and the tail of the pancreas are more insidious and are detected at advanced stages. The preoperative diagnosis, staging, and treatment of pancreatic cancer remains challenging even for experienced clinicians. The prognosis of pancreatic carcinoma is extremely poor, with most patients dying within 2 years of diagnosis. Surgical resection is the only potentially curative approach. Only 3% of newly diagnosed patients will survive 5 years. Pancreaticoduodenectomy improves 5-year survival to more than 20%, with 2% to 3% mortality in carefully selected patients (38). Adverse prognostic factors include histologic grade, lymphatic vessel invasion, perineural invasion, and capsular infiltration.

Acinar cell carcinomas comprise no more than 1% to 2% of all pancreatic cancer, and the prognosis is as poor as for ductal cell carcinoma. Cystic neoplasms can arise in the pancreas, and differentiation of benign from malignant is critical.

Islet cell tumors and other endocrine tumors make up a small fraction of all pancreatic neoplasms and are most often located in the body and tail of the pancreas. They are usually slow-growing tumors and are associated with endocrine abnormalities.

Role and Limitations of Conventional Imaging in the Evaluation of Pancreatic Carcinoma

The suspicion for pancreatic cancer is often raised when a pancreatic mass, or dilatation of the biliary or pancreatic ducts, is detected by US or CT. CT is superior to US, not only in detecting a pancreatic mass but also in its superb ability to assess vascular involvement and invasion of adjacent organs. In a multicenter trial (39), the diagnostic accuracy of CT for staging and assessing resectability was 73% and the positive predictive value for nonresectability was 90%. More-recent studies have reported accuracies in the 85% to 95% range, probably related to technical improvements in CT technology (40, 41). Unfortunately, interpretation of the CT scan is sometimes difficult in the

setting of mass-forming pancreatitis or questionable findings such as enlargement of the pancreatic head without definite signs of malignancy (42, 43). The diagnosis of regional lymph node metastases is also difficult with CT and US. Furthermore, small metastases (less than 1 cm) cannot reliably be differentiated from cysts (44). Therefore, the reported negative predictive value for nonresectability is less than 30%. Other anatomic imaging modalities, including MRI, endoscopic US, and ERCP, have similar limitations. Despite several recent technical improvements in MRI, including MRCP, the diagnostic performance of MRI remains in the same range as CT (45–48). Endoscopic ultrasound offers the possibility of tissue diagnosis with fine-needle biopsy but the field of view is limited (49, 50). The accuracy of ERCP is 80% to 90% for differentiation of benign from malignant pancreatic processes, including differentiation of tumor from chronic pancreatitis, because of the high degree of resolution of ductal structures. The limitations of ERCP include false negatives when the tumor does not originate from the main duct, a 10% rate of technical failure, and up to 8% morbidity, primarily iatrogenic pancreatitis. The principal advantages are the ability to perform fine-needle biopsy and interventional procedures. Although fine-needle biopsy may provide a tissue diagnosis, this technique suffers from significant sampling error (51, 52).

FDG-PET and PET/CT in the Evaluation of Pancreatic Carcinoma

The Role of FDG-PET in the Preoperative Diagnosis of Pancreatic Carcinoma

The difficulty in correctly determining a preoperative diagnosis of pancreatic carcinoma is associated with two types of adverse outcomes. First, less-aggressive surgeons may abort attempted resection because of a lack of tissue diagnosis; this is borne out by the significant rate of "reoperative" pancreaticoduodenectomy performed at major referral centers (53–55). In a recent review of the M.D. Anderson Cancer Center's experience with 29 patients undergoing successful pancreaticoduodenectomy after failure to resect at the time of initial laparotomy, 31% did not undergo resection at the time of the initial procedure because of the lack of tissue confirmation of malignancy (55). A second type of adverse outcome generated by failure to obtain a preoperative diagnosis occurs when more aggressive surgeons inadvertently resect benign disease. As many as 55% of patients who undergo pancreaticoduodenectomy for suspected malignancy without an associated mass on CT scan, are found to have benign disease (56).

To avoid these adverse outcomes, metabolic imaging with FDG-PET may improve the accuracy of the preoperative diagnosis of pancreatic carcinoma. Most malignancies, including pancreatic carcinoma, demonstrate increased glucose utilization due to an increased number of glucose transporter proteins and increased hexokinase and phosphofructokinase activity (57–60). The summary of the literature published in 2001 reported an average sensitivity and specificity of 94% and 90%, respectively (61). All studies report relatively high rates of sensitivity (85%–100%), specificity (67%–99%), and accuracy (85%–93%) for FDG-PET in the differentiation of benign from malignant pancreatic masses, and the majority suggest improved accuracy compared to CT. These results are similar to the findings in the series of Rose et al. (62) at Vanderbilt University, with a sensitivity of 92% and specificity of 85% for FDG-PET compared to 65% and 62%, respectively, for CT imaging One must keep in mind, however, that these studies suffer from biases. For example, the acquisition of the CT data is often not done prospectively and the quality of the CT images may vary among different institutions. Together, these series support the conclusion that FDG-PET imaging may represent a useful adjunctive study in the evaluation of patients with suspected pancreatic cancer, especially when CT imaging results are inconclusive.

As with any imaging modality, FDG-PET has limitations in the evaluation of pancreatic cancer. The high incidence of glucose intolerance and diabetes exhibited by patients with pancreatic pathology represents a potential limitation of this modality in the diagnosis of pancreatic cancer, because elevated serum glucose levels result in decreased FDG uptake in tumors as a result of competitive inhibition. Low SUV values and false-negative FDG-PET scans have been noted in hyperglycemic patients, which has led some investigators to suggest that the SUV be corrected according to serum glucose level (63–66). The true impact of serum glucose levels on the accuracy of FDG-PET in pancreatic cancer and other neoplasms remains controversial. Several studies have demonstrated a lower sensitivity in hyperglycemic compared to euglycemic patients (66–68). For example, in a study of 106 patients with a prevalence of disease of 70%, Zimny et al. (68) found that FDG-PET had a sensitivity of 98%, specificity of 84%, and accuracy of 93% in a subgroup of euglycemic patients compared to 63%, 86%, and 68%, respectively, in a subgroup of hyperglycemic patients. Other investigators (69, 70) noted no variation in the accuracy of FDG-PET based on serum glucose levels. In the studies of Delbeke et al. (71) and Diederichs et al. (65), the presence of elevated serum glucose levels and/or diabetes mellitus may have contributed to false negative interpretations, but correction of the SUV for serum glucose level has not significantly improved the accuracy of FDG-PET in the diagnosis of pancreatic carcinoma. False-negative studies may also occur when the tumor diameter is less than 1 cm (i.e., small ampullary carcinoma).

Both glucose and FDG are substrates for cellular mediators of inflammation. Some benign inflammatory lesions, including chronic active pancreatitis with or without abscess formation, can accumulate FDG and give false-positive interpretations on PET images (72). False-positive studies are frequent in patients with elevated C-reactive protein and/or acute pancreatitis with a specificity as low as 50% (66). Therefore, screening for C-reactive protein has been recommended.

Of interest, studies on a small number of patients suggest that the degree of FDG uptake has prognostic value. Nakata et al. (73) noted a correlation between SUV and survival in 14 patients with pancreatic adenocarcinoma. Patients with an SUV greater than 3.0 had a mean survival of 5 months, compared to 14 months in those with an SUV less than 3.0. Zimny et al. (74) performed a multivariate analysis on 52 patients, including SUV and accepted prognostic factors, to determine the prognostic value of FDG-PET. The median survival of 26 patients with SUV greater than 6.1 was 5 months, compared to 9 months for 26 patients with SUV less than 6.1. The multivariate analysis revealed that SUV and Ca 19-9 were independent factors of prognosis.

In summary, FDG-PET imaging is complementary to CT in the evaluation of patients with pancreatic masses or in whom the diagnosis of pancreatic carcinoma is suspected. In view of the decreased sensitivity seen in patients with hyperglycemia, PET acquisition should be performed under controlled metabolic conditions and in the absence of acute inflammatory abdominal disease.

Role of FDG-PET in Staging of Pancreatic Carcinoma

Stage II disease is characterized by extrapancreatic extension (T stage), stage III by lymph node involvement (N stage), and stage IV by distant metastases (M stage). T staging can only be evaluated with anatomic imaging modalities, which demonstrate best the relationship between the tumor, adjacent organs, and vascular structures. Functional imaging modalities can obviously not replace anatomic imaging in the assessment of local tumor resectability.

As for many other tumors, FDG imaging has not been superior to helical CT for N staging, but is more accurate than CT for M staging. In the study of Delbeke et al. (71), metastases were diagnosed both on CT and PET in 10 of 21 patients with stage IV disease, but PET demonstrated hepatic metastases not identified or equivocal on CT and/or distant metastases unsuspected clinically in 7 additional patients. In 4 patients, neither CT nor PET imaging showed evidence of metastases, but surgical exploration revealed carcinomatosis in 3 and a small liver metastasis in 1 patient. False-positive findings have been reported in the liver of patients with dilated bile ducts and formation of inflammatory granulomas (75).

Impact of FDG-PET on the Management of Patients with Pancreatic Carcinoma

The rate with which FDG-PET results may lead to alterations in clinical management clearly depends on the specific therapeutic philosophy employed by an evaluating surgeon. A common approach is performing pancreatico-duodenectomy only for those patients with potentially curable pancreatic cancer. In these patients, an aggressive approach to resection consist in en bloc retroperitoneal lymphadenectomy and selective resection of the superior mesenteric–portal vein confluence when necessary. Although certain patients with chronic pancreatitis may also benefit from pancreaticoduodenectomy, the majority of patients with nonmalignant biliary strictures are better managed without resection. In a series of 65 patients, the addition of FDG-PET imaging to CT altered the surgical management in 41% of the patients, 27% by detecting CT-occult pancreatic carcinoma and 14% by identifying unsuspected distant metastases, or by clarifying the benign nature of lesions equivocal on CT (71). In this regard, FDG-PET may allow selection of the optimal surgical approach in patients with pancreatic carcinoma.

However, Kalady et al. (76) reviewed the performance of FDG-PET in 54 patients with suspected periampullary malignancy. Despite high sensitivity (90%) and specificity in diagnosing periampullary malignancy, FDG-PET did not change clinical management in most patients previously evaluated by CT. In addition, FDG-PET missed more than 10% of periampullary malignancies and did not provide the anatomic detail necessary to define unresectability.

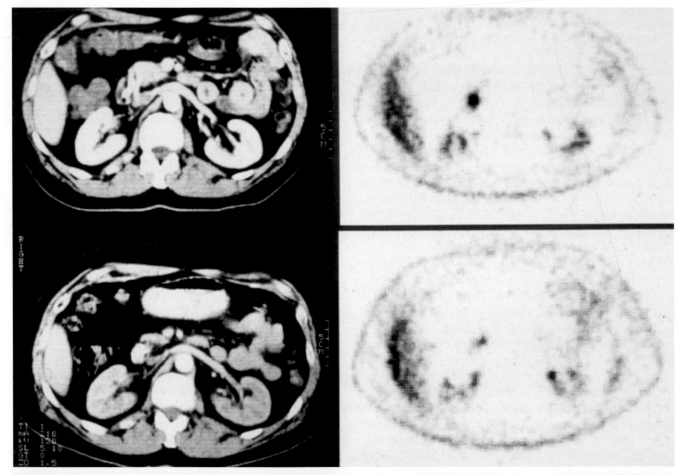

Figure 12.3. A 53-year-old man presented with biliary obstruction. *Upper left:* the CT scan shows prominence of the head of the pancreas but no definite mass. The fine-needle biopsy was inconclusive. *Upper right:* the corresponding FDG-PET image (without attenuation correction) shows marked uptake corresponding to the head of the pancreas, indicating a malignant tumor that was proved to be pancreatic adenocarcinoma with a biopsy obtained at endoscopic ultrasound. The patient underwent neoadjuvant chemoradiotherapy before surgery. *Lower left:* the CT obtained after completion of neoadjuvant chemoradiotherapy shows no pancreatic lesion. *Lower right:* the corresponding FDG-PET image (without attenuation correction) demonstrates uptake but to a lesser degree than on the pretherapy image. The patient underwent extended pancreatoduodenectomy. The specimen demonstrated an adenocarcinoma (2 cm in size). The histologic examination demonstrated 15% to 20% tumor necrosis from chemoradiotherapy. 📖

The Role of FDG-PET in Monitoring Therapy and Assessing Recurrence of Pancreatic Carcinoma

The potentially significant morbidity associated with pancreaticoduodenectomy, which can compromise the delivery of postoperative adjuvant chemoradiation, has led to the development of preoperative adjuvant (neoadjuvant) chemoradiation in these patients. In addition, preliminary studies suggest that neoadjuvant chemoradiation improves the resectability rate and survival of patients with pancreatic carcinoma (77, 78). A preliminary study suggests that FDG-PET imaging is useful for the assessment of tumor response to neoadjuvant therapy and the evaluation of possible recurrent disease following resection (62). Nine patients underwent FDG-PET imaging before and after neoadjuvant chemoradiation therapy. FDG-PET successfully predicted histologic evidence of chemoradiation-induced tumor necrosis in all four patients who demonstrated at least a 50% reduction in tumor SUV following chemoradiation (Figure 12.3). Among these patients, none showed measurable change in tumor diameter as assessed by CT. Three patients showed stable FDG uptake, and two showed increasing FDG uptake indicative of tumor progression. Of the two patients with progressive disease demonstrated by FDG-PET, one showed tumor progression on CT and the other demonstrated stable disease. The four patients who had FDG-PET evidence of tumor response went on to successful resection, all showing 20% to 80% tumor necrosis in the resected specimen. Among the five patients who showed no response by FDG-PET, the disease could be subsequently resected in only two, and only one patient who underwent resection showed evidence of chemoradiation-induced necrosis in the resected specimen. Another pilot study suggests that the absence of FDG uptake at 1 month following chemotherapy is an indicator of improved survival (79). Definitive conclusions regarding the role of FDG-PET in assessing treatment response obviously require evaluation in a larger group of patients. However, given the poor track record of CT in assessing histologic response to neoadjuvant chemoradiation, the potential utility of FDG-PET in this capacity deserves further investigation.

The majority of the prior reports concerning the clinical utilization of FDG-PET scanning for pancreatic malignancy have emphasized the identification of recurrent nodal or distant metastatic disease. In a preliminary study (62), 8 patients were evaluated for possible recurrence because of either indeterminate CT findings or a rise in serum tumor marker levels. All were noted to have significant new regions of FDG uptake, 4 in the surgical bed and 4 in new hepatic metastases. In all patients, metastases or local recurrence was confirmed pathologically or clinically. Another study on 19 patients concluded that FDG-PET added important additional information to clinicians in 50% of the patients, resulting in a change of therapeutic procedure (80); this includes patients with elevated tumor marker levels and no findings on anatomic imaging. Therefore, FDG-PET may be particularly useful when CT identifies an indistinct region of change in the bed of the resected pancreas that is difficult to differentiate from posttreatment changes, for the evaluation of new hepatic lesions that may be too small to biopsy, and in patients with rising tumor marker levels and a negative conventional workup (Figure 12.4). Again, the fusion images aided precise localization of the focus of FDG uptake in a lesion in the pancreatic bed and increased confidence in the diagnosis of local recurrence compared to PET or CT alone.

FDG-PET for Evaluation of Islet Cells and Other Endocrine Tumors of the Pancreas

Most neuroendocrine tumors, including carcinoid, paraganglioma, and islet cell tumors, express somatostatin receptors (SSR) and can, therefore, be imaged effectively with somatostatin analogues such as ^{111}In-octreotide (81). This modality has been reported to be more sensitive than CT for defining the extent of metastatic disease, especially in extrahepatic and extraabdominal sites. However, there may be significant heterogeneity in regard to SSR expression, even in the same patient in adjacent sites, probably related to dedifferentiation of the tumor. Absence of SSR positivity is reported to be a poor prognostic sign, but virtually all these SSR-negative neuroendocrine tumors accumulate FDG and can therefore be imaged with PET (82). More-differentiated SSR-positive tumors do not reliably accumulate significant FDG and may, therefore, be false negative with FDG-PET imaging (83). There is controversy in the literature about the sensitivity of FDG imaging for detection of carcinoid tumors (84), but at least in some reports, ^{111}In-octreotide scintigraphy is more sensitive than FDG-PET imaging; FDG-positive/octreotide-negative tumors tend to be less differentiated and may have a less-favorable prognosis.

Other positron-emitting tracers seem to be more promising. A serotonin precursor 5-hydroxytryptophan (5-HTP) labeled with ^{11}C has shown increased uptake in carcinoid tumors. This uptake seems to be selective, and some clinical evidence has demonstrated that it allows the detection of more lesions with PET than with CT or octreotide scintigraphy (85). Other radiopharmaceuticals in development for PET are ^{11}C L-DOPA and ^{18}F-DOPA, which seem to be useful in visualizing gastrointestinal neuroendocrine tumors (86, 87). A study of 17 patients with 92 carcinoid tumors comparing FDG-PET, ^{18}F-DOPA-PET, and somatostatin-receptor scintigraphy demonstrated the following sensitivities: 29% for FDG-PET, 60% for ^{18}F-DOPA, 57% for somatostatin-receptor scintigraphy, and 73% for morphologic procedures (88).

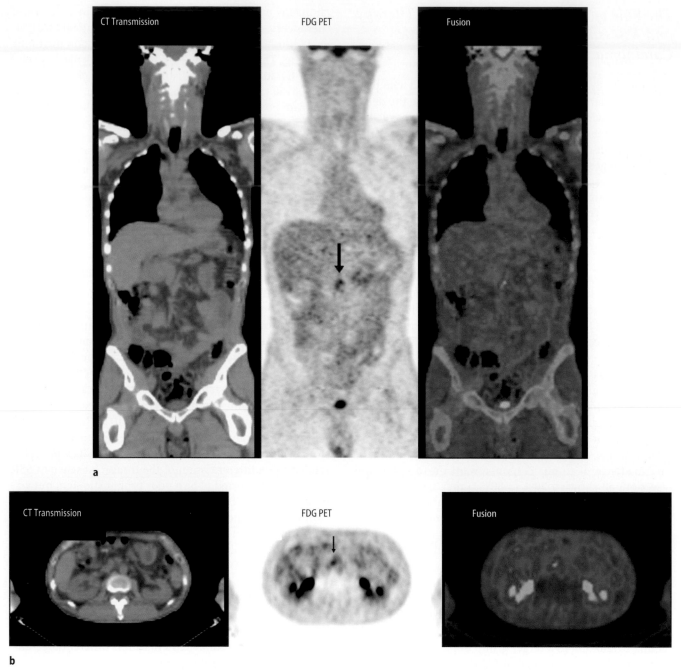

Figure 12.4. A 44-year-old woman who underwent a Whipple procedure for pancreatic cancer 8 months earlier presented with abdominal pain. FDG-PET/CT imaging was performed for restaging. (**a**) A coronal image through the location of the head of the pancreas demonstrated a focus of FDG uptake (*arrow*) suspicious for local recurrence. (**b**) Transaxial views through the focus of FDG uptake in the midabdomen demonstrated that the focus of uptake corresponded to a lesion in the pancreatic bed indicating local recurrence (*arrow*), which was proven at surgery.

An octreotide derivative can be labeled with ^{64}Cu [half-life, 12.7 h; beta+, 0.653 MeV (17.4%); beta–, 0.579 MeV (39%)] and has shown potential as a radiopharmaceutical for PET imaging and radiotherapy. A pilot study in humans has demonstrated that ^{64}Cu-TETA-octreotide (where TETA is 1,4,8,11-tetraazacyclotetradecane-*N,N',N'',N'''*-tetraacetic acid) PET imaging can be used to detect somatostatin-receptor-positive tumors in humans. The high rate of lesion detection and favorable dosimetry and pharmacokinetics of ^{64}Cu-TETA-OC indicates that it

may be a promising radiopharmaceutical for PET imaging of patients with neuroendocrine tumors (89).

Limitations of FDG Imaging

Scintigraphic tumor detectability depends on both the size of the lesion and the degree of uptake, as well as surrounding background uptake and intrinsic resolution of the imaging system. Small lesions may yield false-negative results as a result of partial-volume averaging, leading to underestimation of the uptake in small lesions (less than twice the resolution of the imaging system, for example, small ampullary carcinoma, cholangiocarcinoma of the infiltrating type, and military carcinomatosis) or in necrotic lesions with a thin viable rim, falsely classifying these lesions as benign instead of malignant. The sensitivity of FDG-PET for detection of mucinous adenocarcinoma is lower than for nonmucinous adenocarcinoma (41%–58% versus 92%), probably because of the relative hypocellularity of these tumors (90). Other false negatives include differentiated neuroendocrine tumors and HCC. The high incidence of glucose intolerance and diabetes exhibited by patients with pancreatic pathology represents a potential limitation of this modality in the diagnosis of pancreatic cancer because elevated serum glucose levels result in decreased FDG uptake in tumors due to competitive inhibition. Low SUV values and false-negative FDG-PET scans have been noted in hyperglycemic patients.

In view of the known high uptake of FDG by activated macrophages, neutrophils, fibroblasts, and granulation tissue, it is not surprising that inflammatory tissue demonstrates FDG activity. Mild to moderate FDG activity seen early after radiation therapy, along recent incisions, and in infected incisions, biopsy sites, drainage tubing, and catheters, as well as colostomy sites, can lead to errors in interpretation if the history is not known. Some inflammatory lesions, especially granulomatous ones, may be markedly FDG avid and can be mistaken for malignancies; these include abscesses, acute cholangitis, acute cholecystitis, and acute pancreatitis and chronic active pancreatitis with or without abscess formation. False-positive studies are frequent in patients with elevated C-reactive protein and/or acute pancreatitis with a specificity as low as 50% (68, 91).

Summary

FDG-PET imaging appears helpful in differentiating malignant from benign hepatic lesions, with the exception of false-negative HCC, false-negative infiltrating cholangiocarcinoma, and false-positive inflammatory lesions. It is not helpful to identify HCC in patients with cirrhosis and regenerating nodules. In patients with primary malignant hepatic tumors that accumulate FDG, PET imaging does identify unexpected distant metastases (although miliary carcinomatosis is often false negative) and can help in monitoring response to therapy. FDG-PET imaging seems promising for monitoring patient response to therapy, including regional therapy to the liver, but larger studies are necessary.

FDG-PET imaging is especially helpful for the preoperative diagnosis of pancreatic carcinoma in patients with suspected pancreatic cancer in whom CT fails to identify a discrete tumor mass or in whom biopsy is nondiagnostic.

By providing scintigraphic preoperative documentation of pancreatic malignancy in these patients, laparotomy may be undertaken with a curative intent, and the risk of aborting resection because of diagnostic uncertainty is minimized. FDG-PET imaging is also useful for M staging and restaging by detecting CT-occult metastatic disease and allowing nontherapeutic resection to be avoided altogether in this group of patients. As is true with other neoplasms, FDG-PET can accurately differentiate posttherapy changes from recurrence and holds promise for monitoring neoadjuvant chemoradiation therapy.

FDG-PET imaging is complementary to morphologic imaging with CT; therefore, integrated PET/CT imaging provides optimal images for interpretation.

References

1. Kemmerer SC, Mortele KJ, Ros PR. CT scan of the liver. Radiol Clin N Am 1998;36(2):247–260.
2. Siegelman ES, Outwater EK. MR imaging technique of the liver. Radiol Clin N Am 1998;36(2):263–284.
3. Fulcher AS, Turner MA, Capps GW. MR cholangiography: technical advances and clinical applications. Radiographics 1999;19(1):25–43.
4. del Pilar Fernandez M, Redvanly RD. Primary hepatic malignant neoplasms. Radiol Clin N Am 1998;36(2):333–348.
5. Oppenheim BE. Liver imaging. In: Sandler MP, Coleman RE, Wackers FTJ, et al., editors. Diagnostic Nuclear Medicine. Baltimore: Williams & Wilkins, 1996:749–758.
6. Townsend DW, Beyer T, Bloggett TM. PET/CT scanners: a hardware approach to image fusion. Semin Nucl Med 2003;33(3):193–204.
7. Czernin J, editor. PET/CT: imaging structure and function. J Nucl Med 2004;45(suppl 1):1S–103S.
8. Kinkel K, Lu Y, Both M, Warren RS, Thoeni RF. Detection of hepatic metastases from cancers of the gastrointestinal tract by using noninvasive imaging methods (US, CT, MR imaging, PET): a meta-analysis. Radiology 2002;224(3):748–756.
9. Weber G, Cantero A. Glucose 6 phosphatase activity in normal, precancerous, and neoplastic tissues. Cancer Res 1955;15:105–108.
10. Weber G, Morris HP. Comparative biochemistry of hepatomas. III. Carbohydrate enzymes in liver tumors of different growth rates. Cancer Res 1963;23:987–994.
11. Messa C, Choi Y, Hoh CK, et al. Quantification of glucose utilization in liver metastases: parametric imaging of FDG uptake with PET. J Comput Assist Tomogr 1992;16:684–689.
12. Okazumi S, Isono K, Enomoto D, et al. Evaluation of liver tumors using fluorine-18-fluorodeoxyglucose PET: characterization of tumor and assessment of effect of treatment. J Nucl Med 1992;33:333–339.
13. Torizuka T, Tamaki N, Inokuma T, et al. In vivo assessment of glucose metabolism in hepatocellular carcinoma with FDG PET. J Nucl Med 1995;36:1811–1817.
14. Khan MA, Combs CS, Brunt EM, Lowe VJ, Wolverson MK, Solomon H, Collins BT, Di Bisceglie AM. Positron emission tomography scanning in the evaluation of hepatocellular carcinoma. J Hepatol 2000;32:792–797.
15. Delbeke D, Martin WH, Sandler MP, Chapman WC, Wright JK Jr, Pinson CW. Evaluation of benign vs. Malignant hepatic lesions with positron emission tomography. Arch Surg 1998;133:510–515.
16. Iwata Y, Shiomi S, Sasaki N, Jomura H, Nishigushi S, Seki S, Kawabe J, Ochi H. Clinical usefulness of positron emission tomography with fluorine-18-fluorodeoxiglucose in the diagnosis of liver tumors. Ann Nucl Med 2000;14:121–126.
17. Trojan J, Schroeder O, Raedle J, Baum RP, Herrmann G, Jacobi V, Zeuzem S. Fluorine-18 FDG positron emission tomography for

imaging ofd hepatocellular carcinoma. Am J Gastroenterol 1999;94:3314–3319.

18. Schroder O, Trojan J, Zeuzem S, Baum RP. Limited value of fluorine-18-fluorodeoxyglucose PET for the differential diagnosis of focal liver lesions in patients with chronic hepatitic C virus infection. Nuklearmedizin 1998;37:279–285.

19. Liangpunsakul S, Agarwal D, Horlander JC, Kieff B, Chalasani N. Positron emission tomography for detecting occult hepatocellular carcinoma in hepatitis C cirrhotics awaiting for liver transplantation. Transplant Proc 2003;35:2995–2997.

20. Teefey SA, Hildeboldt CC, Dehdashti F, Siegel BA, Peters MG, Heiken JP, et al. Detection of primary hepatic malignancy in liver transplant candidates: prospective comparison of CT, MR imaging, US, and PET. Radiology 2003;226(2):533–542.

21. Rose AT, Rose DM, Pinson CW, Wright JK, Blair T, Blanke C, et al. Hepatocellular carcinoma outcome based on indicated treatment strategy. Am Surg 1998;64:1122–1135.

22. Wudel LJ, Delbeke D, Morris D, Rice MH, Washington MK, Pinson CW, Chapman WC. The role of FDG-PET imaging in the evaluation of hepatocellular carcinoma. Am Surg 2003;69:117–126.

23. Ho CL, Yu SC, Yeung DW. ^{11}C-Acetate PET imaging in hepatocellular carcinoma and other liver masses. J Nucl Med 2003;44:213–221.

24. Torizuka T, Tamaki N, Inokuma T, et al. Value of fluorine-18-FDG-PET to monitor hepatocellular carcinoma after interventional therapy. J Nucl Med 1994;35:1965–1969.

25. Akuta K, Nishimura T, Jo S, et al. Monitoring liver tumor therapy with [^{18}F]FDG positron emission tomography. J Comput Assist Tomogr 1990;14:370–374.

26. Vitola JV, Delbeke D, Meranze SG, Mazer MJ, Pinson CW. Positron emission tomography with F-18-fluorodeoxyglucose to evaluate the results of hepatic chemoembolization. Cancer (Phila) 1996;78:2216–2222.

27. Langenhoff BS, Oyen WJ, Jager GJ, Strijk SP, Wobbes T, Corstens FH, Ruers TJ. Efficacy of fluorine-18-deoxyglucose positron emission tomography in detecting tumor recurrence after local ablative therapy for liver metastases: a prospective study. J Clin Oncol 2002;20:4453–4458.

28. Anderson GS, Brinkmann F, Soulen MC, Alavi A, Zhuang H. FDG positron emission tomography in the surveillance of hepatic tumors treated with radiofrequency ablation. Clin Nucl Med 2003;28:192–197.

29. Ludwig V, Hopper OW, Martin WH, Kikkawa R, Delbeke D. FDG-PET surveillance of hepatic metastases from prostate cancer following radiofrequency ablation: case report. Am Surg 2003;69:593–598.

30. Wong CY, Salem R, Raman S, Gates VL, Dworkin HJ. Evaluating ^{90}Y-glass microsphere treatment response of unresectable colorectal liver metastases by [^{18}F]FDG PET: a comparison with CT or MRI. Eur J Nucl Med Mol Imaging 2002;29:815–820.

31. Keiding S, Hansen SB, Rasmussen HH, et al. Detection of cholangiocarcinoma in primary sclerosing cholangitis by positron emission tomography. Hepatology 1998;28:700–706.

32. Anderson CA, Rice MH, Pinson CW, Chapman WC, Ravi RS, Delbeke D. FDG PET imaging in the evaluation of gallbladder carcinoma and cholangiocarcinoma. J. Gastrointest Surg 2004;8(1):90–97.

33. Kim YJ, Yun M, Lee WJ, Kim KS, Lee JD. Usefulness of ^{18}F-FDG PET in intrahepatic cholangiocarcinomas. Eur J Nucl Med Mol Imaging 20034;30(11):1467–1472.

34. Drouard F, Delamarre J, Capron JP. Cutaneous seeding of gallbladder cancer after laparoscopic cholecystectomy. N Engl J Med 1991;325:1316.

35. Weiss SM, Wengert PA, Harkavy SE. Incisional recurrence of gallbladder cancer aftyer laparoscopic cholecystectomy. Gastrointest Endosc 1994;40:244–246.

36. Hoh CK, Hawkins RA, Glaspy JA, et al. Cancer detection with whole-body PET using 2-[^{18}F]fluoro-2-deoxy-D-glucose. J Comput Assist Tomogr 1993;17:582–589.

37. Lomis KD, Vitola JV, Delbeke D, Snodgrass SL, Chapman WC, Wright JK, Pinson CW. Recurrent gallbladder carcinoma at laparoscopy port sites diagnosed by PET scan: implications for primary and radical second operations. Am Surg 1997;63:341–345.

38. Gold EB, Goldin SB. Epidemiology of and risk factors for pancreatic cancer. Surg Oncol Clin N Am 1998;7:67–91.

39. Megibow AJ, Zhou XH, Rotterdam H, Francis IR, Zerhouni EA, Balfe DM, et al. Pancreatic adenocarcinoma: CT versus MR imaging in the evaluation of resectability: report of the Radiology Diagnostic Oncology Group. Radiology 1995;195(2):327–332.

40. Diehl SJ, Lehman KJ, Sadick M, Lachman R, Georgi M. Pancreatic cancer: value of dual-phase helical CT in assessing resectability. Radiology 1998;206:373–378.

41. Lu DSK, Reber HA, Krasny RM, Sayre J. Local staging of pancreatic cancer: criteria for unresectability of major vessels as revealed by pancreatic-phase, thin section helical CT. Am J Roentgenol 1997;168:1439–1444.

42. Johnson PT, Outwater EK. Pancreatic carcinoma versus chronic pancreatitis: dynamic MR imaging. Radiology. 1999;212(1):213–218.

43. Lammer J, Herlinger H, Zalaudek G, Hofler H. Pseudotumorous pancreatitis. Gastrointest Radiol 1995;10:59–67.

44. Bluemke DA, Cameron IL, Hurban RH, et al. Potentially resectable pancreatic adenocarcinoma: Spiral CT assessment with surgical and pathologic correlation. Radiology 1995;197:381–385.

45. Bluemke DA, Fishman EK. CT and MR evaluation of pancreatic cancer. Surg Oncol Clin N Am 1998;7:103–124.

46. Catalano C, Pavone P, Laghi A, et al. Pancreatic adenocarcinoma: combination of MR angiography and MR cholangiopancreatography for the diagnosis and assessment of resectability. Eur Radiol 1998;8:428–434.

47. Irie H, Honda H, Kaneko K, et al. Comparison of helical CT and MR imaging in detecting and staging small pancreatic adenocarcinoma. Abdom Imaging 1997;22:429–433.

48. Trede M, Rumstadt B, Wendl et al. Ultrafast magnetic resonance imaging improves the staging of pancreatic tumors. Ann Surg 1997;226:393–405.

49. Hawes RH, Zaidi S. Endoscopic ultrasonography of the pancreas. Gastrointest Endosc Clin N Am 1995;5:61–80.

50. Legmann P, Vignaux O, Dousset B, et al. Pancreatic tumors: comparison of dual-phase helical CT and endoscopic sonography. Am J Roentgenol 1998;170:1315–1322.

51. Brandt KR, Charboneau JW, Stephens DH, Welch TJ, Goellner JR. CT- and US-guided biopsy of the pancreas. Radiology 1993;187:99–104.

52. Chang KJ, Nguyen P, Erickson RA, et al. The clinical utility of endoscopic ultrasound-guided fine-needle aspiration in the diagnosis and staging of pancreatic carcinoma. Gastrointest Endosc 1997;45:387–393.

53. McGuire GE, Pitt HA, Lillemoe KD, et al. Reoperative surgery for periampullary adenocarcinoma. Arch Surg 1991;126:1205–1212.

54. Tyler DS, Evans DB. Reoperative pancreaticoduodenectomy. Ann Surg 1994;219:211–221.

55. Robinson EK, Lee JE, Lowy AM, et al. Reoperative pancreaticoduodenectomy for periampullary carcinoma. Am J Surg 1996;172:432–438.

56. Thompson JS, Murayama KM, Edney JA, Rikkers LF. Pancreaticoduodenectomy for suspected but unproven malignancy. Am J Surg 1994;168:571–575.

57. Flier JS, Mueckler MM, Usher P, Lodish HF. Elevated levels of glucose transport and transporter messenger RNA are induced by ras or src oncogenes. Science 1987;235:1492–1495.

58. Monakhov NK, Neistadt EL, Shavlovskil MM, et al. Physiochemical properties and isoenzyme composition of hexokinase from normal and malignant human tissues. J Natl Cancer Inst 1978;61:27–34.

59. Higashi T, Tamaki N, Honda T, et al. Expression of glucose transporters in human pancreatic tumors compared with increased F-18 FDG accumulation in PET study. J Nucl Med 1997;38:1337–1344.

60. Reske S, Grillenberger KG, Glatting G, et al. Overexpression of glucose transporter 1 and increased F-18 FDG uptake in pancreatic carcinoma. J Nucl Med 1997;38:1344–1348.

61. Gambir SS, Czernin J, Schimmer J, Silverman D, Coleman RE, Phelps ME. A tabulated summary of the FDG PET literature. J Nucl Med 2001;42(suppl):1S–93S.

62. Rose DM, Delbeke D, Beauchamp RD, Chapman WC, Sandler MP, Sharp KW, et al. ^{18}Fluorodeoxyglucose–positron emission tomography (^{18}FDG-PET) in the management of patients with suspected pancreatic cancer. Ann Surg 1998;229:729–738.

63. Wahl RL, Henry CA, Ethrer SP. Serum glucose: effects on tumor and normal tissue accumulation of 2-[F-18]-fluoro-2-deoxy-D-glucose in rodents with mammary carcinoma. Radiology 1992;183:643–647.

64. Lindholm P, Minn H, Leskinen-Kallio S, et al. Influence of the blood glucose concentration on FDG uptake in cancer: a PET study. J Nucl Med 1993;34:1–6.

65. Diederichs CG, Staib L, Glatting G, Beger HG, Reske SN. FDG PET: elevated plasma glucose reduces both uptake and detection rate of pancreatic malignancies. J Nucl Med 1998;39:1030–1033.

66. Diederichs CG, Staib L, Vogel J, et al. Values and limitations of FDG PET with preoperative evaluations of patients with pancreatic masses. Pancreas 2000;20:109–116.

67. Stollfuss JC, Glatting G, Friess H, Kocher F, Berger HG, Reske SN. 2-(Fluorine-18)-fluoro-2-deoxy-D-glucose PET in detection of pancreatic cancer: value of quantitative image interpretation. Radiology 1995;195:339–344.

68. Zimny M, Bares R, Faß J, et al. Fluorine-18 fluorodeoxyglucose positron emission tomography in the differential diagnosis of pancreatic carcinoma: a report of 106 cases. Eur J Nucl Med 1997;24:678–682.

69. Ho CL, Dehdashti F, Griffeth LK, et al. FDG-PET evaluation of indeterminate pancreatic masses. Comput Assist Tomogr 1996;20:363–369.

70. Friess H, Langhans J, Ebert M, et al. Diagnosis of pancreatic cancer by 2[F-18]-fluoro-2-deoxy-D-glucose positron emission tomography. Gut 1995;36:771–777.

71. Delbeke D, Rose M, Chapman WC, Pinson CW, Wright JK, Beauchamp DR, Leach S. Optimal interpretation of F-18 FDG imaging of FDG PET in the diagnosis, staging and management of pancreatic carcinoma. J Nucl Med 1999;40:1784–1792.

72. Shreve PD. Focal fluorine-18 fluorodeoxyglucose accumulation in inflammatory pancreatic disease. Eur J Nucl Med 1998;25:259–264.

73. Nakata B, Chung YS, Nishimura S, et al. ^{18}F-Fluorodeoxyglucose positron emission tomography and the prognosis of patients with pancreatic carcinoma. Cancer (Phila) 1997;79:695–699.

74. Zimny M, Fass J, Bares R, Cremerius U, Sabri O, Buechin P, Schumpelick V, Buell U. Fluorodeoxyglucose positron emission tomography and the prognosis of pancreatic carcinoma. Scand J Gastroenterol 2000;35:883–888.

75. Frolich A, Diederichs CG, Staib L, et al. Detection of liver metastases from pancreatic cancer using FDG PET. J Nucl Med 1999;40:250–255.

76. Kalady MF, Clary BM, Clark LA, Gottfried M, Rohren EM, Coleman RE, et al. Clinical utility of positron emission tomography in the diagnosis and management of periampullary neoplasms. Ann Surg Oncol 2002;9(8):799–806.

77. Yeung RS, Weese JL, Hoffman JP, Solin LJ, Paul AR, Engstrom PF, et al. Neoadjuvant chemoradiation in pancreatic and duodenal carcinoma. A Phase II Study. Cancer (Phila) 1993;72(7):2124–2133.

78. Jessup JM, Steele G Jr, Mayer RJ, Posner M, Busse P, Cady B, et al. Neoadjuvant therapy for unresectable pancreatic adenocarcinoma. Arch Surg 1993;128(5):559–564.

79. Maisey NR, Webb A, Flux GD, Padhani A, Cunningham DC, Ott RJ, Norman A. FDG PET in the prediction of survival of patients with cancer of the pancreas: a pilot study. Br J Cancer 2000;83:287–293.

80. Franke C, Klapdor R, Meyerhoff K, Schauman M. 18-F positron emission tomography of the pancreas: diagnostic benefit in the follow-up of pancreatic carcinoma. Anticancer Res 1999;19:2437–2442.

81. Kaltsas G, Rockall A, Papadogias D, Reznek R, Grossman AB. Recent advances in radiological and radionuclide imaging and therapy of neuroendocrine tumours. Eur J Endocrinol 2004;151(1):15–27.

82. Adams S, Baum R, Rink T, Schumm-Drager PM, Usadel KH, Hor G. Limited value of fluorine-18 fluorodeoxyglucose positron emission tomography for the imaging of neuroendocrine tumors. Eur J Nucl Med 1998;25:79–83.

83. Jadvar H, Segall GM. False-negative fluorine-18-FDG PET in metastatic carcinoid. J Nucl Med 1997;38(9):1382–1383.

84. Foidart-Willems J, Depas G, Vivegnis D, et al. Positron emission tomography and radiolabeled octreotide scintigraphy in carcinoid tumors. Eur J Nucl Med 1995;22:635.

85. Sundin A, Eriksson B, Bergstrom M, Langstrom B, Oberg K, Orlefors H. PET in the diagnosis of neuroendocrine tumors. Ann N Y Acad Sci. 2004;1014:246–257.

86. Eriksson B, Bergstrom M, Sundin A, et al. The role of PET in localization of neuroendocrine and adrenocortical tumors. Ann N Y Acad Sci 2002;970:159–169.

87. Bombardieri E, Maccauro M, De Deckere E, et al. Nuclear medicine imaging of neuroendocrine tumours. Ann Oncol 2001;12(suppl 2):S51–S61.

88. Hoegerle S, Altehoefer C, Ghanem N et al. Whole body ^{18}F dopa PET for detection of gastrointestinal carcinoid tumors. Radiology 2001;220(2):373–380.

89. Anderson CJ, Dehdashti F, Cutler PD, et al. ^{64}Cu-TETA-octreotide as a PET imaging agent for patients with neuroendocrine tumors. J Nucl Med 2001;42:213–221.

90. Whiteford MH, Whiteford HM, Yee LF, Ogunbiyi OA, Dehdashti F, Siegel BA, et al. Usefulness of FDG-PET scan in the assessment of suspected metastatic or recurrent adenocarcinoma of the colon and rectum. Dis Colon Rectum 2000;43(6):759–767; discussion 767–770.

91. Ho CL, Dehdashti F, Griffeth LK, et al. FDG-PET evaluation of indeterminate pancreatic masses. Comput Assist Tomogr 1996;20:363–369.

13

PET and PET/CT Imaging in Breast Cancer

Richard L. Wahl

This chapter reviews the use of positron emission tomography (PET) and PET-computed tomography (CT) in the detection and characterization of primary breast cancers, in locoregional and systemic tumor staging and restaging, in predicting and assessing tumor response to therapy, and in changing patient management. Although PET has been in use for more than 15 years in breast cancer imaging, we continue to learn how best to apply the technique in the management of patients with known and suspected breast cancer.

Similar to many epithelial cancers, breast cancers have a variety of phenotypic deviations from normal breast tissue, which include, but are by no means limited to, the following factors: (1) increased tumor blood flow and increased vascular permeability compared to normal breast tissue [the physiology by which gadolinium contrast magnetic resonance imaging (MRI) of breast cancer appears to produce its signal]; (2) increased levels of glucose metabolism; (3) increased amino acid transport and protein synthesis; (4) increased receptor expression [such as overexpression of the estrogen receptor (ER)]; (5) increased DNA synthetic rates; and (6) decreased oxygen tension (hypoxia). All these processes can be imaged with PET. However, the most extensive clinical work using PET in breast cancer imaging has been performed with the tracer 2[^{18}F]fluoro-2-deoxy-D-glucose (FDG) (1, 2).

We and others have shown that the vast majority of breast cancers in women overexpress the glucose transporter molecule GLUT-1, and that there is a general relationship between the viable cancer cell number and the FDG accumulation in primary breast cancers and a correlation between the GLUT-1 levels and the FDG accumulation in these tumors (3–5). In addition, there is a positive relationship between FDG uptake and the degree of aggressiveness of breast cancers (6).

The first human studies of breast cancer imaging with PET evaluated tumor blood flow, oxygen extraction, and oxygen utilization in nine patients using ^{11}C- and ^{15}O-labeled tracers. In these studies, regional blood flow was higher in the tumors than in surrounding normal breast.

Although oxygen utilization was slightly higher in the tumors than in normal breast, the oxygen extraction ratio was significantly lower (7). A subsequent study in 20 patients showed mean tumor blood flow to be about five to six times higher than flow in normal breast tissue (8). To date, the use of PET to evaluate tumor blood flow has had only modest clinical application, although Wahl and colleagues (9) have preliminarily reported on the use of the generator-produced agent [^{62}C]PTSM for breast cancer imaging. They demonstrated the feasibility of the technique for imaging some primary and systemically metastatic breast cancers. Recently, several other studies of PET to image tumor blood flow were reported. Mankoff et al. (10, 11) reported that changes in tumor blood flow using ^{15}O-water were predictive of response to therapy, and Zasadny et al. (12) have shown that FDG uptake and ^{15}O-water uptake were somewhat correlated. At present, ^{15}O-water, with a 2-min half-life, remains a research tool and not a clinical tool for imaging breast cancer.

PET imaging of ER in breast cancer has also been performed. Thirteen women with primary breast masses were imaged with 16-[^{18}F]fluoro-17-estradiol (FES) (13). PET demonstrated uptake of the radiotracer in the primary breast mass and the axillary nodes in several cases. An excellent correlation ($r = 0.96$) was reported between the estrogen receptor concentration measured in the tumors and the extent of FES uptake into the tumors in this study. A larger follow-up study by the same investigators evaluated 16 patients with 57 known foci of metastatic disease and demonstrated 93% sensitivity for lesion detection (14). Patients who received antiestrogen therapy showed a decline in the fraction of FES reaching the tumors posttreatment (14). Imaging tumors that were estrogen receptor negative was not successful with this approach. Despite the early and promising work with this agent, it has not achieved widespread clinical utilization, in part because it does not target ER-negative tumors and because treatment trails with antiestrogen agents are of low risk. Further, recent data with aromatase inhibitors suggest that such treatments might be quite generally used in breast cancer therapy.

FDG is the most commonly used PET tracer to image breast cancer. This radiopharmaceutical is transported into cancer cells, probably by the facilitative glucose transporter protein GLUT-1 in many breast cancers, and is then phosphorylated by hexokinases in cancer cells (HKII) to FDG-6-phosphate, a polar material that is retained within the cell (2). The intracellular FDG-6-phosphate is detected in tumors by PET imaging. ^{18}F has a half-life of 109 min, so that it is more practical for use in a clinical setting than the very short lived positron-emitting isotopes. Indeed, much of the use of FDG in the United States is at medical centers that are remote from a medical cyclotron, and thus FDG is shipped from a manufacturing site, sometimes over hundreds of miles.

FDG was first used in breast cancer imaging by a planar imaging (non-PET) technique by Minn et al. (15) in 1989. Using a specially collimated gamma camera, the investigators studied 17 patients with breast cancer and were able to detect the tumor in 14 (82%), including 6 of 8 known lymph node metastases. FDG was also able to detect skeletal metastases and was more sensitive for detecting lytic or mixed lesions than purely sclerotic lesions. In assessing treatment response in 10 patients, increase in FDG uptake was consistently associated with disease progression, whereas decline in FDG uptake was often, but not invariably, associated with resolving or stable disease (15). However, planar imaging is a very insensitive technique and images are limited by low resolution and sensitivity.

In preliminary reports in 1989, Wahl et al. (16) and Kubota et al. (17) separately reported on the feasibility of imaging breast cancer using PET with FDG in several patients. Subsequently, Wahl et al. (18) showed the feasibility of imaging primary, regional, and systemic metastases of breast cancer using FDG-PET in a larger series of patients. FDG-PET imaging allowed the detection of 25 of 25 known foci of breast cancer including primary lesions (10/10), soft tissue lesions (5/5), and skeletal metastases (10/10). In addition, PET imaging allowed detection of 4 additional nodal lesions that were not previously identified. Several of the primary cancers were detected in women with radiographically dense breasts. It must be noted, however, that the primary lesions evaluated were all greater than 2 cm in size and therefore larger than most cancers detected by screening mammography.

In preclinical studies, FDG uptake in vitro and in vivo in breast cancer declined with rising glucose levels, suggesting strongly that PET imaging should be performed in the fasted state (19). These results have been confirmed clinically in various human cancers, and all PET centers perform breast cancer imaging in fasting patients (20). Uptake of the tracer is lower in diabetics than in euglycemic patients.

In general terms, the use of FDG-PET in breast cancer imaging can be discussed in terms of evaluation of the primary lesion, evaluation for regional and distant metastatic disease, evaluation of treatment response, and overall in the context of patient management approaches. Unless otherwise stated, the following results were obtained in patients imaged with whole-body dedicated PET scanners and not with dedicated PET breast imaging devices.

Evaluation of Primary Lesions

Anatomic imaging using mammography has long been the standard tool for detecting primary breast cancers. Mammography, however, detects the mass or calcifications produced as a result of the genetic changes that caused the tumor to develop. However, the appearance of benign and malignant masses and calcifications on anatomic imaging overlap with one another in many instances. This overlap in anatomic appearance often requires biopsies be performed to determine the nature of the lesion. In the United States, about 70% of breast biopsies result in benign histologic samples and thus represent false-positive anatomic imaging tests. Functional imaging of tumor biology holds great promise over the simple anatomic imaging approaches to reduce the number of false-positive biopsies and increase the overall accuracy of imaging, as mammography not uncommonly is falsely negative, especially in women with radiodense breast tissues (1).

Although the study of Wahl et al. (18) succeeded in detecting untreated primary lesions of breast cancer by FDG-PET, this small study demonstrating proof of concept was restricted to large primary lesions and did not address the more critical and clinically relevant issue of detection of smaller lesions in the breast. Because imaging devices for PET have been gradually improving, past results may not always be indicative of the current state of the art in PET, as higher-resolution systems are now available so that smaller lesions should be detectable.

Using whole-body PET imaging without attenuation correction, Hoh et al. (21) were able to detect 15 of 17 foci of primary breast carcinoma, as well as regional and systemic metastases. In a study of transverse PET with attenuation correction in 11 patients with primary breast cancer, 10 of 11 primary lesions were identified by FDG-PET (22). Only modest uptake was seen in patients with fibrocystic disease. Tumor to normal tissue uptake ratios were 4.9:1. In a larger prospective study, Adler et al. (23) reported FDG-PET results using attenuation correction and PET in 28 patients with 35 suspicious breast masses. Twenty-six of 27 malignant lesions were identified (sensitivity, 96%) and separated (in retrospect) from the 8 benign breast lesions. The separation was based on FDG uptake levels of the primary lesions. In this study, there was a modest, but significant, correlation between the nuclear grade of the tumors and the quantity of FDG uptake.

Dehdashti et al. (24) evaluated 32 breast masses, of which 24 were malignant, using FDG-PET. They found FDG uptake to be much greater, on average, in the breast cancers than in the benign lesions with a standard uptake value (SUV) of 4.5 ± 2.8 versus 1.05 ± 0.41. Using a SUV cutoff value of 2.0 for diagnosis of malignancy achieved 88% sensitivity and 100% specificity for primary breast lesions.

Bassa et al. (25) reported a sensitivity of 100% for FDG-PET for locally advanced primary breast cancers and a sensitivity of 77% for detection of nodal metastasis before treatment. These results were better than those obtained by anatomic imaging methods. Detection of some primary breast cancers that are locally advanced may be challenging anatomically, but this high sensitivity value must not be considered representative of the performance of PET across a full range of primary lesion sizes. Small lesions will clearly be more difficult to detect than larger lesions for reasons of the size-based resolution and sensitivity limitations of dedicated PET scanners, which typically have approximately 10-mm full with half-maximum (FWHM) reconstructed resolution. Rather, evaluation of the method in patients with suspected but undiagnosed primary lesions of all sizes is more clinically relevant.

Avril et al. (26) initially reported findings in 51 patients with 72 suspicious breast masses, 57% of which proved to be malignant. In this study, malignant lesions had a mean SUV 2.5 times that of benign lesions. By correcting for partial-volume effects, sensitivity was increased from 75% to 92% while specificity decreased from 100% to 97%. Detection of lesions less than 1 cm in diameter was not optimal, however, because of resolution limitations of the whole body PET scanner used in the study. This same group subsequently reported findings in 185 breast cancers using FDG; these included 132 cancers and 53 benign lesions. The overall sensitivity was much lower at 64% with specificity of 94%, using standard visual analysis. When a lower visual threshold of uptake was used for diagnosing malignancies in these same studies, the sensitivity increased to 80% and specificity declined to 76% (27). Nearly two-thirds of invasive lobular cancers were FDG-PET negative, compared to about 24% of invasive ductal cancers. In this study, there was a clear relationship between tumor size and lesion detectability. Using the most sensitive visual detection approach, only 68% sensitivity was obtained in primary tumors less than 2 cm in diameter. Sensitivity was higher for larger lesions, reaching 92% for 2- to 5-cm lesions. All three lesions more than 5 cm in diameter were detected (27). Lobular carcinomas have low GLUT-1 expression, likely accounting for the lower sensitivity of PET in these types of tumors (5, 27).

Similar results to those obtained in this large series were reported by Yutani et al. (28), who found 79% sensitivity for detecting primary breast cancer. They also showed that FDG clearly achieved higher target-to-background ratios than methoxyisobutyl isonitrile (MIBI) (6/1 versus 3.5/1), again indicating that not all tracers are

equivalent for breast imaging (28). Thus, lesion histology and size affected the rate of primary cancer detection using FDG-PET. Certainly, a technique with low sensitivity is not suitable for use as a screening method, so that negative PET findings in lesions less than 2 cm in diameter have limited negative predictive value. However, positive PET results must always be considered of serious concern, with very high and focal FDG uptake having a strong positive predictive value for cancer. Ishimori et al. (29) reported discovery of two incidental breast cancers in patients studied with FDG-PET for recurrent tumors of other types of 1,912 patients studied.

The low sensitivity of current whole-body imaging devices for smaller primary breast cancers makes PET less than optimal for screening, even in high-risk populations. However, the performance characteristics of PET compare very favorably with the characteristics of other noninvasive imaging modalities and are better than those of planar scintillation camera imaging of MIBI. FDG-PET has also been able to detect primary lesions in patients with silicone breast prostheses, which are opaque to standard low-energy mammographic X-rays (30, 31). The small studies directly comparing FDG-PET and MRI have shown that FDG-PET is typically less sensitive than MRI for primary lesion detection but more specific than MRI. As an example, in 42 breast lesions, MRI was 89% sensitive and 63% specific for cancer. By contrast, PET was 74% sensitive but 91% specific (32). Similarly, Goerres et al. (33) showed PET to be 79% sensitive, 94% specific, and 84% accurate versus MRI, which was 94% sensitive, 94% specific, and 88% accurate in the diagnostic setting of detection of possible recurrent tumor in the breast following a known diagnosis of breast cancer. Thus, FDG-PET using whole-body imaging devices appears to be a more-specific method but is less sensitive than MRI as currently interpreted.

The relationship between the MRI signal and the FDG-PET signal was studied in 20 patients with large or locally advanced primary breast cancers. A significant association (P less than 0.05) was observed between the calculated exchange rate constants of both pharmacokinetic models and calculated PET FDG dose SUV; this was a modest relationship accounting for less than half of the findings. Thus, the SUV and MRI enhancement rate are related but not equivalent (34). As with most tracers, delivery does have a role in the total signal detected.

FDG is the dominant tracer in breast cancer imaging currently, as it is in most cancers; however, the use of alternative tracers has been extensively explored to some extent in breast cancer, generally in small studies. L-[Methyl-^{11}C]-methionine PET has been used in the imaging of primary breast cancers and in assessing their response to chemotherapy. In a pilot study of primary and metastatic breast cancers, [^{11}C]-L-methionine uptake in primary breast cancer was less than in the liver but more than in most normal tissues. All tumors greater than 3 cm in size were detected, but three smaller tumors were not detected. Increased uptake of [^{11}C]-L-methionine was associated with a large S-phase fraction (SPF) measured with flow cytometry ($r = 0.77$, $P = 0.01$). These data indicate that [^{11}C]-L-methionine can be used to image breast cancers but weigh against its being useful in small lesions or in the liver due to high background uptake (35). The need for an onsite cyclotron for use of [^{11}C]-L-methionine makes its widespread clinical application even less likely. However, other tracers with higher uptake into breast tumors and lower background activity could be developed, making the PET method more potent for assessing primary lesions even with existing imaging systems.

At present, FDG-PET is an excellent method to help detect primary breast lesions more than 2 cm in diameter and to help characterize such lesions. The exact clinical role of PET in breast imaging remains in evolution. The performance of PET will likely improve as better-resolution scanners increasingly become available. It should be noted that in the United States the role for noninvasive "diagnostic" tests, which help to characterize a lesion as malignant or benign, is a challenging one because a missed diagnosis not uncommonly leads to litigation. For this reason, biopsy and excision remain widely used as the primary approach to diagnosis of breast cancer in the United States following lesion detection by anatomic methods.

Detection of Primary lesions with Positron Cameras Dedicated for Breast Imaging or Positron Emission Mammography

Although FDG is the most common tracer used in breast cancer imaging with PET, it should be noted that most PET scanners used in breast imaging to date have been designed as all-purpose devices to image the entire body, not just the breast. Thus, they represent a compromise in performance as compared to dedicated breast imaging instruments, analogous, perhaps, to using a chest X-ray device to perform mammography, with consequent relative degradation of breast images. Dedicated breast imaging devices carry with them far more potential for successful imaging of small lesions in the breast.

This area is in rapid development and to date only small studies have been performed in limited numbers of patients. In principle, a smaller PET imaging device devoted to imaging the breast would avoid much of the soft tissue attenuation faced by whole-body PET scanners. The devices could be less expensive than whole-body PET scanners, and more sensitive, and could lead to development of radionuclide-guided-biopsies. They are, however, limited by partial-volume effects and the degree of the localization of FDG or other PET tracer in the tumor compared to normal breast (36, 37).

Raylman et al. (37) reviewed issues of detectability with FDG-positron emission mammography (PEM) in phantom studies. In brief, in the absence of breast background, simulated breast lesions as small as 5 mm could be detected. However, as back ground activity was increased to the levels expected in women with more metabolically active breasts, and breast thickness was also increased, only lesions of 12 mm in diameter and larger were detected. This finding indicates that dedicated breast devices likely will do better than dedicated whole-body tomographs and that the role of FDG and PET in evaluating primary breast lesions will require reevaluation as improved instruments are developed (37). Nonetheless, the reliable detection of lesions smaller than 12 mm is essential for such a method to achieve clinical utilization.

Several groups have been carrying out such studies with dedicated high-resolution PET breast imaging devices. Thompson and colleagues (38) have shown the feasibility of PEM devices in patient studies, and more-sensitive devices are being made using different detector materials (39). One of the other potential benefits of dedicated devices is the ability to coregister PEM images with mammographic images to display both anatomic and PET data simultaneously (40). Recently, sensitivity of about 86% and specificity of 91% were reported for detection of breast cancer using a dedicated positron breast imaging device that has less than 2 mm resolution (41).

Using a newer large field of view, high-sensitivity detector system, PEM demonstrated 20 focal abnormalities, of which 18 were malignant and 2 were benign. Three of 20 malignant lesions demonstrated at conventional mammography were not demonstrated at PET mammography. In a small series of patients, the overall sensitivity of PEM for detection of malignancy was 86% with a positive predictive value of 90% (42, 43).

PET must also be considered in the context of other imaging modalities that are used for diagnosis of breast cancer. Because of its high current cost per study, PET is not a realistic candidate for use in cancer screening of low-risk populations, is in contrast to screening mammography, which has been shown to save lives and is the key method for breast cancer detection. Mammography is, however, of limited value for detecting cancers in women with dense breasts (especially younger women), assessing the breast after biopsy, characterizing lesions as malignant or benign, determining whether disease is unifocal or multifocal, and predicting or assessing response to treatment. Mammography cannot determine whether the tumor is localized or metastatic. Thus, it is in these groups of patients that PET may have its greatest potential role for primary lesion assessment.

Detection of Regional Lymph Node Involvement

Clinical studies have supported preclinical findings in rodents of high tumor-to-lymph node ratios, which suggested that FDG may permit detection of lymph node

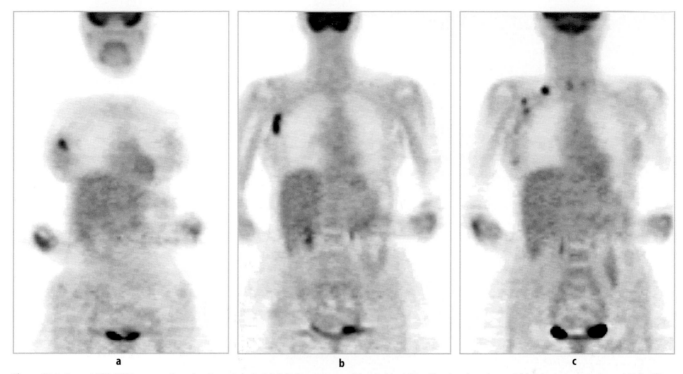

Figure 13.1. Coronal FDG-PET image sections showing uptake in (**a**) right breast cancer, (**b**) palpable right axillary lymph nodes, and (**c**) right supraclavicular and high axillary lymph nodes that were not clinically apparent.

metastases. It was noted that when FDG was injected subcutaneously, there was intense tracer uptake in the lymph nodes proximal to the injection site (44). In a preliminary study, Wahl et al. (45) were able to characterize correctly 8 of 9 axillae with FDG-PET in the preoperative setting using tracer injection in the arm opposite to the side of the tumor. No false-positive findings occurred but there was one false-negative result. Several other reports of imaging axillary nodal metastases from breast cancer with PET have now appeared. In a report by Tse et al. (46) using the whole-body PET technique, 11 of 14 axillae were accurately characterized as to the presence or absence of metastasis. In 5 cases, there was uptake of FDG in the axillary nodal region and in each instance there was tumor involvement of the axilla. In a prospective study, Adler et al. (47) reported 20 patients with newly diagnosed breast cancer who had axillary lymph node dissections following PET scans. In these patients, 9 of 10 patients with nodal metastases were correctly diagnosed whereas 10 of 10 disease free axillae did not show increased uptake.

Figure 13.1 shows a whole body FDG-PET scan in a patient with a primary right breast cancer and palpable right axillary lymph nodes. The PET scan confirms high FDG uptake in the primary tumor and axillary metastasis and also demonstrates high axillary and supraclavicular lymph node metastases, which were not clinically apparent. Figures 13.2 and 13.3 demonstrate the potential usefulness of fusion PET/CT images for anatomic localization of focal uptake in the axillary or chest wall regions. Figure 13.4 shows right internal mammary lymph node metasta-

sis detected by PET. This metastatic lesion was not demonstrable by other imaging techniques.

Subsequent reports have suggested that the sensitivity of PET for detecting nodal metastases is substantially lower than initially reported. Avril and colleagues (48) recently reported sensitivity of 82% for the detection of nodal metastases in breast cancer, with a specificity of 96%. In this series, sensitivity was higher in patients with large primary lesions and lower in those with small primary lesions. Clearly, fewer positive foci of FDG uptake are seen by PET than were seen at pathologic examination, using moderate-resolution whole-body PET devices. Utech et al. (49) reported that 100% of the 44 tumor-positive axillae in 124 patients with primary breast cancer were detected by PET, whereas PET was negative in 75% of the 80 tumor negative axillae. Yutani et al. (50) obtained only 50% sensitivity for detecting axillary metastasis, which was still better than the 38% sensitivity of MIBI. Accuracy values are typically somewhat higher than sensitivity because pathologically negative axillae are more common than positive in most series.

Because of these encouraging but mixed results, a large prospective study was needed to establish finally the false-negative rate for staging the axilla by FDG-PET. In any case, PET can be expected to have a higher false-negative rate than sentinel lymph node imaging and histologic examination of tissue. This expectation is based on the known limitations of PET for detecting lesions smaller than 5 to 10 mm, depending on background tracer activity levels. Crippa et al. (51) in a study of melanoma, which typically shows much higher FDG uptake than breast

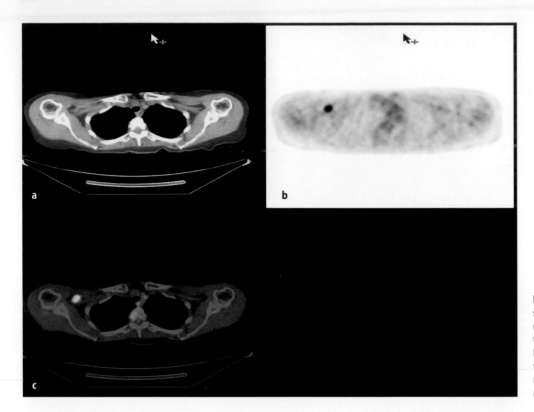

Figure 13.2. PET/CT fusion images showing right axillary lymph node metastasis: (**a**) CT image; (**b**) attenuation-corrected PET image; and (**c**) fused PET/CT image. These images show that the focal activity in the right axillary region corresponds to a small lymph node metastasis. 📖

cancer, found that detectability of nodal metastasis was highly dependent on the size of the lesions. In melanoma, PET succeeded in detecting all nodal metastases more than 1 cm in diameter, 83% of metastases 6 to 10 mm in diameter, and only 23% of those that were 5 mm or less.

A large NCI-sponsored multicenter trial evaluating the accuracy of PET for staging the axilla was completed and reported recently (52). In that study, 360 women older than 18 years of age with newly diagnosed, untreated, invasive breast carcinoma were evaluated. Surgery had to be planned, including axillary dissection, within 30 days of the PET scan. These patients had no prior therapy or intercurrent illnesses. There were three blinded interpreters who examined both the attenuation corrected and the nonattenuation-corrected images. In the overall population, the prevalence of axillary metastases was 35%. For the detection of axillary nodal metastasis, the mean esti-

mated area under the receiver operating characteristic (ROC) curve for the three interpreters was 0.74 (range, 0.70–0.76). When the finding of at least one probably or definitely abnormal axillary focus was the criterion for a positive axilla, the mean (and range) sensitivity, specificity, and positive and negative predictive values for PET were 61% (54%–67%), 80% (79%–81%), 62% (60%–64%), and 79% (76%–81%), respectively. From this study it was concluded that FDG-PET has moderate accuracy for detecting axillary metastasis and that PET commonly fails to detect axillae involved with small and few nodal metastases. In this study, although highly predictive

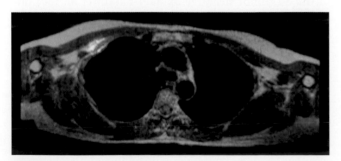

Figure 13.3. Fused PET/CT image shows breast cancer metastatic to a right-sided rib. In the FDG-PET image, it was not clear whether the metastatic focus was located in a high right axillary lymph node or the chest wall. The fused image shows that the focus is situated in a right-sided rib. Fusion was achieved using software methods. 📖

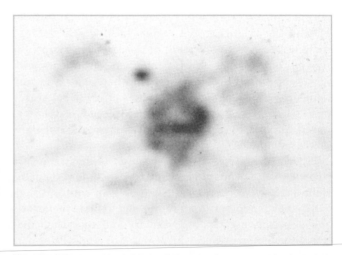

Figure 13.4. Transverse FDG-PET image of the thorax shows a right internal mammary lymph node metastasis. 📖

for tumor involvement when multiple intense foci of tracer uptake were identified in the axillae, FDG-PET could not be recommended for routine staging of the axilla in most patients with newly diagnosed breast cancer. The prognostic value of PET is not yet determined but may still be substantial and warrants further study.

The large multicenter study just described compared PET staging of the axilla to axillary dissection and pathologic assessment of the axilla. Increasingly, sentinel lymph node sampling is replacing axillary dissection because of its lower morbidity and good performance. In some studies, sentinel node sampling has had sensitivity of less than 90% relative to axillary dissections. Thus, the performance of PET versus sentinel node imaging in staging the axilla is relevant to current practice patterns. A study from Canada in 98 patients with early-stage breast cancer compared the sensitivity, specificity, predictive values, and likelihood ratios of PET scanning with standard axillary lymph node dissection and sentinel lymph node biopsy in staging the axilla (53). This study demonstrated a sensitivity of 40% and specificity of 97% for PET compared to axillary dissection and a similar performance compared to sentinel node sampling. PET accuracy was better in patients with high-grade and larger tumors. Increased size and number of positive nodes were also associated with a positive PET scan These findings suggested PET could be useful only if positive as a means of circumventing the need for lymph node dissection (53).

Although the overall sensitivity of PET is too low, especially in small nodal lesions that typically result from small primary tumors, PET is able to detect larger metastases including metastases to the internal mammary lymph nodes. Such nodal metastases are of equivalent prognostic significance to axillary lymph node metastases and are of particular relevance for medially situated primary breast tumors. No careful study of the sensitivity of PET for detecting such metastases has been performed in which all internal mammary lymph nodes were biopsied, but there are cases in which only the internal mammary nodes are positive in which PET could add very useful information to staging and treatment planning. Similarly, a clearly positive PET scan of the axilla may be expected to carry a very high positive predictive value for metastasis. Thus, it is possible to speculate that PET could be performed at diagnosis in larger primary breast cancers to establish the presence of axillary nodal metastases, which could then be treated by neoadjuvant chemotherapy.

In a retrospective study, internal mammary nodal tracer uptake on FDG-PET was compared with standard radiographic imaging and was correlated with putative risk factors for internal mammary lymph node involvement and with clinical patterns of failure. Abnormal FDG uptake in these nodes was seen in 25% (7/28) of women with locally advanced breast carcinoma at presentation. Prospective conventional chest imaging failed to identify internal mammary lymph node metastases in any patient. Internal mammary lymph node uptake on PET was associated with large size of the primary tumor ($P = 0.03$) and with inflammatory disease ($P = 0.04$) (54).

PET may also be useful in the assessment of axillary adenopathy when patients present with axillary nodal tumor and no known cancer in the breast or other systemic metastases. In such situations, PET with FDG has been able to locate primary lesions in the breast that were not clearly detected by other methods (55–58).However, MRI may be superior in this setting. PET/CT should improve on the performance of PET alone in the diagnosis of axillary nodal metastases, but it has not been extensively studied. An example of a positive PET/CT in the axilla is shown in Figure 13.5.

Given the limitations resulting from partial-volume effects with current scanners, PET is unlikely to replace sentinel node dissection in patients with small primary tumors in whom small nodal metastases would be most likely. Rather, PET has the greatest potential role for evaluating the axillae of patients with large primary tumors in whom neoadjuvant chemotherapy is planned and in whom axillary sampling would generally not be performed before surgery. Patients viewed as having high risk for internal mammary lymph node metastases may also benefit from PET imaging.

It is of interest that there appears to be substantial prognostic information available from a baseline PET imaging study. Inoue et al. (59) showed that the prognosis was worse for patients with a high than a lower SUV primary lesion. In addition, if the primary tumor SUV was high and there were axillary lymph nodes visualized, the prognosis was much worse than if the primary tumor had low FDG uptake and no axillary nodes were identified. The 5-year survival was 44.4% if FDG uptake was high in the primary and lymph nodes positive on PET, compared to 96.8% if FDG uptake was low in the primary and lymph nodes negative on PET. However, primary lesion uptake is related to lesion size, so that this prognostic value is, in part, related to lesion size (59).

Evaluation of Systemic Metastatic Disease

Whole-body PET is superior to conventional diagnostic imaging for many but not all systemic metastases of breast cancer (18, 24). Dehdashti et al. (24) reported 89% sensitivity and 100% specificity for the detection of 21 proven metastases using an SUV cutoff value of 2.0. While there is high FDG uptake into untreated primary breast cancers, this uptake is typically lower than in other tumors such as lung cancer (60). Thus, detection of primary and metastatic lesions may be more challenging in breast cancer than some other malignant tumors using FDG-PET.

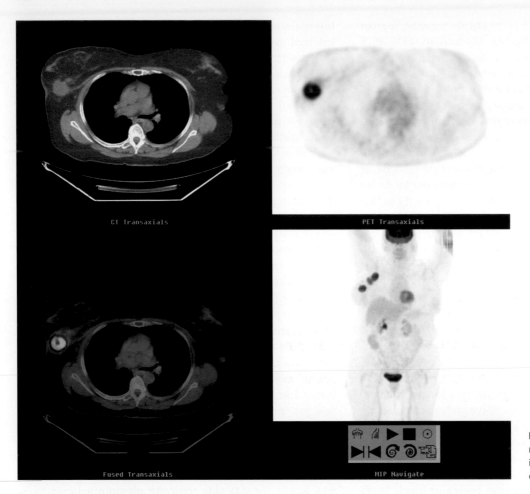

Figure 13.5. Transaxial FDG-PET/CT and maximum intensity projection (MIP) images of a patient with right breast cancer and right axillary metastases.

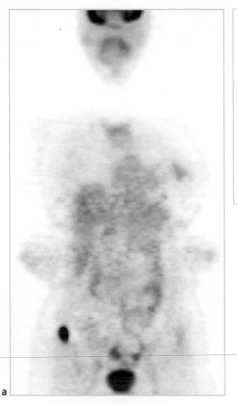

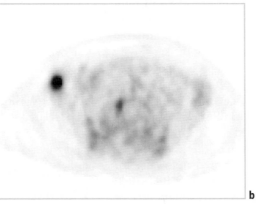

Figure 13.6. Whole-body FDG-PET scan performed in a patient who had undergone lumpectomy 2 weeks, found to be breast cancer on pathologic examination. Staging PET scan was then undertaken, demonstrating a metastasis in the right iliac crest seen in (**a**) coronal and (**b**) transaxial image sections. As a result of this finding, axillary dissection was not undertaken, and the patient was treated by chemotherapy for stage IV disease.

For detection of skeletal metastases, there have been varying results. Two studies showed FDG-PET imaging to be more sensitive overall than ^{99m}Tc-methylene disphosphonate (^{99m}Tc-MDP) because PET imaging was more sensitive for detection of predominantly lytic metastases (61, 62). Cook et al. (61) showed that FDG-PET imaging detected a mean of 14.1 metastatic lesions in the skeleton versus 7.8 by ^{99m}Tc-MDP skeletal scintigraphy. Cook and colleagues also showed that patients with osteolytic lesions that had higher FDG uptake levels had a poorer prognosis. Blastic lesions may be better detected by ^{99m}Tc-MDP skeletal scintigraphy than by FDG-PET imaging.

Figure 13.6 shows a metastasis in the right iliac crest in a patient with primary breast cancer detected by whole-body FDG-PET imaging. Gallowitsch et al. (63) reported in a retrospective series than FDG-PET detected just 61% of bone metastases in breast cancer whereas conventional ^{99m}Tc-MDP skeletal scintigraphy detected 97% (P less than 0. 05). Other reports have shown comparable sensitivity of PET imaging and ^{99m}Tc-MDP skeletal scintigraphy for metastases (95%), but there was lower specificity for the ^{99m}Tc-MDP skeletal scintigraphy than for FDG-PET (94.5% versus 78.7%) (64). An advantage of FDG is that it can potentially monitor tumor response in the skeleton (65).

Because it can be difficult to secure "gold standard" proof for skeletal lesions, direct comparative studies between FDG-PET and ^{99m}Tc-MDP skeletal scintigraphy are limited. A small recent comparison study in 15 breast cancer patients who had both FDG-PET and ^{99m}Tc-MDP single photon emission computed tomography (SPECT) skeletal scintigraphy showed a sensitivity of 85% for SPECT and 17% for PET in the lesion-by-lesion analysis ($n = 900$). Bone SPECT was significantly more sensitive (P less than 0.0001) and accurate (P less than 0.0001) than FDG-PET in this small study. No statistically significant difference was seen with regard to specificity. It seems clear that there are breast cancer metastases to bone that are not seen well with FDG and which can only be seen with bone-avid tracers (66). The overall topic of imaging skeletal metastases with PET has recently been reviewed by Fogelman et al. (67).

FDG is not the only PET agent used for bone imaging. ^{18}F-Fluoride ion can be used for PET imaging of bone itself, and scans with this agent have very high sensitivity for lesion detection compared to ^{99m}Tc MDP skeletal scintigraphy. In a study of 44 patients with breast cancer, 64 skeletal metastases were detected in 17 patients by ^{18}F-PET compared to 29 metastases in 11 patients by ^{99m}Tc-MDP skeletal scintigraphy (68). ^{18}F-Fluoride ion PET imaging is also reasonably sensitive for lytic lesions. This method is not in wide clinical use at present. NaF-PET has also been reported to be more sensitive than ^{99m}Tc-MDP skeletal scintigraphy in the detection of skeletal metastases in a study of 53 patients with lung cancer (69).

PET with FDG is an accurate tool for imaging soft tissue metastases. A study of 109 patients showed FDG-PET to be 89% accurate for characterizing the primary lesion, 91% accurate for characterizing nodal status (with only 10% false negatives), and equal to other conventional diagnostic methods for detecting visceral metastases (19/19 detected) (70). This study must be interpreted with some caution as detectability of lesions depends on lesion size. Metastatic disease in these patients was not necessarily detected at an early stage, possibly reducing the apparent difference between PET and conventional imaging.

Several groups have reported that FDG-PET is a very sensitive method for the detection of brachial plexus metastases of breast cancer, even at a stage when MRI is equivocal (71, 72). A potential challenge in evaluating the upper mediastinum and thorax with FDG-PET is the presence of metabolically active fat, "USA fat" or "brown fat" (73). This fat can cause a high level signal and potentially may obscure the detection of nodal metastases or soft tissue lesions. An example of a patient with a new left primary breast cancer and intense FDG uptake in brown fat is shown in Figure 13.7. Brown fat activity can likely be reduced by warming the patient before FDG injection (74).

Skeletal and visceral metastases are most commonly encountered in the context of tumor recurrence when a patient who has been treated for primary breast cancer presents at a later time with laboratory or clinical evidence of recurrence. FDG-PET is a reasonable alternative to CT for the detection of systemic metastases. Figure 13.8 shows whole body FDG-PET images obtained in a breast cancer patient who was clinically asymptomatic and was found to have an elevated serum marker level. The PET scan showed multiple small metastases in the superior and inferior mediastinum, the preaortic region of the upper abdomen, and in the left acetabulum.

The whole-body FDG-PET scan in Figure 13.9 was obtained in a patient with a history of breast cancer surgery 22 months earlier who was found to have an indefinite opacity in the upper zone of the right lung by CT imaging. PET images showed low-level uptake in the lung lesion, suggesting an inflammatory etiology, but showed definite foci of high uptake in the liver and just below the left lobe of the liver in the upper abdomen. The PET study demonstrated unsuspected hepatic and extrahepatic abdominal tumor. PET/CT images of a patient with an hepatic metastasis and brown fat uptake in the neck are shown in Figure 13.10. The patients shown in Figure 13.11 presented clinically with nodal enlargement, which was shown to be tumor recurrence by biopsy. The PET scan demonstrated high axillary metastases as well as the palpable right supraclavicular lesions. In Figure 13.12, PET/CT images demonstrated extensive involvement of the skeleton. The sagittal view is often very helpful for displaying such lesions, and the CT scan can show sclerotic non-FDG-avid lesions to good advantage. It should be noted that colony-stimulating factors can markedly increase FDG uptake in bone and can potentially obscure the visualization of bone metastases by FDG. We have shown that the granulocyte

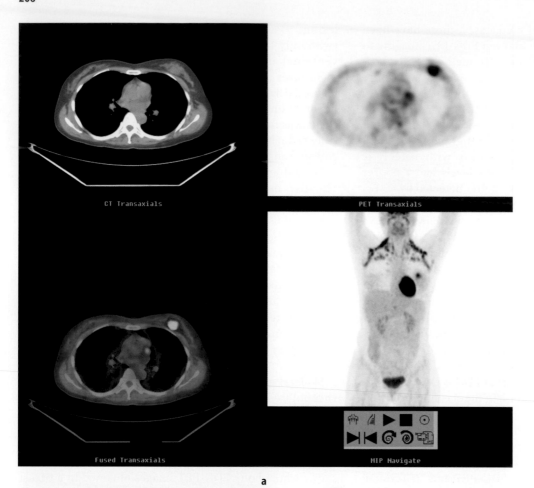

a

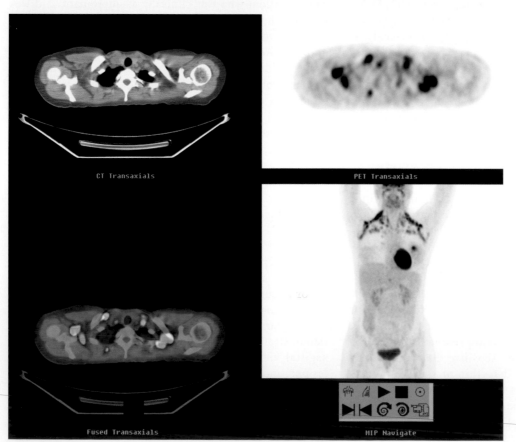

b

Figure 13.7. Transaxial FDG-PET/CT and MIP images of a patient with left breast cancer and brow fat uptake in the neck and supraclavicular regions bilaterally. The fusion images illustrate the superimposition of FDG uptake to the left breast lesion (**a**) and to fatty tissue (**b**).

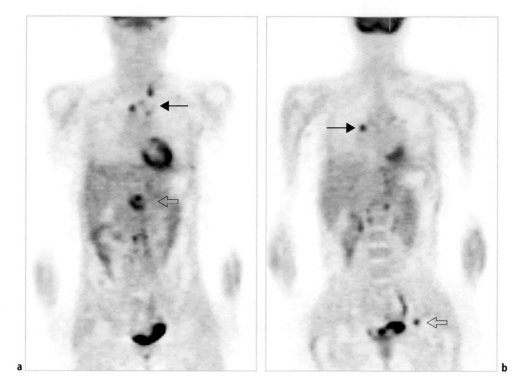

Figure 13.8. Coronal FDG-PET image sections in a patient with a history of breast cancer resection 17 months earlier who was found to have serum CA 3-15 elevation. The patient was asymptomatic. PET demonstrated (**a**) small metastatic foci in the superior mediastinum (*solid arrow*) with larger tumor foci in the preaortic region in the upper abdomen (*open arrow*) and (**b**) small tumor foci in the right lung hilum (*solid arrow*) and the left acetabulum (*open arrow*). 📖

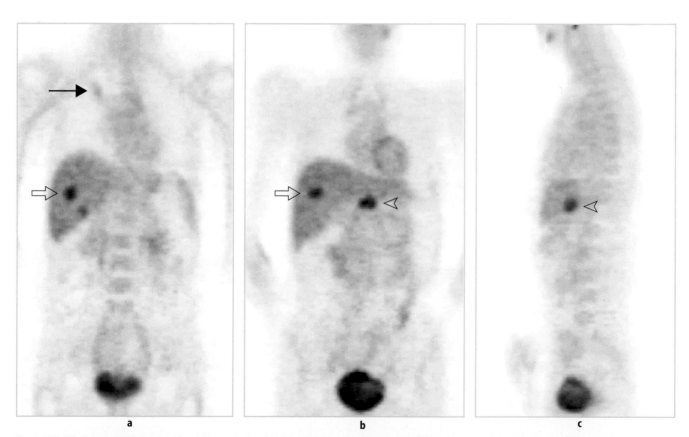

Figure 13.9. Whole-body FDG-PET scan performed in a patient with history of breast cancer resection 22 months earlier who was found to have an ill-defined opacity in the upper zone of the right lung by chest X-ray. (**a**, **b**) Coronal image sections; (**c**) sagittal image section. PET showed low-level activity in the right-side lesion (*arrow*), suggesting an inflammatory etiology, but showed foci of high uptake in the liver (**a**, **b**) (*open arrow*) and in the upper abdomen, immediately anterior to the left lobe of the liver (**b**, **c**) (*arrowheads*). These findings were diagnostic of hepatic and extrahepatic metastasis. 📖

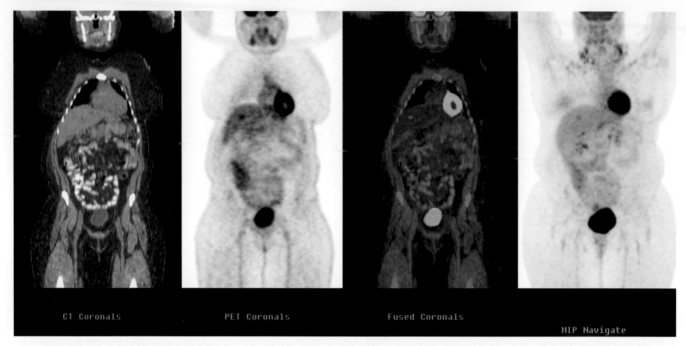

Figure 13.10. Coronal FDG-PET/CT and MIP images of a patient with an hepatic metastasis and brown fat uptake in the neck and supraclavicular regions bilaterally.

colony-stimulating factor (G-CSF) effects can be minimized by delaying the time from G-CSF dosing until PET scanning by at least 2 to 3 weeks (75).

Although PET can assess the entire body for metastases, it has limitations that should be recognized. In addition to failing to detect a moderate number of bone metastases of a predominantly blastic nature, FDG-PET is not optimal for brain metastasis detection because of the high normal background FDG activity of cerebral gray matter, as was shown in a variety of metastatic cancers by Griffeth

et al. (76). Similarly, small pulmonary lesions may escape detection because of the effects on resolution of partial-volume averaging and respiratory motion. For lesions less than 5 mm, PET with FDG often is negative. Thus, PET is a valuable tool for detecting visceral metastases but has limitations related to lesion size, lesion tracer uptake, and background normal tissue activity, as well as physiologic patient motion. In general, whole-body PET can serve as a single imaging session to assess the entire body for the presence or absence of cancer with high accuracy.

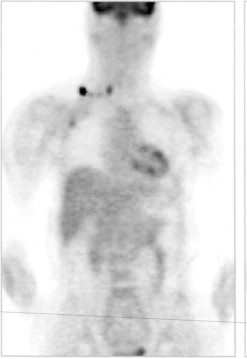
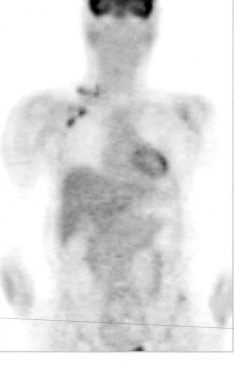

a   b

Figure 13.11. Coronal FDG-PET image sections in a patient who was found to have right supraclavicular lymphadenopathy 3 years after right mastectomy for cancer. PET confirmed metastasis in the supraclavicular lesions (**a**) and demonstrated metastases in right high axillary lymph nodes as well (**b**). 📖

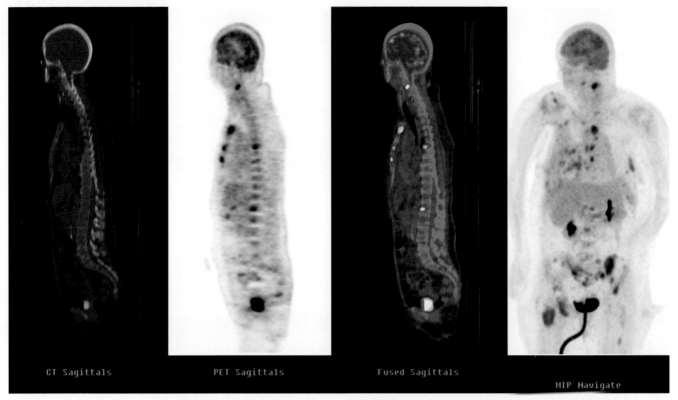

Figure 13.12. Sagittal FDG-PET/CT and MIP images of a patient with extensive skeletal metastases.

Nonetheless, in a meta analysis of 16 studies of 808 subjects with breast cancer, the median sensitivity was 92.7% and the median specificity was 81.6% for metastatic disease (77). Thus, PET with FDG is a useful tool for staging and restaging breast cancer systemically.

Evaluation of Treatment Response

A diagnostic PET scan ideally would predict whether a tumor was likely to respond to a given type of therapy before the therapy was started. Short of this, a PET scan before and again soon after treatment began might be able to provide an early indication of efficacy. Both approaches have been evaluated in breast cancer. Breast cancer estrogen receptor levels can be measured before treatment by PET imaging with estrogen receptor-binding ligands. Similarly, the early efficacy of cancer treatment can be evaluated by assessment of early treatment-induced changes in FDG uptake from baseline levels.

Tumors expressing estrogen receptors have been shown to be more likely to respond to antiestrogen therapy than tumors with lower-level receptor expression. In 43 women with breast cancer, PET scans with FDG-PET and FES-PET were done before treatment. In this study, all ER-negative tumors were negative on FES scan but only 70% of the ER-positive tumors were positive, which may have been related to partial-volume effects in smaller tumors or levels of ER expression. About 76% of the patients re-

sponded to the antiestrogen therapy, as was predicted by the FES scan. Using FDG, all tumors were detected, and in 4 instances additional lesions were seen beyond the lesions seen by anatomic imaging. A rise in FDG uptake in lesions after treatment, which has been referred to as metabolic flare, was associated with effective hormonal therapy in a group of 11 patients. Thus, FDG-PET can in some instances depict the metabolic changes induced downstream following binding of ligands to receptors (78, 79).

Because FDG-PET mainly images viable cancer cells, it is logical to follow the efficacy of cancer therapy using FDG-PET. A prospective evaluation of PET during breast cancer chemohormonotherapy performed several years ago by Wahl et al. (80) demonstrated that, in women treated with a multiagent regimen, there was a rapid and significant decline in the tumor FDG uptake, k_3 kinetic rate constant, and K_i (or influx constant) for FDG as soon as 8 days after treatment was initiated. Further declines in FDG uptake were apparent at 21, 42, and 63 days of treatment in the patients who went on to have a complete or partial response, whereas no significant decline in FDG uptake was seen in the nonresponding patients ($n = 3$) when examined at 63 days after initiation of treatment. This study also showed that the metabolic changes antedated anatomic changes, and that substantial declines in tumor glucose metabolism were apparent in the responding patient population, despite absence of change in tumor size (80). Figure 13.13 shows sequential PET scans obtained before, during, and after chemotherapy of a

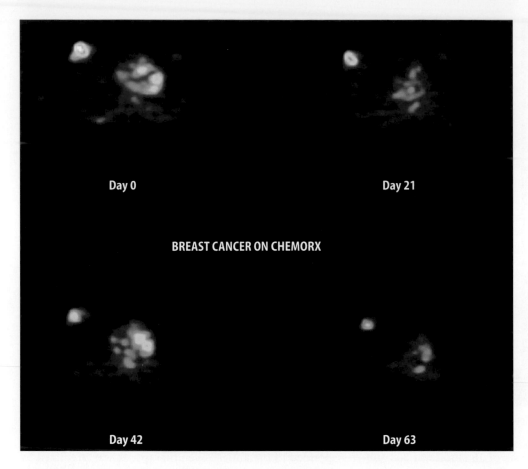

Day 0

Day 21

BREAST CANCER ON CHEMORX

Day 42

Day 63

Figure 13.13. Sequential transverse PET image of the thorax show a large primary right breast cancer that demonstrates progressive reduction in FDG uptake with each cycle of chemotherapy. This progressive decrease in glucose metabolism shows at least a partial response to treatment. Note that myocardial uptake is somewhat variable from study to study.

large right breast cancer that showed progressive decrease in the level of glucose metabolism with treatment. This metabolic response indicated at least a partial tumor response to therapy.

Similar results were reported by Jansson et al. (81) in primary and metastatic breast cancer using both FDG and ^{11}C-L-methionine. Bassa et al. (25) used FDG-PET to follow breast cancer in 16 patients. In 2 of 3 of patients, declines in FDG uptake were seen midway through treatment. PET was not sensitive for residual nodal metastases (42%) and was only 75% sensitive for residual primary tumor. Thus, negative PET findings could not be equated to absence of microscopic disease. As discussed earlier, high FDG uptake after treatment indicates residual tumor whereas low uptake is much more commonly associated with tumor response. It remains to be determined whether the early metabolic changes induced by PET will be predictive of long-term outcome.

Two more recent studies have confirmed the utility of FDG and PET for predicting response to treatment. Smith et al. (82) demonstrated in 30 patients with primary breast cancers more than 3 cm in diameter that PET performed before and after one cycle of chemotherapy was able to predict a complete pathologic remission with 90% sensitivity and 74% specificity. They also showed higher FDG baseline uptake in the tumors that had the best treatment response. Schelling et al. (83) showed similar results in 24 patients with primary breast cancer. Using a retrospec-

tively determined cutoff value of 45% or greater decline in FDG uptake after a single course of treatment, the patients who ultimately achieved elimination of gross residual disease were identified with 100% sensitivity and 85% specificity. This finding has been replicated and shown the ability of FDG-PET to predict the response to chemotherapy (84). These same observations appear to be valid for metastatic tumor foci as well, and PET is now being used in both the primary and metastatic settings clinically. There are confounding situations, such as after hormonal therapy, in which there can be a rise in tracer uptake (flare) response that appears to predict a good outcome (85). A challenge with FDG response studies assessing therapy are exactly what cutoff value in decline in FDG uptake is most predictive; this value likely will vary by treatment type. ROC analysis has been used in some studies, and it seems clear that the larger and more rapid the decline in FDG uptake, the more likely an effective response will be seen, although there is clearly overlap among response groups in these analyses (86).

Other PET tracers can be used to assess treatment response but are less practical and less easy to use than FDG. For example, ^{15}O-water uptake (reflective of tumor blood flow) declines with effective therapy, as shown in a 19-patient study of doxorubicin and docetaxel (87). Similarly, Mankoff et al. (10, 11) have compared FDG and ^{15}O-water PET studies and found that ^{15}O-water flow may be more predictive of response than is the change in FDG uptake in the tumor. Combining PET with FDG and MRI in assessing treatment response may also be more useful than either method alone (88).

Another area of potential utility in planning treatment of breast cancer is the use of labeled chemotherapeutic agents to plan treatment response. Although this area is not yet proven effective, a variety of labeled compounds have been made that may have potential for treatment monitoring. For example, labeled taxanes and labeled Xeloda (capecitabine) have been developed (89). Eventually, such labeled agents might have a role in predicting tumor response to treatments.

Labeling substrates for DNA synthesis may also prove useful as an alternative to assessing glycolysis or protein synthesis. Both FLT and FMAU labeled with ^{18}F have shown early promise in this setting (90, 91). Similarly, assessing breast cancer oxygenation status has shown potential using hypoxia-specific tracers such as ^{18}F-fluoromisonidazole (FMISO). Although FDG uptake and FMISO uptake are correlated in a variety of cancer types, including breast cancer, they are only moderately correlated, and FMISO may provide unique information regarding tumor biology not available from FDG, as hypoxic tumors often are resistant to radiation therapy and chemotherapy (92).

Other PET-Related Approaches in Breast Cancer

The ability of PET to detect breast cancer in radiodense breasts when other methods fail to detect the tumor presents a new set of problems in diagnosis. Specifically, when such lesions are identified with PET, they may be very difficult to locate for biopsy. Raylman and colleagues (93) have explored stereotactic biopsy techniques for breast cancer based on sinogram projections of PET or SPECT images. Based on the location of "hot spots" in multiple angular projections, biopsy needles can be precisely placed into tumor foci. Although this approach has not been explored extensively in clinical trials, it offers an attractive means of obtaining biopsy confirmation of lesions that are seen by radionuclide imaging but not by other methods. An alternative approach, which can be useful, involves the combination of the anatomic information of CT or MRI with the metabolic information of PET into a single "anatometabolic" image, which can precisely localize an FDG-avid focus on an anatomic template for biopsy (94) (see Figures 13.2, 13.3). A variety of methods are under evaluation for such fusions. At present, the use of PET/CT methods is clearly the most attractive and commonly performed of such approaches.

Another area where PET-related methods may prove useful is in the intraoperative localization of tumors. Methods to detect the emitted positron, but not the photon, are capable of precise localization of tumor foci in experimental model systems. We have used such a positron-sensitive probe to locate the precise margin of breast cancer following FDG injection at the time of experimental surgery in rodent models of breast carcinoma. This approach offers the potential for less-deforming surgical procedures in patients with breast cancer by removing all the tumor but less normal tissue (95). This approach is in its earliest stages of clinical evaluation, so it remains an area for future study and opportunity in the application of PET methodologies to improve the management of patients with breast cancer.

Cost-Effectiveness of PET in Breast Cancer and Changes in Management

The cost of PET imaging is much higher than that of a mammogram, so that it is unlikely that PET will have a major role in breast cancer screening in the general patient population unless there are major changes in the technology and its pricing. It is possible that PET could be used in selected very high risk populations and might have the potential to be cost effective in such a setting. Rather, PET is currently being used in problem-solving modes in most medical centers.

To be cost effective, the information from PET must alter management appropriately. Recently, management alterations were shown to occur in about 30% of patients based on PET scan results (96). Although these data were based on a survey instrument supplied to referring physicians and only a minority of the surveys were returned, they were based on actual patient data and clearly indicate that major management decisions are made by the referring physicians using PET data. It should be noted that these management changes were often substantial in nature, such as from chemotherapy to radiation therapy or from surgery to chemotherapy. The economic impact of such changes is expected to be substantial.

In another study of the role of PET in a diverse group of patients (n =165) referred for FDG-PET imaging, distant metastases were diagnosed in about 5% of patients, causing a major change in patient management (97). Thus, as in the rest of clinical PET, if a major diagnostic decision is to be made, accurate imaging information is an important part of the data set used to make such decisions. Thus, PET can have great clinical impact in breast cancer management (98).

Summary

The role of PET in the imaging management of patients with known or suspected breast cancer continues to evolve. For assessing primary lesions, it is sometimes possible with PET to detect cancers not detected by standard methods, and incidental cancers of the breast are detected when whole-body PET studies are performed in large numbers of patients. Size limitations with current whole-body scanners limit the accuracy of the method in detecting small primary tumors. While intense tracer uptake on FDG-PET is usually indicative of cancer and low uptake is generally benign, this differentiation is best addressed by biopsy in most cases.

In women with breast implants, PET may have a growing and important role for assessing the entire breast. PET appears promising for detecting occult lesions in the breast when positive axillary nodes are found without a known primary tumor. Similarly, PET may offer advantages in breasts that show postoperative changes. However, there certainly can also be FDG uptake in inflammatory tissue, and PET is not likely to be able to separate infection from tumor (99). It is likely that PET will come to have a greater role in detecting primary breast lesions as dedicated devices for breast imaging evolve that have superior resolution and performance to current instruments.

Results in FDG-PET imaging for axillary and internal mammary nodal staging have been variable. Small nodal metastases are missed by PET whereas larger ones generally are detected. PET can detect internal mammary nodes; the accuracy of the method in this setting is not known but is likely similar to that in detecting axillary metastases. PET may have a major role in staging the axilla in patients with larger primary breast cancers who then may be treated by neoadjuvant chemotherapy without the need for axillary dissection or preoperative sentinel node biopsy if PET is clearly positive. Similarly, an unequivocally positive FDG-PET study of the axilla may obviate the need for sentinel node surgery and may move the treatment plan to axillary dissection. A negative PET scan cannot be accepted as evidence of absence of axillary metastasis, and PET is not as sensitive as axillary dissection or sentinel lymph node surgery.

Clearly, PET can provide accurate staging for distant metastatic disease. Skeletal scintigraphy with ^{18}F-fluoride is highly sensitive for metastases, and FDG imaging is also effective. However, FDG-PET can miss some blastic skeletal metastases. PET is effective for detecting recurrences in soft tissues and the brachial plexus region in particular.

PET results in planning the treatment of individual patients are very promising. The results must be confirmed in larger studies, but it appears safe to conclude that failure of a chemotherapy regimen to decrease FDG uptake promptly in breast cancer is a bad prognostic sign. This reasoning does not hold true for hormonal therapy. At present, labeled estrogens are not widely available and cannot be recommended for clinical use. Alternative tracers may prove useful in treatment response assessment.

In the author's experience, PET imaging in breast cancer is most commonly applied to solve difficult imaging questions in specific patients, rather than serving as a routine diagnostic and staging procedure in the ap-proximately 200,000 patients per year who develop breast cancer in the United States. However, there is increasing use of PET in monitoring of patients with breast cancer during therapy. Good candidates for baseline PET studies at diagnosis include patients with large or high-risk lesions who may have a greater chance of systemic metastases at presentation, patients with medial lesions, and patients in whom sentinel node sampling or axillary dissection is not planned. Similarly, patients who will be treated primarily with chemotherapy are well suited to early monitoring by PET using a baseline. PET is clearly a useful tool in following disease progression and response to treatment for systemic disease, and in many practices assessing the breast cancer patient for efficacy of treatment of disseminated metastatic disease using PET is a common and important approach to patient management.

References

1. Wahl RL. Clinical oncology update: the emerging role of positron emission tomography. Part 1. In: DeVita VT, Hellman S, Rosenberg S, editors. Principles and Practice of Oncology Updates, vol I. Cedar Knolls, NJ: Lippincott-Raven, 1997:1–18.
2. Wahl RL. Current status of PET in breast cancer imaging, staging and therapy. Semin Roentgenol 2001;36:250–256.
3. Brown RS, Wahl RL. Overexpression of Glut-1 glucose transporter in human breast cancer: an immunohistochemical study. Cancer (Phila) 1993;72(10):2979–2985.
4. Brown RS, Leung JY, Fisher S, Frey KA, Ethier SF, Wahl RL. Intratumoral distribution of tritiated-FDG in breast carcinoma: correlation between glut-1 expression and FDG uptake. J Nucl Med 1996;37(6):1042–1047.
5. Bos R, van Der Hoeven JJ, van Der Wall E, van Der Groep P, van Diest PJ, Comans EF, et al. Biologic correlates of (18) fluorodeoxyglucose uptake in human breast cancer measured by positron emission tomography. J Clin Oncol 2002;20(2):379–387.
6. Buck A, Schirrmeister H, Kuhn T, Shen C, Kalker T, Kotzerke J, et al. FDG uptake in breast cancer: correlation with biological and clinical prognostic parameters. Eur J Nucl Med Mol Imaging 2002;29(10):1317–1323.
7. Beaney RP, Lammertsma AA, Jones T, McKenzie CG, Halnan KE. Positron emission tomography for in-vivo measurement of regional blood /low, oxygen utilisation, and blood volume in patients with breast carcinoma. Lancet 1984;1(8369):131–134.
8. Wilson CB, Lammertsma AA, McKenzie CG, Sikora K, Jones T. Measurements of blood /low and exchanging water space in breast tumors using positron emission tomography: a rapid and noninvasive dynamic method. Cancer Res 1992;52(6):1592–1597.
9. Wahl RL, Zasadny K, Koeppe RA, Mangner T, Mintun MA, Lee KS et al. Initial evaluation of Cu-62PTSM as a tumor blood /low agent in man: comparison studies with O-15 water. J Nucl Med 1993;34(5):224P (abstract).
10. Mankoff DA, Dunnwald LK, Gralow JR, Ellis GK, Charlop A, Lawton TJ, et al. Blood flow and metabolism in locally advanced breast cancer: relationship to response to therapy. J Nucl Med. 2002;43(4):500–509.
11. Mankoff DA, Dunnwald LK, Gralow JR, Ellis GK, Schubert EK, Tseng J, et al. Changes in blood flow and metabolism in locally advanced breast cancer treated with neoadjuvant chemotherapy. J Nucl Med 2003;44(11):1806–1814.
12. Zasadny KR, Tatsumi M, Wahl RL. FDG metabolism and uptake versus blood flow in women with untreated primary breast cancers. Eur J Nucl Med Mol Imaging 2003;30(2):274–280.

13. Mintun MA, Welch MJ, Siegel BA, Mathias CJ, Brodack JW, McGuire AH, et al. Breast cancer: PET imaging of estrogen receptors. Radiology 1988;169:45–48.

14. McGuire AH, Dehdashti F, Siegel BA, Lyss AP, Brodack JW, Mathias CJ, et al. Positron tomographic assessment of 16 alpha-[18F]-fluoro-17 beta-estradiol uptake in metastatic breast carcinoma. J Nucl Med 1991;32(8):1526–1531.

15. Minn H, Soini I. [18F]Fluorodeoxyglucose scintigraphy in diagnosis and follow up of treatment in advanced breast cancer. Eur J Nucl Med 1989;15(2):61–66.

16. Wahl RL, Cody RL, Hutchins GD, Kuhl DE. PET imaging of breast cancer with 18-FDG. Radiology 1989;173:419.

17. Kubota K, Matsuzawa T, Amemiya A, Kondo M, Fujiwara T, Watanuki S, et al. Imaging of breast cancer with [18F]-fluorodeoxyglucose and positron emission tomography. J Comput Assist Tomogr 1989;13(6):1097–1098.

18. Wahl RL, Cody R, Hutchins GD, Mudgett E. Primary and metastatic breast carcinoma: initial clinical evaluation with PET with the radiolabeled glucose analog 2-[18F]-fluorodeoxy-2-D-glucose (FDG). Radiology 1991;179:765–770.

19. Wahl RL, Henry CA, Ethier SF. Serum glucose: effects on tumor and normal tissue accumulation of 2-[18F]-fluoro-2-deoxy-D-glucose in rodents with mammary carcinoma. Radiology 1992;183(3):643–647.

20. Lindholm P, Minn H, Leskinen-Kallio S, Bergman J, Ruotsalainen U, Joensuu H. Influence of the blood glucose concentration on FDG uptake in cancer: a PET study. J Nucl Med 1993;34(1):1–6.

21. Hoh CK, Hawkins RA, Glaspy JA, Dahlbom M, Tse NY, Hoffman EJ, et al. Cancer detection with whole-body PET using 2-[18F]fluoro-2-deoxy-D-glucose. J Comput Assist Tomogr 1993;17(4):582–589.

22. Nieweg OE, Kim EE, Wong WH, Broussard WF, Singletary SE, Hortobagyi GN, et al. Positron emission tomography with fluorine-18-deoxyglucose in the detection and staging of breast cancer. Cancer (Phila) 1993;71(12):3920–3925.

23. Adler LP, Crowe JP, al-Kaisi NK, Sufshine JL. Evaluation of breast masses and axillary lymph nodes with [18F]-2-deoxy-2-fluoro-D-glucose PET. Radiology 1993;187(3):743–750.

24. Dehdashti F, Mortimer JE, Siegel BA, Griffeth LK, Bonasera TJ, et al. Positron tomographic assessment of estrogen receptors in breast cancer: comparison with FDG-PET and in vitro receptor assays. J Nucl Med 1995;36:1766–1774.

25. Bassa P, Kim EE, Inoue T, Wong FCL, Korkmaz M, et al. Evaluation of preoperative chemotherapy using PET with fluorine-18 fluorodeoxyglucose in breast cancer. J Nucl Med 1996;37:931–938.

26. Avril N, Dose J, Janicke DF, Bense S, Ziegler S, et al. Metabolic characterization of breast tumors with positron emission tomography using 18F-fluorodeoxyglucose. J Clin Oncol 1996;14:1848–1857.

27. Avril N, Rose CA, Schelling M, Dose J, Kuhn W, Bense S, et al. Breast imaging with positron emission tomography and fluorine-18-fluorodeoxyglucose: use and limitations. J Clin Oncol 2000;18(20):3495–3502.

28. Yutani K, Shiba E, Kusuoka H, Tatsumi M, Uehara T, Taguchi T, et al. Comparison of FDG-PET with MIBI-SPECT in the detection of breast cancer and axillary lymph node metastasis. J Comput Assist Tomogr 2000;24(2):274–280.

29. Ishimori T, Patel PV, Wahl RL. Detection of unexpected additional primary malignancies with PET/CT. J Nucl Med 2005;46(5):752–757.

30. Wahl RL, Helvie MA, Chang AE, Andersson I. Detection of breast cancer in women after augmentation mammoplasty using positron emission tomography: initial clinical evaluation. J Nucl Med 1994;35:872–875.

31. Noh DY, Yun IJ, Kang HS, Kim YC, Cim IS, Chung JK, et al. Detection of cancer in augmented breasts by positron emission tomography. Eur J Surg 1999;165(9):847–851.

32. Walter C, Scheidhauer K, Goering UJ, Theissen P, Kugel H, Krahe T, Pietrzyk U. Clinical and diagnostic value of preoperative MR mammography and FDG-PET in suspicious breast lesions. Eur Radiol 2003;13(7):1651–1656.

33. Goerres GW, Michel SC, Fehr MK, Kaim AH, Steinert HC, Seifert B, et al. Follow-up of women with breast cancer: comparison between MRI and FDG-PET. Eur Radiol 2003;13(7):1634–1644.

34. Semple SI, Gilbert FJ, Redpath TW, Staff RT, Ahearn TS, Welch AE, et al. The relationship between vascular and metabolic characteristics of primary breast tumours. Eur Radiol 2004;14(11):2038–2045.

35. Leskinen-Kallio S, Nagren K, Lehikoinen P, Ruotsalainen U, Joensuu H. Uptake of 11C-methionine in breast cancer studied by PET. An association with the size of S-phase fraction. Br J Cancer 1991;64(6):1121–1124.

36. Levine EA, Freimanis RI, Perrier ND, Morton K, Lesko NM, Bergman S, et al. Positron emission mammography: initial clinical results. Ann Surg Oncol 2003;10(1):86–91.

37. Raylman RR, Majewski S, Wojcik R, Weisenberger AG, Kross B, Popov V, et al. The potential role of positron emission mammography for detection of breast cancer. A phantom study. Med Phys 2000;27(8):1943–1954.

38. Gagnon JH, Murthy K, Aznar M, Bergman AM, Thompson CI, Robar JL, et al. Positron emission mammographic instrument: initial results. Radiology 2000;215(1):280–285.

39. Doshi NK, Shao Y, Silverman RW, Cherry SR. Design and evaluation of an LSO PET detector for breast cancer imaging. Med Phys 2000;27(7):1535–1543.

40. Bergman AM, Thompson CJ, Murthy K, Robar JL, Clancy RL, English MJ. Technique to obtain positron emission mammography images in registration with X-ray mammograms. Med Phys 1998;25(11):2119–2129.

41. Weinberg IN, Beylin D, Zavarzin V, Yarnall S, Stepanov PY, Anashkin E, et al. Positron emission mammography: high-resolution biochemical breast imaging. Recent studies of 16 patients with 18 lesions. Technol Cancer Res Treat 2005;4(1):55–60.

42. Smith MF, Raylman RR, Majewski S, Weisenberger AG. Positron emission mammography with tomographic acquisition using dual planar detectors: initial evaluations. Phys Med Biol 2004;49(11):2437–2452.

43. Rosen EL, Turkington TG, Soo MS, Baker JA, Coleman RE. Detection of primary breast carcinoma with a dedicated, large-field-of-view FDG-PET mammography device: initial experience. Radiology 2005;234(2):527–534.

44. Wahl RL, Kaminski MS, Ethier SP, Hutchins GD. The potential of 2-deoxy-2[18F]fluoro-D-glucose (FDG) for the detection of tumor involvement in lymph nodes. J Nucl Med 1990;31(11):1831–1835.

45. Wahl RL, Cody RL, August D. Initial evaluation of FDG PET for the staging of the axilla in newly diagnosed breast carcinoma patients. J Nucl Med 1991;32(5):981.

46. Tse NY, Hoh CK, Hawkins RA, Zinner MJ, Dahlbom M, Choi Y, et al. The application of positron emission tomographic imaging with fluorodeoxyglucose to the evaluation of breast disease. Ann Surg 1992;216(1):27–34.

47. Adler LP, Faulhaber PF, Schnur KC, Al-Kasi NL, Shenk RR. Axillary lymph node metastases: screening with [F-18] 2-deoxy-2-fluoro-D-glucose (FDG) PET. Radiology 1997;203:323–327.

48. Avril N, Dose J, Janicke F, Ziegler S, Romer W, Weber W, et al. Assessment of axillary lymph node involvement in breast cancer patients with positron emission tomography using radiolabeled 2-(fluorine-18)-fluoro-2-deoxy-D-glucose. J Natl Cancer Inst 1996;88(17):1204–1209.

49. Utech O, Young CS, Winter PF. Prospective evaluation of fluorine-18 fluorodeoxyglucose positron emission tomography in breast cancer for staging of the axilla related to surgery and immunocytochemistry. Eur J Nucl Med 1996;23(12):1588–1593.

50. Yutani K, Shiba E, Kusuoka H, Tatsumi M, Uehara T, Taguchi T, et al. Comparison of FDG-PET with MIBI-SPECT in the detection of breast cancer and axillary lymph node metastasis. J Comput Assist Tomogr 2000;24(2):274–280.

51. Crippa F, Leutner M, Belli F, Gallino F, Greco M, Pilotti S, et al. Which kinds of lymph node metastases can FDG PET detect? A clinical study in melanoma. J Nucl Med 2000;41(9):1491–1494.

52. Wahl RL, Siegel BA, Coleman RE, Gatsonis CG, PET Study Group. Prospective multicenter study of axillary nodal staging by positron emission tomography in breast cancer: a report of the staging breast cancer with PET Study Group. J Clin Oncol 2004; 22(2):277–285.

53. Lovrics PJ, Chen V, Coates G, Cornacchi SD, Goldsmith CH, Law C, et al. A prospective evaluation of positron emission tomography scanning, sentinel lymph node biopsy, and standard axillary dissection for axillary staging in patients with early stage breast cancer. Ann Surg Oncol 2004;11(9):846–853.

54. Bellon JR, Livingston RB, Eubank WB, Gralow JR, Ellis GK, Dunnwald LK, Mankoff DA. Evaluation of the internal mammary lymph nodes by FDG-PET in locally advanced breast cancer (LABC). Am J Clin Oncol 2004;27(4):407–410.

55. Trampal C, Sorensen J, Engler H, Langstrom B. ^{18}F-FDG whole body positron emission tomography (PET) in the detection of unknown primary tumors. Clin Posit Imaging 2000;3(4):160.

56. Scoggins CR, Vitola JV, Sandler MP, Atkinson JB, Frexes-Steed M. Occult breast carcinoma presenting as an axillary mass. Am Surg 1999;65(1):1–5.

57. Lonneux M, Reffad A. Metastases from unknown primary tumor. PET-FDG as initial diagnostic procedure? Clin Posit Imaging 2000;3(4):137–141.

58. Bohuslavizki KH, Klutmann S, Kroger S, Sonnemann U, Buchert R, Werner JA, et al. FDG PET detection of unknown primary tumors. J Nucl Med 2000;41(5):816–822.

59. Inoue T, Yutani K, Taguchi T, Tamaki Y, Shiba E, Noguchi S. Preoperative evaluation of prognosis in breast cancer patients by [(18)F]2-deoxy-2-fluoro-D-glucose-positron emission tomography. J Cancer Res Clin Oncol 2004;130(5):273–278.

60. Torizuka T, Zasadny KR, Recker B, Wahl RL. Untreated primary lung and breast cancers: correlation between F-18 FDG kinetic rate constants and findings of in vitro studies. Radiology 1998;207(3):767–774.

61. Cook GJ, Fogelman I. Skeletal metastases from breast cancer: imaging with nuclear medicine. Semin Nucl Med 1999;29(1):69–79.

62. Kao CH, Hsieh JF, Tsai SC, Ho YJ, Yen RF. Comparison and discrepancy of ^{18}F-2-deoxyglucose positron emission tomography and Tc-99m MDP bone scan to detect bone metastases. Anticancer Res 2000;20(3B):2189–2192.

63. Gallowitsch HJ, Kresnik E, Gasser J, Kumnig G, Igerc I, Mikosch P, Lind P. F-18 fluorodeoxyglucose positron-emission tomography in the diagnosis of tumor recurrence and metastases in the follow-up of patients with breast carcinoma: a comparison to conventional imaging. Invest Radiol 2003;38(5):250–256.

64. Yang SN, Liang JA, Lin FJ, Kao CH, Lin CC, Lee CC. Comparing whole body (18)F-2-deoxyglucose positron emission tomography and technetium-99m methylene diphosphonate bone scan to detect bone metastases in patients with breast cancer. J Cancer Res Clin Oncol 2002;128(6):325–328.

65. Stafford SE, Gralow JR, Schubert EK, Rinn KJ, Dunnwald LK, Livingston RB, Mankoff DA. Use of serial FDG PET to measure the response of bone-dominant breast cancer to therapy. Acad Radiol 2002;9(8):913–921.

66. Uematsu T, Yuen S, Yukisawa S, Aramaki T, Morimoto N, Endo M, et al. Comparison of FDG PET and SPECT for detection of bone metastases in breast cancer. AJR Am J Roentgenol 2005;184(4):1266–1273.

67. Fogelman I, Cook G, Israel O, Van der Wall H. Positron emission tomography and bone metastases. Semin Nucl Med 2005;35(2):135–142.

68. Schirrmeister H, Guhlmann A, Kotzerke J, Santjohanser C, Kuhn T, Kreienberg R, et al. Early detection and accurate description of extent of metastatic bone disease in breast cancer with fluoride ion and positron emission tomography. J Clin Oncol 1999;17(8):2381–2389.

69. Gabelmann A, Reske SN, Hetzel M. Prospective evaluation of the clinical value of planar bone scans, SPECT, and (18)F-labeled NaF

70. PET in newly diagnosed lung cancer. J Nucl Med 2001;42(12):1800–1804.

70. Rostom AY, Powe J, Kandil A, Ezzat A, Bakheet S, el-Khwsky F, et al. Positron emission tomography in breast cancer: a clinicopathological correlation of results. Br J Radiol 1999;72(863):1064–1068.

71. Wald JJ, Wahl RL. Use of positron emission tomography in evaluation of brachial plexopathy. Neurology 1994;44:A307.

72. Ahmad A, Barrington S, Maisey M, Rubens RD. Use of positron emission tomography in evaluation of brachial plexopathy in breast cancer patients. Br J Cancer 1999;79(3-4):478–482.

73. Cohade C, Osman M, Pannu HK, Wahl RL. Uptake in supraclavicular area fat ("USA-Fat"): description on ^{18}F-FDG PET/CT. J Nucl Med 2003;44(2):170–176.

74. Tatsumi M, Engles JM, Ishimori T, Nicely O, Cohade C, Wahl RL. Intense (18)F-FG uptake in brown fat can be reduced pharmacologically. J Nucl Med 2004;45(7):1189–1193.

75. Sugawara Y, Fisher SJ, Zasadny KR, Kison PV, Baker LH, Wahl RL. Preclinical and clinical studies of bone marrow uptake of fluorine-1-fluorodeoxyglucose with or without granulocyte colony-stimulating factor during chemotherapy. J Clin Oncol 1998;16(1):173–180.

76. Griffeth LK, Rich KM, Dehdashti F, Simpson JR, Fusselman MJ, McGuire AH, et al. Brain metastases from non-central nervous system tumors: evaluation with PET. Radiology 1993;186(1):37–44.

77. Isasi CR, Moadel RM, Blaufox MD. A meta-analysis of FDG-PET for the evaluation of breast cancer recurrence and metastases. Breast Cancer Res Treat 2005;90(2):105–112.

78. Mortimer JE, Dehdashti F, Siegel BA, Katzenellenbogen JA, Fracasso P, Welch MJ. Positron emission tomography with 2-^{18}F]fluoro-2-deoxy-D-glucose and 16-alpha-^{18}F]fluoro-17beta-estradiol in breast cancer: correlation with estrogen receptor status and response to systemic therapy. Clin Cancer Res 1996;2(6):933–939.

79. Dehdashti F, Flanagan FL, Mortimer JE, Katzenellenbogen JA, Welch MJ, Siegel BA. Positron emission tomographic assessment of "metabolic flare" to predict response of metastatic breast cancer to antiestrogen therapy. Eur J Nucl Med 1999;26(1):51–56.

80. Wahl RL, Zasadny KR, Hutchins GD, Helvie M, Cody R. Metabolic monitoring of breast cancer chemohormonotherapy using positron emission tomography (PET): initial evaluation. J Clin Oncol 1993;11(11):2101–2111.

81. Jansson T, Westlin JE, Ahlstrom H, Lilja A, et al. Positron emission tomography studies in patients with locally advanced and/or metastatic breast cancer: a method for early therapy evaluation? J Clin Oncol 1995;13(6):1470–1477.

82. Smith IC, Welch AE, Hutcheon AW, Miller IL, Payne S, Chilcott F, et al. Positron emission tomography using [(18)F]-fluorodeoxy-D-glucose to predict the pathologic response of breast cancer to primary chemotherapy. J Clin Oncol 2000;18(8):1676–1688.

83. Schelling M, Avril N, Nahrig J, Kuhn W, Romer W, Sattler D, et al. Positron emission tomography using [(18)F]fluorodeoxyglucose for monitoring primary chemotherapy in breast cancer. J Cin Oncol 2000;18(8):1689–1695.

84. Burcombe RJ, Makris A, Pittam M, Lowe J, Emmott J, Wong WL. Evaluation of good clinical response to neoadjuvant chemotherapy in primary breast cancer using ^{18}F]-fluorodeoxyglucose positron emission tomography. Eur J Cancer 2002;38(3):375–379.

85. Mortimer JE, Dehdashti F, Siegel BA, Trinkaus K, Katzenellenbogen JA, Welch MJ. Metabolic flare: indicator of hormone responsiveness in advanced breast cancer. J Clin Oncol 2001;19(11):2797–2803.

86. Kim SJ, Kim SK, Lee ES, Ro J, Kang S. Predictive value of ^{18}F]FDG PET for pathological response of breast cancer to neo-adjuvant chemotherapy. Ann Oncol 2004;15(9):1352–1357.

87. Miller KD, Soule SE, Calley C, Emerson RE, Hutchins GD, Kopecky K, et al. Randomized phase II trial of the anti-angiogenic potential of doxorubicin and docetaxel; primary chemotherapy at Biomarker Discovery Laboratory. Breast Cancer Res Treat 2005;89(2):187–197.

88. Chen X, Moore MO, Lehman CD, Mankoff DA, Lawton TJ, Peacock S, et al. Combined use of MRI and PET to monitor response and assess residual disease for locally advanced breast cancer treated with neoadjuvant chemotherapy. Acad Radiol 2004; 11(10):1115–1124.

89. Fei X, Wang JQ, Miller KD, Sledge GW, Hutchins GD, Zheng QH. Synthesis of [18F]Xeloda as a novel potential PET radiotracer for imaging enzymes in cancers. Eur J Nucl Med Mol Imaging 2005;32(1):15–22.

90. Sun H, Sloan A, Mangner TJ, Vaishampayan U, Muzik O, Collins JM, et al. Imaging DNA synthesis with [18F]FMAU and positron emission tomography in patients with cancer. Eur J Nucl Med Mol Imaging 2005;32(1):15–22.

91. Smyczek-Gargya B, Fersis N, Dittmann H, Vogel U, Reischl G, Machulla HJ, et al. PET with [18F]fluorothymidine for imaging of primary breast cancer: a pilot study. Eur J Nucl Med Mol Imaging 2004;31(5):720–724.

92. Rajendran JG, Mankoff DA, O'Sullivan F, Peterson LM, Schwartz DL, Conrad EU, et al. Hypoxia and glucose metabolism in malignant tumors: evaluation by [18F]fluoromisonidazole and [18F]fluorodeoxyglucose positron emission tomography imaging. Clin Cancer Res 2004;10(7):2245–2252.

93. Raylman RR, Ficaro EP, Wahl RL. Stereotactic coordinates from ECT sinograms for radionuclide-guided breast biopsy. J Nucl Med 1996;37(9):1562–1567.

94. Wahl RL, Quint LE, Cieslak RD, Aisen AM, Koeppe RA, Meyer CR. "Anatometabolic" tumor imaging: fusion of FDG PET with CT or MRI to localize foci of increased activity. J Nucl Med 1993;34(7):1190–1197.

95. Raylman RR, Fisher S, Brown RS, Ethier SP, Wahl RL. Fluorine-18-fluorodeoxyglucose-guided breast cancer surgery with a positron-sensitive probe: validation in preclinical studies. J Nucl Med 1995;36(10):1869–1874.

96. Yap CS, Seltzer MA, Schiepers C, Gambhir SS, Rao J, Phelps ME, et al. Impact of whole-body 18F-FDG PET on staging and managing patients with breast cancer: the referring physician's perspective. J Nucl Med 2001;42(9):1334–1337.

97. Weir L, Worsley D, Bernstein V. The value of FDG positron emission tomography in the management of patients with breast cancer. Breast J 2005;11(3):204–209.

98. Eubank WB, Mankoff DA. Evolving role of positron emission tomography in breast cancer imaging. Semin Nucl Med 2005;35(2):84–99.

99. Bakheet SM, Powe J, Kandil A, Ezzat A, Rostom A, Amartey J. F-18 FDG uptake in breast infection and inflammation. Clin Nucl Med 2000;25(2):100–103.

14

PET and PET/CT Imaging in Testicular and Gynecologic Cancers

Sharon F. Hain

Tumors of the testes and gynecologic malignancies are important cancers with significant morbidity, and the role of imaging is vital. Although testicular cancer is not common, it is increasing in incidence and is a cancer of relatively young men. Ovarian, uterine, and cervical cancer are among the most common tumors in women, with high morbidity and mortality rates. There are early screening tests for cervical cancer but, despite this, mortality from this tumor remains significant. Ovarian cancer is a problem as it often presents in late stage. In all three, outcome is dependent on stage at diagnosis and imaging plays a crucial role in assessment. Following initial treatment, appropriate posttreatment evaluation and timely additional therapy can decrease morbidity from unnecessary further treatment or can lead to cure when residual tumor is detected. Imaging techniques have been fundamental in this process, and positron emission tomography (PET) is developing an important role in pretreatment assessment and in assessing early response to therapy. This chapter considers the role of imaging in the management of these tumors.

Testicular Cancer

Testicular cancer (seminoma and nonseminoma, NSGCT) is a relatively rare tumor affecting only 1% of men, but it is the commonest tumor in young males (aged 15–35) and its incidence is increasing (1). The two tumor types differ in their biologic behavior and potential for metastases so that treatment also differs. The overall prognosis for these tumors is good.

Diagnosis and Tumor Staging

Most testicular cancers present as an asymptomatic lump in the testes, and urgent orchidectomy is performed. The basic histologic distinction is between seminoma and nonseminomatous germ cell tumor (NSGCT), although the histology can be complex and 10% of patients have mixed tumors (2). The tumors spread to the paraaortic region initially, although hematogenous spread is more common in NSGCT and metastases are seen in the lung, brain, liver, and bone. At diagnosis all patients should be staged by clinical examination and computed tomography (CT) scans of the chest and abdomen and pelvis. Tumor markers should be measured as they provide prognostic information, allow monitoring of treatment response and assessment of recurrence (3, 4). In addition, histologic factors in the tumor such as the presence of blood vessel invasion by the primary tumor, the percentage of embryonal cancer, and involvement of the rete testes (in seminoma) provide prognostic information. All these data are used to stage the patient. There are many staging systems. One of the most common is the Marsden System (Table 14.1) (5).

On the basis of staging investigations, patients with NSGCT are classified as low or high risk for metastatic disease, and treatment regimens are based on this stratification. Low-risk patients may be observed or treated with two courses of chemotherapy followed by surveillance, and both these regimens have good cure rates (4, 6). The factors influencing staging, however, have proven to be unreliable for absolute determination of risk for metastases in any individual. Because of advances in chemotherapy, cure is now possible for the majority of patients with minimal metastatic disease. Accordingly, there has been a need to reevaluate the treatment strategies that are currently being used to minimize chemotherapy toxicity by accurately differentiating the patients who have metastatic disease from those who do not. Further concerns have been raised over the long-term effects of chemotherapy on the cardiac system and the precipitation of second malignancies (7). If tumor spread could be reliably assessed, some patients with NSGCT stage I (no evidence of metastases) could be clinically observed rather than undergo prophylactic chemotherapy. Of the patients initially classified as stage I, 20% to 30% have lymph

Table 14.1. Staging of testicular cancer.

Stage	Description
I	No evidence of metastases
IM	Rising concentrations of serum markers with no other evidence of metastases
II	Abdominal node metastases
A	=2 cm in diameter
B	2–5 cm in diameter
C	=5 cm in diameter
III	Supradiaphragnatic nodal metastasis
M	Mediastinal
N	Supraclavicular, cervical, or axillary
0	No abdominal node metastasis
ABC	Node stage as defined in stage II
IV	Extralymphatic metastasis
Lung	
L1	=3 metastases
L2	=3 metastases, all =2 cm in diameter
L3	=3 metastases, one or more =2 cm in diameter
H+, Br+, Bo+	Liver, brain, or bone

Source: From Horwich A. Testicular Cancer. In: Horwich A, ed. Oncology: A Multidisciplinary Textbook. London: Chapman & Hall 1995:485–498.

nodes at retroperitoneal lymph node dissection and therefore need further treatment. Overall, as many as 50% are understaged and 25% overstaged by currently available techniques. Eventually, 20% of these patients will relapse whereas 50% of patients who have been classified as high risk on the basis of prognostic factors and therefore sent for treatment do not relapse (4).

In seminoma, conventional practice has been to perform retroperitoneal radiotherapy even in stage I disease, and about 15% of patients at presentation have disease confined to the abdomen. Retroperitoneal and pelvic radiotherapy is a common practice and has a good rate of achieving local control. Even so, 15% of clinical stage I relapse and 20% to 30% of patients with seminoma will relapse at distant sites (8).

Imaging Procedures in Tumor Staging

Anatomic staging techniques including CT, ultrasound, and lymphangiography have all been used to stage testicular cancer. The most widely used now is CT, which is routinely performed as part of the initial staging protocol. All staging procedures have limitations, and even for CT false-negative rates of 59% have been reported. The false-negative rates for lymphangiography and ultrasound are 64% and 70%, respectively (9). The diagnosis of nodal metastases by CT is based on detection of nodal enlargement, with a 1-cm upper limit for normal lymph node size. Before nodal enlargement, the entire volume of a

lymph node may be replaced by malignant cells, whereas a large lymph node may contain only benign reactive cells. As a result, the false-positive rate of CT is also high at 40% (10). This inaccuracy has led to search for more-accurate imaging methods including metabolic imaging with PET.

Because of the ability of ^{18}F-2-deoxy-D-glucose (FDG)-PET to detect metabolically active disease without reliance on size criteria, FDG-PET imaging has the potential to identify small-volume disease in a lymph node that is normal in size; this may have a direct effect on patient management. In stage I tumor, more accurate classification of patients as high or low risk would avoid unnecessary treatment and morbidity. In stage II and III tumor, where prognosis is based on many factors including nonvisceral disease and tumor markers, accurate classification may determine whether radiotherapy/ chemotherapy or surgery should be used to treat patients.

At diagnosis, FDG-PET can clearly identify more sites of disease in patients with established metastatic disease than seen on CT (11, 12). This finding will have minimal effect on initial management if the patient is to receive chemotherapy based on traditional staging. In a few cases, it has identified unsuspected visceral or bone disease and therefore altered management (11).

There are only a few studies that have addressed the issue of improving the initial staging using PET, and these include between 31 and 50 patients (11–15). The sensitivity ranged between 70% and 87% and the specificity between 94 and 100%. The three major initial studies (11, 13, 14) confirmed overall better sensitivity, positive predictive value (PPV), and negative predictive value (NPV) for PET than for CT. Both CT and PET missed small (approximately 1 cm) retroperitoneal lymph node metastases (13).

One limitation of these studies was that not all patients had histologic confirmation of findings on PET and assessment of true negativity or positivity. Albers et al. (14) had histopathology results in some patients and found 70% sensitivity and a 100% specificity in a study of 37 patients. They concluded that PET was useful in stage II tumors to correctly diagnose the false positives that were seen on CT but was not useful in stage I tumor because, among 15 patients with stage I NSGCT, PET identified only 4 of 6 patients with pathologically involved lymph nodes negative on CT. Even with the limitation of low sensitity for detection of small-volume disease, the use of PET could significantly improve management of stage I NSGCT patients. Although there are no studies regarding the performance of integrated PET/CT imaging for staging of testicular tumors, precise localization of FDG-avid nodes can guide biopsy to these specific lymph nodes, even if normal by size criteria, and further improve staging.

Understaging with imaging is of most concern in NSGCT because patients with true stage I tumor could be followed by surveillance only. Lassen et al. (15) performed a prospective analysis of 46 patients with stage I tumor after a normal CT and tumor markers who underwent FDG PET imaging. All patients underwent routine follow-up with repeated CT and tumor marker evaluation; 22% (10/46) relapsed, of whom 7 of 10 had positive initial PET scans, which gave a negative predictive value of 92% for PET compared to that of 78% for conventional imaging. The authors concluded that PET was superior to conventional staging ($P = 0.06$) in stage I NSGCT with the potential to improve patient management.

To fully define if PET may be useful in staging, large-scale prospective studies are needed in which patients are followed until relapse or until abnormalities found on PET and CT are biopsied. If it can be confirmed that PET has the suggested improved sensitivity for occult disease, then it could significantly improve the current management of patients with stage I NSGCT. Studies have been started including the Medical Research Council (MRC)-funded TE22 Trial in the U.K. to further address this problem.

Tumor Recurrence

Patients with metastatic disease must be monitored to detect relapse at a time when salvage treatment would have the best chance of cure or disease control. Patients must be followed for several years clinically, biochemically with serum tumor markers, and radiologically.

Following treatment, patients with metastatic disease frequently have residual masses and the treatment of these remains difficult. When the mass contains persistent tumor, immediate further treatment is indicated. Often the mass contains merely necrosic or fibrous tissue, in which case no further treatment is indicated and the patient may be observed. In the case of NSGCT, such masses may also contain mature teratoma differentiated (MTD), which is a benign tumor. Such masses need to be removed because there is a risk of malignant tumor recurrence, but the procedure can be delayed and undertaken later at a time when the patient is less likely to experience surgical morbidity from recent chemotherapy. CT imaging is the standard method of monitoring these patients, and by this means it is not possible to determine whether the residual mass contains any active tumor.

Determination of tumor markers is the second important procedure in the follow-up of patients with testicular cancer because a detectable level of such markers may indicate residual or recurrent tumor. Serum tumor markers are not sufficiently sensitive or specific to determine tumor presence or absence reliably and cannot indicate the anatomic site of any tumor that is present (16, 17).

PET Imaging in Tumor Recurrence

Functional imaging techniques, including imaging with radiolabeled antibodies ([67]gallium and [201]thallium), have been used to assess disease presence but with variable results (18–20). FDG-PET potentially has the ability to detect small-volume tumors in solitary residual masses, to identify a specific mass as the site of relapse in patients with multiple masses, to detect other unsuspected sites of tumor, and to determine the site and extent of disease in patients with raised tumor markers.

Residual Masses

Several studies have focused on the ability of PET to determine which patients should undergo resection of residual masses following treatment (12, 21–23). Initially, Stephens et al. (21) studied 30 patients with NSGCT who had residual masses after chemotherapy. PET was able to differentiate viable tumor from fibrosis/necrosis or MTD but could not differentiate these nonmalignant lesions from each other. Hain et al. (22) evaluated 70 patients with FDG-PET posttreatment (47 for assessment of residual masses). FDG-PET had sensitivity and specificity of 88% and 95%, respectively, for detecting residual tumor in masses, as well as high PPV and NPV, 90% and 96%, respectively. Most studies confirmed the improved PPV and accuracy of PET over CT (23). Hain et al. (22) and Cremerius et al. (23) each had one case of histologically proven MTD that had low uptake; both these masses were in the chest. Cremerius postulated that this may have been the result of tissue attenuation differences between the chest and other organs. However, not all intrathoracic MTDs were found to be positive in the study by Hain et al. (21).

The sensitivity, specificity, and predictive values of PET for active tumor allow a sufficient degree of certainty in distinguishing active tumor from nonactive tumor to allow early intervention in patients with active disease. However, PET could not differentiate MTD from fibrosis or necrosis. MTD has the potential to become malignant and must eventually be resected. Resection may be delayed and performed electively at a later time when the patient has recovered from the effects of chemotherapy/radiotherapy.

Analysis of PET data by standard uptake value (SUV) determination has been used in an effort to improve diagnosis of residual masses. Stephens et al. (21) found that recurrent NSGCT had a higher SUV (mean, 8.81) than MTD (mean, 3.07) and necrosis/fibrosis (mean, 2.86). Patients with an SUV greater than 5 were 75 times more likely to have persistent tumor. Cremerius et al. (23) also considered SUV analysis and found seminoma to have a higher SUV than NSGCT, although their results for NSGCT were affected by their considering MTD as tumor in their analysis. MTD has been successfully differentiated from fibrosis or necrosis with PET using kinetic rate constants, and this approach needs further evaluation (24).

In seminoma, detection of active disease is even more important because treatment is more difficult. The value of surgical removal is uncertain, and there is no advantage to be gained from radiotherapy after chemotherapy (25). Cremerius et al. (23) found, in a study of 42 posttreatment PET scans (seminoma and NSGCT), that FDG-PET had a 90% sensitivity for detection of residual tumor in seminoma, and these high values were confirmed by Hain et al. (22). De Santis et al. (26) published the largest study, including 51 patients with seminomas and postchemotherapy residual masses. PET detected residual tumor in all masses greater than 3 cm and in 95% of masses less than 3 cm, with PPVs and NPVs of 100% and 96%, respectively, for PET versus 37% and 92%, respectively, for CT. They concluded that PET was the best predictor of residual tumor and should be used as a standard investigation in this group of patients with seminomas and residual masses after therapy.

Two problems emerged in the studies of residual masses. First, FDG-PET can miss some small-volume active disease. Hain et al. (22) had two false-negative studies, both of which are of concern as they involved small numbers of malignant cells in large masses otherwise containing MTD. Cremerius et al. (23) had one patient with a 15-mm lymph node containing tumor that was missed on PET and CT, but they did not indicate whether the nodes contained only a small number of malignant cells or were entirely replaced by tumor. FDG-PET may therefore have an undefined detection limit that varies with tumor type. It is also possible that the uptake time may be important for detecting malignant testicular cancer, as is the case for sarcoma and breast cancer (27, 28). Cremerius et al. and Hain et al. (22, 23) both found false-negative studies within 2 weeks of chemotherapy, suggesting that it would be appropriate to wait more than 2 weeks after chemotherapy to perform a PET study.

Overall, the numbers of false-negative PET studies were small and the NPV was high.

Rising Tumor Markers

The presence of tumor markers is an important prognostic factor in testicular cancer, and tumor markers are used in the routine monitoring and follow-up of patients. Rising markers may be the first indicator of disease recurrence (29). However, they are neither sensitive nor specific for tumor detection, and marker-negative relapse may occur even where the initial tumor was marker positive. Also, some patients with residual masses posttreatment may show modest elevation of markers even though the masses contain no active tumor (30), and a return of markers to normal posttreatment does not guarantee disease remission (16, 17).

There are two groups of patients with recurrent or residual disease in whom elevated tumor markers are diagnostically problematic. In patients with elevated markers and no residual masses, the anatomic location of recurrent disease may be difficult to determine, and in patients with elevated markers and residual masses, it may not be apparent which, if any, of the known masses contain active disease.

Cremerius et al. (23) found that adding PET imaging to tumor marker determination improved sensitivity and NPV of markers, but adding the tumor marker determination to PET imaging was of little value. However, tumor marker determination was not available for all patients. Hain et al. (22) had marker information for all 70 patients who underwent PET imaging. They found that in patients with raised tumor markers, including those with a resid-

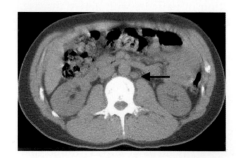

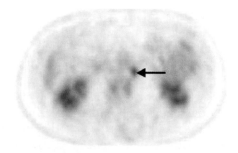

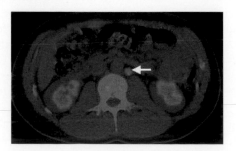

Figure 14.1. A 33-year-old man with a history of testicular cancer 18 months previously presented with rising tumor markers. There was no evidence of disease on conventional imaging. PET/CT imaging revealed a ¹⁸F-2-deoxy-D-glucose (FDG)-avid lymph node lying close to the L2 vertebra, as shown on the transaxial images of CT (*top left*), PET (*top right*), and PET/CT fusion images (*lower*).

ual mass, PET identified the site of disease in all patients but 1. In the group with raised tumor markers and no residual masses, PET demonstrated the tumor in all patients (Figure 14.1). In the group with raised tumor markers and residual masses, there was one false positive. Negative PET scans in the presence of raised tumor markers presented more of a problem as there were 5 patients with false-negative findings in this group. In 3 of these cases, all imaging was normal and subsequent PET scans were the first studies to identify the site of recurrence. This finding suggests that, in the presence of raised tumor markers and negative imaging findings, the most appropriate follow-up procedure is repeating the PET study.

These findings have important implications for the management of patients. Hain et al. (22) found that the ability of PET to find unsuspected disease resulted in management changes in 57% of patients. Management changes involved changes from local therapy–radiotherapy/surgery to chemotherapy or surveillance (Figure 14.2). Many of their patients had had multiple recurrences and had chemotherapy-resistant tumor, and here local control of active sites may be the only chance of cure. In the first relapse, determination of whether there are one or multiple sites will help to determine the type of consolidation treatment.

Predicting Response to Treatment

In other tumors, for example, lymphoma and breast (31–33), PET can predict the response to chemotherapy early during the course and predict long-term outcome. Bokemeyer et al. (34) have evaluated the value of FDG-PET imaging compared to tumor markers and CT/MR in 23 patients with relapsed testicular cancer after two or three cycles of induction chemotherapy before high-dose chemotherapy. The outcome of high-dose chemotherapy was correctly predicted by PET/CT scan/serum tumor marker in 91%, 59%, and 48% of patients, respectively. In those patients who showed response to induction chemotherapy according to CT scans or serum tumor marker evaluation, a positive PET study correctly predicted treatment failure. In addition, PET identified patients most likely to achieve a favorable response to subsequent high-dose chemotherapy. It was suggested that FDG-PET is a valuable addition to the prognostic model of low-, intermediate-, and high-risk patients, particularly in the low and intermediate groups, for further selection of patients who would benefit from high-dose chemotherapy (Figure 14.3).

Cervical Cancer

Cervical carcinoma is one of the most common cancers in women (35). About 80% are squamous cell carcinomas and about 20% adenocarcinomas, in addition to other rare types of tumors. Cervical carcinoma can be detected early in the course of the disease with screening Pap smears and biopsy when appropriate. The International

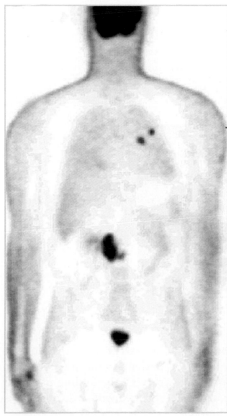
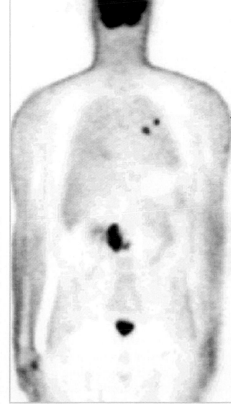

Figure 14.2. FDG-PET image in a patient with a history of testicular cancer and a right para-aortic residual mass on CT and rising markers. He was being considered for laparotomy as definitive treatment if this was the only site of disease. CT of the abdomen and chest was otherwise normal although he previously had lung metastases. He was referred for a PET scan to exclude other sites of disease and thereby enable surgery. The images show increased uptake in the known mass as well as disease in the lungs and mediastinum, which directly altered the patient's management. 📖

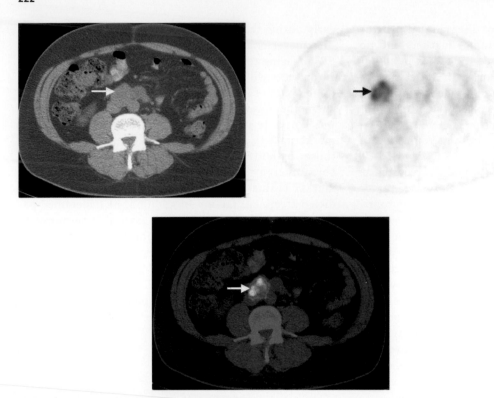

a

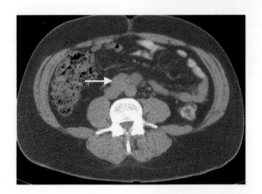

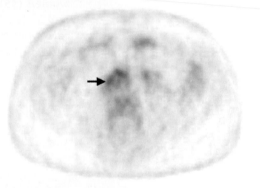

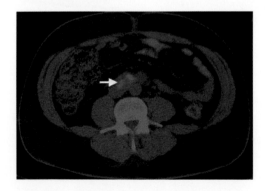

b

Figure 14.3. (a) A patient with known metastatic testicular cancer had an FDG-PET scan performed before chemotherapy. The transaxial images of CT (*top left*), PET (*top right*), and PET/CT fusion image (*lower*) showed an FDG-avid nodal lesion (*arrows*). **(b)** One week later, following chemotherapy, there has been a rapid decrease in uptake in the tumor (*arrows*), indicating an early response to chemotherapy. The patient responded well to the current course of chemotherapy.

Federation of Gynecology and Obstetrics (FIGO) (36) has defined a staging system for carcinoma of the cervix that uses a combination of clinical and radiologic findings (Table 14.2). Although prognosis is related to stage, other factors are important in determining prognosis, including the extent of lymph node involvement. For example, in patients with stage IB tumor (tumor confined to the cervix), the finding of tumor-positive lymph nodes is associated with a decrease in survival, from 85% to 95% to 45% to 55% (37–40).

Table 14.2. Staging of cervical cancer.

Stage	Description
I	Confined to the uterus
IA	Invasion carcinoma diagnosed only by microscopy
IB	Clinically visible lesions confined to the cervix or lesions more than 5 mm and with 7 mm or greater horizontal spread
II	Tumor invades beyond the uterus but not to pelvic wall or lower third of vagina
IIA	Without parametrial invasion
IIB	With parametrial invasion
III	Tumor extends to pelvic wall and/or involves lower one-third of vagina and/or causes hydronephrosis or nonfunctioning kidney
IIIA	Tumor involves lower third of the vagina, no extension to pelvic wall
IIIB	Tumor extends to pelvic wall, and/or causes hydronephrosis or nonfunctioning kidney
IV	Cancer has spread to the bladder, rectum, or outside the pelvis
IVA	Tumor invades mucosa of bladder or rectum and/or extends beyond true pelvis
IVB	Distant metastases

Source: Reprinted from FIGO Committee on Gynaecological Oncology: Benedet JL, Bender H, Jones H III, Ngan HYS, Pecorelli S. FIGO staging classifications and clinical practice guidelines in the management of gynecologic cancers. Int J Gynecol Obstet 2000;70:209–262. Copyright © 2000, with permission from International Federation of Gynecology and Obstetrics.

Tumor Staging

Imaging Procedures in Tumor Staging

Definition of lymph node status is important. Currently, the best methods for defining the status of lymph nodes are lymphangiography, CT, and MRI scanning. As these procedures are not widely available, FIGO has not included them to the staging criteria (36). Lymphangiography has been widely used in the past for assessment of lymph nodes but suffers from technical problems and may have high false-positive rates (41–43), as well as false-negative findings because of nonopacification in lymph nodes that are totally replaced by tumor (44). The problem in lymph node staging by anatomic imaging using CT or MRI is the need to use anatomic criteria for tumor detection. As with other tumors, small (less than 1 cm) lymph nodes may contain tumor and enlarged (greater than 1 cm) lymph nodes may be reactive; this results in sensitivities as low as 34% (45) for detection of lymph node metastases by CT and sensitivity and specificity for MR of 38% to 89% and 78% to 99%, respectively (44–47). Attempts have been made to improve the results in MR by contrast enhancement and by the use of circular polarized phased-array coil, but without success (47). A meta-analysis of studies that evaluated lymphangiography, CT, and MR showed that all perform similarly with regard to detection of lymph node metastases. The less-invasive nature of CT/MR as well as the extra anatomic information provided on tumor extent has made the imaging procedures preferable (44).

These problems with conventional imaging have led to evaluation of FDG-PET as an alterative for the staging of cervical carcinoma and for the evaluation of lymph nodes in particular. Several studies that have examined the performance of PET imaging compared either CT or MRI for initial staging of patients with cervical cancer diagnosed both at early and at more-advanced stages (48–53). Overall, the sensitivity for FDG-PET imaging ranged from 83% to 100% and the specificity from 89% to 100%. This sensitivity for CT/MR in the same studies ranged from 50% to 73%. In studies where PET was evaluated in patients with negative CT/MR (48, 50, 51), the sensitivity and specificity of PET for detection of metastases were 83.3% to 85.7% and 94.4% to 96.7%, respectively. There was some difference between studies in early- and late-stage disease.

Rose et al. (48) studied a group of 32 patients with stage IIb–IVA cervical cancer presurgery and used PET to evaluate pelvic and paraaortic lymph nodes that were negative by CT. In 17 patients who underwent pelvic nodal resection, PET detected nodal metastases in 11 of 17 patients, with no false-positive or false-negative results. Rose et al. (48) analyzed the data according to identification of pelvic and paraortic disease. FDG-PET was less effective in the paraaortic region. Overall, PET in the paraaortic region had a PPV of 75% and NPV of 92%, with positive PET indicating a relative risk of 0.9 for paraaortic metastasis.

Reinhardt et al. (49) compared PET and MRI in patients with stage IB–IIA disease. On a patient-by-patient basis, the PPV for nodall staging by PET and MRI was 100% and 67%, respectively. However, this difference was not found to be statistically significant. These investigators also evaluated the detection of nodal metastases on a site-by-site basis and found PPV for PET and MRI as 90% and 64%, respectively (P less than 0.05; Fischer exact test). Unlike Rose et al. (48), they found some false-negative PET results in the pelvis and speculated that the difference in the two studies may have resulted from patient selection, with Rose et al. (48) studying more patients with advanced disease.

On the basis of the difference in PPV for detection of metastatic lymph nodes by PET and MRI (90% and 64%, respectively), FDG-PET was considered useful for planning patient management. Any patient with positive lymph nodes by PET, regardless of findings by MRI, should undergo treatment of the involved nodal region, either by extended-field radiotherapy or surgical clearance (49). On the basis of this, it can be concluded that more patients would benefit from the use of FDG-PET in the staging procedure.

Both Rose et al. (48) and Reinhardt et al. (49) reported cases of enlarged lymph nodes by conventional imaging that were merely reactive, and a negative PET finding in these patients was also useful to the managing clinican. As in testicular cancer, such negative results may mean that more-limited treatment is required, thereby avoiding unnecessary morbidity.

Miller and Grisgsby (54) have evaluated the usefulness of tumor volume measurement with PET in patients with advanced cervical cancer treated by radiation therapy. They concluded the following: (1) tumor volume can be accurately measured by PET; (2) tumor volume separates patients with a good prognosis from those with a poorer prognosis; (3) a subset of patients with relatively small tumors and no lymph node involvement does remarkably well; and (4) tumor volume does not correlate with the presence of lymph node disease.

The same group of investigators (55) have evaluated a treatment planning method for dose escalation to the paraaortic lymph nodes based on PET with intensity-modulated radiotherapy (IMRT) for cervical cancer patients with paraaortic lymph node involvement. They subsequently determined the guidelines regarding the selection of appropriate treatment parameters (56).

Miller et al. (57) developed a simple, rapid, and highly reproducible system for visual grading of characteristics of the primary tumor in patients with cervical cancer at the time of diagnosis. This approach allowed separation of patients with a poor prognosis from those who will do well, thus providing a new tool for accurate estimation of prognosis. In addition, as this did not correlate with lymph node status, it provides a potentially independent predictor of outcome. Singh et al. (58) evaluated the outcome of patients with FIGO clinical stage IIIb cervical carcinoma as a function of site of initial regional lymph node metastasis as detected by FDG-PET. They concluded that the cause-specific survival in this group was highly dependent on the extent of lymph node metastasis as identified on FDG-PET.

The performance of FDG-PET imaging for evaluation of patients with cervical carcinoma pre- and postradiotherapy allows predicting reponse to treatment and overall outcome. Grisgby et al. (59) studied 152 patients with cervical cancer who underwent radiotherapy and/or chemotherapy with pre- and posttreatment FDG-PET imaging. The 5-year survival of patients with negative FDG-PET posttherapy was 80%, whereas patients with positive FDG-PET scans (at previous or new sites) had a 5-year survival of 32%. They concluded that persistent or new FDG uptake on the posttherapy scan was the most significant prognostic factor for developing metastatic disease and for predicting death from cervical cancer.

There are some technical aspects of PET scanning for cervical cancer that need to be considered. Because of the close proximity to the bladder and the excretion of tracer in the urine, it has been thought by some investigators that problems in interpretation could arise from activity in the ureters and image artifact from the bladder. Most reported studies used continuous bladder irrigation and some also used preimaging hydration. Some have used

vigorous hydatrion and furosemide as well (60). Others investigators have not found these patient manipulations to be necessary (49). Sugawara et al. (52) imaged patients pre- and postvoid without patient interventions and found 100% sensitivity for tumor detection on postvoid images. With integrated PET/CT imaging, precise localization of FDG uptake should facilitate the differentiation of physiologic urine activity from pathologic activity in tumor.

Tumor Recurrence

Imaging Procedures in Tumor Recurrence

There are limited data on the place of FDG-PET imaging in recurrent cervical cancer. Studies have focused particularly on patients without radiologic evidence of recurrence (also called asymptomatic or disease free). Three studies (61–63) have shown sensitivity ranging from 80% and 90.3%, with specificity from 76.1% to 100%, which is of particular value when other imaging is normal or equivocal. Ryu et al. (62) evaluated PET in different body regions and found that the sensitivity of PET imaging was high in mediastinal, hilar, and scalene lymph nodes and in liver and spine, but was relatively low in lungs and retrovesical and paraaortic lymph nodes.

In patients with documented (also called symptomatic) recurrence, PET imaging may also be of value (61, 64). Unger et al. (61) evaluated 21 PET scans in women symptomatic of recurrence and found PET to have a sensitivity of 100% and a specificity of 85.7%. Lai et al. (64) prospectively examined 44 patients with recurrent disease and compared restaging with PET and CT/MR. PET imaging was superior to CT/MR in the overall detection of lesions and in the identification of metastatic disease. In addition, PET altered management in 55% of patients. They concluded that in patients with recurrence who are candidates for salvage treatment by CT/MR crtiteria, adding PET imaging significantly reduced unnecessary salvage therapies. Yen et al. (65) defined a prognostic scoring system using PET for patients with recurrent cervical cancer that allowed ranking patients into three groups. They concluded that, using this risk score, FDG-PET may offer maximal benefits by selecting appropriate recurrent cervical cancer patients for salvage therapy with precise restaging information.

As with other tumors, timing of the acquisition of the PET images postadministration of FDG may be important, and a distribution period of 1 h for FDG may not be optimal (27, 28). Ma et al. (66) studied patients with primary and recurrent cervical cancer at 40 min and 3 h after injection of FDG. On the 3-h-delayed images, additional lower paraaortic lymph nodes were detected, and some FDG-avid lesions at 40 min demonstrated decreased FDG uptake at 3 hours, allowing classifying them as benign.

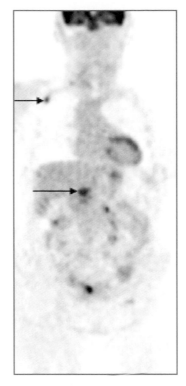
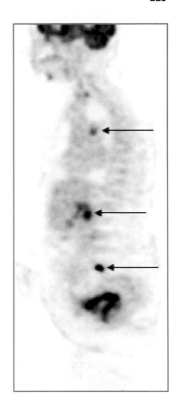

Figure 14.4. A 79-year-old women with carcinoma of the cervix and known involvement of the left ureter was treated with surgery and radiotherapy and considered to be cured. Nine years later, she presented with gross hematuria. CT showed a complex mass at the vaginal vault, and the biopsy was negative. A PET was performed to assess local and possible metastatic disease. The coronal (*left*) and sagittal (*right*) FDG-PET images showed widespread metastases (*arrows*) including the right pectoral region, the lung hilum, the porta hepatis, and retroperitoneal lymph nodes in the abdomen. Biopsy confirmed metastatic disease.

When cervical cancer recurs, PET imaging may also be useful in differentiating tumor from fibrosis after surgery or radiotherapy, a common problem with anatomic imaging. The ability to coregister images from different imaging modalities using software package or integrated PET/CT systems (Figure 14.4) provides fusion images allowing identification of FDG-avid anatomic lesions and guidance of biopsies. Further studies need to be done to define the role of integrated PET/CT imaging compared to PET alone role in patients with cervical cancer.

[11]C-Methionine has been used to successfully to image cervical carcinoma although, because of physiologic uptake of tracer, correlation with anatomic imaging is required (67). It is possible that this procedure may find a place imaging the difficult-to-manage patient.

Ovarian Carcinoma

Ovarian cancer has a high mortality. The tumor may be asymptomatic until quite late (68), and there are no readily available screening tests similar to those that are available for detection of cervical cancer. At primary diagnosis, accurate staging is an important in determinig the prognosis (69), but no imaging modality has as yet provided accurate staging information. The FIGO staging system is presented in Table 14.3 (36). In recurrent disease, early detection and detection of peritoneal carcinomatosis are important for appropriate selection of patients for second-line salvage therapy. Similar to testicular

Table 14.3. Staging of ovarian cancer.

Stage	Description
I	Tumor confined to ovaries
IA	Tumor limited to one ovary, with the capsule, no tumor on surface, no malignant cells in ascites or peritoneal washings
IB	Tumor limited to both ovaries, but otherwise as above
IC	Tumor limited to one or both ovaries with any of the following capsule ruptured, tumor on surface, positive malignant cells in ascites or peritoneal washings
II	Tumor involves one or both ovaries with pelvic extension
IIA	Extension/implants in uterus and/or tubes, no malignant cells in ascites or peritoneal washings
IIB	Extension to other pelvic organs, no malignant cells in ascites or peritoneal washings
IIC	Either IIA or IIB, with malignant cells in ascites or peritoneal washings
III	Tumor involves one or both ovaries with microscopically confirmed peritoneal metastases outside the pelvis and/or regional lymph node metastases
IIIA	Microscopic peritoneal metastases beyond the pelvis
IIIB	Macroscopic peritoneal metastases beyond the pelvis 2 cm or less in diameter
IIIC	Peritoneal metastasis beyond the pelvis more than 2 cm in diameter and/or regional lymph node metastases
IV	Distant metastases beyond the peritoneal cavity

Source: Reprinted from FIGO Committee on Gynaecological Oncology: Benedet JL, Bender H, Jones H III, Ngan HYS, Pecorelli S. FIGO staging classifications and clinical practice guidelines in the management of gynecologic cancers. Int J Gynecol Obstet 2000;70:209–262. Copyright © 2000, with permission from International Federation of Gynecology and Obstetrics.

cancer, ovarian cancer does have a biochemical marker in CA-125, and this has been useful in the monitoring of patients after treatment. CA-125 levels can, however, be falsely elevated in a number of conditions, and in early-stage disease it is elevated in fewer than half the patients (70).

Ultrasound, CT, and MR imaging have all been used in the assessment of ovarian cancer, both for assessment of adnexal masses and for detection of recurrent disease. Ultrasound has been used in primary disease but lacks specificity (71). CT and MR are not accurate for defining small-volume primary disease. Even in recurrent tumor, the sensitivity of CT has been reported to be 40% to 93% (72–74) and, because of its low NPV (45%–50%) (75, 76), CT cannot replace second-look laparotomy. Much recent work had focused on FDG uptake in both primary and recurrent disease.

Diagnosis and Tumor Staging

Imaging Procedures in Tumor Staging

In the evaluation of primary disease, FDG-PET can identify tumor with high sensitivity and specificity (both, 90%). However, PET fails to detect small stage I cancers and tumors that are histologically borderline for malignancy (77, 78). False-positive FDG uptake may be seen in some benign lesions, including ovarian endometriosis and corpus luteal cysts (79). SUV evaluation of lesions has been performed and, although an SUV greater than 7.9 is a strong indicator of malignancy (79), it does not improve diagnostic accuracy, especially in the differentiation of benign from borderline tumors (77). These studies have all been performed at early imaging times, and it remains to be seen whether delayed scanning permits better differentiation of benign from malignant, as seen in other tumors (27, 28). ^{11}C-Methionine was found to be less dependent on tumor type (80). Physiologic uptake was a limiting factor for staging, but there may be potential for imaging with this tracer, especially when combined with FDG.

PET, MRI, and ultrasound have been compared for the evaluation of adnexal masses in several studies (78, 81, 82) as well as MR versus PET after identification of a mass on ultrasound (83). Generally, ultrasound was more sensitive than PET or MR but less specific (77, 80). Grab et al. (78) concluded that the best specificity without loss of sensitivity was obtained when all three were combined. However, because negative PET and MRI findings did not exclude stage I or borderline disease, ultrasound was the most appropriate screening tool in the diagnosis of ovarian cancer. Furthermore, it was concluded that any mass that was suspected of being malignant on ultrasound evaluation required surgical evaluation.

Adding an FDG-PET study to the preoperative assessment of patients at diagnosis improves the accuracy of staging. Yoshida et al. (84) found that in 15 patients CT agreed with pathology in 53% whereas CT and FDG-PET agreed in 87%.

Tumor Recurrence

Imaging Procedures in Tumor Recurrence

FDG-PET may be more useful in recurrent disease and restaging than in primary ovarian cancer. Many small studies have combined groups with suspected recurrence and unsuspected recurrence and compared PET versus conventional imaging and tumor markers. The overall sensitivity and specificity of PET was in the range of 80% to 100% (85–89). PET did, however, miss lesions less than 1cm and microscopic metastatic disease (77). PET was most valuable when conventional imaging was negative or equivocal and tumor markers were raised, with a sensitivity ranging from 87% to 96% (85, 86) (Figure 14.5). The abnormalities on PET preceded those on conventional imaging where markers were raised by as much as 6 months, which reflects a similar situation as described by Hain et al. (22) in testicular cancer and Valk et al. (90) for raised CEA in colorectal cancer. PET may have an important role in monitoring these patients after initial treatment.

Integrated PET/CT imaging is becoming more available, and Makhija et al. (91) reported findings on eight patients, six with ovarian and two with fallopian tube tumors, after cytoreductive therapy. Five of the eight patients had recurrrent disease identified on PET/CT imaging whereas CT alone was negative. Bristow et al. (92) studied a larger group of 22 patients more than 6 months after initial therapy. All patients presented with rising CA-125 and negative or equivocal CT and underwent PET/CT imaging followed by surgery. They found that combined PET/CT imaging was valuable, with overall patient-based sensitivity, specificity, and accuracy of 83%, 75%, and 82%, respectively, and lesion-based sensitivity, specificity, and accuracy of 60%, 95%, and 72%, respectively. Although promising, as suggested by Pannu et al. (93) following their study, a larger trial is necessary to judge the impact of PET/CT on clinical practice.

The treatment of ovarian cancer is limited by the inability to diagnose early peritoneal tumor spread. Many patients undergo second-look laparotomy, which is not without morbidity, because imaging is not sufficiently sensitive for detection of peritoneal tumors. The appropriate use of salvage chemotherapy also depends on accurate assessment of disease extent. Anatomic imaging modalities have been poor at imaging peritoneal carcinomatosis. Results with PET for detection of peritoneal carcinomatosis have been conflicting. Rose et al. (94) found that PET had a sensitivity and specificty of 10% and 42%, respectively, suggesting little value. Later studies (95, 96) have been more favorable. Kim et al. (96) found that there

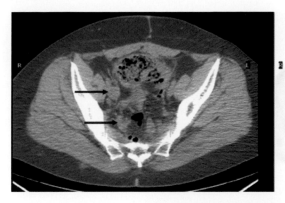
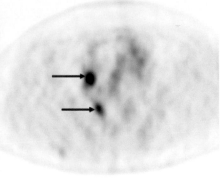

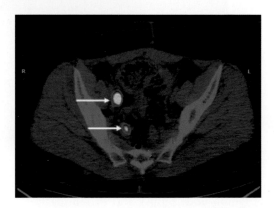

Figure 14.5. A 40-year-old woman was diagnosed with ovarian cancer 20 months earlier and was treated with radical surgery including hysterectomy, bilateral salpingo-oophorectomy, and anterior resection with end-colostomy. Postoperatively she had a right pelvic abscess that had resolved. She was found on routine follow-up to have a rising CA-125 tumor marker. Magnetic resonance imaging (MRI) was stable without evidence of recurrence. The transaxial images of CT (*top left*), PET (*top right*), and PET/CT fusion image (*lower*) showed FDG-avid lesions in the pelvis that were proven to be recurrent disease.

was no statistical signficance in the progression and disease-free interval between patients evaluated with PET or second-look laparotomy. These data suggest that PET imaging could replace second-look laporotomy for evaluation of patients with ovarian cancer, especially those at high risk. One early study (97) evaluated the cost-effectiveness of PET in managing patients with recurrent ovarian cancer. Using Monte Carlo simulation analysis, FDG-PET imaging was found to reduce unnecessary invasive surgical staging, leading to cost savings of $US 1,941–11,766 per patient.

In assessing metastatic disease, the use of half-body scans may also facilitate identification of extrapelvic suspected or unsuspected disease. Some authors have speculated on the use of PET to monitor response to chemotherapy (77), an application that has been demonstrated in other tumors, for example, breast and testicular cancers (32–34).

Uterine Cancer

Cancer of the uterine body (endometrial) is the most common gynecologic cancer but, in contrast to cervical cancer, it occurs predominantely in postmenopausal women. Endometrial cancer metastasizes predominently by hematogenous spread and is associated with the tumor marker CA-125. Endometrial cancer has been shown to be FDG avid (Figure 14.6) but has been less studied with FDG-PET than cervical cancer. There are no published

data on the use of PET in initial staging. Two studies (98, 99) that examined its use in recurrent disease found that PET altered management in 30% to 35% of patients, often by detecting clinically and radiologically unsuspected extrapelvic metastases. Saga et al. (99) compared the performance of PET, CT/MRI, and tumor markers in the follow-up of postoperative patients with endometrial cancer. They reported a sensitivity, specificity, and accuracy of 100%, 88%, and 93% for PET, 85%, 86%, and 85% for CT/MRI, and 100%, 71%, and 83% for tumor markers.

Endometrial cancer has been imaged with ^{11}C-methionine with success. The physiologic accumulation of methionine in the pelvis provided some confusion and therefore correlative imaging would always be required (100). The use of FDG-PET for diagnosing uterine sarcoma has been reported (101). Given the avidity of FDG in soft tissue sarcoma (27), this result is not surprising.

Vulval Cancer

Primary vulvar cancer is less common than the other gynecologic tumors. It commonly spreads via the groin lymph nodes, and their status is important for treatment decisions and prognosis. There is one study evaluating FDG-PET imaging for staging of vulvar cancer. All the primaries were FDG avid, and PET was sensitive for detection of extranodal metastases but relatively insensitive for detection of inguinal lymph node involvement (102). This limitation is related to the inabilty of PET imaging to

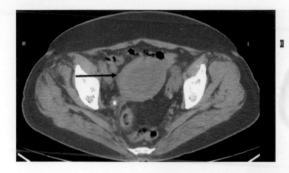

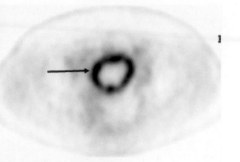

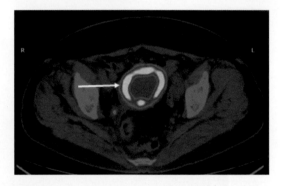

Figure 14-6. A 61-year-old women with stage IV endometrial cancer had completed chemotherapy and radiotherapy and was being considered for surgery. PET/CT imaging was performed to exclude metastatic disease. There was no evidence of disease outside the uterus. The transaxial images of CT (*top left*), PET (*top right*), and PET/CT fusion image (*lower*) showed persistent FDG uptake in the primary endometrial tumor indicating residual tumor.

detect microscopic metastases, as reported in studies comparing sentinel lymph node biopsy and PET in patients with melanoma and head and neck cancer (103, 104).

Conclusions

In testicular cancer, PET is particularly useful for detection of viable tumor in posttreatment residual masses and for localization of marker-positive relapse. In both testicular and gynecologic cancer, PET can also be used to identify primary tumors and metastatic disease, and it can be used in surveillance for tumor recurrence. Investigations regarding the role of PET in these tumors have been increasing, but larger studies are needed to define the role of PET in the staging of cancer of the testis and ovary and in the evaluation of cancer of the cervix.

The role of FDG-PET is likely to be extended with the use of integrated PET/CT to assist in localizing disease with a view to guide biopsy. Further assessment of its role in determining and predicting response to treatment may have direct effects and benefits for individual patient management.

References

1. Mead GM. Testis. In: Price P, Sikora K, editors. Treatment of Cancer, 3rd ed. Chapman & Hall, London, 1995:627–645.
2. Dearnley DP, Huddart RA, Horwich A. Managing testicular cancer. Br Med J 2001;322:1583–1588.
3. Wishnow KI, Johnson DE, Swanson, et al. Identifying patients at low risk clinical stage I testicular tumoirs who should be treated by surveillance. Urology 1989;34:339–343.
4. Read G, Stenning SP, Cullen MH, et al. Medical Research Council prospective study of surveillance for Stage I testicular teratoma. J Clin Oncol 1992;10:1762–1768.
5. Horwich A. Testicular cancer. In: Horwich A, editor. Oncology: A Multidisciplinary Textbook. London: Chapman & Hall 1995:485–498.
6. Cullen MH. Adjuvant chemotherapy in high risk stage I non-seminomatous germ cell tumours of the testis. In: Horwich A, editor. Testicular Cancer: Investigation and Management. London: Chapman & Hall 1996:181–191.
7. Huddart R, Norman A, Shahidi M, et al. cardiovascular disease as a long-term complication of treatment for testicular cancer. J Clin Oncol 2003;21;1513–1523.
8. Warde P, Jewett MA. Surveillance for stage I testicular seminoma. Is it a good option? Urol Clin N Am 1998;25:425–433.
9. McLeod DG, Weiss RB, Stablein DM, et al. Staging relationships and outcome in early stage testicular cancer: a report from the Testicular Cancer Intergroup Study. J Urol 1991;145:1178–1183.
10. Fernandez EB, Moul JW, Foley JP, Colon E, McLeod DG. Retroperitoneal imaging with third and fourth generation computed axial tomography in clinical stage I nonseminomatous germ cell tumours. Urology 1994;44:548–552.
11. Hain SF, O'Doherty MJ, Timothy , et al. Fluorodeoxyglucose PET in the initial staging of germ cell tumours. Eur J Nucl Med 2000;27:590–594.
12. Tsatalpas P, Beuthien-Baumann B, Kropp J, et al. Diagnostic value of ¹⁸F-FDG positron emission tomography for detection and treatment control of malignant germ cell tumors. Urol Int 2002;68:157–163.
13. Cremerius U, Wildberger JE, Borchers H, et al. Does positron emission tomography using 18-fluoro-2-deoxyglucose improve clinical staging of testicular cancer? Results in 50 patients. Urology 1999;54:900–904.
14. Albers P, bender H, Yilmaz H, et al. Positron emission tomography in the clinical staging of patients with stageI and II testicular germ cell tumours. Urology 1999;53:808–811.

15. Lassen U, Daugaard G, Eigtved A, et al. Whole body FDG-PET in patients with stage I non-seminomatous germ cell tumours. Eur J Nucl Med Mol Imaging 2003;30:396–402.

16. Mostofi FK, Spaander P, Grigor K, Parkinson CM, Skakkekaek NE, Oliver RTD. Consensus on pathological classifications of testicular tumours. In: Prostate Cancer and Testicular Cancer. EORTC Genitourinary Group Monograph 7. New York: Wiley-Liss, 1990:267–277.

17. Javadpour N. Current status of tumor markers in testicular cancer. A practical review. Eur Urol 1992;21:34–36.

18. Kalofonos HP, Kosmas C, Pawlikowska TR, Bamias A, Snook D, Dhokia B, et al. Immunolocalisation of testicular tumours using radiolabelled monoclonal antibody to placental alkaline phosphatase. J Nucl Med Allied Sci 1990;34:294–298.

19. Warren GP. Gallium scans in the evaluation of residual masses after chemotherapy for seminoma. J Clin Oncol 1995;13:2784–2788.

20. Uchiyama M, Kantoff PW, Kaplan WD. Gallium-67 citrate imaging extragonadal and gonadal seminomas: relationship to radiologic findings. J Nucl Med 1994;35:1624–1630.

21. Stephens AW, Gonin R, Hutchins GD, Einhorn LH. Positron emission tomography of residual radiological abnormalities in postchemotherapy germ cell tumour patients. J Clin Oncol 1996;14:1637–1641.

22. Hain SF, O'Doherty MJ, Timothy AR, et al. Fluorodeoxyglucose positron emission tomography in the evaluation of germ cell tumours at relapse. Br J Cancer 2000;83:863–869.

23. Cremerius U, Effert PJ, Adam G, Sabri O, Zimmy M, Wagenkneckt GG, et al. FDG PET for detection and therapy control of metastatic germ cell tumour. J Nucl Med 1998;39:815–822.

24. Sugawara Y, Zasadny KR, Grossman HB, et al. Germ cell tumour: differentiation of viable tumor, mature teratoma and necrotic tissue with FDG PET and kinetic modelling. Radiology 1999;211:249–256.

25. Duchesne GM, Stenning SP, Aass N, Mead GM, Fossa SD, Oliver RTD. Minimal benefit from radiotherapy after chemotherapy for metastatic seminoma: analysis of Medical Research Council (MRC-UK) database. Proc ASCO 1997;16:318.

26. De Santis M, Becherer A, Bokemeyer C, et al. 2-[18]Fluoro-2-deoxy-D-glucose positron emission tomography is a reliable predictor for viable tumor in postchemotherapy seminoma: an update of the prospective multicentric SEMPET trial. J Clin Oncol 2004;22:1034–1039.

27. Lodge MA, Lucas JD, Marsden PK, et al. A PET study of [18]FDG uptake in soft tissue masses. Eur J Nucl Med 1999;26:22–30.

28. Boerner AR, Weckesser M, Herzog H, et al. Optimal scan time for fluorine-18 fluorodeoxyglucose positron emission tomography in breast cancer. Eur J Nucl Med 1999;26:226–230.

29. Rathmall AJ, Brand IR, Carey BM, Jones WG. Early detection of relapse after treatment for metastatic germ cell tumour of the testis: an exercise in medical audit. Clin Oncol R Coll Radiol 1993;5:34–38.

30. Coogan CL, Foster RS, Rowland RG, et al. Postchemotherapy retroperitoneal lymph node dissection is effective therapy in selected patients with elevated tumor markers after primary chemotherapy alone. Urology 1997;50:957–962.

31. Mikhaeel NG, Timothy AR, O'Doherty MJ, Hain S, et al. 18-FDG-PET as a prognostic indicator in the treatment of aggressive non-Hodgkin's lymphoma: comparison with CT. Leuk Lymphoma 2000;39:543–553.

32. Schelling M, Avril N, Nahrig J, Kuhn W, et al. Positron emission tomography using [(18)F]fluorodeoxyglucose for monitoring primary chemotherapy in breast cancer. J Clin Oncol 2000;18:1689–1695.

33. Smith IC, Welch AE, Hutcheon AW, Miller ID, et al. Positron emission tomography using [(18)F]-fluorodeoxy-D-glucose to predict the pathologic response of breast cancer to primary chemotherapy. J Clin Oncol 2000;18:1676–1688.

34. Bokemeyer C, Kollmannsberger C, Oechsle K, Dohmen BM, et al. Early prediction of treatment response to high-dose salvage chemotherapy in patients with relapsed germ cell cancer using [(18)F]FDG PET. Br J Cancer 2002;86:506–511.

35. Parkin DM, Laara E, Muir CS. Estimates of worldwide frequency of sixteen major cancers in 1990. Int J Cancer 1998;80:827–841.

36. FIGO Committee on Gynaecological Oncology: Benedet JL, Bender H, Jones H III, Ngan HYS, Pecorelli S. FIGO staging classifications and clinical practice guidelines in the management of gynecologic cancers. Int J Gynecol Obstet 2000;70:209–262.

37. Kjorstad KE, Kolbenstvedt A, Strickert T. The value of complete lymphadenectomy in radical treatment of cancer of the cervix, stage IB. Cancer (Phila) 1984;54:2215–2219.

38. Alvarez RD, Potter ME, Soong SJ, Gay FL, Hatch KD, Partridge EE, Shingleton HM. Rationale for using pathologic tumor dimensions and nodal status to subclassify surgically treated stage IB cervical cancer patients. Gynecol Oncol 1991;43:108–112.

39. Delgado G, Bundy B, Zaino R, Sevin BU, Creasman WT, Major F. Prospective surgical-pathological study of disease-free interval in patients with stage IB squamous cell carcinoma of the cervix: a Gynecologic Oncology Group study. Gynecol Oncol 1990;38:352–357.

40. Steren A, Nguyen HN, Averette HE, Estape R, Angioli R, Donato DM, et al. Radical hysterectomy for stage IB adenocarcinoma of the cervix: the University of Miami experience. Gynecol Oncol 1993;48:355–359.

41. Piver MS, Barlow JJ. Para-aortic lymphadenectomy, aortic node biopsy, and aortic lymphangiography in staging patients with advanced cervical cancer. Cancer (Phila)1973;32:367–370.

42. Brown C, Buchsbaum HJ, Tewfik HH, et al. Accuracy of lymphangiography in the diagnosis of paraaortic lymph node metastases from carcinoma of the cervix. Obstet Gynecol 1979;54:571–575.

43. Pendlebury SC, Cahill S, Crandon AJ, et al. Role of bipedal lymphangiogram in radiation treatment planning for cervix cancer. Int J Radiat Oncol Biol Phys 1993;27:959–962.

44. Scheidler J, Hrick H, Yu KK, et al. Radiological evaluation of lymph node metastases in patients with cervical cancer. A meta-analysis. JAMA 1997;278:1096–1101.

45. Heller PB, Malfetano JH, Bundy BN, et al. Clinical-pathologic study of stage IIB, III, and IVA carcinoma of the cervix. Extended diagnostic evaluation for paraaortic node metastasis: a Gynecologic Oncology Group study. Gynecol Oncol 1990;38:425–430.

46. Kim SH, Kim SC, Choi BI, Han MC. Uterine cervical carcinoma: evaluation of pelvic lymph node metastasis with MR imaging. Radiology 1994;190:807–811.
Kim SH, Choi BI, Han JK, Kim HD, Lee HP, Kang SB, et al. Preoperative staging of uterine cervical carcinoma: comparison of CT and MRI in 99 patients. J Comput Assist Tomogr 1993;17(4):633–640.

47. Hawighorst H, Schoenberg SO, Knapstein PG, Knopp MV, Schaeffer U, Essig M, van Kaick G. Staging of invasive cervical carcinoma and of pelvic lymph nodes by high resolution MRI with a phased-array coil in comparison with pathological findings. J Comput Assist Tomogr 1998;22:75–81.

48. Rose PG, Adler LP, Rodriguez M, et al. PET for evaluating para-aortic nodal metastases in locally advanced cervical cancer before surgical staging: a surgicopathological study. J Clin Oncol 1999;17:41–45.

49. Reinhardt MJ, Ehritt-Braun C, Vogelgesang D, et al. Metastatic lymph nodes in patients with cervical cancer; detection with MR imaging and FDG-PET. Radiology 2001;218:776–782.

50. Yeh LS, Hung YC, Shen YY, Kao CH, Lin CC, Lee CC. Detecting para-aortic lymph nodal metastasis by positron emission tomography of ^{18}F-fluorodeoxyglucose in advanced cervical cancer with negative magnetic resonance imaging findings. Oncol Rep 2002;9:1289–1292.

51. Lin WC, Hung YC, Yeh LS, Kao CH, Yen RF, Shen YY. Usefulness of (18)-F-fluorodeoxyglucose positron emission tomography to

detect para-aortic lymph nodal metastasis in advanced cervical carcinoma with negative computed tomography findings. Gynecol Oncol 2003;89:73–76.

52. Sugawara Y, Eisbruch A, Kosuda S, Recker BE, Kison PV, Wahl RL. Evaluation of FDG PET in patients with cervical cancer. J Nucl Med 1999;40:1125–1131.

53. Narayan K, Hicks RJ, Jobling T, Bernshaw D, McKenzie AF. A comparison of MRI and PET scanning in surgically staged loco-regionally advanced cervical cancer: potential impact on treatment. Int J Gynecol Cancer 2001;11:263–271.

54. Miller TR, Grigsby PW. Measurement of tumor volume by PET to evaluate prognosis in patients with advanced cervical cancer treated by radiation therapy. Int J Radiat Oncol Biol Phys 2003;53:353–359.

55. Esthappan J, Mutic S, Malyapa RS, et al. Treatment planning guidelines regarding the use of CT/PET guided IMRT for cervical carcinoma with positive paraaortic lymph nodes. Int J Radiat Oncol Biol Phys 2004;58:1289–1297.

56. Mutic S, Malyapa RS, Grigsby PW, et al. PET-guided IMRT for cervical carcinoma with positive para-aortic nodes: a dose escalation treatment planning strategy. Int J Radiat Oncol Biol Phys 2003;56:489–493.

57. Miller TR, Pinkus E, Dehdashti F, Grigsby PW. Improved prognostic value of ^{18}F-FDG PET using a simple visual analysis of tumor characteristics in patients with cervical cancer. J Nucl Med 2003;44(2):192–197.

58. Singh AK, Grigsby PW, Dehdashti F, Herzog TJ, Siegel BA. FDG-PET lymph node staging and survival of patients with FIGO stage IIIb cervical carcinoma. Int J Radiat Oncol Biol Phys 2003;56(2):489–493.

59. Grigsby PW, Siegel BA, Dehdashti F, Rader J, Zoberi I. Posttherapy ^{18}F] fluorodeoxyglucose positron emission tomography in carcinoma of the cervix: response and outcome. J Clin Oncol 2004;22:2167–2171.

60. Leisure GP, Vesselle HJ, Faulhaber PF, et al. Technical improvements in fluorine-18-FDG PET imaging of the abdomen and pelvis. J Nucl Med Technol 1997;25:115–119.

61. Unger JB, Ivy JJ, Connor P, et al. Detection of recurrent cervical cancer by whole-body FDG PET scan in asymptomatic and symptomatic women. Gynecol Oncol 2004;94:212–216.

62. Ryu SY, Kim MH, Choi SC, Choi CW, Lee KH. Detection of early recurrence with ^{18}F-FDG PET in patients with cervical cancer. J Nucl Med 2003;44:347–352.

63. Havrilesky LJ, Wong TZ, Secord AA, Berchuck A, Clarke-Pearson DL, Jones EL. The role of PET scanning in the detection of recurrent cervical cancer. Gynecol Oncol 2003;90:186–190.

64. Lai Ch, Huang KG, See LC, et al. Restaging of recurrent cervical carcinoma with dual-phase ^{18}F]-fluoro-2-deoxy-D-glucose positron emission tomography. Cancer (Phila) 2004;100:544–552.

65. Yen T-C, See L-C, Chang T-C, et al. Defining the priority of using ^{18}F-FDG PET for recurrent cervical cancer. J Nucl Med 2004;45:1632–1638.

66. Ma SY, See LC, Lai CH, et al. Delayed (18)F FDG PET for the detection of paraaortic lymph node metastases in cervical cancer patients. J Nucl Med 2004;44:1775–1783.

67. Lapela ML, Leskinen-Kallio S, Varpula M, et al. Imaging of uterine carcinoma by carbon-11-methionine and PET. J Nucl Med 1994;35:1618–1623.

68. Cannistra SA. Cancer of the ovary. N Engl J Med 1993;329:1550–1559.

69. Omura GA, Brady MF, Homesley HD, et al. Long-term follow-up and prognostic factor analysis in advanced ovarian carcinoma: the Gynecological Oncology Group experience. J Clin Oncol 1991;9:1138–1150.

70. Jacobs I, Bast RC Jr. The CA 125 tumour-associated antigen: a review of the literature. Hum Reprod 1989;4:1–12.

71. Outwater EK, Dunton CJ. Imaging of ovary and adnexa: clinical issues and applications of MR imaging. Radiology 1995;194:1–18.

72. Warde P, Ridenour DF, Herman S. Computed tomography in advanced ovarian cancer: inter- and intraobserver reliability. Invest Radiol 1986;21:31–33.

73. Buy J-N, Ghossian MA, Schiot C, Bazot M, et al. Epithelial tumours of the ovary: CT findings and correlation with US. Radiology 1991;178:811–818.

74. Nelson BE, Rosenfield A, Schwartz PE. Preoperative abdominopelvic computed tomographic prediction of optimal cytoreduction in epithelial ovarian carcinoma. J Clin Oncol 1993:166–172.

75. Lund B, Jacobsen K, Rasch I. Correlation of abdominal ultrasound and computed tomography scans with second or third look laparotomy in patients with ovarian carcinoma. Gynecol Oncol 1990;37:279–283.

76. Pestasides D, Kayianni H, Facou A, et al. Correlation of abdominal computed tomography scanning and second-look operation findings in with ovarian cancer patients Am J Clin Oncol 1991;14:457–462.

77. Schroder W, Zimny M, Rudlowski C, Bull U, Rath W. The role of 18F-fluoro-deoxyglucose positron emission tomography (18F-FDG PET) in diagnosis of ovarian cancer. Int J Gynecol Cancer 1999;9:117–122.

78. Grab D, Flock F, Stohr I, et al. Classification of asymptomatic adnexal masses by ultrasound, magnetic resonance imaging and positron emission tomography. Gynecol Oncol 2000;77:454–459.

79. Lerman H, Metser U, Grisaru D, Fishman A, Lievshitz G, Even-Sapir E. Normal and abnormal 18F-FDG endometrial and ovarian uptake in pre- and postmenopausal patients: assessment by PET/CT. J Nucl Med 2004;45:266–271.

80. Lapela ML, Leskinen-Kallio S, Varpula M, et al. Metabolic imaging of ovarian tumours with carbon-11-methionine: A PET study. J Nucl Med 1995;36:2196–2200.

81. Reiber A, Nussle K, Stohr I, et al. Preoperative diagnosis of ovarian tumours with MR imaging: comparison with transvaginal sonography, positron emission tomography and histological findings. Am J Roentgenol 2001;177:123–129.

82. Fenchel S, Grab D, Nuessle K, et al. Asymptomatic adnexal masses: correlation of FDG PET and histopathological findings. Radiology 2002 223:780–788.

83. Kawahara K, Yoshida Y, Kurokawa T, et al. Evaluation of positron emission tomography with tracer 18-fluorodeoxyglucose in addition to magnetic resonance imaging in the diagnosis of ovarian cancer in selected women after ultrasonography. J Comput Assist Tomogr 2004;28:505–516.

84. Yoshida Y, Kurokawa T, Kawahara T, et al. Incremental benefit of FDG positron emission tomography over CT alone for the preoperative staging of ovarian cancer. Am J Roentgenol 2004;182:227–233.

85. Zimny M, Siggelkow W, Schroder W, et al. 2-[Fluorine-18]-fluoro-2-deoxy-D-glucose positron emission tomography in the diagnosis of recurrent ovarian cancer. Gyncol Oncol 2001;83:310–315.

86. Torizuka T, Nobezawa S, Kanno T, et al. Ovarian cancer recurrence: role of whole body positron emission tomography using 2-[fluorine-18]-fluoro-2-deoxy-D-glucose. Eur J Nucl Med Mol Imaging 2002;29:797–803

87. Jimenez-Bonilla J, Maldonado A, Morales S, et al. Clinical impact of 18F-FDG-PET in the suspicion of recurrent ovarian carcinoma based on elevated tumour serum marker levels. Clin Posit Imaging 2000;3:231–236.

88. Yen RF, Sun SS, Shen YY, Changlai SP, Kap A. Whole body positron emission tomography with 18F-fluoro-2-deoxyglucose for the detection of recurrent ovarian cancer. Anticancer Res 2001;21:3691–3694.

89. Nakamoto Y, Saga T, Ishimori T, et al. Clinical value of positron emission tomography with FDG for recurrent ovarian cancer. Am J Roentgenol 2001;176:1449–1454.

90. Valk PE, Abella-Columna E, Haseman MK, et al. Whole-body PET imaging with ^{18}F]fluorodeoxyglucose in management of recurrent colorectal cancer. Arch Surg 1999;134:503–511.

91. Makhija S, Howden N, Edwards R, Kelley J, Townsend DW, Meltzer CC. Positron emission tomography/computed tomography imaging for the detection of recurrent ovarian and fallopian tube carcinoma: a retrospective review. Gynecol Oncol 2002;85:53–58

92. Bristow RE, del Carmen MG, Pannu HK, et al. Clinically occult recurrent ovarian cancer: patient selection for secondary cytoreduction surgery using combined PET/CT. Gynecol Oncol 2003;90:519–528.

93. Pannu HK, Cohade C, Bristow RE, Fishman EK, Wahl L. PET/CT detection of abdominal recurrence of ovarian cancer: radiologic-surgical correlation. Abdom Imaging 2004;29:398–403.

94. Rose PG, Faulhaber P, Miraldi F, Abdul-Karim FW. Positive emission tomography for evaluating a complete clinical response in patients with ovarian or peritoneal carcinoma: correlation with second-look laparotomy. Gynecol Oncol 2001;82:17–21.

95. Turlakow A, Yeung HW, Salmon AS, Macapinlac HA, Larson SM. Peritoneal carcinomatosis: role of (18)F-FDG PET. J Nucl Med 2003;44:1407–1412.

96. Kim S, Chung JK, Kang SB, et al. [¹⁸F]FDG PET as a substitute for second look laparotomy in patients with advanced ovarian carcinoma. Eur J Nucl Med Mol Imaging 2004;31:196–201.

97. Smith GT, Hubner KF, McDonald T, Thie JA. Cost analysis of FDG PET for managing patients with ovarian cancer. Clin Posit Imaging 1999;2:63–70.

98. Belochine T, De Barsy C, Hustinx R, Willems-Foidart J. Usefulness of (18)F-FDG-PET in the post-therapy surveillance of endometrial carcinoma. Eur J Nucl Med Mol Imaging 2002;29:1132–1139.

99. Saga T, Higashi T, Ishimori T, et al. Clinical value of FDG-PET in the follow up of postoperative patients with endometrial cancer. Ann Nucl Med 2003;17:197–303.

100. Lapela ML, Leskinen-Kallio S, Varpula M, et al. Imaging of uterine carcinoma by carbon-11-methionine and PET. J Nucl Med 1994;35:1618–1623.

101. Umesaki N, Tanaka T, Miyama M, et al. Positron emission tomography with (18)F-fluorodeoxyglucose of uterine sarcoma: a comparison with magnetic resonance imaging and power Doppler imaging. Gynecol Oncol 2001;80(3):372–377.

102. Cohn DE, Dehdashti F, Gibb RK, et al. Prospective evaluation of positron emission tomography for the detection of groin node metastases from vulvar cancer. Gynecol Oncol 2002;85:179–184.

103. Acland KM, Healy C, Calonje E, et al. Comparison of positron emission tomography scanning and sentinel node biopsy in the detection of micrometastases of primary cutaneous malignant melanoma. J Clin Oncol 2001;19:2674–2678.

104. Hyde NC, Prvulovich E, Newman L, Waddington WA, Visirkis D, Ell P. A new approach to pre-treatment assessment of the N0 neck in oral squamous cell carcinoma: the role of sentinel node biopsy and positron emission tomography. Oral Oncol 2003;39:350–360.

15

PET and PET/CT Imaging in Melanoma

George M. Segall and Susan M. Swetter

Epidemiology

The incidence of melanoma has tripled in the United States during the past 40 years. In the year 2005, more than 59,000 Americans were expected to develop melanoma (1). The lifetime risk is now 1 in 68, and is estimated to rise to 1 in 50 by the year 2010. Melanoma is rare among black, Asian, Hispanic, and Native American people. Approximately 7,800 people die of the disease each year. Melanoma is the ninth most common cancer in the United States (2). It is the most common cancer in women who are 25 to 29 years old, and is second only to breast cancer in the 30- to 34-year-old age group. It accounts for 3.5% of cancer cases and 1.4% of cancer deaths. Despite the increased incidence of melanoma, the case-based fatality rate has decreased over the past 50 years because of earlier detection and treatment of thinner melanomas, which lack the biologic potential to metastasize (3). The average 5-year survival was 50% in 1950; by 1990, 5-year survival had increased to 90%.

Risk Factors

Melanoma occurs at sites of intermittent, intense sun exposure, in contrast to other skin cancers such as basal cell carcinoma and squamous cell carcinoma, which occur at sites of high cumulative sun exposure. The trunk is the most common site in men, whereas the back and legs are the most common sites in women. There is an increased incidence near the equator worldwide, suggesting a causal relationship with exposure to ultraviolet radiation. Other major risk factors include increased age (particularly age over 60 years), fair complexion, family history of melanoma in a first-degree relative, and the presence of atypical or dysplastic moles.

Melanoma frequently arises from precursor lesions, which include congenital melanocytic nevi (particularly "giant" varieties), common acquired nevi, dysplastic nevi,

and the in situ forms of melanoma (lentigo maligna, superficial spreading melanoma in situ, and acral lentiginous melanoma in situ). Changes in the appearance of a pigmented lesion may herald the development of melanoma; however, precursor lesions are estimated to be present in only 30% to 50% of melanomas.

Tumor Histology

There are four clinical-histologic subtypes of primary cutaneous melanoma: superficial spreading, lentigo maligna, acral lentiginous, and nodular melanoma. The first three subtypes have an in situ phase, or a period of radial, horizontal growth, where the tumor is growing laterally but not invading the dermis . Tumors in this phase have not yet acquired the potential to metastasize and are cured with excision alone. Primary cutaneous melanoma can occur anywhere on the body, but lentigo maligna melanoma occurs primarily on the face and other sun-exposed sites, whereas acral lentiginous melanoma occurs primarily on the palms, soles, and nail beds.

Superficial spreading melanoma is the most common type, accounting for 70% of cases. Most patients are between 30 and 50 years old at the time of diagnosis. A change in a preexisting mole is commonly reported. Lesions may be flat or elevated, are usually irregular with asymmetrical borders, and have variegated pigmentation. Focal regression within the lesion may cause visible hypopigmentation in spots.

Nodular melanoma is the second most common type and is seen in 15% to 30% of cases. As with superficial spreading melanoma, the legs and trunk are the most common sites. Nodular melanoma is characterized by rapid growth over weeks to months because it lacks an in situ or radial growth phase. Nodular melanomas usually present as deeper lesions at the time of diagnosis and may elude early detection, but the prognosis is similar to superficial spreading melanoma when matched for tumor depth.

Lentigo maligna melanoma occurs in 4% to 15% of cases but is the most common type of melanoma in some populations (4). Similar to basal cell carcinoma, and squamous cell carcinoma, this type of melanoma occurs on sun exposed areas such as the face and is linked to cumulative sun exposure. Acral lentiginous melanoma only accounts for 2% to 8% of cases. It is the least common form in light-skinned individuals, but it is the most common form in dark-skinned individuals, accounting for 29% to 72% of melanoma in black, Asian, and Hispanic peoples. The most common site is the soles, but it also occurs on the palms and in the nail beds, where it arises from the nail. Unclassified types of melanoma and rare types account for less than 5% of all cutaneous melanomas.

Diagnosis and Treatment of Melanoma

Diagnosis of Primary Tumor

Appropriate biopsy is essential for diagnosis and histologic microstaging. Excisional biopsy with narrow (1–3 mm) margins is recommended so as not to disrupt lymphatic drainage before sentinel node mapping. Multiple punch or deep saucerization biopsies are acceptable for large lesions. Important histologic features include tumor depth in millimeters, level of invasion, ulceration, mitoses, angiolymphatic invasion, microsatellites, and host response.

Tumor thickness, also known as the Breslow depth, is the most important histologic determinant of prognosis, and is measured in millimeters from the top of the granular layer to the deepest part of the tumor in the dermis or subcutaneous tissue. The anatomic level of tumor penetration is known as the Clark level (level I, intraepithelial or in situ; level II and III, papillary dermis; level IV, reticular dermis; level V, subcutaneous tissue). This measurement has been largely superseded by the Breslow depth, which more accurately predicts prognosis. Clark level has predictive value in thin melanomas that are less than 1 mm deep and is incorporated into melanoma staging in this subgroup only (5).

There are also some clinical features that have prognostic value (6). In general, older patients do worse than younger patients; men do worse than women; and patients with truncal melanomas do worse than patients with lesions on the extremities. Six percent of all melanomas in women are diagnosed during pregnancy, but there is no convincing evidence that pregnancy increases risk. There is no difference in prognosis when melanoma is confined to the skin, but pregnancy makes the prognosis worse if there are nodal or distant metastases. This observation is true whether the melanoma is diagnosed before, during, or after the pregnancy.

Table 15.1. Simplified version of the American Joint Cancer Committee and European Joint Cancer Committee (AJCC/UICC) staging system for melanoma.

Stage 1	Cutaneous melanoma without metastases
1A	< 1 mm without ulceration
1B	< 1 mm with ulceration; 1.01–2.0 mm without ulceration
Stage 2	
II A	1.01–2.0 mm with ulceration; 2.01–4.0 mm without ulceration
II B	2.01–4.0 mm with ulceration; >4.0 mm without ulceration
II C	>4.0 mm with ulceration
Stage 3	Cutaneous melanoma with lymph node metastases
III A	Any thickness without ulceration, and 1–3 microscopic lymph node metastases
III B	Any thickness with ulceration, and 1–3 microscopic lymph node metastases; or any thickness without ulceration, and 1–3 macroscopic lymph node metastases; or any thickness with or without ulceration, and in transit/satellite metastases without metastatic regional nodes
III C	Any thickness with ulceration, and 1–3 macroscopic lymph node metastases; or =4 metastatic nodes, matted nodes, or in transit metastases/satellites with metastatic regional nodes
Stage 4	Cutaneous melanoma with distant metastases
	Distant skin, subcutaneous, nodal, visceral, skeletal or central nervous system metastases

Source: Updated from Valk PE, Bailey DL, Townsend DW, Maisey MN. Positron Emission Tomography: Basic Science and Clinical Practice. Springer-Verlag London Ltd 2003, p. 627.

Staging of Primary Tumor

Staging has changed dramatically over the past 15 years. The most commonly used staging system was formulated by the American Joint Cancer Committee and European Joint Cancer Committee (AJCC/UICC) and underwent major revisions in 2002 (Table 15.1).

The revisions to the AJCC/UICC staging system in the year 2002 included the presence of ulceration to upstage disease to the next worse prognostic level (7). New tumor depth cutoff points were set at 1 mm, 2 mm, and 4 mm because of more-recent data on the biologic behavior of thin melanomas (less than 1 mm). Microscopic or macroscopic satellite lesions, as well as in-transit lymph node metastases, are now grouped into stage 3, regardless of the distance from the primary tumor. In addition, the number of metastatic lymph nodes has been taken into account, rather than the size of lymph nodes, because the number of metastatic lymph nodes is a more important prognostic factor. Regional nodal tumor burden is determined microscopically by sentinel lymph node biopsy in patients without clinically palpable disease.

Treatment of Primary Melanoma

Excisional surgery is the primary therapy for localized cutaneous melanoma (stages I and II using the AJCC/UICC system). The definition of adequate surgical margins has evolved significantly over time. A National Institute of Health (NIH) Consensus Conference in 1992 stated that 0.5-cm margins were adequate for melanoma in situ and 1.0-cm margins were adequate for thin melanomas to 1 mm in depth (8). Longitudinal studies have shown that such margins were curative for in situ melanoma and resulted in greater than 95% 8-year survival for thin melanomas.

The surgical approach at Stanford University is to obtain 1-cm margins for tumors less than 1 mm deep, 1- to 2-cm margins for tumors 1 to 2 mm deep (typically 1 cm in certain anatomic locations, such as the face, to achieve primary closure and a good cosmetic result), 2-cm margins for tumors 2 to 4 mm deep, and at least 2-cm margins for tumors more than 4 mm deep. There are no prospective studies validating this approach in this last subgroup; however, a retrospective study of patients with high-risk primary melanomas showed that margins greater than 2 cm did not improve local relapse, disease-free survival, or overall survival in patients with thick (greater than 4 mm) primary melanoma (9).

The role of elective regional lymph node dissection for stages 1 and 2 primary cutaneous melanoma has been controversial. Lymph node dissection is not recommended for thin lesions (less than 1 mm) because of the low risk of metastatic spread, or for deep lesions (greater than 4 mm), which have a propensity to metastasize by the hematogenous route (10). Elective lymph node dissection was originally thought to decrease the risk of metastases in intermediate-thickness lesions (1–4 mm), based on nonrandomized studies. Randomized trials such as the WHO Melanoma Trial, however, have shown no benefit in patients with extremity lesions, and marginal, if any, benefit in patients with truncal lesions (11). Subgroup analysis, however, suggested that removal of micrometastatic disease in patients with nonpalpable lymph nodes improved overall survival.

This observation led to the development of preoperative and intraoperative lymph node mapping and of sentinel node biopsy as a means of assessing high risk (12). If the sentinel lymph node, or first lymph node draining the site of the primary lesion, is free of metastasis, this finding predicts the histology of the remaining lymph nodes in the same lymph node basin (13). When metastasis is present, 90% of the time the sentinel lymph node is the only lymph node involved. Identification of the sentinel lymph node is accomplished by preoperative lymphoscintigraphy with intraoperative localization using a gamma probe, as well as intraoperative intradermal injection of vital blue dye.

The addition of lymphoscintigraphy to vital blue dye increases the rate of identification of the sentinel lymph node up to 99% by identifying lymphatic basins at risk in ambiguous drainage sites (such as the head and neck, and trunk), as well as identifying in-transit lymph nodes between the primary lesion and regional lymph node basin. Wide excision of the primary lesion before sentinel node assessment can disturb lymphatic drainage and reduces the likelihood of identifying the sentinel node. All patients without palpable regional lymph nodes are candidates for sentinel lymph node biopsy if the primary lesion is 1 mm or more deep, has reached Clark level IV, is ulcerated, or shows signs of regression. Multiple studies have shown that sentinel lymph node status is the most important prognostic factor for recurrence and the most powerful predictor of survival. Sentinel lymph node biopsy also identifies patients who may benefit from lymph node dissection and adjuvant therapies.

Melanoma with regional lymph node metastases is treated by resection. Adjuvant high-dose interferon-alpha therapy and experimental vaccine therapies are options for patients with stage III melanoma. Involved-field radiation is frequently used if there is extracapsular lymph node extension. Distant metastases, if limited in number and spread, are treated by resection, which has been shown to improve survival.

Treatment of Metastatic and Recurrent Melanoma

Metastatic melanoma may occur locally by intralymphatic spread around the melanoma scar, in regional lymph node basins, or in remote locations, including subcutaneous, lymph node, visceral, skeletal, or central nervous system sites. Recurrence most commonly occurs within 3 years of treatment of the primary melanoma, initially within the skin, subcutaneous tissue, or regional lymph nodes. Visceral metastasis is observed most often in the lungs, liver, brain, bone, and gastrointestinal (GI) tract, in order of frequency.

Regional nodal disease is classified as stage III in the 2002 AJCC melanoma staging system. Five-year overall survival in stage III disease ranges from 13% with the worst combination of predictive factors (multiple macroscopic metastatic nodes, gross extracapsular extension, ulcerated primary melanoma) to 69% for the lowest risk combination (single regional nodal micrometastasis, nonulcerated primary melanoma). Diagnosis of recurrent nodal disease is most often made in the setting of palpable adenopathy, where confirmatory needle aspiration or nodal biopsy is performed before complete regional lymphadenectomy. The number of nodes is directly correlated with prognosis.

Therapeutic lymph node dissection is the treatment of choice for regional nodal disease whether synchronously diagnosed with the primary melanoma or occurring as metachronous recurrent disease; however, 5-year survival remains less than 50%. Adjuvant high-dose interferon-alpha therapy has been shown to improve both relapse-free and overall survival in stage III melanoma, and immunomodulatory therapy with melanoma vaccines is being actively investigated in these high-risk patients.

Involved-field radiation is frequently used if there is microscopic or gross extracapsular lymph node extension to improve local control in the regional nodal basin. Complete metastatic staging is recommended for appropriate patient management following the diagnosis of nodal metastasis.

Distant metastasis is classified as stage IV disease with 5-year survival ranging from 7% to 19%, depending on the site(s) of metastasis. In general, patient with soft tissue, nodal, and isolated lung metastases fare better than those with other visceral metastases. Resection of limited metastatic tumor in the skin, subcutis, nodal basins, and lung has shown to improve survival, although survival is generally measured in months, rather than years, in stage IV disease. Surgical debulking should also be considered for palliation of visceral metastasis.

Chemotherapy is the mainstay of treatment for stage IV disease despite response rates less than 20%. Furthermore, most responses tend to be short lived, and

therapy is not currently effective for central nervous system metastasis. Studies have shown that combination cytotoxic chemotherapy is no more effective than single agents such as dacarbazine and temozolomide. Radiation therapy is largely palliative, particularly for brain and skeletal metastasis. Biochemotherapy employing standard chemotherapeutic agents with biologic response modifiers such as granulocyte-macrophage colony-stimulating factor (GM-CSF), interleukin 2, and alpha-interferon has had limited success in the management of unresectable stage IV melanoma, as has treatment with high-dose interleukin 2 alone. As with regional nodal metastasis, numerous trials are underway investigating the use of melanoma vaccines, with or without biologic response modifiers, in the treatment of disseminated disease. Again, total-body and brain imaging of patients with advanced melanoma is important for accurate metastatic staging and is used to guide both surgical and medical treatment decisions.

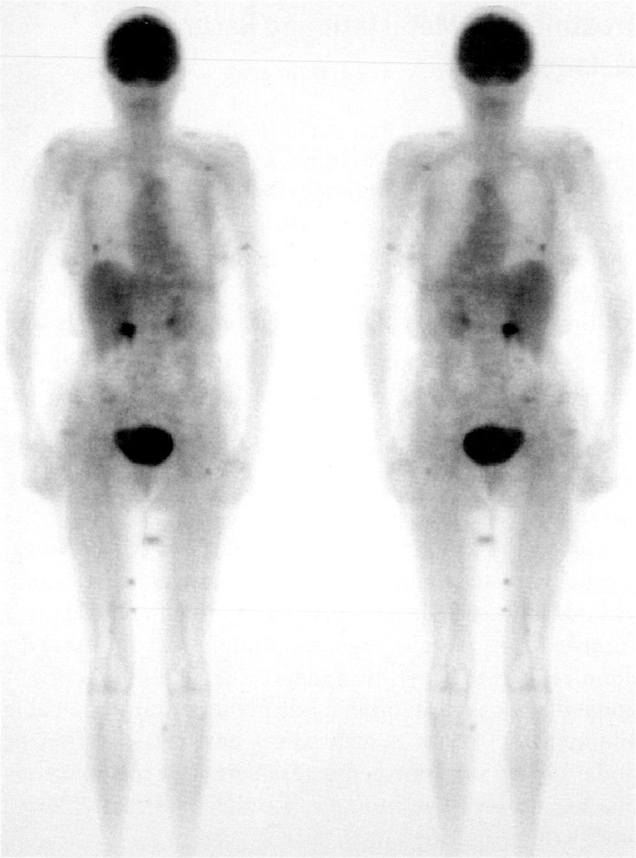

Positron Emission Tomography

Overview

Positron emission tomography (PET) with 2-[^{18}F]fluoro-2-deoxy-D-glucose (FDG) is a highly sensitive technique for the detection of melanoma. Clinical FDG-PET imaging of melanoma followed experiments using human tumor xenografts in nude mice that showed high intracellular concentration of FDG by melanoma cells (14).

Whole-body PET imaging has a number of advantages that are significant in evaluating melanoma. The whole body can be easily imaged, which is important because melanoma metastasizes widely. Skin, muscle, and bone metastases are easily detected (Figure 15.1).

PET is also very sensitive for detection of metastases to the bowel, omentum, and mesentery, which are difficult to image by computerized tomography (CT) (Figure 15.2). PET can detect metastatic disease in lymph nodes smaller than 1 cm, which would not be considered abnormal by CT criteria. This factor has particular importance in the mediastinum where PET is more accurate than CT. The supraclavicular area is also easier to evaluate by PET, as this area is anatomically complex and small lymph nodes are difficult to detect radiographically.

PET imaging also has some limitations. Evaluation of the brain is limited by high glucose uptake in the normal gray matter, and contrast-enhanced magnetic resonance imaging (MRI) is more accurate for detection of metastases to the cerebrum or cerebellum. The relatively low resolution of PET prevents detection of pulmonary lesions smaller than 5 mm, which are more accurately detected by CT. However, pulmonary nodules less than 10 mm are difficult to biopsy, and PET may be used to determine the likelihood of malignancy in a 5- to 10-mm nodule that is

Figure 15.1. Anterior (*left*) and posterior (*right*) FDG-PET whole-body images of a 76-year-old woman with melanoma in situ removed from the forehead 7 years before the PET scan. She had two local recurrences, and a pulmonary metastasis to the right middle lobe, which were resected on three separate occasions. She began to develop superficial and subcutaneous nodules on her trunk and extremities 2 months before the PET scan. Biopsy of a nodule on the left breast was positive for melanoma. New nodules were appearing at an alarming rate when she was referred for a PET scan to evaluate the extent of metastatic disease. The PET scan shows widespread metastatic melanoma with cutaneous, subcutaneous, and muscular lesions throughout the trunk and extremities. Metastases are also seen in the lower lobe of the right lung.

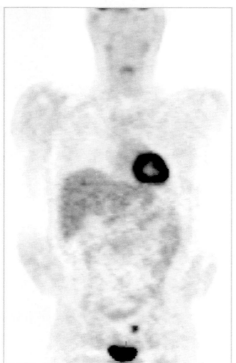
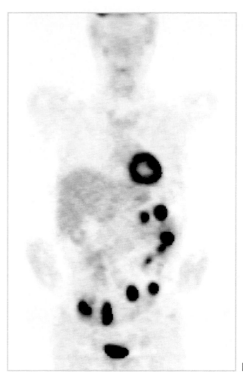

Figure 15.2. Coronal PET images obtained in a 65-year old man, 1 month after resection of a Clark's level III melanoma from the right thigh, show a focus of increased uptake in the left pelvis (**a**). The patient was asymptomatic, and CT scan of the pelvis was negative. A follow-up CT 5 months later also showed no pelvic abnormality. One year after the PET study, the patient presented with gastrointestinal (GI) bleed and was found to have a mass in the gastric mucosa, which proved to be recurrent melanoma on biopsy. Repeat PET scan after the biopsy showed multiple tumor masses in the abdomen and pelvis (**b**).

a

b

detected by CT. PET may also detect metastases in lymph nodes that are not enlarged, but it cannot substitute for lymph node biopsy for detection of micrometastases.

Imaging Protocol

General protocols for whole-body oncologic imaging are described in detail elsewhere. Fasting for 4 h or more before tracer injection is recommended because elevation of blood sugar reduces tissue FDG uptake by competitive inhibition and lowers the sensitivity of tumor detection. The entire body should be imaged because of the propensity of melanoma to metastasize widely. Images from the base of the skull to the upper thighs may be sufficient in patients with truncal melanoma. The legs should also be imaged when the primary lesion is on the feet, to detect in-transit or regional lymph nodes as well as subcutaneous nodules.

The recent introduction of hybrid PET/CT cameras has the potential to improve diagnostic accuracy by providing combined anatomic as well as metabolic information (Figure 15.3). Equivocal lesions on PET alone or CT alone are more accurately characterized as being benign or malignant by PET/CT, and metastases are more easily identified and located for directed biopsy or resection. At this time, data on the utility of PET/CT are preliminary. There are no published reports comparing PET/CT to either modality alone in patients with melanoma, but evidence for the improved accuracy of PET/CT is mounting in other cancers.

Indications for FDG-PET Imaging

PET is indicated for determining the extent of disease in patients with recurrent disease that is thought to be limited and surgically resectable. PET may demonstrate occult metastases that are not resectable, thereby contraindicating surgery. PET may also be used to characterize nonspecific radiographic abnormalities that are not amenable to biopsy or those where biopsy has yielded an indeterminate result. On the other hand, PET may correctly indicate resectability in patients who are thought to have nonresectable tumor on the basis of multiple indeterminate radiographic findings.

PET is generally not useful for initial staging in primary melanoma because of its low sensitivity for detecting micrometastatic disease in regional lymph nodes. Patients with clinical stage 3 disease might benefit from PET, but the incidence of distant metastatic disease in this subgroup of patients is low.

Whole-body imaging and the high sensitivity for detection of distant metastases make PET a potentially useful tool for posttreatment surveillance in patients with high-risk lesions and normal physical findings. Approximately 72% of patients with high-risk primary lesions will develop metastases, the majority within the first 2 to 5 years following diagnosis. Early detection of limited recurrent disease by PET may permit surgical resection, which has been shown to prolong survival. At the present time, however, there is no evidence of the utility and cost-effectiveness of PET in this setting.

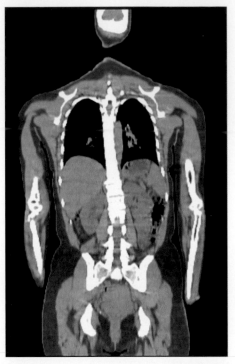
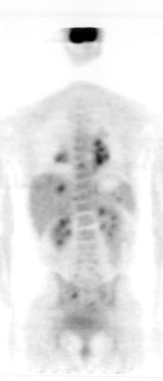
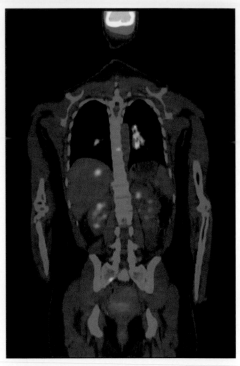

Figure 15.3. Coronal images from a PET/CT scan of a 42-year-old woman with widespread metastatic melanoma. The CT image on the *left* is added to the PET image in the *middle* to produce the fused PET/CT image on the *right*. Anatomic information provided by CT makes it possible to localize the hypermetabolic lesions seen on PET. The PET/CT scan shows metastatic disease in the hila, liver, and right sacrum.

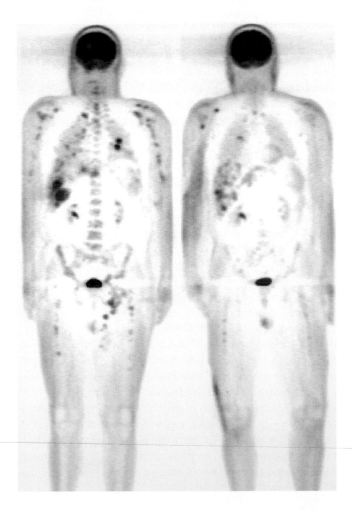

PET is an effective technique for evaluating the response to chemotherapy or immunotherapy when there are widespread lesions throughout the body (Figure 15.4). The capability of PET to scan the entire body makes it possible to evaluate the total burden of disease, rather than the response of a few index lesions. Also, metabolic response to chemotherapy is seen earlier than decrease in lesion size and can be used for early assessment of tumor response.

Accuracy of Tumor Detection in Regional Lymph Nodes

The presence or absence of regional lymph node metastases is an important prognostic factor in patients with melanoma. Sentinel node biopsy accurately predicts the presence or absence of metastases in the regional lymph

Figure 15.4. Coronal FDG PET images of a 54-year-old man with primary cutaneous melanoma of the left heel diagnosed 7 months before the first PET scan. Recurrent disease that was nonresectable was found in the left inguinal region 5 months after initial diagnosis. CT of the thorax, abdomen, and pelvis showed multiple bilateral small pulmonary nodules and hepatic lesions that were suspicious for metastatic disease. He was referred for a PET scan to evaluate the extent of metastatic disease before starting an intensive program of immunotherapy and chemotherapy. The first PET scan (*left*) shows widespread metastatic lesions that are too numerous to count throughout the axial and proximal appendicular skeleton, lungs, liver, and scattered lymph nodes. The second PET scan (*right*) done 4 months later after two cycles of therapy shows a marked reduction in the burden of disease, although there are a few new lesions, and some old lesions are more prominent.

Table 15.2. Accuracy of PET and sentinel node biopsy (SNB) for detecting regional lymph node metastases in patients with primary melanoma.

First author	Number (patients)	Sensitivity		Specificity		
		PET	SNB	PET	SNB	
Hafner 2004 (22)	100	8%	—	88%	—	(patients)
Havenga 2003 (21)	53	15%	—	88%	—	(patients)
Acland 2001 (20)	50	0%	100%	—-	—	(patients)
Crippa 2000 (19)	38	95%	—	84%	—	(basins)
		66%	—	99%		(nodes)
Wagner 1999 (18)	70	17%	94%	96%	100%	(basins)
Macfarlane 1998 (16)	22	85%	—	92%	—	(basins)
Wagner 1997 (17)	11	100%	—	100%	—	(patients)
Blessing 1995 (15)	20	74%	—	93%	—	(nodes)

Source: Updated from Valk PE, Bailey DL, Townsend DW, Maisey MN. Positron Emission Tomography: Basic Science and Clinical Practice. Springer-Verlag London Ltd 2003, p. 633.

node basin (13), but biopsy is invasive and does not provide any information about possible distant metastases. For this reason, PET imaging has been evaluated for regional nodal staging.

Early reports of the accuracy of PET in evaluating regional lymph nodes were largely obtained in patients in whom there was clinical or imaging suspicion of nodal spread (Table 15.2). In one study of 20 patients with clinically palpable lymph nodes, the sensitivity of PET for individual lymph node metastases was 28 of 38 (74%) and the specificity was 42 of 45 (93%) (15). The low sensitivity was mostly the result of failure to detect nonenlarged lymph nodes with minimal tumor burden. In a second study of 22 patients, some of whom had clinically abnormal or pathologically proven lymph node metastases, PET had a sensitivity of 11 of 13 (85%) and a specificity of 10 of 11 (92%) for the evaluation of regional lymph node basins (16). More-recent studies have consistently shown that the sensitivity of PET is very low for detecting regional lymph node metastases compared to sentinel node biopsy in patients without palpable adenopathy (see Table 15.2).

Havenga et al. (21) studied 55 patients with primary cutaneous melanoma greater than 1.0-mm Breslow thickness and no palpable regional lymph nodes. All patients had PET before to sentinel lymph node biopsy. Thirteen patients (24%) had positive sentinel lymph node biopsies but PET only identified 2 patients with metastatic disease. The diameter of the lymph node metastasis was 7 and 8 mm in the 2 patients identified by PET, whereas the diameter was less than 6 mm in 10 of 11 false-negative PET studies.

Hafner et al. (22) prospectively studied 100 patients with primary cutaneous melanoma less than 1-mm Breslow thickness. Twenty-six patients (26%) had a positive sentinel lymph node biopsy. Only 3 patients had lymph node metastases detected by PET, including 2 pa-

tients with palpable disease measuring 8 and 12 mm. Twenty-two patients had metastases less than 2.0 mm, which were not detected.

The sensitivity of PET can also be correlated with tumor volume. Wagner et al. (23) showed that PET detected 14% of nodal metastases when tumor volume was less than 78 mm^3 and 90% of metastases when tumor volume was greater than 78 mm^3. Mijnhout et al. (24) used morphometric analysis to predict the usefulness of PET for detecting lymph node metastases. They examined the tumor volume in positive sentinel lymph nodes from 59 of 108 patients with melanoma. Median tumor volume was only 0.15 mm^3. Based on these data and the assumption that the spatial resolution of a state-of-the-art PET scanner is 5 mm, only 15% to 49% of positive sentinel lymph nodes would be detected by PET.

The conclusion from these studies is that PET is sensitive for detecting disease in clinically suspicious lymph nodes but is not sufficiently sensitive to replace sentinel lymph node biopsy in patients with nonpalpable nodes.

Accuracy of Tumor Detection in Distant Metastasis

There have been multiple reports of the utility of PET for evaluating the extent of regional and distant metastatic disease in patients with melanoma (Table 15.3) (25–34). Most studies compared PET to conventional diagnostic methods, and results were reported on the basis of individual lesions or anatomic regions of involvement, whereas some studies reported results on a patient-by-patient basis.

Table 15.3. Accuracy of PET and conventional diagnostics (CD) for detecting metastatic disease in patients with melanoma.

First author	Number (patients/scans)	Sensitivity		Specificity		
		PET	CD	PET	CD	
Fuster 2004 (25)	156/184	74%	86%	58%	45%	(patients)
Finkelstein 2004 (26)	18/18	79%	76%	87%	87%	(lesions)
Swetter 2002 (27)	104/157	84%	58%	97%	70%	(lesions)
	53/66	81%	57%	—	—	(lesions)
Acland 2000 (28)	44/62	87%	—	78%	—	(patients)
Tyler 2000 (29)	95/106	87%	—	44%	—	(areas)
Eigtved 2000 (30)	38/—	97%	62%	56%	22%	(lesions)
Holder 1998 (31)	76/103	94%	55%	83%	84%	(patients)
Rinne 1998 (32)	100/100	100%	85%	96%	68%	(patients)
	100/100	92%	58%	94%	45%	(lesions)
Damian 1996 (33)	100/100	93%	—	—	—	(lesions)
Steinert 1995 (34)	33/33	92%	—	77%	—	(lesions)

Source: Updated from Valk PE, Bailey DL, Townsend DW, Maisey MN. Positron Emission Tomography: Basic Science and Clinical Practice. Springer-Verlag London Ltd 2003, p. 634.

Sensitivity for detecting individual metastatic lesions or abnormal regions ranged from 84% to 97%. Sensitivity for identifying individual patients with metastatic disease ranged from 74% to 100%.

Metastatic lesions that were missed were usually smaller than 10 mm. Six studies (25–27, 30–32) compared PET to conventional diagnostic imaging, including CT, MRI, ultrasound, and bone scans. PET was found to be more sensitive than conventional diagnostic imaging in five studies (25, 27, 30–32), especially for metastases to musculocutaneous sites, lymph nodes, abdomen, and bone. Finkelstein et al. (26) found that sensitivity was equivalent, although the combination of PET with conventional diagnostics was more sensitive and specific than either approach alone.

In studies that compared PET and conventional diagnostics, PET sensitivity on the basis of individual lesions was 79% to 97%, compared to 57% to 76% for conventional diagnostics. On a patient-by-patient basis, PET and conventional diagnostic sensitivities were 74% to 100% and 55% to 85%, respectively.

Some studies have reported CT to be more sensitive than PET for detecting small pulmonary metastases (25, 32, 35, 36). This finding is expected, given the higher resolution of CT and the high contrast between pulmonary tumor and the surrounding air-containing lung.

In three studies that compared PET and conventional diagnostics on a patient-by-patient basis (25, 31, 32), specificity was 83% to 96% and 45% to 84%, respectively. False-positive results were mostly the result of FDG uptake in surgical wounds, inflammatory sites, or benign tumors. All three studies reported a substantial decrease in the rate of false-positive findings when PET scans were interpreted with knowledge of the clinical history.

We conducted a retrospective study to determine the accuracy of PET in 157 scans and 104 patients with melanoma (27). Seventy patients also had CT imaging. The sensitivity and specificity of PET for individual metastases were 84% and 97%, respectively, compared to 58% and 70% for CT. A subgroup of 53 patients had 66 PET studies that were performed within 3 weeks of a CT study. CT and PET were ordered at the same time to eliminate the possibility of referral bias. The overall sensitivity of PET versus CT in this subgroup was 81% versus 57%. Exclusion of body areas that were not evaluated by CT in these patients (head, neck, and extremities) increased the sensitivity of CT to 69%. PET was more sensitive than CT in detecting metastatic disease to subcutaneous sites, peripheral lymph nodes, mediastinal and hilar lymph nodes, and intraabdominal regions. PET was equivalent to CT in detecting lung and skeletal metastases.

The observations and conclusions of most of the published studies are limited by pretest selection bias or posttest evaluation bias or both, which tend to cause overestimation of sensitivity and underestimation of specificity. Histologic confirmation of PET findings is difficult to obtain in patients with widespread disease and patients with normal imaging results. Such results can only be validated by systematic patient follow-up, which was not done in most reported studies. Despite these limitations, the results do demonstrate that PET is a sensitive technique for detection of metastatic melanoma, and is superior to CT, which is limited by a smaller field-of-view and low sensitivity for small lesions in areas other than the lungs.

Impact on Patient Management

A question that is not clearly answered by published results is when to use PET in patients with melanoma. Most studies have included mixed populations, including patients at the time of initial diagnosis, at the time of known or suspected recurrence, or during immunochemotherapy. It is difficult to generalize the results of these studies to specific populations. Tyler et al. (29) specifically evaluated PET in patients with clinical stage III melanoma (involvement of regional lymph nodes) who were undergoing pretreatment staging. Management was changed by PET findings in 16 of 106 studies (15%) because of the finding of previously unsuspected distant metastatic disease. Similarly, Acland et al. (28) reported that 28% of patients with clinical stage III disease were found to have distant metastases by PET, which significantly altered management.

Gulec et al. (37) used PET to evaluate extent of disease in 49 patients with known (46 patients) or suspected (3 patients) locoregional or distant metastases. Each patient's treatment options were originally based on history, physical examination, and conventional imaging. All patients then had PET, and the results were compared with conventional imaging to confirm or change the original treatment plan. PET demonstrated a greater extent of metastatic disease in 27 of the 49 patients (55%) and the same extent in 18 patients (37%). PET did not identify two brain metastases, one skeletal metastasis, and one subcutaneous metastasis in 4 patients (8%). PET confirmed the original treatment plan in 25 patients (51%) and prompted a change in 24 patients (49%). In 12 patients the planned operative procedure was canceled because of additional metastases identified by PET. In 6 patients an additional operation was performed because PET identified limited or otherwise unrecognized disease. In the remaining 6 patients, PET resulted in the addition of biochemotherapy, radiation therapy, or experimental immunotherapy because additional metastases were identified.

In another study, Valk et al. (38) found that planned tumor resection was deemed inappropriate in 5 of 45 patients (11%) following PET. An additional 2 of 45 patients (4%) who were thought to have nonresectable tumor based on CT were shown to have only limited disease by PET. The addition of PET to the diagnostic algorithm for presurgical evaluation of melanoma resulted in a savings-to-cost ratio of 2:1 because unnecessary procedures were avoided.

The general consensus based on these studies is that PET is indicated for presurgical staging of patients with known or suspected metastatic disease that is thought to be resectable in extent. Less often, PET indicates limited disease that is potentially resectable in patients erroneously believed to have widespread metastases based on conventional diagnostics.

Summary

PET is a sensitive and specific technique for the detection of melanoma, but micrometastases and lesions smaller than 10 mm may not be detected. PET is more sensitive than CT for detection of metastases in subcutaneous sites, lymph nodes, abdomen, and skeleton, but CT is equivalent to, or more sensitive than, PET for detecting small pulmonary lesions. Contrast-enhanced MRI remains the preferred method for detection of brain metastases. False-positive PET findings may be seen at surgical sites and in inflammatory lesions and may also be seen in some benign tumors. Clinical correlation significantly improves the specificity of PET.

PET is indicated for determining the extent of known metastatic disease, especially if patients are potentially operable. Approximately 4% to 24% of patients who are thought to have resectable limited disease based on physical examination and conventional diagnostic tests are found to have nonresectable disease by PET. Conversely, in patients who are erroneously considered to have nonresectable disease by conventional diagnostics, PET may show limited disease that is potentially operable.

References

1. American Cancer Society. Surveillance research. Estimated new cancer cases for selected cancer sites by state, US. Atlanta, GA: American Cancer Society, Inc., 2005.
2. Howe HL, Wingo PA, Thun MJ, Ries LA, Rosenberg HM, Feigal EG, Edwards BK. Annual report to the nation on the status of cancer (1973 through 1998), featuring cancers with recent increasing trends. J Natl Cancer Inst 2001;93:824–842.
3. McKean-Cowdin R, Feigelson HS, Ross RK, Pike MC, Henderson BE. Declining cancer rates in the 1990s. J Clin Oncol 2000;18:2258–2268.
4. Swetter SM, Jung S, Harvell JD, Egbert BM. Increased proportion of lentigo maligna and lentigo maligna melanoma subtypes in the Veterans Affairs Palo Alto Health Care System and Stanford University Medical Center. J Invest Dermatol 2002;119:245.
5. Buzaid AC, Ross MI, Balch CM, Soong S, McCarthy WH, Tinoco L, et al. Critical analysis of the current American Joint Committee on Cancer staging system for cutaneous melanoma and proposal of a new staging system. J Clin Oncol 1997;15:1039–1051
6. Chang AE, Karnell LH, Menck HR. The National Cancer Data Base report on cutaneous and noncutaneous melanoma: a summary of 84,836 cases from the past decade. The American College of Surgeons Commission on Cancer and the American Cancer Society. Cancer (Phila)1998;83:1664–1678.
7. Balch CM, Buzaid AC, Soong SJ, Atkins MB, Cascinelli N, Coit DG, et al. Final version of the American Joint Committee on Cancer Staging system for cutaneous melanoma. J Clin Oncol 2001;19:3635–3648.
8. NIH Consensus conference. Diagnosis and treatment of early melanoma. JAMA 1992;268:1314–1319.
9. Heaton KM, Sussman JJ, Gershenwald JE, Lee JE, Reintgen DS, Mansfield PF, Ross MI. Surgical margins and prognostic factors in patients with thick (>4mm) primary melanoma. Ann Surg Oncol 1998;5:322–328.
10. Mraz-Gernhard S, Sagebiel RW, Kashani-Sabet M, Miller JR III, Leong SP. Prediction of sentinel lymph node micrometastasis by histological features in primary cutaneous malignant melanoma. Arch Dermatol 1998;134:983–987.
11. Cascinelli N, Morabito A, Santinami M, MacKie RM, Belli F. Immediate or delayed dissection of regional nodes in patients with melanoma of the trunk: a randomised trial. WHO Melanoma Programme. Lancet 1998;351:793–796.
12. Morton DL, Wen DR, Wong JH, Economou JS, Cagle LA, Storm FK, et al. Technical details of intraoperative lymphatic mapping for early stage melanoma. Arch Surg 1992;127:392–399.
13. White RR, Tyler DS. Management of node-positive melanoma in the era of sentinel node biopsy. Surg Oncol 2000;9:119–125.
14. Wahl RL, Hutchins GD, Buchsbaum DJ, Liebert M, Grossman HB, Fisher S. Feasibility studies for cancer imaging with positron-emission tomography. Cancer (Phila) 1991;67:1544–1550.
15. Blessing C, Feine U, Geiger L, Carl M, Rassner G, Fierlbeck G. Positron emission tomography and ultrasonography. A comparative retrospective study assessing the diagnostic validity in lymph node metastases of malignant melanoma. Arch Dermatol 1995;131:1394–1398.
16. Macfarlane DJ, Sondak V, Johnson T, Wahl RL. Prospective evaluation of 2-[^{18}F]-2-deoxy-D-glucose positron emission tomography in staging of regional lymph nodes in patients with cutaneous malignant melanoma. J Clin Oncol 1998;16:1770–1776.
17. Wagner JD, Schauwecker D, Hutchins G, Coleman JJ III. Initial assessment of positron emission tomography for detection of nonpalpable regional lymphatic metastases in melanoma. J Surg Oncol 1997;64:181–189.
18. Wagner JD, Schauwecker D, Davidson D, Coleman JJ III, Saxman S, Hutchins G, et al. Prospective study of fluorodeoxyglucose-positron emission tomography imaging of lymph node basins in melanoma patients undergoing sentinel node biopsy. J Clin Oncol 1999;17:1508–1515.
19. Crippa F, Leutner M, Belli F, Gallino F, Greco M, Pilotti S, Cascinelli N, Bombardieri E. Which kinds of lymph node metastases can FDG-PET detect? A clinical study in melanoma. J Nucl Med 2000;41:1491–1494.
20. Acland KM, Healy C, Calonje E, O'Doherty M, Nunan T, Page C, Higgins E, Russell-Jones R. Comparison of positron emission tomography scanning and sentinel node biopsy in the detection of micrometastases of primary cutaneous malignant melanoma. J Clin Oncol 2001;19:2674–2678.
21. Havenga K, Cobben DCP, Oyen WJG, Nienhuijs S, Hoekstra HJ, Ruers TJM, Wobbes TH. Fluorodeoxyglucose-positron emission tomography and sentinel lymph node biopsy in staging primary cutaneous melanoma. Eur J Surg Oncol 2003;29:662–664.
22. Hafner J, Schmid MH, Kempf W, Burg G, Kunzi W, Meuli-Simmen C, et al. Baseline staging in cutaneous malignant melanoma. Br J Dermatol 2004;150:677–686.
23. Wagner JD, Schauwecker DS, Davidson D, Wenck S, Jung SH, Hutchins G. FDG-PET sensitivity for melanoma lymph node metastases is dependent on tumor volume. J Surg Oncol 2001;77:237–242.
24. Mijnhout GS, Hoekstra OS, van Lingen A, van Diest PJ, Ader HJ, Lammertsma AA, et al. How morphometric analysis of metastatic load predicts the (un)usefulness of PET scanning: the case of lymph node staging in melanoma. J Clin Pathol 2003;56(4):283–286.
25. Fuster D, Chiang S, Johnson G, Schuchter LM, Zhuang H, Alavi A. Is 18F-FDG PET more accurate than standard diagnostic proce-

dures in the detection of suspected recurrent melanoma? J Nucl Med 2004;45:1323–1327.

26. Finkelstein SE, Carrasquillo JA, Hoffman JM, Galen B, Choyke P, White DE, et al. A prospective analysis of positron emission tomography and conventional imaging for detection of stage IV metastatic melanoma in patients undergoing metastasectomy. Ann Surg Oncol 2004;11:731–738.

27. Swetter SM, Carroll LA, Johnson DL, Segall GM. Positron emission tomography is superior to computed tomography for metastatic detection in melanoma patients. Ann Surg Oncol 2002;9:646–653.

28. Acland KM, O'Doherty MJ, Russell-Jones R. The value of positron emission tomography scanning in the detection of subclinical melanoma. J Am Acad Dermatol 2000;42:606–611.

29. Tyler DS, Onaitis M, Kherani A, Hata A, Nicholson E, Keogan M, et al. Positron emission tomography scanning in malignant melanoma. Cancer (Phila) 2000;89:1019–1025.

30. Eigtved A, Andersson AP, Dahlstrom K, Rabol A, Jensen M, Holm S, et al. Use of fluorine-18 fluorodeoxyglucose positron emission tomography in the detection of silent metastases from malignant melanoma. Eur J Nucl Med 2000;27:70–75.

31. Holder WD Jr, White RL Jr, Zuger JH, Easton EJ Jr, Greene FL. Effectiveness of positron emission tomography for the detection of melanoma metastases. Ann Surg 1998;227:764–769; discussion 769–771.

32. Rinne D, Baum RP, Hor G, Kaufmann R. Primary staging and follow-up of high risk melanoma patients with whole-body 18F-fluorodeoxyglucose positron emission tomography. Cancer (Phila) 1998;82:1664–1670.

33. Damian DL, Fulham MJ, Thompson E, Thompson JF. Positron emission tomography in the detection and management of metastatic melanoma. Melanoma Res 1996;6:325–329.

34. Steinert HC, Huch Boni RA, Buck A, Boni R, Berthold T, Marincek B, et al. Malignant melanoma: staging with whole-body positron emission tomography and 2-[F-18]-fluoro-2-deoxy-D-glucose. Radiology 1995;195:705–709.

35. Krug B, Dietlein M, Groth W, Stutzer H, Psaras T, Gossmann A, et al. Fluor-18-fluorodeoxyglucose positron emission tomography (FDG-PET) in malignant melanoma. Diagnostic comparison with conventional imaging methods. Acta Radiol 2000;41:446–452

36. Gritters LS, Francis IR, Zasadny KR, Wahl RL. Initial assessment of positron emission tomography using 2-fluorine-18-fluoro-2-deoxy-D-glucose in the imaging of malignant melanoma. J Nucl Med 1993;34:1420–1427

37. Gulec SA, Faries MB, Lee CC, Kirgan D, Glass C, Morton DL, Essner R. The role of fluorine-18 deoxyglucose positron emission tomography in the management of patients with metastatic melanoma: impact on surgical decision making. Clin Nucl Med 2003;28:961–965.

38. Valk PE, Pounds TR, Tesar RD, Hopkins DM, Haseman MK. Cost-effectiveness of PET imaging in clinical oncology. Nucl Med Biol 1996;23:737–743.

16

PET and PET/CT Imaging in Urologic Tumors

Paul D. Shreve

Urologic applications of positron emission tomography (PET) have centered primarily in oncology using ^{18}F-2-deoxy-D-glucose (FDG), although a variety of tracers and specialized applications have been approached in limited studies. The primary clinical application of diagnostic imaging in urology relates to the management of the chief urologic malignancies including renal cancer, bladder cancer, prostate cancer, and testicular cancer. Testicular cancer is covered in a separate chapter of this volume (Chapter 14). Upper urinary tract obstruction, relative size and function of kidneys, presence of infection, and urinary tract congenital abnormalities are still assessed by anatomic imaging, contrast fluoroscopy, ultrasound, magnetic resonance imaging (MRI), computed tomography (CT), and existing radionuclide studies (1). Renal blood flow (2, 3) and blood flow and metabolism (4) as well as renal angiotensin receptor distribution (5) have been quantitatively measured using PET techniques; however, practical clinical applications of these studies have not yet emerged.

The applications in urologic malignancies parallels other oncologic applications in the body, that is, static imaging using FDG or other tracers preferentially accumulated by the malignant tissue for purposes of diagnosis and staging. The relevance of FDG-PET in the clinical management of urologic malignancies is, as with any malignancy, inextricably related to the need for diagnostic imaging in the management of the disease. Hence, experience with FDG PET in urologic malignancies must be viewed in the context of the contemporary management of these diseases.

Renal Malignancy

Renal malignancies include renal cell carcinoma (RCC), transitional cell carcinoma, squamous cell carcinoma, lymphoma, and metastatic neoplasm, usually lung cancer or melanoma. Renal cell carcinoma, the most common malignant neoplasm of the kidney, is primarily a surgically managed disease. Surgical extirpation is the only effective means of curing RCC. In patients with advanced disease, medical treatment options currently offer little in terms of improved survival or palliation. Hence, most renal masses that are not unequivocally benign on anatomic imaging are removed in attempt to provide a surgical cure, providing that there is no diagnostic imaging evidence the neoplasm has spread beyond the local compartment of Gerota's fascia. The prognosis for patients with RCC is most strongly influenced by stage at the time of surgery (6), although the histology and nuclear grade of the primary neoplasm influence long-term survival (7, 8) among RCC of the same grade.

The role of diagnostic imaging in the management of renal malignancy, then, is twofold: (1) characterization of the renal mass, that is, determining whether it is a malignant neoplasm requiring extirpation, and (2) staging, that is, determining the presence of locoregional and distant metastases.

Historically, diagnostic imaging evaluation of a renal mass was performed as part of evaluation of hematuria, flank pain, or a palpable mass. Increasingly, as a result of the widespread use of cross-sectional imaging modalities such as CT and ultrasound, detection of a renal mass is serendipitous (9), and consequently RCC is discovered at a much earlier stage. In the absence of clear-cut evidence of metastases on CT or ultrasound, the principal task of diagnostic imaging of a renal mass is distinguishing cysts (always benign) from complex cysts (sometimes malignant) from solid masses (usually malignant). Simple cysts are well characterized on contrast CT and ultrasound. More-complex cystic renal masses are placed in categories first described by Bosniak (10). Bosniak category I, the simple cysts, require no further evaluation. Category II complex cysts can be managed by directed anatomic imaging follow-up. The category III and IV cystic masses, however, typically require surgical exploration or extirpation, respectively, as there is currently no reliable method of excluding a malignancy (11). Hence, PET potentially could aid in the noninvasive evaluation of Bosniak category III and IV renal masses.

Solid renal masses are generally considered malignant because 80% to 90% prove to be so when resected. Differentiating RCC from other malignancies such as lymphoma or metastases is generally not an issue, as such are an uncommon cause of a solid renal mass, and urothelial malignancy is commonly differentiated an anatomic imaging. Benign solid masses include angiomyolipoma, which are usually reliably identified by demonstration of fat components on CT or ultrasound (12), and oncocytomas, which are not reliably distinguished from RCC on anatomic imaging (13). Preoperative needle biopsy of solid renal masses is generally eschewed because of inherent sampling errors and the possibility of spread of neoplastic cells along the needle tract (14). Hence, solid renal masses without evidence of metastatic disease are generally resected; there is little role for additional imaging characterization of a solid renal mass greater than 3 cm in diameter. Characterization of the small renal mass, however, is taking on greater importance for reasons of the growing incidental detection of solid masses less than 3 cm in diameter and the conservative management of such masses, particularly in settings where minimally invasive surgery and nephron-sparing surgery are contemplated (15–17). PET potentially could aid noninvasive characterization of the malignant potential of small renal masses.

Elevated FDG uptake in renal malignancy was described in RCC by Wahl et al. (18) in an animal tumor model and a limited number of human subjects more than a decade ago. Characterization of a solid renal mass has been described in initial limited series (19, 20) in which surgically proven RCC renal masses were identified as positive on FDG-PET. However, the renal masses in these series were large, in many cases readily identifiable in the PET images by sheer bulk even if the tumor was isointense with the nonmalignant adjacent renal parenchyma. False-negatives tended to be small renal masses where evaluation on FDG-PET is, at least in part, limited by the urinary excretory route of FDG (21, 22). The series reported by Goldenberg et al. (19) did include 11 Bosniak category III renal masses; however, 10 of 11 were benign and the 1 containing RCC was false negative on FDG-PET. In one series, 68 renal masses with histopathologic proof reported by Shreve et al. (23) found the average standardized uptake value (SUV) for RCC to be 4.6, thus on average only slightly higher than adjacent nonmalignant renal parenchyma between 1 and 2 h after tracer injection. FDG uptake in oncocytomas fell within the range of RCC. Further, no significant correlation with histologic classification or Fuhrman nuclear grade and SUV was evident. In a retrospective chart review by Kang et al., (24) among 66 patients with known or suspected RCC the sensitivity of FDG-PET for the primary tumor was 60% compared to 92% for contrast-enhanced CT. In a series of 35 patients, Aide et al. (25) reported a sensitivity of 47% and specificity of 80% for FDG-PET characterization of renal masses. Hence, there may be inherent limitations in using the tracer FDG for evaluation of RCC in general, and renal masses in particular, because of the relatively modest FDG avidity of a significant fraction of malignant renal tumors.

Renal cell carcinoma can be quite FDG-avid, and when FDG-avid is readily evident as a renal mass or metastatic deposit (Figure 16.1). Non- or relatively low FDG-avid RCC can be entirely occult on FDG-PET, however, even when obvious on anatomic imaging (Figure 16.2). In the detection of locoregional and distant metastases, FDG-PET sensitivity and specificity were 63% to 77% and 75% to 100%, respectively (24–27). Positive predictive value appears to high, in excess of 90% (28), whereas generally the negative predictive value is too low to be clinically useful (for example, a negative study does not exclude malignancy). At least in one series (26), true-positive FDG-PET metastases were anatomically abnormal by size criteria (greater than 1.7 cm), whereas the false-negative metastases ranged in size from 0.7 to 1.4 cm, consistent

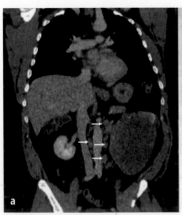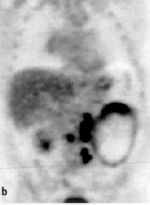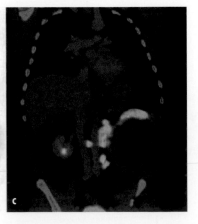

Figure 16.1. FDG avid renal cell carcinoma. Contrast-enhanced (CT) (**a**), FDG-PET with attenuation correction (**b**), and fusion image (**c**) demonstrate a large left renal cell carcinoma with central necrosis. FDG-PET demonstrates glucose metabolism along the peripheral rim of tumor and in left paraaortic and aortocaval lymph nodes (*arrows*).

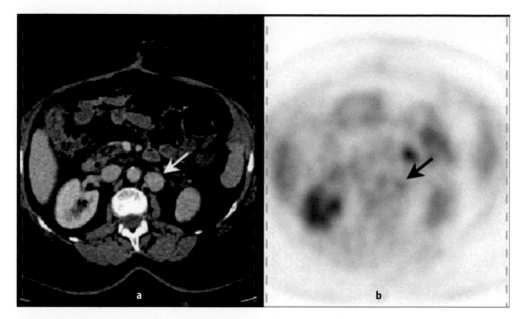

Figure 16.2. Non-FDG-avid lymph node metastasis of renal cell carcinoma. Contrast-enhanced CT (**a**) depicts enlarged left paraaortic lymph node that is not FDG avid on the FDG-PET image with attenuation correction (**b**) (*arrows*).

with a relatively modest, on average, FDG tracer uptake by RCC metastases. Regarding osseous metastases, whereas FDG-PET can detect lytic metastases occult on conventional bone scan (29), in a retrospective chart review of 66 patients, sensitivity and specificity of FDG-PET were 77% and 100%, respectively, in comparison to 94% and 87%, respectively, for combined CT and bone scan (24). Hence, FDG PET alone is probably not suited for detection of recurrent or metastatic RCC. Given the high positive predictive value of a positive FDG-PET finding, however, combined FDG-PET/CT may well have advantages over CT alone in the staging of RCC and the detection of recurrence.

Given the limitations of FDG as a tracer for renal malignancies, primarily the modest and variable FDG uptake and urinary excretory route, other tracers have been investigated, including amino acids and amino acid analogues (30). Acetate is also retained by RCC but rapidly cleared from the renal parenchyma as CO_2, and with no urinary excretion (4). Higher average SUV and tumor to renal cortex values are obtained within 10 min of tracer injection compared to FDG at 1 h post injection, and the highest acetate tracer accumulation was found in granulocytic tumors (31). Although such tracers of amino acid transport or lipid-related metabolism may have a role in characterizing a small renal mass or response to therapy, detecting RCC in complex renal masses and metastatic disease requires high consistent tracer uptake in the renal neoplasm, and such has not yet been demonstrated with this tracer.

Bladder Cancer

Bladder and related urothelial malignancies account for approximately 4% of clinical malignancies. Of the three

major urothelial malignancies, transitional cell carcinoma, squamous cell carcinoma, and adenocarcinoma, transitional cell carcinoma is by far the most common in both the bladder and upper urinary tract. In contemporary clinical practice, bladder cancer is diagnosed primarily by cystoscopy and upper urinary tract urothelial malignancy by CT urography or retrograde contrast ureteropylography. The most important prognostic factor in bladder cancer is the development of, and degree of, bladder wall invasion (32). Diagnostic imaging is most helpful in assessing the depth of muscle invasion and degree of perivesical involvement. Such distinctions require, in addition to high contrast between neoplasm and nonneoplastic tissue, high spatial resolution. PET would be expected to have limited utility in determining bladder wall invasion and adjacent spread. It is possible to visualize primary bladder cancer on FDG-PET when appropriate maneuvers are taken to minimize urinary FDG tracer activity (Figure 16.3).

Locoregional nodal staging is critical for proper management of bladder cancer patients. As with N staging elsewhere, the size criteria of nodal involvement used with anatomic imaging is of limited accuracy (33). Hence, PET with FDG or other tumor-specific tracers could provide increased accuracy in N stage of bladder cancer, as has been demonstrated with several other malignancies. Distant metastatic disease, most commonly osseous, pulmonary, and hepatic metastases, is important in patients with invasive bladder cancer. It is possible PET could offer some improvement in detection of osseous or hepatic metastases analogous to that observed in other malignancies such as lung or esophageal cancer.

Metastases not closely associated with the upper urinary tract and bladder are readily detected due to relatively good apparent FDG avidity of transitional cell

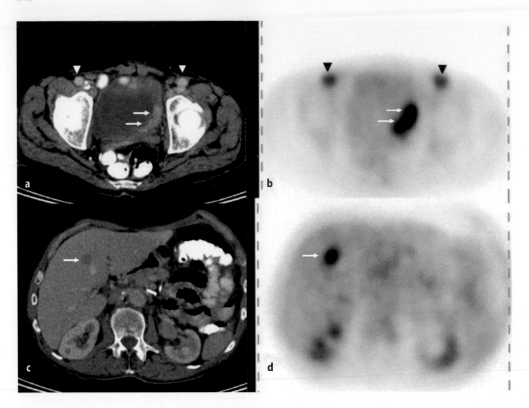

Figure 16.3. FDG-avid left bladder wall thickening (**a** and **b**, *arrows*) corresponding to a primary urothelial neoplasm. A solitary hepatic metastasis is also intensely FDG-avid (**c** and **d**, *arrows*). The bilateral FDG uptake in the groins is physiologic in aortofemoral bypass graft (**a** and **b**, *arrowheads*).

carcinoma of the bladder (Figure 16.3). Limited pilot studies (34, 35) have demonstrated that metastatic bladder cancer is FDG-avid and that involved local lymph nodes as small as 9 mm could be detected, whereas smaller involved nodes (less than 5 mm) were false negative. In a series of 64 patients, Bachor et al. (36) reported a sensitivity of 67% and a specificity of 86% for FDG-PET detection of pelvic lymph node metastases of bladder cancer. Osseous metastases of bladder cancer are readily detected on FDG-PET, but the relative accuracy of FDG-PET versus conventional bone scintigraphy has yet to be fully addressed. Bladder cancer appears to have relatively consistent avidity for FDG, and adding FDG-PET to conventional anatomic evaluation of locoregional and distant spread of bladder cancer such as with PET/CT may well prove clinically valuable.

As elsewhere in the urinary tract, alternative tracers that do not undergo urinary excretion, or can be imaged before the arrival of the excreted urinary tracer activity, have been investigated in an attempt to obviate the confounding effects of urinary tract excretion. ^{11}C-L methionine was used in a limited series to investigate the PET detection of primary bladder cancer (37). T4, most T3, and 2 of 4 of T2 primary bladder cancers were detected. The T staging was not superior to anatomic imaging, and there were insufficient proven cases of nodal metastases to evaluate accuracy of local lymph node metastases. In a preliminary series, de Jong et al. (38) reported detection of 10 of 18 primary bladder cancers with ^{11}C-choline PET. In 2 patients, pelvic lymph node metastases were visualized; however, again there were insufficient proven cases of

nodal metastases in the series to evaluate the accuracy for local lymph node metastases.

In addition to locoregional lymph node staging, differentiating postradiation therapy scar from recurrent tumor in patients treated for locally advanced disease and assessment of neoadjuvant therapy response are areas warranting further investigation of PET with FDG and other tracers.

Prostate Cancer

The role of diagnostic imaging in the management of prostate cancer is both as rapidly evolving and as controversial as the clinical management of the disease. Although two decades ago staging before prostatectomy with bone scintigraphy was common, today the management of prostate cancer is varied, with far less reliance on surgery and the routine use of serum markers (prostate-specific antigen) to assess disease progression and response to therapy. Because the prostate itself is easily accessed via the rectal vault, very high resolution anatomic imaging by ultrasound or MRI is possible (39). Biopsy of all sectors of the prostate gland, either randomly or assisted by ultrasound guidance, is routine and hence tumor histologic grade is readily obtained at initial diagnosis. The potential roles for diagnostic imaging of prostate cancer include diagnosis of primary disease, determination of extracapsular spread, and detection of locoregional lymph node metastases and distant metastatic spread. In addition, emerging roles include guidance of local therapy in patients with organ-confined disease and assessment of tumor response to systemic therapy in patients with advanced metastatic disease.

As with other neoplasms, the value of PET in clinical management is dependent on the avidity of prostate cancer for a given tracer such as FDG and the utility of competing modalities or procedures. Early observations with prostate cancer found relatively low avidity in untreated metastatic disease of sufficient bulk to be readily detected by PET. SUV generally were less than 4, and although soft tissue disease that was anatomically abnormal was detected, the sensitivity of FDG-PET relative to bone scintigraphy was poor (40, 41). Although the reason for the low relative uptake is not fully understood, the relatively slow growth of most prostate cancer may relate to low glucose metabolism. It is of interest, however, that advanced prostate cancer refractory to systemic therapy demonstrates consistently moderately high FDG uptake (42). In such patients, the metastatic lesions in both bone and soft tissue are well demonstrated on FDG-PET (Figure 16.4).

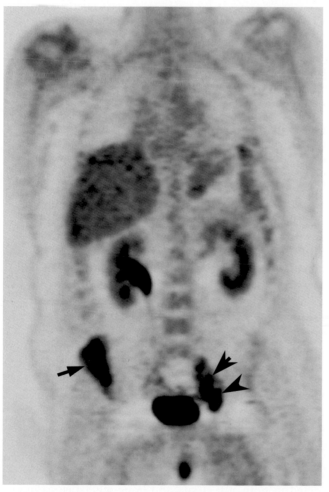

Figure 16.4. FDG-PET of advanced metastatic prostate cancer. Whole-body anterior projection attenuation-corrected FDG-PET image of a patient with advanced metastatic prostate cancer, refractory to hormone therapy. Substantial FDG uptake is present in the right iliac wing (*arrow*) and left iliac lymph nodes (*arrowheads*). 📖

The relatively low avidity of untreated prostate cancer for FDG and the confounding effect of adjacent bladder urinary tracer activity would suggest FDG-PET is not likely to be useful in diagnosis of organ-confined disease. In a series of 24 patients with organ-confined prostate cancer in which urinary tracer activity in the bladder was minimized, only 1 (4% sensitivity) was detected (43). Tumor volume ranged from 1.2 to 10.4 mL with a mean of 6.9 mL. The failure of detection most likely reflects the low tumor to background achieved with FDG. Similar disappointing results were reported for the detection of local recurrence of prostate cancer in patients treated by prostatectomy (44), also attributed to the relatively low avidity of prostate cancer for FDG.

Capsular penetration is a key milestone in stage of prostate cancer, separating traditionally surgically resectable (stage A and B of the Jewett–Whitmore classification) from nonresectable disease (stage C) and a poor prognostic indicator associated with metastatic disease. The distinction between involvement of the capsule of the prostate gland from penetration of the capsule requires exceeding high resolution; it is not a task for which scintigraphic imaging is well suited. Both transrectal ultrasound and endorectal coil MR offer the level of imaging spatial resolution to potentially address this important distinction, with best results reported to date with endorectal coil MR (45). Applications of PET to assess capsular penetration have not been reported.

In addition to capsular penetration, staging of locoregional lymph nodes remains an important task for imaging in patients considered for prostatectomy who are at high risk for nodal spread based on Gleason score and serum prostate-specific antigen (PSA). The limitations of anatomic imaging in the diagnosis of lymph node metastatic involvement of obturator, iliac, and retroperitoneal lymph nodes in the setting of prostate cancer are well established (46, 47). Again, likely reflecting the relatively low FDG avidity of prostate cancer, preliminary assessment of FDG-PET detection of pelvic lymph node metastases on FDG-PET was disappointing (48). In one series, detection of abdominal or pelvic lymph node metastases was no better than anatomic assessment (49). Reported sensitivity for pelvic lymph node metastases in a series of 24 patients with rising serum prostate-specific antigen was, however, 75% with a specificity of 100% (50). It does appear that detection of both locoregional and distant metastatic prostate cancer by FDG-PET is most feasible in patients with untreated, or in particular, progressive, disease (42, 51).

Distant metastatic disease includes bone, abdominal and thoracic lymph nodes, and liver. Detection of lymph node and liver metastases has been reported (40), but no substantial series assessing the accuracy of FDG-PET in the detection of distant soft tissue metastases has been published. Compared to bone scan, FDG-PET appeared to be substantially inferior in the detection of osseous metastases, with sensitivity relative to bone scan ranging from

18% to 65% (40, 41). The osteoblastic nature of osseous prostate cancer metastases and the relatively low FDG avidity of prostate cancer likely underlie these initial findings. In patients with progressive metastatic prostate cancer, however, FDG-PET-positive osseous lesions correlated with active foci of metastatic disease (42). In untreated prostate cancer, high FDG tracer uptake correlated with significantly higher rates of relapse postprostatectomy (52). In series of FDG-PET scans of patients undergoing chemotherapy, progression of FDG tracer uptake, as measured by SUV increase greater than 33%, predicted disease progression in more than 90% of patients who were subsequently found to have disease progression by clinical criteria (53).

The generally low FDG avidity of prostate cancer and the confounding effect of the urinary excretory route of FDG has fostered interest in other positron tracers. Exogenous choline is used by cells in the synthesis of phosphatidylcholine, the most abundant phospholipid in cell membranes. ^{11}C-Choline was first investigated as a tumor-imaging tracer in primary brain tumors (54) and subsequently in several cancers including prostate cancer (55, 56). Both soft tissue and bone metastases were identified, with SUVs ranging from 2.5 to 9, with a mean in the 4.5 to 5.0 range, whereas normal prostate was in the 2 to 3 range. Blood pool clears within minutes, and tracer is generally retained in the tumor with little change. Pancreas, liver, and renal parenchyma had high uptake and retention, with little urinary tract activity. The absence of urinary tracer activity in the early dynamic phase of imaging allows for some utility in the detection of locally recurrent prostate cancer in the setting of rising serum prostate-specific antigen, although detection of the foci of primary cancer within the prostate gland is limited by tracer uptake in benign prostatic hypertrophy (57–59). ^{11}C-Choline PET was 80% sensitive and 96% specific in the staging of pelvic lymph node metastases in a prospective series of 67 patients (60).

[18]F-Labeled choline derivatives have subsequently been synthesized and tested, including ^{18}F-fluoromethyl choline (61) and ^{18}F-fluoroethylcholine (62). Fluromethylcholine most closely matches the in vivo phosphorylation rate of choline and appears to be the preferred ^{18}F-labeled choline analogue for PET imaging (63). Both soft tissue and bone metastases are readily identified with fluorocholine, with SUVs ranging from 2.5 to as high as 10, but on average roughly 4.5 in untreated prostate cancer (Figure 16.5). In addition to high liver, pancreas, and bowel activity, fluorocholine undergoes urinary excretion. The rapid tumor uptake and blood pool clearance, however, does permit early imaging of the prostate and adjacent tissues before arrival of urinary tracer in the bladder. In comparison to FDG-PET, ^{18}F-fluorocholine PET was generally better in detection of primary lesions and osseous and soft tissue metastases on initial clinical evaluation (64). Detection of foci of primary cancer within the prostate gland was, similar to ^{11}C-choline, limited by

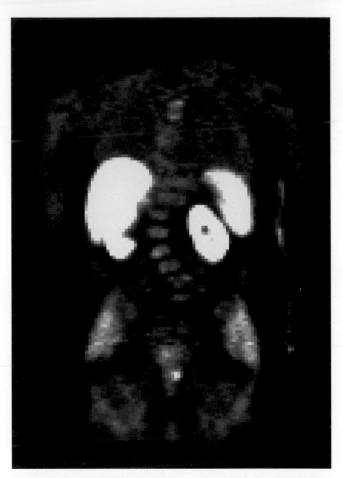

Figure 16.5. Fluorocholine PET of recurrent prostate cancer. Whole-body attenuation-corrected coronal image located posterior to the bladder acquired 10–40 min post [^{18}F]fluoromethylcholine administration. Focal uptake in the prostate bed is suggestive of recurrent disease in patient with serum PSA of 15.7 mg/dL and rising. 📖

tracer uptake in benign prostatic hypertrophy and prostatitis (65, 66).

Altered anabolic metabolism in cancers can be detected on PET using ^{11}C-acetate. First used to assess oxidative metabolism in myocardium, radioacetate has been shown to accumulate in tissues with high levels of anabolic metabolism such as the pancreas (67) and certain cancers including renal cell carcinoma (4, 31), nasopharyngeal and ovarian cancer (68), and prostate cancer (69). With ^{11}C-acetate PET, blood pool tracer activity clears within 2 min and tumor visualization nears maximum within about 5 min after tracer injection, with very slow clearance of retained tracer thereafter. Pancreas is the only abdominal organ with consistently high uptake, with variable moderate uptake in liver and portions of bowel. Renal parenchymal activity clears within 10 min (following oxidation to carbon dioxide), and there is no appreciable urinary excretion. Untreated prostate cancer is readily detected in both soft tissue and bone metastases (Figure 16.6), with SUVs ranging from 2.5 to more than 10 but averaging about 4.5 (70, 71). The absence of urinary excretion permits unencumbered visualization of the prostate

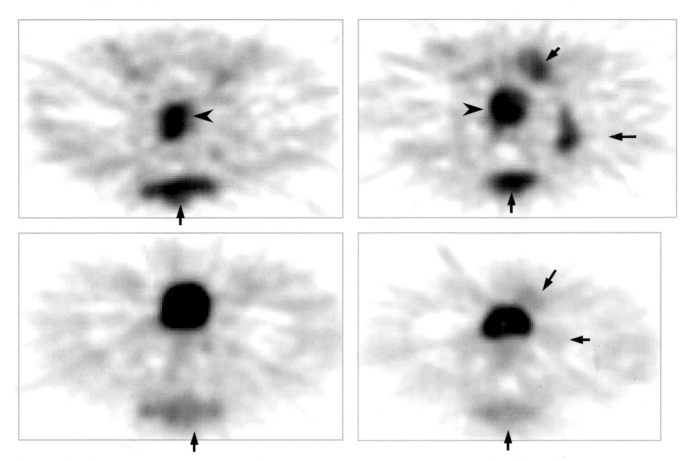

Figure 16.6. Acetate PET of primary and metastatic prostate cancer. Transaxial images of pelvis above and at the base of the bladder 10 min after ¹¹C-acetate administration (*upper images*) demonstrate osseous metastasis (*arrows*) in the sacrum, posterior acetabulum, and pubic ramus, as well as primary prostate cancer/prostate tissue (*arrowheads*). Note absence of urinary tracer activity in the bladder. Comparison images at the same transaxial levels 1 h after FDG administration (*lower images*) demonstrate very low FDG uptake (*arrows*) in the prostate cancer metastases.

and adjacent structures. Normal prostate tissue is, however, associated with tracer uptake with SUVs ranging from 1.1 to 4.5 with a mean of 2.8 (72), limiting potential assessment of primary prostate cancer. In a limited series of 15 patients comparing ¹¹C-acetate PET with FDG-PET, locoregional metastases were visualized slightly better on ¹¹C-acetate PET, whereas distant metastatic lesions were slightly more conspicuous on FDG-PET (73). A limited series of 12 patients found the degree of ¹¹C-acetate and ¹¹C-choline tracer uptake in prostate cancer or its metastases to be nearly identical (74).

As with acetate and choline PET tracers, amino acid tracers have been investigated both as a probe of an alternate metabolic pathway and as a strategy to avoid the confounding effects of urinary tracer in the bladder. In patients with progressive metastatic prostate cancer, ¹¹C-L-methionine uptake in metastatic lesions was consistently higher than FDG uptake and demonstrated progressing metastatic lesions more consistently than did FDG on PET imaging (75, 76). Some success was also reported in using ¹¹C-L-methionine PET to direct biopsy in patients with rising serum prostate-specific antigen and negative routine biopsies (77).

Although there has been considerable enthusiasm for ¹¹C-choline, ¹⁸F-fluorocholine, ¹¹C-acetate, and ¹¹C-L-methionine as solutions to the limitations of FDG in PET imaging of prostate cancer, it should be noted that the average uptake of these tracers in prostate cancer, although higher than FDG, is still modest on average, with SUV less than 5.0. Such uptake is about one-half of the average values observed with FDG in the cancers, such as lung cancer, where PET has shown superior accuracy over anatomic imaging. Thus, although selected cases with high uptake of these tracers (SUV 8 or higher) suggest high potential diagnostic accuracy, the experience with moderate or average uptake in FDG-PET, such as with breast cancer, indicates the applications of these tracers to prostate cancer may be more limited. One potential application where detection of small deposits of tumor is not critical is assessment of tumor response to therapy in patients with advanced prostate cancer. Serum prostate-specific antigen is widely used for determining response to therapy of prostate cancer but is not always reliable, particularly in advanced hormone-refractory prostate cancer (78). There is some evidence of prognostic value of FDG (glucose metabolism) and ¹¹C-methionine (amino acid transport and metabolism) PET imaging in patients with progressing metastatic prostate cancer (76). Similarly, other metabolic pathways such as depicted by choline and acetate tracers, as well as androgen receptor status tracers (79, 80), could be valuable in the management of advanced prostate cancer.

References

1. Pollack HM, McClennan BL, editors. Clinical Urography. Philadelphia: Saunders, 2000.

2. Nitzche EU, Choi Y, Killion D, et al. Quantification and parametric imaging of renal cortical blood flow in vivo based on Patlak graphical analysis. Kidney Int 1993;44:985–996.

3. Chen BC, Germano G, Huang S-C, et al. A new noninvasive quantification of renal blood flow with N-13 ammonia, dynamic positron emission tomography and a two-compartment model. J Am Soc Nephrol 1992;3:1295–1306.

4. Shreve P, Chiao P-C, Humes HD, Schwaiger M, Gross MD. Carbon-11 acetate PET imaging of renal disease. J Nucl Med 1995;36:1595–1601.

5. Szabo Z, Speth RC, Brown PR, et al. Use of positron emission tomography to study AT1 receptor regulation in vivo. J Am Soc Nephrol 2001;12:1350–1358.

6. Robson CJ, Churchill BM, Anderson W. Results of radical nephrectomy for renal cell carcinoma. J Urol 1969;101:297–301.

7. Skinner DG, Colvin RB, Vermillion CD, et al. Diagnosis and management of renal cell carcinoma: a clinical and pathological study of 309 cases. Cancer (Phila) 1971;28:1165.

8. Fuhrman SA, Lasky LC, Limus C. Prognostic significance of morphologic parameters in renal cell carcinoma. Am J Surg Pathol 1982;6:665.

9. Smith SJ, Bosniak MA, Megibow AJ, et al. Renal cell carcinoma: earlier discovery and increased detection. Radiology 1989;170:699–703.

10. Bosniak MA. The current radiographic approaches to renal cysts. Radiology 1986;158:1–10.

11. Wolfe JS Jr. Evaluation and management of cystic renal masses. J Urol 1998;159:1120–1133.

12. Bosniak MA, Megibow AJ, Hulnick DH, et al. CT diagnosis of renal angiomyolipoma: the importance of detecting small amounts of fat. AJR 1988;151:497–501.

13. Licht MR. Renal adenoma and oncocytoma. Semin Urol Oncol 1995;13:262–266.

14. Wehle MJ, Grebstald H. Contraindications to needle aspiration of a solid renal mass: tumor dissemination by renal needle aspiration. J Urol 1986;136:446–448.

15. Bosniak MA, Rofsky NM. Problems in the detection and characterization of small renal masses. Radiology 1996;198:638–641.

16. Licht MR, Novick AC. Nephron sparing surgery for renal cell carcinoma. J Urol 1993;149:1–7.

17. Butler BP, Novick AC, Miller DP, et al. Management of small unilateral renal cell carcinomas: radical versus nephron-sparing surgery. Urology 1995;45:34–41.

18. Wahl RL, Harney J, Hutchins G, Grossman HB. Imaging of renal cancer using positron emission tomography with 2-deoxy-2-(18F)-fluoro-D-glucose: pilot animal and human studies. J Urol 1991;146:1470.

19. Goldenberg MA, Mayo-Smith WW, Papanicolaou N, Fischman AJ, Lee MJ. FDG-PET characterization of renal masses: preliminary experience. Clin Radiol 1997;52:510–515.

20. Ramdave S, Thomas GW, Berlangieri SU, et al. Clinical role of F-18 fluorodeoxyglucose positron emission tomography for detection and management of renal cell carcinoma. J Urol 2001;166:825–830.

21. Shreve PD, Anzai Y, Wahl RL. Pitfalls in oncologic diagnosis with FDG PET imaging: physiologic and benign variants. RadioGraphics 1999;19:61–77.

22. Zhuang H, Duarte PS, Pourdehnad M, et al. Standardized uptake value as a unreliable index of renal disease on fluorodeoxyglucose PET imaging. Clin Nucl Med 2000;25:358–360.

23. Shreve PD, Miyauchi T, Wahl RL. Characterization of primary renal cell carcinoma by FDG PET. Radiology 1998;209;P94.

24. Kang DE, White RL Jr, Zuger JH, Sasser HC, Teigland CM. Clinical use of fluorodeoxyglucose F-18 positron emission tomography for detection of renal cell carcinoma. J Urol 2004;17:1806–1809.

25. Aide N, Cappele O, Bottet P, Bensadoun H, Regeasse A, Comoz F, et al. Efficiency of [(18)F]FDG PET in characterizing renal cancer and detecting distant metastases: a comparison with CT. Eur J Nucl Med Mol Imaging 2003;30:1236–1245.

26. Majhail NS, Urbain J-LCP, Olencki TE, et al. F-18 fluorodeoxyglucose positron emission tomography in the evaluation of distant metastases from renal cell carcinoma. J Clin Oncol 2003;21:3995–4000.

27. Safaei A, Figlin R, Hoh CK, Silverman DH, Seltzer M, Phelps ME, Czernin J. The usefulness of F-18 deoxyglucose whole-body positron emission tomography (PET) for re-staging of renal cell cancer. Clin Nephrol 2022; 57:56–62.

28. Jadvar H, Kherbache HM, Pinski JK, Conti PS. Diagnostic role of [F-18]-FDG positron emission tomography in restaging renal cell carcinoma. Clin Nephrol 2003;60:395–400.

29. Seto E, Segall GM, Terris MK. Positron emission tomography detection of osseous metastases of renal cell carcinoma not identified on bone scan. Urology 2000;55:286P.

30. Borner AR, Langen K-J, Herzog H. Whole-body kinetics and dosimetry of cis-4[^{18}F]fluoro-L-proline. Nucl Med Biol 2001;28:287–292.

31. Shreve PD, Wahl RL. Carbon-11 acetate PET imaging of renal cell carcinoma. J Nucl Med 1999;40:257P.

32. Catalona WJ. Bladder carcinoma. J Urol 1980;123:35–36.

33. Walsh JW, Amendola MA, Konerding KF, et al. Computed tomography detection of pelvic and inguinal lymph node metastases from primary and recurrent pelvic malignant disease. Radiology 1980;137:157–166.

34. Kosuda S, Kison PV, Greenough R, Grossman HB, Wahl RL. Preliminary assessment of fluorine-18 fluorodeoxyglucose positron emission tomography in patients with bladder cancer. Eur J Nucl Med 1997;24:615–620.

35. Heicappell R, Muller-Mattheis V, Reinhardt M, et al. Staging of pelvic lymph nodes in neoplasms of the bladder and prostate by positron emission tomography with 2-[^{18}F]-2-deoxy-D-glucose. Eur Urol 1999;36:582–587.

36. Bachor R, Kotzerke J, Reske SN, Hautmann R. Lymph node staging of bladder carcinoma with positron emission tomography. Urologe A 1999;38:46–50.

37. Ahlstrom H, Malmstrom P-U, Letocha H, Andersson J, Langstrom B, Nilsson S. Positron emission tomography in the diagnosis and staging of urinary bladder cancer. Acta Radiol 1996;37:180–185.

38. de Jong IJ, Pruim J, Elsinga PH, Jongen MM, Mensink HJ, Vaalburg W. Visualisation of bladder cancer using C-11-choline PET: first clinical experience. Eur J Nucl Med Mol Imaging 2002;29:1283–1288.

39. Yu KK, Hricak H. Imaging prostate cancer. Radiol Clin N Am 2000;38:59–85.

40. Shreve PD, Grossman HB, Gross MD, Wahl RL. Metastatic prostate cancer: initial finding of PET with 2-deoxy-2-[F-18]fluoro-D-glucose. Radiology 1996;199:751–756.

41. Effert PJ, Bares R, Handt S, Wolff JM, Bull U, Jakse G. Metabolic imaging of untreated prostate cancer by positron emission tomography with ^{18}F-fluorine-labeled deoxyglucose. J Urol 1996;155:994.

42. Morris MJ, Akhurst T, Osman I, Nunez R, Macapinlac H, Siedlechi K, et al. Fluorinated deoxyglucose positron emission tomography

imaging in progressive metastatic prostate cancer. Urology 2002;59:913–918.

43. Liu IJ, Zafar MB, Segall GM, Terris MK. Fluorodeoxyglucose positron emission tomography studies in diagnosis and staging of clinically organ-confined prostate cancer. Urology 2001;57:108–111.

44. Hofer C, Laubenbacher C, Block T, Breul J, Hartung R, Schwaiger M. Fluorine-18-fluorodeoxyglucose positron emission tomography is useless for the detection of local recurrence after radical prostatectomy. Eur Urol 1999;36:31–35.

45. Yu KK, Schidler J, Hricak H, et al. Prostate cancer: prediction of extracapsular extension by endorectal MR imaging and 3D H-MR spectroscopic imaging. Radiology 1999;213:481–488.

46. Oyen RH, Van Poppel HP, Ameye FE, Van de Voorde WA, Baert AL, Baert AL. Lymph node staging of localized prostate carcinoma with CT and CT-guided fine-needle aspiration biopsy: prospective study of 285 patients. Radiology 1994;190:315–322.

47. Tempany CM, Zhou X, Zerhouni EA, et al. Staging of prostate cancer: results of Radiology Diagnostic Oncology Group project comparison of three MR imaging techniques. Radiology 1994;192:47–54.

48. Sanz G, Robles JE, Gimenez M, et al. Positive emission tomography with 18-fluorine-labelled deoxyglucose: utility in localized and advanced prostate cancer. BJU Int 1999;84:1028–1031.

49. Seltzer MA, Barbaric Z, Belldegrun A, et al. Comparison of helical computerized tomography, positron emission tomography and monoclonal antibody scans for evaluation of lymph node metastases in patients with prostate specific antigen relapse after treatment for localized prostate cancer. J Urol 1999;162:1322–1328.

50. Chang CH, Wu HC, Tsai JJ, Shen YY, Changlai SP, Kao A. Detecting metastatic pelvic lymph nodes by ^{18}F-2-deoxyglucose positron emission tomography in patients with prostate-specific antigen relapse after treatment for localized prostate cancer. Urol Int 2003;70:311–315.

51. Sung J, Espiritu JI, Segall GM, Terris MK. Fluorodeoxyglucose positron emission tomography studies in the diagnosis and staging of clinically advanced prostate cancer. BJU Int 2003;92:24–27.

52. Oyama N, Akino H, Suzuki Y, Kanamaru H, Miwa Y, Tsuka H, et al. Prognostic value of 2-deoxy-2-[F-18]fluoro-D-glucose positron emission tomography imaging for patients with prostate cancer. Mol Imaging Biol 2002;4:99–104.

53. Morris MJ, Akhurst T, Larson SM, et al. Fluorodeoxyglucose positron emission tomography as an outcome measure for castrate metastatic prostate cancer treated with antimicrotubule chemotherapy. Clin Cancer Res 2005;11:3210–3216.

54. Hara T, Kosaka N, Shinora N, Kondo T. PET imaging of brain tumor with [methyl-^{11}C]choline. J Nucl Med 1997;38:842–847.

55. Hara T, Koska N, Kishi H. PET imaging of prostate cancer using carbon-11 choline. J Nucl Med 1998;39:990–995.

56. de Jong IJ, Pruim J, Elsinga PH, Vaalburg W, Mensink HJ. Visualization of prostate cancer with ^{11}C-choline positron emission tomography. Eur Urol 2002;42:18–23.

57. de Jong IJ, Pruim J, Elsinga PH, Vaalburg W, Mensink HJ. ^{11}C-Choline positron emission tomography for the evaluation after treatment of localized prostate cancer. Eur Urol 2003;44:32–38.

58. Yamaguchi T, Lee J, Uemura H, et al. Prostate cancer: a comparative study of (11)C-choline PET and MR imaging combined with proton MR spectroscopy. Eur J Nucl Med Mol Imaging 2005;32(7):742–748.

59. Yoshida S, Nakagomi K, Goto S, Futatsubashi M, Torizuka T. C-Choline positron emission tomography in prostate cancer: primary staging and recurrent staging. Urol Int 2005;74:214–220.

60. de Jong IJ, Pruim J, Elsinga PH, Vaalburg W, Mensink HJ. Preoperative staging of pelvic lymph nodes in prostate cancer by ^{11}C-choline PET. J Nucl Med 2003;44:331–335.

61. Hara T, Yuasa M. Automated synthesis of fluorine-18 labeled choline analogue: 2-fluoroethyl-dimethyl-2-oxytheylammonium. J Nucl Med 1997;38:44P.

62. DeGrado TR, Coleman RE, Wang S, et al. Synthesis and evaluation of ^{18}F-labeled choline as an oncologic tracer for positron emission tomography: initial findings in prostate cancer. Cancer Res 2001;61:110–117.

63. DeGrado TR, Baldwin SW, Wang S, et al. Synthesis and evaluation of ^{18}F-labeled choline analogs as oncologic PET tracers. J Nucl Med 2001;42:1805–1814.

64. Price DT, Coleman RE, Liano RP, Robertson CN, Polascik TJ, DeGrado TR. Comparison of [^{18}F]fluorocholine and [^{18}F]fluorodeoxyglucose for positron emission tomography of androgen dependent and androgen independent prostate cancer. J Urol 2002;168:273–280.

65. Kwee SA, Coel MN, Lim J, Ko JP. Prostate cancer localization with ^{18}fluorine fluorocholine positron emission tomography. J Urol 2005;173:252–255.

66. Schmid DT, John H, Zwwefel R, Cservenyak T, Westera G, Goerres GW, et al. Fluorocholine PET/CT in patients with prostate cancer: initial experience. Radiology 2005;235:623–628.

67. Shreve PD, Gross MD. Imaging of the pancreas and related diseases with PET carbon-11 acetate. J Nucl Med 1997;38:1305–1310.

68. Liu RS, Yuan CC, Chang CP, Chou, KL, Chang CW, Ng HT, Yeh SH. Positron emission tomography (PET) with [C-11] acetate (ACE) in detecting malignant gynecologic tumors. Eur J Nucl Med 1998;25:963P.

69. Shreve PD. Carbon-11 acetate PET imaging of prostate cancer. J Nucl Med 1999;40:60P.

70. Oyama N, Akino H, Kanamaru H, Suzuki Y, Muramoto S, Yonekura Y, et al. ^{11}C-Acetate PET imaging of prostate cancer. J Nucl Med 2002;43:181–186.

71. Kotzerke J, Volkmer BG, Neumaier B, Gschwend JE, Hautmann RE, Reske SN. Carbon-11 acetate positron emission tomography can detect local recurrence of prostate cancer. Eur J Nucl Med Mol Imaging 2002;29:1380–1384.

72. Kato T, Tsukamoto E, Kuge Y, Takei T, Shiga T, Shinohara N, et al. Accumulation of [^{11}C]acctate in normal prostate and benign prostatic hyperplasia: comparison with prostate cancer. Eur J Nucl Med Mol Imaging 2002;29:1492–1495.

73. Fricke E, Machtens S, Hofmann M, van den Hoff J, Bergh S, Brunkhorst T, et al. Positron emission tomography with ^{11}C-acetate and ^{18}F-FDG in prostate cancer patients. Eur J Nucl Med Mol Imaging 2003;30:607–611.

74. Kotzerke J, Volkmer BG, Glatting G, van der Hoff J, Gschwend JE, Messer P, et al. Intraindividual comparison of [^{11}C]acetate and [^{11}C]choline PET for detection of metastases of prostate cancer. Nuklearmedizin 2003;42:25–30.

75. Macapinlac HA, Humm JL, Akhurst T, Osman I, Pentlow K, Shangde C, et al. Differential metabolism and pharmacokinetics of L-[1-(11)C]-methionine and 2-[(18)F] fluoro-2-D-glucose (FDG) in androgen independent prostate cancer. Clin Posit Imaging 1999;2:173–181.

76. Nunez R, Macapinlac HA, Yeung HW, Akhurst T, Cai S, Osman I, et al. Combined ^{18}F-FDG and ^{11}C-methionine PET scans in patients with newly progressive metastatic prostate cancer. J Nucl Med 2002;43:46–55.

77. Toth G, Lengyel Z, Balkay L, Salah MA, Tron L, Toth C. Detection of prostate cancer with ^{11}C-methionine positron emission tomography. J Urol 2005;173:66–69.

78. Eisenberger MA, Nelson WG. How much can we rely on the level of prostate-specific antigen as an end point for evaluation of clinical trials? A word of caution! J Natl Cancer Inst 1996;88:779–781.

79. Larson SM, Morris M, Gunther I, et al. Tumor localization of 16beta-18F-fluoro-5alpha-dihydrotestoserone versus 18F-FDG in patients with progressive, metastatic prostate cancer. J Nucl Med 2004;45:366–373.

80. Dehdashti F, Picus J, Michalski JM, Dence CS, Siegel BA, Katzenellenbogen JA, Welch MJ. Positron tomographic assessment of androgen receptors in prostatic carcinoma. Eur J Nucl Med Mol Imaging 2005;32:344–350.

17

PET and PET/CT in Sarcoma

Michael J. O'Doherty and Michael A. Smith

The management of patients with cancer is becoming increasingly important, with between one in three and one in four members of the population developing cancer in their lifetime. The rare tumors present a number of difficulties to clinicians. As with other tumors, positron emission tomography (PET) has a role in distinguishing benign from malignant disease, staging the extent of disease, monitoring response to treatment, and evaluating local recurrence or distant relapse.

The role of imaging in soft tissue sarcomas (STS) is of particular interest because the presence of benign soft tissue masses is common and the ability to distinguish between these and malignant tumors is essential. Diagnosis is made by surgical biopsy. The outcome is dependent on the stage of the tumor, an essential part of which is the grade. Imaging can play a fundamental role in all these processes. Imaging plays a similar role in the grading and staging of bone tumors. In this group of tumors, in contrast to STS, adjuvant therapy, especially preoperatively, can play a key role in the treatment. Such treatment may mean the difference between limb salvage and amputation. PET has the potential to play a key role in the assessment and evaluation of these patients.

This chapter discusses the role of imaging in the management of both soft tissue sarcomas and osteogenic sarcomas.

Pathology

Soft Tissue Sarcomas

Soft tissue sarcomas are a rare heterogeneous group of tumors with an incidence of between 1.5 and 2 per 100,000 of the population in the United States and the U.K. (1). To more accurately define the biologic potential of a soft tumor, the World Health Organization (WHO) classification of tumors has recommended four categories:

benign, intermediate (locally aggressive), intermediate (rarely metastasize), and malignant (Tables 17.1, 17.2) (2).

The incidence increases with age. Soft tissue sarcomas can occur throughout the body, but more than 70% occur in the limbs or limb girdles. Certain factors are associated with malignancy. A mass with a history of rapid growth, a mass situated deep to the deep fascia, and a mass of a size greater than 5 cm should alert the treating clinician. To establish the diagnosis, appropriate investigations, coupled with a well-planned biopsy, are necessary (3).

The histologic diagnosis is fundamental and emphasizes further the importance of the biopsy. Tumor grade is dependent on histologic features; probably the most widely accepted scheme is that proposed by Trojani et al. (4). It is based on three factors: cell differentiation, mitotic rate, and extent of necrosis. The huge number and variety of tumors underlines the importance of the quality of the biopsy for the histopathologist to make the diagnosis. The process can be made even more difficult if there has been preoperative adjuvant therapy.

Osteogenic Sarcomas

Lichenstein (5) advocated a classification of bone tumors based on the cytologic features of the tumor cells and of the tissue they produce. In most instances there are benign and malignant examples (see Table 17.2); most present as single lesions. Diaphyseal aclasis is the only significant benign exception. Malignant exceptions include myeloma, lymphoma, and, most importantly, metastatic bone disease, which is the most common form of malignant bone tumor. Metastasis can present as an isolated tumor, and this presentation is a feature of certain malignancies such as renal carcinoma. For a proportion of malignancies, the presenting feature is a single metastatic lesion rather than a primary tumor.

Other conditions that affect bone and must be considered in the differential diagnosis include sepsis, fibrous dysplasia, benign bone cysts, hyperparathyroidism, and

Table 17.1. Adaptation of the World Health Organization (WHO) classification of malignant soft tissue tumors.

Adipocytic
- Atypical lipomatous/well-differentiated liposarcoma
- De-differentiated liposarcoma
- Myxoid liposarcoma
- Pleomorphic liposarcoma
- Mixed-type liposarcoma

Fibroblastic/myofibroblastic
- Low-grade myofibroblastic sarcoma
- Myxoinflammatory fibroblastic sarcoma
- Infantile fibrosarcoma
- Adult fibrosarcoma
- Myxofibrosarcoma
- Low-grade fibromyxoid sarcoma
- Sclerosing epithelioid fibrosarcoma

So-called fibrohistiocytic
- Pleomorphic malignant fibrous histocytoma (MFH)
- Giant cell MFH
- Inflammatory MFH

Smooth muscle
- Leiomyosarcoma

Skeletal muscle
- Rhabdomyosarcoma
  1. Embryonal
  2. Alveolar
  3. Pleomorphic

Vascular
- Kaposiform haemangioendothelioma
- Retiform haemangioendothelioma
- Papillary intralymphatic angioendothelioma
- Composite haemangioendothelioma
- Kaposi sarcoma
- Epitheliod haemangioendothelioma
- Angiosarcoma of soft tissue

Chondro-osseous
- Extraskeletal osteosarcoma

Tumors of uncertain differentiation
- Synovial sarcoma
- Epithelioid sarcoma
- Alveolar soft part sarcoma
- Clear cell sarcoma of soft tissue
- Extraskeletal myxoid chondrosarcoma
- Malignant mesenchymoma
- Desmoplastic small round cell tumor
- Extrarenal rhabdoid tumor
- Intimal sarcoma

Table 17.2. Adaptation of WHO classification of malignant bone tumors.

Cartilage tumors
- Chondrosarcoma
- Dedifferentiated chondrosarcoma
- Mesenchymal chondrosarcoma
- Clear cell chondrosarcoma

Osteogenic tumors
- Conventional osteosarcoma
- Telangiectatic osteosaroma
- Small cell osteosarcoma
- Low-grade central osteosarcoma
- Secondary osteosarcoma
- Parosteal osteosarcoma
- Periosteal osteosarcoma
- High-grade surface osteosarcoma

Fibrogenic tumors
- Fibrosarcoma of bone

Fibohysticytic tumors
- Malignant fibrous histiocytoma of bone

Ewing sacoma/primitive neuroectodermal tumors (PNET)
- Ewing sarcoma tumor/PNET

Haematopoietic tumors
- Plasma cell myeloma
- Malignant lymphoma

Giant cell tumors
- Malignancy in giant cell tumors

Notochordal tumors
- Chordoma

Vascular tumors
- Angiosarcoma

Myogenic lipogenic neural and epithelial tumors
- Leiomyosarcoma of bone
- Liposarcoma of bone
- Adimantinoma
- Metastases in bone

Paget's disease. In the latter, malignant change can occur as a complication, particularly in the elderly. Thus, investigations to establish the diagnosis and stage the tumor are an essential part of the workup of any bone tumor.

Osteogenic sarcomas are tumors of malignant connective tissue that produces osteoid. These tumors have osteoblastic components and may have fibroblastic or chondroblastic features. The WHO published a revised classification with tumors grouped according to where they arise, either central or on surface of bone (6).

Diagnosis depends on the biopsy. A needle core biopsy is preferred. The soft tissue element of the bone sarcoma, which can account for up to 90% of the tumor, provides sufficient tissue for the pathologist. To make the diagnosis, there must be a clinical and radiologic correlation with the histologic findings, which requires close cooperation between the clinician and the pathologist. Ideally, the treating surgeon should do the biopsy as the method and placement of the biopsy scar are critical. If malignancy is suspected on clinical and radiologic grounds, referral to a specialized center is to be advocated.

Staging of Sarcomas

Soft Tissue Sarcomas

Staging is used for two main reasons: (1) staging is a guide to prognosis, and (2) staging is the most accurate and reproducible available measure of the disease for monitoring response to treatment.

Factors that are known to have a bearing on prognosis include size, site (usually meaning depth in relation to the deep fascia), grade, and the presence or absence of metastases. No one system appears to include all the factors satisfactorily, and a number of systems have been proposed. The American Joint Committee on Cancer (AJCC) staging is based on the TNM system and includes size of the tumor and three grades but omits the site (7). The Musculoskeletal Society system (8) assesses two grades and the site in relation to the compartment but omits size. The Memorial Sloan-Kettering Hospital assesses size, depth, and grade but not metastases (9).

There are additional factors to be considered. Children's sarcomas behave differently and rhabdomyosarcomas, for example, have a special staging system of their own. Preoperative adjuvant therapy has been advocated by certain centers, affecting the staging in terms of size and grading, especially with the degree of necrosis.

There are a number of limitations associated with tumor grading as part of the staging process, including interpathologist variability, the significance of mitoses in various tumor types, and determination of the extent of these abnormalities within a particular tumor. These difficulties highlight the potential of PET imaging to provide a general overview of the entire tumor and its grade, stage, and future behaviour (see following), eliminating sampling variation. The staging of a particular STS is the best predictor of its prognosis. Accurate staging is not only essential at the time of diagnosis but also subsequently in managing recurrences should they occur. In many instances, tumor that is diagnosed as recurrent may in fact represent residual tumor that has become evident since the time of initial treatment.

Between 10% and 23% of patients have metastases, and 33% of these are in the lung. Skeletal, hepatic, and cerebral metastases account for approximately 40%, and the other sites—regional lymph nodes, retroperitoneum, and soft tissues—make up the remaining 25% (3, 10).

Bone Sarcomas

The most common sites of osteosarcoma in children and adolescents are in the metaphyseal region of the femur (44%), tibia (17%), and humerus (15%). In older patients, the axial skeleton is frequently involved. Approximately 15% to 20% of patients have metastases at presentation.

Osteogenic sarcomas show a linear increase in bone metastases, approximately 1% per month, between 6 and 30 months after the primary diagnosis. Pulmonary metastases remain more common than skeletal metastases but skeletal and pulmonary metastases tend to develop simultaneously (11–13). A similar course of metastatic development has been identified with Ewing's sarcoma (11, 12). Other tissues involved with metastases include the liver, lymph nodes, kidneys, brain, soft tissue, and heart, but these are uncommon.

Surgical Management

Soft Tissue Sarcomas

The early and prompt diagnosis of soft tissue sarcomas depend on clinical awareness. Sixty percent occur in the limbs, the lower more commonly than the upper. In the majority of instances they are situated deep to the deep fascia. Thus, any mass lying within muscle and more than 5 cm in diameter, and especially where there is a recent history of rapid growth, must be regarded as malignant until proven otherwise. Soft tissue sarcomas occur at all ages, although each histologic subtype tends to fall within set age ranges.

Plain radiography can be helpful in the diagnosis. Soft tissue calcification is found in certain benign conditions, such as myositis ossificans and hamangiomas, but also in certain malignancies, most notably liposarcoma, synovial sarcoma, and soft tissue osteosarcoma. The essential investigations, if malignancy is suspected from the history and examination, are magnetic resonance imaging (MRI) or computed tomography (CT). Both methods give anatomic definition and may provide a clue to the assumed tissue of origin and its benignity or malignancy. The findings are seldom diagnostic.

The definitive management of any STS is dependent upon adequate tissue being obtained for definitive histologic examination. The management of STS is difficult and often suboptimal (14). The initial diagnosis is hampered by the rarity of the tumors, their clinical, radiologic, and histologic similarity to benign soft tissue masses, the variety of surgical specialties to which these patients are referred (14), and the frequent failure to perform an adequate initial biopsy (14). If an inadequate sample is taken or there is a misclassification of the tumor preoperatively, then an inappropriate operation may be performed, requiring further and often more debilitating surgery; this may lead to increased patient morbidity and in some cases mortality (14, 15). The biopsy itself may not define the true malignant grade of the tumor and therefore the selection of techniques to investigate such masses needs to be considered. Any method that can improve the identification of the most malignant site within the tumor mass, and which can accurately assess the body for distant metastases, would improve greatly the management of these tumors.

There are several methods described to obtain a histologic diagnosis, and biopsy is not without hazard (16, 17). Ideally a biopsy should be performed by the surgeon treating the patient, in a specialized center following consultation with the pathologist. Needle biopsy has the disadvantage of providing a very small fragment of tissue. It is often not representative of the tumor and is seldom sufficient if preoperative adjuvant therapy is to be used. The preferred method is an incisional biopsy with due regard to the definitive treatment; this should guarantee obtaining a tumor specimen of sufficient size that it is representative and will satisfy the pathologists' requirements. Where excision of the lump is a comparable procedure to incisional biopsy, then excision with meticulous regard to hemostasis is acceptable. Frozen section is seldom if ever indicated, other than to confirm that there is tumor tissue in the specimen.

Surgery is the mainstay of treatment for primary STS. Wide local excision with margins of at least 2 cm where possible should be performed. However, meeting this requirement may not be possible in certain regions of the body, particularly in the upper limbs. Preservation of vital structures is preferred provided the excision is not seriously compromised. Where clearance is necessarily marginal, for example, when the mass is adjacent to a main nerve or bone, adjuvant radiotherapy can compensate. The results are said to be comparable to surgery with satisfactory margins. Adjuvant therapy has been advocated in certain centers; both preoperative and postoperative radiotherapy and chemotherapy have been described regardless of the extent of the surgery. However, in the majority of primary STS, there is little if any effect and any response is unpredictable. There are notable exceptions, such as embryonal rhabdomyosarcoma in children.

As already stated, postoperative adjuvant radiotherapy of the primary site can be used to sterilize areas where margins are less than 2 cm. Care must be taken, particularly with preoperative radiotherapy, because surgical morbidity and complications may be increased. Postoperative chemotherapy may have a role in the management of metastatic disease, and there is some evidence to support this. Single-agent chemotherapy appears to be as efficacious as multiple-agent chemotherapy, but again the response is variable and unpredictable. There is evidence that metastases can be delayed, but overall mortality remains unchanged.

Follow-up must be undertaken on a regular basis. Knowledge of the natural history of the individual STS is important to time the appropriate investigations. MRI of the site of the primary, 2 to 3 months following surgery, provides a baseline for subsequent scans. CT of the chest is required for the higher-grade STS, but as the chest only accounts for up to 40% of metastases, whole-body PET can provide a more-comprehensive evaluation, which is particularly important in higher-grade tumors.

A follow-up of 5 years is a reasonable limit and is sufficient to be able to ensure a cure in most tumors. There are, however, notable exceptions, such as alveolar soft part tumor and epithelioid tumors, which can recur or metastasize after many years.

Bone Sarcoma

Primary malignant bone tumors are rare, and in terms of numbers are a relatively minor group of tumors. However, the majority affect children and young adults (of 5 to 25 years), and there is a smaller peak incidence in the fifth and sixth decades (18). Optimal management of these tumors requires multidisciplinary specialist collaboration. Because of these factors, malignant bone tumors should be managed in specialized centers.

Pain, swelling, and a degree of loss of function are the classical features of a primary malignancy of bone, but these are not unique. Clinical awareness is essential in making an early diagnosis. The majority occur at the end of long bones, more commonly in the lower than upper limbs. There is a wide age spectrum depending on the tumor type. Osteosarcomas are the most common primary bone tumor and occur mainly in children. Chondrosarcomas occur in young adults with a second peak in the fifth and sixth decades. Because of the rarity of these tumors, their diagnosis is often significantly delayed.

Management depends on the tumor type and the staging (19). Surgical ablation with suitable margins is the mainstay of treatment and if successful is associated with the best results. Limb salvage is advocated where possible; to this end prostheses are increasingly being used. Of concern is their increasing failure with time. This form of reconstructive surgery can be augmented with local radiotherapy either preoperatively, postoperatively, or combined. Adjuvant chemotherapy is given for most primary bone tumors in an attempt to control metastases. A 5-year survival of 55% has been reported for patients presenting without apparent metastases (20), and more recently, 5-year survival rates of 68% with multiagent chemotherapy given both preoperatively and postoperatively (21).

Increasing the chemotherapeutic dose does not increase survival in osteosarcoma (22). For Ewing's, where there is believed to be histologic evidence of response to chemotherapy, that is, more than 90% necrosis following preoperative adjuvant therapy, there is a 60% relapse-free survival at 5 years (23). However, there is a significant complication rate in the longer-term survivors. This increase in survival compares favorably to a 22% 5-year survival some 25 years ago. Improvement may be the result, at least in part, of earlier and more accurate diagnosis with consequently more effective surgery (24, 25) rather than adjuvant chemotherapy.

Imaging

Soft Tissue Sarcomas

Accurate staging at presentation is difficult to achieve, but a baseline chest X-ray and MRI or CT of the mass should be obtained before the biopsy. The size, location, and relationship of the tumor to surrounding tissue are well delineated by MRI scanning (26–29). Following the confirmation of diagnosis of malignancy, CT of the chest will satisfactorily detect thoracic metastases. Other sites in the body are more difficult to assess and are often ignored until symptoms lead to further investigation.

Nuclear imaging of the whole body has been attempted with a variety of tracers, such as ^{67}Ga (30–32), ^{201}Tl (33), and ^{99m}Tc-MIBI [^{99m}Tc-hexakis(2-methoxyisobutylisonitrile)] (34) or pentavalent ^{99m}Tc-DMSA [(V)-dimercaptosuccinic acid], (35) with variable success. The variable and lowly success of these imaging agents for STS has resulted in the application of PET imaging to this tumor group.

Bone Sarcomas

Plain radiography can be diagnostic. MRI is essential as it is more sensitive and accurate in delineating the extent of bone involvement and the presence of skip lesions. It is also crucial in identifying and defining any soft tissue extension.

This tumor group is best visualized with ^{99m}Tc-diphosphonate imaging. The various tumor types may have vari-

able uptake, with a suggestion by McLean and Murray (36) that the appearance of the skeletal scintigraphy may give a clue as to the type of tumor. Osteogenic sarcoma and Ewing's sarcoma tend to have high uptake but the former is more heterogeneous than the latter. Chondrosarcoma tends to have an intermediate grade of uptake with focal high-activity areas. Unfortunately, none of these changes is specific enough to differentiate the tumor types. Both ^{67}Ga (37) and ^{201}Tl have been used to differentiate tumors from benign disease (with absent uptake). Chondrosarcomas also show variable uptake and therefore make the interpretation of images difficult. ^{99m}Tc-sestamibi also has variable uptake in benign and malignant lesions resulting in the technique not being helpful in individual cases (38). Skeletal scintigraphy remains the imaging modality of choice to assess both local and distant disease.

MRI provides the most accurate assessment of the extent of the primary disease. Metastatic disease has been identified with a variety of agents including ^{201}Tl, ^{67}Ga, and ^{99m}Tc-MIBI, also with variable success. More recently, FDG-PET has been used to identify primary and metastatic disease and to assess tumor response to treatment.

PET Imaging

FDG-PET

FDG-PET has been used to image STS for a number of years, both to delineate the nature of the primary lesion and to detect distant metastases. The variability of FDG uptake by different primary tumors (Figure 17.1) has led the interest by a number of groups in investigating quantitative measures of uptake to establish the benign or malignant nature of the growth and, if malignant, its grade.

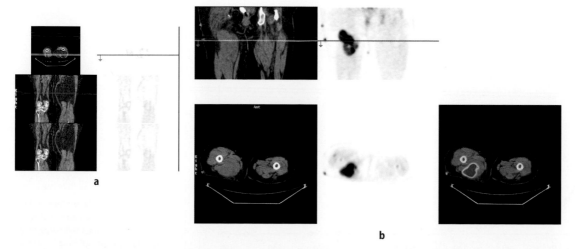

Figure 17.1. FDG-PET/CT images illustrate low uptake in a benign tumor (i) and the higher, more-heterogeneous uptake in a high-grade sarcoma (ii). The benign tumor has uptake that is less than the normal muscle uptake of the opposite popliteal fossa. The high-grade tumor has marked increased uptake compared to the surrounding soft tissue and compared to the opposite thigh. There is particularly high uptake in the proximal aspect of the tumor, which corresponded to the most malignant part of the tumor. The heterogeneous uptake in the distal part of the tumor corresponded to necrotic areas. These findings have implications for the biopsy site.

Grading Primary Malignancy

Soft Tissue Sarcoma

Adler et al. (39, 40) used semiquantitative standardized uptake value (SUV) analysis to distinguish high-grade malignancies from low-grade or benign tumors. Similar results have been found by a number of workers, demonstrating a separation between benign and malignant tumors using SUV (41–43). This approach, however, has not been found to be useful by others investigators (44–46). Nieweg et al. (46) determined the metabolic rate of glucose and were able to separate benign lesions from high-grade malignancies but not from intermediate- or low-grade malignancies. Similarly, Eary et al. (47, 48) reported that metabolic rate measurements correlated well with tumor grade. Their actual results show a marked overlap, ranging from 0.4 to 15 ?mol/min/g in low-grade tumors, to 2.0 to 5.3 ?mol/min/g in intermediate-grade tumors, and to 3.1 to 38.7 ?mol/min/g in high-grade tumors. On the basis of these results, it would be difficult to determine the grade of malignancy. Dimitrakopoulou-Strauss et al. (49) also looked at the use of quantification using a variety of rate constants and SUV, and similarly showed quite marked overlap between the various tumor types and benign disease, with the best separation in the highest grade tumors. These results were acquired over the first 60 min after injection of tracer.

The workers used a number of different methods to quantify the tumor metabolic rate, including Patlak graphic analysis and the SUV approach. The SUV should be related to the metabolic rate, assuming that the tracer concentration has reached a plateau at the time of measurement. This assumption is now known not to be the case. Lodge et al. (45) demonstrated that significant differences were observed in the time–activity response of benign and high-grade tumors. High-grade sarcomas were found to reach a peak activity concentration approximately 4 h after injection (Figure 17.2) whereas benign lesions reached a maximum within 30 min. Therefore, the two tumor types could be differentiated by comparing SUVs determined from images acquired at early and late postinjection times (Figure 17.3). An SUV measured 4 h postinjection was found to be as useful an index of tumor malignancy as the metabolic rate of FDG, determined using Patlak or nonlinear regression techniques. Each of these indices had a sensitivity and specificity of 100% and 76%, respectively, for the discrimination of high-grade sarcomas from benign tumors.

The use of delayed imaging has been applied to the assessment of patients with neurofibromatosis 1 (50) and in painful plexiform neurofibromas associated with neurofibromatosis type 1 to detect malignant change (51). This latter group found that 3-h-delayed imaging appeared to separate benign from malignant disease if SUV values exceed 3.3. Studies in other types of tumor indicate that delayed plateaus are observed in breast cancer (52). A simulation study in lung cancer by Hamberg et al. (53) has shown that tumor concentration did not reach a plateau within a 90-min study period. In the case of breast cancer this plateau was not reached for 3 h, and in lung cancer a simulation projected a 5-h plateau.

Recently it has been shown that the tumor-to-background ratios measured at 60 min also have similar overlap between benign, low-, intermediate-, and high-grade tumors (54–56). Given this failure to demonstrate

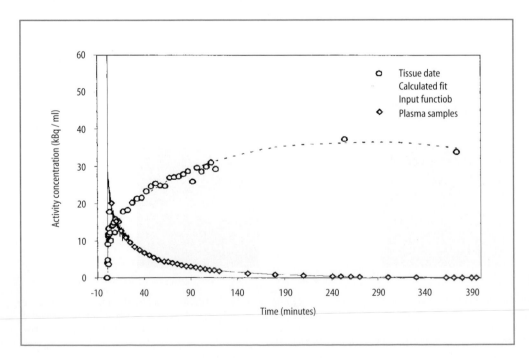

Figure 17.2. Characteristic time–activity curves for a high-grade malignant sarcoma and a benign soft tissue mass. The data are expressed in units of standardized uptake value (SUV). (Reprinted from Lodge MA, Lucas JD, Marsden PK, et al. A PET study of ^{18}FDG uptake in soft tissue masses. Eur J Nucl Med 1999;26:22–30.)

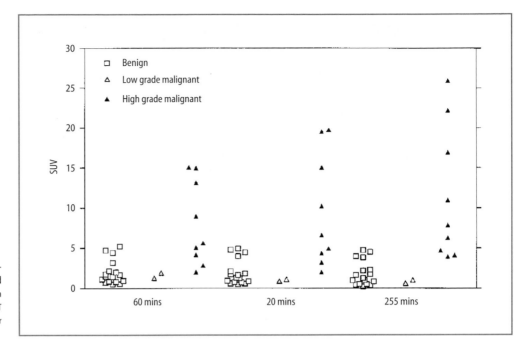

Figure 17.3. Plot of SUV versus malignancy for SUVs measured 60, 120, and 255 min postinjection. (Reprinted from Lodge MA, Lucas JD, Marsden PK, et al. A PET study of ^{18}FDG uptake in soft tissue masses. Eur J Nucl Med 1999;26:22–30.)

clear differences between the various grades of tumor, other imaging modalities have been evaluated, including L-1-[^{11}C]-tyrosine (57) and, more recently, ^{18}F-?-methyltyrosine (56). In both groups, it was concluded that FDG-PET was better at grading STS. ^{11}C-Tyrosine may be more useful in monitoring tumor response to therapy. The investigators suggested that ^{18}F-?-methyltyrosine was superior to FDG in distinguishing benign from malignant disease, but again the SUV measurements showed a high degree of overlap.

Bone Sarcomas

The published data on bone sarcomas suggest that there is no clear separation of malignant disease from benign disease. Kole et al. (58) examined the glucose consumption of a variety of bone tumors and SUV measurements and found that there was a large overlap of MRglc and SUV between benign and malignant lesions. Schulte et al. (59, 60) and Watanabe et al. (55) also found that there was marked overlap between benign and malignant causes of bone abnormalities and in particular benign and malignant tumors. All lesions were visualized. These data disagreed with the findings of Dehdashti et al. (61), who showed a clear separation between benign and malignant tumors. This group, however, only included three primary bone tumors; the rest were metastatic lesions, which are almost all by definition high grade. Aoki et al. (62) also found no overlap in a larger group of benign chondromatous lesions and chondrosarcomas; this contrasts with the study by Lee et al. (63) where grade I

chondrosarcomas could not be separated from benign lesions, whereas there was adequate separation of grade II and III chondrosarcomas. Using an SUV of 2.3 for grade II and III chondrosarcomas, the positive predictive value was 0.82 and the negative predictive value 96%. The largest study, by Schulte et al. (59), examined 202 patients and found a sensitivity of 93% and a specificity of 66% for malignancy, using a tumor-to-background (T/B) ratio exceeding 3 at 45 to 60 min postinjection. The study found a high degree of overlap, with some infections or indeed fibromas having an uptake T/B ratio of 18 to 24 and osteosarcomas and Ewing's tumors having T/B ratios of 3.3 to 33.2. The presence of very low T/B ratios did not exclude malignancy because chondrosarcomas ranged from 1.4 to 11.6. Despite this range of quantitative data, Eary et al. (64) demonstrated that the outcome in a range of sarcomas was predicted by the maximum SUV, and this was more significant than conventional tumor grading. Further work needs to be performed or alternative tracers used to separate the groups. Absolute quantification of uptake may be needed to assess tumor response to chemotherapy as ratios are dependent on the background changes as well as tumor changes.

Staging Disease

Soft Tissue Sarcomas

The use of FDG to assess the whole body for metastatic disease has a lower sensitivity than CT for detecting pul-

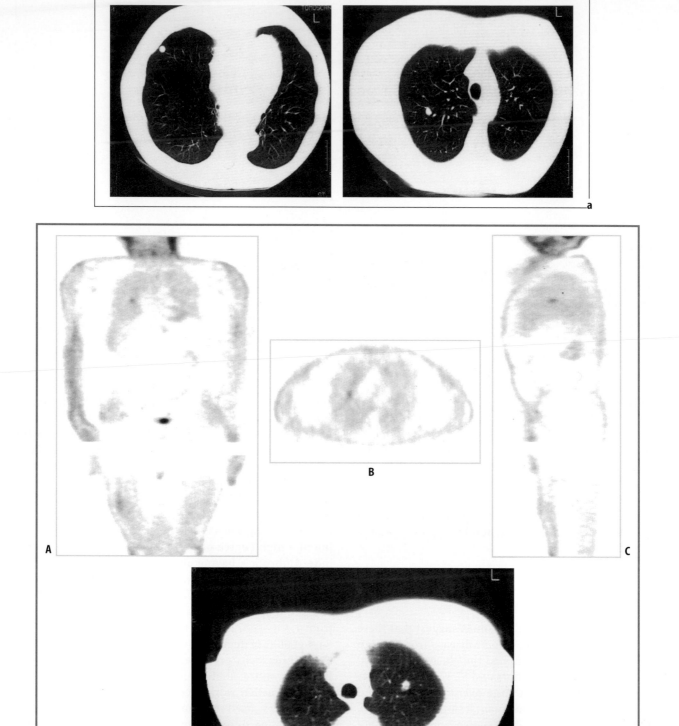

Figure 17.4. CT (**a**) and FDG-PET (**b**) images without attenuation correction in a patient with multiple pulmonary metastases from a high-grade sarcoma. FDG uptake is seen only in some of the CT lesions. The patient was undergoing follow-up for a high-grade soft tissue sarcoma.

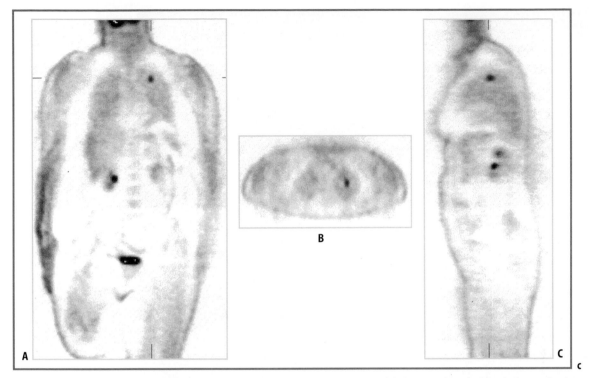

Figure 17.4. Contd. An area of increased uptake was seen in the left upper lobe (c). The CT scan appearance and the history suggested a second primary rather than a metastasis, and an adenocarcinoma of the lung was identified by biopsy. 📖

monary metastases (86.7% versus 100%) (Figure 17.4) and a lower sensitivity than MRI for detecting recurrent local disease (73.7% versus 88.2%) but a similar specificity (94.3% versus 96%) (65, 66). FDG-PET does, however, detect metastases at other sites (Figure 17.5) and therefore should be used in the primary assessment of patients with these tumors. The role of PET in patients with primary STS is to define preoperatively the grade of malignancy, detect distant metastases, and therefore guide the operative approach. It is also likely that, in patients with a heterogeneous mass, the most malignant area within that tumor mass can be identified with metabolic imaging. With image coregistration, the most appropriate site can be identified for biopsy, and it is likely that PET/CT will improve the registration to MRI, allowing interventional MR scanners to be used to direct biopsies.

Bone Sarcomas

There are few data on the identification of distant disease. Tse et al. (67) demonstrated that FDG-PET can visualize pulmonary metastases in a single patient with osteosarcoma, and Shulkin et al. (68) further demonstrated metastases from a Ewing's sarcoma in a child using FDG. Schulte et al. (59) identified pulmonary metastases in 4 patients using FDG-PET. Franzius et al. (69) examined 70 patients with primary bone tumors and found that 21 had metastases; 54 osseous metastases were identified by other imaging modalities. FDG-PET showed a sensitivity of

90%, a specificity of 96%, and an accuracy of 95% for detection of these lesions, compared to skeletal scintigraphy values of 71%, 92%, and 88%, respectively. The superiority to skeletal scintigraphy was found in patients with Ewing's sarcoma, where FDG-PET had a sensitivity of 100% and a specificity of 96%. The number of patients with osteosarcoma metastases was too small to draw meaningful conclusions. In this study, the failure to identify lung metastases was also demonstrated when compared with CT. The sensitivity was 50% for FDG-PET but 75 % with CT with very similar specificity and accuracy. It is likely that the introduction of integrated PET CT imaging will change these conclusions because the combined modality imaging would probably allow detection of metastases on the CT transmission images. However, this concept needs to be tested in a prospective evaluation.

Other sites of metastases have not been clinically evaluated with PET. There is no reason to believe that osseous metastases will be any less well visualized than the primary tumor because these are high-grade tumors.

Recurrent Disease

Soft Tissue Sarcoma

Recurrent STS occurs locally in between 15% and 47% of patients after initial surgery (70). MRI has been shown to be the investigation of choice to demonstrate local recur-

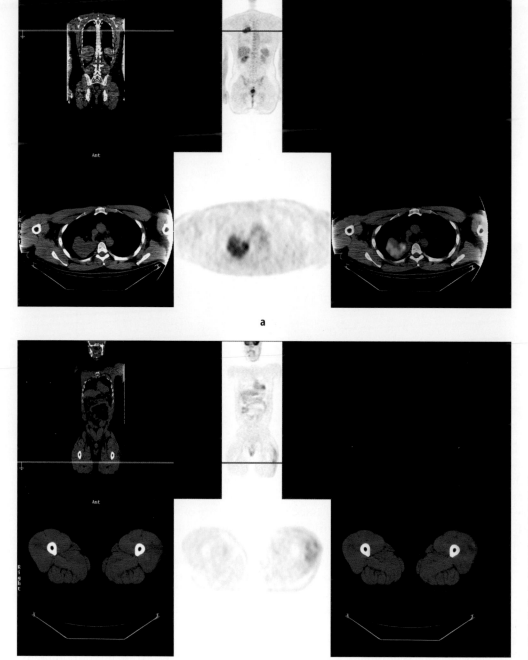

Figure 17.5. FDG-PET/CT images in a patient with metastases from a synovial sarcoma 14 years after excision from the anterior abdominal wall. The FDG-PET/CT images demonstrate not only a pulmonary metastasis (**a**) but also a metastasis in the thigh (**b**). Whole-body PET/CT imaging offers the possibility of detection of metastases in all organ systems and precise localization of the anatomic site. The pulmonary metastasis was identified on a diagnostic CT, but the additional site in the thigh was not known.

rence. It has a sensitivity as high as 96% when a mass is seen on MRI with high signal intensity on gadolinium-enhanced or T_2-weighted images (26, 27). Recurrence, however, can be difficult to identify after surgery, especially if there has been local radiotherapy (26, 27). False-positive and -negative results can occur if the characteristics and signal intensity of the primary tumor are unknown or not available for comparison (29). Recurrence is often the result of inadequate surgery because the primary tumor can extend beyond the anatomic boundaries. It is for this reason that a minimum margin of clearance at surgery is essential.

Detection of local recurrence by FDG-PET imaging has been investigated by a number of groups. Kole et al. (71) examined FDG-PET scans in 17 patients undergoing evaluation for recurrence of STS and found that PET identified 14 of 15 confirmed recurrences. Tumors as small as 0.5 cm were as easily identified as larger lesions (up to 20 cm). One low-grade liposarcoma recurrence was missed, and 2 of 17 patients had benign causes for an abnormal MRI. MRI and CT yielded false-negative results in 3 patients and false-positive findings in 2. Similar results were found by Schwarzbach et al. (43) in a smaller number of recurrences. Lucas et al. (66) examined a larger group of 60 patients; FDG-PET had a sensitivity of 73.7% and a specificity of 94.3%. There were 5 false-negative results, of which MRI and CT failed to identify 3 with recurrence. There were 3 false-positive results with FDG-PET. MRI had a sensitivity of 88.2% and a specificity of 96%. Schwarzbach et al. (42) examined 28 patients with suspected local recurrence and showed a sensitivity of 88% and specificity of 92%. The minor differences between the study by Schwarzbach et al. (42) and the study by Lucas et al. (66) are likely to be related to the tumor types. The results confirm that the use of a combination of imaging modalities is likely to result in the highest detection rate. At the end of the day, a tissue diagnosis is required to confirm the presence or otherwise of tumor.

Recurrent disease may result in the need for amputation, and the FDG-PET appearances in the stump need to be recognized. Hain et al. (72) showed that there are a range of physiologic appearances in the amputee stump and that recurrence can be identified by the specific features:

1. Diffuse uptake was found in stumps for up to 18 months postsurgery without any evidence of disease recurrence, this was noticed to be greater in lower limb amputations than upper limb.
2. Focal areas of uptake either were associated with known pressure areas with skin breakdown that could be seen clinically, or, in the absence of localized clinical changes, represented a recurrence and needed a biopsy.

Distant metastatic disease has been identified previously.

Bone Sarcomas

The identification of local recurrence has been demonstrated by Franzius et al. (73). In their study, FDG-PET imaging detected six local recurrences and one false-positive case, whereas MRI detected six recurrences but had two false-positive cases. The additional value from PET imaging is detection of distant metastases during the same imaging session because of the whole-body nature of the procedure. There are potential problems with both imaging modalities if prosthetic implants are in place, but these are less of a problem with FDG-PET than with MRI.

Monitoring Therapy Response

Soft Tissue Sarcoma

Individual randomized trials have not demonstrated a benefit from adjuvant chemotherapy in patients with localized and respectable STS. A recent meta-analysis (74) suggested that doxorubicin may extend recurrence-free intervals and showed a trend to improved overall survival. The quality of life during the treatment periods

was not assessed, and serious side effects were reported. A variety of new methods of therapy are being tried, and combinations of chemotherapy or new targeted therapies may prove to be successful. An effective method of assessing tumor response is needed for evaluation of new therapies.

Jones et al. (75) assessed the value of FDG-PET in nine patients who underwent either chemotherapy or combined radiotherapy and hyperthermia. In areas of radiotherapy treatment, very low uptake corresponded to areas of radiation-induced necrosis, and a peripheral high uptake was seen with a fibrous pseudocapsule forming, making differentiation from viable tumor difficult. Following chemotherapy, there was a more homogeneous reduction in FDG uptake, but the uptake that remained was often associated with benign therapy-related fibrous tissue. The timing of the second scans is therefore likely to be a crucial factor in disease response assessment.

van Ginkel et al. (65) used ^{11}C-tyrosine to assess response of sarcomas to isolated limb perfusion and thought that inflammatory changes did not interfere with residual viable tumor assessment. They found marked reduction of ^{11}C-tyrosine uptake in the responders. Further work has also been performed using ^{11}C-thymidine (see following).

Bone Sarcomas

Over the past 20 years, survival in bone sarcomas has improved from 20% to more than 60%, predominantly because of the use of adjuvant therapy and improved surgical techniques. The utilization of chemotherapy preoperatively, sometimes with radiotherapy, has facilitated the use of limb-sparing surgery. The risk with limb-sparing procedures was an increase in local recurrence rate, which was approximately 10% compared with 2% for amputation. The higher risk was related to poor response to chemotherapy and narrow resection margins.

Established methods for assessing response to chemotherapy have included ^{67}Ga and ^{201}Tl imaging, which were found to be superior to ^{99m}Tc-diphosphonates (76–79). Recently, the role of dynamic MRI assessment in predicting tumor response (80) and the use of magnetic resonance spectroscopy to monitor therapy (81) have raised interest. FDG-PET imaging has undergone assessment in neoadjuvant therapy using tumor-to-background ratios (TBR). Schulte et al. (60), using TBR measurements, identified all responders to chemotherapy, all these responders had TBR ratios of less than 0.6 and 8 of 10 nonresponders had TBR greater than 0.6.

These results are surprising, given the crudeness of the observations and the potential effect of chemotherapy on the background regions. Hawkins et al. (82) demonstrated a good correlation between the change in maximum SUV prechemotherapy and postchemotherapy with the extent of necrosis and histologic response. However, the absolute

SUV postchemotherapy and the change in SUV failed to identify a good response and an unfavorable response in 16% to 27% of patients. The authors discuss one possible explanation, which is the use of maximum SUV rather than an average change over the tumor. A maximum SUV, although predictive of the most aggressive area of the tumor and therefore the overall behavior of the tumor, does not necessarily reflect the heterogeneous tumor mass response. The other possible problem is the timing of the posttherapy study with regard to macrophage infiltration.

Other Positron Emission Tracers

The primary tumor can be detected with a number of radiotracers by means of determination of blood flow (43), DNA turnover using ^{11}C-thymidine (83) or ^{18}F-fluorothymidine (FLT), amino acid turnover (75, 84), and tumor hypoxia (85–87), as well as glucose metabolism. This combination of approaches could possibly be used for metabolic staging of tumors, which might have a predictive value comparable to histologic techniques. These tracers could also provide information to direct and monitor response to neoadjuvant therapy (60, 74), possibly by assessment of the delivery of therapy with the flow and hypoxia images and the response to therapy using FDG and [11]C- thymidine and/or FLT. An interesting study of only two patients with sarcoma showed that, using FDG and ^{11}C-thymidine uptake as markers of tumor glucose metabolism and DNA biosynthesis, respectively, the response to treatment could be followed (83). One patient who was unresponsive had an increase in FDG uptake and essentially no change in thymidine flux, whereas the responder had a decline in both measurements. The changes suggest that either agent or both could prove to be of value in the assessment of novel therapies before larger clinical trials. Furthermore, in bone sarcomas, the failure of therapy is likely to be related to multidrug-resistant gene expression, and it is possible this expression could be evaluated before therapy and following therapy using PET tracers. The role of ^{99m}Tc-MIBI should also be explored in that regard. The use of ^{18}F-fluoride for skeletal imaging in bone sarcomas also needs further investigation in terms of identifying skip lesions and metastatic disease.

Hypoxia-Cell Tracers

Hypoxia in tumors can result in areas of necrosis leading to diagnostic and therapeutic problems for the clinician. Necrotic areas related to severe hypoxia can lead to non-diagnostic or inaccurate biopsies. Hypoxic and ischaemic tissue may account, in part, for resistance to radiation therapy (88) and chemotherapy (86, 89), and it has been suggested that hypoxia itself may promote the develop-

ment of multidrug resistance. The ability to identify tumor hypoxia will allow identification of appropriate sites for image-guided biopsy and may identify tumors that will have a poor response to radiotherapy and/or chemotherapy. Nitroimidazole compounds, initially developed as radiosensitizing agents, have been of great interest because of their accumulation in hypoxic tumors (90) and areas of ischemia (91). Several compounds have been developed for imaging hypoxia (92). The PET tracer ^{18}F-fluoromisonidazole (FMISO) has been studied in a variety of tumors with models produced for analyzing tumor hypoxia (85). This technique has been used to determine the tumor hypoxic fraction both before and during radiotherapy of lung cancer and other cancers (87). There are now a number of tracers under development to identify hypoxic tissue, including [62]Cu- and ^{64}Cu-labeled compounds (93). The use of such imaging in STS may predict tumor responsiveness to radiation. Rajendren et al. (94) examined FMISO and FDG-PET imaging in soft tissue and bone sarcomas and found that the majority of patients had regions of hypoxia. There was no correlation between hypoxic regions and FDG uptake, vascular endothelial growth factor, or the size of the tumor. It has been suggested that FDG can be used as a surrogate marker for hypoxia, but this study would suggest this is not the case. It is disappointing that tumor size could not be related to the hypoxic fraction because this would have been expected. The study included both soft tissue and bone sarcomas, however, and inclusion of different types of tumors may have masked correlation in selected types.

Conclusion

Two recent meta-analyses have been performed to investigate the role of FDG-PET imaging in soft tissue sarcomas (95, 96), one of which included bone sarcomas (95). Each study reached slightly different conclusions despite similar search strategies. The pooled sensitivity and specificity for detection of malignant lesions were 91% and 85%, respectively (95), and 87% and 73% (using SUV greater than 2.0) after a 60-min imaging time for FDG (96). The conclusions from both studies were that FDG-PET imaging can discriminate between benign and low-grade tumors and intermediate- and high-grade tumors. Ioannidis et al. (96) suggested that FDG-PET imaging is useful for detection of tumor recurrence and may be helpful for grading tumors. Both groups highlight the difficulty in using PET to distinguish low-grade tumors from benign tumors.

FDG-PET imaging has a role in the grading and staging of primary tumors of soft tissue and bone. FDG-PET imaging has also a role for surveillance of disease recurrence and for detection of metastatic disease, and this role is likely to be enhanced by the use of integrated PET/CT

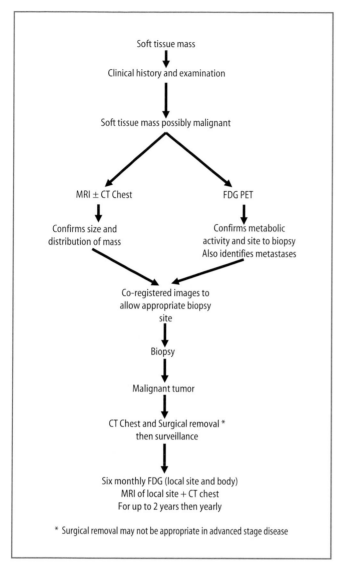

Figure 17.6. Algorithm for the management of soft tissue sarcomas.

imaging. There does appear to be variable uptake in pulmonary metastases, and therefore the combination of FDG-PET and CT of the chest seems a useful combination of imaging to assess the whole body. The advent of integrated PET/CT imaging may therefore obviate the need for separate CT of the chest. The use of PET imaging in diagnosis and follow-up of patients with sarcomas is summarized in Figure 17.6.

The role in the initial assessment of soft tissue sarcomas is likely to be extended. The use of MRI and FDG-PET image coregistration will allow selection of the most appropriate site within the tumor to biopsy (97, 98), avoiding areas of low metabolic activity. Furthermore, the development of novel therapies may benefit from a means of rapid, noninvasive assessment using a combination of PET tracers to assess ischemia, glucose metabolism, protein metabolism, and cell proliferation.

References

1. Mack TM. Sarcomas and other malignancies of soft tissue, retroperitoneum, peritoneum, pleura, heart, mediastinum, and spleen. Cancer (Phila) 1995;75(1 suppl): 211–244.
2. Fletcher CDM, Uni KK, Meteus F, editors. World Health Organisation Classification of Tumors. Pathology and Genetics of Tumors of Soft Tissue and Bone. Lyon: IARC Press, 2002.
3. Enzinger FM, Weiss SW, editors. Soft Tissue Tumors, 3rd ed. St. Louis: Mosby, 1995.
4. Trojani M, Contesso G, Oindrej M, et al. Soft tissue sarcomas of adults: study of pathological and prognostic variables and definition of a grading system. Int J Cancer 1984;33: 37–42.
5. Lichenstein L. Classification of primary tumors of bone. Cancer (Phila) 1951;4:335–341.
6. Schwajowicz F, Sissons H, Sobin L. The World Health Organization's histologic classification of bone tumors. Cancer (Phila) 1995;75:1208–1214.
7. Russell WD, Cohen J, Enzinger FM, et al. A clinical and pathological staging system for soft tissue sarcomas. Cancer (Phila) 1977;40:1562–1570.
8. Enneking WF, Spanier SS, Goodman MA. A system for the surgical staging of musculo-skeletal sarcoma Clin Orthop 1980;153:105–120.
9. Hajdu SI. Pathology of soft tissue tumors. Philadelphia: Lea & Febiger, 1979.
10. Lawrence W Jr, Donegan WL, Natarajan N, Mettlin C, Beart R, Winchester D. Adult soft tissue sarcoma. A pattern of care survey of the American College of Surgeons. Ann Surg 1996;205(4):349–359.
11. Goldstein H, McNeil BJ, Zufall E, Jaffe N, Treves S. Changing indications for bone scintigraphy in patients with osteosarcoma. Radiology 1980;135:177–180.
12. Murray IPC, Ellison BS. Radionuclide bone imaging for primary bone malignancy. Clin Oncol 1986;5:141–158.
13. McKillop JH, Etcubanas E, Goris ML. The indications for and limitations of bone scintigraphy in osteogenic sarcoma:a review of 55 patients. Cancer (Phila) 1981;48(5):1133–1138.
14. Clasby R, Tilling K, Smith MA, Fletcher CDM. Variable management of soft tissue sarcoma: regional audit with implications for specialist care. Br J Surg 1998;84:692–696.
15. Goodlad JR, Fletcher CDM, Smith MA. Surgical resection of primary soft tissue sarcoma: incidence of residual tumor in 95 patients needing re-excision after local resection. J Bone Joint Surg 1996;78B:658–661.
16. Mankin HJ, Lange TA, Spanier SS. The hazards of biopsy in patients with malignant primary bone and soft tissue tumors. J Bone Joint Surg 1982;64A:1121–1127.
17. Springfield DS, Rosenberg A. Biopsy: complicated and risky. J Bone Joint Surg 1996;78A:639–643.
18. Brenner W, Bohuslavizki KH, Eary JF. PET imaging of osteosarcoma. J Nucl Med 2003;44:930–942.
19. Dome JS, Schwartz CL. Osteosarcoma. In: Waterhouse DO, Cohn SC, editors. Diagnostic and Therapeutic Advances in Paediatric Oncology. Boston: Kluwer, 1997:215–251.
20. Myers PA, Gorlick R, Heller G, et al. Intensification of preoperative chemotherapy for osteogenic sarcoma: results of the Memorial Sloan-Kettering (T12 protocol). J Clin Oncol 1998;16:2452–2458.
21. Bielack SS, Kempf-Bielack B, Delling G, et al. Prognostic factors in high-grade osteosarcoma of the extremities or the trunk: an analysis of 1,702 patients treated on neoadjuvant cooperative osteosarcoma study group protocols. J Clin Oncol 2002;20:776–790.
22. Souhami RL, Craft AW, Van Dereiojken JW, et al. Randomised trial of two regimes of chemotherapy in operable osteosarcoma. A study of the European Osteosarcoma Group. Lancet 1997;350:911–917.
23. Craft AW, for UK Children's Cancer Study Group. Long term results from the first UKCCSG Ewing's tumor study (ET-1). Eur J Cancer 1997;33:1061–1069.

24. van Trommel MF, Kroon HM, Bloem JL, Hogendoorn PC, Taminiau AH. MR imaging based strategies in limb salvage surgery for osteosarcoma of the distal femur. Skeletal Radiol 1997;26(11):636–641.

25. Safram MR, Codie MH, Nambar S, et al. 151 Endoprosthetic reconstructions for patients with primary tumors involving bone. Contemp Orthop Addit 1994;29:15–25.

26. van der Woude HJ, Bloem JL, Hogendoorn PC. Preoperative evaluation and monitoring chemotherapy in patients with high-grade osteogenic and Ewing's sarcoma: review of current imaging modalities. Skeletal Radiol 1998;27(2):57–71.

27. van der Woude HJ, Verstraete KL, Hogendoorn PC, Taminiau AH, Hermans J, Bloem JL. Musculoskeletal tumors: does fast dynamic contrast-enhanced subtraction MR imaging contribute to the characterization? Radiology 1998;208(3):821–828.

28. van der Woude HJ, Bloem JL, Pope TL Jr. Magnetic resonance imaging of the musculoskeletal system. Part 9. Primary tumors. Clin Orthop 1998;347:272–286.

29. Bloem JL, van der Woude HJ, Geirnaerdt M, Hogendoorn PC, Taminiau AH, Hermans J. Does magnetic resonance imaging make a difference for patients with musculoskeletal sarcoma? Br J Radiol 1997;70(832):327–337.

30. Southee AE, Kaplan WD, Jochelson MS, et al. Gallium imaging in metastatic and recurrent soft tissue sarcoma. J Nucl Med 1992;33:1594–1599.

31. Finn HA, Simon MA, Martin WB, Darakjain H. Scintigraphy with gallium-67 citrate in staging soft tissue sarcoma of the extremity. J Bone Joint Surg 1987;69A:886–891.

32. Schwarz HS, Jones CK. The efficacy of gallium scintigraphy in detecting malignant soft tissue neoplasms. Ann Surg 1992;215:78–82.

33. Sato O, Kawai A, Ozaki T, Kunisada T, Danura T, Inoue H. Value of thallium-201 scintigraphy in bone and soft tissue tumors. J Orthop Sci 1998;3(6):297–303.

34. Garcia JR, Kim EE, Wong FCL, et al. Comparison of fluorine-18-FDG-PET and technetium-99m-MIBI SPECT in evaluation of musculoskeletal sarcomas. J Nucl Med 1996;37:1476–1479.

35. Ohta H, Endo K, Fujita T, et al. Clinical evaluation of tumor imaging using 99mTc(V) dimercaptosuccinic acid, a new tumor seeking agent. Nucl Med Commun 1988;9:105–116.

36. McLean RG, Murray IPC. Scintigraphic patterns in certain primary bone malignancies. Clin Radiol 1984;35:379–384.

37. Simon MA, Kirchener PT. Scintigraphic evaluation of primary bone tumors: comparison of technetium-99m phosphonate and gallium citrate imaging. J Bone J Surg 1980;62A:758–764.

38. Caner B, Kitapci M, Aras T, Erbengi G, Ugur O, Bekdik C. Increased accumulation of hexakis (2-methoxyisobutylisonitrile) technetium (99m) in osteosarcoma and it metastatic lymph nodes. J Nucl Med 1991;32:1977–1978.

39. Adler LP, Blair HF, Makley JT, et al. Noninvasive grading of musculoskeletal tumors using PET. J Nucl Med 1991;32:1508–1512.

40. Adler LP, Blair HF, Williams RP, et al.. Grading liposarcomas with PET using [18F]FDG. J Comput Assist Tomogr 1990;14:960–962.

41. Griffeth LK, Dehdashti F, McGuire AH, et al. PET evaluation of soft tissue masses with fluorine-18 fluoro-2-deoxy-d-glucose. Radiology 1992;182:185–194.

42. Schwarzbach MHM, Dimitrakopolou-Strauss A, Willeke F, et al. Clinical value of [18-F] fluorodeoxyglucose positron emission tomography imaging in soft tissue sarcomas. Ann Surg 2000;231:380–386.

43. Schwarzbach M, Willeke F, Dimitrakopoulou Strauss A, et al. Functional imaging and detection of local recurrence in soft tissue sarcomas by positron emission tomography. Anticancer Res 1999;19:1343–1349.

44. Lucas JD, O'Doherty MJ, Cronin BF, et al. Prospective evaluation of soft tissue masses and sarcomas using fluorodeoxyglucose positron emission tomography. Br J Surg 1999;86:550–556.

45. Lodge MA, Lucas JD, Marsden PK, Cronin BF, O'Doherty MJ, Smith MA. A PET study of [18]FDG uptake in soft tissue masses. Eur J Nucl Med 1999;26:22–30.

46. Nieweg OE, Pruim J, van Ginkel RJ, et al. Fluorine-18-fluorodeoxyglucose PET imaging of soft-tissue sarcoma. J Nucl Med 1996;37(2):257–261.

47. Eary JF, Conrad EU, Bruckner JD, et al. Quantitative [F-18]fluorodeoxyglucose positron emission tomography in pretreatment and grading of sarcoma. Clin Cancer Res 1998;4(5):1215–1220.

48. Eary JF, Mankoff DA. Tumor metabolic rates in sarcoma using FDG-PET. J Nucl Med 1998;39:250–254.

49. Dimitrakopolou-Strauss A, Strauss LG, Schwarzbach M, et al. Dynamic PET ^{18}F-FDG studies in patients with primary and recurrent soft tissue sarcomas: impact on diagnosis and correlation with grading. J Nucl Med 2001;42:713–720.

50. Kaplan AM, Chen K, Lawson MA, Wodrich DL, Bonstelle CT, Reiman EM. Positron emission tomography in children with neurofibromatosis. 1. J Child Neurol 1997;12:499–506.

51. Ferner R, Lucas JD, O'Doherty MJ, et al. Evaluation of [18]fluorodeoxyglucose positron emission tomography ([18]FDG-PET) in the detection of malignant peripheral nerve sheath tumors arising from within plexiform neurofibromas in neurofibromatosis. 1. J Neurol Neurosurg Psychiatry 2000;68:353–357.

52. Boerner AR, Weckesser M, Herzog H, et al. Optimal scan time for fluorine-18 fluorodeoxyglucose positron emission tomography in breast cancer. Eur J Nucl Med 1999;26:226–230.

53. Hamberg LM, Hunter GJ, Alpert NM, Choi NC, Babich JW, Fischman AJ. The dose uptake ratio as an index of glucose metabolism: useful parameter or oversimplification? J Nucl Med 1994;35:1308–1312.

54. Schulte M, Brecht Krauss D, Heymer B, et al. Fluorodeoxyglucose positron emission tomography of soft tissue tumors: is a non-invasive determination of biological activity possible? Eur J Nucl Med 1999;26:599–605.

55. Watanabe H, Shinozaki T, Yanagawa T. Glucose metabolic analysis of musculoskeletal tumors using fluorine-18-FDG-PET as an aid to preoperative planning. J Bone Joint Surg [Br] 2000;82:760–767.

56. Watanabe H, Inoue T, Shinozaki T, et al. PET imaging of musculoskeletal tumors with fluorine-18 a-methyl tyrosine: comparison with fluorine-18 fluorodeoxyglucose PET. Eur J Nucl Med 2000;27:1509–1517.

57. Kole AC, Plaat BBC, Hoekstra HJ, Vaalburg W, Molenaar WM. FDG and L-1-[11C]-tyrosine imaging of soft-tissue tumors before and after therapy. J Nucl Med 1999;40:381–386.

58. Kole AC, Nieweg OE, Hoekstra HJ, van Horn JR, Koops HS, Vaalburg W. Fluorine-18-fluorodeoxyglucose assessment of glucose metabolism in bone tumors. J Nucl Med 1998;39(5):810–815.

59. Schulte M, Brecht Krauss D, Heymer B, et al. Grading of tumors and tumor-like lesions of bone: evaluation by FDG PET. J Nucl Med 2000;41:1695–1701.

60. Schulte M, Brecht Krauss D, Werner M, et al. Evaluation of neoadjuvant therapy response of osteogenic sarcoma using FDG PET. J Nucl Med 1999;40:1637–1643.

61. Dehdashti F, Siegel BA, Griffeth LK, et al. Benign versus malignant intraosseous lesions: discrimination by means of PET with 2-[F-18] fluoro-2-deoxy-D-glucose. Radiology 1996;200:243–247.

62. Aoki J, Watanabe H, Shinozaki T, Tokunaga M, Inoue T, Endo K. FDG-PET in differential diagnosis and grading of chondrosarcomas. J Comput Assist Tomog 1999;23:603–608.

63. Lee FY-I, Yu J, Chang S-S, Fawwaz R, Parisien MV. Diagnostic value and limitations of fluorine-18 fluorodeoxyglucose positron emission tomography for cartilaginous tumors of bone. J Bone Joint Surg 2004;86(12):2677–2685.

64. Eary JF, O'Sullivan F, Powitan Y, et al. Sarcoma tumor FDG uptake measured by PET and patient outcome: a retrospective analysis. Eur J Nucl Med 2002;29:1149–1154.

65. van Ginkel RJ, Kole AC, Nieweg OE, et al. L-[1-11C]-Tyrosine PET to evaluate response to hyperthermic isolated limb perfusion for locally advanced soft-tissue sarcoma and skin cancer. J Nucl Med 1999;40(2):262–267.

66. Lucas JD, O'Doherty MJ, Wong JCH et al. Evaluation of the role of fluoro-deoxyglucose in the follow up management of soft tissue sarcomas. J Bone Joint Surg 1998;80B:441–447.

67. Tse N, Hoh C, Hawkins R, Phelps M, Glaspy J. Positron emission tomography diagnosis of pulmonary metastases in osteogenic sarcoma. Am J Clin Oncol 1994;17:22–25.

68. Shulkin BL, Mitchell DS, Ungar DR, et al. Neoplasms in a pediatric population: 2-[F-18]-fluoro-2-deoxy-D-glucose PET studies. Radiology 1995;194(2):495–500.

69. Franzius C, Sciuk J, Daldrup-Link HE, Jurgens H, Schober O. FDG-PET for detection of osseous metastases from malignant primary bone tumors: comparison with bone scintigraphy. Eur J Nucl Med 2000;27:1305–1311.

70. Potter DA, Glenn J, Kinsella T, et al. Patterns of recurrence in patients with high grade soft tissue sarcomas. J Clin Oncol 1985;3:353–366.

71. Kole AC, Nieweg OE, van Ginkel RJ, et al. Detection of local recurrence of soft-tissue sarcoma with positron emission tomography using [18F]fluorodeoxyglucose. Ann Surg Oncol 1997;4:57–63.

72. Hain SF, O'Doherty MJ, Lucas JD, et al. Fluorodeoxyglucose PET in the evaluation of amputations for soft tissue sarcoma. Nucl Med Commun 1999;20:845–848.

73. Franzius C, Daldrup-Link HE, Wagner-Bohn A, et al. FDG PET for detection of recurrences from malignant primary bone tumors: comparison with conventional imaging. Ann Oncol 2002;13:157–160.

74. Sarcoma Meta-analysis Collaboration. Adjuvant chemotherapy for localised resectable soft tissue sarcoma in adults. Cochrane Library 2000;3:1–26.

75. Jones DN, McCowage GB, Sostman HD, et al. Monitoring of neoadjuvant therapy response of soft-tissue and musculoskeletal sarcoma using fluorine-18-FDG PET. J Nucl Med 1996;37(9):1438–1444.

76. Ramanna L, Waxman A, Binney G, Waxman S, Mirra J, Rosen G. Thallium-201 scintigraphy in bone sarcoma: comparison with gallium-67 and MDP in the evaluation of therapeutic response. J Nucl Med 1990;31:567–572.

77. Knop J, Delling G, Heise U, Winkler K. Scintigraphic evaluation of tumor regression during preoperative chemotherapy of osteosarcoma. Skeletal Radiol 1990;19:165–172.

78. Estes DN, Magill HL, Thompson EI, Hayes FA. Primary Ewing sarcoma: follow up with Ga-67 scintigraphy. Radiology 1990;177:449–453.

79. Ohtomo K, Terui S, Yokoyama R, et al. Thallium-201 scintigraphy to assess effect of chemotherapy in osteosarcoma. J Nucl Med 1996;37:1444–1448.

80. Erlemann R, Sciuk J, Bosse A, et al. Response of osteosarcoma and Ewing sarcoma to preoperative chemotherapy: assessment with dynamic and static MR imaging and skeletal scintigraphy. Radiology 1990;175:791–796.

81. Ross B, Helsper JT, Cox IJ. Osteosarcoma and other neoplasms of bone: magnetic resonance spectroscopy to monitor therapy. Arch Surg 1987;122:1464–1469.

82. Hawkins DS, Rajendran JG, Conrad EU, Bruckner JD, Eary JF. Evaluation of chemotherapy response in paediatric bone sarcomas by [F-18]-fluorodeoxy-D-glucose positron emission tomography. Cancer (Phila) 2002;94:3277–3284.

83. Shields AF, Mankoff DA, Link JM, et al. Carbon-11-thymidine and FDG to measure therapy response. J Nucl Med 1998;39(10):1757–1762.

84. Plaat B, Kole A, Mastik M, Hoekstra H, Molenaar W, Vaalburg W. Protein synthesis rate measured with L-[1-11C]tyrosine positron emission tomography correlates with mitotic activity and MIB-1 antibody-detected proliferation in human soft tissue sarcomas. Eur J Nucl Med 1999;26(4):328–332.

85. Casciari JJ, Graham MM, Rasey JS. A modeling approach for quantifying tumor hypoxia with [F-18] fluoromisonidazole PET time activity data. Med Phys 1995;22:1227–1139.

86. Moulder JE, Rockwell S. Tumor hypoxia: its impact on cancer therapy. Br J Radiol 1987;26:638–648.

87. Rasey JS, Koh WJ, Evans ML, et al. Quantifying regional hypoxia in human tumors with positron emission tomography of [F-18]fluoromisonidazole: a pretherapy study of 37 patients. Int J Radiat Biol Phys 1996;36:417–428.

88. Peters LJ, Withers HR, Thames HD, Fletcher GH. Keynote address. The problem: tumor radioresistance in clinical radiotherapy. Int J Radiat Oncol Biol Phys 1982;8:101–108.

89. Rockwell S. Effect of some proliferative and environmental factors on the toxicity of mitomycin C to tumor cells in vitro. Int J Cancer 1986;38:229–235.

90. Adams GE, Flockhart IR, Smithen CE, et al. Electron-affinic sensitisation. VII. A correlation between structures, one electron reduction potentials and efficiencies of nitroimidazoles as hypoxic cell sensitisers. Radiat Res 1976;67:9–20.

91. Chapman JD, Baer K, Lee J. Characteristics of the metabolism-induced binding of misonidazole to hypoxic mammalian cells. Cancer Res 1983;43:1523–1528.

92. Nunn A, Linder K, Strauss HW. Nitroimidazoles and imaging hypoxia. Eur J Nucl Med 1995;22:265–280.

93. Lewis JS, Sharp TL, Laforest R, Fujibayashi Y, Welch MJ. Tumor uptake of copper-diacetyl-bis(N4-methylsemicarbazone): effect of changes in tissue oxygenation. J Nucl Med 2001; 42:655–661.

94. Rajendran JG, Wilson DC, Conrad EU, et al. [18F]FMISO and [18F]FDG PET imaging in soft tissue sarcomas: correlation of hypoxia metabolism and VEGF expression. Eur J Nucl Med Mol Imaging 2003;30:695–704.

95. Bastiaannet E, Groen H, Jager PL, et al. The value of FDG-PET in the detection, grading and response to therapy of soft tissue and bone sarcomas: a systematic review and meta-analysis. Cancer Treat Rev 2004;30(1):83–101.

96. Ioannidis JP, Lau J. 18F-FDG PET for the diagnosis and grading of soft-tissue sarcoma: a meta-analysis. J Nucl Med 2003; 44(5):717–724.

97. Somer EJ, Marsden PK, Benatar NA, Goodey J, O'Doherty MJ, Smith MA. PET-MR image fusion in soft tissue sarcoma: accuracy, reliability and practicality of interactive point-based and automated mutual information techniques. Eur J Nucl Med Mol Imaging 2003;30(1):54–62.

98. Hain SF, O'Doherty MJ, Bingham J, Chinyama C, Smith MA. Can FDG PET be used to successfully direct preoperative biopsy of soft tissue tumors? Nucl Med Commun 2003;24(11):1139–1143.

18

PET and PET/CT Imaging in Thyroid and Adrenal Diseases

I. Ross McDougall

I first met Peter Valk when I was on a fellowship to Stanford University from 1972 to 1974. He was a fellow at the Donner Laboratory in Berkeley. Although we were not collaborators, our geographic proximity during the next three decades meant that our paths crossed at local and regional meetings and even rarely by chance in San Francisco. His early entry into clinical positron emission tomography (PET) resulted in my frequently being in the audience listening to, and admiring, his clinical and research activities. I was proud to be invited to contribute to the first edition of his textbook devoted to PET imaging and am equally proud but greatly saddened as I rewrite this chapter for the memorial edition.

The major role for PET in thyroid disease is in the management of patients with cancer, and ^{18}F-fluorodeoxyglucose (FDG) is the radiopharmaceutical of choice for tumor imaging. The major part of this chapter concerns FDG in the management of thyroid cancer. The role of PET in differentiating benign from malignant thyroid nodules is described briefly. There have been quantitative studies of thyroid function using ^{124}I for thyroid imaging and dosimetry, which are included. In addition, the importance of recognizing abnormalities in the thyroid in FDG scans in patients with diseases of nonthyroidal origin is discussed. Finally, the limited role of PET in the investigation of adrenal disorders is addressed.

Thyroid Glands

The Normal Thyroid on PET Imaging

Although some textbooks state that the normal thyroid can be seen distinctly on FDG scan, I disagree with that statement. In many patients, there is no concentration of FDG in normal glands, as is well demonstrated with PET/CT imaging where computed tomography (CT) demonstrates the anatomy of the thyroid and surround-

ing structures (Figure 18.1) Symmetrical intense uptake of FDG is seen in Grave's or autoimmune thyroid disease, in particular Hashimoto's thyroiditis (Figure 18.2).

Thyroid Cancer

Cancers of the thyroid can arise from the follicular or parafollicular cells (1). Rarely, they come from malignant transformation of lymphocytes. Metastases to the thyroid can occur from primary lung, breast, or kidney cancer and melanoma, and although present at autopsy in patients dying of widespread metastases, they are rare as isolated lesions (2).

Pathologically, cancers of the follicular cells are classified as papillary, follicular, or anaplastic. The first two are usually grouped together as differentiated thyroid cancers because they retain structural and physiologic features of normal thyroid cells. They account for about 90% of thyroid cancers and, in countries where the intake of iodine is large, there is a preponderance of papillary cancers. In contrast, in regions of iodine deficiency the prevalence of follicular cancers increases, and these can become the majority. Differentiated cancers usually retain the ability to trap iodine and secrete thyroglobulin (Tg). There are variants of these cancers such as mixed papillary-follicular, follicular variant of papillary, tall and clear cell variants, and Hürthle cell cancer.

The principles of treatment are the same for all types. The thyroid should be removed by total, or near-total, thyroidectomy, and in patients with larger, or invasive, or metastasizing cancers, whole-body imaging with radioiodine and treatment of functioning tissue with ^{131}I is advised (3). When therapy has been successful, there should be no abnormal uptake on whole-body ^{131}I or ^{123}I scan, or measurable Tg. These tests are concordant in about 75% to 80% of patients, and in the remainder, the usual disparity is an abnormal Tg but a negative radioiodine scan. Disparate results are more common in patients with papillary cancers. Metastases from Hürthle cell

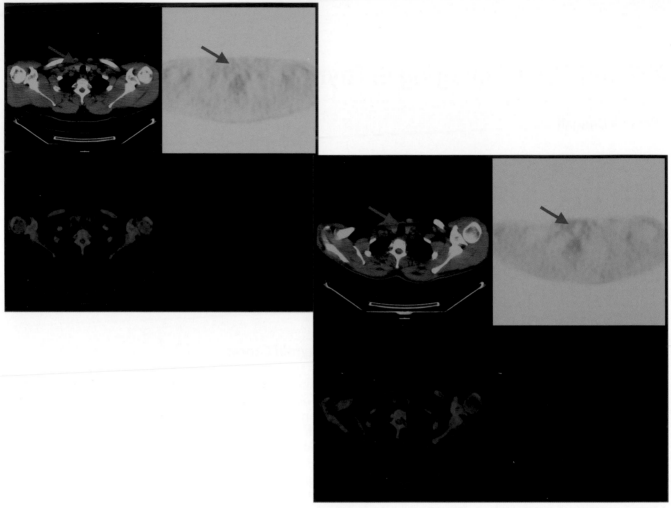

Figure 18.1. PET and combined PET/CT in two patients who have normal thyroid structure and function. There is no uptake of ¹⁸FDG 1 h after intravenous injection of 555 MBq. (arrows)

cancer, which is usually classified as follicular carcinoma, are almost always Tg positive and radioiodine negative.

Anaplastic cancers that also derive from follicular cells are classified as undifferentiated because they have no structural or functional characteristics of normal thyroid. The prognosis is dismal, with most patients dying in 3 to 6 months. ¹³¹I imaging, or therapy and measurement of Tg,

are of no value in the management of patients with this cancer.

Cancers arising from the parafollicular cells are called medullary cancers and secrete calcitonin, which is a valuable tumor marker (4). They account for about 5% of thyroid cancers. The majority of medullary cancers are isolated lesions, but 25% to 33% are hereditary cancers:

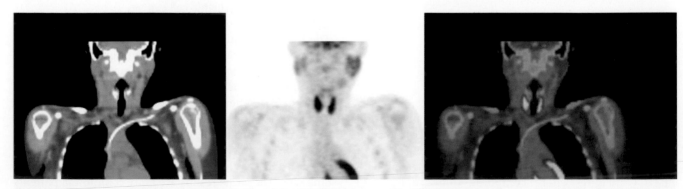

Figure 18.2. PET and combined PET/CT in patient who has Hashimoto's thyroiditis. There is intense uptake of ¹⁸FDG symmetrically throughout the gland 1 h after intravenous injection of 555 MBq.

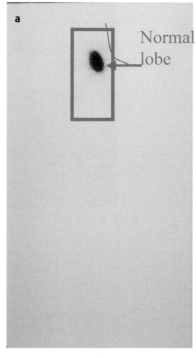

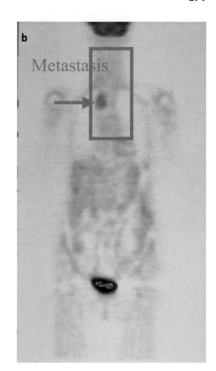

Iodine-131

F-18 FDG

Figure 18.3. (a) Whole-body ^{131}I scan in a 69-year-old woman referred to Stanford Medical Center. She had a right lobectomy several years previously for papillary cancer, and she had external radiation for a local recurrence. She had not had the left lobe removed. Chest roentgenogram and CT showed a right superior mediastinal mass. [131]I scan showed the residual left lobe, (arrow) whereas ^{18}F-FDG-PET scan (b) showed uptake in the chest mass (arrow) but not the left lobe. At surgery, the mass was metastatic papillary cancer and the left lobe was normal, demonstrating that cancer is seen on PET scan but normal thyroid is not.

these can be classified as familial medullary cancer, or as part of multiple endocrine neoplasias (MEN IIa, and MEN IIb). Medullary cancers do not trap iodine. The management problem is when a patient has had apparently curative surgery but calcitonin values remain elevated.

The major role of PET is to identify those cancers that cannot be imaged using radioiodine, and its place in the management of these individual cancers is discussed next.

PET Imaging in Patients with Differentiated Thyroid Cancers

It is generally acknowledged that PET imaging is unnecessary in the majority of patients with differentiated thyroid cancer because radioiodine scintigraphy, either diagnostic or posttherapeutic, together with serum Tg measurements, usually define the presence or absence of disease. In addition, it has been shown that FDG-PET imaging has a lower sensitivity in well differentiated cancers that retain the ability to trap radioiodine (Figure 18.3). The main role for FDG-PET imaging is in patients with negative ^{123}I or ^{131}I scan but measurable Tg. There are several reports addressing this topic (5–22), the first by Joensuu and Ahonen in 1987 (5). Table 18.1 provides a synopsis of the data. The sensitivity of PET imaging for identifying the source of Tg varies from 60% to 90%. In these reports, imaging was conducted with a dedicated PET camera. Most of the lesions were in the cervical region, usually in the thyroid bed or in adjacent lymph nodes, but some lesions were distant. The demonstration of lesions can lead to a change in therapy, including surgery or external

radiation instead of radioiodine. Figure 18.4 shows examples of PET and radioiodine images in patients who were suspected of having tumor but the extent of disease was not known.

The approach to the Tg-positive/radioiodine-negative patients has been reported by Desai et al. and Karwowski

Table 18.1. Frequency of radioiodine-negative/positron emission tomography (PET)-positive scans in patients with suspected recurrent or metastatic disease.

Study	Number ^{131}I negative	Number [18]FDG positive	Comments
Chung et al.	23	17	High thyroglobulin
Chung et al.	15	14	Normal thyroglobulin
Wang et al.	19, 13 true negative	18, 14 true positive	High thyroglobulin in 18
Jadvar et al.[a]	10	5	High thyroglobulin
Altenvoerde et al.	12	6	High thyroglobulin
Dietlin et al.	11	7	High thyroglobulin
Feine	21	19	High thyroglobulin or abnormal exam
McDougall[b]	42	24	Mostly high Tg
Alnafasi et al.	11	11	Posttherapy ^{131}I negative, high Tg
Muros et al.	10	6	Posttherapy [131]I
Grunwald et al.	13	11	

Tg, thyroglobulin.
[a]Early and [b]later reports from Stanford.
Source: From Valk PE, Bailey DL, Townsend DW, Maisey MN. Positron Emission Tomography: Basic Science and Clinical Practice. Springer-Verlag London Ltd 2003, p. 671.

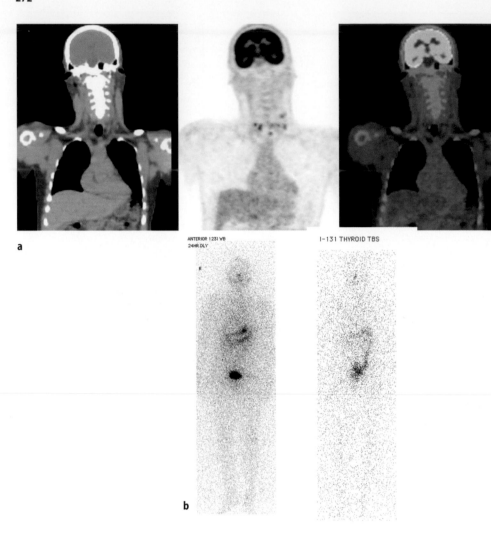

ANTERIOR 123I WB
24HR DLY

I-131 THYROID TBS

Figure 18.4. (**a**) PET/CT obtained 1 h after injection of 555 MBQ FDG. There are several foci of abnormal uptake in the low neck corresponding to metastases in lymph nodes. The patient had an elevated Tg and negative ^{123}I and ^{131}I whole-body scans (**b**).

et al. (23, 24). When the Tg is significantly elevated, PET (now PET/CT) imaging is performed. When local lesions are identified, ultrasound (US) of that specific region is undertaken and an US-guided fine-needle biopsy (FNA) obtained. When recurrent thyroid cancer is diagnosed the patient undergoes surgery, and intraoperative US has helped ensure the cancer is localized and fully excised.

Several alternative nuclear medicine procedures have been evaluated, including whole-body scan with [201]Tl (25), ^{99m}Tc-sestamibi (26), ^{99m}Tc-tetrafosmin (27), or ^{111}In-octreotide (28, 29), and there are several reports that present sensitivity, specificity, and accuracy of these tests. PET imaging, when available, became the imaging modality of choice because the better resolution of the images and superior sensitivity for detection of disease.

Most of the studies with PET were in patients with negative ^{131}I scans and measurable Tg values, and in most published reports the results are seldom compared with other imaging procedures in significant numbers of patients. The largest number of patients in a single publication is the report of a multicenter study from seven PET centers in Germany. The investigators evaluated 222 patients (30). Of these, 166 had negative ^{131}I scans, but many of these were believed to be true negatives. The investiga-

tors were able to calculate overall sensitivity and specificity as well as these percentages in ^{131}I-negative and ^{131}I-positive patients, as well as in relation to the Tg and thyrotropin (TSH) levels at the time of scanning (Table 18.2).

Table 18.2. Sensitivity and specificity of PET scan extracted from Grunwald et al. (30).

Category[a]	Sensitivity (%)	Specificity (%)
All patients (222)	75	90
[131]I positive (56)	65	100
[131]I negative (166)	85	90
Tg less than 5 ng/mL (109)	87	61
Tg 5 ng/ml or more (107)	100	100
TSH less than 5 mu/L (46)	91	74
TSH ≥ 5 mu/l (141)	67	94
Papillary cancer (134)	73	86
Follicular cancer (80)	78	100
Hurthle cell cancer (8)	87	100

TSH, thyrotropin.
[a]Number of patients in parentheses.
Source: From Valk PE, Bailey DL, Townsend DW, Maisey MN. Positron Emission Tomography: Basic Science and Clinical Practice. Springer-Verlag London Ltd 2003, p. 672.

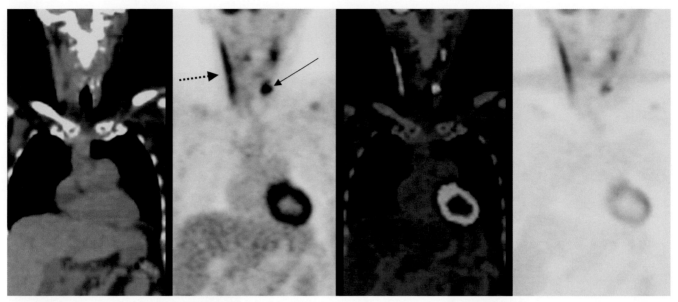

Figure 18.5. PET and combined PET/CT in patient who has had multiple operations, ^{131}I treatments, and external radiation for recurrent thyroid cancer. There is intense uptake of FDG in metastases shown by *arrow* after intravenous injection of 555 MBq. Uptake in muscle in the right neck is indicated by *dotted arrow*.

In addition, it has been demonstrated that FDG-PET can define the site of disease in patients who have negative posttreatment scans. Alnafisi et al. (20) showed positive results in 11 patients, all of whom had negative posttreatment scans. Muros et al. (19) reported a sensitivity of 90% in this situation (9 positive PET in 10 negative posttherapy scans).

In contrast, van Tol et al. (18) reported a high incidence of false-positive results in 10 patients, but this finding brings up two important issues. First, is it correct to label a PET abnormality in a Tg-positive patient as a false positive? Second, in some patients, how important is it to find microscopic differentiated thyroid cancer? The patients described by van Tol had Tg values from 3.4 to 30 ng/ml. The FDG-positive lesions were in the lungs or cervical nodes, and none became clinically apparent after follow-up of 17 to 33 months. The Tg value increased in 8 patients and decreased in 2. Therefore, it is not correct to conclude that FDG imaging gives a high incidence of false-positive results; rather, it provides early or preclinical positive information. The availability of Tg measurement also allows identifying residual or recurrent disease at a very early stage. We know from older follow-up reports, before Tg measurement was available, that most patients with differentiated thyroid cancer have an excellent prognosis (31). We also know that whole-body radioiodine scan and clinical examination accurately determine the extent of disease in most patients.

False-Positive and False-Negative PET Scans

False-positive scans do occur (32). Uptake of FDG in actively contracting muscles is one cause, as shown in Figure 18.5. Muscle relaxants are of value in tense patients (33). More recently, with integrated PET/CT imaging, it became evident that metabolically active (brown fat) is FDG avid (Figure 18.6). The role of PET/CT in correctly identifying muscle and fat is important. The muscles of speech can be imaged and can be misinterpreted as residual thyroid or nodal metastases (34, 35). PET/CT imaging shows the anatomic–physiologic relationship. Recurrent laryngeal nerve paralysis results in asymmetry of FDG uptake in laryngeal muscles, and this is easier to identify correctly using PET/CT imaging (Figure 18.7). After administration of FDG, the patient should be resting in a quiet environment and should not eat, talk, or exercise. The diffuse uptake of FDG in Hashimoto's thyroiditis should not be misinterpreted as cancer (36). Infectious and inflammatory conditions can cause FDG uptake, which can be misinterpreted as a metastasis. Identification of focal disease in the neck is usually an indication for surgery; therefore, to exclude a false-positive result, histologic confirmation of the malignant nature of the FDG-avid lesion is recommended before surgical reexploration.

False-negative FDG-PET images can occur in well-differentiated cancers. They are most likely in patients with very small volumes of cancer because the resolution of PET is about 4 to 6 mm full-width half-maximum (FWHM). Another factor that can potentially cause false-negative results is intake of food before injection of FDG; this can cause a redistribution of FDG into muscle, which can obscure small lesions.

More than 100 patients with thyroid cancer underwent FDG-PET imaging at Stanford University Medical Center, most of whom had ^{131}I-negative and Tg-positive findings. Most patients who were ^{131}I negative and FDG-PET negative had no clinical evidence of disease on follow-up of 12

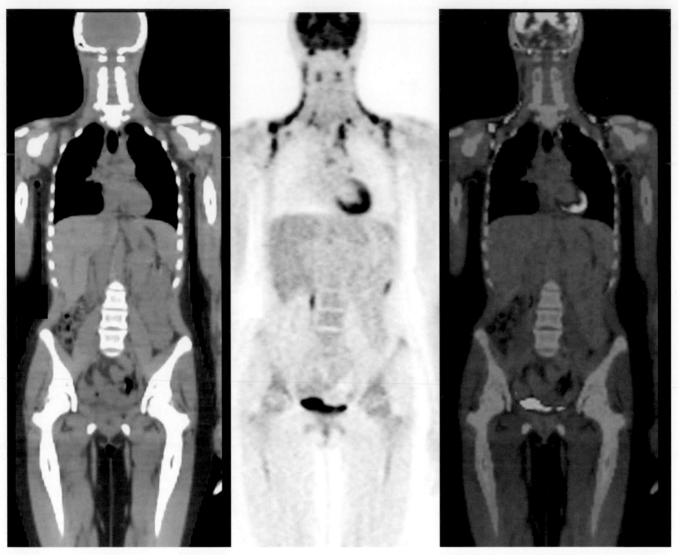

Figure 18-6. PET/CT in a patient with uptake of FDG in the neck that could be confused as extensive metastases to cervical lymph nodes. The CT shows this corresponds to fat ("brown" fat).

to 36 months. The reason was probably a small volume of cancer that does not grow when TSH is suppressed.

In patients with positive findings, the therapeutic approach was changed to include surgery or external radiation and in some a combination of treatments. At Stanford University Medical Center, patients that have a rising level of Tg but negative radioiodine images are seldom treated with large therapy doses of ^{131}I as recommended by some investigators (37, 38).

The Role of Thyrotropin in PET Imaging

For radioiodine imaging, it is necessary to have an elevated serum TSH level. TSH stimulates follicular cells to trap and to organify iodine. FDG imaging is based on glucose uptake by cancers, a direct result of the relatively elevated metabolic rate of the tumor. It would be anticipated that the metabolism of cancers would be increased by high levels of TSH, which stimulates both

growth and function. However, when the elevated TSH is caused by hypothyroidism, the result of withdrawal of thyroid hormone, the patient has reduced metabolism. Data on the role of an elevated TSH in increasing glucose uptake are limited and not uniform. Wang et al. (39) found no clinical difference in 4 patients who had paired studies with normal and elevated TSH values. Others have reported an increased uptake in the hypothyroid condition. Recently, recombinant human TSH (rhTSH) was shown to increase the number of both FDG-positive patients (19 versus 9) and FDG-positive lesions (82 versus 45) (40). Chin et al. (41) compared scans with and without rhTSH in 7 patients; 1 patient was only positive with rhTSH stimulation. The role of recombinant human TSH in stimulating uptake of FDG is an interesting and important area for study. If the patient is metabolically normal, exogenous TSH may augment uptake of glucose.

Our experience in this area is limited, although a lesion was seen in one hypothyroid patient that was not seen

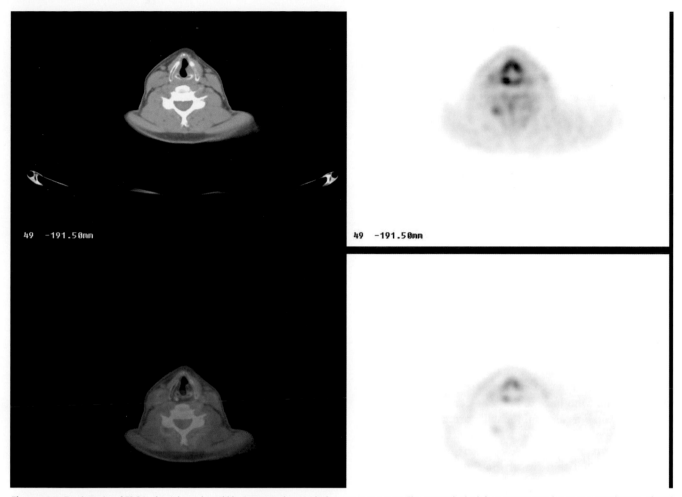

Figure 18.7. Focal uptake of FDG in the right neck could be interpreted as residual or recurrent cancer. The patient had a left recurrent nerve lesion as a complication of total thyroidectomy, and the FDG uptake is in the laryngeal muscles on the right.

when he was euthyroid. In other patients with negative PET scans, we found no advantage.

PET Imaging in Patients with Hürthle Cell Cancers

Hürthle cells are follicular cells that appear microscopically to be hyperactive. Their cytoplasm shows an increase in mitochondria and mitochondrial DNA. In the United States, Hürthle lesions are not common and the main concern is to differentiate a benign from malignant neoplasm., which requires excision of the lesion for pathologic examination to determine whether there is vascular or capsular invasion. The presence of either or both of these features classifies the neoplasm as malignant. Although Hürthle cell cancers are traditionally grouped pathologically with follicular cancers, these cancers seldom trap iodine and should be considered separately. Therefore, when there is residual or recurrent disease in a patient with Hürthle cell cancer that cannot be identified by whole-body radioiodine imaging, alternative whole-body imaging tests should be considered. There are reports of [201]Tl (26), ^{111}In-octreotide (42, 43), and ^{99m}Tc-sestamibi (26, 44) demonstrating lesions. FDG-PET imaging has a similar role, but the number of reports is limited (30, 45). In the multicenter study from Germany, the sensitivity of PET for Hürthle cell cancer was 87%, but only 8 patients were evaluated (30). Plotkin et al. (46) imaged 17 patients, of whom 13 had elevated Tg values. The sensitivity of FDG-PET imaging was 92% and the specificity 80% (1 false positive among 4 patients). Lowe et al. (47) imaged 14 patients with FDG, and these images provided additional important clinical information in 7 patients. The lesions are usually intense and, because Hürthle cell cancers are almost always radioiodine negative, FDG provides an important imaging tool for the follow-up of patients.

PET Imaging in Patients with Anaplastic Cancer of the Thyroid

The clinical diagnosis of anaplastic cancer is sadly easy, and the diagnosis usually predicts the patient's death within months. The cancers are inevitably radioiodine negative and there is seldom a role for diagnostic nuclear medicine imaging. However, in patients where the extent of disease is not obvious and in whom there could be an attempt for curative treatment, FDG has a role. Anaplastic cancers are usually intense on FDG. Combined PET/CT imaging gives added information of the extent of invasion of structures by the cancer.

FDG-PET Imaging in Patients with Medullary Cancers of the Thyroid

Medullary cell cancers arise from parafollicular cells and are best treated by total thyroidectomy with a central and lateral dissection of lymph nodes. The hope is that serum calcitonin becomes undetectable after surgery. When calcitonin remains measurable after surgery, there is residual or recurrent disease. In most patients the cancer is in the neck. Less commonly, the metastases are in the liver, lung, or skeleton. However, the lesions can be small and difficult to detect by anatomic or functional imaging. Because medullary cancer is uncommon and imaging is usually only undertaken in patients in whom calcitonin is elevated, the number of patients studied with FDG is small. Patients who have undergone curative surgery and have undetectable calcitonin usually are not investigated. Therefore, the sensitivity of the test has been determined only in patients in whom disease is suspected but not yet identified, rather than in all patients with medullary cancer. There are no data on specificity.

In a study of 20 patients, Brandt-Mainz et al. (48) found that the sensitivity was 76%. Seventeen scans were positive and cancer was identified in 13. There were 4 false-negative and 1 true-negative results. Lesions were correctly identified in cervical and mediastinal nodes as well as the lung and skeleton. A report by Conti et al. (10) included cancers from follicular and parafollicular origin. They identified lesions in 6 patients with medullary cancer whose calcitonin values ranged from 104 to 9,500 ng/mL. Musholt et al. (49) found FDG to be superior to CT and magnetic resonance imagine (MRI) in 10 patients. Anatomic imaging demonstrated 11 lesions and FDG demonstrated 31.

In Europe, ^{99m}Tc-pentavalent dimercaptosuccinate (DMSA) has been used in this situation. Adams et al. (50) compared DMSA and ^{111}In-octreotide with FDG in 8 patients with medullary cancer. FDG was positive in seven patients and showed 38 lesions. ^{99m}Tc-DMSA was positive in two patients and showed only 1 lesion in each patient. ^{111}In-Octreotide showed 1 lesion in one patient, and seven scans were false negative. ^{99m}Tc-DMSA has not been approved for use in the United States. Diehl et al. (51) evaluated FDG in 85 patients and reported a sensitivity of 78% and specificity of 79%. A recent report validates the role of PET/CT imaging (52). PET appears to have a role in calcitonin-positive patients, but more studies are needed to determine the calcitonin cutoff value below which the scan would be ineffective.

Recent reports have demonstrated the value of ^{18}F-DOPA in medullary cancer, and the results were superior to FDG (53, 54).

PET Imaging in Patients with Lymphoma of the Thyroid

FDG is very accurate for identifying lymphoma and is used widely for staging and for determining the success of therapy (55). Primary lymphoma of the thyroid usually occurs in older women who have had preexisting Hashimoto's thyroiditis. It is rare, accounting for about 2% to 5% of cases of primary thyroid cancers. It is also a rare presentation of lymphoma, accounting for about 2% of these tumors (56). The diagnosis is made clinically by finding a rapidly expanding thyroid mass in an older woman with a preexisting goiter. Biopsy demonstrates a monoclonal population of lymphocytes. The role of FDG in localized thyroid lymphoma is small because the diagnosis is usually made pathologically. Because autoimmune thyroid disorders such as Hashimoto's show elevated FDG uptake, it is not possible to differentiate benign from malignant lymphomatous infiltrate. FDG-PET has an important role in staging of lymphoma, and in primary thyroid lymphoma it is valuable for ensuring that the lesion is indeed restricted to the thyroid gland.

Use of Gamma Camera Versus PET Scanner

Although the number of centers capable of PET scanning is increasing, the high cost of the equipment and the radiopharmaceuticals has resulted in a nonuniform availability of the procedure throughout the United States as well as the world in general. In an effort to overcome this limitation, some manufacturers of whole-body gamma cameras have implemented designs that allow coincidence imaging (57). Dedicated PET scanners usually have a ring detector, whereas the bifunctional gamma camera has two planar detectors in opposition. There has been concern that the gamma camera lacks the resolution of a true PET scanner. To date there are very few published reports comparing these procedures in patients with thyroid cancer. One study did demonstrate 4- and 5-mm lesions with a coincidence gamma camera (58). Goshen et al. (59) studied 20 patients with dual-headed camera and FDG; 15 patients had positive scans and 14 were true positives. A third study (in abstract form) compared PET and coincidence counting in 19 patients (60). Coincidence scanning identified 58 of the 84 lesions (69%) seen by PET. Lesions smaller than 1.5 cm were missed. The tumor stage in 3 patients would have been incorrect using only coincidence imaging. Therefore, PET imaging using a dedicated PET camera is superior to a hybrid gamma camera. Integrated PET/CT imaging that provides function and anatomy allows greater comfort when interpreting the causes of abnormal uptake of FDG in the neck.

Role of PET in Differentiating Benign from Malignant Thyroid Nodule

Thyroid nodules are very common and can be identified in 5% of healthy adults. With high-resolution ultrasound imaging, up to 50% of normal adults have a thyroid nodule. Most thyroid nodules are benign; however, both the patient and physician are concerned about the presence of cancer, which needs to be detected early and treated surgically. Many studies have shown that standard nuclear medicine imaging using ^{123}I or $^{99m}TcO_4$ has a very low specificity and poor positive predictive value, and the same applies to nonnuclear methods such as ultrasound, CT, or MRI. It was hoped that radiopharmaceuticals such as ^{201}Tl, ^{99m}Tc-MIBI, or ^{99m}Tc-tetrafosmin would differentiate malignant lesions by showing higher uptake, but this has not been the case. The small number of patients studied by PET does not support widespread clinical use (61–63). The most cost-effective approach is fine-needle aspiration of the nodule for cytologic interpretation. Will FDG be useful in the cytopathologically indeterminate thyroid nodule, where now excisional biopsy is required to differentiate follicular adenoma from follicular carcinoma? It would be hoped that FDG-positive lesions would all be malignant and FDG-negative lesions all benign adenomas (64). Unfortunately, there are reports of benign adenomas that are FDG positive.

Thyroid Uptake of FDG on PET Imaging Performed for Nonthyroidal Investigations

Most whole-body FDG scans are for the management of patients with nonthyroidal cancers. There is a significant body of information on its value in cancers of the lung, breast, and head and neck as well as melanoma and lymphoma. Rarely in the course of these studies is thyroidal uptake noted. Intense diffuse uptake suggests the presence of autoimmune thyroid disease such as Graves' hyperthyroidism or Hashimoto's thyroiditis. This occurrence has been described in "healthy" individuals who had annual PET imaging as part of a preventive health screening program (65, 66). In contrast, focal uptake has been associated with a significant likelihood of thyroid cancer (67–69). Therefore, when this is recognized and a thyroid nodule palpated, a fine-needle aspiration is recommended for pathologic diagnosis. When there is no palpable lesion, an ultrasound should be considered, and when a nodule of 1 cm or greater identified it should be aspirated for cytopathologic interpretation (Figure 18.8).

Use of Other Positron Emitters: Iodine-124 (124I)

One of the major shortcomings of treatment with radioiodine is the uncertainty of the volume of tissue being treated. However, because the treatments are so successful using empiric doses of ^{131}I, detailed knowledge of volume of cancer has been considered less important. For detailed dosimetry, it is necessary to know the percentage uptake of the radiopharmaceutical, its volume and uniformity of

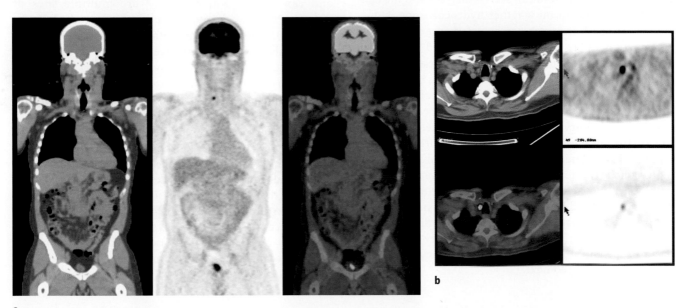

Figure 18.8. Coronal (**a**) and transraxial (**b**) PET/CT images demonstrate. A small focus of uptake of FDG in the right thyroid bed. The patient was being imaged because he had a history of melanoma. Ultrasound of the thyroid showed an 8-mm lesion, and ultrasound-guided fine-needle aspiration (FNA) biopsy demonstrated papillary cancer. The patient has had a thyroidectomy and has undetectable thyroglobulin (Tg) on a dose of thyroid hormones designed to keep his thyrotropin (TSH) at the lower limit of normal.

distribution, its effective half-life, and the type and energies of the radioactive emissions from the therapeutic radioisotope. PET imaging using [124]I allows volumetric measurements. [124]I is a positron emitter with a half-life of 4.2 days. Diagnostic doses of [124]I have been used to determine the volume of the thyroid in patients with Graves' hyperthyroidism and shown to be superior to planar nuclear imaging and to ultrasound (70–74).

In practice, precise dosimetry for treating Graves' disease is not advantageous, but it is conceivable that dosimetry may be useful some patients with thyroid cancer. The patient would still need to be prepared by a low-iodine diet and elevated TSH, either by withdrawal of thyroid hormone or by injection of recombinant human TSH. [124]I could be used for imaging and for determining the volume of cancer and measurement of the effective half-life within the gland. There are reports indicating the value of [124]I in this setting combined with anatomic imaging (75, 76). [124]I allows accurate volumetric imaging, and the 4.2-day half-life is advantageous for sophisticated dosimetry incorporating both the volume of tissue to be treated and the effective half-life of radioiodine (77). The radiation to thyroid tissues would be less than with diagnostic ^{131}I and concern about stunning (if the concept of stunning is accepted) would be lessened (78, 79).

Adrenal Glands

PET scanning has a limited role in imaging primary cancers of the adrenals. Adrenal tumors are uncommon; they are best characterized as arising in the cortex, or the medulla. Cortical tumors can secrete steroids and the clinical features depend on whether hydrocortisone, aldosterone, or sex hormones are secreted in excess. Some cortical tumors are nonsecretory and found fortuitously on imaging studies such as CT or MRI; these have been termed incidentalomas. The word tumor has been used in the text consciously because adrenal cortical lesions are usually benign but can be occasionally be cancerous. Adrenal medullary tumors are called pheochromocytomas. Most of these are benign but about 10% are malignant; 10% are bilateral and 10% extraadrenal.

In contrast, metastases to the adrenals are more common, in particular from primary lung cancer. Noncancerous and nonfunctioning masses in the adrenals are also common, and these have to be differentiated from metastasis in a patient with a primary cancer that has a propensity to spread to the adrenals. The role of FDG-PET is discussed next.

Normal Adrenals

The normal adrenal glands have an approximate volume of 4 mL. They have a rich vascular supply and are meta-

bolically active. However, normal adrenal glands are not usually identified on FDG-PET scan.

Adrenal Cortical Lesions

In most patients, careful evaluation of peripheral steroid values in blood and urine as well as pituitary adrenocorticotrophic hormone (ACTH), plus anatomic imaging, define whether an adrenal lesion is causing Cushing's or Conn's syndrome. Maurea et al. (80) demonstrated that FDG-PET is negative in nonsecretory adrenal tumors that were detected on anatomic imaging. Malignant adrenal lesions and their metastases were identified by intense uptake.

Adrenal Medullary Lesions

Once a diagnosis of pheochromocytoma is made, anatomic imaging usually defines the site of the tumor, which is adrenal in about 90% of patients. [131]I- ^{123}I-Metaiodobenzylguanidine (MIBG) scanning has a role in confirming the abnormal gland, demonstrating whether there is one or more lesions and whether there are metastases. A diagnostic MIBG scan can help determine whether therapeutic MIBG is appropriate. There are considerable technical aspects concerning MIBG studies, including ensuring that the patient is not ingesting medications which can interfere with the uptake of MIBG. ^{123}I-MIBG, which is optimal for diagnosis, is not widely available. There are well-documented false-negative MIBG studies. As a result, there is a role for an alternative nuclear medicine procedure that covers the entire body and can answer whether there are multiple sites of disease. PET is such a test. Shulkin et al. studied 29 patients (81); 15 of 17 patients with malignant pheochromocytoma showed intense concentration of FDG in the lesions, compared to 7 of 12 benign tumors. Malignant lesions in general showed more intense uptake. MIBG-negative tumors could be identified using FDG; however, when lesions concentrated both MIBG and FDG, the images of the former were judged equal or better in 88% of cases. PET could be used when other tests have failed to identify the tumor. Yun et al. (82) interpreted uptake in the adrenal equal to liver uptake as abnormal. Using this criterion, they calculated a sensitivity of 100% for cancerous lesions and a specificity of 94%.

Two recent reports from investigators at the National Institutes of Health (NIH) indicate that 6-^{18}F-fluorodopamine is valuable in identification of primary and metastatic pheochromocytoma (83, 84).

Metastases to the Adrenals

Lung cancer can metastasize to the adrenals. PET scan is both sensitive (100%) and specific (80%–100%). One study evaluated PET in 113 adrenal masses detected by CT or MRI (85). The investigators used uptake in the gland equal to or greater than hepatic uptake to diagnose metastases and lower uptake to exclude that . PET was positive in 71 glands and negative in 42. The sensitivity was 93% and specificity 90%, which are similar to the values of 94.4% and 91.6%, respectively, reported by Gupta et al. (86). Therefore, high adrenal uptake of FDG would indicate with confidence that a primary lung cancer should not be approached surgically. Clearly, not all adrenal masses are metastases, and in another study of FDG-PET in non-small cell lung cancer, 3 of 120 patients were downstaged by the finding of no uptake in adrenal masses seen on CT (87).

References

1. Schlumberger M, Baudin E, Travagli JP. Papillary and follicular cancers of the thyroid. Presse Med 1998;27(29):1479–1481.
2. Nakhjavani, M, Gharib H, Goellner JR, Van Heerden JA. Metastasis to the thyroid gland. A report of 43 cases. Cancer (Phila) 1997;79:574–578.
3. Mazzaferri E. Radioiodine and other treatment and outcomes. In: Braverman LE, Utiger RD, editors. Werner and Ingbar's The Thyroid: A Fundamental and Clinical Text, 8th ed. Philadephia: Lippincott Williams & Wilkins, 2000:904–929.
4. Ball D, Baylin SB, De Bustros AC. Medullary Thyroid Carcinoma. In: Braverman LE, Utiger RD, editors. Werner and Ingbar's The Thyroid: A Fundamental and Clinical Text, 8th ed. Philadephia: Lippincott Williams & Wilkins, 2000:930–943.
5. Joensuu H, Ahonen A. Imaging of metastases of thyroid carcinoma with fluorine-18 fluorodeoxyglucose. J Nucl Med 1987;28:910–914.
6. Grunwald F, Menzel C, Bender H, Palmedo H, Willkomm P, Ruhlmann J, et al. Comparison of [18]FDG-PET with [131]iodine and ^{99m}Tc-sestamibi scintigraphy in differentiated thyroid cancer. Thyroid 1997;7(3):327–335.
7. Fiene U. Fluoro-18-deoxyglucose positron emission tomography in differentiated thyroid carcinoma. Eur J Endocrinol 1998;138:492–496.
8. Dietlein M, Scheidhauer K, Voth E, Theissen P, Schicha H. Fluorine-18 fluorodeoxyglucose positron emission tomography and iodine-131 whole-body scintigraphy in the follow-up of differentiated thyroid cancer. Eur J Nucl Med 1997;24(11):1342–1348.
9. Dietlein M, Moka D, Scheidhauer K, Schmidt M, Theissen P, Voth E, et al. Follow-up of differentiated thyroid cancer: comparison of multiple diagnostic tests. Nucl Med Commun 2000;21(11):991–1000.
10. Conti PS, Durski JM, Bacqai F, Grafton ST, Singer PA. Imaging of locally recurrent and metastatic thyroid cancer with positron emission tomography. Thyroid 1999;9(8):797–804.
11. Wang W, Macapinlac H, Finn RD, et al. PET scanning with [^{18}F] 2-fluoro-2 deoxy-D-glucose (FDG) can localize residual differentiated thyroid cancer in patients with negative [^{131}I]-iodine whole-body scans. J Clin Endocrinol Metab 1999;84:2291–2302.
12. Altenvoerde G, Lerch H, Kuwert T, Matheja P, Schafers M, Schober O. Positron emission tomography with F-18-deoxyglucose in patients with differentiated thyroid carcinoma, elevated thyroglobulin levels, and negative iodine scans. Langenbecks Arch Surg 1998;383:160–163.
13. Jadvar H, McDougall IR, Segall GM. Evaluation of suspected recurrent papillary thyroid carcinoma with [^{18}F] fluorodeoxyglucose positron emission tomography. Nucl Med Commun 1998; 19:547–554.
14. Huang T, Chieng PU, Chang CC, Yen RF. Positron emission tomography for detecting iodine-131 nonvisualized metastasis of well-dif-

ferentiated thyroid carcinoma: two case reports. J Endocrinol Invest 1998;21:392–398.
15. Lips P, Comans EF, Hoekstra OS, van der Poest Clement E, van Mourik JC, Teule GJ. Positron emission tomography for the detection of metastases of differentiated thyroid carcinoma. Neth J Med 2000;57(4):150–156.
16. Macapinlac HA. Clinical usefulness of FDG PET in differentiated thyroid cancer. J Nucl Med 2001;42(1):77–78.
17. McDougall IR, Davidson J, Segall GM. Positron emission tomography of the thyroid, with an emphasis on thyroid cancer. Nucl Med Commun 2001;22(5):485–492.
18. van Tol KM, Jager PL, Dullaart RP, Links TP. Follow-up in patients with differentiated thyroid carcinoma with positive ^{18}F-fluoro-2-deoxy-D-glucose-positron emission tomography results, elevated thyroglobulin levels, and negative high-dose ^{131}I posttreatment whole body scans. J Clin Endocrinol Metab 2000;85(5):2082–2083.
19. Muros MA, Llamas-Elvira JM, Ramirez-Navarro A, Gomez MJ, Rodriguez-Fernandez A, Muros T, et al. Utility of fluorine-18-fluorodeoxyglucose positron emission tomography in differentiated thyroid carcinoma with negative radioiodine scans and elevated serum thyroglobulin levels. Am J Surg 2000; 179(6):457–461.
20. Alnafisi N, Driedger AA, Coates G, Moote DJ, Raphael SJ. FDG PET of recurrent or metastatic ^{131}I-negative papillary thyroid carcinoma. J Nucl Med 2000;41:1010–1015.
21. Al-Nahhas AM. FDG PET and alternative imaging in the management of thyroid carcinoma. Nucl Med Rev Cent East Eur 2003;6(2): 139–145.
22. Chung J-K, So Y, Lee JS, et al. Value of FDG PET in papillary thyroid carcinoma with negative ^{131}I whole-body scan. J Nucl Med 1999;40:486–492.
23. Desai D, Jeffrey RB, McDougall IR, Weigel RJ. Intraoperative ultrasonography for localization of recurrent thyroid cancer. Surgery (St. Louis) 2001;129(4):498–500.
24. Karwowski J, Jeffrey RB, McDougall IR, Weigel RJ. Intraoperative ultrasonography improves identification of recurrent thyroid cancer. Surgery (St. Louis) 2002;132:924–928.
25. Nakada K, Katoh C, Kanegae E, et al. Thallium-201 scintigraphy to predict therapeutic outcome of iodine-131 therapy of metastatic thyroid carcinoma. J Nucl Med 1998;39:807–810.
26. Yen T, Lin HD, Lee CH, Chang SL, Yeh SH. The role of technetium-99m sestamibi whole-body scans in diagnosing metastatic Hurthle cell carcinoma of the thyroid gland after total thyroidectomy: a comparison with iodine-131 and thallium-201 whole-body scans. Eur J Nucl Med 1994;21:980–983.
27. Gallowitsch H, Mikosch P, Kresnik E, Unterweger O, Gomez I, Lind P. Thyroglobulin and low-dose iodine-131 and technetium-99m-tetrofosmin whole-body scintigraphy in differentiated thyroid carcinoma. J Nucl Med 1998;39:870–875.
28. Baudin E, Schlumberger M, Lumbroso J, Travagli JP, Caillou B, Parmentier C. Octreotide scintigraphy in patients with differentiated thyroid carcinoma; contribution for patients with negative radioiodine scans. J Clin Endocrinol Metab 1996;81:2541–2544.
29. Krenning E, Kwekkeboom DJ, Bakkar WH, et al. Somatostatin receptor scintigraphy with [^{111}In-DTPA-D-Phe1]- and [^{123}I-Tyr3]- octreotide: the Rotterdam experience with more than 1000 patients. Eur J Nucl Med 1994;20:716–731.
30. Grunwald F, Kalicke T, Feine U, Lietzenmayer R, Scheidhauer K, Dietlein M, et al. Fluorine-18 fluorodeoxyglucose positron emission tomography in thyroid cancer: results of a multicentre study. Eur J Nucl Med 1999;26(12):1547–1552.
31. Mazzaferri E, Young RL, Oertel JE, et al. Papillary thyroid carcinoma. The impact of therapy on 576 patients. Medicine (Baltim) 1977;56:171–196.
32. Cook G, Maisey MN, Fogelman I. Normal variants, artefacts and interpretative pitfalls in PET with 18-fluoro-2-deoxyglucose and carbon-11 methionine. Eur J Nucl Med 1999;26:1363–1378.
33. Barrington S, Maisey MN. Skeletal muscle uptake of fluorine-18 FDG: effect of oral diazepam. J Nucl Med 1996;37:1127–1129.

34. Zhu Z, Chou C, Yen TC, Cui R. Elevated F-18 FDG uptake in laryngeal muscles mimicking thyroid cancer metastases. Clin Nucl Med 2001;26(8):689–691.

35. Igerc I, Kumnig G, Heinisch M, Kresnik E, Mikosch P, Gallowitsch HG, et al. Vocal cord muscle activity as a drawback to FDG-PET in the followup of differentiated thyroid cancer. Thyroid 2002;12(1):87–89.

36. Schmid DT, Kneifel S, Stoeckli SJ, Padberg BC, Merrill G, Goerres GW. Increased 18F-FDG uptake mimicking thyroid cancer in a patient with Hashimoto's thyroiditis. Eur Radiol 2003;13(9):2119–2121.

37. Pineda J, Lee T, Ain K, Reynolds JC, Robbins J. Iodine-131 therapy for thyroid cancer patients with elevated thyroglobulin and negative diagnostic scan. J Clin Endocrinol Metab 1995;80:1488–1492.

38. Schlumberger M, Mancusi F, Baudin E, Pacini F. ^{131}I therapy for elevated thyroglobulin levels. Thyroid 1997;7:273–276.

39. Wang W, Larson SM, Fazzari M, Tickoo SF, Kolbert K, Sgouros G, et al. Prognostic value of [^{18}F]fluorodeoxyglucose positron emission tomographic scanning in patients with thyroid cancer. J Clin Endocrinol Metab 2000;85(3):1107–1113.

40. Boerner AR, Petrich T, Weckesser E, Fricke H, Hofmann M, Otto D, et al. Monitoring isotretinoin therapy in thyroid cancer using ^{18}F-FDG PET. Eur J Nucl Med Mol Imaging 2002;29(2):231–236.

41. Chin BB, Patel P, Cohade C, Ewertz M, Wahl R, Ladenson P. Recombinant human thyrotropin stimulation of fluoro-D-glucose positron emission tomography uptake in well-differentiated thyroid carcinoma. J Clin Endocrinol Metab 2004;89(1):91–95.

42. Wilson C, Woodroof JM, Girod DA. First report of Hurthle cell carcinoma revealed by octreotide scanning. Ann Otol Rhinol Laryngol 1998;107: 847–850.

43. Gulec SA, Serafini AN, Sridhar KS, Peker KR, Gupta A, Goodwin WJ, et al. Somatostatin receptor expression in Hurthle cell cancer of the thyroid. J Nucl Med 1998;39(2):243–245.

44. Fisher C, Vehec A, Kashlan B, et al. Incidental detection of a malignant hurthle cell carcinoma by Tc-99m sestamibi cardiac imaging. Clin Nucl Med 2000;25:469–470.

45. Wiesner W, Engel H, von Schulthess GK, Krestin GP, Bicik I. FDG PET-negative liver metastases of a malignant melanoma and FDG PET-positive Hurthle cell tumor of the thyroid. Eur Radiol 1999;9(5):975–978.

46. Plotkin M, Hautzel H, Krause BJ, Schmidt D, Larisch R, Mottaghy FM, et al. Implication of 2-18-fluoro-2-deoxyglucose positron emission tomography in the follow-up of Hurthle cell thyroid cancer. Thyroid 2002;12(2):155–161.

47. Lowe VJ, Mullan BP, Hay ID, McIver B, Kasperbauer JL. ^{18}F-FDG PET of patients with Hurthle cell carcinoma. J Nucl Med 2003;44(9):1402–1406.

48. Brandt-Mainz K, Muller SP, Gorges R, Saller B, Bockisch A. The value of fluorine-18 fluorodeoxyglucose PET in patients with medullary thyroid cancer. Eur J Nucl Med 2000;27(5):490–496.

49. Musholt TJ, Musholt PB, Dehdashti F, Moley JF. Evaluation of fluorodeoxyglucose-positron emission tomographic scanning and its association with glucose transporter expression in medullary thyroid carcinoma and pheochromocytoma: a clinical and molecular study. Surgery (St. Louis) 1997;122(6):1049–1060; discussion 1060–1061.

50. Adams S, Baum R, Rink T, Schumm-Drager PM, Usadel KH, Hor G. Limited value of fluorine-18 fluorodeoxyglucose positron emission tomography for the imaging of neuroendocrine tumours. Eur J Nucl Med 1998;25(1):79–83.

51. Diehl M, Risse JH, Brandt-Mainz K, Dietlein M, Bohuslavizki KH, Matheja P, et al. Fluorine-18 fluorodeoxyglucose positron emission tomography in medullary thyroid cancer: results of a multicentre study. Eur J Nucl Med 2001;28(11):1671–1676.

52. Bockisch A, Brandt-Mainz K, Gorges R, Muller S, Stattaus J, Antoch G. Diagnosis in medullary thyroid cancer with [^{18}F]FDG-PET and improvement using a combined PET/CT scanner. Acta Med Aust 2003;30(1):22–25.

53. Hoegerle S, Altehoefer C, Ghanem N, Brink I, Moser E, Nitzsche E. ^{18}F-DOPA positron emission tomography for tumour detection in patients with medullary thyroid carcinoma and elevated calcitonin levels. Eur J Nucl Med 2001;28(1):64–71.

54. Gourgiotis L, Sarlis NJ, Reynolds JC, VanWaes C, Merino MJ, Pacak K. Localization of medullary thyroid carcinoma metastasis in a multiple endocrine neoplasia type 2A patient by 6-[^{18}F]-fluorodopamine positron emission tomography. J Clin Endocrinol Metab 2003;88(2):637–641.

55. Kostakoglu L, Goldsmith SJ. Fluorine-18 fluorodeoxyglucose positron emission tomography in the staging and follow-up of lymphoma: is it time to shift gears? Fluorine-18 fluorodeoxyglucose positron emission tomography in the staging and follow-up of lymphoma: is it time to shift gears? Eur J Nucl Med 2000;27:1564–1578.

56. Schlumberger M, Pacini F. Unusual thyroid tumors. In: Schlumberger M, Pacini F, editors. Thyroid Tumors. Paris: Nucleon, 1999:301–307.

57. Patton J, Turkington TG. Coincidence imaging with a dual-head scintillation camera. J Nucl Med 1999;40:432–441.

58. Stokkel MP, Hoekstra A, van Rijk PP. The detection of small carcinoma with 18F-FDG using a dual head coincidence camera. Eur J Radiol 1999;32(3):160–162.

59. Goshen E, Cohen O, Rotenberg G, Oksman Y, Karasik A, Zwas ST. The clinical impact of 18F-FDG gamma PET in patients with recurrent well differentiated thyroid carcinoma. Nucl Med Commun 2003;24(9):959–961.

60. Berger F, Knesewitsch P, Tausig A, et al. [^{18}F]Fluorodeoxyglucose hybrid PET in patients with differentiated thyroid cancer: comparison with dedicated PET. Eur Assoc Nucl Med Congr Paris 2000.

61. Adler LP, Bloom AD. Positron emission tomography of thyroid masses. Thyroid 1993;3(3):195–200.

62. Uematsu H, Sadato N, Ohtsubo T, et al. Fluorine-18-fluorodeoxyglucose PET versus thallium-201 scintigraphy evaluation of thyroid tumors. J Nucl Med 1998;39:453–459.

63. Park CH, Lee EJ, Kim JK, Joo HJ, Jang JS. Focal F-18 FDG uptake in a nontoxic autonomous thyroid nodule. Clin Nucl Med 2002;27(2):136–137.

64. Reimer S, Adler LP, Bloom AD. Prospective evaluation of PET-FDG in FNA indeterminate thyroid nodules. J Nucl Med 1998;39:123.

65. Yasuda S, Shohtsu A. Cancer screening with whole-body 18F-fluorodeoxyglucose positron-emission tomography. Lancet 1997;350:1819.

66. Yasuda S, Shohtsu A, Ide M, et al. Chronic thyroiditis: diffuse uptake of FDG at PET. Radiology 1998;207:775–778.

67. Cohen MS, Arslan N, Dehdashti F, Doherty GM, Lairmore TC, Brunt LM, Moley JF. Risk of malignancy in thyroid incidentalomas identified by fluorodeoxyglucose-positron emission tomography. Surgery (St. Louis) 2001;130(6):941–946.

68. Davis PW, Perrier ND, Adler L, Levine EA. Incidental thyroid carcinoma identified by positron emission tomography scanning obtained for metastatic evaluation. Am Surg 2001;67(6):582–584.

69. Ramos CD, Chisin R, Yeung HW, Larson SM, Macapinlac HA. Incidental focal thyroid uptake on FDG positron emission tomographic scans may represent a second primary tumor. Clin Nucl Med 2001;26(3):193–197.

70. Frey P, Townsend D, Jeavons A, Donath A. In vivo imaging of the human thyroid with a positron camera using I-124. Eur J Nucl Med 1985;10:472–476.

71. Frey P, Townsend D, Flattet A, De Gautard R, Widgren S, Jeavons A, et al. Tomographic imaging of the human thyroid using ^{124}I. J Clin Endocrinol Metab 1986;63(4):918–927.

72. Flower, M., Irvine AT, Ott RJ, et al., Thyroid imaging using positron emission tomography: a comparison with ultrasound imaging and conventional scintigraphy in thyrotoxicosis. Br J Radiol 1990;63:325–330.

73. Crawford D, Flower MA, Pratt BE, et al. Thyroid volume measurement in thyrotoxic patients: comparison between ultrasonography

and iodine-124 positron emission tomography. Eur J Nucl Med 1997;24:1470–1478.

74. Flower M, Al-Saadi A, Harmer CL, et al. Dose response study on thyrotoxic patients undergoing positron emission tomography and radioiodine therapy. Eur J Nucl Med 1994;21:531–536.

75. Freudenberg LS, Antoch G, Gorges R, Knust J, Pink R, Jentzen W, et al. Combined PET/CT with iodine-124 in diagnosis of spread metastatic thyroid carcinoma: a case report. Eur Radiol 2003;13 Suppl 4:L19–L23.

76. Freudenberg LS, Antoch G, Jentzen W, Pink R, Knust J, Gorges R, et al. Value of (^{124})I-PET/CT in staging of patients with differentiated thyroid cancer. Eur Radiol 2004;14(11):2092–2098.

77. Sgouros G, Kolbert KS, Sheikh A, Pentlow KS, Mun EF, Barth A, et al. Patient-specific dosimetry for ^{131}I thyroid cancer therapy using ^{124}I PET and 3-dimensional-internal dosimetry (3D-ID) software. J Nucl Med 2004;45(8):1366–1372.

78. Park H-M, Perkins OW, Edmondson JW, et al. Influence of diagnostic radioiodines on the uptake of ablative dose of iodine-131. Thyroid 1994;4:49–54.

79. McDougall I. 74 MBq radioiodine ^{131}I does not prevent the uptake of therapeutic doses of ^{131}I (i.e., it does not cause stunning in differentiated thyroid cancer). Nucl Med Commun 1997;18:505–512.

80. Maurea S, Mainolfi C, Bazzicalupo L, et al. Imaging of adrenal tumors using FDG PET: comparison of benign and malignant lesions. AJR 1999;173:25–29.

81. Shulkin BL, Shapiro B, Francis IR, Sisson JC. Pheochromocytomas: imaging with 2-[fluorine-18]fluoro-2-deoxy-D-glucose PET. Radiology 1999;212:235–241.

82. Yun M, Kim W, Alnafisi N, Lacorte L, Jang S, Alavi A. ^{18}F-FDG PET in characterizing adrenal lesions detected on CT or MRI. J Nucl Med 2001;42(12):1795–1799.

83. Pacak K, Eisenhofer G, Carrasquillo JA, Chen CC, Whatley M, Goldstein DS. Diagnostic localization of pheochromocytoma: the coming of age of positron emission tomography. Ann N Y Acad Sci 2002;970:170–176.

84. Ilias I, Yu J, Carrasquillo JA, Chen CC, Eisenhofer G, Whatley M, et al. Superiority of 6-[^{18}F]-fluorodopamine positron emission tomography versus [^{131}I]-metaiodobenzylguanidine scintigraphy in the localization of metastatic pheochromocytoma. J Clin Endocrinol Metab 2003;88(9):4083–4087.

85. Kumar R, Xiu Y, Yu JQ, Takalkar A, El-Haddad G, Potenta S, et al. 18F-FDG PET in evaluation of adrenal lesions in patients with lung cancer. J Nucl Med 2004;45(12):2058–2062.

86. Gupta NC, Graeber GM, Tamim WJ, Rogers JS, Irisari L, Bishop HA. Clinical utility of PET-FDG imaging in differentiation of benign from malignant adrenal masses in lung cancer. Clin Lung Cancer 2001;3(1):59–64.

87. Brink I, Schumacher T, Mix M, Ruhland S, Stoelben E, Digel W, et al. Impact of [(18)F]FDG-PET on the primary staging of small-cell lung cancer. Eur J Nucl Med Mol Imaging 2004;31(12):1614–1620.

19

PET and PET/CT Imaging in Multiple Myeloma, Solitary Plasmacytoma, MGUS, and Other Plasma Cell Dyscrasias

Ronald C. Walker, Laurie B. Jones-Jackson, Erik Rasmussen, Marisa Miceli, Edgardo J.C. Angtuaco, Frits Van Rhee, Guido J. Tricot, Joshua Epstein, Elias J. Anaissie, and Bart Barlogie

Multiple myeloma (MM), also called plasma cell myeloma, is a malignancy of terminally differentiated B cells (plasma cells) typically associated with secretion by these plasma cells of a complete and/or partial (light-chain) monoclonal immunoglobulin protein (M protein). MM comprises about 1% of all malignancies and 10% of hematologic malignancies, representing the second most common hematologic malignancy in the United States (1, 2).

Currently, about 50,000 people live with MM, with approximately 15,000 new cases diagnosed and 11,000 patients dying each year. The median age at diagnosis is approximately 65 years, although occasionally MM even occurs in teenagers. MM occurs more commonly in men than women and in African-American males than in the general population. Although the etiology of MM is unknown, several cytogenetic abnormalities associated with the disease define biologic and prognostically distinct entities, in particular with chromosome 13 deletions conferring poor prognosis. Most cases of MM evolve from a premalignant condition known as monoclonal gammopathy of undetermined significance (MGUS), with a conversion to MM of about 1% annually, although some cases of MM probably develop de novo (2, 3).

Tumor Biology

MM belongs to a spectrum of disorders known as *plasma cell dyscrasias*. The tumor growth is typically restricted to the bone marrow until late in tumor evolution. The myeloma plasma cells induce changes in the bone marrow microenvironment that are critical for tumor cell survival, growth, and microenvironment-mediated drug resistance, as well as osteolysis. These signals eventually contribute to disease manifestations (e.g., bone disease).

The disease causes clinical symptoms through a variety of mechanisms, including direct tumor mass effect (such as cord compression), cytokine production (leading to secondary anemia, diffuse systemic osteoporosis, and focal bone destruction), and immune system dysfunction (producing immunosuppression associated with increased infections, secondary malignancies, and autoimmune disorders). The clinical course depends on the individual biologic behavior of a given cell line, which can vary from relatively indolent to highly aggressive disease.

Other plasma cell dyscrasias include solitary plasmacytoma and rare conditions such as Castleman's disease, alpha-heavy-chain disease, and Waldenström's macroglobulinemia. Solitary plasmacytoma exists in two varieties, solitary extramedullary plasmacytoma occurring exclusively in the soft tissues and with a high potential for cure, and the more ominous and more common skeletal-associated entity, solitary bone plasmacytoma. Solitary bone plasmacytoma is frequently a harbinger for future development of MM, with a 3% annual conversion rate to MM. If more than one focal lesion is detected in a patient with suspected solitary disease, then the diagnosis is typically MM or, less likely, the uncommonly recognized condition "multiple solitary plasmacytomas" (4–11).

Diagnosis and Conventional Staging

Radiographic and laboratory findings form the basis of diagnosis. The common clinical presentations are fatigue and bone pain (back or ribs) with or without associated fractures or infection. About 15% to 30% of patients present with hypercalcemia and renal insufficiency because of precipitation of monoclonal light chains in the collecting tubules. Another 10% of patients present with other symptoms, such as hyperviscosity syndrome, compression of the spinal cord, radicular pain, soft tissue deposits, or abnormal bleeding. In patients with asymptomatic disease, laboratory findings of anemia or hyperproteinemia lead to the diagnosis.

Criteria for diagnosing MM are summarized in Table 19.1. The hallmark of MM is the detection in blood and/or urine of a monoclonal protein, called M protein, produced by the malignant plasma cells. Serum protein electrophoresis reveals a monoclonal peak in 80% of patients.

Table 19.1. Criteria for diagnosis of multiple myeloma.[a]

Major criteria:

Plasmacytomas on tissue biopsy

Marrow plasmacytosis with >30% plasma cells

Monoclonal globulin spike on serum electrophoresis >3.5 g/dL for IgG or >2.0 g/dL for IgA; 1.0 g/24 h of κ or λ light-chain excretion on urine electrophoresis in the absence of amyloidosis

Minor criteria:

Marrow plasmacytosis 10%–30%

Monoclonal globulin spike present, but less than the levels defined above

Lytic bone lesions

Normal IgM <0.05 g/dL, IgA <0.1 g/dL, or IgG <0.6 g/dL

[a]The diagnosis of plasma cell myeloma is confirmed when at least one major and one minor criterion or at least *three minor criteria* are documented in *symptomatic* patients with *progressive* disease. The presence of features not specific for the disease supports the diagnosis, particularly if of recent onset: anemia, hypercalcemia, azotemia, bone demineralization, or hypoalbuminemia.
Source: Reprinted with permission from Barlogie B, Shaughnessy J, Sanderson R, Epstein J, Walker R, Anaissie E, Tricot G. Plasma cell myeloma. In: Lichtman MA, Beutler E, Kaushansky K, Kipps TJ, editors. *Williams Hematology*, 7th edition. Copyright © 2006 chapter 106, pp 1279–1304. McGraw-Hill Professional.

Table 19.2. Durie–Salmon PLUS Staging System (DS+).

Classification	PLUS	New imaging: MRI and/or FDG-PET
MGUS		All negative
Stage IA*(smoldering or indolent)		Can have single plasmacytoma and/or limited disease on imaging
Multiple myeloma stage IB*		Fewer than 5 focal lesions; mild diffuse disease
Multiple myeloma stage IIA/B*		5–20 focal lesions; moderate diffuse disease
Stage III A/B*		More than 20 focal lesions; severe diffuse disease
	*A Serum creatinine <2.0 mg/dL and No extramedullary disease	
	*B Serum creatinine >2.0 mg/dL or Extramedullary disease	

Source: As adapted from Durie BGM, Kyle RA, Belch A, Bensinger W, et al. Myeloma management guidelines: a consensus report from the Scientific Advisors of the International Myeloma Foundation. *The Hematology Journal* 2003;4(6):379–398.

The remaining 20% of patients have either hypogamma-globulinemia or a normal-appearing (nonsecretory) electrophoresis profile. By using more-sensitive techniques, M protein (in serum or/and urine) will be detected in 99% of patients. Once MM is suspected, a radiographic skeletal survey and bone marrow aspiration and biopsy are performed for histologic examination, cytogenetic analyses, and determination of plasma cell labeling index. Minimal criteria for establishing a diagnosis of MM is the detection of at least 10% abnormal plasma cells in a random bone marrow biopsy specimen and M protein in either the serum (usually more than 3 g/dL) and/or urine (usually more than 1 g/24-h collection). Osteolytic lesions are seen on skeletal survey in 80% or more of patients (3).

Most plasma cell dyscrasias result from the expansion of a single clone of cells, with resultant monoclonal protein secretion. A small fraction (3% at diagnosis) will be hyposecretory or nonsecretory, either with very low levels or without the secretion of the hallmark M protein. Serial M protein levels can be very useful to determine response to treatment and the hyposecretory or nonsecretory status. Transformation to hyposecretory or nonsecretory status frequently develops during treatment, is problematic for clinical management, and also reflects dedifferentiation of the tumor, typically associated with a more-aggressive form of the disease. Sensitive assays for serum free light-chain protein allow quantifiable detection of free kappa/lambda proteins in approximately 70% of patients with nonsecretory disease and/or minimally secretory disease. Positron emission tomography (PET) using ^{18}F-fluorodeoxyglucose (FDG) as a tracer is also useful in this clinical setting because FDG uptake is not affected by the secretory status of the tumor (3–11).

Staging at diagnosis is based on the classic Durie–Salmon staging (D-S staging) (Table 19.2). To incorporate advances in medical technology (especially in imaging) in the interval since the D-S staging system (11) was initially adopted in 1975, Durie and others have recently proposed the Durie–Salmon Plus Staging system (D-S Plus Staging), which is currently being incorporated into new protocols.

The D-S Plus Staging system incorporates the number of either magnetic resonance imaging (MRI)- or FDG-PET-defined focal lesions into staging. Increasing numbers of focal lesions on either MRI and/or FDG-PET result in upstaging of the patient. Importantly, an increasing number of MRI- or FDG-PET-defined focal lesions is inversely related to prognosis, both independently and, in a multivariate analysis, with or without the presence of cytogenetic abnormalities. Increasing numbers of baseline focal lesions of bone confers a statistically significant poor prognosis in terms of both event-free survival (EFS) and overall survival (OS), independent of standard prognostic factors [i.e., elevated lactate dehydrogenase (LDH), M protein, C-reactive protein, creatinine, or ß-2 microglobulin, or abnormally depressed levels of hemoglobin or platelets] (4, 8, 12) (Figure 19.1).

Current Treatment

Autologous peripheral blood stem cell transplant supported by high-dose melphalan is now considered standard therapy for symptomatic patients with MM up to 70 years of age, with tandem autotransplants achieving an almost 50% complete remission rate, and with median survival ex-

Relative disadvantages of MRI compared to FDG-PET are time and expense for a thorough examination of the skeletal system (necessitating separate studies of the calvarium, vertebral column, pelvis, shoulder girdle, and sternum), MRI's limited field of view to the region under examination, and its contraindicated use in some patients (such as patients with pacemakers, aneurysm clips, cochlear implants).

The relative advantages of MRI compared to FDG-PET and integrated PET/CT imaging are the superior spatial and contrast resolution of MRI (typically 2 mm for a high-field system) and its more widespread availability at this time. The high spatial resolution of MRI is important in

the definition of [
cord compression[
tures. Avascular n[
from glucocortico[
of myeloma are ea[
well seen on FDG[
soft tissue contras[
ticularly helpful to[
skull base and face[
PET and PET/CT[
normal FDG distr[
(Figure 19.3).

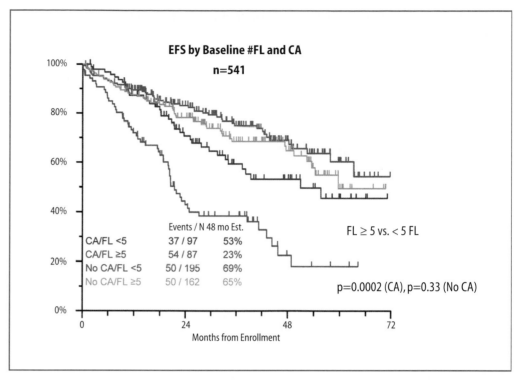

a

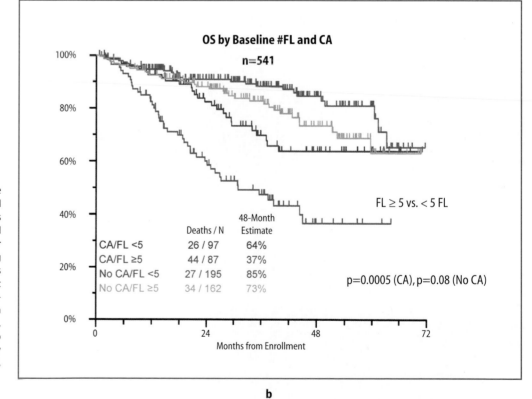

b

a   b   c

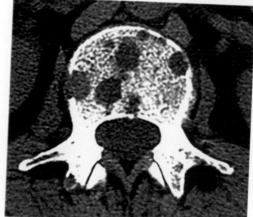

d

Figure 19.1. Impact of the presence of cytogenetic abnormalities and number of focal lesions at diagnosis on event-free survival (EFS) and overall survival (OS). (**a**) Kaplan–Meier (KM) curves for EFS; (**b**) OS correlating significance of number of focal lesions of bone at baseline with or without cytogenetic abnormalities with outcome. (Reproduced with permission from Walker RC, Jones-Jackson L, Rasmussen E, et al. Presentation to the American Society of Hematology Annual Meeting, December 4–7, 2004, San Diego, CA.)

Figure 19.3. Severe diffuse an[
nance imaging (MRI) T₁- (**a**) an[
weighted (**b**) and FDG-PET imagi[
demonstrating classic osteolytic [
pression fracture with neural arc[
FDG-PET.

tending beyond 6 to 7 years. "Mini-allotransplants" are less toxic than standard therapy, exploiting a "graft-versus-myeloma" effect. New active drugs include immunomodulatory agents, such as thalidomide and CC-5013 (Revimid; Celgene, Warren, NJ, USA), and the proteasome inhibitor

Velcade (Millenium, Cambridge, MA, USA), all of which not only target myeloma cells directly but which also exert an indirect effect by suppressing growth and survival signals elaborated by the bone marrow microenvironment interaction with the myeloma cells.

Cytogenetic abnormalities, present in one-third of patients at diagnosis, identify a particularly poor prognostic subgroup with a median survival not exceeding 2 to 3 years. In the absence of cytogenetic abnormalities, the use of tandem autotransplantation can yield 4-year survival rates of 80% to 90% (13).

Imaging of Multiple Myeloma

Role of X-Ray

One or more osteolytic lesions on X-ray identify specific classic Durie–Salmon stages of MM at baseline and on restaging. The use of X-ray is a mainstay of the classic Durie–Salmon staging established in 1975 but has less importance in the new Durie–Salmon Plus system, which incorporates newer imaging technology (MRI and PET). In contrast to osteolytic lesions associated with other malignancies that respond to treatment, osteolytic lesions from MM usually do not heal (3, 4, 14) (Figure 19.2).

Role of MRI

MRI is an impor...
tary plasmacytor...
ument the exter...
This MRI patter...
microfocal ("spe...
disease (5 mm or...
combinations. M...
focal areas of dise...
into the soft tissu...
still occur in the c...
marrow microen...
sively soft tissue o...
tumor population...
marrow microenv...
therapy is not only...
directed at interr...
marrow interactio...
the marrow micro...
in need of aggressi...

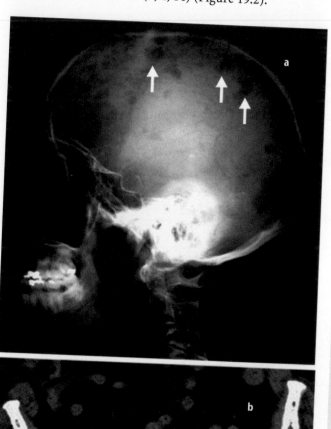

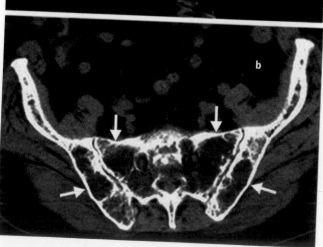

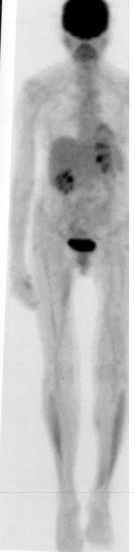

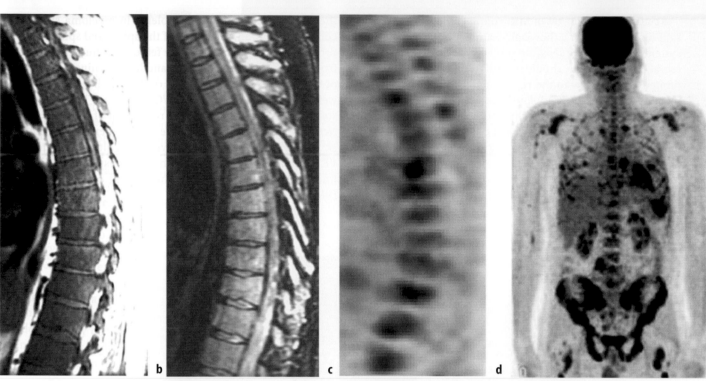

Figure 19.4. "Masking" of focal lesions on MRI from the signal produced by diffuse, intense tumor infiltration on T_1- (**a**) and STIR (**b**) -weighted sagittal thoracic spine images. The many focal lesions as well as areas of diffuse infiltration are shown clearly on sagittal (**c**) and on anterior 3D maximum intensity projection (MIP) (**d**) FDG-PET images.

If severe diffuse marrow infiltration with tumor is present, focal lesions from MM can be "masked" on MRI. Diffuse tumor infiltration of the marrow has the same signal characteristics as focal disease [high signal on short-T_1 inversion recovery (STIR-wt) weighted and low signal on T_1-weighted (T1-wt)] image sequences. Severe diffuse marrow infiltration with MM will also enhance diffusely after gadolinium injection to a similar degree as focal disease. Thus, severe diffuse tumor infiltration can obscure the presence of focal disease on MRI on all sequences. FDG-PET or integrated PET/CT, however, can still visualize these focal lesions against the diffuse infiltration because of the frequent greater metabolic rate of the focal lesions (Figure 19.4).

MRI and FDG-PET and integrated PET/CT are complementary. MRI and FDG-PET or integrated PET/CT both track response to treatment of the diffuse infiltration from MM in essential real time. However, the focal lesions seen on MRI will take, typically, months to years to resolve, and, indeed, may never resolve. MRI can detect relapse in these focal lesions by increasing size or number of focal lesions or the presence of enhancement with gadolinium. These MRI-defined focal lesions can still harbor microscopic viable tumor cells and thus can be sites for relapse months to years after remission. FDG-PET or integrated PET/CT demonstrate essentially real-time response of both diffuse and focal lesions, in both secretory or, importantly, nonsecretory disease. FDG-PET or integrated PET/CT will demonstrate relapse on treatment as areas of persisting and/or recurrent metabolic uptake on treatment. Thus, MRI is ideally used to document "complete-

ness" of ultimate response whereas FDG-PET or integrated PET/CT is ideally used for monitoring short-term response (Figure 19.5) (7).

Whole-body MRI is an emerging technology with a currently undefined role in this disease.

Clinical PET and PET/CT Imaging Protocols

The FDG-PET imaging protocol for MM and related plasma cell dyscrasias is designed to image the entire marrow space and the soft tissues. The patient is instructed to begin a high-protein, low-carbohydrate diet and to limit physical activity, beginning 24 h before the study. Because imaging of the marrow space of the entire body is very sensitive to interference from muscle uptake, this restriction of physical activity is very important. As for other FDG-PET imaging procedures, insulin-dependent diabetic patients are instructed to withhold insulin for a minimum of 6 h before administration of FDG if possible. Measuring and documenting the serum glucose level at time of injection of the FDG is important. Because of frequent use of high-dose glucocorticoids in MM and related diseases, high serum glucose levels are common, frequently between 100 and 200 mg/dL. This hyperglycemia interferes to some degree with the quality of the PET images; and the degree of uptake by tumors is reduced by competitive inhibition, decreasing measured standard uptake values (SUV). The SUV is a measure of the degree of uptake in a region of interest normalized by the dose administered to the patient and is dependent of

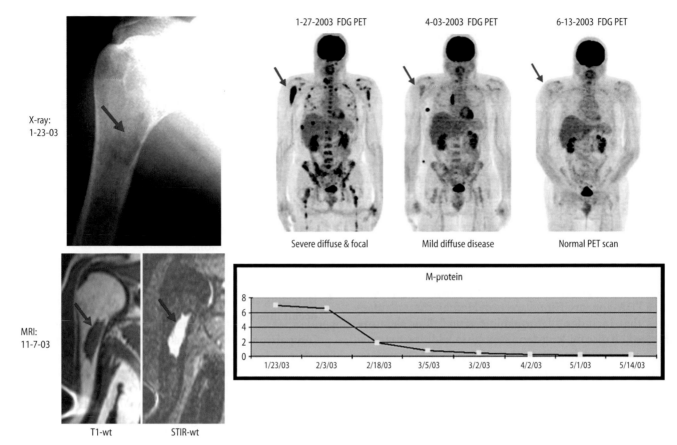

Figure 19.5. PET normalization of diffuse and focal disease components tracks clinical response. X-ray is abnormal permanently. MRI–focal lesion normalization lags behind monoclonal immunoglobulin protein (M protein) and FDG-PET for months to years. Patient is in complete remission since April 2003.

many other parameters that have to be carefully controlled if the SUV is used clinically. Importantly, high-dose glucocorticoid medications can produce false-negative examinations in some patients from transient inhibition of still-viable tumor (3).

The radiopharmaceutical should be administered in a peripheral vein rather than in central line, to allow detection of infected thrombi that may form at the tip of a central venous catheter. If direct venous access is not technically feasible, flushing the central catheter with 200 mL sterile normal saline help to diminish residual activity in the central catheter and improve the image appearance in general, although detection of infection associated with the central catheter will be less reliable (15–18).

A dose of FDG 555 MBq (15 mCi) of FDG is typically administered intravenously 60 min before beginning image acquisition. During the distribution time, the patient should rest quietly in a semirecumbent position. The patient is instructed to void immediately before imaging.

Protocols for PET image acquisition and processing are identical to the technical settings used for other malignancies on a given system, such as melanoma or lymphoma, although specific settings vary from manufacturer to manufacturer. The "whole-body" protocol for MM is similar to that for melanoma, imaging from vertex to feet.

At the University of Arkansas Medical Center, PET/CT imaging is performed using an integrated PET/CT system (CTI Reveal 16 "Hi Res" system from CTI Medical Systems, Knoxville, TN) with 4-mm LSO detectors and a 16-slice helical CT scanner; this is a "3D-only" system. PET/CT imaging is performed from vertex to feet in a continuous manner, with whole-body CT attenuation correction.

The protocol is the following: a CT topogram is performed with 1-mm helical slices using 120 KV and 50 mA for the full length of the body. If the patient exceeds the single-pass length capability of the system, the knees to feet section of the body is imaged separately. The protocol for CT image acquisition includes effective mAs of 190 and KV of 120, 1.5-mm "fast" slice with 18-mm feed/rotation. A general "whole-body" kernel filter is used, B.31f, with medium smoothing with a 500-mm field of view, and with reconstruction increments of 3.4 mm.

For the PET acquisition of the PET/CT examination, emission images are acquired for 4 min per bed position, although if the patient is obese, the time is extended on a "floating scale" to obtain sufficient count statistics for reliable imaging and quantitation. The raw data are transferred to a "Wizard" workstation (CTI Medical Systems, Knoxville, TN) for reconstruction to save time on the system for scanning the next patient. The PET images are

recor
gram
tions
imag
work

Wi
recor
becau
"who
using
both
ated
512 x
mm
168 x
matri

Im
(CTI
etary
turer
archi
fer o
Wher
imag
const

Wl
the A
USA)
of Ar
Atter
imag
sion
midtl
midtl
min
per l
obese

Th
filter
subse
form

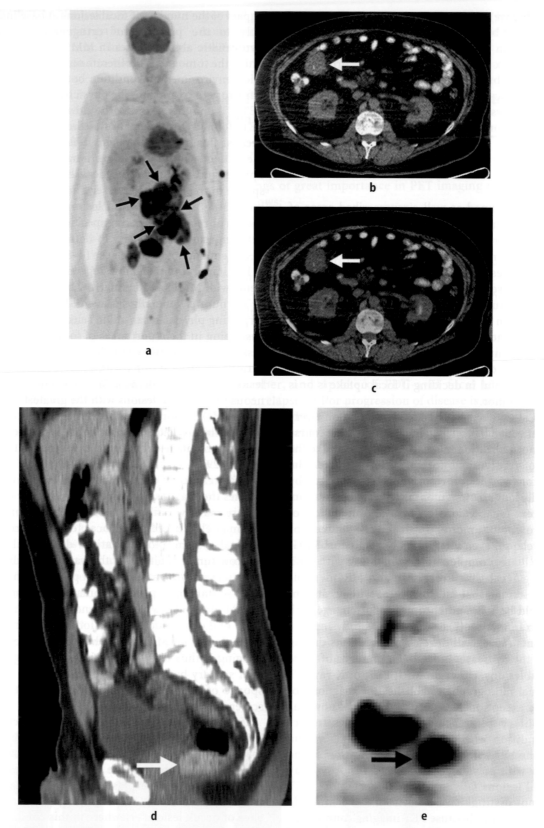

Figure 19.7. Various examples of extramedullary disease. (**a**) 3D MIP FDG-PET image of a patient with newly diagnosed multiple myeloma with severe extramedullary disease of the retroperitoneum (*arrows*). Smaller foci of extramedullary disease are in the soft tissues of the left lower extremity. Injection artifact in the left arm. (**b, c**) Occult extramedullary disease from multiple myeloma discovered with PET/CT (*arrows*). (**d, e**) Extramedullary multiple myeloma of the rectum on sagittal CT (*left*) and FDG-PET (*right*) (*arrows*).

**Figure
the cor
microe

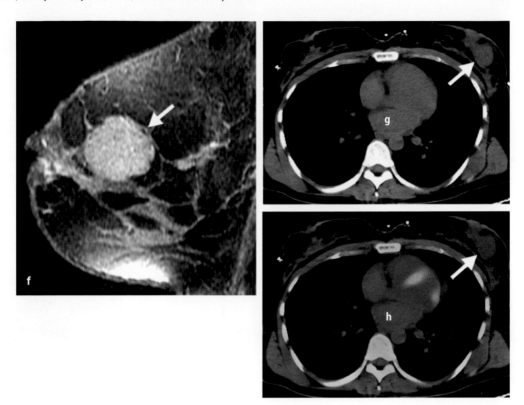

Figure 19.7. Contd. (f–h) Extramedullary multiple myeloma. A solid mass is found on RODEO MRI (**f**). Axial CT (**g**) and fused PET (**h**) demonstrate a focal mass in the left breast with slight FDG uptake. Needle biopsy demonstrated extramedullary multiple myeloma.

reliably distinguished from MM by FDG-PET or PET/CT imaging (12, 31, 32).

Extramedullary Disease

It is of great importance that the PET or PET/CT interpreting physician distinguishes between intramedullary and extramedullary disease. The presence of extramedullary disease defines a high-risk population with poor survival in need of close monitoring and aggressive treatment. "Breakout" lesions (focal areas of tumor that expand beyond the confines of the bone from which they originate, via erosion/permeation through the cortex) do not have the same biologic significance as true extramedullary disease as the breakout lesions are still growing in association with (and presumed dependency upon) the bone marrow microenvironment. Extramedullary disease represents a clinical entity that is less differentiated, often nonsecretory, and more aggressive, capable of growing without dependency on the marrow microenvironment. Because much of the treatment of MM is directed not only at the tumor but also at the tumor–bone marrow interaction and dependency, development of extramedullary disease indicates development of resistance to many therapeutic agents. Median survival for relapsing or refractory patients with extramedullary disease is less than 1 year (7) (Figure 19.7).

Diffuse Uptake

Attention to diffuse FDG uptake in the hematopoietic marrow is essential. Diffuse increased uptake can be caused by tumor infiltration, especially if the pattern is irregular or heterogeneous. Other conditions, for example, bone marrow-stimulating medication and bone marrow hyperplasia following chemotherapy, can cause diffuse metabolic stimulation of the hematopoietic marrow (Figure 19.8a, b).

Severe diffuse increased uptake can hide or "mask" the presence of underlying focal lesions that have a similar metabolic rate. As treatment results in rapid suppression of the diffuse component, on subsequent PET imaging the focal lesions may become visible against the now less-active diffuse background activity, producing the spurious appearance of interval development of new focal lesions ("unmasking"). Thus, it is important to document in the report for each examination the SUV of a representative region of the hematopoietic marrow that is not involved with focal uptake. If, on follow-up PET examination the diffuse component SUV has dramatically decreased whereas the number of focal lesions appears to have increased, "unmasking" of focal lesions may be present, representing tumor responding to treatment.

At the University of Arkansas, the maximum SUV of the L3 or L4 vertebral body is typically reported, assuming there are no compression fractures, vertebroplasty, focal

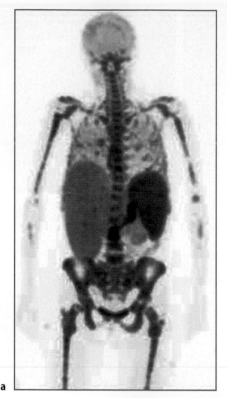

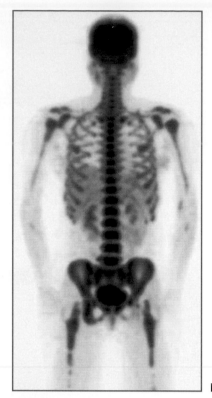

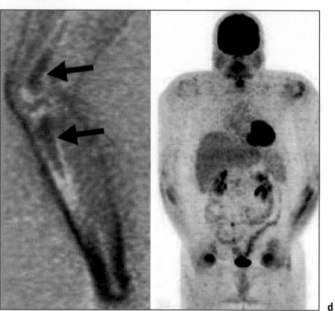

Figure 19.8. (a) Anterior 3D MIP FDG-PET of a "Superscan" appearance from severe intramedullary and extramedullary (stomach, liver, and spleen) multiple myeloma. Notice the brain "see-through" effect with the massive tumor "steal" of the FDG. **(b)** Anterior 3D MIP FDG-PETof a "Superscan" appearance from bone marrow-stimulating medication (G-CSF, granulocyte colony-stimulating factor). **(c, d)** Patients in remission after stem cell transplantation can demonstrate more uptake in the peripheral marrow than typically seen in the middle-aged to elderly adult (**c**, *arrows*); this is probably caused by the redistribution of hematopoietic marrow. Anterior 3D attenuation-corrected MIP image (**d**) demonstrated no evidence of active disease. The patient was in complete remission.

lesion, or other abnormality present. If an abnormality is present at these levels, the SUV of the first normal-appearing cephalad vertebral body is reported. All SUV values are measured from the transaxial images. Although nonspecific, comparison between SUV values of diffuse uptake between examinations can thus provide suggestive information, with careful correlation with clinical information, as to possible "unmasking" versus progression of disease on treatment, with the latter indicating a need for rapid intervention.

Skeletal disease in MM typically mirrors the distribution of hematopoietic marrow. The hematopoietic marrow in the typical late middle aged to elderly patient with MM typically extends from the calvarium to the pelvis and includes the sternum, the ribs, portions of the scapulae (the glenoid and coracoid regions and inferior scapular tips), and the proximal portions of the humeri and femora. In more aggressive or in later stages of the disease, spread to more peripheral locations occurs.

After bone marrow transplantation, alteration of the "normal" hematopoietic distribution can be significant, with FDG uptake extending to the peripheral skeleton, sometimes with an irregular or "patchy" appearance of the more central marrow (this irregular pattern can be caused by repopulation of a scarred marrow space from antecedent large, focal osteolytic defects). Thus, active marrow in the distal femurs or proximal tibias is sometimes seen (Figure 19.8c, d).

Because MM does rarely occur as young as the teenage years, the age of the patient must be considered when evaluating the extent and intensity of diffuse skeletal uptake. In the adolescent to young adult, the hematopoietic marrow normally extends further into the extremities than in the older adult and is more metabolically active than in the typical, otherwise healthy older adult (3).

Detection of Infection

Infection continues to represent a diagnostic and therapeutic challenge in the management of immunosuppressed patients, including those with MM, in whom timely diagnosis and treatment is often delayed because of the absence of the typical signs and symptoms of infection (19, 20).

Because FDG accumulates nonspecifically in cells with increased glucose metabolism, including both neoplastic and inflammatory cells, it represents a potentially important tool for the diagnosis of infection in this patient population (21–25).

A recent study demonstrated the ability of FDG-PET scan to identify a wide range of infections caused by various pathogens (bacteria, fungi, and/or viruses) at a variety of sites, including pneumonia, sinusitis, discitis, osteomyelitis, septic arthritis, cellulitis, colitis, diverticulitis, esophagitis, abdominal abscess, and periodontal abscess, even in the presence of severe neutropenia (15–17). Additionally, FDG-PET imaging was particularly useful for the diagnosis of vascular infections such as septic thrombophlebitis, either spontaneous or due to infection of implantable catheters. FDG-PET imaging also contributed to the management of patients with infection, particularly those involving the vascular system, skeleton, and joints, including detection of infectious foci that were silent clinically and were not identified by conventional diagnostic methods, determination of the extent of infection, and modification of the diagnostic workup and therapeutic strategy (15, 17).

Other nuclear imaging techniques have a significant role in the diagnosis of infection, for example, using tracers such as [111]In- or ^{99m}Tc-labeled white blood cell scans or ^{67}Ga citrate. Compared to these tracers, FDG is less labor intensive, allows earlier diagnosis because of earlier imaging time (less than 1 h versus 4–24 h), and provides less radiation dose to the patient because of a shorter half-life (less than 2 h versus 6–67 h). In addition, FDG-PET imaging can be diagnostic in the setting of severe neutropenia and is also useful for staging the underlying cancer. With PET/CT imaging, the foci of abnormal FDG uptake can be precisely localized anatomically (15–17, 19–27).

A potential limitation of FDG-PET imaging for diagnosing infections is the inability to discriminate between malignancy and infection. This potential difficulty is largely circumvented by discriminative interpretation of the findings on imaging in the context of the typical anatomic distribution of the underlying malignancy. For instance, foci of FDG uptake in certain locations would be extremely unlikely to represent metastatic spread, such as foci in the vascular system, dentition, sinuses, disks, and joints. The interpreter has to be familiar with the usual pattern of spread of the malignancy affecting the patient. For example, MM rarely affects the respiratory and gastrointestinal tracts; thus, abnormal FDG uptake at these sites should raise suspicious for infection. Conversely, if the patient has a primary malignancy known to involve the respiratory or gastrointestinal tract, such as lung, colon, or breast cancer, malignant involvement should be considered. Evaluation of response of the underlying malignancy by parameters other than imaging (for example, clinical examination, serum markers such as C-reactive protein, bone marrow aspirate, and biopsy) can also assist in the evaluation of the etiology of abnormal FDG uptake at these extramedullary sites.

The radiopharmaceutical should be administered in a peripheral vein rather than through a central catheter, which does improve reliability of SUV measurements by decreasing the amount of isotope that is not circulating in the body as almost invariably some radiopharmaceutical remains in the central catheter. More importantly, focal uptake at the catheter tip following peripheral venous injection of the radiopharmaceutical strongly suggests infection of the thrombus at the catheter tip from hematogenous microorganism seeding. Similarly, FDG uptake along the catheter tract or associated with the reservoir of an indwelling catheter strongly suggests infection, unless the catheter was placed within the last 2 weeks. If the radiopharmaceutical is injected via the central catheter, even if the catheter is flushed liberally with saline, the reliability of these findings for detection of infection is decreased.

FDG-PET imaging should be considered in immunosuppressed patients with suspected infection who have a negative diagnostic evaluation to detect and localize a source of infection. FDG-PET imaging should similarly be considered to evaluate the significance of persistent findings of infection by conventional radiologic tests following a recently treated infection, such as a persistent lung infiltrate on chest X-ray or chest CT scan. A negative FDG-PET study, in the absence of other clinical or laboratory markers for infection, suggests resolution of the infection, hence allowing safe resumption of antineoplastic chemotherapy. FDG-PET imaging should also be obtained when an intravascular infection is suspected, including spontaneous septic thrombophlebitis and infections of implantable catheters.

Finally, abnormal areas of FDG uptake at extramedullary sites should prompt an evaluation for infection, particularly when the location of the abnormal uptake is not typical of organ system involvement by the

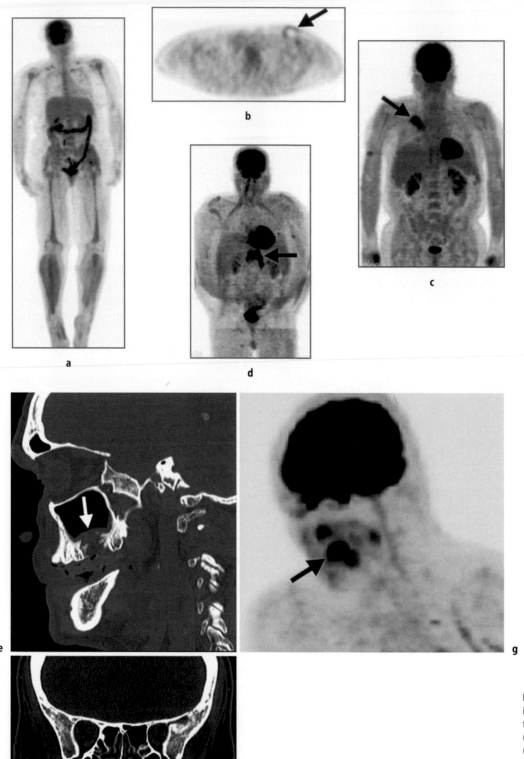

Figure 19.9. (**a**–**d**) Proven examples of infections discovered on FDG-PET in patients with hematologic malignancies. (**e**–**g**) Sagittal (*upper left*) and coronal (*lower left*) CT images and oblique 3D MIP FDG-PET (*upper left*) demonstrate osteomyelitis of the maxilla (*arrows*) in a patient with multiple myeloma and a severe periodontal abscess. Maxillary or mandibular infections in this population can be bacterial or fungal (for example, actinomycosis), and their incidence may be increased by the use of bisphosphonate medications, which may also predispose to osteonecrosis of the mandible.

underlying malignancy. Detecting and properly treating such infections before immunosuppressive therapy may avoid serious clinical consequences (15–17).

Because detection of infection in MM and related malignancies frequently results in significant changes in patient management, rapid communication of the presence of suspected infection to the referring physician is essential (Figure 19.9).

Pitfalls, Pearls, and Caveats

Vertebroplasty

Much of the morbidity of MM comes from the bone disease resulting from inhibition of osteoblasts and stimulation of osteoclasts from various cytokines released from the tumor–marrow microenvironmental interactions. Systemic osteoporosis, with or without coexisting focal osteolytic lesions, is a frequent cause of compression fractures of the vertebral bodies.

Vertebroplasty is now a common procedure performed for acute to subacute compression fractures, often with dramatic and rapid relief of pain. The vertebroplasty cement is radioopaque. Depending on the PET or PET/CT manufacturer's imaging algorithm and method of attenuation correction, the levels of vertebroplasty can appear as areas of increased or decreased uptake on the attenuation-corrected PET or PET/CT images. Additionally, because compression fractures often occur at levels with concurrent osteolytic focal lesions, careful attention and comparison to prior examinations is needed to determine if uptake is from the compression fracture alone, from attenuation correction artifact (by reviewing the images without attenuation correction), or is also associated with focal tumor uptake (18) (Figure 19.10).

Biopsies

Bone marrow biopsy is a mainstay for diagnosis and restaging of MM. This procedure is typically done in the posterior iliac crest but can be performed in the sternum and occasionally elsewhere (for example, a specific focal bone lesion or area of extramedullary disease). As with other surgical interventions, biopsies can produce FDG uptake from the induced inflammatory response in the acute to subacute time frame. The interpreting physician should recognize focal uptake in these typical locations as possible biopsy sequelae, especially for new areas of uptake. PET/CT imaging is particularly helpful in this regard, often showing the needle track. On PET-only images, extension of uptake from the area of suspected biopsy to the skin is often visible, particularly in the region of the iliac wing (18) (Figure 19.11).

False-Negative Results: Chemotherapy Effect

Any partially effective treatment, such as dexamethasone for MM, can produce false-negative FDG images despite the presence of extensive disease in the bone marrow on both a diffuse and a focal basis. As in other malignancies, it is important to coordinate the timing of the treatment and performance of PET imaging following treatment to avoid false-negative examinations (5) (Figure 19.12).

Other Plasma Cell Dyscrasias

Monoclonal Gammopathy of Undetermined Significance

Most authorities consider monoclonal gammopathy of undetermined significance (MGUS) the usual precursor of MM, being indistinguishable from MM based on gene expression profiling and having the same genomic instability. There is progression to MM in approximately 1% of patients per year. PET imaging in MGUS is usually normal. As focal lesions or extramedullary disease are not seen in MGUS, PET or PET/CT imaging revealing such findings demonstrate conversion to MM. Because of the whole-body nature of FDG-PET imaging, this procedure

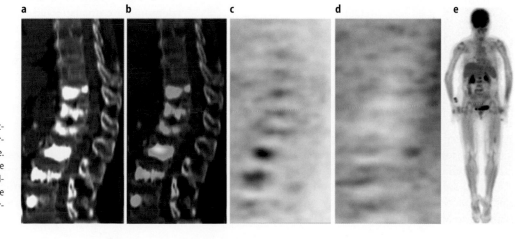

Figure 19.10. (a–e) Attenuation correction artifact producing spurious appearance of focal lesions of the lumbar spine. These images are from a multiple myeloma patient in remission with multilevel vertebroplasty. Attention to the sagittal image without attenuation correction demonstrated no focal lesions.

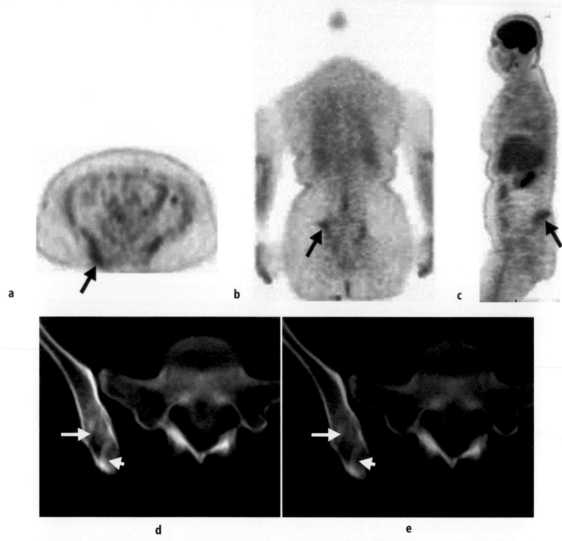

Figure 19.11. (**a–c**) Recent bone marrow needle biopsy of the iliac wing. This should not been mistaken for a focal lesion of multiple myeloma. Clues to the true etiology include typical location (for example, posterior iliac wing or sternum), answers to a patient questionnaire about recent surgeries or biopsies, and/or extension of uptake to the skin as a result of inflammatory response. FDG-PET changes are seen on axial (**a**), coronal (**b**), and sagittal (**c**) images. On integrated PET/CT imaging (**d, e**), the bony defect from the biopsy needle is often visible. (**d, e**) Focal uptake from a bone marrow biopsy the previous day, not to be mistaken from a focal lesion from multiple myeloma (*long arrow*). The linear track is seen on the CT image. An old, healed biopsy site with sclerotic margins is also visible (*short arrow*).

is an excellent modality to screen for conversion to MM, compared to conventional imaging (X-ray, CT, and MRI) (11) (Figure 19.13).

Solitary Plasmacytoma

Solitary plasmacytoma occurs in two varieties, solitary extramedullary plasmacytoma (occurring exclusively in the soft tissues) and solitary bone plasmacytoma (occurring in concert with the skeletal system). Solitary extramedullary plasmacytoma has a high likelihood of cure with radiation and is unlikely to progress to MM. Solitary extramedullary plasmacytoma appears as a nonspecific FDG-avid soft tissue mass shown on biopsy to be a plasmacytoma. Solitary bone plasmacytoma is more likely to progress to MM, with a conversion rate of about 3% of pa-

tients annually. Solitary bone plasmacytoma has the same osteolytic appearance on X-ray as a plasmacytoma associated with MM. FDG-PET and PET/CT imaging can demonstrate unsuspected multiple lesions, upgrading the diagnosis to MM. FDG-PET and PET/CT can also demonstrate response of solitary extramedullary plasmacytoma or solitary bone plasmacytoma to treatment (11, 31) (Figure 19.14).

Castleman's Disease

Benjamin Castleman first described Castleman's disease in 1956. It was later divided into two subtypes, unicentric (with a relatively favorable prognosis and often diagnosed incidentally) and multicentric. The multicentric variant has a relatively poor prognosis, representing a polyclonal

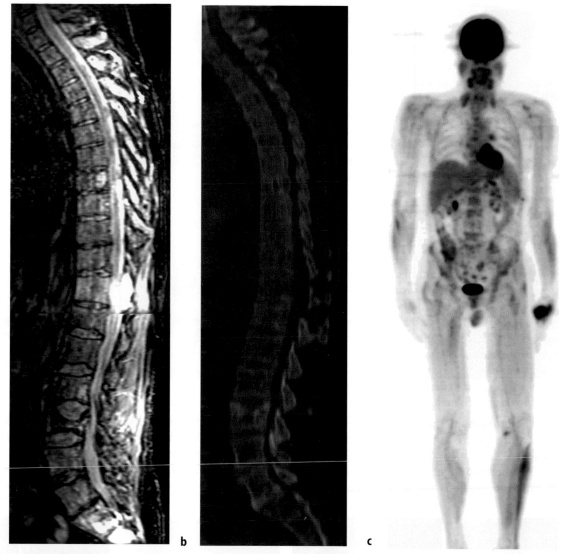

Figure 19.12. False-negative PET/CT scan in a patient with known extensive disease. The MRI examination was obtained 1 day before the FDG-PET/CT. Administration of dexamethasone to the patient resulted in a rapid transient suppression of tumor metabolism with no areas of increased uptake seen on the PET/CT study (fused PET/CT, **b**) and anterior 3D MIP, **c**), despite the focal mass with epidural involvement and focal lesions on bone seen on the STIR-weighted MRI images (**a**).

lymphoproliferative disorder characterized by recurrent fevers, lymphadenopathy, hepatosplenomegaly, and autoimmune phenomena, sometimes progressing to malignant lymphoma. The plasma cell variant of multicentric Castleman's disease frequently occurs in the setting of underlying human immunodeficiency virus (HIV) infection and appears to be associated with HIV-8 infection.

FDG-PET imaging can demonstrate active disease in Castleman's disease in an analogous manner to lymphoma, demonstrating active regions of disease and response to treatment. FDG-PET/CT imaging is particularly helpful because, as in lymphoma, the CT imaging allows demonstration of lymphadenopathies that may not be FDG avid, allowing comparison on follow-up examination for both anatomic and physiologic changes (33–35) (Figure 19.15).

Waldenström's Macroglobulinemia

Waldenström's macroglobulinemia is a variant of non-Hodgkin's lymphoma, a lymphoid neoplasm characterized by a monoclonal lymphoplasmacytic expansion accompanied by secretion of a serum monoclonal immunoglobulin M (IgM). Jan G. Waldenström originally described the disease in 1944, reporting two patients with epistaxis, anemia, lymphadenopathy, and hypergammaglobulinemia. The monoclonal protein often results in a characteristic hyperviscosity syndrome. Current therapeutic modalities include alkylator agents, purine nucleoside analogues, and rituximab.

Waldenström's macroglobulinemia is uncommon disorder, comprising 6% of all B-cell lymphoproliferative

References

1. Barlogie B, Shaughnessy J, Epstein J, Sanderson R, Anaissie R, Walker R, Tricot G. Plasma cell myeloma. In: Williams Hematology, 7th ed. New York: McGraw-Hill, 2005: .

2. Shaughnessy J, Barlogie B. Interpreting the molecular biology and clinical behavior of MM in the context of global gene expression profiling. Immunol Rev 2003;194:140–163.

3. Angtuaco EJC, Fassas AB, Walker RC, et al. Multiple myeloma: clinical review and diagnostic imaging. Radiology 2004; 231:11–23.

4. Durie, B, Kyle, R, Belch, A, et al. Myeloma management guidelines: a consensus report from the Scientific Advisors of the International Myeloma Foundation. Hematol J 2003;4:379–398.

5. Walker R, Jones-Jackson L, Miceli M, et al. FDG PET functional imaging in multiple myeloma: clinically important caveats, pitfalls, and pearls. In: ASH Meeting 2004, San Diego. Blood 2004;104:11 (abstract).

6. Walker R, Jones-Jackson L, Rasmussen E, et al. Rapid response to treatment of multiple myeloma detected with FDG PET scanning in multiple myeloma: early results from total therapy III. In: ASH Meeting 2004, San Diego. Blood 2004;104:11 (abstract).

7. Walker R, Jones-Jackson L, Rasmussen E, et al. Diagnostic imaging of multiple myeloma: FDG PET and MRI complementary for tracking short vs. long term tumor response. In: ASH Meeting 2004, San Diego. Blood 2004;104:11 (abstract).

8. Walker R, Jones-Jackson L, Rasmussen E, et al. MRI-detectable focal lesions (FL) in multiple myeloma (MM) at relapse frequently involve novel sites not involved at diagnosis. In: ASH Meeting 2004, San Diego. Blood 2004;104:11 (abstract).

9. Orchard K, Barrington S, Buscombe A. et al. Fluoro-deoxyglucose positron emission tomography imaging for the detection of occult disease in MM. Br J Haematol 2002;117:133–135.

10. Schirrmeister H, Bommer M, Buck AK, et al. Initial results in assessment of MM using ^{18}F-FDG PET. Eur J Nucl Med 2002;29:361–366.

11. Durie B, Waxman A, D'Agnolo A, et al. Whole-body ^{18}F-FDG PET identifies high-risk myeloma. J Nucl Med 2002;43:1457–1463.

12. Durie BG, Salmon SE. A clinical staging system for MM: correlation of measured myeloma cell mass with presenting clinical features, response to treatment and survival. Cancer (Phila) 1975;36:842–854.

13. Barlogie B, Shaughnessy J, Tricot G, et al. Treatment of multiple myeloma. Blood 2004;103:20–32.

14. Roodman GD. Pathogenesis of myeloma bone disease. Blood Cells Mol Dis 2004;32:290–292.

15. Miceli MH, Atoui R, Walker R, et al. Diagnosis of deep septic thrombophlebitis in cancer patients by fluorine-18 fluorodeoxyglucose positron emission tomography scanning: a preliminary report. J Clin Oncol 2004;22:1949–1956.

16. Miceli MH, Jones-Jackson LB, Walker RC, et al. Diagnosis of infection of implantable central venous catheters by fluorine-18 fluorodeoxyglucose positron emission tomography. Nucl Med Commun 2004;25:813–818.

17. Anaissie E, Miceli M, Stroud S, et al. ^{18}F-Fluorodeoxyglucose (FDG) positron emission tomography (PET): an important tool in the management of infection in patients with hematological cancer. In: ASH Meeting 2004, San Diego. Blood 2004;104:11 (abstract).

18. Walker RC. Personal communication, unpublished data. University of Arkansas for Medical Sciences, Little Rock, AR, 2005.

19. O'Brien SN, Blijlevens NM, Mahfouz TH, et al. Infections in patients with hematological cancer: recent developments. Hematology (Am Soc Hematol Educ Program) 2003;438–472.

20. Crawford J, Dale DC, Lyman GH. Chemotherapy-induced neutropenia: risks, consequences, and new directions for its management. Cancer (Phila) 2004;100:228–237.

21. Sugawara Y. Infection and inflammation. In: Wahl R, Buchanam J, editors. Principles and Practice of Positron Emission Tomography, 1st ed. Philadelphia Lippincott Williams & Wilkins, 2002:381–394.

22. Stumpe KD, Dazzi H, Schaffner A, et al. Infection imaging using whole-body FDG-PET. Eur J Nucl Med 2000;27:822–832.

23. Zhuang H, Alavi A. 18-Fluorodeoxyglucose positron emission tomographic imaging in the detection and monitoring of infection and inflammation. Semin Nucl Med 2002;32:47–59.

24. Rennen H, Boerman O, Oyen W, et al. Imaging infection/inflammation in the New Millennium. Eur J Nucl Med 2001;28:241–252.

25. De Winter F, Vogelaers D, Gemmel F, et al. Promising role of 18-F-fluoro-D-deoxyglucose positron emission tomography in clinical infectious diseases. Eur J Clin Microbiol Infect Dis 2002;21:247–257.

26. Gibel LJ, Hartshorne MF, Tzamaloukas AH. Indium-111 oxime leukocyte scan in the diagnosis of peritoneal catheter tunnel infections. Perit Dial Int 1998;18:234–235.

27. Mettler F, Guiberteau M. Inflammation and infection imaging. In: Mettler F, editor. Essentials of Nuclear Medicine Imaging, 4th ed. Philadelphia: Saunders, 1998:387–403.

28. Wahl R. Principles of cancer imaging with fluorodeoxyglucose. In: Wahl R, editor. Positron Emission Tomography, 1st ed. Philadelphia: Lippincott Williams & Wilkins, 2002:100–110.

29. Avva R, Vanhemert R, Barlogie B, et al. CT-guided biopsy of focal lesions in patients with multiple myeloma may reveal new and more aggressive cytogenetic abnormalities. Am J Neuroradiol 2001;22:781–785.

30. Callander NS, Roodman GD. Myeloma bone disease. Semin Hematol 2001;38:276–285.

31. Kyle R, Child JA, Anderson K, et al. Criteria for the classification of monoclonal gammopathies, MM and related disorders: a report of the International Myeloma Working Group. Br J Haematol 2003;121:749–757.

32. Migliorati CA. Bisphosphonates and oral cavity avascular bone necrosis. J Clin Oncol 2003;21:4253–4254.

33. Marx RE. Pamidronate (Aredia) and zoledronate (Zometa) induced avascular necrosis of the jaws: a growing epidemic. J Oral Maxillar Surg 2003;61:1115–1118.

34. van Rhee F, Alikhan M, Munshi N, et al. Anti-IL6 antibody (ab) based strategies improve the management of HIV negative Castleman's disease. In: ASH Meeting 2004, San Diego. Blood 2004;104:11 (abstract).

35. Volberding P, Baker K, Levine A. Human immunodeficiency virus hematology. Hematology 2003;294–313 (www.hematology.org).

36. Brown, J, Skarkin A. Clinical mimics of lymphoma. Oncologist 2004;9:406–416.

37. Ghobrial IM, Gertz MA, Fonseca R. Waldenström macroglobulinaemia. Lancet Oncol 2003;4:679–685.

20

Evolving Role of FDG-PET Imaging in the Management of Patients with Suspected Infection and Inflammatory Disorders

Hongming Zhuang and Abass Alavi

Timely identification and localization of infectious and inflammatory processes is a critical step toward appropriate treatment of patients with known or suspected of such disorders. Radiologic techniques including computed tomography, magnetic resonance imaging, and ultrasonography have been frequently utilized for this purpose. However, these techniques rely solely on structural changes, and therefore discrimination of active infectious and inflammatory processes from alterations following surgery or other intervention remain difficult with these modalities. Also, infectious and inflammatory disorders cannot be detected in the early stages of their development because of the lack of substantial structural derangements, which render anatomic techniques insensitive for early diagnosis. Furthermore, these techniques, based on current standards, provide information on only a limited part of the body. However, because of advance made in recent years, it becomes feasible to scan the entire body within a short period of time.

Conventional nuclear medicine modalities, such as ^{67}Ga imaging and labeled leukocyte studies, have been useful in diagnosing some infectious disease and have contributed significantly to the management of such patients during the past three decades. However, these techniques suffer from many shortcomings. Recent developments in the field have substantially improved the ability of the radiotracer imaging techniques to detect infectious disease. These new methods are based on utilizing radiolabeled chemostatic peptides (1), radiolabed liposomes (2), avidin-mediated agents (3–5), radiolabeled antibodies (6, 7), radiolabeled antibiotics (8), and positron-emitting compounds, such as [^{18}F]fluorodeoxyglucose (FDG). In this chapter, we discuss the evolving role of positron emission tomography (PET) in this setting.

FDG is transported into cells by glucose that is transported and phosphorylated by hexokinase to FDG-6-phosphate. Phosphorylated FDG is trapped inside cells for a prolonged period of time. The deoxy-substitution prevents further metabolism of FDG and, therefore, over time monophosphorylated product accumulates in the tissue (9). Multiple factors influence the degree of uptake and intracellular accumulation of FDG. High level of expression of glucose transporters is known to be an important factor for facilitating FDG uptake in tumor cells. Tumor cells show increased expression of glycolytic enzymes, including hexokinase, and decreased expression of glucose-6-phosphatase, compared to normal cells (10, 11). Consequently, high levels of FDG-6-phosphate are produced and retained in malignant tissues.

FDG is not a tumor-specific tracer. Due to enhanced glycolytic activity, activated inflammatory cells such as neutrophils and macrophages also have increased FDG uptake, which results in high FDG concentration of this agent at sites of inflammation and infection (12–15). FDG rapidly accumulates at sites of bacterial infection and in reactive lymph nodes and results in high contrast between the affected and noninvolved tissues (16). There is evidence that inflammatory cells also have increased expression of glucose transporters when they are activated (17, 18). Acute human immunodeficiency virus (HIV) infection leads to increased expression of glucose transporters at high levels and a corresponding increase in glucose metabolic activity (19). It is possible that the metabolism of glucose and FDG uptake by inflammatory cells is more complicated than in malignant cells. For example, there is evidence that numerous cytokines and growth factors, whose levels are often increased during infection, may dramatically affect glucose uptake by inflammatory cells (20–24).

The hierarchy for the degree of FDG uptake by resting inflammatory cells is as follows: neutrophils greater than macrophages greater than lymphocytes (25), which would indicate that infectious or inflammatory sites where neutrophils or macrophages dominate are more likely to be visualized by FDG-PET than those dominated by lymphocytes. However, in spite of low levels of FDG uptake by lymphocytes in the resting state, glucose metabolic activity of these cells can increase dramatically upon stimulation. Using a mouse skin transplantation model, Heelan et al. (26) found that FDG uptake was 1.5 to 2 times higher in allografts than in syngeneic grafts and that the increase in uptake correlated with the levels of T-cell infiltrate seen histologically. Furthermore, FDG-PET images of simian immunodeficiency virus (SIV)-infected animals were readily distinguishable from those of an uninfected control cohort and revealed a pattern consistent with widespread lymphoid tissue activation. Areas of elevated FDG uptake on the PET images correlated well with productive SIV infection using in situ hybridization as a test for virus replication. These data suggest that the FDG-PET method can be used to evaluate the presence and the state of inflammatory reaction in infected tissues in vivo, thereby avoiding invasive procedures such as biopsy (27).

FDG-PET imaging has been used with great success in the management of patients with a variety of malignancies. In contrast, reports of the utility of FDG-PET in detecting and characterizing infectious diseases are limited in the literature. FDG accumulates in a variety of infectious and inflammatory sites, including abscesses in abdomen (28), brain (29), lung (30), kidney (31), tubo-ovarian (32, 33), inflammatory pancreatic disease (34), lobar pneumonia (35, 36), sarcoidosis (37, 38), osteomyelitis (7, 39), tuberculosis (40, 41), mastitis (42), and infectious mononucleosis (43). The accumulation of FDG in sites of infection or inflammation is one of the major causes of false-positive results when this imaging technique is used in oncology. On the other hand, high accumulation of FDG at sites of infection provides an opportunity to use FDG-PET imaging for the management of patients with suspected or known infection. In an early report with a total of 24 patients with various infectious lesions, FDG-PET achieved a sensitivity of 92% and a specificity of 100% (44).

Imaging Protocols for Infection and Inflammation

Currently, there is no consensus about an optimal FDG-PET imaging protocol for the assessment of infectious and inflammatory processes because of limited experience with this application. Many centers employ the same protocol that is in place for assessment of patients with cancer. However, because of inherent biologic differences between the two processes, imaging procedures may need to be adjusted for optimal results.

The optimal time interval between the injection of FDG and data acquisition is not well established when FDG-PET is used to examine patients with a malignant disease. Most centers initiate imaging 60 min following the administration of FDG. Some authors believe that a short 30-min dynamic data acquisition is as effective as longer

acquisition time in the diagnosis of lung cancer (45). On the other hand, many publications suggest that delayed imaging up to several hours from the time of injection of FDG would result in increase of sensitivity of the technique in the diagnosis and management of patients with several types of cancer (46–59). In contrast, in infection imaging, there is strong evidence that the interval between injection and data acquisition can be substantially shortened without an adverse effect on the sensitivity of the technique in detecting the site of abnormalities. Early experiments in animal models demonstrated that FDG uptake by the inflammatory cells increases gradually until 60 min following injection, and then it reaches a plateau and then declines thereafter. This observation has been confirmed in later studies. Therefore, the time interval between the administration of FDG and data acquisition could be shortened to less than 60 min for this indication, but this protocol requires validation with future studies.

The effect of serum glucose levels on the sensitivity of FDG-PET imaging in the diagnosis of infection is poorly defined at this time. It is well known that hyperglycemia results in significantly reduced FDG uptake by malignant lesions (60). As a result, hyperglycemia can lead to false-negative findings in a variety of malignancies, including bronchial carcinoma (61), head and neck cancer (62), and pancreatic neoplasias (63). To reduce serum glucose levels before to FDG-PET imaging, patients with cancer are instructed to fast for several hours. Also, in these patients, their serum glucose concentration is assessed before FDG is administered to make certain that the levels are optimal for this purpose. Based on recent data generated in our laboratory, inflammatory cells differ significantly from malignant cells with regard to the time course of FDG uptake and its retention, as measured by in vitro techniques. There is evidence that hyperglycemia may not adversely affect FDG uptake by inflammatory lesions, which is in contrast to malignant lesions (64). Animal experiments have shown that during septic shock glucose metabolism in macrophage-rich tissues remains elevated regardless of serum glucose level (65). In a clinical study in patients with chronic pancreatitis, Diederichs et al. (63) found that the average standardized uptake value (SUV) of the pancreatic tissue obtained in hyperglycemic patients was slightly higher than that in euglycemic patients, although the difference was not statistically significant. Most tumor cells have been shown to exhibit increased glycolytic activity and, therefore, have low glycogen storage and fully depend on extracellular glucose supply. In contrast, inflammatory cells are capable of mobilizing their intracellular glycogen storage to produce glucose during periods of low plasma glucose. One example is the process of granulocyte phagocytosis, which often occurs in a hypoglycemic environment and depends heavily on glycogen storage (66). Therefore, the effects of glucose concentration on FDG uptake appear to be negligible in FDG-PET imaging to detect inflammation.

In addition, cytokines released during the process of inflammation and infection also modulate FDG uptake by inflammatory cells. One example is platelet-activating factor, which is an important cytokine involved in a variety of inflammatory and infectious disorders (24, 67, 68). In animal experiments, it has been shown that platelet-activating factor can cause hyperglycemia (69, 70). In addition, platelet-activating factor increases glucose uptake by polymorphonuclear leukocytes and monocytes in such a hyperglycemic environment (70). Similarly, granulocyte colony-stimulating factor (G-CSF)-treated rats revealed enhanced glucose uptake by several organs, which was not affected by changes in plasma glucose or insulin concentrations (20). This regulatory effect of cytokines and growth factors on FDG uptake would be very small in malignant cells because of their relatively autonomous nature. Accordingly, fasting may not be necessary when FDG-PET is used for the assessment of infection. If this phenomenon can be confirmed in clinical settings, it would simplify patient preparation considerably when FDG-PET imaging employed to detect inflammatory processes. This change would be of especially great importance in examining diabetic patients, who are prone to have frequent infections.

Clinical Applications

Chronic Osteomyelitis

When bone architecture has been altered by previous trauma, surgery, or soft tissue infection, it is often difficult to exclude chronic osteomyelitis and to differentiate soft tissue infection from osseous infection by imaging techniques (71). Bone scintigraphy plays an important role in the diagnosis of osteomyelitis, especially in an otherwise intact bone. Three-phase bone scanning has good sensitivity for detecting infection, but its specificity is relatively low (72). ⁶⁷Gallium imaging in conjunction with bone scintigraphy has been used to distinguish infections from other processes, but this combination study has an accuracy of only 70% (73). When combined with bone scanning, labeled-leukocyte imaging has a good accuracy for diagnosing chronic osteomyelitis in the appendicular skeleton (71, 73). However, in chronic osteomyelitis of the axial skeleton and other sites with high concentration of red marrow, labeled white blood cell (WBC) images of infection frequently appears as areas of decreased activity, which is nonspecific and is seen in inactive process such as fibrosis, surgical changes, fractures, necrosis, degenerative disease, and hematologic disorders (74–76). For these reasons, WBC imaging of sites of infection in the axial skeleton and in areas with hematopoietic activity has an accuracy of 53% to 76% and therefore is of limited value (77).

FDG-PET imaging has proven to be useful in the diagnosis of osteomyelitis (Figures 20.1, 20.2), with a sensitivity of 95% to 100% and specificity of 85.7% to 95% in a relatively small number of patients (7, 39, 78). According to these reports, FDG-PET is able to correctly diagnose chronic osteomyelitis when both MRI and antigranulocyte antibody imaging appear negative. In a study of 51 patients, Guhlmann et al. (7) noted that FDG-PET imaging was superior to combined 99mTc-labeled monoclonal antigranulocyte antibody imaging and bone scintigraphy in the diagnosis of chronic osteomyelitis. Meller et al. compared the efficacy of labeled-leukocyte imaging and FDG-PET with coincident camera in the evaluation of chronic osteomyelitis. They reported that labeled-leukocyte imaging was true positive in 2 of 18 regions, true negative in 8 of 18 regions, false negative in 7 of 18 regions,

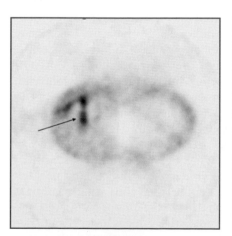

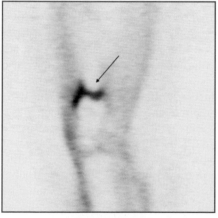

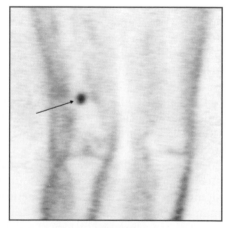

TRANSVERSE   SAGITTAL   CORONAL

Figure 20.1. FDG-PET images of the femur of a 41-year-old man who was involved in a motor vehicle accident and was suspected of suffering from osteomyelitis of the right distal femur. Increased glucose metabolism in the marrow space of the right distal femur is seen (*arrow*). The site of inflammatory reaction extends outward from the distal femur to the lateral fascia of the right thigh. [From Zhuang H, Duarte PS, Pourdehand M, et al. Exclusion of chronic osteomyelitis with F-18 fluorodeoxyglucose positron emission tomographic imaging. Clin Nucl Med 2000;25(4):281–284.]

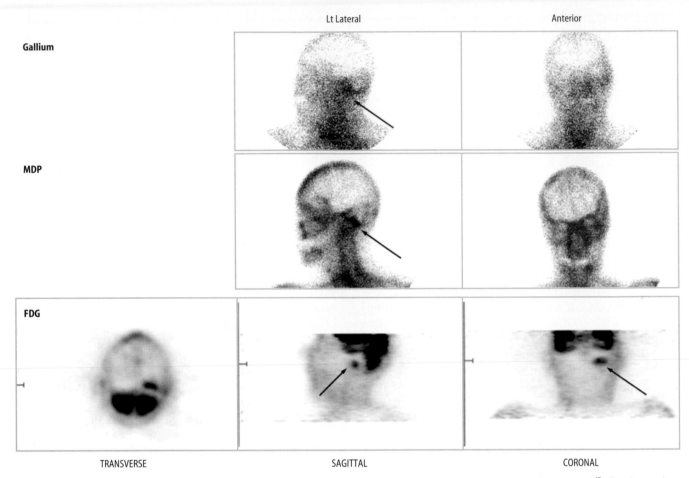

Figure 20.2. FDG-PET images of a 33-year-old man with left ear malignant otitis (*arrows*) caused by *Pseudomonas aeruginosa*. Corresponding bone scan and [67]gallium images show an extensive area of abnormal activity whereas FDG-PET localizes the infection to a considerably smaller area.

and false positive in 1 of 18 regions. In contrast, FDG imaging was true positive in 11 of 11 regions and true negative in 23f of 25 regions (79). FDG-PET was found to be especially useful for the evaluation of the axial skeleton. In a study of 32 patients with suspected vertebral osteomyelitis, De Winter et al. (80) reported sensitivity, specificity, accuracy, and interobserver agreement of 100%. When the results of FDG-PET imaging were compared with those of the combined bone and white blood cell scintigraphy, FDG-PET was significantly more accurate.

Particularly, FDG-PET imaging can correctly distinguish chronic osteomyelitis from a healing bone reaction (81). It is known that there is frequently increased FDG accumulation in the acute fracture site (82, 83) (Figure 20.3). However, the increased FDG activity in the fracture sites is usually transient. It is uncommon to note significant FDG activity at the fracture sites after a few months if no complications have followed the incident (84, 85). Therefore, the history of fracture several months earlier is unlikely to affect the diagnosis of osteomyelitis by FDG-PET. This argument was confirmed by an animal experiment by Koort et al. (86) in which they found that uncomplicated bone healing in rabbits was associated

with a temporary increase in FDG uptake at 3 weeks, which mostly disappeared by 6 weeks. In the experimental animals, localized osteomyelitis resulted in an intense continuous uptake of FDG, the degree of which was higher than that of healing bones at both 3 weeks and at 6 weeks. This pattern differs from that of bone scintigraphy, where a positive result may indicate either an active infection or a reparative process following successful treatment (87)].

The high spatial and contrast resolution of FDG-PET imaging may also permit differentiation of osteomyelitis from infection of soft tissue surrounding the bone. FDG-PET imaging can provide definitive evidence and the exact location of sites of infection within a few hours of tracer administration, compared to 24 to 48 h for ^{67}Ga or labeled-leukocyte imaging. Because of the high sensitivity of FDG-PET imaging in detecting infection, a negative study essentially rules out osteomyelitis (39). In contrast, caution should be exercised in the interpretation of a positive FDG-PET, as its predictive value is lower than that of a negative result. FDG-PET is a cost-effective alternative to the combined use of three-phase bone scanning, leukocyte scintigraphy, and bone marrow imaging in this setting and should be considered as the study of choice for optimal management of such patients.

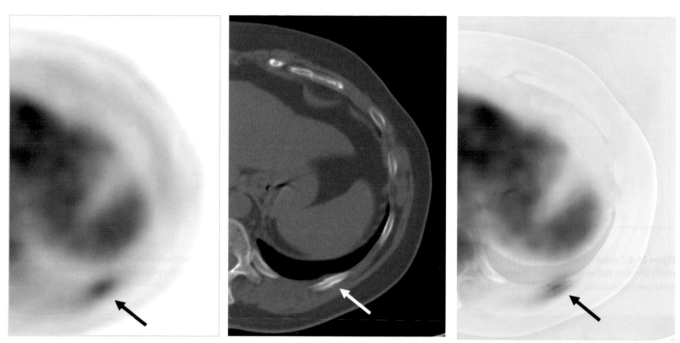

Figure 20.3. FDG-PET/CT images of a patient who sustained a severe trauma to the chest wall 3 weeks previously. Increased FDG activity is shown in the left posterior chest wall, and CT shows a recent fracture (*arrows*). These findings were interpreted to represent a recent fracture with inflammatory reaction resulting from the healing process.

Infections Associated with Prostheses

The success of prosthetic joint implant surgery has greatly improved the quality of life for individuals suffering from degenerative and other joint diseases. The number of lower limb arthroplasties performed in the United States has increased dramatically over the past two decades. More than 10% of patients will develop discomfort or pain at some time point following arthroplasty. The majority of these symptoms are the result of mechanical failure or loosening, while only a small fraction is caused by superimposed infection. Infection occurs in fewer patients following initial lower limb arthroplasty, but the incidence can be as high as 30% following prosthesis revisions (88, 89). It is crucial that the diagnosis of infected prosthesis is established before the patient is subjected to further surgical intervention. Multiple tests, which often include X-ray radiography, bone scanning, labeled-leukocyte scintigraphy, measurement of C-reactive protein, and joint aspiration followed by bacterial culture, are frequently required to establish the diagnosis of infection (90). Among nuclear medicine studies, bone and ^{67}Ga scans were first used to detect periprosthetic infections but did so with relatively low sensitivity (91, 92). Some authors have suggested that labeled-leukocyte scanning (93) can provide high sensitivity and specificity in this setting, especially when it is combined with ^{99m}Tc-sulfur colloid bone marrow imaging (94,95). However, the reported accuracy of this technique has varied and no consensus has been established concerning its utility. In a study by Scher et al. (96) in which 153 labeled-leukocyte scans were performed on 143 patients with painful prostheses, the authors re-

ported a sensitivity of 77%, a specificity of 86%, and positive and negative predictive values of 54% and 95%. They concluded that labeled-leukocyte scan should not be recommended routinely for evaluating periprosthetic infection because of the complexity of the procedure, the cost associated with this study, the need for two separate visits separated by at least 24 h, and unsatisfactory accuracy.

FDG-PET has proved to be an effective modality in the diagnosis of infection associated with lower limb arthroplasty. Clinical studies have demonstrated an overall sensitivity of 90% to 100% and a specificity of 81% to 89% for FDG-PET (97, 98). Based on some preliminary data, it appears that this technique has a higher accuracy in the diagnosis of painful hip prostheses than that in knee prostheses (98). In a study of 36 knee prostheses and 38 hip prostheses, the sensitivity, specificity, and accuracy of FDG-PET for detecting infection associated with knee prostheses were 90.2%, 72.0%, and 77.8%, respectively. In comparison, the sensitivity, specificity, and accuracy of FDG-PET for detecting infection associated with hip prostheses were 90.0%, 89.3%, and 89.5% (98). In contrast to computed tomography (CT) and magnetic resonance imaging (MRI), FDG-PET is not hindered by metallic implants that are commonly used for orthopedic procedures (99, 100). In addition, bone marrow uptake of FDG is minimal in the elderly population who are candidates for this surgical procedure. The higher spatial resolution of FDG-PET compared to conventional nuclear medicine imaging modalities enable small and subtle lesions to be detected (Figure 20.4).

Accurate diagnosis of periprosthetic infection would require defining appropriate criteria for distinguishing in-

skeleton, it has the advantage of demonstrating all metastatic sites, whether they be in soft tissue or bone.

With regard to imaging of skeletal infection and inflammation, uptake of FDG is likely to be largely associated with activated white cells. As this mechanism of uptake is not specific to the skeleton, there is the potential to also image the soft tissue component of infection associated with the skeleton.

[18]F-Fluoromisonidazole

[18]F-Fluoromisonidazole is one of a number of labeled compounds that have been synthesized which incorporate a 2-nitroimidazole moiety as a bioreductive molecule (9). The nitro group undergoes one-electron reduction in viable cells to produce a radical anion. In hypoxic cells, this intermediate is further reduced to species that react with cellular components and are trapped within the cell. In normoxic conditions, reoxidation rapidly takes place and the compound eventually diffuses out of the cell. These compounds have been most extensively evaluated in the context of tumor and myocardial hypoxia but may have a role in the investigation of anaerobic infection of bone (10, 11).

Malignant Skeletal Disease

Skeletal Metastases

High image contrast between normal and diseased bone is achievable by PET imaging as early as 1 h following injection of ^{18}F-fluoride, and this method has been evaluated in the investigation of bone metastases (12–17).

In breast cancer, both sclerotic and lytic bone metastases have been reported to show increased uptake of ^{18}F-fluoride (15), as is also seen with conventional nuclear medicine bone tracers such as ^{99m}Tc-methylene diphosphonate (MDP). Lytic metastases, more frequently encountered in most cancers, are caused by osteoclastic bone resorption that is stimulated by tumor-derived cytokines, and this is nearly always accompanied by local reactive bone formation and hence uptake of bone tracers. Sclerotic metastases occur when newly formed bone is laid down without prior resorption, and these are characteristically associated with marked uptake of bone tracers.

Although PET shows superior quantitative accuracy over planar or single photon emission computed tomography (SPECT) gamma camera imaging, it is unlikely that quantification of uptake of ^{18}F-fluoride uptake, as a nonspecific bone agent, would be able to differentiate benign from malignant focal skeletal lesions. Using a relatively crude index of uptake (lesion to normal ratios), it has not been possible to differentiate benign from malignant lesions (13). Dynamic measurements of regional skeletal ^{18}F-fluoride activity, including determination of K_i, the rate constant of plasma clearance of ^{18}F-fluoride by bone mineral, have shown 3- to 10-fold-higher activity in metastases compared to normal bone (12, 15). However, comparisons to benign lesions have not been reported. It is also possible to generate parametric images of regional indices from dynamic acquisitions, a process that may facilitate measurement of regional kinetic indices (12).

Because of better spatial resolution and the routine acquisition of tomographic images, ^{18}F-fluoride PET imaging has potential advantages over conventional bone scintigraphy in detecting bone metastases. Schirrmeister and colleagues (17) have compared ^{18}F-fluoride PET with ^{99m}Tc-MDP in 44 patients with varied primary cancers (prostate, lung, and thyroid), using computed tomography (CT), magnetic resonance imaging (MRI), and ^{131}I-scintigraphy as reference methods. It was found that all known metastases were detected by ^{18}F-fluoride PET and nearly twice as many benign and malignant lesions were identified by PET than by ^{99m}Tc-MDP bone scans. It was also possible to correctly classify a larger number of lesions as benign or malignant with ^{18}F-fluoride (97%) compared to ^{99m}Tc-MDP (80.5%) because of the superior spatial localization of the former, particularly in the spine. In a further study of patients with breast cancer, the greater accuracy of ^{18}F-fluoride PET led to a change in management in 4 of 34 patients compared to conventional ^{99m}Tc-MDP bone scans (16).

It is possible that some of the benefit of ^{18}F-fluoride PET derives from the better spatial resolution and tomographic images compared to planar bone scintigraphy. A further study by Schirrmeister et al. (18) in patients with lung cancer showed a significant advantage in diagnostic accuracy for ^{18}F-fluoride PET compared to planar bone scintigraphy but only a small, nonstatistically significant advantage compared to bone scintigraphy augmented with SPECT. ^{18}F-Fluoride PET led to a change in management in 11% of patients compared to planar bone scintigraphy. In a subsequent study, ^{18}F-fluoride PET was found to be more effective than SPECT but was associated with higher incremental costs (19). It remains to be seen whether routine PET bone scanning by ^{18}F-fluoride PET would be cost effective across a broader spectrum of malignancies.

With the advent of combined PET/CT, it is likely that the anatomic and structural information available from the CT component will enhance the diagnostic accuracy compared to ^{18}F-fluoride PET alone, as has been confirmed in an initial study by Even-Sapir et al. in which the specificity of PET/CT was significantly greater than PET (97% versus 72%; P less than 0.001) (Figure 21.1) (20).

Because the uptake of FDG is within tumor cells and not on bone mineral surface, measurement of quantitative indices of FDG and ^{18}F-fluoride uptake is attractive, as there is the potential to grade lesions, infer prognosis, and monitor response to treatment. This potential has not been explored so far.

The possible role of FDG in the evaluation of skeletal metastases has been explored in a number of studies. On a patient-by-patient basis, the sensitivity for detecting metastases was found to be the same as ^{99m}Tc-MDP in non-small cell lung cancer, but FDG-PET correctly confirmed the absence of bone metastases in a larger number of cases [specificity of 98% (87/89) compared to 61% (54/89)] (21). The most likely explanation for these observations is that uptake of FDG is more specific for tumor and does not accumulate in coincidental benign skeletal disease such as osteoarthritis. Subsequent studies in lung cancer have shown advantages of FDG-PET over bone scintigraphy with regard to overall accuracy (22, 23).

In a similar study of 145 patients with a variety of cancers, FDG-PET demonstrated higher sensitivity and specificity on a lesion-by-lesion basis (24). This improved sensitivity may result from detection of metastases at an earlier stage, when only bone marrow is involved, as well as the better spatial and contrast resolution offered by PET.

FDG-PET is not more sensitive than bone scintigraphy for the detection of all bone metastases. In breast cancer, a higher false-negative rate has been described in the skeleton compared to other sites (25, 26). In a group of patients with breast cancer and progressive disease, a significantly higher number of metastases was reported overall with FDG-PET (Figure 21.2), but it was noted that a subgroup with predominantly sclerotic disease showed fewer lesions by PET than by conventional ^{99m}Tc-MDP bone scintigraphy (27). Sclerotic metastases showed lower uptake than lytic metastases in terms of standardized uptake value (SUV) (Figure 21.3). It has also been noted in a number of reports that FDG was less accurate in assessing skeletal disease in prostate cancer, a tumor that produces predominantly sclerotic bone metastases (28, 29).

A number of possible explanations have been advanced for the lower uptake of FDG in sclerotic bone metastases. It may be that this type of metastasis is metabolically less active. This explanation would be supported by the finding of a better prognosis in breast cancer patients with sclerotic bone metastasis than those with lytic metastasis (27). Sclerotic metastases are relatively less cellular (30), so that smaller tumor volumes may account for the lower uptake of FDG. In addition, more aggressive lytic disease might be expected to outstrip its blood supply, rendering the metastasis relatively hypoxic, a property that is associated with increased FDG uptake in some cell lines (7). In a different study, the sensitivity for detection of bone metastases was no different between FDG-PET and bone scintigraphy, but FDG-PET showed a higher specificity (97.6% versus 80.9%) (31).

In lymphoma, where skeletal involvement is predominantly marrow based rather than in cortical bone, FDG-PET has shown greater sensitivity than conventional bone scintigraphy (32) (Figure 21.4). As bone marrow staging of lymphoma is limited to iliac crest biopsy, which involves obvious sampling errors, it has also been suggested that FDG-PET might replace bone marrow biopsy as a staging procedure, with the added advantage that other tissues will be imaged at the same time (33, 34). For assessing bone marrow response to treatment, the situation may be more difficult with FDG, because it is known that chemotherapy and granulocyte colony-stimulating factors cause benign, diffuse increase in marrow activity (35, 36) (Figure 21.5). Similarly, in other malignancies where bone scintigraphy characteristically has a lower sensitivity as a result of predominantly lytic skeletal disease, including multiple myeloma and renal cell carcinoma, FDG-PET has been reported as showing greater accuracy in assessing the extent of skeletal involvement (37–39).

It would be anticipated that MRI would also be sensitive to marrow-based metastatic disease compared to bone scintigraphy. In a study of 39 children and young adults with Ewing's sarcoma, osteosarcoma, lymphoma, rhabdomyosarcoma, melanoma, and Langerhan's cell histiocytosis, sensitivity for the detection of bone metastases for FDG-PET was 90%, for whole-body MRI, 82%, and for bone scintigraphy, 71%, although FDG-PET showed the highest number of false-positive results (40).

Early data using FDG with combined PET/CT for evaluation of skeletal metastases in the spine have shown that having CT as an anatomic and structural correlate results in a higher specificity (41). PET/CT also allows the assessment of soft tissue extension of potential neurologic significance (Figures 21.6, 21.7).

Skeletal metastases are notoriously difficult to assess with regard to response to systemic therapies using current standard criteria depending on changes in radiographic appearances, a method that is accepted as being relatively insensitive compared to standard methods for monitoring soft tissue metastases. Assessing response with conventional bone scintigraphy is also limited by a slow change in bone activity or by the flare response. Direct measurement of skeletal tumor viability with FDG-PET is therefore of potential interest. At the present time there are few data to support the use of FDG-PET in this role, but a preliminary study reported by Stafford and colleagues (42) showed a correlation in response as measured by tumor markers and that measured by serial FDG-PET.

Primary Bone Tumors

There are fewer published data regarding the role of PET in primary bone tumors than bone metastases. As in many other tumor types, there is a relationship between the uptake of FDG, whether measured by semiquantitative indices such as SUV or kinetic parameters such as the metabolic rate of FDG and tumor grade in soft tissue and bone sarcomas (43–45). Because of low uptake in low-grade sarcomas, FDG-PET may not be an adequate tool to differentiate low-grade sarcomas from benign lesions (46, 47). However, uptake of FDG does not generally reach a plateau for 3 to 4 h in malignant lesions whereas benign lesions show maximal uptake long before this (48). There

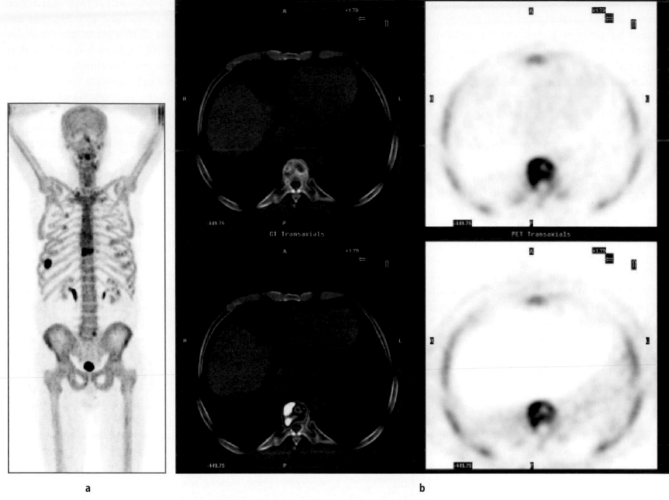

Figure 21.1. ^{18}F-Fluoride PET/CT images of a 73-year-old man with hormone refractory prostate cancer [Gleason 9; prostate-specific antigen (PSA), 37 ng/mL]. A sclerotic T9 metastasis and a pathologic rib fracture with an associated abnormal soft tissue mass are demonstrated on the PET (**a**) and PET/CT (**b, c**) images. (Courtesy Dr. Einat Even-Sapir, Souraski Medical Center, Tel-Aviv, Israel.)

is therefore the potential for better differentiation of benign from malignant sarcomas by scanning later than the conventional 1 h postinjection or by measuring changes in uptake between two time points.

In chondrosarcomas, a correlation between SUV and histopathologic stage has been described with an improvement in identifying patients at risk of local relapse and metastatic disease with higher SUVs (49). A further improvement in prognostic power was found when histologic tumor grade and SUV information were combined. Similarly, high FDG activity has adverse prognostic significance in osteosarcoma (50). For detecting skeletal metastases, it would appear that FDG-PET is more sensitive than conventional bone scintigraphy in Ewing's sarcoma but relatively insensitive in osteosarcoma (51).

The value of FDG in the monitoring response of bone tumors to chemotherapy is unclear. Certainly there appears to be a reduction in uptake corresponding to tumor necrosis following neoadjuvant therapy (52), but uptake into benign reactive fibrous tissue has been described that may cause difficulty in differentiating complete from partial response (53). However, FDG-PET may

be complementary to MRI in differentiating residual viable tumor from treatment effects (54). Good correlation between serial FDG-PET and histopathologic response has also been observed in pediatric bone sarcomas (55).

FDG-PET has also been used in the evaluation of residual or recurrent musculoskeletal tumors, and results suggest that it may be more sensitive than other tumor agents such as ^{99m}Tc-sestamibi. In addition, it is possible that additional information may be gained by using both FDG and ^{99m}Tc-sestamibi, where the former gives information on tumor viability and the latter gives information on the presence of multidrug resistance (56).

Benign Skeletal Disease

[18]F-Fluoride

Outside oncology, most of the reported work using ^{18}F-fluoride is in quantifying regional skeletal kinetics. The

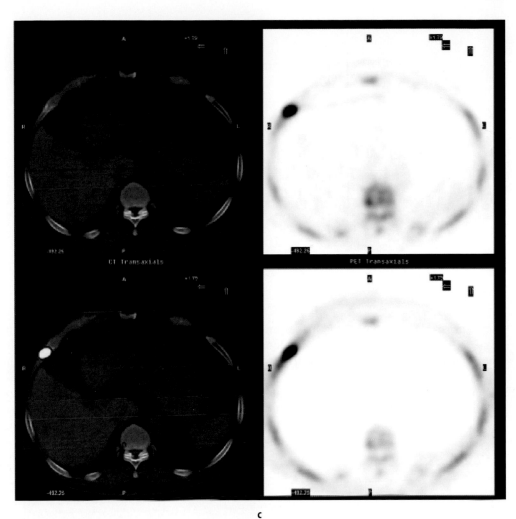

Figure 21.1. Contd. (c) ^{18}F-Fluoride PET/CT images of a 73-year-old man with hormone refractory prostate cancer [Gleason 9; prostate-specific antigen (PSA), 37 ng/mL]. A sclerotic T9 metastasis and a pathologic rib fracture with an associated abnormal soft tissue mass are demonstrated on the PET/CT.

c

method was pioneered by Hawkins and colleagues (12) and has subsequently been used to quantify a number of indices of regional skeletal metabolism in normal subjects (12, 57, 58), metabolic bone diseases (59, 60), avascular necrosis of the hip (61), and surgical bone grafts of the maxilla and hip (62, 63), as well as measuring differences between trabecular and cortical sites within the same subject (57).

By measuring plasma arterial activity and acquiring dynamic PET data over a skeletal region of interest, it is possible to calculate a number of parameters related to different aspects of bone metabolism. Hawkins et al. (12) described a three-compartment model and with the use of nonlinear regression analysis was able to determine the rate constants describing exchange of tracer between compartments as well as the macro-constant K_i, where

$$K_i = K_1 k_3/(k_2 + k_3) \tag{1}$$

It was found that a three-compartment model, consisting of a plasma compartment, a central extravascular compartment, and a bone mineral compartment, fitted the data better than a two-compartment model consisting of plasma and bone only. Subsequently, this model has been validated by using model-independent methods of

data analysis. It would appear that although the central compartment is an oversimplification, that is, it contains both bone marrow and bone extracellular fluid (ECF) as well as bone marrow cellular elements, this has little effect on parameter estimation, particularly K_i (58). This finding suggests that bone marrow ^{18}F-fluoride is available for uptake into bone mineral, even if on a longer time scale than true bone ECF.

The two most important parameters are K_i (net plasma clearance of fluoride by bone mineral; units, ml min^{-1} ml^{-1}) and K_1 (plasma clearance of fluoride by total bone tissue; units, ml min^{-1} ml^{-1}). The parameters k_2, k_3, and k_4 are simple rate constants (units of min^{-1}). K_i has been shown to correlate with biochemical and histomorphometric parameters of bone formation and mineralization. K_1 is related to blood flow (Q) and E by the equation:

$$K_1 = Q \cdot E \tag{2}$$

If E is close to unity, as has been deduced from measurements on rabbit hindlimb (3), K_1 represents regional bone blood flow. The Renkin–Crone model of capillary diffusion (64) predicts that E is reduced at high rates of blood flow, as might be seen in trabecular sites. K_1 therefore underestimates true regional blood flow in these cir-

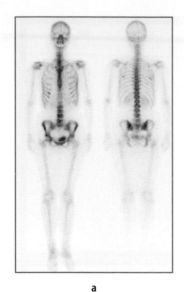

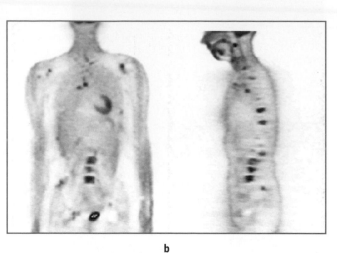

a

b

Figure 21.2. ^{99m}Tc-MDP) bone scan (**a**) and FDG-PET scan (**b**) in a patient with breast cancer. The scans were performed 2 weeks apart with no intervening therapy. The bone scan is essentially normal, but the PET scan shows a number of skeletal metastases.

cumstances; this is supported by work from Piert and colleagues (65) who compared vertebral blood flow in pigs measured with ^{15}O-labeled water with K_1 for ^{18}F-fluoride determined as just described.

A number of potential clinical applications have been described for dynamic, quantitative ^{18}F-fluoride PET. One clinical problem in bone disease is the differentiation of

renal osteodystrophy patients with accelerated turnover from those with adynamic bone. Bone biopsy may be the only way of resolving this problem, and thus alternative noninvasive methods to determine bone turnover are attractive. Messa and colleagues (59), by measuring K_i in vertebrae of patients with renal osteodystrophy, were able to differentiate those with high turnover from those with

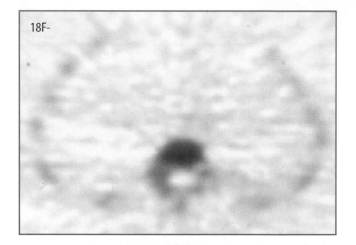

a

b

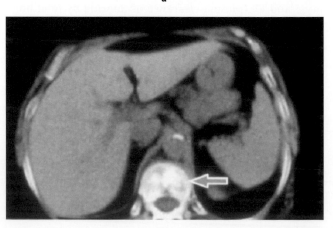

c

Figure 21.3. Transaxial ^{18}F-fluoride (**a**), FDG (**b**), and CT (**c**) scans demonstrate focally increased bone turnover in a sclerotic metastasis (*arrows*) but with no abnormal FDG activity.

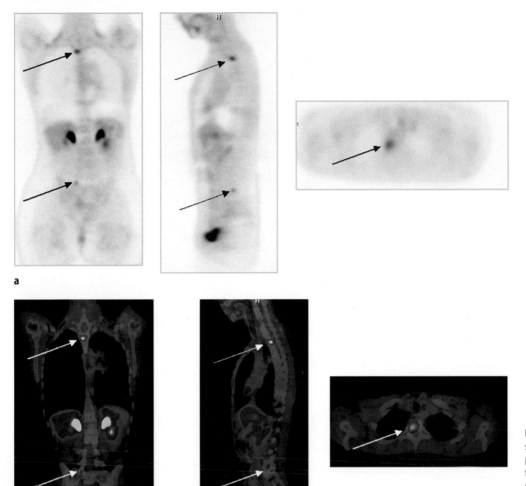

a

b

Figure 21.4. FDG-PET (**a**) and PET/CT (**b**) scans in a patient with recurrent lymphoma. On the FDG-PET scan, abnormal focal uptake is seen (*arrows*), but it is not possible to confidently localize these lesions to the skeleton. The combined PET/CT scan allows accurate skeletal localization of the lesions, which had been unsuspected and changed management.

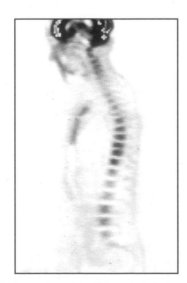

Figure 21.5. Diffuse increase in uptake of FDG seen in the bone marrow of a patient 2 weeks after completing chemotherapy is consistent with reactive changes rather than active lymphoma.

adynamic bone and were also able to show correlations between K_i and histomorphometric and biochemical indices of bone formation.

Schiepers et al. (60) studied the axial skeleton in a small number of subjects with a variety of metabolic bone diseases, including Paget's disease and osteoporosis, by this method. Two subjects with Paget's disease showed an expected increase in both K_i and K_1, reflecting an increase in both regional mineralization and blood flow. We have noted similar findings in seven patients with pagetic vertebrae (66). It was also noted that k_4, describing the release of ^{18}F-fluoride from bone mineral back to the extravascular compartment, was lower in pagetic vertebrae compared to adjacent normal vertebrae, suggesting that ^{18}F-fluoride remains more tightly bound to bone mineral in Paget's disease, an observation also recorded by Fogelman and colleagues using the ^{99m}Tc-MDP retention method (67).

Osteoporotic subjects show low values of vertebral K_i (expressed in units of ml min^{-1} ml^{-1}) (58, 60), a finding

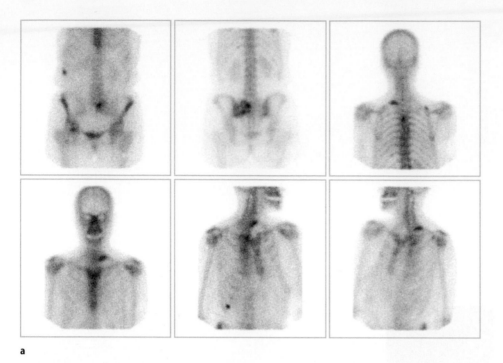

a

b

Figure 21.6. A 47-year-old woman with metastatic breast cancer treated with chemotherapy 2 years previously presented with increasing serum tumor markers. Planar bone scintigraphy (**a**) shows areas of abnormal uptake at the sacrum, T5, first left rib, a right anterior rib, and the right femoral neck. FDG-PET coronal slices (**b**) demonstrate pathologic uptake in the sacrum, left first rib, and T5.

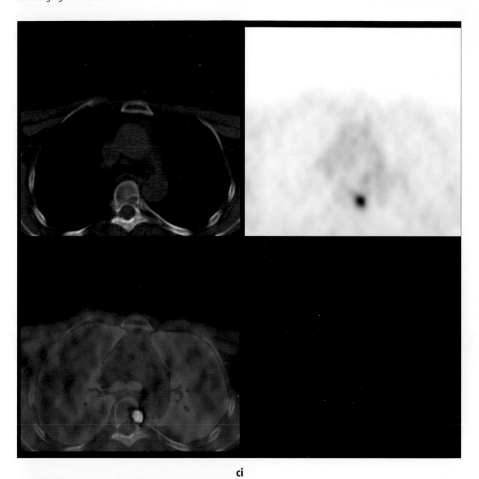

ci

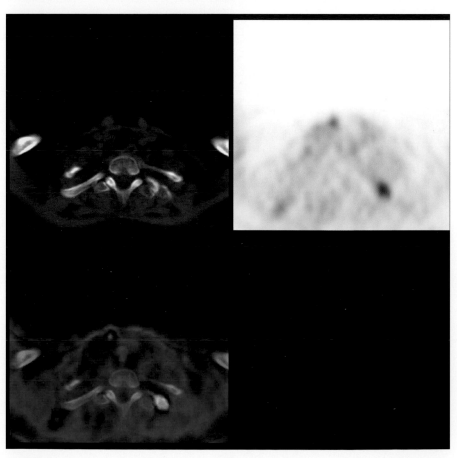

cii

Figure 21.6. Contd. PET/CT images (**c**) show an active lytic metastasis in the pedicle of T4 (i), an active mixed sclerotic and lytic metastasis in the left first rib (ii)

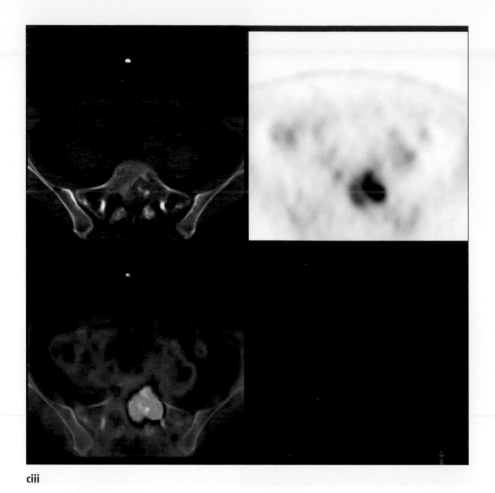

ciii

Figure 21.6. Contd. (**c**) and an active lytic metastasis in the sacrum (iii).

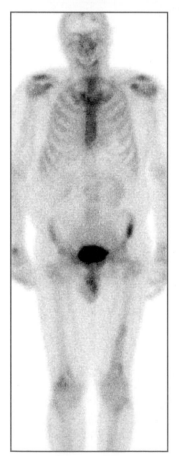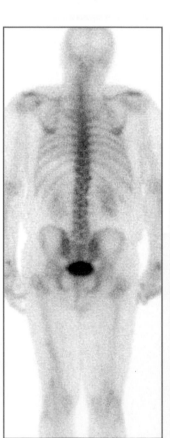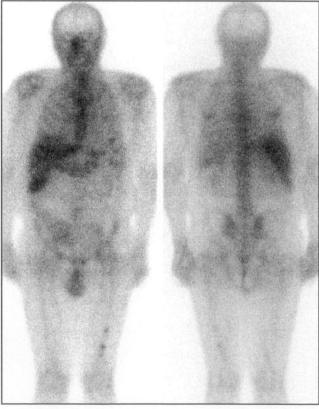

a   b

Figure 21.7. A 75-year-old man with diffuse large B-cell lymphoma. Bone scintigraphy (**a**) shows slight increased uptake in the left iliac wing and distal femora. Planar 48-h ⁶⁷Ga scintigraphy (**b**) demonstrates mildly increased uptake in the left distal femur. FDG-PET coronal slices

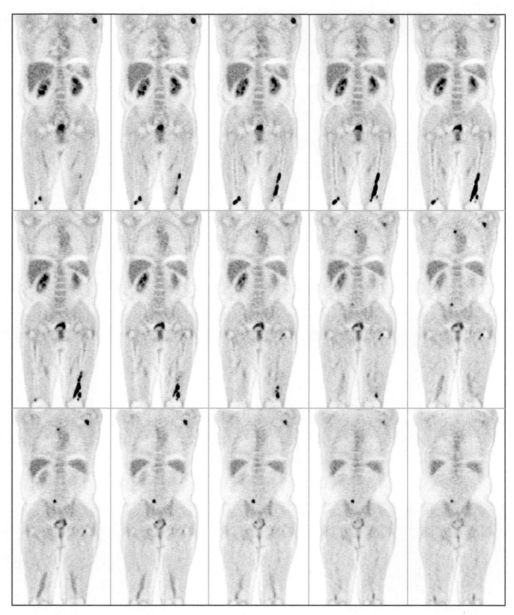

Figure 21.7. Contd. (c) demonstrate pathologic uptake in both femora, the left proximal humerus, T4, and L5 (the left iliac crest is not visible on these slices).

c

that is at odds with reports of increased bone turnover as measured by other methods (68). It is possible that there is a dissociation between global skeletal turnover and that measured in the predominantly trabecular site of the lumbar spine (58). It is clear that further work, perhaps with correlation with histomorphometric measurements, is required in osteoporotic subjects, before the full significance of kinetic parameters derived by ^{18}F-fluoride PET is understood. However, using this method a direct metabolic effect of antiresorptive therapy on skeletal kinetics at the clinically important site of the lumbar spine has been observed in osteoporotic women (69).

Quantification of regional skeletal metabolic indices also has potential clinical applications in orthopedics. A pilot study has suggested that by measuring K_1 as an index of blood flow in the femoral head following trauma, it may be possible to predict which patients will require surgical intervention rather than a conservative approach (61). Two studies have also employed this method in assessing bone grafts, suggesting that it may be a useful method to monitor graft metabolism and incorporation (62, 63).

^{18}F-Fluorodeoxyglucose

Accumulation of FDG is not specific to malignant cells, and observation of increased uptake into activated inflammatory cells (70, 71) has led to its use in the detec-

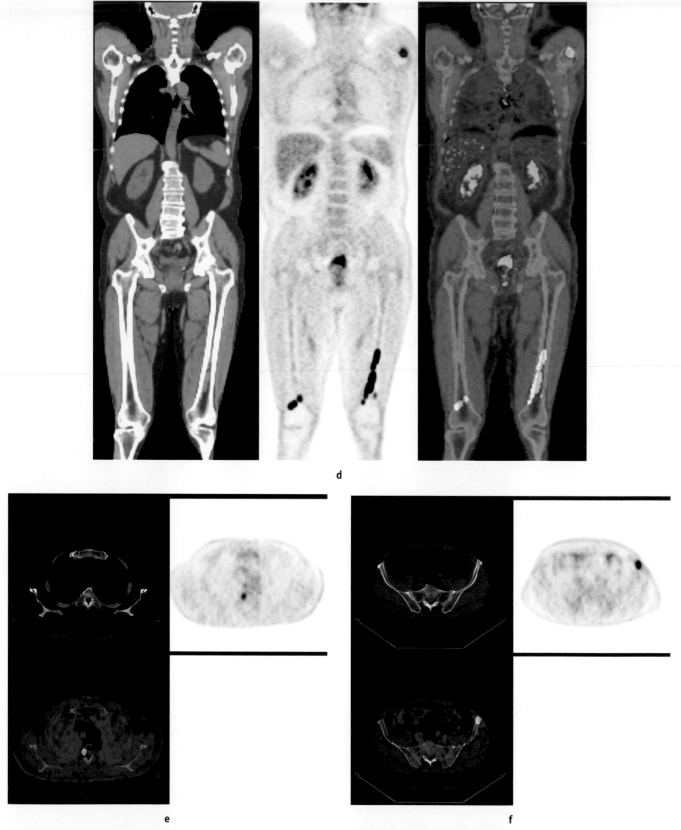

Figure 21.7. Contd. FDG-PET/CT images show the intramedullary location of sites of disease in the femora and left humerus (**d**), an active lytic lesion in the right pedicle of T4 (**e**), and an intramedullary lesion in the left anterior iliac crest (**f**)

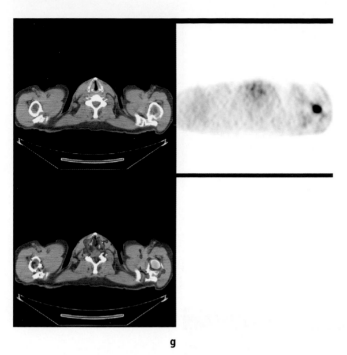

g

Figure 21.7. Contd. and the left humerus (**g**).

tion of infection in humans (72). Early reports suggest that this application may be particularly successful in the skeleton (73, 74). Increased uptake of FDG occurs in both acute and chronic osteomyelitis (Figure 21.8), but care is required in image interpretation following surgical intervention when nonspecific uptake may be seen (75–78).

There is increasing evidence that FDG-PET may be helpful in assessing painful prostheses, although there is the potential for false-positive interpretation for infection if close attention is not paid to the pattern of uptake in painful hip prostheses (79–81). Here it has been suggested that a pattern of uptake in the interface between the prosthesis and bone is more predictive of infection than uptake around the head or neck of the prosthesis (82). Accuracy has been reported as being better for hip prosthetic infection compared to knee prostheses (83) (Figure 21.9).

It has been noted that increased uptake of FDG may also be seen in Paget's disease, particularly in patients with more active disease as measured by serum alkaline phosphatase levels (84) (Figure 21.10). As the majority of patients with Paget's disease are asymptomatic and may be unaware of the disease, there is the possibility of an occasional false-positive scan in patients being staged for cancer. Although osteosarcoma is only a rare complication of Paget's disease, the efficacy of FDG-PET for differ-

entiating this tumor from benign pagetic changes may be diminished.

Conversely, it does not appear that benign degenerative disease in the spine or elsewhere causes uptake of FDG that would be mistaken for metastatic deposits, but uptake may be seen in inflammatory joint disease, particularly in the shoulder where uptake is assumed to be caused by inflammatory capsulitis or tendonitis. FDG-PET has been found to be very reliable in differentiating degenerative and infective vertebral endplate abnormalities detected on MRI, with no false positives or false negatives reported in one series of 30 patients (85). In the detection of bone metastases a potential false positive is caused by recent fractures (86–89). It has been suggested that measurement of SUV may help differentiate benign from pathologic malignant fractures. In a series of 20 patients, it was found that the mean SUV in those with benign fractures was 1.36 ± 0.49 whereas those with pathologic fractures had a mean SUV of 4.46 ± 2.12 (88). It has been noted that it is rare for FDG uptake to persist longer than 3 month after trauma (89).

^{18}F-Fluoromisonidazole

There is a limited literature on the potential use of this hypoxia selective tracer in skeletal infection. Its accumulation has been described in situations that commonly involve infection by anaerobic organisms, such as diabetic foot infection and osteomyelitis, as well as in odontogenic disease (10, 11).

Conclusion

The role for PET imaging of the skeleton is evolving. It is possible that ^{18}F-fluoride PET may have incremental value in detecting bone metastases over conventional bone scintigraphy, but this has yet to be established. A role may also exist for quantitative studies using ^{18}F-fluoride PET for research or clinical applications where there is a need to quantify regional skeletal metabolism.

FDG-PET is highly sensitive for detecting skeletal metastases in most cancers, and it is possible that the use of conventional bone scintigraphy may diminish as FDG-PET is used more routinely for cancer staging. There is a developing role for the use of this method also in primary bone tumors and in the detection of infection related to the skeleton.

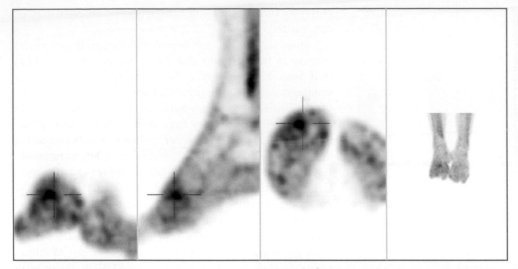

a

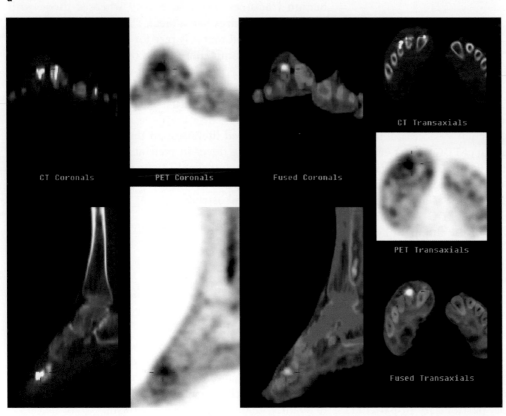

b

Figure 21.8. A 40-year-old woman with a previous history of osteomyelitis of the right foot complained of recurrent swelling and pain in the foot. FDG-PET (**a**) shows a focus of abnormal uptake in the right foot, localized to the second right metatarsal on the PET/CT images (**b**), consistent with recurrent osteomyelitis.

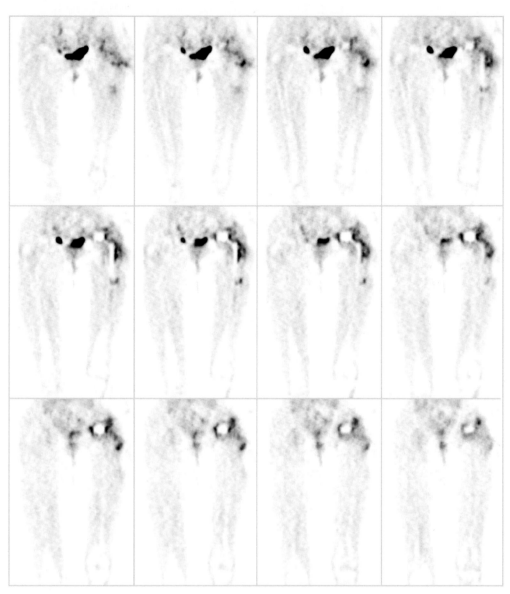

Figure 21.9. A 28-year-old woman with a left total hip replacement following an osteogenic sarcoma in childhood complained of pain related to the prosthesis. Coronal FDG-PET images with attenuation correction (**a**),

a

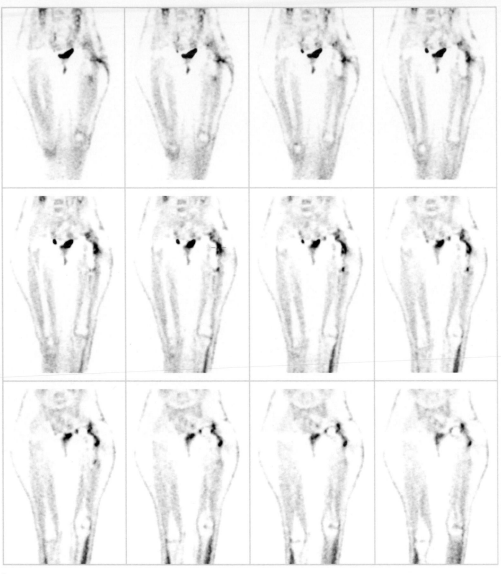

b

Figure 21.9. Contd. and without attenuation correction (**b**) show foci of abnormal uptake surrounding both the acetabular and femoral components of the prosthesis, consistent with infection.

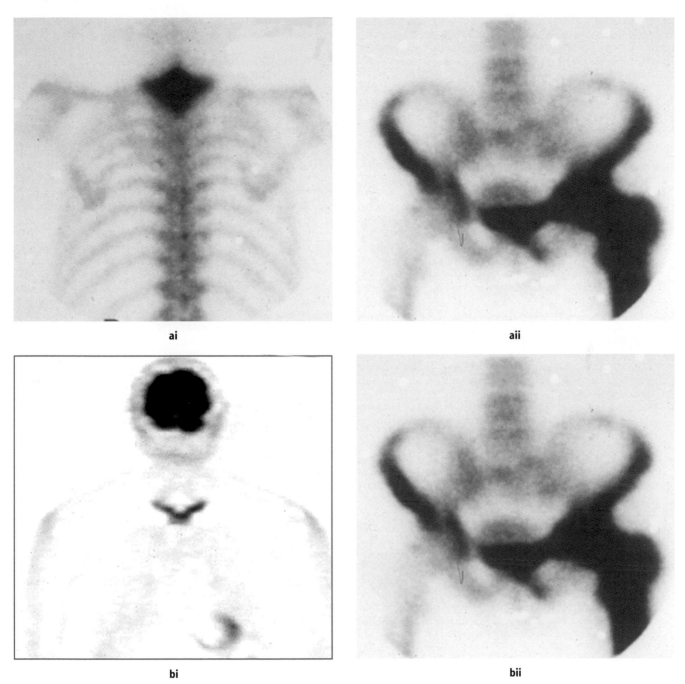

ai

aii

bi

bii

Figure 21.10. ^{99m}Tc-MDP bone scan (**a**) demonstrates Paget's disease in the left femur and in the upper thoracic spine; increased activity is also shown on a FDG-PET scan (**b**).

References

1. Blau M, Nagler W, Bender MA. A new isotope for bone scanning. J Nucl Med 1962;3:332–334.
2. Bang S, Baud CA. Topographical distribution of fluoride in iliac bone of a fluoride-treated osteoporotic patient. J Bone Miner Res 1990;5:S87–S89.
3. Wootton R, Dore C. The single-passage extraction of ^{18}F in rabbit bone. Clin Phys Physiol Meas 1986;7:333–343.
4. Piert M, Zittel TT, Machulla, HJ et al. Blood flow measurements with [(15)O]H$_2$O and [18F]fluoride ion PET in porcine vertebrae. J Bone Miner Res 1998;13:1328–1336.
5. Blake GM, Park-Holohan SJ,Cook GJR, Fogelman I. Quantitative studies of bone using ^{18}F-fluoride and ^{99m}Tc-methylene diphosphonate. Semin Nucl Med 2001;31:28–49.
6. Warburg O. On the origin of cancer cells. Science 1954;123:306–314.
7. Clavo AC, Brown RS, Wahl RL. Fluorodeoxyglucose uptake in human cancer cell lines is increased by hypoxia. J Nucl Med 1995;36:1625–1632.
8. Yamamoto T, Seino Y, Fukumoto H, et al. Overexpression of facilitative glucose transporter genes in human cancer. Biochem Biophys Res Commun 1990;170:223–230.
9. Nunn A, Linder K, Strauss HW. Nitroimidazoles and imaging hypoxia. Eur J Nucl Med 1995;22:265–280.

10. Liu RS, Chu LS, Yen SH, et al. Detection of odontogenic infections by fluorine-18 fluoromisonidazole. Eur J Nucl Med 1996;23:1384–1387.

11. Liu RS, Chu LS, Chang CP, et al. Assessment of the outcome of diabetic foot lesions by [F-18]fluoromisonidazole PET scan. J Nucl Med 1996;37:27P.

12. Hawkins RA, Choi Y, Huang SC, et al. Evaluation of the skeletal kinetics of fluorine-18-fluoride ion with PET. J Nucl Med 1992;33:633–642.

13. Hoh CK, Hawkins RA, Dahlbom M, et al. Whole body skeletal imaging with [18F]fluoride ion and PET. J Comput Assist Tomogr 1993;17:34–41.

14. Hoegerle S, Juengling F, Otte A, et al. Combined FDG and [F-18]fluoride whole-body PET: a feasible two-in-one approach to cancer imaging? Radiology 1998;209:253–258.

15. Petren-Mallmin M, Andreasson I, Ljunggren, et al. Skeletal metastases from breast cancer: uptake of 18F-fluoride measured with positron emission tomography in correlation with CT. Skeletal Radiol 1998;27:72–76.

16. Schirrmeister H, Guhlmann A, Kotzerke J, et al. Early detection and accurate description of extent of metastatic bone disease in breast cancer with fluoride ion and positron emission tomography. J Clin Oncol 1999;17:2381–2389.

17. Schirrmeister H, Guhlmann A, Elsner K, et al. Sensitivity in detecting osseous lesions depends on anatomic localization: planar bone scintigraphy versus 18F PET. J Nucl Med 1999;40:1623–1629.

18. Schirrmeister H, Glatting G, Hetzel J, et al. Prospective evaluation of the clinical value of planar bone scans, SPECT and (18)F-labeled NaF PET in newly diagnosed lung cancer. J Nucl Med 2001;42:1800–1804.

19. Hetzel M, Arslandemir C, Konig HH, et al. F-18 NaF PET for detection of bone metastases in lung cancer: accuracy, cost-effectiveness and impact on patient management. J Bone Miner Res 2003;18:2206–2214.

20. Even-Sapir E, Metser U, Flusser G, et al. Assessment of malignant skeletal disease: initial experience with 18F-fluoride PET/CT and comparison between 18F-fluoride PET and 18F-fluoride PET/CT. J Nucl Med 2004;45:272–278.

21. Bury T, Barreto A, Daenen F, Barthelemy N, Ghaye B, Rigo P. Fluorine-18 deoxyglucose positron emission tomography for the detection of bone metastases in patients with non-small cell lung cancer. Eur J Nucl Med 1998;25:1244–1247.

22. Cheran SK, Herndon JE, Patz EF. Comparison of whole-body FDG-PET to bone scan for detection of bone metastases in patients with a new diagnosis of lung cancer. Lung Cancer 2004;44:317–325.

23. Hsia TC, Shen YY, Yen RF, et al. Comparing whole body 18F-2-deoxyglucose positron emission tomography and technetium-99m methylene diphsophonate bone scan to detect bone metastases in patients with non-small cell lung cancer. Neoplasma 2002;49:267–271.

24. Chung JK, Kim YK, Yoon JK, et al. Diagnostic usefulness of F-18 FDG whole body PET in detection of bony metastases compared to 99mTc-MDP bone scan. J Nucl Med 1999;40:96P.

25. Moon DH, Maddahi J, Silverman DH, Glaspy JA, Phelps ME, Hoh CK. Accuracy of whole-body fluorine-18-FDG PET for the detection of recurrent or metastatic breast carcinoma. J Nucl Med 1998;39:431–435.

26. Gallowitsch HJ, Kresnik E, Gasser J, et al. F-18 fluorodeoxyglucose positron emission tomography in the diagnosis of tumour recurrence and metastases in the follow-up of patients with breast carcinoma: a comparison to conventional imaging. Invest Radiol 2003;38:250–256.

27. Cook GJ, Houston S, Rubens R, Maisey MN, Fogelman I. Detection of bone metastases in breast cancer by FDG PET: differing metabolic activity in osteoblastic and osteolytic lesions. J Clin Oncol 1998;16:3375–3379.

28. Shreve PD, Grossman HB, Gross MD, et al. Metastatic prostate cancer: initial findings of PET with 2-deoxy-2-[F-18]fluoro-D-glucose. Radiology 1996;199:751–756.

29. Yeh SD, Imbriaco M, Larson SM, et al. Detection of bony metastases of androgen-independent prostate cancer by PET-FDG. Nucl Med Biol 1996;23:693–697.

30. Galasko CSB. Skeletal Metastases. London: Butterworths, 1986.

31. Ohta M, Tokuda Y, Suzuki Y, et al. Whole body PET for the evaluation of bony metastases in patients with breast cancer: comparison with 99mTc-MDP bone scintigraphy. Nucl Med Commun 2001;22:875–879.

32. Moog F, Kotzerke J, Reske SN. FDG PET can replace bone scintigraphy in primary staging of malignant lymphoma. J Nucl Med 1999;40:1407–1413.

33. Carr R, Barrington SF, Madan B, et al. Detection of lymphoma in bone marrow by whole-body positron emission tomography. Blood 1998;91:3340–3346.

34. Moog F, Bangerter M, Kotzerke J, et al. 18-F-Fluorodeoxyglucose-positron emission tomography as a new approach to detect lymphomatous bone marrow. J Clin Oncol 1998;16:603–609.

35. Cook GJ, Fogelman I, Maisey MN. Normal physiological and benign pathological variants of 18-fluoro-2-deoxyglucose positron-emission tomography scanning: potential for error in interpretation. Semin Nucl Med 1996;26:308–314.

36. Hollinger EF, Alibazoglu H, Ali A, et al. Hematopoietic cytokine-mediated FDG uptake simulates the appearance of diffuse metastatic disease on whole-body PET imaging. Clin Nucl Med 1998;23:93–98.

37. Jadvar H, Conti PS. Diagnostic utility of FDG PET in multiple myeloma. Skeletal Radiol 2002;31:690–694.

38. Orchard K, Barrington S, Buscombe J, et al. Fluoro-deoxyglucose positron emission tomography for the detection of occult disease in multiple myeloma. Br J Haematol 2002;117:133–135.

39. Wu HC, Yen RF, Shen YY, et al. Comparing whole body 18F-2-deosxyglucose positron emission tomography and technetium-99m methylene diphosphonate bone scan to detect bone metastases in patients with renal cell carcinomas: a preliminary report. J Cancer Res Clin Oncol 2002;128:503–506.

40. Daldrup-Link HE, Franzius C, et al. Whole-body MR imaging for detection of bone metastases in children and young adults: comparison with skeletal scintigraphy and FDG PET. Am J Roentgenol 2001;177:229–236.

41. Metser U, Lerman H, Blank A, et al. Malignant involvement of the spine: assessment by 18F-FDG PET/CT. J Nucl Med 2004;45:279–284.

42. Stafford SE, Gralow JR, Schubert EK, et al. Use of serial FDG PET to measure the response of bone-dominant breast cancer to therapy. Acad Radiol 2002;9:913–921.

43. Eary JF, Conrad EU, Bruckner JD, et al. Quantitative [F-18]fluorodeoxyglucose positron emission tomography in pretreatment and grading of sarcoma. Clin Cancer Res 1998;4:1215–1220.

44. Eary JF, Mankoff DA. Tumor metabolic rates in sarcoma using FDG PET. J Nucl Med 1998;39:250–254.

45. Kern KA, Brunetti A, Norton JA, et al. Metabolic imaging of human extremity musculoskeletal tumors by PET. J Nucl Med 1988;29:181–186.

46. Adler LP, Blair HF, Makley JT, et al. Noninvasive grading of musculoskeletal tumors using PET. J Nucl Med 1991;32:1508–1512.

47. Kole AC, Nieweg OE, Hoekstra HJ, et al. Fluorine-18-fluorodeoxyglucose assessment of glucose metabolism in bone tumors. J Nucl Med 1998;39:810–815.

48. Lodge MA, Lucas JD, Marsden PK, et al. A PET study of 18FDG uptake in soft tissue masses. Eur J Nucl Med 1999;26:22–30.

49. Brenner W, Conrad EU, Eary JF. FDG PET imaging for grading and prediction of outcome in chondrosarcoma patients. Eur J Nucl Med Mol Imaging 2004;31:189–195.

50. Franzius C, Bielack S, Flege S, Sciuk J, Jurgens H, Schober O. Prognostic significance of (18)F-FDG and (99m)Tc-methylene diphosphonate uptake in primary osteosarcoma. J Nucl Med 2002;43:1012–1017.

51. Franzius C, Sciuk J, Daldrup-Link HE, Jurgens H, Schober O. FDG-PET for detection of osseous metastases from malignant primary

bone tumours: comparison with bone scintigraphy. Eur J Nucl Med 2000;27:1305–1311.

52. Schulte M, Brecht-Krauss D, Werner M, et al. Evaluation of neoadjuvant therapy response of osteogenic sarcoma using FDG PET. J Nucl Med 1999;40:1637–1643.

53. Jones DN, McCowage GB, Sostman HD, et al. Monitoring of neoadjuvant therapy response of soft-tissue and musculoskeletal sarcoma using fluorine-18-FDG PET. J Nucl Med 1996 37:1438–1444.

54. Bredella MA, Caputo GR, Steinbach LS. Value of FDG positron emission tomography in conjunction with MR imaging for evaluating therapy response in patients with musculoskeletal sarcomas. AJR Am J Roentgenol 2002;179:1145–1150.

55. Hawkins DS, Rajendran JG, Conrad EU, Bruckner JD, Eary JF. Evaluation of chemotherapy response in pediatric bone sarcomas by [F-18]-fluorodeoxy-D-glucose positron emission tomography. Cancer (Phila) 2002;94:3277–3284.

56. Garcia R, Kim EE, Wong FC, et al. Comparison of fluorine-18-FDG PET and technetium-99m-MIBI SPECT in evaluation of musculoskeletal sarcomas. J Nucl Med 1996;37:1476–1479.

57. Cook GJR, Lodge MA, Blake GM, Marsden PK, Fogelman I. Differences in skeletal kinetics between vertebral and humeral bone measured by ^{18}F-fluoride PET in postmenopausal women. J Bone Miner Res 2000;15:763–769.

58. Frost ML, Fogelman I, Blake GM, Marsden PK, Cook GJR. Dissociation between global markers of bone formation and direct measurement of spinal bone formation in osteoporosis. J Bone Miner Res 2004;19:1797–1804.

59. Messa C, Goodman WG, Hoh CK, et al. Bone metabolic activity measured with positron emission tomography and [^{18}F]fluoride ion in renal osteodystrophy: correlation with bone histomorphometry. J Clin Endocrinol Metab 1993;77:949–955.

60. Schiepers C, Nuyts J, Bormans G, et al. Fluoride kinetics of the axial skeleton measured in vivo with fluorine-18-fluoride PET. J Nucl Med 1997;38:1970–1976.

61. Schiepers C, Broos P, Miserez M, et al. Measurement of skeletal flow with positron emission tomography and ^{18}F-fluoride in femoral head osteonecrosis. Arch Orthop Trauma Surg 1998;118:131–135.

62. Berding G, Burchert W, van den Hoff J, et al. Evaluation of the incorporation of bone grafts used in maxillofacial surgery with [^{18}F]fluoride ion and dynamic positron emission tomography. Eur J Nucl Med 1995;22:1133–1140.

63. Piert M, Winter E, Becker GA, et al. Allogeneic bone graft viability after hip revision arthroplasty assessed by dynamic [^{18}F]fluoride ion positron emission tomography. Eur J Nucl Med 1999;26:615–624.

64. Renkin EM. Transport of potassium-42 from blood to tissue in isolated mammalian skeletal muscles. Am J Physiol 1959;197:1205–1210.

65. Piert M, Zittell TT, Machulla HJ, et al. Blood flow measurements with [^{15}O]H$_2$O and [^{18}F]fluoride ion PET in porcine vertebrae. J Bone Miner Res 1998;13:1328–1336.

66. Cook GJ, Maisey MN, Fogelman I. Fluorine-18-FDG PET in Paget's disease of bone. J Nucl Med 1997;38:1495–1497.

67. Fogelman I, Martin W. Assessment of skeletal uptake of 99mTc-diphosphonate over a five day period. Eur J Nucl Med 1983;8:489–490.

68. Garnero P, Sornay-Rendu E, Chapuy MC, Delmas PD. Increased bone turnover in late postmenopausal women is a major determinant of osteoporosis. J Bone Miner Res 1996;11:337–349.

69. Frost ML, Cook GJ, Blake GM, Marsden PK, Benatar NA, Fogelman I. A prospective study of risedronate on regional bone metabolism and blood flow at the lumbar spine measured by ^{18}F-fluoride positron emission tomography. J Bone Miner Res 2003;18:2215–2222.

70. Yamada S, Kubota K, Kubota R, et al. High accumulation of fluorine-18-fluorodeoxyglucose in turpentine-induced inflammatory tissue. J Nucl Med 1995;36:1301–1306.

71. Kubota R, Kubota K, Yamada S, et al. Methionine uptake by tumor tissue: a microautoradiographic comparison with FDG. J Nucl Med 1995;36:484–492.

72. Sugawara Y, Braun DK, Kison PV, et al. Rapid detection of human infections with fluorine-18 fluorodeoxyglucose and positron emission tomography: preliminary results. Eur J Nucl Med 1998;25:1238–1243.

73. Zhuang H, Yu JQ, Alavi A. Applications of fluorodeoxyglucose-PET imaging in the detection of infection and inflammation and other benign disorders. Radiol Clin N Am 2005;43:121–134.

74. Crymes WB, Demos H, Gordon L. Detection of musculoskeletal infection with ^{18}F-FDG PET: review of the current literature. J Nucl Med Technol 2004;32:12–15.

75. Zhuang T, Duarte PS, Pourdehand M, et al. Exclusion of chronic osteomyelitis with F-18 fluorodeoxyglucose PET imaging. Clin Nucl Med 2000;25:281–284.

76. Kalicke T, Schmitz A, Risse JH, et al. Fluorine-18 fluorodeoxyglucose PET in infectious bone diseases: results of histologically confirmed cases. Eur J Nucl Med 2000;27:524–528.

77. Guhlmann A, Brecht-Krauss D, Suger G, et al. Fluorine-18-FDG PET and technetium-99m antigranulocyte antibody scintigraphy in chronic osteomyelitis. J Nucl Med 1998;39:2145–2152.

78. Guhlmann A, Brecht-Krauss D, Suger G, et al. Chronic osteomyelitis: detection with FDG PET and correlation with histopathologic findings. Radiology 1998;206:749–754.

79. Stumpe KD, Notzli HP, Zanetti M, et al. FDG PET for differentiation of infection and aseptic loosening in total hip replacements: comparison with conventional radiography and three-phase bone scintigraphy. Radiology 2004;231:333–341.

80. Vanquickenborne B, Maes A, Nuyts J, et al. The value of (18)FDG-PET for the detection of infected hip prosthesis. Eur J Nucl Med Mol Imaging 2003;30:705–715.

81. Schiesser M, Stumpe KD, Trentz O, Kossmann T, Von Schulthess GK. Detection of metallic implant-associated infections with FDG PET in patients with trauma: correlation with microbiologic results. Radiology 2003;226:391–398.

82. Chacko TK, Zhuang H, Stevenson K, Moussavian B, Alavi A. The importance of the location of fluorodeoxyglucose uptake in periprosthetic infection in painful hip prostheses. Nucl Med Commun 2002;23:851–855.

83. Zhuang H, Duarte PS, Pourdehnad M, et al. The promising role of ^{18}F-FDG PET in detecting infected lower limb prosthesis implants. J Nucl Med 2001;42:44–48.

84. Cook GJ, Maisey MN, Fogelman I. Fluorine-18-FDG PET in Paget's disease of bone. J Nucl Med 1997;38:1495–1497.

85. Stumpe KD, Zanetti M, Weishaupt D, Hodler J, Boos N, Von Schulthess GK. FDG positron emission tomography for differentiation of degenerative and infectious endplate abnormalities in the lumbar spine detected on MR imaging. AJR 2002;179:1151–1157.

86. Fayad LM, Cohade C, Wahl RL, Fishman EK. Sacral fractures: a potential pitfall of FDG positron emission tomography. AJR 2003;181:1239–1243.

87. Shon IH, Fogelman I. F-18 FDG positron emission tomography and benign fractures. Clin Nucl Med 2003;28:171–175.

88. Kato K, Aoki J, Endo K. Utility of FDG-PET in differential diagnosis of benign and malignant fractures in acute to subacute phase. Ann Nucl Med 2003;17:41–46.

89. Zhuang H, Sam JW, Chacko TK, et al. Rapid normalization of osseous FDG uptake following traumatic or surgical fractures. Eur J Nucl Med Mol Imaging 2003;30:1096–1103.

22

PET Imaging in Pediatric Disorders

Hossein Jadvar, Leonard P. Connolly, Frederic H. Fahey, and Barry L. Shulkin

Development and validation of applications for positron emission tomography (PET) in pediatrics has proceeded more slowly than in adult medicine, partly because diseases to which PET has been most widely applied in adults are uncommon in pediatrics. Only about 2% of all cancers, for example, occur before 15 years of age. Experience has therefore been gained more slowly in pediatrics. Accrual of experience has been further slowed because there are few PET units in pediatric hospitals. The labor-intensive nature of imaging sick children has limited the ability of adult-oriented PET centers, which are faced with a shortage of available imaging slots, to take on a substantial number of time-consuming, challenging pediatric cases. Despite these considerations, PET, and more recently the hybrid positron emission tomography-computed tomography (PET/CT) imaging systems, are emerging as important tools in pediatric nuclear medicine. In this chapter, we review the clinical applications of PET in pediatrics with an emphasis on the more-common applications in epilepsy and oncology. General considerations in patient preparation and radiation dosimetry are also discussed.

Patient Preparation

Preparation of children and parents for nuclear medicine imaging has been thoroughly reviewed elsewhere (1, 2). As with any imaging study, gaining the trust and allaying the fears of both the patient and the parents are essential before attempting to image. Because parental attitudes and anxieties are readily conveyed to children, it is essential that the parents be well informed and cooperative if their child's trust is to be gained. Patient cooperation, once achieved, may be assisted by relatively simple methods. Sheets wrapped around the body, sandbags, and/or special holding devices are often sufficient for immobilization. Parents may accompany their child during the course of a study to provide emotional support.

However, on occasion, an anxious parent may actually impair the performance of the study.

Sedation is indicated when, on the basis of careful consideration, it is anticipated that simpler approaches will be inadequate. Sedation protocols, particularly regarding the recommended medications and the level of sedation required for an imaging procedure, vary from institution to institution. Guidelines such as those advanced by the Society of Nuclear Medicine (3) are useful in developing an institutional sedation program and a sedation formulary.

Also important to consider is the potential effect of sedatives on ^{18}F-2-deoxy-D-glucose (FDG) distribution. Many sedatives may affect cerebral metabolism. When performing FDG-PET of the brain, it is best if sedatives are not administered for 30 min after FDG administration because it is during this interval that the majority of cerebral FDG uptake occurs. Sedatives are not known to cause significant changes in tumoral metabolism and can be administered at any time relative to FDG administration for studies of tumors outside the central nervous system (4). During this time, the patient should be kept quiet and inactive to reduce regional stimulation of the cerebral cortex, which might confuse interpretation of the results.

Other important technical issues specific to performing PET studies in pediatric patients (e.g. consent, intravenous access, bladder catheterization) have recently been reviewed (5, 6). The recent introduction of PET/CT imaging systems also present unique issues that will need to be addressed. Kaste et al. (7) has reviewed the experience with implementing PET/CT at a tertiary pediatric hospital. Issues such as physical location of the PET/CT unit, the roles of CT and nuclear medicine technologists, and the methodology for study interpretation are discussed. Additional important considerations deliberated are the use of intravenous and sugar-free oral contrasts for the CT portion of the PET/CT examinations and the management of hyperglycemia.

Table 22.1. Radiation dosimetry for ^{18}F-2-deoxy-D-glucose (FDG).

	1 year	5 years	10 years	15 years	Adult
Mass (kg)	9.8	19.0	32.0	55.0	70.0
Administered activity (MBq)	54.5	105.6	177.8	305.6	389.0
Bladder	32.1	33.8	49.8	64.2	62.2
Brain	2.6	3.6	5.3	8.6	10.9
Heart	19.1	21.1	21.3	24.8	24.1
Kidneys	5.2	5.7	6.4	7.6	8.2
Red marrow	3.3	3.4	3.9	4.3	4.3
Effective dose (mSv)	5.2	5.3	6.4	7.6	7.4

The doses are reported in mSv (8). Patient masses represent the 50% percentile for that age (9).

Radiation Dosimetry

Several factors affect the dosimetry of positron emitters relative to single photon imaging agents. On the one hand, the energy per photon is higher (511 keV as compared to 140 keV for ^{99m}Tc), and there are two photons emitted per disintegration, which leads to much higher energy per unit activity than with most single photon agents. However, the higher photon energy also leads to a smaller fraction of the photons being absorbed within the patient. Table 22.1 summarizes the dosimetry of FDG for selected organs as well as the effective dose in the pediatric population based upon the administered activity of 5.55 kBq/kg (0.15 "Ci/kg).

Because the administered activities are scaled by body weight, the doses are similar across the age range, being slightly higher in adults. The effective dose is 5.2 and 7.4 mSv for the 1-year-old child and the adult, respectively. The critical organ is the bladder wall, with the dose being six to eight times higher than the effective dose. Table 22.2 compares the effective dose from FDG to a number of commonly used single photon imaging agents.

Table 22.2 illustrates that the radiation-absorbed dose to the patient from an FDG-PET study is very similar to the dose received from other nuclear medicine imaging procedures (4, 10, 11).

In pediatric imaging, the parents of the patient often prefer to remain with the patient during the procedure. The exposure rate constants for ^{18}F and ^{99m}Tc are 0.0154 and 0.00195 mR/h per MBq at 1 m, respectively. The difference is primarily due to the higher photon energy for ^{18}F as compared to ^{99m}Tc and the fact that two photons are emitted per disintegration. It is therefore prudent to consider the radiation exposure to the parent during these procedures. As shown in Table 22.1, pediatric patients receive a range of administered activities depending on patient size. The following assumptions are made: the patient receives 260 MBq and is considered to be a point source with no self-absorption. The patient sits in a preparatory room for 60 min during uptake and then is imaged for 60 min. These assumptions are considered quite conservative; that is, these will probably lead to an overestimation of the radiation dose to the parent. Table 22.3 estimates the total exposure to the parent during both the uptake and imaging periods, provided the parent maintains the specified distance from the patient.

Even if the parent stayed within 1 m of the patient during the entire uptake and imaging periods, the exposure to the parent would be no more than 5.5 mR. Therefore, the parents can be allowed to stay with the patient during the procedure but are instructed to stay as far from the patient as they feel comfortable.

Hybrid PET/CT scanners use the CT portion of the examination for attenuation correction. The dose to the patient from CT can vary greatly, depending on the tube voltage and current and the size of the patient. Table 22.4 summarizes the radiation dose to patients from the CT transmission as a function of tube voltage (based on a phantom study using phantoms of various sizes).

Table 22.2. Effective dose in pediatric patients for a variety of radiopharmaceuticals.

Radiopharmaceutical	Maximum administered activity (MBq)	1 year	5 years	10 years	15 years	Adult
FDG	389	5.2	5.3	6.4	7.6	7.4
[67]Ga-Citrate	222	19.9	19.9	20.3	22.7	22.2
[99m]Tc HMPAO	740	5.1	5.4	5.8	6.4	6.9
[99m]Tc MDP	740	2.8	2.8	3.7	4.1	4.2
[99m]Tc SestaMIBI	740	4.7	4.6	5.4	5.8	5.8

The doses are reported in mSv. The maximum administered activity is that which would be administered to a 70-kg adult. The pediatric dose administered is scaled by the patient's weight as in Table 22.1 (8, 9). Patient masses represent the 50% percentile for that age (9).

Table 22.3. Total exposure to the parent from a patient receiving 260 MBq of ^{18}F for a positron emission tomography (PET) study.

Distance from patient during uptake period (m)	Distance from patient during imaging period (m)	Total exposure to parent (mR)
1	1	5.5
1	2	4.0
2	2	1.4
2	3	1.1

It is assumed that the parent stayed with the patient during a 60-min uptake period and a 60-min imaging period.

Table 22.4. Radiation dose from computed tomography (CT) transmission as a function of tube voltage.

Kvp	Newborn	1 year	5 years	10 years	Medium adult
80	7.0	5.7	4.5	3.8	1.5
100	13.5	11.3	9.0	7.9	3.5
120	21.4	18.2	14.9	12.9	6.0
140	30.1	25.8	21.8	18.9	9.0

All doses are reported in mGy. All data were obtained at 130 mA and a pitch of helical 1.5:1 (12).

Smaller patients receive a substantially higher dose than adults from the same CT acquisition parameters. For example, the radiation dose to a 10-year-old is approximately twice that to a medium-sized adult for the same CT acquisition parameters. An alternative to using CT for attenuation correction is to use rotating rod sources. With a total activity in the rods of 370 MBq, the dose to the patient is between 0.05 and 0.2 mGy for a 15-min acquisition, based on a phantom study using phantoms of different sizes (12). Thus, the dose to the patient from a CT scan used for attenuation correction is substantially higher than that associated with the rotating rods sources. However, the CT scan provides anatomic correlation to the functional images, a feature that is not available using the rod sources.

Comparing the values in Tables 22.1 and 22.4, the dose to the patient from the CT portion of the scan can be equal to, if not higher than, the dose received from the radiopharmaceutical. Thus, the acquisition parameters for the CT portion of the scan should be tailored to the patient's size. For diagnostic CT, reduction of exposure by 30% to 50% relative to an adult has been suggested (13). Reducing the milliamps (mA) proportionately decreases the absorbed radiation dose without significant loss in the information provided. In addition, there is the potential to further reduce the tube voltage and current without adversely affecting the quality of the attenuation correction in those cases where anatomic correlation is not required.

PET in Pediatric Neurology

Normal Brain Development

An understanding of the normal brain development and evolution of cerebral glucose utilization is important when FDG-PET is considered as a diagnostic functional imaging study. Functional maturation proceeds phylogenetically. Glucose metabolism is initially high in the sensorimotor cortex, thalamus, brainstem, and cerebellar vermis. During the first 3 months of life, glucose metabolism gradually increases in the basal ganglia and in the parietal, temporal, calcarine, and cerebellar cortices. Maturation of the frontal cortex, which proceeds from lateral to medial, and of the dorsolateral cortex occur during the second 6 months of life. Cerebral FDG distribution in children after the age of 1 year resembles that of adults (Figure 22.1).

Childhood Epilepsy

Epilepsy is a relatively common and potentially devastating neurologic condition during childhood. Its incidence in children and adolescents is between 40 and 100 per 100,000 (17). The 1990 National Institutes of Health Consensus Conference on Surgery for Epilepsy estimated that 10% to 20% of epilepsy cases prove medically intractable and that 2,000 to 5,000 epilepsy patients per year can benefit from surgical resection of the seizure focus (18).

Accurate preoperative localization of the epileptogenic region is an essential but difficult task that is best accomplished by finding a concordance between results obtained with clinical examination, electroencephalography (EEG), neuropsychologic evaluation, and imaging studies. Computed tomography (CT) and magnetic resonance imaging (MRI) are used to detect anatomic lesions that may cause the seizures. However, structural lesions occur in a relatively small percentage of patients with epilepsy and, when such lesions are detected, they may not necessarily correspond to the epileptogenic region (19). Ictal or interictal single photon emission tomography (SPECT) evaluation of regional cerebral blood flow (rCBF) with tracers such as ^{99m}Tc-hexamethylpropylene (^{99m}Tc-HMPAO) and ^{99m}Tc-ethyl cysteinate dimer (^{99m}Tc-ECD) can localize the epileptogenic region in the presence or absence of structural abnormalities. The characteristic appearance of an epileptogenic region is relative zonal hyperperfusion on ictal SPECT and relative zonal hypoperfusion on interictal SPECT. The sensitivity of ictal rCBF tracer SPECT may approach 90% whereas that of interictal SPECT is in the range of 50% (20). The utility of ictal SPECT is somewhat reduced by the difficulty in coordinating tracer administration with seizures. Noninvasive evaluation is often unsuccessful in precisely localizing an epileptogenic region. As a result, surgical placement of

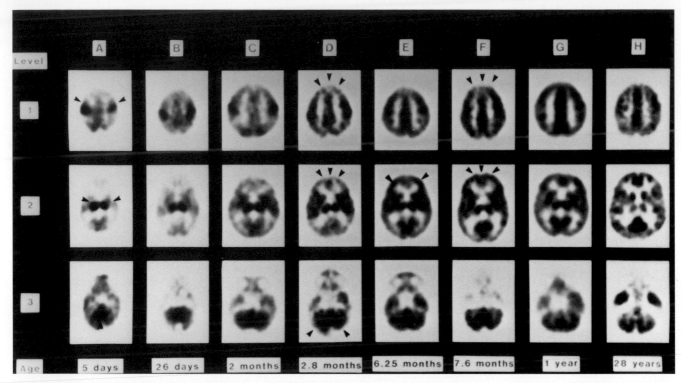

Figure 22.1. Normal brain maturation. Positron emission tomography (PET) scans show ontogenic changes in local cerebral glucose metabolism of the normal human infant. (**a**) In the 5-day-old infant, glucose metabolism is highest in the sensorimotor cortex, thalamus, cerebellar vermis (*arrows*), and brainstem (not shown). (**b–d**) Glucose metabolism increases gradually in the parietal, temporal, and calcarine cortices, basal ganglia, and cerebellar cortex (*arrows*), especially during the second and third months. (**e**) In the frontal cortex, glucose metabolism increases first in the lateral prefrontal regions by approximately 6 months. (**f**) By about 8 months, glucose metabolism also increases in the medial aspects of the frontal cortex (*arrows*) as well as the dorsolateral occipital cortex. (**g**) By 1 year of age, the glucose metabolic pattern resembles that of a normal adult, although the metabolic rates are two- to threefold elevated in comparison to values expected in normal adults. (Photo kindly provided by Harry Chugani, MD. From Chugani HT. Positron emission tomography. In: Berg BO, editor. Principles of Child Neurology. New York: McGraw-Hill, 1996:113–128.)

electrode grids on the brain surface or insertion of depth electrodes becomes necessary.

FDG-PET has proven useful in preoperative localization of the epileptogenic region (Figure 22.2) (21). FDG-PET is generally performed following an interictal injection. Although metabolic alterations might be localized better ictally than interictally, the relatively short half-life of ^{18}F limits the window of opportunity during which it can be administered ictally. Even when FDG can be administered at seizure onset, the approximately 30-min brain uptake time of FDG means that the study may depict periictal as

well as ictal FDG distribution. Ictal FDG studies may also show areas of seizure propagation that could be confused with the actual seizure focus. Despite these considerations, some favorable results have been reported employing ictal PET for patients with continuous or frequent seizures (22).

For interictal PET, FDG should be administered in a setting such as a quiet room with dim lights where environmental stimuli are minimal during the 30 min following FDG administration. It is best to have the child remain awake with minimal parental interaction during this period. EEG monitoring is essential to identify seizure ac-

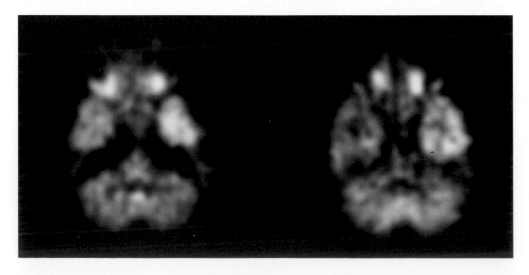

Figure 22.2. Transverse brain FDG-PET scan of a 14-year-old girl shows a hypometabolic epileptogenic right temporal lobe. (Reprinted with permission from Jadvar H, Connolly LP, Shulkin BL, Treves ST, Fischman AJ. Positron-emission tomography in pediatrics. Nuclear Medicine Annual 2000;53–83.)

tivity that might affect FDG distribution. The sensitivity of interictal FDG-PET approaches that of ictal rCBF SPECT in localizing the epileptogenic region, which is detected as regional hypometabolism. Importantly, the hypometabolism may predominantly affect cortex bordering the epileptogenic focus. Epileptic activity may originate in cortical areas bordering the hypometabolic regions rather than the hypometabolic region itself (23).

Incorporation of FDG-PET into preoperative evaluation of epilepsy patients reduces the need for intracranial EEG monitoring and the cost of preoperative evaluation (24). The best results have been obtained in temporal lobe epilepsy, for which metabolic abnormalities may be evident in as many as 90% of surgical candidates (24, 25). Extratemporal epileptogenic regions are more difficult to identify, but some success has been achieved in children with intractable frontal lobe epilepsy and normal CT or MRI studies (26).

FDG-PET has been reported as being of particular help in the evaluation of infantile spasms, a subtype of seizure disorder. This condition, which has an incidence of 2 to 6 per 10,000 live births, consists of a characteristic pattern of infantile myoclonic seizures and is frequently associated with profound developmental delay despite medical treatment (16, 27). Before the availability of FDG-PET, surgical intervention was attempted and successful in only isolated instances. Incorporation of FDG-PET into the evaluation of children with infantile spasms has resulted in identification of a substantial number of children who benefit from cortical resection. FDG-PET has revealed marked focal cortical glucose hypometabolism associated with malformative or dysplastic lesions that are not evident on anatomic imaging. There is marked decline in seizure frequency and in some patients reversal of developmental delay when a single metabolic abnormality that correlates with EEG findings is shown by FDG-PET. Patients with bitemporal hypometabolism on FDG-PET have a poor prognosis and typically are not candidates for resective surgery (27–31).

In addition to FDG, PET tracers that assess altered abundance or function of receptors, enzymes, and neurotransmitters in epileptogenic regions have been applied to localizing the epileptogenic region. Among alterations that have been observed are relatively reduced uptake of ^{11}C-flumazenil, a central benzodiazepine receptor antagonist, and ^{11}C-labeled (S)-(N-methyl) ketamine, which binds to the N-methyl-D-aspartate receptor-gated ion channel (32–36). Relative increases in uptake of ^{11}C-carfentanil, a selective mu opiate receptor agonist, and ^{11}C-deuterium-deprenyl, an irreversible inhibitor of monoamine oxidase type B (MAO-B) have also been described (37, 38).

Other Neurological Applications

PET with ^{15}O-water has also been investigated in infants with intraventricular hemorrhage and hemorrhagic infarction and in infants with hypoxic–ischemic encephalopathy (39, 40). Cerebral blood flow was markedly reduced not only in the hemorrhagic areas but also in the remainder of the involved hemisphere, suggesting that neurologic deficits may be caused by ischemia rather than the presence of blood within the brain parenchyma or cerebral ventricles (39). In full-term infants with perinatal asphyxia, diminished blood flow to the parasagittal cortical regions suggested that injury to the brain in these infants was also ischemic in etiology (40).

PET has also been employed to study the pathophysiology of many other childhood brain disorders such as autism (41), attention deficit hyperactivity disorder (42), schizophrenia (43), sickle cell encephalopathy (44), anorexia and bulimia nervosa (45, 46), Rasmussen's syndrome (47), Krabbe's disease (48), Sturge–Weber syndrome (49), and cognitive impairment in Duchenne muscular dystrophy (50). However, the exact role of PET in these clinical settings remains unclear. Further experience may result in expanded role of PET in many childhood neurologic disorders.

PET in Pediatric Cardiology

Currently, PET plays a relatively minor role in pediatric cardiology. Quinlivan et al. (51) have reviewed the cardiac applications of PET in children. PET with ^{13}N-ammonia has been employed to measure myocardial perfusion in infants after anatomic repair of congenital heart defects and after Norwood palliation for hypoplastic left heart syndrome (52). Infants with repaired heart disease had higher resting blood flow and less coronary flow reserve than previously reported for adults. Infants with Norwood palliation also had less perfusion and oxygen delivery to the systemic ventricle than the infants with repaired congenital heart lesion, explaining in part the less favorable outcome for patients with Norwood palliation. Evaluation of myocardial perfusion with ^{13}N-ammonia PET in infants following a neonatal arterial switch operation has demonstrated that patients with myocardial perfusion defects may have a more complicated postoperative course (53).

A major application of PET in adult cardiology is the assessment of myocardial viability with FDG as a tracer for glucose metabolism. A recent study evaluated regional glucose metabolism and contractile function by gated FDG-PET in seven infants and seven children after arterial switch operation and suspected myocardial infarction (54). Gated FDG-PET was found to contribute pertinent information to guide additional therapy including high-risk revascularization procedures. Recent reports have also provided evidence for the utility of PET in the assessment of myocardial perfusion and viability in infants and children with coronary abnormalities (55, 56). In another study in children with Kawasaki disease, PET with ^{13}N-ammonia and FDG showed abnormalities in about 60% of patients during the acute and subacute stages and about 40% of patients in the convalescent stage of disease (57). PET was valuable in assessing immunoglobulin therapy response at differing doses and administration schedules. ^{13}N-ammonia PET may also reveal reduction of coronary flow reserve in children with Kawasaki disease and angiographically normal epicardial coronary arteries (58). Beyond the more common assessment of myocardial perfusion and oxidative metabolism, PET has been used to study such fundamental functional abnormalities as mitochondrial dysfunction in children with hypertrophic or dilated cardiomyopathy (59). Dynamic PET with ^{11}C-acetate demonstrated reduced myocardial Kreb cycle activity (i.e., decreased oxidative metabolism) in children with cardiomyopathy despite normal myocardial perfusion. The diminished oxidative metabolism was associated with compensatory increased glycolytic activity as demonstrated on FDG-PET.

PET in Pediatric Oncology

The incidence of cancer is estimated to be 133.3 per million children in the United States (60). Although cancer is much less common in children than in adults, it is still an important cause of mortality in pediatrics. Approximately 10% of deaths during childhood are attributable to cancer, making it the leading cause of childhood death from disease (61).

Childhood cancers often differ from those encountered in adults, as is illustrated in Table 22.5, which delineates the estimated incidences of the more commonly encountered cancers in American children. Of the adult cancers to which FDG-PET has been most widely applied, only lymphomas and brain tumors occur with an appreciable incidence in children. However, the diagnostic utility of

FDG-PET and its impact on patient management have been reported for many pediatric cancers (62–64). In decreasing order of frequency, PET led to important changes in clinical management of lymphoma (32%), brain tumors (15%), and sarcomas (13%) (63).

Before reviewing the applications of PET in pediatric oncology, it is important to consider potential causes of confusion on FDG-PET that relate to physiologic FDG distribution in children. High FDG uptake in thymus (65–67) and in skeletal growth centers, particularly the long bone physes, are two important physiologic variations in FDG distribution encountered in children (Figures 22.3, 22.4). With the introduction of PET/CT imaging systems, it has been recognized that elevated FDG uptake in the normal brown adipose tissue may also be a source of false-positive findings (68, 69). The common anatomic areas involved include the neck and shoulder region, axillae, mediastinum, and the paravertebral and perinephric regions. Neck brown fat hypermetabolism is seen more commonly in the pediatric than in the adult population (15% versus 2%; P less than 0.01) and appears to be stimulated by cold temperatures (68, 69). Recent data have shown that brown fat metabolic activity may be suppressed pharmacologically (e.g., propranolol) (70).

Other potential pitfalls, which also apply to imaging adults, include variable FDG uptake in active skeletal muscles, the myocardium, the thyroid gland, and the gastrointestinal tract, as well as accumulation of FDG excreted into the renal pelves and bladder, and possible tracer accumulation in draining lymph nodes from extravasated tracer at the time of injection (71). Diffuse high bone marrow and splenic FDG uptake following adminis-

Table 22.5. Cancer incidence rates per million children younger than 15 years of age in the United States as derived from the Surveillance, Epidemiology, and End Results (SEER) and reported in reference.

Histology	Total[a] No.	Rate	(%)	Males No.	Rate	Females No.	Rate	Male:Female	Whites No.	Rate	Blacks No.	Rate	White:Black
All histologic types	10,555	133.3	100.0	5711	140.9	4844	125.1	1.13	8756	139.5	1064	108.3	1.29
Acute lymohoid leukemia	2484	30.9	23.2	1383	33.7	18.0	1.20	2092	32.9	169	16.9	1.95	
All central nervous system	2205	27.6	1195	29.3	1010	26.0	1.13	1847	29.3	239	23.8	1.23	
Astrocytomas and gliomas	1329	16.8	12.6	692	17.1	637	16.2	1.06	1130	17.9	144	14.3	1.25
Primitive neuroectodermal	532	6.6	5.0	311	7.7	221	5.6	1.38	433	6.8	58	5.9	1.15
Other CNS	344	4.3	3.2	192	4.6	152	3.9	1.18	284	4.6	37	3.6	1.28
Neuroblastoma	74	9.7	7.3	389	9.8	365	9.6	1.02	632	10.2	78	7.8	1.31
Non-Hodgkin's lymphoma	666	8.4	6.3	484	12.0	182	4.6	2.61	578	9.1	53.	5.4	1.69
Wilm's tumor	638	8.1	6.1	287	6.9	351	8.9	0.78	520	8.3	94	9.4	0.88
Hodgkin's disease	511	6.6	5.0	295	7.4	216	5.6	1.32	451	7.3	46	4.7	1.55
Acute mycloid leukemia	454	5.6	4.2	224	5.5	230	6.0	0.92	358	5.8	47	4.8	1.21
Rhabdomyosarcoma	354	4.5	3.4	211	5.2	143	3.6	1.44	294	4.7	40	4.1	1.15
Retinoblastoma	306	3.9	2.9	144	3.6	162	4.2	0.86	234	3.9	44	4.5	0.87
Osteosarcoma	262	3.4	2.6	130	3.3	132	3.4	0.97	197	3.4	38	3.9	0.87
Ewing's sarcoma	208	2.8	2.1	109	2.8	99	2.6	1.08	194	3.3	3	0.3	11.00
All other histologic types	1713	21.8	16.4	860	21.4	853	22.6	0.95	1359	21.3	213	22.7	0.94

[a] Rates are standardized to the 1980 SEER population and reported per million children per year.
Source: From Bailey DL, Townsend DW, Valk PE, Maisey MN. Positron Emission Tomography: Basic Sciences. Springer-Verlag London Ltd 2005, p.759.

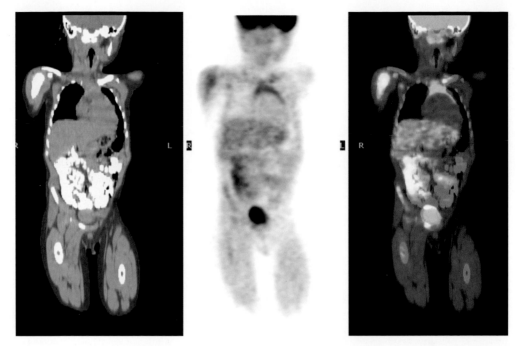

Figure 22.3. Five-year-old boy with history of T-cell large cell anaplastic lymphoma. Coronal images from PET/CT obtained before bone marrow transplantation. *Left panel:* CT of neck, chest, abdomen, pelvis, and thighs. *Center panel:* FDG-PET of same area shows normal-appearing thymic activity, left greater than right, in the superior mediastinum. *Right panel:* Fused PET/CT image overlays the metabolic information from the FDG-PET study onto the anatomic map provided by the CT images.

tration of hematopoietic stimulating factors may also resemble disseminated metastatic disease (Figure 22.5) (72, 73). Elevated bone marrow FDG uptake has been observed in patients as long as 4 weeks after completion of treatment with granulocyte colony-stimulating factor (G-CSF) (72). This observation is probably reflective of increased bone marrow glycolytic metabolism in response to hematopoietic growth factors.

Another important issue specific to PET imaging of pediatric patients is the choice of measurement parameter for the standardized uptake value (SUV), which is commonly used as a semiquantitative measure of the degree of FDG uptake in a region of interest. The calculation of SUV based on body surface area appears to be a more-uniform parameter than that calculated based on body weight in pediatric patients (74).

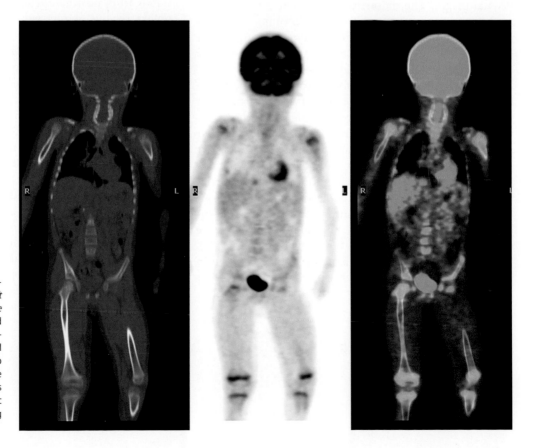

Figure 22.4. Coronal images of a 4-year-old boy with history of Wilms tumor. *Left panel:* CT images bone window. *Middle panel:* FDG-PET images show increased uptake representing growth plate activity in the proximal humerii, proximal femurs, and about both knees, similar to activity found in these areas on bone scan. *Right panel:* Fused FDG-PET images on CT show more precisely the anatomic location of the FDG activity representing growth plates.

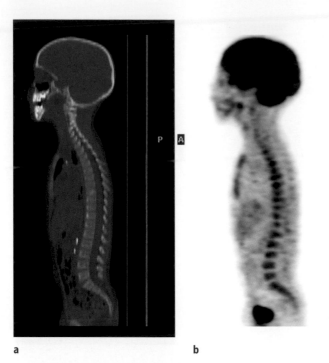
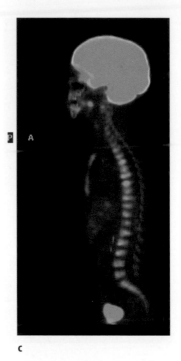

Figure 22.5. Sagittal images of PET/CT from 4-year-old child undergoing chemotherapy for metastatic Wilms tumor and receiving granulocyte colony-stimulating factor (G-CSF). (**a**) CT images bone window. (**b**) FDG-PET images show uniform increased uptake in the spine and sternum. (**c**) Fusion images of FDG-PET and CT.

a   b   c

Central Nervous System Tumors

Taken as a group, tumors of the central nervous system (CNS) are the most common nonhematologic tumors of childhood. They account for about 20% of all pediatric malignancies. The grouping includes many histologically diverse tumors of both neuroepithelial and nonneuroepithelial origin. The majority of pediatric brain tumors arise from neuroepithelial tissue. CNS tumors are subclassified histopathologically by cell type and graded for degree of malignancy using criteria that include mitotic activity, infiltration, and anaplasia (75, 76).

To grasp the wide variety of pediatric CNS tumors, one need only consider the distribution of the most common tumors according to the major anatomic compartment involved. In the posterior fossa, medulloblastoma, cerebellar astrocytoma, ependymoma, and brainstem gliomas are most common. Tumors about the third ventricle include tumors that arise from suprasellar, pineal, and ventricular tissue. The most common neoplasms about the third ventricle are optic and hypothalamic gliomas, craniopharyngiomas, and germ cell tumors. Supratentorial tumors are most often astrocytomas, many of which are low grade (76).

MRI and CT are the main imaging modalities used in staging and following children with CNS tumors. Their main limitation is the inability to distinguish viable recurrent or residual tumor from abnormalities resulting from surgery or radiation. SPECT with ^{201}Tl-chloride and ^{99m}Tc-methoxyisobutylisonitrile (MIBI) have proven valuable for this determination in a number of pediatric brain tumors (77–80). Use of FDG-PET in brain tumors has been widely reported in series that predominantly include adult patients for whom FDG-PET has helped distinguish viable tumor from posttherapeutic changes (81–83). High FDG uptake relative to adjacent brain indicates residual or recurrent tumor whereas low or absent FDG uptake is observed in areas of necrosis. This distinction is most readily made with high-grade tumors that show high uptake of FDG at diagnosis. Even with high-grade tumors, the presence of microscopic tumor foci is not excluded by an FDG-PET study that does not show increased uptake; this is particularly true after intensive radiation therapy, in which case FDG-PET results may not accurately correlate with tumor progression (84). Furthermore, in the immediate posttherapy period, elevated FDG uptake may persist (85).

FDG-PET has been applied to tumor grading and prognostic stratification. Higher-grade aggressive tumors typically have higher FDG uptake than do lower-grade tumors (Figures 22.6, 22.7) (86). Among low-grade tumors, some show insufficient FDG uptake to be distinguished from adjacent brain and some appear hypometabolic. The development of hypermetabolism as evidenced by increased

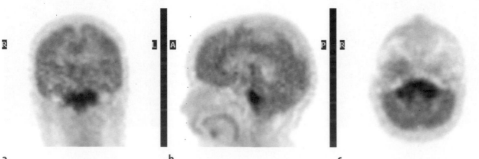

Figure 22.6. FDG-PET brain scan from a 10-year-old boy with anaplastic pontine astrocytoma shows considerably increased FDG uptake within the tumor compared to surrounding gray matter and extracerebral soft tissue. (**a**) coronal images; (**b**) sagittal images; (**c**) transverse images.

a   b   c

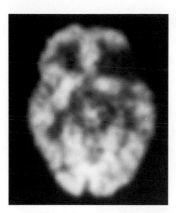

Figure 22.7. Transverse FDG-PET brain image from a 4.5-year-old boy with generalized tonic–clonic seizure demonstrates hypometabolism in the left frontal lobe. This area was resected and shown to be a low-grade glioma. (Reprinted with permission from Jadvar H, Connolly LP, Shulkin BL, Treves ST, Fischman AJ. Positron-emission tomography in pediatrics. Nuclear Medicine Annual 2000;53–83.)

FDG uptake in a low-grade tumor that appeared hypometabolic at diagnosis indicates degeneration to a higher grade (87). The biologic behavior of high-grade tumors may be reflected in their appearance on FDG-PET. Shorter survival times have been reported for patients whose tumors show the highest degree of FDG uptake (88). Data, although limited, suggest that FDG-PET findings also correlate well with pathology and clinical outcome in children (89–91). A potential pediatric application of this approach entails a reported excellent correlation between FDG-PET findings and clinical outcome in children affected by neurofibromatosis who have low-grade astrocytomas (92). In that series, high tumoral glucose metabolism shown by FDG-PET was a more-accurate predictor of tumor behavior than was histologic analysis. Combining FDG-PET imaging and MRI in the planning of stereotactic brain biopsies has been reported to improve the diagnostic yield in infiltrative, poorly defined lesions and to reduce sampling in high-risk functional areas (93).

Another positron-emitting radiotracer that has been used to study pediatric brain tumors is ^{11}C-methionine (^{11}C-Met), which localizes to only a minimal degree in normal brain. Uptake of this labeled amino acid reflects transmethylation pathways that are present in some tumors. However, as with FDG, some low-grade gliomas may escape detection (94, 95). ^{11}C-Met-PET has been reported to be useful in differentiating viable tumor from treatment-induced changes (94, 96). It is worth noting, however, that ^{11}C-Met is not tumor specific as it has been shown to accumulate in some nontumoral CNS diseases, probably as a result of blood–brain barrier disruption (97). Both FDG-PET and ^{11}C-Met-PET have been shown to be independent predictors of event-free survival (98). ^{11}C-Met, because of the short 20-min half-life of the ^{11}C label, must be produced locally for administration and is not commercially available.

Lymphoma

Lymphomas of non-Hodgkin's and Hodgkin's types account for between 10% and 15% of pediatric malignan-cies. Non-Hodgkin's lymphoma occurs throughout childhood. Lymphoblastic and small cell tumors, including Burkitt's lymphoma, are the most common histologic types. The disease is usually widespread at diagnosis. Mediastinal and hilar involvement are common with lymphoblastic lymphoma. Burkitt's lymphoma most often occurs in the abdomen. Hodgkin's disease has a peak incidence during adolescence. Nodular sclerosing and mixed cellularity are the most common histologic types. The disease is rarely widespread at diagnosis and the majority of cases have intrathoracic nodal involvement (60, 99).

^{67}Ga-citrate scintigraphy has proven useful in staging and monitoring therapeutic response of patients with non-Hodgkin's and Hodgkin's lymphomas (100–104). In numerous studies, which have included predominantly adult patient populations, FDG has been shown to accumulate in non-Hodgkin's and Hodgkin's lymphoma tissue (Figure 22.8) (71, 105–123). Similarly to ^{67}Ga-citrate, FDG uptake is generally greater in higher- than in lower-grade lymphomas (112, 114). FDG-PET has been shown to reveal sites of nodal and extranodal disease that are not detected by conventional staging methods, resulting in upstaging of disease (110–117, 124). FDG-PET, when performed at the time of initial evaluation, has also been recently shown to change disease stage and treatment in up to 10% to 23% of children with lymphoma (125, 126). Identification of areas of intense FDG uptake within the bone marrow can be particularly useful in directing the site of biopsy or even eliminating the need for biopsy at staging (110, 123). FDG-PET is also useful for assessing residual soft tissue masses shown by CT after therapy. Absence of FDG uptake in a residual mass is predictive of remission whereas high uptake indicates residual or recurrent tumor (117). A negative FDG-PET scan after completion of chemotherapy, however, does not exclude the presence of residual microscopic disease (127). The potential role of FDG-PET in radiation treatment planning for pediatric oncology including lymphoma has also been recently described (128–130).

FDG-PET has been compared to ^{11}C-Met-PET in a relatively small series of 14 patients with non-Hodgkin's lymphoma. ^{11}C-Met-PET provided superior tumor-to-background contrast, whereas FDG-PET was superior in distinguishing between high- and low-grade lymphomas (108).

Neuroblastoma

Neuroblastoma is the most common extracranial solid malignant tumor in children. The mean age of patients at presentation is 20 to 30 months, and it is rare after the age of 5 years (99).

The most common location of neuroblastoma is the adrenal gland. Other sites of origin include the paravertebral and presacral sympathetic chain, the organ of Zuckerkandl, posterior mediastinal sympathetic ganglia, and cervical sympathetic plexuses. Gross or microscopic calcification is often present in the tumor. Two related

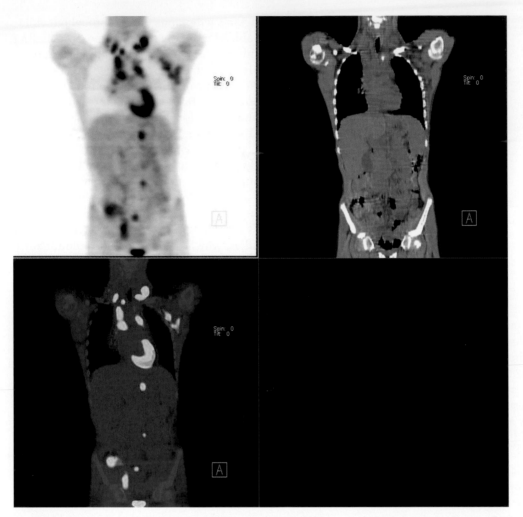

Figure 22.8. Coronal images of FDG-PET/CT of 11-year-old child with newly diagnosed nodular sclerosing Hodgkins disease following left axillary lymph node biopsy. *Upper left panel:* FDG-PET scan shows increased uptake of FDG in bilaterally in the lower neck and mediastinum, the left axilla, the right hilum, the left mid- and lower abdomen, and centrally within the pelvis. *Upper right panel:* CT image corresponding with FDG-PET scan. *Lower left panel:* Fusion image of these FDG-PET and CT images.

neural crest tumors, ganglioneuroma and ganglioneuroblastoma, have been described. Some neuroblastomas spontaneously regress or mature into ganglioneuroma, which is benign. However, the unpredictability and apparent infrequency of spontaneous regression and maturation, and the consequences of delaying therapy, require that treatment be instituted at diagnosis in most cases. Ganglioneuroblastoma is a malignant tumor that contains both undifferentiated neuroblasts and mature ganglion cells.

Disseminated disease is present in up to 70% of neuroblastoma cases at diagnosis and most commonly involves cortical bone and bone marrow. Less frequently, there is involvement of liver, skin, or lung. A primary tumor is not detected in up to 10% of children with disseminated neuroblastoma (131). The primary tumor may also go undetected in patients who present with paraneoplastic syndromes such as infantile myoclonic encephalopathy.

Surgical excision is the preferred treatment of localized neuroblastoma. When local disease is extensive, intensive preoperative chemotherapy may be utilized. When distant metastases are present, surgical removal is not likely to improve survival. The prognosis in these cases is poor, but high-dose chemotherapy, total-body irradiation, and bone marrow reinfusion are beneficial for some children with this presentation.

Delineation of local disease extent is achieved with MRI, CT, and scintigraphic studies (132). These tests are also utilized in localizing the primary site in children who present with disseminated disease or with a paraneoplastic syndrome. Metaiodobenzylguanidine (MIBG) and ^{111}In-pentetreotide scintigraphy have been employed in these settings with a sensitivity of greater than 85% for detecting neuroblastoma. Uptake of MIBG, which is an analogue of guanethidine and norepinephrine, into neuroblastoma is by a neuronal sodium- and energy-dependent transport mechanism. The localization of ^{111}In-pentetreotide in neuroblastoma reflects the presence of somatostatin type 2 receptors on some neuroblastoma cells (133).

Bone scintigraphy has been most widely used for detection of skeletal involvement for staging. MIBG and, to a lesser extent, ^{111}In-pentetreotide imaging have also been increasingly used for detecting skeletal involvement (134).

Patients with residual unresected primary tumors are periodically evaluated with MRI or CT. However, these studies cannot distinguish viable tumor from treatment-related scar or tumor that has matured into ganglioneuroma. Specificity in establishing residual viable tumor can be improved with MIBG or ^{111}In-pentetreotide imaging when the primary tumor had been shown to ac-

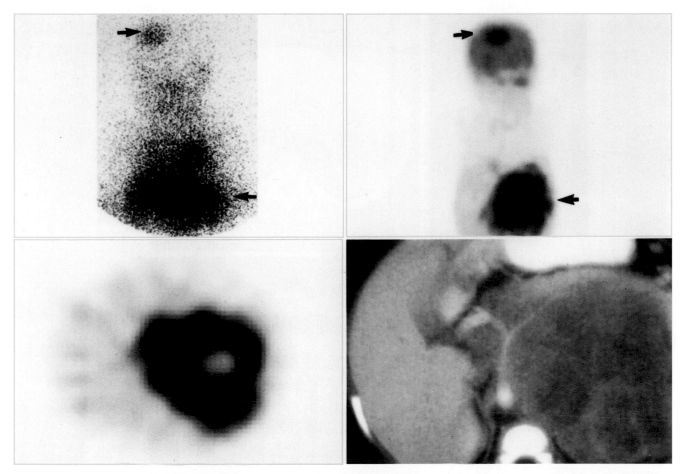

Figure 22.9. Images of a 2-year-old girl with neuroblastoma at presentation. *Top left:* Anterior planar view of the head, neck, chest, and upper abdomen 48 h following [131I]MIBG injection. An area of abnormal uptake is present in the right superior aspect of the skull (*upper arrow*). A large focus of abnormal uptake in the primary tumor (*lower arrow*) extends from the upper left abdomen medially and inferior to the liver. *Top right:* Anterior projection of the head, neck, chest, and upper abdomen approximately 1 h following FDG injection. Increased FDG uptake in the skull (*upper arrow*) corresponding to the site of abnormal [131I]MIBG uptake is well visualized against the normal brain uptake of FDG. Increased FDG uptake is present in the primary tumor (*lower arrow*). Note the paucity of background activity in the chest and abdomen. *Bottom left:* Transaxial image through the midabdomen shows marked FDG uptake in the primary tumor. A central area of decreased FDG accumulation probably represents necrosis. *Bottom right:* A large mass in the left abdomen is shown in a transaxial CT image. This set of images demonstrates that neuroblastomas are metabolically active. The primary tumor and metastases can be well visualized with FDG-PET. (Reprinted with permission from Jadvar H, Connolly LP, Shulkin BL, Treves ST, Fischman AJ. Positron-emission tomography in pediatrics. Nuclear Medicine Annual 2000;53–83.)

cumulate one of these agents. These agents are also useful in assessing residual skeletal disease in patients with MIBG- or ^{111}In-pentetreotide-avid skeletal metastases. Bone scintigraphy, however, is unable to distinguish active disease from bony repair on the basis of tracer uptake.

Neuroblastomas are metabolically active tumors. Neuroblastomas or their metastases avidly concentrated FDG before chemotherapy or radiation therapy in 16 of 17 patients studied with FDG-PET and MIBG imaging (135).Uptake after therapy was variable but tended to be lower. FDG and MIBG results were concordant in most instances (Figures 22.9, 22.10). However, there were occasions that one agent accumulated at a site of disease and the other did not (Figure 22.11). MIBG imaging was overall considered superior to FDG-PET, particularly in delineation of residual disease. As the patients in this series had aggressive tumors and poor prognoses, the value of FDG-PET for assessing therapeutic response could not be determined. An advantage of FDG-PET is the initiation of imaging 30 to 60 min after FDG administra-

tion whereas MIBG imaging is performed 1 or more days following tracer administration.

FDG-PET imaging may be of limited value for the evaluation of the bone marrow involvement of neuroblastoma for reason of the mild FDG accumulation by the normal bone marrow (135). Pitfalls resulting from physiologic FDG uptake in the bowel and the thymus are additional factors that may limit the role for FDG-PET in neuroblastoma. A recent study has reported that once the primary tumor is resected, PET and bone marrow examination suffice for monitoring neuroblastoma patients at high risk for progressive disease in soft tissue, bone, and bone marrow (136). Currently, however, the primary role of FDG-PET in neuroblastoma is in the evaluation of known or suspected neuroblastomas that do not demonstrate MIBG uptake. We have also noted variable uptake of FDG in ganglioneuromas (Figure 22.12). This finding suggests that FDG may not reliably differentiate between neuroblastomas and ganglioneuromas.

[11]C-Hydroxyephedrine (^{11}C-HED), an analogue of norepinephrine, and ^{11}C-epinephrine PET have also been

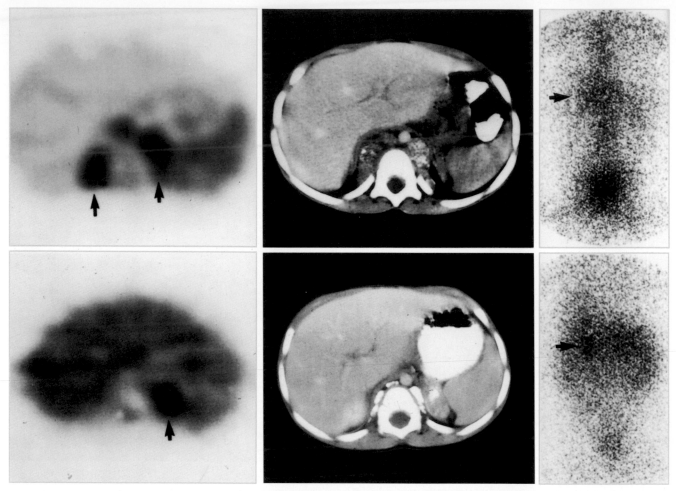

Figure 22.10. Three-year-old boy who presented with leg and abdominal pain: *top panel*, images from the time of diagnosis; *bottom panel*, images obtained at reevaluation 6 months later following chemotherapy. Following reevaluation, the patient underwent surgical exploration, and a tumor consisting of ganglioneuroblastoma was removed from the sites identified by MIBG scintigraphy and PET-FDG scanning the adrenal bed, and ganglioneuroma from the adjacent paraspinal lymphadenopathy, which no longer concentrated FDG. *Top left panel:* Transverse PET image of the upper abdomen acquired 60 min after injection. There is extensive paraspinal uptake of FDG (*arrows*). *Top middle panel:* CT scan demonstrates scattered calcifications within the paraspinal lymphadenopathy, which corresponds well with the increased metabolic activity demonstrated by FDG-PET scanning. *Top right panel:* Posterior image of the chest–abdomen–pelvis 48 h after [131I]MIBG injection. There is a large area of abnormal uptake in the upper abdomen–lower thorax (*arrow*) and extensive uptake of MIBG within the skeleton, consistent with metastatic disease in the bone marrow. *Lower left panel:* Transverse PET image of the upper abdomen acquired at 60 min after injection shows persistent activity to the left of the spine, consistent with residual neoplasm (*arrow*). Uptake in paraspinal lymphadenopathy is no longer present. *Lower middle panel:* CT scan shows considerable calcification and reduction in the size of the paraspinal tumor, but a partially calcified mass persists in the region of the left adrenal gland. *Lower right panel:* Posterior image of the chest–abdomen 48 h after [131I]MIBG injection. Slightly increased activity remains in the area of the left adrenal gland (*arrow*). (Reprinted with permission from Shulkin BL, Hutchinson RJ, Castle VP, Yanik GA, Shapiro B, Sisson JC. Neuroblastoma: positron emission tomography with 2-[fluorine-18]-fluoro-2-deoxy-D-glucose compared with metaiodobenzylguanidine scintigraphy. Radiology 1996;199:743–750.)

used in evaluating neuroblastoma (Figure 22.13). All seven neuroblastomas studied showed uptake of ^{11}C-HED (137), and four of five neuroblastomas studied showed uptake of ^{11}C-epinephrine (138). Uptake of these tracers is demonstrated within minutes after administration, which is an advantage over MIBG imaging. Limitations regarding cost and the need for onsite synthesis of short lived [11]C (half-life, 20 min) suggest that neither ^{11}C-HED nor ^{11}C-epinephrine PET is likely to replace MIBG imaging. These tracers may prove useful adjuncts in difficult cases where a primary tumor is difficult to identify with more readily available agents and to further the understanding of this disease. Compounds labeled with ^{18}F, such as fluoronorepinephrine, fluorometaraminol, and fluorodopamine, may also be useful tracers. PET using 4-[fluorine-18]-fluoro-3-iodobenzylguanidine

(139) and [124]I-labeled MIBG (140) has also been described.

Wilms Tumor

Wilms tumor is the most common renal malignancy of childhood. Wilms tumor is predominantly seen in younger children and is uncommonly encountered after the age of 5 years (60). Bilateral renal involvement occurs in about 5% of all cases and can be identified synchronously or metachronously (97, 114). An asymptomatic abdominal mass is the typical mode of presentation. Nephrectomy with adjuvant chemotherapy is the treatment of choice. Radiation therapy is used in selected cases when resection is incomplete.

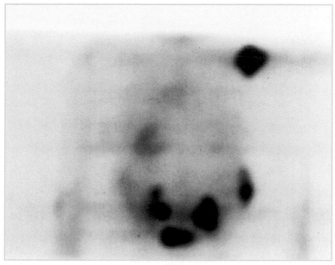

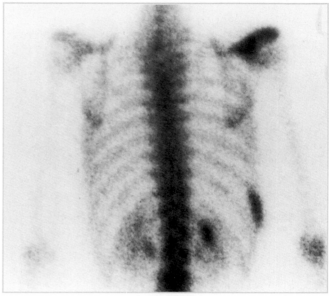

Figure 22.11. Neuroblastoma recurrence post bone marrow transplantation. *Left panel:* Posterior projection image of PET scan obtained 1 h following FDG injection. There is marked accumulation of FDG in the right shoulder (*arrow*), a right lower rib (*arrow*), and an abnormal focus of activity within the abdomen (*arrow*) between the kidneys, which most likely represented metastatic disease. *Right panel:* Posterior image of the chest and abdomen from a bone scan acquired 2 h following injection of 25 mCi [⁹⁹ᵐᵀᶜ]MDP shows abnormal uptake in the right shoulder (*arrow*) and ninth right rib (*arrow*), confirming that two of the sites of abnormal FDG uptake involve bone. The patient declined further evaluation and died of widely metastatic disease. (Reprinted with permission from Shulkin BL, Hutchinson RJ, Castle VP, Yanik GA, Shapiro B, Sisson JC. Neuroblastoma: positron emission tomography with 2-[fluorine-18]-fluoro-2-deoxy-D-glucose compared with metaiodobenzylguanidine scintigraphy. Radiology 1996;199:743–750.)

Scintigraphy has not played an important role in imaging of Wilms tumor. Radiography, ultrasonography, CT, and MRI are commonly employed in anatomic staging and detection of metastases, which predominantly involve lung, occasionally liver, and only rarely other sites. Anatomic imaging, however, is of limited value in the assessment for residual or recurrent tumor (141). Uptake of FDG by Wilms tumor (Figure 22.14) has been described (142), but a role for FDG-PET in Wilms tumor has not been established. Normal excretion of FDG through the kidney is a limiting factor. However, careful correlation with anatomic cross-sectional imaging allows distinction of tumor uptake from normal renal FDG excretion.

Bone Tumors

Osteosarcoma and Ewing's sarcoma are the two primary bone malignancies of childhood; of the two, osteosarcoma is the more common and predominantly affects adolescents and young adults. A second peak affects older adults,

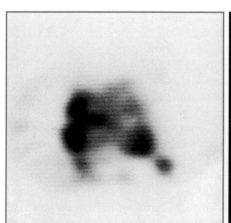

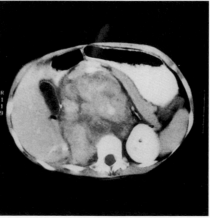

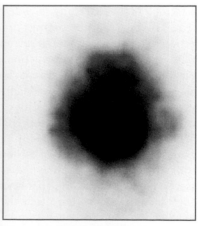

Figure 22.12. Uptake of FDG within ganglioneuroma. The patient presented at age 2 years with intractable diarrhea and was found to have a nonresectable 10-cm abdominal mass containing mostly stroma and rare elements of neuroblastoma. She received radiotherapy, chemotherapy, and subsequent surgical debulking. MIBG scans showed persistent abdominal uptake. The patient eventually received therapy with [¹³¹I]MIBG. Surgical debulking was again repeated. Multiple resected portions of the tumor contained only mature stroma and no neuroblastoma. She did well and returned for reevaluation 16 months later. *Left panel:* Transverse image of PET scan obtained 60 min following injection of FDG shows heterogeneous increased uptake of FDG in midabdomen. *Middle panel:* CT scan of the abdomen demonstrates the very large abdominal mass. *Right panel:* Anterior image of [¹³¹I]MIBG scan 48 h after injection shows extensive, intense uptake in the abdomen, corresponding to the tumor. (Reprinted with permission from Shulkin BL, Hutchinson RJ, Castle VP, Yanik GA, Shapiro B, Sisson JC. Neuroblastoma: positron emission tomography with 2-[fluorine-18]-fluoro-2-deoxy-D-glucose compared with metaiodobenzylguanidine scintigraphy. Radiology 1996;199:743–750.)

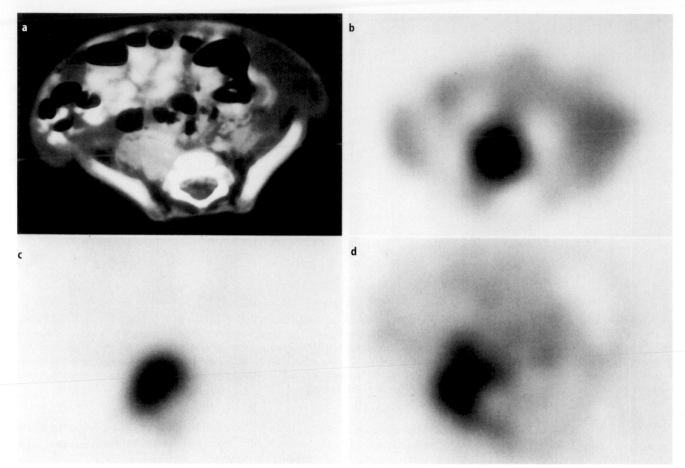

Figure 22.13. (**a**) CT scan of the pelvis of a 6-month-old boy following surgical debulking of abdominal–pelvic neuroblastoma. There is abnormal soft tissue with speckled calcification in the right posterior pelvis. (**b**) MIBG-SPECT scan at 24 h shows uptake into the tumor. (**c**) ¹¹C-Hydroxyephedrine (HED)-PET scan at 20 min following injection. There is excellent uptake within the neuroblastoma, and the image appears similar to the MIBG-SPECT examination. (**d**) FDG-PET scan at 50 min shows moderate accumulation of FDG within the mass relative to surrounding background. However, the tumor appears better delineated with the more specific adrenergic tumor imaging agent HED. (Reprinted by permission of the Society of Nuclear Medicine from Shulkin BL, Wieland DM, Baro ME, et al. PET hydroxyephedrine imaging of neuroblastoma. J Nucl Med 1996;37:18, Figure 2.)

predominantly individuals with a history of prior radiation to bone or Paget's disease. This tumor rarely affects children younger than 7 years of age. Osteosarcoma is typically a lesion of the long bones. The treatment of choice for osteosarcoma of an extremity is wide resection and limb-sparing surgery. Limb-sparing procedures entail the resection of tumor with a cuff of surrounding normal tissue at all margins, skeletal reconstruction, and muscle and soft tissue transfers. Employing current chemotherapeutic regimens pre- and postoperatively and imaging to define tumor extent and tumor viability preoperatively, limb-sparing procedures can be appropriately performed in 80% of patients with osteosarcoma (143).

Almost all cases of Ewing's sarcoma occur between the ages of 5 and 30 years with the highest incidence being in the second decade of life. In patients younger than 20 years, Ewing's sarcoma most often affects the appendicular skeleton. Beyond that age, pelvic, rib, and vertebral lesions predominate. The tumor is believed to be of neuroectodermal origin and, along with the primitive neuroectodermal tumor (PNET), to be part of a spectrum of a single biologic entity (144). Ewing's sarcoma is considered an undifferentiated variant and PNET a more differentiated peripheral neural tumor. Therapy for Ewing's sarcoma involves multi-

agent chemotherapy for eradication of microscopic or overt metastatic disease and irradiation and/or surgery for control of the primary lesion. Because late recurrence is not uncommon, resection of the primary tumor is gaining favor for local disease control (145).

MRI is used to define the local extent of osteosarcoma and Ewing's sarcoma in bone and soft tissue. However, signal abnormalities caused by peritumoral edema can result in an overestimation of tumor extension (146). Scintigraphy has been used primarily to detect skeletal metastases of these tumors at diagnosis and during follow-up. With osteosarcoma, skeletal scintigraphy occasionally demonstrates extraosseous metastases, most often pulmonary, as a result of osteoid production by the metastatic deposits.

Determination of preoperative chemotherapeutic response is important in planning limb salvage surgery. Because of the nonspecific appearance of viable tumor on MRI, variable results have been reported for assessing chemotherapeutic response (147–152). Scintigraphy with ²⁰¹Tl has been shown to be useful for assessing therapeutic response in osteosarcoma (153–158) and perhaps Ewing's sarcoma (154, 155). Marked decrease in tumoral ²⁰¹Tl uptake indicates a favorable response to

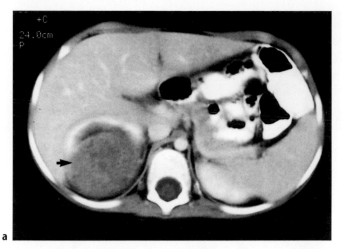

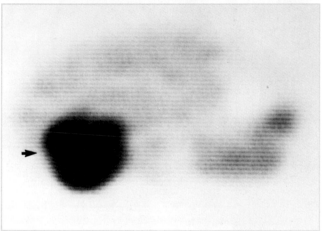

Figure 22.14. CT scan (**a**) PET scan (**b**). Markedly increased uptake of FDG (*arrow*) is present within the right-sided mass seen on CT (*arrow*). The mass was surgically removed and confirmed histologically to represent Wilms tumor. (Copyright 1997. From Shulkin BL, Chang E, Strouse PJ, et al. PET FDG studies of Wilms tumors. J Pediatr Hem/Onc 1997;19:334–338. Reproduced by permission of Taylor & Francis, Inc., http://www.taylorandfrancis.com.)

chemotherapy. When tumoral ^{201}Tl uptake does not decrease within weeks of chemotherapy, a therapeutic change may be needed. ^{99m}Tc-MIBI may also be useful in osteosarcoma but seemingly not with Ewing's sarcoma (159, 160).

The exact role of FDG-PET in osteosarcoma and Ewing's sarcoma has not yet been determined (Figures 22.15, 22.16). However, early experience suggests that in patients with Ewing's sarcoma FDG-PET may play a role

in monitoring response to therapy (161–167). When compared to bone scintigraphy, FDG-PET may be superior for detecting osseous metastases from Ewing's sarcoma but may be less sensitive for those from osteosarcoma (168). A second potential role is in assessing patients with suspected or known pulmonary metastasis, which is particularly common with osteosarcoma.

Soft Tissue Tumors

Rhabdomyosarcoma is the most common soft tissue malignancy of childhood. The peak incidence occurs between 3 and 6 years of age. Rhabdomyosarcomas can develop in any organ or tissue but, contrary to what the name implies, this tumor does not usually arise in muscle. The most common anatomic locations are the head, particularly the orbit and paranasal sinuses, the neck, and the genitourinary tract. CT or MRI is important for establishing the extent of local disease. Radiography and CT are used for detecting pulmonary metastases, and skeletal scintigraphy is employed for identifying osseous metastases. Radiation therapy and surgery are utilized for local disease control, and chemotherapy is employed for treatment of metastases. Rhabdomyosarcomas show variable degrees of FDG accumulation. Cases showing the clinical utility of FDG-PET have been described, but the exact role of FDG-PET in rhabdomyosarcoma is yet to be determined (Figure 22.17) (4, 161).

Conclusion

FDG-PET is being applied increasingly to study diseases of childhood, especially tumors. Because the tumors encountered are relatively rare, it will be difficult to perform well-designed, prospective clinical trials at single institutions. The recent merger of the CCG (Children's Cancer Group) and the POG (Pediatric Oncology Group) to form COG (Children's Oncology Group) brings an opportunity to examine the use of FDG-PET in the management of childhood tumors in multiinstitutional, cooperative efforts. We expect that future data will show that FDG-PET does contribute unique, valuable information for the care of childhood tumors.

Figure 22.15. A 16-year-old girl with osteosarcoma. Images obtained following four courses of chemotherapy to assess for residual disease. (**a**) Coronal CT images show sclerotic lesions in the distal right tibia. (**b**) FDG-PET images show elevated FDG uptake at the periphery of the sclerotic lesions, indicating viable tumor. (**c**) Fusion images of CT and FDG-PET.

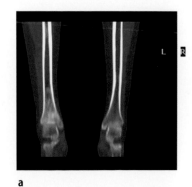

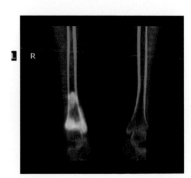

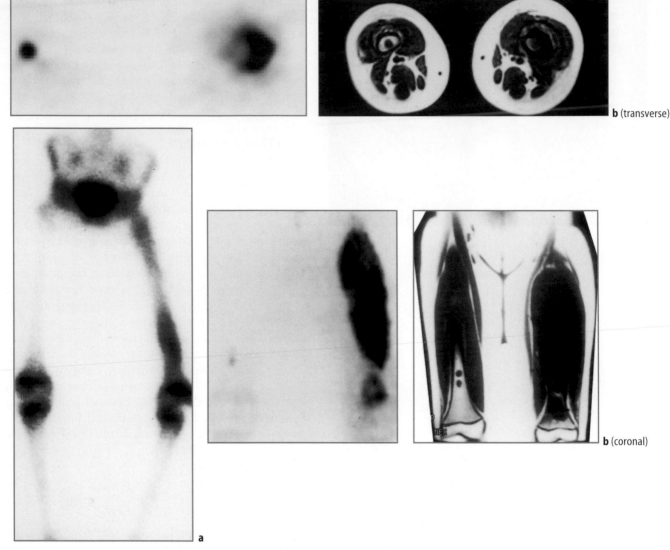

b (transverse)

b (coronal)

a

Figure 22.16. An 11-year-old girl who complained of pain and swelling in the left thigh. Plain film radiographs showed findings suggestive of Ewing's sarcoma, subsequently confirmed on biopsy. (**a**) *Left panel:* Bone scan obtained 2 h following injection of ^{99m}Tc-MDP. Abnormal accumulation of tracer is noted throughout the left femur. The right femur is unremarkable. *Center panel:* Anterior projection image from PET study shows intense, irregular uptake of FDG within the soft tissues of the left thigh. Two small foci of activity are seen in the region of the distal right femur that are not present on bone scanning. *Right panel:* T$_1$-weighted coronal MR image shows soft tissue mass in the left thigh, replacement of normal marrow of the left femur, and two focal lesions in the distal right femur. (**b**) *Left panel:* Transverse and coronal image from PET scan at the level of the distal femurs. Intense uptake of FDG is present in the soft tissues surrounding the left femur and focally within the right femur. *Right panel:* T$_1$-weighted MR image at same level shows the soft tissue mass surrounding the left femur, replacement of marrow of the left femur, and a lesion in the center of the right femur. (Reprinted with permission from Shulkin BL, Mitchell DS, Ungar DR, et al. Neoplasms in a pediatric population: 2-[F-18]-fluoro-2-deoxy-D-glucose PET studies. Radiology 1995;194:495–500.)

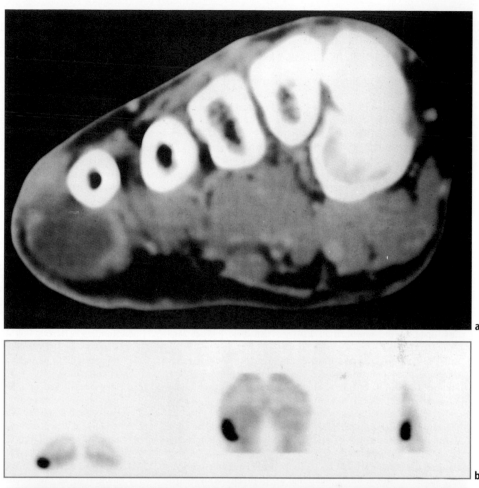

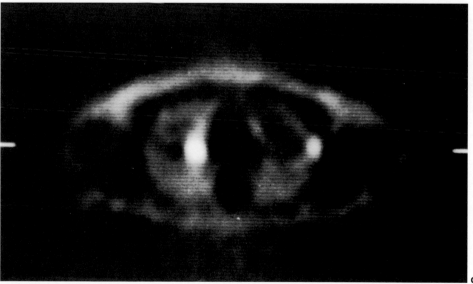

Figure 22.17. Rhabdomyosarcoma. (**a**) CT of the right foot of an 18 year-old girl following chemotherapy shows abnormal residual soft tissue mass below the right fifth metatarsal. (**b**) Depicted *left* to *right* are transverse, coronal, and sagittal images. (**c**) Nonattenuation-corrected images of the chest show metastases in the medial right lung and lateral left lung. (Reprinted with permission from Shulkin BL. PET applications in pediatrics. Q J Nucl Med 1997;41:281–291.)

Acknowledgments

Supported in part by grant NCI 54216 (BS). The authors thank Alan Fischman, M.D., Ph.D., for assistance with illustrations.

References

1. Gordon I. Issue surrounding preparation, information, and handling the child and parent in nuclear medicine. J Nucl Med 1998;39:490–494.

2. Treves ST. Introduction. In: Treves ST, editor. Pediatric Nuclear Medicine, 2nd ed. New York: Springer-Verlag, 1995:1–11.

3. Mandell GA, Cooper JA, Majd M, et al. Procedure guidelines for pediatric sedation in nuclear medicine. J Nucl Med 1997;38:1640–1643.

4. Shulkin BL. PET applications in pediatrics. Q J Nucl Med 1997;41:281–291.

5. Borgwardt L, Larsen HJ, Pedersen K, et al. Practical use and implementation of PET in children in a hospital PET center. Eur J Nucl Med Mol Imaging 2003;30:1389–1397.

6. Roberts EG, Shulkin BL. Technical issues in performing PET studies in pediatric patients. J Nucl Med Technol 2004;32:5–9.

7. Kaste SC. Issues specific to implementing PET/CT for pediatric oncology: what we have learned along the way. Pediatr Radiol 2004;34:205–213.

8. ICRP Report 80. Radiation dose to patients from radiopharmaceuticals. Stockholm: International Commission on Radiation Protection, 1998:49–110.

9. ICRP Report 56. Age-dependent doses to members of the public from intake of radionuclides: Part 1. Stockholm: International Commission on Radiation Protection, 1989:4.

10. Jones SC, Alavi A, Christman D, et al. The radiation dosimetry of 2-[^{18}F]fluoro-2-deoxy-D-glucose in man. J Nucl Med 1982;23:613–617.

11. Ruotsalainen U, Suhonen-Povli H, Eronen E, et al. Estimated radiation dose to the newborn in FDG-PET studies. J Nucl Med 1996;37:387–393.

12. Fahey F, Palmer M, Strauss K, et al. Image quality and dosimetry using CT-based attenaution correction. (In preparation.)

13. Brenner D, Elliston C, Hall E, et al. Estimated risks of radiation-induced fatal cancer from pediatric CT. Am J Radiol 2001;176:289–296.

14. Chugani HT, Phelps ME. Maturational changes in cerebral function in infants determined by ^{18}FDG positron emission tomography. Science 1986;231:840–843.

15. Chugani HT, Phelps ME, Mazziotta JC. Positron emission tomography study of human brain functional development. Ann Neurol 1987;22:487–497.

16. Chugani HT. Positron emission tomography. In: Berg BO, editor. Principles of Child Neurology. New York: McGraw-Hill, 1996:113–128.

17. Hauser W. Epidemiology of epilepsy in children. Neurosurg Clin N Am 1995;6:419–428.

18. National Institutes of Health Consensus Development Conference Statement: surgery for epilepsy. Epilepsia ;31:806–812.

19. Kuzniecky R, Suggs S, Gaudier J, et al. Lateralization of epileptic foci by magnetic resonance imaging in temporal lobe epilepsy. J Neuroimaging 1991;1:163–167.

20. Treves ST, Connolly LP. Single photon emission computed tomography in pediatric epilepsy. Neurosurg Clin N Am 1995;6:473–480.

21. Snead OC III, Chen LS, Mitchell WG, et al. Usefulness of [^{18}F]fluorodeoxyglucose positron emission tomography in pediatric epilepsy surgery. Pediatr Neurol 1996;14:98–107.

22. Meltzer CC, Adelson PD, Brenner RP, et al. Planned ictal FDG PET imaging for localization of extratemporal epileptic foci. Epilepsia 2000;41(2):193–200.

23. Juhasz C, Chugani DC, Muzik O, et al. Is epileptogenic cortex truly hypometabolic on interictal positron emission tomography? Ann Neurol 2000;48(1):88–96.

24. Cummings TJ, Chugani DC, Chugani HT. Positron emission tomography in pediatric epilepsy. Neurosurg Clin N Am 1995;6:465–472.

25. Engel J Jr, Kuhl DE, Phelps ME. Patterns of human local cerebral glucose metabolism during epileptogenic seizures. Science 1982;218:64–66.

26. da Silva EA, Chugani DC, Muzik O, et al. Identification of frontal lobe epileptic foci in children using positron emission tomography. Epilepsia 1997;38:1198–1208.

27. Hrachovy R, Frost J. Infantile spasms. Pediatr Clin N Am 1989;36:311–329.

28. Chugani HT, Shields WD, Shewmon DA, et al. Infantile spasms: I. PET identifies focal cortical dysgenesis in cryptogenic cases for surgical treatment. Ann Neurol 1990;27:406–413.

29. Chuagni HT, Shewmon DA, Shields WD, et al. Surgery for intractable infantile spasms: neuroimaging perspectives. Epilepsia 1993;34:764–771.

30. Chugani HT, Da Silva E, Chugani DC. Infantile spasms: III. Prognostic implications of bitemporal hypometabolism on positron emission tomography. Ann Neurol 1996;39:643–649.

31. Chugani HT, Conti JR. Etiologic classification of infantile spasms in 140 cases: role of positron emission tomography. J Child Neurol 1996;11:44–48.

32. Savic I, Svanborg E, Thorell JO. Cortical benzodiazepine receptor changes are related to frequency of partial seizures: a positron emission tomography study. Epilepsia 1996;37:236–244.

33. Arnold S, Berthele A, Drzezga A, et al. Reduction of benzodiazepine receptor binding is related to the seizure onset zone in extratemporal focal cortical dysplasia. Epilepsia 2000;41(7):818–824.

34. Richardson MP, Koepp MJ, Brooks DJ, et al. ^{11}C-Flumanezil PET in neocortical epilepsy. Neurology 1998;51:485–492.

35. Debets RM, Sadzot B, van Isselt JW, et al. Is ^{11}C-flumazenil PET superior to ^{18}FDG PET and ^{123}I-iomazenial SPECT in presurgical evaluation of temporal lobe epilepsy? J Neurol Neurosurg Psychiatry 1997;62:141–150.

36. Kumlien E, Hartvig P, Valind S, et al. NMDA-receptor activity visualized with (S)-[N-methyl-11-C]ketamine and positron emission tomography in patients with medial temporal epilepsy. Epilepsia 1999;40:30–37.

37. Mayberg HS, Sadzot B, Meltzer CC, et al. Quantification of mu and non-mu opiate receptors in temporal lobe epilepsy using positron emission tomography. Ann Neurol 1991;30:3–11.

38. Kumlien E, Bergstrom M, Lilja A, et al. Positron emission tomography with [C-11]deuterium deprenyl in temporal lobe epilepsy. Epilepsia 1995;36:712–721.

39. Volpe JJ, Herscovitch P, Perlman JM, et al. Positron emission tomography in the newborn: extensive impairment of regional cerebral blood flow with intraventricular hemorrhage and hemorrhagic intracerebral involvement. Pediatrics 1983; 72(5):589–601.

40. Volpe JJ, Herscovitch P, Perlman JM, et al. Positron emission tomography in the asphyxiated term newborn: parasagittal impairment of cerebral blood flow. Ann Neurol 1985;17(3):287–296.

41. Zilbovicius M, Boddaert N, Belin P, et al. Temporal lobe dysfunction in childhood autism: a PET study. Am J Psychiatry 2000;157(12):1988–1993.

42. Ernst M, Zametkin AJ, Matochik JA, et al. High midbrain [^{18}F]DOPA accumulation in children with attention deficit hyperactivity disorder. Am J Psychiatry 1999;156(8):1209–1215.

43. Jacobson LK, Hamburger SD, Van Horn JD, et al. Cerebral glucose metabolism in childhood onset schizophrenia. Psychiatry Res 1997;75(3):131–144.

44. Reed W, Jagust W, Al-Mateen M, et al. Role of positron emission tomography in determining the extent of CNS ischemia in patients with sickle cell disease. Am J Hematol 1999;60(4):268–272.

45. Delvenne V, Lotstra F, Goldman S, et al. Brain hypometabolism of glucose in anorexia nervosa: a PET scan study. Biol Psychiatry 1995;37(3):161–169.

46. Delvenne V, Goldman S, Simon Y, et al. Brain hypometabolism of glucose in bulimia nervosa. Int J Eat Disord 1997;21(4):313–320.

47. Lee JS, Asano E, Muzik O, et al. Sturge-Weber syndrome: correlation between clinical course and FDG PET findings. Neurology 2001;57:189–195.

48. Al-Essa MA, Bakheet SM, Patay ZJ, et al. Clinical and cerebral FDG PET scan in a patient with Krabbe's disease. Pediatr Neurol 2000;22:44–47.

49. Lee JS, Juhasz C, Kaddurah AK, et al. Patterns of cerebral glucose metabolism in early and late stages Rasmussen's syndrome. J Child Neurol 2001;16:798–805.

50. Lee JS, PfundZ, Juhasz C, et al. Altered regional brain glucose metabolism in Duchenne muscular dystrophy: a PET study. Muscle Nerve 2002;26:506–512.

51. Quinlivan RM, Robinson RO, Maisey MN. Positron emission tomography in pediatric cardiology. Arch Dis Child 1998;79(6):520–522.

52. Donnelly JP, Raffel DM, Shulkin BL, et al. Resting coronary flow and coronary flow reserve in human infants after repair or palliation of congenital heart defects as measured by positron emission tomography. J Thorac Cardiovasc Surg 1998;115(1):103–110.

53. Yates RW, Marsden PK, Badawi RD, et al. Evaluation of myocardial perfusion using positron emission tomography in infants following a neonatal arterial switch operation. Pediatr Cardiol 2000;21(2):111–118.

54. Rickers C, Sasse K, Buchert R, et al. Myocardial viability assessed by positron emission tomography in infants and children after the arterial switch operation and suspected infarction. J Am Coll Cardiol 2000;36(5):1676–1683.

55. Singh TP, Muzik O, Forbes TF, et al. Positron emission tomography myocardial perfusion imaging in children with suspected coronary abnormalities. Pediatr Cardiol 2003;24:138–144.

56. Hernandez-Pampaloni M, Allada V, Fishbein MC, et al. Myocardial perfusion and viability by positron emission tomography in infants and children with coronary abnormalities: correlation with echocardiography, coronary angiography, and histopathology. J Am Coll Cardiol 2003;41:618–626.

57. Hwang B, Liu RS, Chu LS, et al. Positron emission tomography for the assessment of myocardial viability in Kawasaki disease using different therapies. Nucl Med Commun 2000;21(7):631–636.

58. Huaser M, Bengel F, Kuehn A, et al. Myocardial blood flow and coronary flow reserve in children with "normal" epicardial coronary arteries after the onset of Kawasaki disease assessed by positron emission tomography. Pediatr Cardiol 2004;25:108–112.

59. Litvinova I, Litvinov M, Loeonteva I, et al. PET for diagnosis of mitochondrial cardiomyopathy in children. Clin Posit Imaging 2000;3(4):172.

60. Gurney JG, Severson RK, Davis S, et al. Incidence of cancer in children in the United States. Cancer (Phila) 1995;75:2186–2195.

61. Robison L. General principles of the epidemiology of childhood cancer. In: Pizzo P, Poplack D, editors. Principles and Practice of Pediatric Oncology. Philadelphia: Lippincott-Raven, 1997:1–10.

62. Franzius C, Schober O. Assessment of therapy response by FDG PET in pediatric patients. Q J Nucl Med 2003;47:41–45.

63. Wegner EA, Barrington SF, Kingston JE, et al. Eur J Nucl Med Mol Imaging 2005;32(1):23–30.

64. Shulkin BL. PET imaging in pediatric oncology. Pediatr Radiol 2004;34:199–204.

65. Patel PM, Alibazoglu H, Ali A, et al. Normal thymic uptake of FDG on PET imaging. Clin Nucl Med 1996;21:772–775.

66. Weinblatt ME, Zanzi I, Belakhlef A, et al. False positive FDG-PET imaging of the thymus of a child with Hodgkin's disease. J Nucl Med 1997;38:888–890.

67. Brink I, Reinhardt MJ, Hoegerle S, et al. Increased metabolic activity in the thymus gland studied with 18F-FDG PET: age dependency and frequency after chemotherapy. J Nucl Med 2001;42:591–595.

68. Yeung HW, Grewal RK, Gonen M, et al. Patterns of (18F)-FDG uptake in adipose tissue and muscles: a potential source of false-positives for PET. J Nucl Med 2003;44:1789–1796.

69. Cohade C, Mourtzikos KA, Wahl RL. "USA-fat": prevalence is related to ambient outdoor temperature: evaluation with ^{18}F-FDG PET/CT. J Nucl Med 2003;44:1267–1270.

70. Tatsumi M, Engles JM, Ishimori T, et al. Intense (18)F-FDG uptake in brown fat can be reduced pharmacologically. J Nucl Med 2004;45:1189–1193.

71. Delbeke D. Oncological applications of FDG PET imaging: colorectal cancer, lymphoma, and melanoma. J Nucl Med 1999;40:591–603.

72. Sugawara Y, Fisher SJ, Zasadny KR, et al. Preclinical and clinical studies of bone marrow uptake of fluorine-1-fluorodeoxyglucose with or without granulocyte colony-stimulating factor during chemotherapy. J Clin Oncol 1998;16:173–180.

73. Hollinger EF, Alibazoglu H, Ali A, et al. Hematopoietic cytokine-mediated FDG uptake simulates the appearance of diffuse metastatic disease on whole-body PET imaging. Clin Nucl Med 1998;23:93–98.

74. Yeung HW, Sanches A, Squire OD, et al. Standardized uptake value in pediatric patients: an investigation to determine the optimum measurement parameter. Eur J Nucl Med Mol Imaging 2002;29:61–66.

75. Kleihues P, Burger P, Scheithauer B. The new WHO classification of brain tumours. Brain Pathol 1993;3:255–268.

76. Robertson R, Ball WJ, Barnes P. Skull and brain. In: Kirks D, editor. Practical Pediatric Imaging. Diagnostic Radiology of Infants and Children. Philadelphia: Lippincott-Raven, 1997:65–200.

77. Maria B, Drane WB, Quisling RJ, et al. Correlation between gadolinium-diethylenetriaminepentaacetic acid contrast enhancement and thallium-201 chloride uptake in pediatric brainstem glioma. J Child Neurol 1997;12:341–348.

78. O'Tuama L, Janicek M, Barnes P, et al. Tl-201/Tc-99m HMPAO SPECT imaging of treated childhood brain tumors. Pediatr Neurol 1991;7:249–257.

79. O'Tuama L, Treves ST, Larar G, et al. Tl-201 versus Tc-99m MIBI SPECT in evaluation of childhood brain tumors. J Nucl Med 1993;34:1045–1051.

80. Rollins N, Lowry P, Shapiro K. Comparison of gadolinium-enhanced MR and thallium-201 single photon emission computed tomography in pediatric brain tumors. Pediatr Neurosurg 1995;22:8–14.

81. Valk PE, Budinger TF, Levin VA, et al. PET of malignant cerebral tumors after interstitial brachytherapy. Demonstration of metabolic activity and correlation with clinical outcome. J Neurosurg 1988;69:830–838.

82. Di Chiro G, Oldfield E, Wright DC, et al. Cerebral necrosis after radiotherapy and/or intraarterial chemotherapy for brain tumors: PET and neuropathologic studies. Am J Radiol 1988;150:189–197.

83. Glantz MJ, Hoffman JM, Coleman RE, et al. Identification of early recurrence of primary central nervous system tumors by [^{18}F]fluorodeoxyglucose positron emission tomography. Ann Neurol 1991;29:347–355.

84. Janus T, Kim E, Tilbury R, et al. Use of [^{18}F]fluorodeoxyglucose positron emission tomography in patients with primary malignant brain tumors. Ann Neurol 1993;33:540–548.

85. Rozental JM, Levine RL, Nickles RJ. Changes in glucose uptake by malignant gliomas: preliminary study of prognostic significance. J Neurooncol 1991;10:75–83.

86. Schifter T, Hoffman JM, Hanson MW, et al. Serial FDG-PET studies in the prediction of survival in patients with primary brain tumors. J Comput Assist Tomogr 1993;17:509–561.

87. Francavilla TL, Miletich RS, Di Chiro G, et al. Positron emission tomography in the detection of malignant degeneration of low-grade gliomas. Neurosurgery 1989;24:1–5.

88. Patronas NJ, Di Chiro G, Kufta C, et al. Prediction of survival in glioma patients by means of positron emission tomography. J Neurosurg 1985;62:816–822.

89. Molloy PT, Belasco J, Ngo K, et al. The role of FDG PET imaging in the clinical management of pediatric brain tumors. J Nucl Med 1999;40:129P (abstract).

90. Holthof VA, Herholz K, Berthold F, et al. In vivo metabolism of childhood posterior fossa tumors and primitive neuroectodermal tumors before and after treatment. Cancer (Phila) 1993;1394–1403.

91. Hoffman JM, Hanson MW, Friedman HS, et al. FDG-PET in pediatric posterior fossa brain tumors. J Comput Assist Tomogr 1992;16:62–68.

92. Molloy PT, Defeo R, Hunter J, et al. Excellent correlation of FDG PET imaging with clinical outcome in patients with neurofibromatosis type I and low-grade astrocytomas (abstract). J Nucl Med 1999;40:129P.

93. Pirotte B, Goldman S, Salzberg S, et al. Combined positron emission tomography and magnetic resonance imaging for the planning of stereotactic brain biopsies in children: experience in 9 cases. Pediatr Neurosurg 2003;38:146–155.

94. O'Tuama LA, Phillips PC, Strauss LC, et al. Two-phase [¹¹C]L-methionine PET in childhood brain tumors. Pediatr Neurol 1990;6:163–170.

95. Mosskin M, von Holst H, Bergstrom M, et al. Positron emission tomography with ¹¹C-methionine and computed tomography of intracranial tumours compared with histopathologic examination of multiple biopsies. Acta Radiol 1987;28:673–681.

96. Lilja A, Lundqvist H, Olsson Y, et al. Positron emission tomography and computed tomography in differential diagnosis between recurrent or residual glioma and treatment-induced brain lesion. Acta Radiol 1989;38:121–128.

97. Mineura K, Sasajima T, Kowada M, et al. Indications for differential diagnosis of nontumor central nervous system diseases from tumors. A positron emission tomography study. J Neuroimaging 1997;7:8–15.

98. Utriainen M, Metsahonkala L, Salmi TT, et al. Metabolic characterization of childhood brain tumors: comparison of ¹⁸F-fluorodeoxyglucose and ¹¹C-methionine positron emission tomography. Cancer (Phila) 2002;95:1376–1386.

99. Cohen MD. Imaging of Children with Cancer. St. Louis: Mosby Yearbook, 1992.

100. Nadel HR, Rossleigh MA. Tumor imaging. In: Treves ST, editor. Pediatric Nuclear Medicine, 2nd ed. New York: Springer-Verlag, 1995:496–527.

101. Rossleigh MA, Murray IPC, Mackey DWJ. Pediatric solid tumors: evaluation by gallium-67 SPECT studies. J Nucl Med 1990;31:161–172.

102. Howman-Giles R, Stevens M, Bergin M. Role of gallium-67 in management of paediatric solid tumors. Aust Paediatr J 1982;18:120–125.

103. Yang SL, Alderson PO, Kaizer HA, et al. Serial Ga-67 citrate imaging in children with neoplastic disease: concise communication. J Nucl Med 1979;20:210–214.

104. Sty JR, Kun LE, Starshak RJ. Pediatric applications in nuclear oncology. Semin Nucl Med 1985;15:17–200.

105. Barrington SF, Carr R. Staging of Burkitt's lymphoma and response to treatment monitored by PET scanning. Clin Oncol 1995;7:334–335.

106. Bangerter M, Moog F, Buchmann I, et al. Whole-body 2-[¹⁸F]-fluoro-2-deoxy-D-glucose positron emission tomography (FDG PET) for accurate staging of Hodgkin's disease. Ann Oncol 1998;9:1117–1122.

107. Jerusalem G, Warland V, Najjar F, et al. Whole-body ¹⁸F-FDG PET for the evaluation of patients with Hodgkin's disease and non-Hodgkin's lymphoma. Nucl Med Commun 1999;20:13–20.

108. Leskinen-Kallio S, Ruotsalainen U, Nagren K, et al. Uptake of carbon-11-methionine and fluorodeoxyglucose in non-Hodgkin's lymphoma: a PET study. J Nucl Med 1991;32:1211–1218.

109. Moog F, Bangerter M, Kotzerke J, et al. 18-F-Fluorodeoxyglucose positron emission tomography as a new approach to detect lymphomatous bone marrow. J Clin Oncol 1998;16:603–609.

110. Moog F, Bangerter M, Diederichs CG, et al. Extranodal malignant lymphoma: detection with FDG PET versus CT. Radiology 1998;206:475–481.

111. Moog F, Bangerter M, Diederichs CG, et al. Lymphoma: role of whole-body 2-deoxy-2-[F-18]fluoro-D-glucose (FDG) PET in nodal staging. Radiology 1997;203:795–800.

112. Okada J, Yoshikawa K, Imazeki K, et al. The use of FDG-PET in the detection and management of malignant lymphoma: correlation of uptake with prognosis. J Nucl Med 1991;32:686–691.

113. Okada J, Yoshikawa K, Itami M, et al. Positron emission tomography using fluorine-18-fluorodeoxyglucose in malignant lymphoma: a comparison with proliferative activity. J Nucl Med 1992;33:325–329.

114. Rodriguez M, Rehn S, Ahlstrom H, et al. Predicting malignancy grade with PET in non-Hodgkin's lymphoma. J Nucl Med 1995;36:1790–1796.

115. Paul R. Comparison of fluorine-18-2-fluorodeoxyglucose and gallium-67 citrate imaging for detection of lymphoma. J Nucl Med 1987;28:288–292.

116. Newman JS, Francis IR, Kaminski MS, et al. Imaging of lymphoma with PET with 2-[F-18]-fluoro-2-deoxy-D-glucose: correlation with CT. Radiology 1994;190:111–116.

117. de Wit M, Bumann D, Beyer W, et al. Whole-body positron emission tomography (PET) for diagnosis of residual mass in patients with lymphoma. Ann Oncol 1997;8(suppl 1):57–60.

118. Cremerius U, Fabry U, Neuerburg J, et al. Positron emission tomography with 18-F-FDG to detect residual disease after therapy for malignant lymphoma. Nucl Med Commun 1998;19:1055–1063.

119. Hoh CK, Glaspy J, Rosen P, et al. Whole-body FDG PET imaging for staging of Hodgkin's disease and lymphoma. J Nucl Med 1997;38:343–348.

120. Romer W, Hanauske AR, Ziegler S, et al. Positron emission tomography in non-Hodgkin's lymphoma: assessment of chemotherapy with fluorodeoxyglucose. Blood 1998;91:4464–4471.

121. Stumpe KD, Urbinelli M, Steinert HC, et al. Whole-body positron emission tomography using fluorodeoxyglucose for staging of lymphoma: effectiveness and comparison with computed tomography. Eur J Nucl Med 1998;25:721–728.

122. Lapela M, Leskinen S, Minn HR, et al. Increased glucose metabolism in untreated non-Hodgkin's lymphoma: a study with positron emission tomography and fluorine-18-fluorodeoxyglucose. Blood 1995;86:3522–3527.

123. Carr R, Barrington SF, Madan B, et al. Detection of lymphoma in bone marrow by whole-body positron emission tomography. Blood 1998;91:3340–3346.

124. Hudson MM, Krasin MJ, Kaste SC. PET imaging in pediatric Hodgkin's lymphoma. Pediatr Radiol 2004;34:190–198.

125. Montravers F, McNamara D, Landman-Parker J, et al. [(18)F]FDG in childhood lymphoma: clinical utility and impact on management. Eur J Nucl Med Mol Imaging 2002;29:1155–1165.

126. Depas G, De Barsy C, Jerusalem G, et al. ¹⁸F-FDG PET in children with lymphomas. Eur J Nucl Med Mol Imaging 2005;32(1):31–38.

127. Lavely WC, Delbeke D, Greer JP, et al. FDG PET in the follow-up of management of patients with newly diagnosed Hodgkin and non-Hodgkin lymphoma after first-line chemotherapy. Int J Radiat Oncol Biol Phys 2003;57:307–315.

128. Swift P. Novel techniques in the delivery of radiation in pediatric oncology. Pediatr Clin N Am 2002;49:1107–1129.

129. Korholz D, Kluge R, Wickmann L, et al. Importance of F18-fluorodeoxy-D-2-glucose positron emission tomography (FDG-PET) for staging and therapy control of Hodgkin's lymphoma in childhood and adolescence: consequences for the GPOH-HD 2003 protocol. Onkologie 2003;26:489–493.

130. Krasin MJ, Hudson MM, Kaste SC. Positron emission tomography in pediatric radiation oncology: integration in the treatment-planning process. Pediatr Radiol 2004;34:214–221.

131. Bousvaros A, Kirks DR, Grossman H. Imaging of neuroblastoma: an overview. Pediatr Radiol 1986;16:89–106.

132. Kushner BH. Neuroblastoma: a disease requiring a multitude of imaging studies. J Nucl Med 2004;45:1172–1188.

133. Briganti V, Sestini R, Orlando C et al. Imaging of somatostatin receptors by indium-111-pentetreotide correlates with quantitative determination of somatostatin receptor type 2 gene expression in neuroblastoma tumor. Clin Cancer Res 1997;3:2385–3291.

134. Shulkin BL, Shapiro B, Hutchinson RJ. ¹³¹I-MIBG and bone scintigraphy for the detection of neuroblastoma. Presented at the Fifth Biennial Congress of the South African Society of Nuclear Medicine, Capetown, South Africa, September, 1992. S Afr Med J 1993;83:53.

135. Shulkin BL, Hutchinson RJ, Castle VP, et al. Neuroblastoma: positron emission tomography with 2-[fluorine-18]-fluoro-2-deoxy-D-glucose compared with metaiodobenzylguanidine scintigraphy. Radiology 1996;199:743–750.

136. Kushner BH, Yeung HW, Larson SM, et al. Extending positron emission tomography scan utility to high-risk neuroblastoma:

fluorine-18 fluorodeoxyglucose positron emission tomography as sole imaging modality in follow-up of patients. J Clin Oncol 2001;19:3397–3405.

137. Shulkin BL, Wieland DM, Baro ME, et al. PET hydroxyephedrine imaging of neuroblastoma. J Nucl Med 1996;37:16–21.

138. Shulkin BL, Wieland DM, Castle VP, et al. Carbon-11 epinephrine PET imaging of neuroblastoma. J Nucl Med 1999;40:129P (abstract).

139. Vaidyanathan G, Affleck DJ, Zalutsky MR. Validation of 4-[fluorine-18]fluoro-3-iodobenzylguanidine as a positron-emitting analog of MIBG. J Nucl Med 1995;36:644–650.

140. Ott RJ, Tait D, Flower MA, et al. Treatment planning for ^{131}I-mIBG radiotherapy of neural crest tumors using ^{124}I-mIBG positron emission tomography. Br J Radiol 1992;65:787–791.

141. Barnewolt CE, Paltiel HJ, Lebowitz RL, et al. Genitourinary system. In: Kirks DR, editor. Practical Pediatric Imaging. Diagnostic Radiology of Infants and Children, 3rd edition. Philadelphia: Lippincott-Raven, 1997:1009–1170.

142. Shulkin BL, Chang E, Strouse PJ, et al. PET FDG studies of Wilms tumors. J Pediatr Hematol/Onccol 1997;19:334–338.

143. McDonald DJ. Limb salvage surgery for sarcomas of the extremities. Am J Radiol 1994;163:509–513.

144. Triche TJ. Pathology of pediatric malignancies. In: Pizzo PA, Poplack DG, editors. Principles and Practice of Pediatric Oncology, 2nd ed. Philadelphia: Lippincott, 1993:115–152.

145. O'Connor MI, Pritchard DJ. Ewing's sarcoma. Prognostic factors, disease control, and the reemerging role of surgical treatment. Clin Orthop 1991;262:78–87.

146. Jaramillo D, Laor T, Gebhardt M. Pediatric musculoskeletal neoplasms. Evaluation with MR imaging. MRI Clin N Am 1996;4:1–22.

147. Frouge C, Vanel D, Coffre C, et al. The role of magnetic resonance imaging in the evaluation of Ewing sarcoma: a report of 27 cases. Skeletal Radiol 1988;17:387–392.

148. MacVicar AD, Olliff JFC, Pringle J, et al. Ewing sarcoma: MR imaging of chemotherapy-induced changes with histologic correlation. Radiology 1992;184:859–864.

149. Lemmi MA, Fletcher BD, Marina NM, et al. Use of MR imaging to assess results of chemotherapy for Ewing sarcoma. Am J Radiol 1990;155:343–346.

150. Erlemann R, Sciuk J, Bosse A, et al. Response of osteosarcoma and Ewing sarcoma to preoperative chemotherapy: assessment with dynamic and static MR imaging and skeletal scintigraphy. Radiology 1990;175:791–796.

151. Holscher HC, Bloem JL, Vanel D, et al. Osteosarcoma: chemotherapy-induced changes at MR imaging. Radiology 1992;182:839–844.

152. Lawrence JA, Babyn PS, Chan HS, et al. Extremity osteosarcoma in childhood: prognostic value of radiologic imaging. Radiology 1993;189:43–47.

153. Connolly LP, Laor T, Jaramillo D, et al. Prediction of chemotherapeutic response of osteosarcoma with quantitative thallium-201 scintigraphy and magnetic resonance imaging. Radiology 1996;201(P):349 (abstract).

154. Lin J, Leung WT. Quantitative evaluation of thallium-201 uptake in predicting chemotherapeutic response of osteosarcoma. Eur J Nucl Med 1995;22:553–555.

155. Menendez LR, Fideler BM, Mirra J. Thallium-201 scanning for the evaluation of osteosarcoma and soft tissue sarcoma. J Bone Joint Surg 1993;75:526–531.

156. Ramanna L, Waxman A, Binney G, et al. Thallium-201 scintigraphy in bone sarcoma: comparison with gallium-67 and technetium-99m MDP in the evaluation of chemotherapeutic response. J Nucl Med 1990;31:567–572.

157. Rosen G, Loren GJ, Brien EW, et al. Serial thallium-201 scintigraphy in osteosarcoma. Correlation with tumor necrosis after preoperative chemotherapy. Clin Orthop 1993;293:302––306.

158. Ohtomo K, Terui S, Yokoyama R, et al. Thallium-201 scintigraphy to assess effect of chemotherapy to osteosarcoma. J Nucl Med 1996;37:1444–1448.

159. Bar-Sever Z, Connolly LP, Treves ST, et al. Technetium-99m MIBI in the evaluation of children with Ewing's sarcoma. J Nucl Med 1997;38:13P (abstract).

160. Caner B, Kitapel M, Unlu M, et al. Technetium-99m-MIBI uptake in benign and malignant bone lesions: a comparative study with technetium-99m-MDP. J Nucl Med 1992;33:319–324.

161. Lenzo NP, Shulkin B, Castle VP, et al. FDG PET in childhood soft tissue sarcoma. J Nucl Med 2000;41(suppl 5):96P (abstract).

162. Abdel-Dayem HM. The role of nuclear medicine in primary bone and soft tissue tumors. Semin Nucl Med 1997;27:355–363.

163. Shulkin BL, Mitchell DS, Ungar DR, et al. Neoplasms in a pediatric population: 2-[F-18]-fluoro-2-deoxy-D-glucose PET studies. Radiology 1995;194:495–500.

164. Jadvar H, Connolly LP, Shulkin BL, et al. Positron-emission tomography in pediatrics. Nucl Med Annual 2000;53–83.

165. Franzius C, Sciuk J, Brinkschmidt C, et al. Evaluation of chemotherapy response in primary bone tumors with F-18 FDG positron emission tomography compared with histologically assessed tumor necrosis. Clin Nucl Med 2000;25:874–881.

166. Hawkins DS, Rajendran JG, Conrad EU III, et al. Evaluation of chemotherapy response in pediatric bone sarcomas by [F-18]-fluorodeoxy-D-glucose positron emission tomography. Cancer (Phila) 2002;94:3277–3284.

167. Brisse H, Ollivier L, Edeline V, et al. Pediatr Radiol 2004;34:595–605.

168. Franzius C, Sciuk J, Daldrup-Link HE, et al. FDG-PET for detection of osseous metastases from malignant primary bone tumors: comparison with bone scintigraphy. Eur J Nucl Med 2000;27:1305–1311.

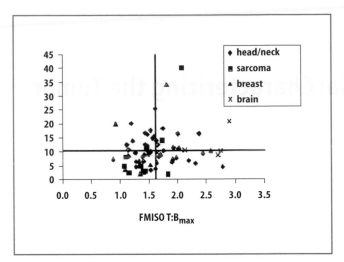

Figure 23.1. Graph of the distribution of ^{18}F-fluorodeoxyglugose (FDG) and fluoromisonidazole (FMISO) uptake (divided into four quadrants based on the median uptake values) in patients with several types of cancer shows the heterogeneous correlation between hypoxia and glycolysis. *SUV*, standardized uptake value. [Reprinted with permission from Rajendran JG, Mankoff DM, O'Sullivan F, et al. Hypoxia and glucose metabolism in malignant tumors: evaluation by FMISO and FDG PET imaging. Clin Cancer Res 2004;10(7):2245–2252.]

(VEGF) and signaling molecules such as interleukin 1 (IL-1), tumor necrosis factor-alpha (TNF-a), and transforming growth factor-beta (TGF-β), and selection of cells with mutant p53 (21, 22). Several consequences of this genetic adaptation are relevant to treatment as well as imaging. For example, hypoxic cells do not readily undergo death by apoptosis (23) or arrest in the G_1 phase of the cell cycle in response to sublethal deoxyribonucleic acid (DNA) damage induced by radiation (24, 25). Increased glucose transporter (GLUT) activity is responsible for much of the increased glucose uptake associated with hypoxia, which

can be as high as twofold (26, 27). Hexokinase levels can also be elevated (28).

Hypoxia-Inducible Factor

The mechanistic aspects of tissue oxygen sensing and the response to hypoxia are the focus of active research. The primary cellular oxygen-sensing mechanism appears to be mediated by a heme protein that uses O_2 as a substrate to catalyze hydroxylation of proline in a segment of hypoxia-inducible factor (HIF1α) this leads to rapid degradation of HIF1α by ubiquitination under normoxic conditions (29). In the absence of O_2, HIF1α accumulates and forms a heterodimer with HIF1β that is transported to the nucleus and promotes *hypoxia-responsive* genes, resulting in a cascade of genetic and metabolic events in an effort to mitigate the effects of hypoxia on cellular energetics (30, 31). Stabilization of HIF1α has been shown to occur early in the process of tumor development, even before the invasive stage (32). Identification of overexpressed HIF1α in tissues by immunocytochemical (IHC) staining has been used as an indirect measure of hypoxia (33–35), but its heterogeneous expression within a tumor and the nonspecific nature of HIF1α expression limit the prognostic value of IHC.

Angiogenesis

Angiogenesis in tumors has been aggressively investigated by many groups, both from an imaging standpoint and, more importantly, for finding a therapeutic strategy (36,

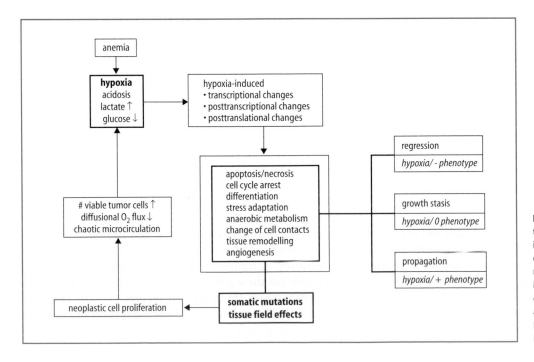

Figure 23.2. Hypoxia-induced proteomic changes in cancer cells influencing propagation of cancer. The net result of these effects is manifested by growth, regression, or stable disease. [From Hockel M, Vaupel P. Tumor hypoxia: definitions and current clinical, biologic, and molecular aspects. J Natl Cancer Inst 2001;93(4):266–276. Reprinted by permission of Oxford University Press.]

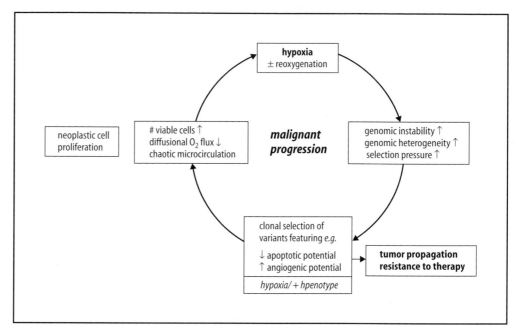

Figure 23.3. Progressive genomic changes in a tumor resulting from hypoxia. [From Hockel M, Vaupel P. Tumor hypoxia: definitions and current clinical, biologic, and molecular aspects. J Natl Cancer Inst 2001;93(4):266–276. Reprinted by permission of Oxford University Press.]

37). Angiogenesis, the formation of new blood vessels, is an important aspect of the tumor phenotype and can be considered as a failure of the balance between proangiogenic and antiangiogenic signals. New blood vessels are essential for the delivery of nutrients to sustain tumor growth, invasion, and metastatic spread. It has been found that tumors do not grow beyond a size of 1 to 2 mm without inducing angiogenesis (38). Angiogenesis is further accentuated in hypoxia as a direct result of transcription factors such as VEGF. However, this process of tumor angiogenesis is chaotic and can exhibit significant functional deficiencies compared to normal vasculature (12, 39). Although angiogenesis is a frequent consequence of hypoxia, some tumors develop extensive angiogenesis without the presence of hypoxia. Severely hypoxic tumors do not promote significant angiogenesis. Imaging of angiogenesis has been attempted by several methods including optical, magnetic resonance imaging (MRI), and PET. Further details can be obtained from recent references (40–44).

Tumor Hypoxia and Clinical Outcome: What Is New?

The negative association of hypoxia in tumor biology and clinical prognosis has long been known. Radiobiologic experiments have overwhelmingly established oxygen as important for *fixing* (making permanent) the radiation-induced cytotoxic products in tissues. In the absence of oxygen, the free radicals formed by ionizing radiation recombine without producing the anticipated cellular damage (45–47). Clinical as well as preclinical experience indicates that it takes three times as much photon radiation dose to cause the same lethality in hypoxic cells as compared to normoxic cells (13, 47–50).

While these are all established concepts, what is new in our understanding is that hypoxia results in the development and selection of an aggressive phenotype, poor response, and poor outcome because of increased metastatic potential (23, 51–53). Hypoxia can also promote resistance to chemotherapeutic agents by any one of several mechanisms. (i) Hypoxia can result in slowing of cellular proliferation, and the effectiveness of chemotherapy frequently correlates with S-phase function. (ii) Changes in perfusion associated with hypoxia may impede delivery of chemotherapy drugs. (iii) Gene amplification results in the induction of numerous stress proteins that are factors in limiting response, for example, apoptosis (53, 54) (Figure 23.4).

However, cancer treatment schemes designed over the years to circumvent the cure-limiting consequences of hypoxia have led to disappointing results (55). Hyperbaric oxygen, neutron therapy, hyperfractionation, and the use of oxygen-mimetic radiation sensitizers have not produced the expected benefit in the clinic. Although several reasons have been postulated for this situation, the most significant for this review is the need for a noninvasive assay to identify patients with hypoxic tumors. Such an assay would help characterize the tumor and would also provide a basis for rational patient selection and for guiding treatment. Optimal time, dose, and fractionation schemes to overcome hypoxia are being actively investigated, but further discussion of that topic is left to the radiation oncology literature.

The negative association of hypoxia with treatment response and patient outcome implies that evaluating hypoxia will help identify tumors with a high hypoxic fraction, so that hypoxia-directed treatments can be im-

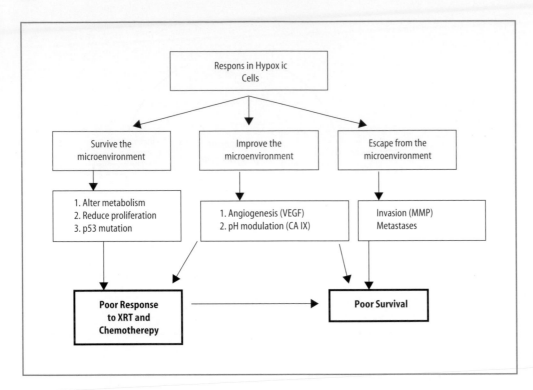

Figure 23.4. Possible biologic and clinical implications of hypoxia response.

Evaluation of Tumor Hypoxia

plemented and treatments that depend on the presence of oxygen can be avoided. If careful selection of patients with hypoxic tumors improved the outcome in patients treated with hypoxia-specific approaches, this would benefit the individual patient. Contrary to expectations, there is abundant evidence that tumor hypoxia does not correlate with tumor size, grade, and extent of necrosis or blood hemoglobin status (56–61). Moreover, most of the commonly used clinicopathologic measures of tumor hypoxia are not strong indicators of prognosis.

The nature of hypoxic cells makes them attractive targets for hypoxia-activated prodrugs (62, 63). Although focal hypoxia in a tumor can be treated with boost radiation using intensity-modulating radiation therapy (IMRT) (64, 65), more-diffuse hypoxia will benefit from hypoxic cell toxins or sensitizers. Newer hypoxia-activated prodrugs (66) with less toxicity and more effectiveness than their early counterparts have been synthesized (67). The most widely tested drug has been SR 4233, now called tirapazamine (TPZ). In addition to direct cytotoxic effects, these agents exhibit synergistic toxicity with radiation and chemotherapy. Tirapazamine is being tried in a number of malignancies with encouraging results (63, 68, 69). Nuclear imaging with hypoxia-specific tracers should be an important tool for selecting patients who might benefit from this treatment. When some early results were analyzed (70), it was obvious that TPZ was much more effective in the presence of hypoxia, indicating that identification of hypoxia before treatment would be beneficial in patient selection (70, 71).

During the past several decades, tumor oxygenation has been evaluated by several methods, and assays of tumor hypoxia have been evaluated to predict patient outcome in cancers of the uterine cervix (72), lung (58), head and neck (64, 73–75), and glioma (76, 77). These techniques have identified widespread heterogeneity in tumor hypoxia within a tumor, between tumors, and between patients with the same tumor type (78). Hypoxia usually resolves as the tumor shrinks with treatment, either by radiotherapy or by chemotherapy, but it may show paradoxical results in some tumors, perhaps because of relative sparing of hypoxic cells by the treatment (79). Because of the changing dynamics in the tumor secondary to treatment, the optimal radiation boost needs to be directed to the hypoxic subvolume from imaging after a few treatments rather than to the baseline information.

Current hypoxia assays can be categorized as in vivo (both invasive and noninvasive) or ex vivo (invasive biopsy) (5, 80, 81). A useful assay should distinguish normoxic regions from hypoxic regions at a level of oxygen relevant to cancer, pO_2 in the range of 3 to 5 mmHg. Experience has shown that regional levels of hypoxia should be measured for individual patients and individual tumor sites. To be maximally successful, hypoxia-directed imaging and treatment should target both chronic hypoxia and hypoxia resulting from transient interruption of blood flow (79). The assay should reflect intracellular pO_2 rather than blood flow or some consequence of the O_2 level on downstream biochemical changes in, for example, thiols or NADH. The observed temporal heterogeneity in tissue pO_2 suggests that a secondary effect, such as intracellular redox status, will not be as relevant to cancer treatment outcome as the intracellular partial pressure of O_2.

Other desirable characteristics for an ideal clinical hypoxia assay include (a) simple and noninvasive method, (b) nontoxic, (c) rapid and easy to perform with consistency between laboratories, and (d) ability to quantify without the need for substantial calibration of the detection instrumentation. Location of the tumor in a patient should not be a limiting factor for the assay. Last, spatial heterogeneity in the distribution of hypoxia dictates that the ideal assay must provide a complete locoregional evaluation of the tumor. All these requirements suggest an important role for imaging in evaluating hypoxia.

Polarographic Electrode Measurements

Early measurements of oxygen levels in tumors were largely based on direct measurement of pO_2 levels using very fine polarographic electrodes. Because this assay can be calibrated in units of mmHg, it has been described as a gold standard for pO_2. Heterogeneity of hypoxia within a tumor, which shows a gradient toward the center of the tumor, presents a challenge for accurately mapping regional pO_2 (57, 82). The electrodes do not provide full maps of a tumor area, but do provide a histogram of the distribution of the electrode's reading as it is inserted and then slowly withdrawn. Interlaboratory variations in calibration of the electrodes further plague the results (83), especially when pO_2 values are compared between laboratories. Because polarographic electrodes measure oxygen tension in a group of cells, the readings can be influenced by the presence of blood in the interstitial or vascular spaces in tumors (84).

Although image-guided sampling can be used to select the path and depth of electrode deployment to avoid blood vessels (85), anatomic imaging methods are notoriously limited in identifying areas of viable tissue within a tumor. For example, a hypodense area visualized within a tumor on CT is considered necrotic, but it can have measurable levels of oxygen (84). Selection of close needle entry points can reduce sampling error (83) but can compromise patient comfort and compliance. Other limitations of electrode measurements include the need for accessible tumor location and difficulties associated with serial measurements. An accurate value of pO_2 may be less informative than once expected, because different tissues have different respiratory demand and may not exhibit a hypoxic stress response at the same level of oxygenation. Normal cells, for example, cardiomyocytes, experience

stress at relatively high pO_2 (86). Moreover, oxygen extraction capacities of cells might be impaired in the whole tumor or regionally within a tumor. Electrode studies commonly report the percentage of readings that fall below some cutoff value that may range from 2.5 to 10 mmHg, depending on the tumor site and standards for a laboratory. An absolute pO_2 value may be less useful and less robust than accurate assessment of the volume or fraction of tumor cells that are hypoxic. In addition, O_2 consumption continues during electrode assay. The distribution of electrode measurements in a tumor can be bimodal in some instances (87). The fraction of cells in the hypoxic peak may be much more important than the mean pO_2 or the pO_2 for the nadir separating hypoxia and normoxia. These limitations provided the impetus to search for a noninvasive method that could be done serially to characterize and quantify tumor hypoxia in patients. Imaging methods for hypoxia provide a complete map of relative oxygenation level in tumor regions with good spatial resolution in a microenvironment that tends to be highly heterogeneous.

PET Hypoxia Imaging

Hypoxia imaging presents the special challenge of making a positive image from the relative absence of O_2. Chemists have developed two different classes of imaging agents to address this problem: bioreductive alkylating agents that are O_2 sensitive and metal chelates which are apparently sensitive to the intracellular redox state that develops as a consequence of hypoxia.

Nitroimidazole Compounds

Misonidazole, an azomycin-based hypoxic cell sensitizer introduced in clinical radiation oncology three decades ago, binds covalently to intracellular molecules at levels that are inversely proportional to intracellular oxygen concentration below about 10 mmHg. It is a lipophilic 2-nitroimidazole derivative whose uptake in hypoxic cells is dependent on the sequential reduction of the nitro group on the imidazole ring (88). This uptake mechanism requires viability of the cells; reduction depends on electron transport to provide electrons for the bioreduction step. In the absence of mitochondrial electron transport, the tracer is not reduced and thus not accumulated in the tissues. The one-electron reduction product is an unstable radical anion that will either give up its extra electron to O_2 or pick up a second electron. In the presence of O_2, the nitroimidazole simply goes through a futile reduction cycle and is returned to its initial nitromidazole state. In the absence of a competitive electron acceptor, the nitroimidazole continues to accumulate electrons to form the hydroxylamine alkylating agent and become trapped

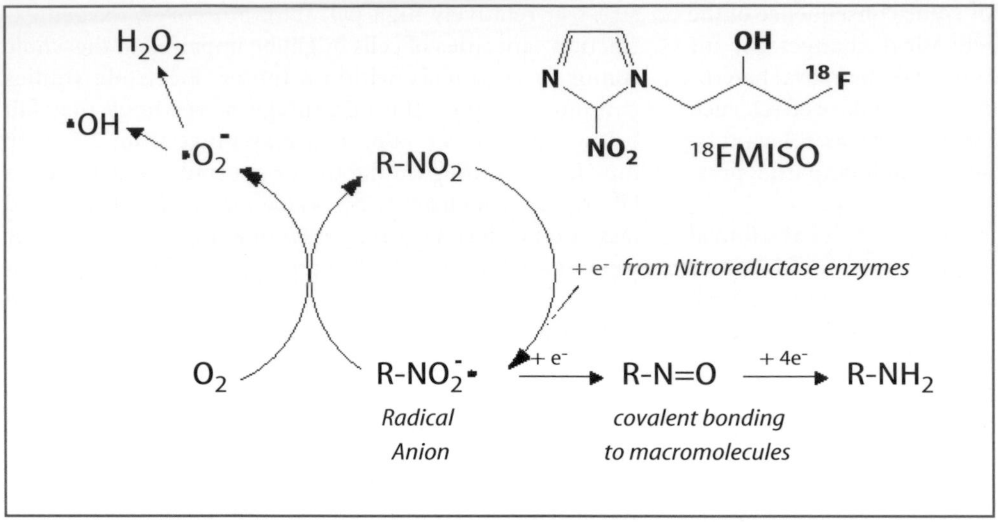

Figure 23.5. Structure of misonidazole showing the mechanism of action in the presence and absence of oxygen.

within the viable but O_2-deficient cell (Figure 23.5). The uptake and retention of ^{18}F-fluoromisonidazole (FMISO) is inversely related to the instantaneous oxygen tension within the cell. In vitro studies have shown that reoxygenated cells exposed to a new batch of the tracer will not accumulate the 2-nitroimidazole compounds.

FMISO is an imaging agent derived from misonidazole, one of the earliest investigated radiosensitizers in clinical radiation therapy. It has a high *hypoxia-specific factor* (HSF) of 20 to 50, defined as the ratio of uptake in hypoxic cells compared to normoxic cells. This factor determines the uptake specificity in vitro (89). Prodrug imaging agents such as FMISO are bioreductively activated in hypoxic tissue but the process is reversible in the presence of oxygen in tissues. The result is a positive image of the absence of O_2.

FMISO is a highly stable and robust radiopharmaceutical that can be used to quantify tissue hypoxia using PET imaging (59, 90). Its easy synthesis and optimal safety profile are responsible for its acceptance in the clinic. After extensive clinical validation, FMISO has maintained its role as the most commonly used PET hypoxia tracer (16, 58, 59, 91–95). It has biodistribution and dosimetry characteristics ideal for PET imaging (96). The partition coeffecient of FMISO is 0.41 (97), similar to that of the blood flow agent iodoantipyrine, so that initially after injection the tissue distribution reflects blood flow. After about an hour the distribution reflects its partition coefficient; it is homogeneously distributed with no tissue specificity (86).

The distribution of pixel uptake values after about 90 min is narrowly dispersed, which has led to a simple analysis of FMISO PET images by scaling the pixel uptake to plasma concentration. The mean value for this ratio in all tissues is close to unity and almost all normoxic pixels have a value less than 1.2. The regional concentration of the intermediate radical anion product parallels nitroreductase levels, which vary only slightly, so this factor does not affect the imaging analysis of hypoxic volume (93). The optimal time for imaging appears to be between 90

and 150 min, which can be adjusted to fit the clinic schedule and has a logistic analogy to a diphosphonate bone scan. The modest tumor-to-background ratio reflects the continuum between normoxia and hypoxia and does not compromise image interpretation. Hypoxia images can be interpreted by either qualitative or quantitative methods. Qualitative interpretations have been used with a scoring system to grade the uptake in a tumor in relation to normal tissue (70). We use a simple but accurate quantitation method using a venous blood sample to calculate a tissue-to-blood ratio that is the result of extensive validation studies (4, 98). The tumor to plasma ratio has proved useful to estimate the degree of hypoxia in a number of studies, as has the tumor-to-muscle ratio (92). The FMISO ratio image provides a reliable and reproducible analysis that can be readily introduced in clinical practice.

Fractional hypoxic volume (FHV), defined as the proportion of pixels within the imaged tumor volume that have a ratio above some cutoff value, has been used (93) but this requires accurate delineation of tumor margins to define the denominator. This parameter is a conceptual extrapolation of the FHV from radiobiology laboratory experiments. We prefer the tumor hypoxic volume (HV) parameter, which is the total number of pixels with a tissue-to-blood ratio of 1.2 or more. Expressed in milliliters (mL), it is a measure of the extent of hypoxic tissue and obviates the need for stringent demarcation of the tumor boundaries (4, 99). Mathematical models with more detailed analysis are useful research tools but are not likely to find acceptance in routine clinical imaging (100).

A typical protocol for PET scanning with FMISO uses an intravenous administration of a dose of 3.7 MBq (0.1 mCi)/kg, which results in an effective total-body dose equivalent of 0.0126 mGy/MBq (96). Scanning typically begins at about 120 min and lasts for 20 min with blood sampling during the scan. A transmission scan is used for attenuation correction of emission data. Typically one axial field of view (AFOV) of 15 cm in the craniocaudal dimension is acquired. An FDG scan of the entire torso in-

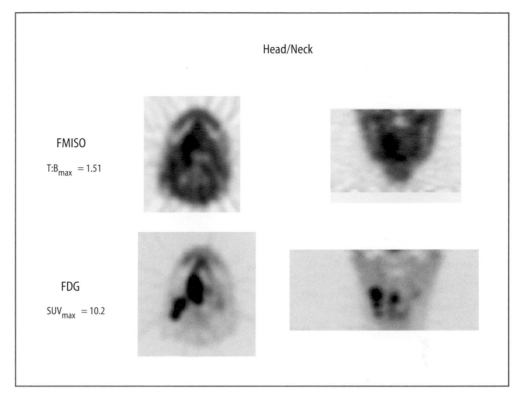

Figure 23.6. Corresponding transaxial and coronal FMISO (*top*) and FDG (*bottom*) images of a patient with cancer of the posterior tongue with metastatic lymph nodes.

clusive of the same region is obtained for some patients, with care taken to reposition the patient between images. Addition of FDG imaging data increases the sensitivity of FMISO imaging by indicating the full extent of tumor and helps in correlating metabolic activity and hypoxia in tumor as well as in staging (16) (Figures 23.6–23.10).

Alternative Azomycin Imaging Agents

Some research groups have developed alternative azomycin (nitroimidazole) radiopharmaceuticals for hypoxia imaging by attempting to manipulate the rate of blood clearance to improve image contrast (101–103). EF-

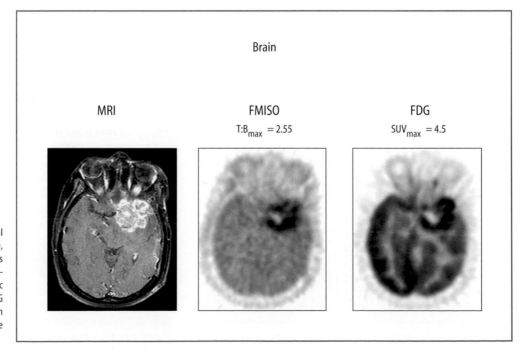

Figure 23.7. Corresponding transaxial magnetic resonance imaging (MRI) (*left*), FMISO (*middle*), and FDG (*right*) images of a patient with glioblastoma multiforme in the left temporal lobe. Hypoxic region generally parallels the FDG uptake, but note the differences in uptake in the anteromedial aspect of the tumor.

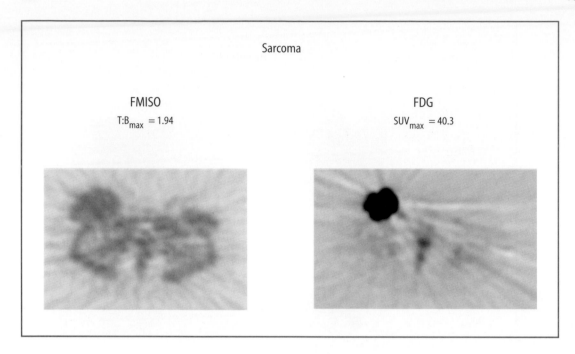

Sarcoma

FMISO
$T{:}B_{max} = 1.94$

FDG
$SUV_{max} = 40.3$

Figure 23.8. Corresponding FMISO (*left*) and FDG (*right*) images of a patient with soft tissue sarcoma of the pelvis show uniform uptake of the tracers in the tumor.

1 was initially developed because of the availability of an antibody stain to verify the distribution in tissue samples (104). Fluoroerythronitroimidazole (FETNIM) was developed as a more-hydrophilic derivative of misonidazole that might have more-rapid plasma clearance as this could be an imaging advantage. Fluoroetanidazole has binding characteristics similar to FMISO but has been reported to have less retention in liver and fewer metabolites in animals (105, 106). It appears to be the best of the alternative azomycins; however, the advantages have not been sufficient to carry these derivatives to wide clinical testing.

Single photon emission computed tomography (SPECT)-based hypoxia imaging compounds have been introduced with the hope of taking this technology to gamma camera imaging (107). The Cross Cancer Center group pioneered the development of iodinated derivatives of nitroimidazoles. Direct halogenation of the imidazole ring does not lead to a stable radiopharmaceutical, so the general approach has been to place sugar residues between the nitroimidazole and the radioiodine to stabilize the labeled molecule. Introduction of the sugar results in a more water soluble molecule than misonidazole. These products exhibit minimal deiodination, and two derivatives have been evaluated in mice and in patients, iodoazomycin arabinoside and iodoazomycin galactoside. The sugar has two consequences: the hydrophilic product clears more rapidly, but its clearance and therefore its background distribution in normoxic tissues are somewhat dependent on blood flow. The resulting images have higher contrast than FMISO when imaging is typically initiated 110 min after injection.

The success with positron-labeled radioiodinated azomycins led to attempts to develop technetium derivatives of 2-nitroimidazole and to take advantage of a ^{99m}Tc label, well known in the nuclear medicine community. Ready availability at low cost, convenient half-life for hypoxia measurements, and versatile chemistry are all benefits of ^{99m}Tc-compounds. Two different approaches have been evaluated: both BMS181321 and HL91 were synthesized and evaluated as hypoxia-based agents. Although both these molecules involve ligands with potential hypoxia-specific binding characteristics, the reduction chemistry of the metal core is also subject to redox chemistry that can result in lability of the Tc=O core from the ligand (101, 108). The BMS compound was so

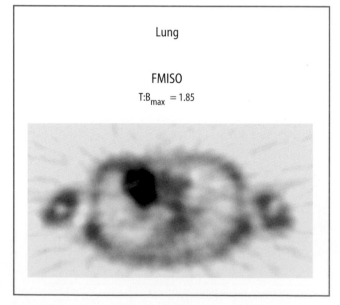

Lung

FMISO
$T{:}B_{max} = 1.85$

Figure 23.9. FMISO image of a patient with a right non-small cell lung cancer. Note the central area of photopenia corresponding to necrosis in the tumor.

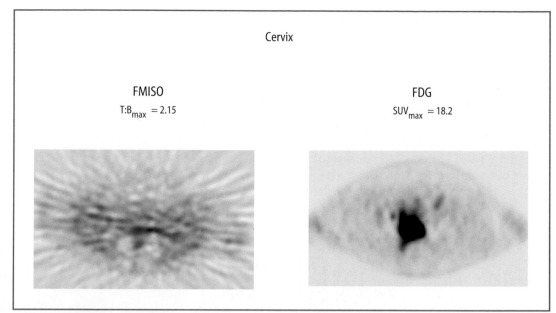

Figure 23.10. Corresponding FMISO (*left*) and FDG (*right*) images of a patient with cervical cancer. Although the FMISO uptake in the primary tumor is not as pronounced as the FDG uptake, this tumor was found to have a FMISO tissue-to-blood ratio (*T:B*) of 2.15 and a hypoxic volume (HV) of 68.4 mL, which is significant.

lipophilic that its background activity remained high, especially in the abdomen. A less-lipophilic derivative, BMS194796, has been developed with better clearance properties, especially from the liver (109). Both the BMS compounds involve a nitroimidazole group, although it is probably not the dominant influence in determining the biodistribution kinetics of the radiopharmaceutical or its specific localization in hypoxic tissues. The HL91 molecule ^{99m}Tc-BnAO does not include a nitroimidazole; the ^{99m}Tc–ligand coordination chemistry is directly reduced and retained in hypoxic environments (108). It requires a much lower level of O_2 for its reduction and uptake, raising concerns for routine clinical applications (110). The lack of specificity for hypoxia has led to abandonment of this molecule as a tracer for imaging hypoxia.

SPECT radiopharmacuticals, both the iodinated compounds (e.g., ^{123}I-IAZA) (111) and the ^{99m}Tc-based agents already noted, suffer from lower image contrast and have less potential for quantification than the PET agents (102). The absence of a gold standard for hypoxia evaluation complicates validation of all hypoxia markers, including FMISO (89). There is an urgent need for treatment outcome studies to provide convincing evidence for the clinical value of hypoxia imaging in the oncology clinic.

Other Imaging Agents

Copper bis(thiosemicarbazones) are a class of molecules that has been evaluated as freely diffusible blood flow tracers that are retained once they enter cells. The altered redox environment associated with hypoxia results in increased retention of some copper chelates so that they produce high-contrast images. The ^{64}Cu-labeled acetyl derivative of pyruvaldehyde bis[N4-methylthiosemicarbazonato]–copper (II) complex, Cu-ATSM, has the

potential advantage of a longer half-life for practical clinical use (112–114) although the mechanism of retention is less well validated than FMISO. Intracellular retention is related to the copper reduction chemistry, Cu^{2+} to Cu^+, which has a redox potential of –297 mV for Cu-ATSM (115). Several other biologic systems have comparable redox potentials: –315 mV for NADH and –230 mV for glutathione.

Retention of Cu-ATSM in hypoxic regions and rapid washout from normal regions have been documented (115). This radiopharmaceutical has rapid washout from normoxic areas. It is a useful imaging agent for identifying regions of tissue that have higher levels of reducing agents, for example, NADH, as a consequence of hypoxia. This mechanism is distinct from that for the nitroimidazoles in that the copper agents reflect a consequence of hypoxia rather than simply the level of pO_2. This mechanistic difference might limit the role of Cu-ATSM for measuring a prompt reoxygenation response because the increased levels of NADH secondary to hypoxia persist after reoxygenation. Diffusion of NADH and other related reducing equivalents might also make Cu-ATSM less reflective of the spatial heterogeneity of hypoxia. However, the same characteristics make the Cu agents preferable for imaging chronic hypoxia where levels of NADH can increase by severalfold (113, 116), resulting in substantial signal amplification.

Summary

The increasing utilization of PET imaging in the clinic makes it an attractive choice for evaluating tumor hypoxia. One can evaluate the entire tumor and regional lymph nodes noninvasively at the same time in a snapshot fashion, and PET imaging is not limited by some of the procedural difficulties of polarographic electrodes. In addition, the noninvasiveness and safety profile provide the ability for repeat imaging (117). The main advantage of PET is its ability to accurately quantify tissue uptake of the hypoxia tracer, independent of anatomic location of the tumor. Widespread availability of PET scanners (and now PET/CT scanners) and distribution of ^{18}F-labeled hypoxia tracers in the community will make this procedure within reach of community nuclear medicine and radiation oncology centers. Although the level of pretherapy hypoxia is an important parameter, its change with treat-

ment will provide an even better understanding of the effectiveness of treatment.

Hypoxia imaging can be combined with other indicators of tissue hemodynamics and oxygenation, such as perfusion imaging using ^{15}O-water, and tissue markers of proteomic response to hypoxia, such as vascular density and HIF1? expression using immunocytochemistry, to provide complementary information. Availability of PET/CT scanners provides the opportunity to combine anatomic imaging and functional information; this will increase the accuracy of hypoxia imaging and will also facilitate incorporation of the images into radiation treat-

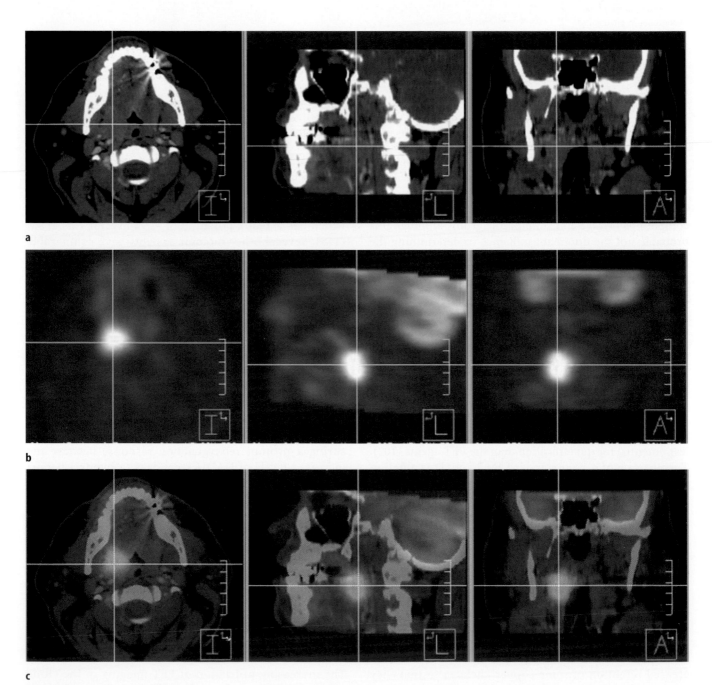

Figure 23.11. (a) Computed tomography (CT) images, (b) FDG PET images, and (c) fused FDG-PET/CT images reconstructed in transaxial, sagittal, and coronal planes with crosshairs centered on the tumor in the right tonsilar region.

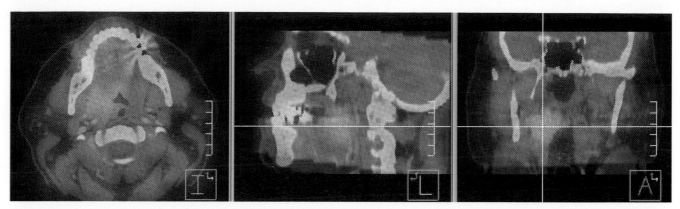

Figure 23.12. Fused FMISO PET/CT images reconstructed in transaxial, sagittal, and coronal planes with crosshairs centered on the hypoxic subvolume in the right tonsilar primary tumor.

ment plans to explore the feasibility of IMRT-based boosts to hypoxic subvolumes (64, 65, 118) (Figures 23.11–23.14).

Rapid advances in technology and expanding sophistication of cancer treatment necessitate noninvasive molecular imaging methods to characterize the tumor and guide treatment decisions. The problem of tumor hypoxia, first realized during the last century, can now be effectively ad-

dressed with newer drugs and advanced radiation techniques in treatment planning and delivery. An effective hypoxia assay would guide treatment selection and help monitor treatment results. However, each of the currently available imaging procedures has at least one disadvantage. The limited quantitative potential of SPECT and MRI will always limit imaging based on these detection systems. PET imaging with FMISO and Cu-ATSM both

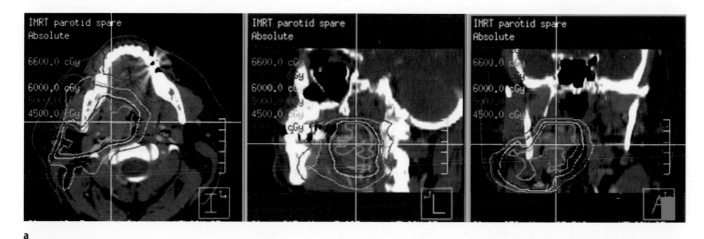

a

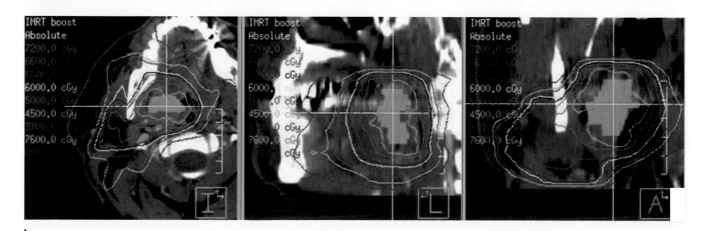

b

Figure 23.13. (a) Isodose distributions for intensity-modulating radiation therapy (IMRT) planning for the gloss target volume (GTV) derived from CT scan. **(b)** Isodose distributions for IMRT planning for additional boost dose to the hypoxic subvolume derived from FMISO images.

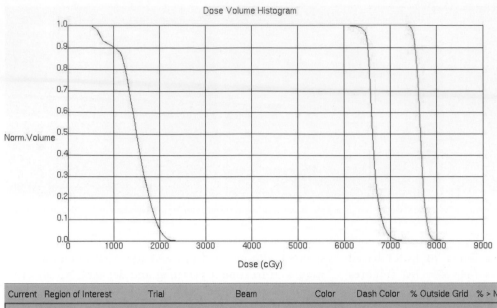

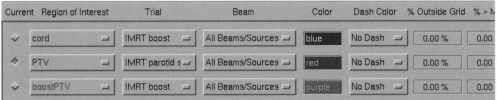

a

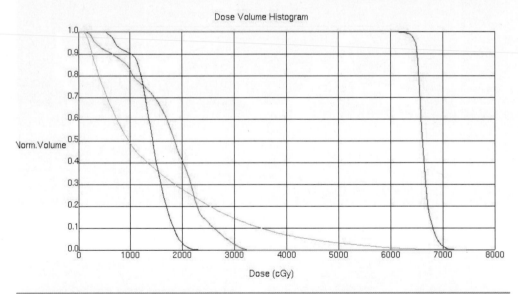

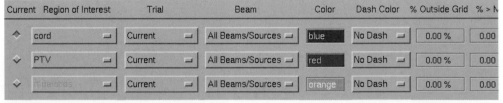

b

Figure 23.14. (**a**) Dose–volume histograms (DVH) for boost to hypoxia subvolume based on FMISO. (**b**) Dose–volume histograms (DVH) for dose to the primary tumor GTV based on CT scan.

have initial clinical experience and need to be studied in a multicenter trial of the importance of hypoxia as a cure-limiting factor.

However, PET, for the foreseeable future, will retain disadvantages associated with cost and availability of the tracers. Modern techniques in radiation treatment planning, such as IMRT, provide the ability to customize radiation delivery based on physical conformity (64, 65, 119, 120). By incorporating information on focal hypoxia, imaging can be used to create a biologic profile of the tumor to direct radiation therapy (121, 122) (see Figures 23.11, 23.12). Also, patients with widespread hypoxia in the tumor will benefit from systemic therapy with hypoxic cell cytotoxin (123). Assessing hypoxia will help development and clinical evaluation of novel hypoxia-directed drugs (62, 124–126). Hypoxia imaging will provide complementary information that would be useful in clinical management of cancer patients.

A comparison of the prominent hypoxia imaging agents FMISO, IAZA, and Cu-ATSM is needed. The Cu-ATSM images show the best contrast early after injection, but these images are confounded by blood flow, and their mechanism of localization is one step removed from the intracellular O_2 concentration. The most common criticism of FMISO is the "less than optimal" contrast on images, but its uptake after 2 h is perhaps the purest reflection of regional pO_2 at the time of radiopharmaceutical administration. For SPECT, the only option is now IAZA, as the ^{99m}Tc-agents are not yet ready for clinical evaluation. The time is ripe for the nuclear medicine community to inform the oncologists of the availability of advanced techniques for hypoxia imaging and impress upon them the need for large-scale multicenter trials of hypoxia imaging to direct treatment.

Acknowledgments

We appreciate the following individuals for the help they provided: Eric Ford, Ph.D., for help with IMRT planning, Lanell M. Peterson, B.A., and Holly Pike for help with the manuscript preparation, and Hubert S. Vesselle, M.D. for help with the oxygen electrode studies. This study was supported in part by NIH grants P01 CA42045 and S10RR17229 and a Pilot grant from the Seattle Cancer Consortium.

References

1. Wahl RL. Anatomolecular imaging with 2-deoxy-2-[18F]fluoro-D-glucose: bench to outpatient center. Mol Imaging Biol 2003;5:49–56.
2. Herschman HR. Molecular imaging: looking at problems, seeing solutions. Science 2003;302:605–608.
3. Chapman JD, Bradley JD, Eary JF, Haubner R, Larson SM, Michalski JM, et al. Molecular (functional) imaging for radiotherapy applications: an RTOG symposium. Int J Radiat Oncol Biol Phys 2003;55:294–301.
4. Rajendran J, Muzi M, Peterson LM, Diaz AZ, Spence AM, Schwartz DS, Krohn KA. Analyzing the results of [F-18] FMISO PET hypoxia imaging: what is the best way to quantify hypoxia? J Nucl Med 2002;43:102P.
5. Peters L, McKay M. Predictive assays: will they ever have a role in the clinic? Int J Radiat Oncol Biol Phys 2001;49:501–504.
6. Rowland DJ, Lewis JS, Welch MJ. Molecular imaging: the application of small animal positron emission tomography. J Cell Biochem Suppl 2002;39:110–115.
7. Maclean D, Northrop JP, Padgett HC, Walsh JC. Drugs and probes: the symbiotic relationship between pharmaceutical discovery and imaging science. Mol Imaging Biol 2003;5:304–311.
8. Collier TL, Lecomte R, McCarthy TJ, Meikle S, Ruth TJ, Scopinaro F, et al. Assessment of cancer-associated biomarkers by positron emission tomography: advances and challenges. Dis Markers 2002;18:211–247.
9. Gambhir SS. Molecular imaging of cancer with positron emission tomography. Nat Rev Cancer 2002;2:683–693.
10. Hanahan D, Weinberg RA. The hallmarks of cancer. Cell 2000;100:57–70.
11. Simon LM, Robin ED, Theodore J. Differences in oxygen-dependent regulation of enzymes between tumor and normal cell systems in culture. J Cell Physiol 1981;108:393–400.
12. Bhujwalla ZM, Artemov D, Aboagye E, Ackerstaff E, Gillies RJ, Natarajan K, Solaiyappan M. The physiological environment in cancer vascularization, invasion and metastasis. Novartis Found Symp 2001;240:23–38; discussion 38–45, 152–153.
13. Thomlinson RH, Gray LH. The histological structure of some human lung cancers and the possible implications for radiotherapy. Br J Cancer 1955;9:537–549.
14. Folkman J. Tumor angiogenesis. Adv Cancer Res 1974;19:331–358.
15. Kourkourakis MI, Giatromanolaki A. Cancer vascularization: implications in radiotherapy? Int J Radiat Oncol Biol Phys 2000;48:545s–553s.
16. Rajendran JG, Mankoff DM, O'Sullivan F, Peterson LM, Schwartz DL, Conrad EU, et al. Hypoxia and glucose metabolism in malignant tumors: evaluation by FMISO and FDG PET imaging. Clin Cancer Res 2004;10:2245–2252.
17. Scandurro AB, Weldon CW, Figueroa YG, Alam J, Beckman BS. Gene microarray analysis reveals a novel hypoxia signal transduction pathway in human hepatocellular carcinoma cells. Int J Oncol 2001;19:129–135.
18. Villaret DB, Wang T, Dillon D, Xu J, Sivam D, Cheever MA, Reed SG. Identification of genes overexpressed in head and neck squamous cell carcinoma using a combination of complementary DNA subtraction and microarray analysis. Laryngoscope 2000;110:374–381.
19. Agani F, Semenza GL. Mersalyl is a novel inducer of vascular endothelial growth factor gene expression and hypoxia-inducible factor 1 activity. Mol Pharmacol 1998;54:749–754.
20. Bae MK, Kwon YW, Kim MS, Bae SK, Bae MH, Lee YM, et al. Identification of genes differentially expressed by hypoxia in hepatocellular carcinoma cells. Biochem Biophys Res Commun 1998;243:158–162.
21. Dachs GU, Tozer GM. Hypoxia modulated gene expression: angiogenesis, metastasis and therapeutic exploitation. Eur J Cancer 2000;36:1649–1660.
22. Eisma RJ, Spiro JD, Kreutzer DL. Vascular endothelial growth factor expression in head and neck squamous cell carcinoma. Am J Surg 1997;174:513–517.
23. Hockel M, Schlenger K, Hockel S, Vaupel P. Hypoxic cervical cancers with low apoptotic index are highly aggressive. Cancer Res 1999;59:4525–4528.
24. Guillemin K, Krasnoq MA. The hypoxic response: huffing and HIFing. Cell 1997;89:9–12.
25. Jiang BH, Semenza GL, Bauer C, Marti HH. Hypoxia-inducible factor 1 levels vary exponentially over a physiologically relevant range of O_2 tension. Am J Physiol 1996;271:C1172–C1180.
26. Clavo AC, Wahl RL. Effects of hypoxia on the uptake of tritiated thymidine, L-leucine, L-methionine and FDG in cultured cancer cells. J Nucl Med 1996;37:502–506.
27. Burgman P, Odonoghue JA, Humm JL, Ling CC. Hypoxia-induced increase in FDG uptake in MCF7 cells. J Nucl Med 2001;42:170–175.
28. Hwang DY, Ismail-Beigi F. Glucose uptake and lactate production in cells exposed to CoCl(2) and in cells overexpressing the Glut-1 glucose transporter. Arch Biochem Biophys 2002;399:206–211.
29. Ivan M, Kondo K, Yang H, Kim W, Valiando J, Ohh M, et al. HIFalpha targeted for VHL-mediated destruction by proline hydroxylation: implications for O_2 sensing. Science 2001;292:464–468.
30. Huang LE, Arany Z, Livingston DM, Bunn HF. Activation of hypoxia-inducible transcription factor depends primarily upon redox-sensitive stabilization of its alpha subunit. J Biol Chem 1996;271:32253–32259.
31. Guillemin K, Krasnow MA. The hypoxic response: huffing and HIFing. Cell 1997;89:9–12.

32. Bos R, Zhong H, Hanrahan CF, Mommers EC, Semenza GL, Pinedo HM, et al. Levels of hypoxia-inducible factor-1 alpha during breast carcinogenesis. J Natl Cancer Inst 2001;93:309–314.

33. Zhong H, De Marzo AM, Laughner E, Lim M, Hilton DA, Zagzag D, et al. Overexpression of hypoxia-inducible factor 1-alpha in common human cancers and their metastases. Cancer Res 1999;59:5830–5835.

34. Marxsen JH, Schmitt O, Metzen E, Jelkmann W, Hellwig-Burgel T. Vascular endothelial growth factor gene expression in the human breast cancer cell line MX-1 is controlled by O_2 availability in vitro and in vivo. Ann Anat 2001;183:243–249.

35. Yaziji H, Gown AM. Immunohistochemical analysis of gynecologic tumors. Int J Gynecol Pathol 2001;20:64–78.

36. Folkman J. Angiogenesis and its inhibitors. Important Adv Oncol 1985;1985:42-62.

37. Boehm-Viswanathan T. Is angiogenesis inhibition the Holy Grail of cancer therapy? Curr Opin Oncol 2000;12:89–94.

38. Folkman J. Tumor angiogenesis: therapeutic implications. N Engl J Med 1971:285:1182–1186.

39. Gullino PM. Angiogenesis and neoplasia. N Engl J Med 1982;305:884–885.

40. Costouros NG, Diehn FE, Libutti SK. Molecular imaging of tumor angiogenesis. J Cell Biochem Suppl 2002;39:72–78.

41. Blankenberg FG, Eckelman WC, Strauss HW, Welch MJ, Alavi A, Anderson C, et al. Role of radionuclide imaging in trials of antiangiogenic therapy. Acad Radiol 2000;7:851–867.

42. Haubner R, Wester HJ, Weber WA, Mang C, Ziegler SI, Goodman SL, et al. Noninvasive imaging of alpha(v)beta3 integrin expression using 18F-labeled RGD-containing glycopeptide and positron emission tomography. Cancer Res 2001;61:1781–1785.

43. Leong-Poi H, Christiansen JP, Klibanov AL, Kaul S, Lindner JR. Noninvasive assessment of angiogenesis by contrast ultrasound imaging with microbubbles targeted to alpha-V integrins. J Am Coll Cardiol 2003;41:430–431.

44. Brasch R, Pham C, Shames D, Roberts T, van Dijke K, van Bruggen N, et al. Assessing tumor angiogenesis using macromolecular MR imaging contrast media. J Magn Reson Imaging 1997;7:68–74.

45. Hall EJ. Radiobiology for the Radiologist. Philadelphia: Lippincott Williams & Wilkins, 2000.

46. Marples B, Greco O, Joiner MC, Scott SD. Molecular approaches to chemo-radiotherapy. Eur J Cancer 2002;38:231–239.

47. Overgaard J, Horsman MR. Modification of hypoxia-induced radioresistance in tumors by the use of oxygen and sensitizers. Semin Radiat Oncol 1996;6:10–21.

48. Fowler JF. Eighth annual Juan del Regato lecture. Chemical modifiers of radiosensitivity–theory and reality: a review. Int J Radiat Oncol Biol Phys 1985;11:665–674.

49. Frommhold H, Guttenberger R, Henke M. The impact of blood hemoglobin content on the outcome of radiotherapy. The Freiburg experience. Strahlenther Onkol 1998;174(suppl 4):31–34.

50. Evans SM, Koch CJ. Prognostic significance of tumor oxygenation in humans. Cancer Lett 2003;195:1–16.

51. Koong AC, Denko NC, Hudson KM, Schindler C, Swiersz L, Koch C, et al. Candidate genes for the hypoxic tumor phenotype. Cancer Res 2000;60:883–887.

52. Blancher C, Moore JW, Talks KL, Houlbrook S, Harris A.L. Relationship of hypoxia-inducible factor (HIF)-1-alpha and HIF-2-alpha expression to vascular endothelial growth factor induction and hypoxia survival in human breast cancer cell lines. Cancer Res 2000;60:7106–7113.

53. Sutherland RM. Tumor hypoxia and gene expression: implications for malignant progression and therapy. Acta Oncol 1998;37:567–574.

54. Amellem O, Pettersen EO. Cell inactivation and cell cycle inhibition as induced by extreme hypoxia: the possible role of cell cycle arrest as a protection against hypoxia-induced lethal damage. Cell Prolif 1991;24:127–141.

55. Moulder JE, Rockwell S. Tumor hypoxia: its impact on cancer therapy. Cancer Metastasis Rev 1987;5:313–341.

56. Brown MJ. The hypoxic cell: a target for selective cancer therapy. Eighteenth Bruce F. Cain memorial award lecture. Cancer Res 1999;59:5863–5870.

57. Hockel M, Schlenger K, Knoop C, Vaupel P. Oxygenation of carcinoma of the uterine cervix: evaluation by computerized oxygen tension measurements. Cancer Res 1991;51:6098–6102.

58. Koh WJ, Bergman KS, Rasey JS, Peterson LM, Evans ML, et al. Evaluation of oxygenation status during fractionated radiotherapy in human non-small cell lung cancers using [F-18]fluoromisonidazole positron emission tomography. Int J Radiat Oncol Biol Phys 1995;33:391–398.

59. Rajendran JG, Wilson D, Conrad EU, Peterson LM, Bruckner JD, Rasey JS, et al. F-18 FMISO and F-18 FDG PET imaging in soft tissue sarcomas: correlation of hypoxia, metabolism and VEGF expression. Eur J Nucl Med 2003;30:695–704.

60. Rajendran JG, Ng P, Peterson LM, Schwartz DL, Scharnhrost J, Conrad EU, et al. F-18 FMISO PET tumor hypoxia imaging: investigating the tumor volume–hypoxia connection. J Nucl Med 2003;44:1340, 1376P.

61. Adam M, Gabalski EC, Bloch DA, Ochlert JW, Brown JM, Elsaid AA, et al. Tissue oxygen distribution in head and neck cancer patients. Head Neck 1999;21:146–153.

62. Blancher C, Harris AL. The molecular basis of the hypoxia response pathway: tumour hypoxia as a therapy target. Cancer Metastasis Rev 1998;17:187–194.

63. Brown JM. Exploiting the hypoxic cancer cell: mechanisms and therapeutic strategies. Mol Med Today 2000;6:157–162.

64. Rajendran JG, Meyer J, Schwartz DL, Kinahan PE, Cheng P, Hummel SM, et al. Imaging with F-18 FMISO-PET permits hypoxia directed radiotherapy dose escalation for head and neck cancer. J Nucl Med 2003;44:415, 127P.

65. Chao KS, Bosch WR, Mutic S, Lewis JS, Dehdashti F, Mintun MA, et al. A novel approach to overcome hypoxic tumor resistance: Cu-ATSM-guided intensity-modulated radiation therapy. Int J Radiat Oncol Biol Phys 2001;49:1171–1182.

66. Lee DJ, Moini M, Giuliano J, Westra WH. Hypoxic sensitizer and cytotoxin for head and neck cancer. Ann Acad Med Singap 1996;25:397–404.

67. Sartorelli AC, Hodnick WF. Mitomycin C: a prototype bioreductive agent. Oncol Res 1994;6:501–508.

68. Denny WA, Wilson WR. Tirapazamine: a bioreductive anticancer drug that exploits tumour hypoxia. Expert Opin Invest Drugs 2000;9:2889–2901.

69. von Pawel J, von Roemeling R, Gatzemeie, U, Boyer M, Elisson LO, Clark P, et al. Tirapazamine plus cisplatin versus cisplatin in advanced non-small-cell lung cancer: a report of the international CATAPULT I study group. Cisplatin and tirapazamine in subjects with advanced previously untreated non-small-cell lung tumors. J Clin Oncol 2000;18:1351–1359.

70. Rischin D, Peters L, Hicks R, Hughes P, Fisher R, Hart R, et al. Phase I trial of concurrent tirapazamine, cisplatin, and radiotherapy in patients with advanced head and neck cancer. J Clin Oncol 2001;19:535–542.

71. Vordermark D, Brown JM. Endogenous markers of tumor hypoxia predictors of clinical radiation resistance? Strahlenther Onkol 2003;179:801–811.

72. Hockel M, Schlenger K, Knoop C, Vaupel P. Oxygenation of carcinomas of the uterine cervix: evaluation by computerized O_2 tension measurements. Cancer Res 1991;51:6098–6102.

73. Brizel DM, Sibley GS, Prosnitz LR, Scher RL, Dewhirst MW. Tumor hypoxia adversely affects the prognosis of carcinoma of the head and neck. Int J Radiat Oncol Biol Phys 1997;38:285–289.

74. Lartigau E, Lusinchi A, Weeger P, Wibault P, Luboinski B, Eschwege F, Guichard M. Variations in tumour oxygen tension (pO_2) during accelerated radiotherapy of head and neck carcinoma. Eur J Cancer 1998;34:856–861.

75. Ng P, Rajendran JG, Peterson LM, Schwartz DL, Scharnhrost J, Krohn KA. Can F-18 fluoromisonidazole PET imaging predict

treatment response in head and neck cancer? J Nucl Med 2003;44:416, 128P.

76. Muzi M, Spence AM, Rajendran JG, Grierson JR, Krohn KA. Glioma patients assessed with FMISO and FDG: two tracers provide different information. J Nucl Med 2002;44:878, 243P.

77. Valk P, Mathis C, Prados M, Gilbert J, Budinger T. Hypoxia in human gliomas: demonstration by PET with fluorine-18-fluoromisonidazole. J Nucl Med 1992;33:2133–2137.

78. Rajendran J, Lanell P, Schwartz DS, Muzi M, Scharnhorst JD, Eary JF, Krohn KA. [F-18] FMISO PET hypoxia imaging in head and neck cancer: heterogeneity in hypoxia. Primary tumor vs. lymph nodal metastases. J Nucl Med 2002;43:73P.

79. Rasey JS, Casciari JJ, Hofstrand PD, Muzi M, Graham MM, Chin LK. Determining hypoxic fraction in a rat glioma by uptake of radiolabeled fluoromisonidazole. Radiat Res 2000;153:84–92.

80. Stone HB, Brown JM, Phillips TL, Sutherland RM. Oxygen in human tumors: correlations between methods of measurement and response to therapy. Summary of a workshop held November 19–20, 1992, at the National Cancer Institute, Bethesda, Maryland. Radiat Res 1993;136:422–434.

81. Hockel M, Vaupel P. Tumor hypoxia: definitions and current clinical, biologic, and molecular aspects. J Natl Cancer Inst 2001;93:266–276.

82. Vaupel P, Kelleher DK, Hockel M. Oxygen status of malignant tumors: pathogenesis of hypoxia and significance for tumor therapy. Semin Oncol 2001;28:29–35.

83. Nozue M, Lee I, Yuan F, et al. Interlaboratory variation in oxygen tension measurement by Eppendorf "Histograph" and comparison with hypoxic marker. J Surg Oncol 1997;66:30–38.

84. Lartigau E, Le Ridant AM, Lambin P, Weeger P, Martin L, Sigal R, et al. Oxygenation of head and neck tumors. Cancer (Phila) 1993;71:2319–2325.

85. Brizel DM, Rosner GL, Harrelson J, Prosnitz LR, Dewhirst MW. Pretreatment oxygenation profiles of human soft tissue sarcomas. Int J Radiat Oncol Biol Phys 1994;30:635–642.

86. Martin GV, Caldwell JH, Graham MM, Grierson JR, Kroll K, Cowan MJ, et al. Noninvasive detection of hypoxic myocardium using fluorine-18 fluoromisonidazole and positron emission tomography. J Nucl Med 1992;33:2202–2208.

87. Rajendran JG, Krohn KA. Imaging hypoxia and angiogenesis in tumors. Radiol Clin N Am 2005;43:169–187.

88. Prekeges JL, Rasey JS, Grunbaum Z, Krohn KH. Reduction of fluoromisonidazole, a new imaging agent for hypoxia. Biochem Pharmacol 1991;42:2387–2395.

89. Chapman JD, Engelhardt EL, Stobbe CC, Schneider RF, Hanks GE. Measuring hypoxia and predicting tumor radioresistance with nuclear medicine assays. Radiother Oncol 1998;46:229–237.

90. Grierson JR, Link JM, Mathis CA, Rasey JS, Krohn KA. Radiosynthesis of of fluorine-18 fluoromisonidazole. J Nucl Med 1989;30:343–350.

91. Bentzen L, Keiding S, Horsman MR, Falborg L, Hansen SB, Overgaard J. Feasibility of detecting hypoxia in experimental mouse tumours with 18F-fluorinated tracers and positron emission tomography: a study evaluating [18F]fluoro-2-deoxy-D-glucose. Acta Oncol 2000;39:629–637.

92. Yeh SH, Liu RS, Wu LC, Yang DJ, Yen SH, Chang CW, et al. Fluorine-18 fluoromisonidazole tumour to muscle retention ratio for the detection of hypoxia in nasopharyngeal carcinoma. Eur J Nucl Med 1996;23:1378–1383.

93. Rasey JS, Koh WJ, Evans ML, Peterson LM, Lewellen TK, Graham MM, Krohn KA. Quantifying regional hypoxia in human tumors with positron emission tomography of [18F]fluoromisonidazole:a pretherapy study of 37 patients. Int J Radiat Oncol Biol Phys 1996;36:417–428.

94. Liu RS, Chu LS, Yen SH, Chang CP, Chou KL, Wu LC, et al. Detection of anaerobic odontogenic infections by fluorine-18 fluoromisonidazole. Eur J Nucl Med 1996;23:1384–1387.

95. Read SJ, Hirano T, Abbott DF, Markus R, Sachinidis JI, Tochon-Danguy HJ, et al. The fate of hypoxic tissue on 18F-fluoromisonidazole positron emission tomography after ischemic stroke. Ann Neurol 2000;48:228–235.

96. Graham MM, Peterson LM, Link JM, Evans ML, Rasey JS, Koh WJ, et al. Fluorine-18-fluoromisonidazole radiation dosimetry in imaging studies. J Nucl Med 1997;38:1631–1636.

97. Rasey JS, Koh WJ, Grierson JR, Grunbaum Z, Krohn KA. Radiolabelled fluoromisonidazole as an imaging agent for tumor hypoxia. Int J Radiat Oncol Biol Phys 1989;17:985–991.

98. Koh WJ, Rasey JS, Evans ML, Grierson JR, Lewellen TK, Graham MM, et al. Imaging of hypoxia in human tumors with [F-18]fluoromisonidazole. Int J Radiat Oncol Biol Phys 1992;22:199–212.

99. Dubois L, Landuyt W, Haustermans K, Dupont P, Bormans G, Vermaelen P, et al. Evaluation of hypoxia in an experimental rat tumour model by [(18)F]fluoromisonidazole PET and immunohistochemistry. Br J Cancer 2004;91:1947–1954.

100. Casciari JJ, Graham MM, Rasey JS. A modeling approach for quantifying tumor hypoxia with [F-18]fluoromisonidazole PET time-activity data. Med Phys 1995;22:1127–1139.

101. Chapman JD, Schneider RF, Urbain JL, Hanks GE. Single-photon emission computed tomography and positron-emission tomography assays for tissue oxygenation. Semin Radiat Oncol 2001;11:47–57.

102. Nunn A, Linder K, Strauss HW. Nitroimidazoles and imaging hypoxia. Eur J Nucl Med 1995;22:265–280.

103. Biskupiak JE, Krohn KA. Second generation hypoxia imaging agents [editorial; comment]. J Nucl Med 1993;34:411–413.

104. Kachur AV, Dolbier WR Jr, Evans SM, Shiue CY, Shiue GG, Skov KA, et al. Synthesis of new hypoxia markers EF1 and [18F]-EF1. Appl Radiat Isot 1999;51:643–650.

105. Tewson TJ. Synthesis of [18F]fluoroetanidazole: a potential new tracer for imaging hypoxia. Nucl Med Biol 1997;24:755–760.

106. Markus R, Reutens DC, Kazui S, Read S, Wright P, Pearce DC, et al. Hypoxic tissue in ischaemic stroke: persistence and clinical consequences of spontaneous survival. Brain 2004;127:1427–1436.

107. Wiebe LI, Stypinski D. Pharmacokinetics of SPECT radiopharmaceuticals for imaging hypoxic tissues. Q J Nucl Med 1996;40:270–284.

108. Siim BG, Laux WT, Rutland MD, Palmer BN, Wilson WR. Scintigraphic imaging of the hypoxia marker (99m)technetium-labeled 2,2'-(1,4-diaminobutane)bis(2-methyl-3-butanone) dioxime (99mTc-labeled HL-91; prognox): noninvasive detection of tumor response to the antivascular agent 5,6-dimethylxanthenone-4-acetic acid. Cancer Res 2000;60:4582–4588.

109. Rumsey WL, Kuczynski B, Patel B, Bauer A, Narra RK, Eaton SM, et al. SPECT imaging of ischemic myocardium using a technetium-99m-nitroimidazole ligand. J Nucl Med 1995;36:1445–1450.

110. Zhang X, Melo T, Ballinger JR, Rauth AM. Studies of 99mTc-BnAO (HL-91): a non-nitroaromatic compound for hypoxic cell detection. Int J Radiat Oncol Biol Phys 1998;42:737–740.

111. Stypinski D, Wiebe LI, McEwan AJ, Schmidt RP, Tam YK, Mercer JR. Clinical pharmacokinetics of 123I-IAZA in healthy volunteers. Nucl Med Commun 1999;20:559–567.

112. Shelton ME, Green MA, Mathias CJ, Welch MJ, Bergmann SR. Assessment of regional myocardial and renal blood flow with copper-PTSM and positron emission tomography. Circulation 1990;82:990–997.

113. Lewis JS, McCarthy DW, McCarthy TJ, Fujibayashi Y, Welch MJ. Evaluation of Cu-64-ATSM in vitro and in vivo in a hypoxic model. J Nucl Med 1999;40:177–183.

114. Dehdashti F, Mintun MA, Lewis JS, Bradley J, Govindan R, Laforest R, et al. In vivo assessment of tumor hypoxia in lung cancer with 60Cu-ATSM. Eur J Nucl Med Mol Imaging 2003;30:844–850.

115. Fujibayashi Y, Taniuchi H, Yonekura Y, Ohtani H, Konishi J, Yokoyama A. Copper-62-ATSM: a new hypoxia imaging agent with high membrane permeability and low redox potential. J Nucl Med 1997;38:1155–1160.

116. Ballinger JR. Imaging hypoxia in tumors. Semin Nucl Med 2001;31:321–329.

117. Gabalski EC, Adam M, Pinto H, Brown JM, Bloch DA, Terris DJ. Pretreatment and midtreatment measurement of oxygen tension levels in head and neck cancers. Laryngoscope 1998;108:1856–1860.

118. Klabbers BM, Lammertsma AA, Slotman BJ. The value of positron emission tomography for monitoring response to radiotherapy in head and neck cancer. Mol Imaging Biol 2003;5:257–270.

119. Alber M, Paulsen F, Eschmann SM, Machulla HJ. On biologically conformal boost dose optimization. Phys Med Biol 2003;48:N31–N35.

120. Buatti J, Yao M, Dornfeld K, Skwarchuk M, Hoffman HT, Funk GF, et al. Efficacy of IMRT in head and neck cancer as monitored by post-RT PET scans. Int J Radiat Oncol Biol Phys 2003;57:S305.

121. Tome WA, Fowler JF. Selective boosting of tumor subvolumes. Int J Radiat Oncol Biol Phys 2000;48:593–599.

122. Ling CC, Humm J, Larson S, Amols H, Fuks Z, Leibel S, Koutcher JA. Towards multidimensional radiotherapy (MD-CRT): biological imaging and biological conformality. Int J Radiat Oncol Biol Phys 2000;47:551–560.

123. Peters LJ. Targeting hypoxia in head and neck cancer. Acta Oncol 2001;40:937–940.

124. Solomon B, McArthur G, Cullinane C, Zalcberg J, Hicks R. Applications of positron emission tomography in the development of molecular targeted cancer therapeutics. BioDrugs 2003;17:339–354.

125. Klimas MT. Positron emission tomography and drug discovery: contributions to the understanding of pharmacokinetics, mechanism of action and disease state characterization. Mol Imaging Biol 2003;4:311–337.

126. Hammond LA, Denis L, Salman U, Jerabek P, Thomas CR Jr, Kuhn JG. Positron emission tomography (PET): expanding the horizons of oncology drug development. Invest New Drugs 2003;21:309–340.

24

Labeled Pyrimidines in PET Imaging

Anthony F. Shields

Cellular proliferation is a requirement of life and is needed as animals develop, replace tissues, and repair damage. On the other hand, uncontrolled proliferation leads to cancer. Because of the intense interest in cell growth, many different aspects of proliferation have been evaluated over the years. One of the basic requirements of proliferating cells is to make a new complement of cellular DNA before division. Because excess cellular growth is a problem and cause in both cancer and some immunologic diseases, DNA synthesis is a common target of therapy. Many drugs and radiotherapy can directly alter DNA and interfere with cell replication, resulting in immediate or delayed (programmed) cell death. It is generally thought that impairment of DNA synthesis by therapy better reflects cell injury than do changes seen in glycolysis, which relies on more robust organelles such as mitochondria (1). Therefore, nucleoside utilization may be a more sensitive and specific measure of early response to therapy because cells may maintain their oxidative and glycolytic function for some time after therapy. In contrast to other positron emission tomography (PET) tracers that measure energetic or biosynthetic functions (for example, glucose and amino acids), nucleosides such as thymidine are precursors of DNA synthesis (2) (Figure 24.1). Imaging pathways of proliferation may be preferable in situations where one needs to rapidly assess physiologic response or noncytotoxic therapy is being employed. Imaging with nucleosides may be useful in diagnosis and staging of some tumor types, particularly in tissues and organs with high physiologic uptake of ^{18}F-fluorodeoxyglucose (FDG).

Thymidine is taken up by cells and used in DNA synthesis or degraded. The pyrimidine basis of thymidine, thymine, is not used in RNA synthesis, unlike adenine, cytidine, and guanine. Thus, thymidine is routinely used of thymidine uptake as a measure of cell growth, and thymidine labeled with ^{14}C and ^{3}H has been used in the laboratory for half a century, producing extensive data for understanding uptake, retention, and degradation of thymidine. After delivery via the bloodstream, thymidine

is transported across the cell membrane, in a variety of cell types, by a number of both active and passive carrier-mediated transporters (3). Transport is rapid in both directions and does not appear to be a rate-limiting step under most circumstances (4). Once in the cell, thymidine can be used for DNA synthesis or degraded. It is important to realize that the major source of thymidine used in DNA synthesis is endogenous synthesis by the cell from uracil. Uracil, in the form of uridine monophosphate, is methylated from folate donors to produce thymidine. The uptake of exogenous thymidine, while accounting for a minority of that used in DNA synthesis, can still be used as a marker of cellular proliferation, based on the fact that the relative utilization of the two pathways depends on the concentration of thymidine outside the cell (5). If this concentration is relatively constant, then the fraction used within the cell is predictable.

Once in the cell, thymidine can be degraded or utilized in DNA synthesis (Figure 24.2). Its utilization is first mediated by its phosphoryation by thymidine kinase. One form of this enzyme, thymidine kinase 1 (TK1), is a cytosolic enzyme and used by the cell as it processes thymidine that is employed in DNA within the nucleus. The cell also possesses a second thymidine kinase (TK2) that is located within the mitochondria and is not involved in cell replication. The regulation of TK1 is closely coordinated with the other enzymes involved in DNA synthesis, and its concentration within the cell increases severalfold as the cells begin to make new DNA. Another important enzyme in the DNA synthetic pathway is thymidylate synthase, the enzyme responsible for the endogenous synthesis of thymidine from uracil monophosphate. Similarly, DNA polymerase is rapidly increased in cells as they enter the DNA synthetic or "S" phase of the cell cycle.

Before the development of PET, a number of parameters dependent on thymidine assays have been defined to help quantitate and describe cell growth (6, 7). Unfortunately, these tests often require repeated specimen examinations and are usually not suitable for routine clinical use. The labeling index (LI) is probably the easiest

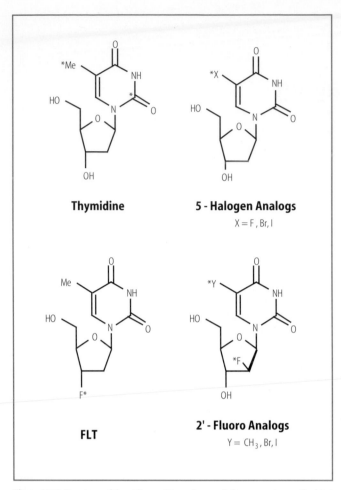

Figure 24.1. Structures of thymidine and its analogues. Halogens can be used to label the 5-position and produce iodo-, bromo-, and fluoro-deoxyuridine. FLT (3′-deoxy-3′-fluorothymidine) is readily labeled with ^{18}F. Analogues of 2′-fluoro compounds can be methyl-, bromo-, or iodo labeled in the 5-position or with ^{18}F in the sugar. *Asterisks* indicate position of possible labels. 📖

thymidine-based assay of cell kinetics to obtain. Before biopsy, labeled thymidine is injected into the animal or patient being studied. The fraction of cells with labeled nuclei is then quantitated with autoradiography. The LI, therefore, is a measure of the fraction of cells in S phase at a given time in a single sample of cells, which may not be representative of the tissue or tumor. The potential advantage of PET is that it could provide a means of noninvasively obtaining repeated measurements from multiple sites in a tumor and, when combined with kinetic modeling, provides an accurate and consistent measure of uptake. Similar approaches using unlabeled 5-bromodeoxyuridine (BUdR) have also been developed where cellular retention is detected using monoclonal antibodies, but the requirement for tissue sampling remains (8).

^{11}C-Thymidine Imaging

^{11}C-Labeled thymidine was considered an attractive tracer to image cellular proliferation in vivo, but a practical syn-

thesis was needed to make this approach feasible. The introduction of ^{11}C with its 20-min half-life into thymidine and the need to synthesize quantities of about 20 mCi for patient studies is a formidable task. ^{11}C-Thymidine was first synthesized with the label in the 5-methyl position by Christman et al., and the method was improved by Sundoro-Wu et al. (9, 10). Subsequently, a synthesis of ^{11}C in the ring 2-position was also described (11). The advantage of labeled thymidine is that it is a native compound that it is readily incorporated into DNA (12). A detailed understanding of its in vivo kinetics is needed to interpret PET images, but the pathways of its uptake, retention, utilization, and degradation are well understood. Preliminary studies in animals with both implanted and spontaneous tumors demonstrated that the retention in tumors was sufficient to make imaging feasible (13). After intravenous injection in vivo, the labeled thymidine is primarily retained in DNA (14), but rapid degradation has also been shown to occur (15, 16).

The initial human studies using 5-methyl-labeled ^{11}C-thymidine were carried out in patients with non-Hodgkin's lymphoma (17). This study first demonstrated that ^{11}C-thymidine uptake was rapid enough to make tumor imaging clinically feasible. The relative retention of thymidine was found to correlate with tumor grade. Patients with low-grade and high-grade lymphoma had a mean uptake of 0.29% and 0.71% dose per 100 mL, respectively. Uptake was much higher in tumors than muscle, with a mean lymphoma-to-muscle ratio of 11.8. In a study of 13 patients with head and neck cancer, the tumor was visualized in each case, although the standard uptake values (SUVs) were relatively modest (mean, 1.3; range, 0.8–2.0) (18). High activity was also noted in the kidney (mean SUV, 13.7), liver (SUV, 18.0), and heart muscle (SUV, 3.2). Retention in the liver is the result of degradation of methyl-labeled ^{11}C-thymidine, and heart retention appears to be mediated by trapping involving mitochondrial TK2. In a study of patients with various brain lesions using methyl-labeled thymidine, the lesion was detected in 8 of 10 patients with cancer. In 10 patients with nonmalignant abnormalities, none had detectable lesion uptake. Although some of the retention of thymidine in the tumors may be attributable to incorporation into DNA, the concentration of metabolites in lesions with a disrupted blood–brain barrier may also contribute to the uptake (19). Methyl-labeled thymidine was also found to be useful in differentiating benign from lymphomatous lesions in the brain in patients with AIDS (20).

Thymidine was labeled for use with ^{11}C in the ring 2-position to take advantage of the alternate degradation found when compared to the methyl-labeled thymidine (Figure 24.3). The degradation of the methyl-labeled thymidine results in the production of numerous labeled metabolites, some of which are retained in the liver, producing a high background in that organ. Ring-2-labeled thymidine undergoes degradation to labeled CO_2. Within 8 eight min after injection, 65% of the blood activity is

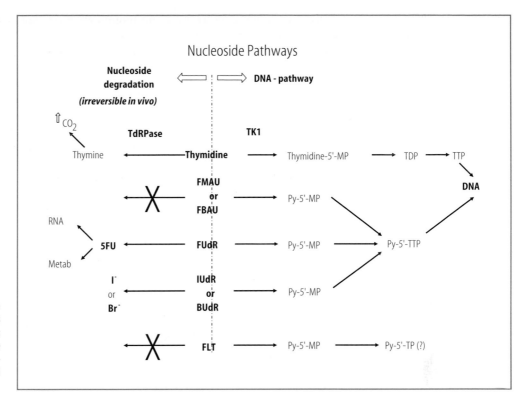

Figure 24.2. Pathways of incorporation and degradation of thymidine and its analogues. *TK1*, thymidine kinase 1; *TdRPase*, thymidine phosphorylase, the initial step in the degradation of thymidine; *Py*, the appropriate phosphorylated pyrimidine. 📖

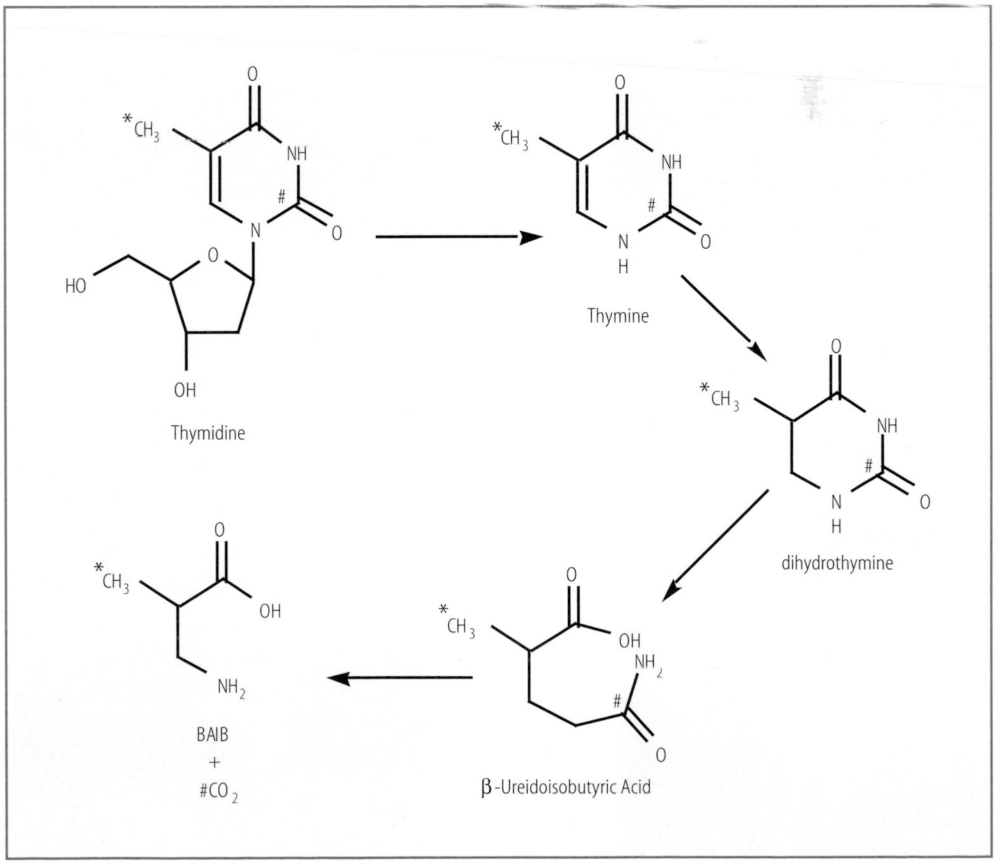

Figure 24.3. Degradation pathway of methyl and ring-2-labeled forms of thymidine. The methyl-labeled form (*) leads to β-amino-isobutyric acid (BAIB) and other metabolites; the ring-2-labeled form (#) generates CO_2. 📖

present as labeled CO_2 (16). Because CO_2 is not concentrated in the liver, one can noninvasively image liver regeneration, as well as tumor within the liver, using 2-[^{11}C]thymidine (21). Although labeled CO_2 is not preferentially retained in any organ, it is widely distributed in the body and is not rapidly exhaled (22). Therefore, kinetic modeling must also take into account the presence of such metabolites.

Patients with brain tumors have been imaged using 2-[^{11}C]thymidine, with results compared to those obtained with ^{18}F-fluorodeoxyglucose (FDG). In a study by Vander Borght and colleagues, 20 patients with intracranial tumors were studied (23). Increased thymidine retention was noted in 11 of 14 and 5 of 6 patients with untreated and recurrent tumors, respectively. The SUV of tumors ranged from 0.68 to 6.94 (mean, 1.7). Normal areas of the cortex have a mean SUV of 0.85. There was no correlation with uptake and tumor grade. High SUV values were seen in the 2 patients with lymphoma (2.29 and 2.62). Although this could be expected, it was unexpected that meningiomas also had very high retention with a mean of 3.6 (range, 2.3–6.9). With thymidine, the tumor-to-normal cortex uptake ratio was 2.0 (range, 0.8–7.2). Because of the high retention of FDG in the normal brain, the tumor-to-cortex ratio was only 0.7 (range, 0.2–3.4) with this tracer. FDG was able to demonstrate the tumor in 11 of 20 cases, compared to 16 of 20 with thymidine. In another study of 13 patients,

brain tumors were visible in all but 1 case with thymidine and all but 2 cases with FDG (24). Even when seen with both tracers, the tumor was better visualized with thymidine in 2 cases and FDG in 1 case, demonstrating the complementary nature of both tracers. Because most of the ring-2 thymidine is degraded to CO_2, imaging with [^{11}C]CO_2 was done independently and these results were used to assess contribution of [^{11}C]CO_2 to the images, to calculate the DNA flux, and understand the contribution of the blood–brain barrier disruption. This approach was used to produce images of thymidine flux, which enhanced the ability to image tumors and demonstrated that DNA synthesis is being measured in the tumor, not metabolites. 2-[^{11}C]Thymidine has also been used in a pilot study to measure response to chemotherapy in patients with small cell lung cancer and sarcoma (25). Rapid declines in thymidine retention were noted within one week of starting therapy (Figure 24.4). Kinetic modeling, allowing for the removal of the metabolite contribution, was more robust for tumor detection than simply measuring changes in SUV. Declines in FDG were also noted, but these were not as dramatic or complete. In patients with locally advanced non-small cell lung cancer receiving chemoradiotherapy, thymidine uptake also appears to decline more rapidly then FDG retention. FDG retention takes about 3 months to decline after successful therapy because of slow loss of both tumor and inflammatory cells (26).

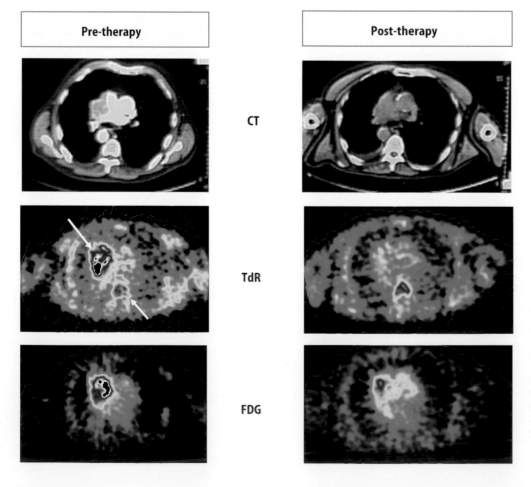

Pre-therapy   **Post-therapy**

CT

TdR

FDG

Figure 24.4. Pretherapy images and those obtained 6 days after treatment. Thymidine (*TdR*) and ^{18}F-2-deoxy-D-glucose (*FDG*) images both demonstrated declines in tumor activity, but the changes were more pronounced with thymidine. *CT*, computed tomography. (Reprinted by permission of the Society of Nuclear Medicine from Shields AF, Mankoff DA, Link JM, et al. [^{11}C]Thymidine and FDG to measure therapy response. J Nucl Med 1998;39:1757–1762, Figure 1.)

5-Halogenated Analogues of Thymidine

The 5-halogenated analogues of thymidine have been of interest from both a therapeutic and an imaging standpoint. Much of the work on alternative labeled nucleosides has focused on halogenated analogues of thymidine, reflecting the potential for imaging with PET using [76]Br, [124]I/^{125}I, and ^{18}F, and the steric match of the halogen atom with the thymidine methyl group. Single photon emission computed tomography (SPECT) imaging is also possible using [77]Br and ^{131}I radionuclides. BUdR was explored initially because the van der Waals radius of bromine is the closest match to that of the methyl group of thymidine. BUdR is problematic for clinical use because of its rapid dehalogenation, resulting in 90% of plasma activity being present as free bromide after 10 min in humans (27). BUdR is readily incorporated into DNA, as shown in cell culture and animal studies; however, one must take into account the distribution of free bromide (28). Attempts have been made to account for the generation of free bromide and to limit its retention using infusions of saline and forced diuresis (29). Despite such efforts, investigators have found that the slow washout of bromide and low specific retention in DNA has made BUdR unsuitable for imaging proliferation in animals and patients (30, 31). It is of note that BUdR is suitable for in vitro, animal, and biopsy studies. For example, a number of studies have been done in patients where unlabeled BUdR was injected intravenously and a biopsy was obtained several hours later. The cells that contained BUdR that had been incorporated into DNA could then be detected using a fluorescent antibody. In such a situation, the free bromine was not a limiting factor in the assay. Unfortunately, noninvasive imaging detects both the intact and degraded BUdR, making it unsuitable for in vivo studies.

5-Idoxurindinum (IUdR), a similarly structured compound incorporated into DNA, has been studied for imaging proliferation in brain tumors using ^{131}I and SPECT or [124]I and PET. Studies in tumor-bearing rats have shown that changes in IUdR retention correlated with change of the S-phase fraction after hormonal treatment of mammary carcinoma (32). Labeled IUdR has limitations similar to those of BUdR because it also undergoes rapid dehalogenation. Once incorporated into DNA, IUdR can still be dehalogenated, but this occurs over a period of days and should not interfere with imaging (33). IUdR has a half-life of only 1.6 min in humans, but the free iodide is cleared with a half-life of about 6.4 h, which is more rapid than the clearance of bromide (34). Waiting 24 h before imaging to allow for iodide washout has proven to be useful in improving the relative retention in the tumors (35) (Figure 24.5). Even after such a delay, a variable fraction of the tumor nonincorporated activity (15%–93%) remained (36). This finding has led to suggestions that increased hydration may improve the quality of such images. The addition of unlabeled IUdR to the labeled agent was also shown to increase retention in murine glioblastoma cell lines studied in vitro and in vivo (37). A similar effect was seen when using 5-fluoro-2'-deoxyuridine (38). The mechanism remains under investigation, but it was not due to a decline in degradation of the labeled IUdR.

Replacement of the methyl group in thymine with fluorine has been studied extensively for both therapeutic and imaging uses. 5-Fluorouracil (5-FU) was the first rationally designed antineoplastic agent. Its synthesis was first reported in 1957 (39), and today it is still extensively used in the treatment of a wide variety of tumors includ-

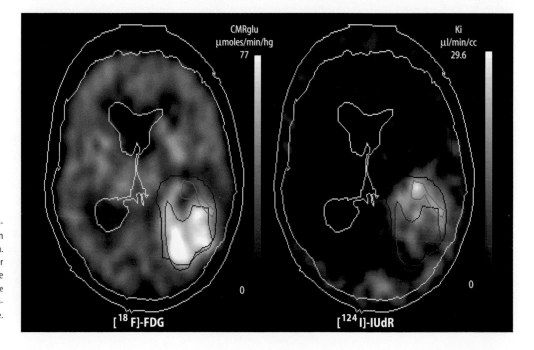

Figure 24.5. FDG (*left*) and [124]I-5-idoxurindinum ([124]I-IUdR) (*right*) images from a patient with a glioblastoma multiform. The IUdR images were obtained 24 h after injection to allow for the clearance of free iodine. (Reprinted from Blasberg RG, Roelcke U, Weinreich R, et al. Imaging brain tumor proliferative activity with [[124]I]iododeoxyuridine. Cancer Res 2000;60(3):624–635.)

ing breast and gastrointestinal malignancies. 5-FU has been labeled with ^{18}F to assist in tracking its retention in tumors and its metabolism (40). It was found that tumor retention was increased in patients who responded to therapy (41). Although labeled 5-FU is very useful in understanding the pharmacokinetics and pharmacodynamics of the drug, it is not useful as an agent for imaging general cellular proliferation. One problem is that 5-FU is almost completely degraded by the liver within several minutes of intravenous injection (42). In addition, it can be incorporated into both RNA and DNA, depending if it is linked to ribose or deoxyribose, respectively.

5-Fluoro-2'-deoxyuridine (FUdR) has also been used as a therapeutic and imaging compound. Therapeutically, it is commonly used in the treatment of tumors within the liver and is administered via intrahepatic infusion. The agent is rapidly cleared by the liver with a first-pass extraction of 69% to 92% (43), which results in a high drug exposure to tumors within the liver but limited systemic exposure and toxicity. One advantage of FUdR over 5-FU is that it is one step closer to incorporation into DNA than 5-FU. A number of studies have explored the use of FUdR as an imaging agent and compared it to other tracers. An advantage of FUdR over the iodine- or bromine-labeled counterparts is more-limited dehalogenation, as the primary path of degradation is to alpha-fluoro-beta-alanine (44). One of the initial studies was done by Abe and colleagues in tumor-bearing rats (45). They demonstrated that, at 120 min after injection, the tumor-to-blood activity ratio was approximately 18 for FUdR and 6 for 5-FU and 5-fluoro-2'-uridine (FUR). In rabbits, they were able to produce images of an implanted tumor. Because of metabolism and excretion, ^{18}F activity present in the liver and kidneys obscures imaging in these organs. A recent study in mice with pancreatic cancer found that the tumor could be well imaged (46). Using flow cytometry, the investigators compared the proportion of cells in S phase with the uptake of FUdR and found a strong correlation ($r = 0.82$, P less than 0.0001). A trial of ^{18}F-FUdR in 22 patients with gliomas found increased retention in those with high-grade compared with low-grade tumors (mean SUV, 0.64 and 0.21, respectively) (47). The relative merits of FUdR imaging compared to some of the other halogentated tracers should be studied.

A comparison of ^{3}H-thymidine to ^{18}F labeled FUdR, FUR, and 5-FU was conducted in tumor-bearing mice and rats (48). Of the fluorinated pyrimidines, FUdR showed the highest incorporation into DNA whereas FUR has the lowest. Although the distribution of FUdR and thymidine was slightly different in tumors, both appeared to provide similar tumor detection capacity. The differences that were seen between FUdR and thymidine were attributed in part to possible uptake of FUdR into organs with high RNA, as well as to DNA synthesis. The pathway for such incorporation involve removal of the deoxyribose sugar from FUdR with subsequent replacement by ribose before use in RNA synthesis (49).

Fluorinated Deoxyribose Analogues of Thymidine

The analogues of thymidine with substitutions on the ribose sugar have been extensively studied because of possible antiviral and antineoplastic properties. 3'-Azidothymidine (AZT) was originally synthesized as a possible antineoplastic agent but has subsequently found extensive use in the treatment of human immunodeficiency virus (HIV). AZT acts as a chain terminator during the synthesis of DNA. The open 5'-hydoxyl on the sugar allows covalent linkage to the growing DNA chain, but the presence of the azido group on the 3'-position prevents further elongation. Replacement of the azido group by a fluorine results in 3'-deoxy-3'-fluorothymidine (FLT), which was also made as a possible antineoplastic compound, but subsequently tested against HIV (50). One distinct advantage of the 3'-fluorine is that it also stabilizes the glycosidic linkage, inhibits degradation of the compound, and allows for the production of an ^{18}F-labeled imaging agent (51). Investigators have also placed the fluorine in the 2'-position, as has been done with 1-(2'-deoxy-2'-fluoro-β-D-arabinofuranosyluracil)thymine (FMAU) (52, 53). Although FLT and FMAU are no longer being used clinically because of hepatotoxicity and neurotoxicity, respectively, at therapeutic doses (54), these and other similar analogues are being explored for their potential as imaging agents for cellular proliferation.

FLT, similar to thymidine, is readily taken up by cells and transported across the cell membrane, which involves a carrier-mediated transport mechanism (55), and then undergoes phosphorylation by thymidine kinase. FLT is a preferential substrate for cytosolic TK1 rather than the mitochondrial isozyme (TK2) (56), which is important as TK2 is maintained at low basal levels in all cells whereas TK1 is increased up to 10 fold in proliferating cells compared to those in a resting state (57). TK1 can readily monophosphorylate FLT at a rate 25% to 50% of the rate of natural thymidine (58). FLT can undergo further phosphorylation to di- and triphosphates, but at a rate that is about 25 fold lower than thymidine (59). FLT can be incorporated into DNA, where it terminates the growing strand of the DNA, but less than 0.1% of intracellular FLT is incorporated into DNA (55). The uptake and retention of FLT, therefore, are dependent on phosphorylation by thymidine kinase, but not on incorporation into DNA. Phosphorylated FLT is cleared from the cells with a phase 1 half-life of about 1.2 h (55). In studies in dogs and humans, uptake at 60 min is high in proliferating organs and tumors, with little evidence of tracer loss from the cell over that time period (60, 61). Although little FLT is incorporated into DNA, the uptake of FLT reflects the distribution of thymidine kinase (62) and this mirrors cell proliferation when studied in cell lines, lung, and colon cancer (63–66). In some tumor cell lines, however, TK does not appear to rapidly increase as the cell enters S phase (67); FLT retention may not well image tumors that seem to rely more on endogenous thymidine synthesis and, therefore, do not readily incorporate FLT through the exogenous, TK-dependent pathway. The conclusions of these studies indicate that FLT deserve to be further investigated.

In studies in rats, dogs, monkeys, and humans, little degradation of FLT was found to occur (55, 60, 61, 68), so that labeled catabolites were not a problem after administration of ^{18}F-FLT. The only other expected labeled compound in the blood is the polar glucuronide, as found in studies of AZT, which will not enter tissues and is cleared by the kidneys (69). In monkeys, about 7% to 31% of the FLT is excreted as the glucuronide, depending on the species studied (55, 70). In humans, about 15% to 35% of FLT is present as the glucuronide 60 min after injection (71, 72). For kinetic analysis, this must be taken into account when determining the flux of FLT (72, 73).

FLT produces images of high contrast with many types of tumors when used in patients imaging (74–76). Its use is limited for imaging the liver because of retention of FLT during glucuronidation and is limited in the kidney and bladder by excretion of FLT. The marrow takes up FLT because of its normal proliferative activity. On the other hand, retention of FLT in brain tumors allows for easy imaging of high-grade tumors because little FLT is transported across the intact blood–brain barrier, producing very little normal cerebral uptake. In the chest, there is little uptake in normal organs, aside from the bone marrow (Figure 24.6). Although thymidine is readily re-

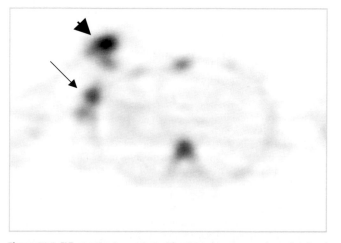

Figure 24.6. FLT retention in a patient with primary breast cancer (*arrowhead*) and lymph node metastases (*arrow*). FLT retention in normal vertebral and sternal marrow is also visible. 📖

tained in the heart, due to the presence of TK2, the specificity of FLT for TK1 means that the heart has little uptake. FLT imaging, therefore, allows for the detection of lung, esophageal, and breast cancers.

Studies have started to compare FLT and FDG for the detection of tumors, including a study of 26 patients with lung lesions (77). Of the 18 patients with malignant lesions, FDG only missed 1 (carcinoma in situ), whereas FLT missed 2 additional lesions. On the other hand, all 8 benign lesions were normal with FLT-PET, but 4 were false positive with FDG. Another study comparing FLT and FDG in pulmonary lesions found that FLT had a sensitivity and specificity of 86% and 100%, respectively, for detecting cancer compared to 100% and 73%, respectively, for FDG (78). A study comparing both tracers in patients with colorectal cancer found that both had a similar sensitivity for detection disease outside the liver, although the degree of FLT retention was less than that of FDG. Within the liver, FLT missed many lesions because of the physiologic retention of FLT within normal liver. FLT may be more specific than FDG for detection of malignancy because of less retention in inflammatory cells. A study in mice with a sterile inflammatory lesion caused by turpentine demonstrated inflamed muscle-to-normal muscle uptake ratio of 4.8 with FDG and 1.3 with FLT (79). However, a study comparing FLT and FDG imaging in patients with head and neck cancer demonstrated that both gave comparable results for detection of malignancies but also inflammatory lesions (80). In another study using FLT-PET to evaluate patients with metastatic melanomas, 2 patients were FLT positive in the groin, but no cancer was found. (81).

Although FLT may have a greater specificity than FDG, owing to its retention on proliferating tissues, its greatest value may be in monitoring response to treatment. The two tracers were compared in a study of mice bearing fibrosarcoma treated with a single injection of 5-FU (82). The animals were imaged in an animal PET device 24 and 28 h after treatment. The FLT and FDG retention declined by 72.9% and 48.2%, respectively.

FMAU also had potential advantages for imaging cell proliferation. Similar to FLT, the placement of fluorine in the sugar, now in the 2'-position, stabilizes the molecule but still allows for chain elongation during DNA synthesis. Labeled and unlabeled FMAU is primarily excreted unchanged in the urine in mice (70%–86%) and in rats and dogs (more than 90%) (53, 83, 84). Studies in mice with ^{14}C-FMAU found tumor-to-blood ratios of 4.2 and 7.7 at 1 and 2 h, respectively. As was seen with FLT, uptake varies in different types of tumors, possibly reflecting the variable reliance on the exogenous pathway. Studies have shown that more than 90% of both labeled thymidine and FMAU were incorporated into DNA of prostate cancer cell lines (85). FMAU has been labeled with ^{11}C in the 5-methyl position for use in PET imaging

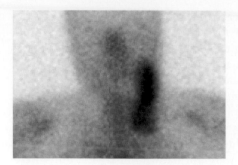

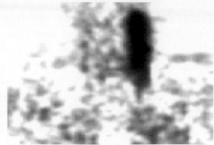

Figure 24-7. ^{11}C-1-(2'-Deoxy-2'-fluoro-β-D-arabinofuranosyluracil)thymine (FMAU) (*right*), FDG (*center*), and gallium (*left*) images of in a patient with non-Hodgkin's lymphoma in the neck. 📖

(86) (Figure 24.7), and recently, FMAU has also been produced and labeled with ^{18}F on the sugar (87, 88). Pilot imaging studies have been done with both labeled tracers in patients with cancer. Marrow and tumors have been found to concentrate the tracer, permitting PET imaging. Similar to the results with FLT, increased retention is seen in the liver, kidney, and bladder. Because FMAU is also phosphorylated by TK2, it is retained in the heart as well. The advantage of FMAU is that it is incorporated into DNA, in contrast to FLT, which does not proceed beyond phosphorylation. The relative merits of both FMAU and FLT need to be further explored.

Another potential imaging tracer is 1-(2'-deoxy-2'-fluoro-β-D-arabinofuranosyluracil) bromouracil (FBAU) (89, 90). Given the similar sizes of the methyl and bromo moieties, it behaves very much like FMAU. Furthermore, it can be labeled with ^{76}Br with a 16-h half-life for use with PET. FBAU can also be labeled with ^{18}F with only slight modification of the synthesis used for FMAU (88). Studies in rats have shown that it is retained in rapidly proliferating organs such as the spleen and small intestine. Extraction of the incorporated radioactivity demonstrated that about 90% is incorporated into DNA and that this was blocked by the DNA synthesis inhibitor hydroxyurea. As with FMAU, there was little catabolism of the compound, and more than 95% of the urine activity present was intact FBAU. In summary, this also may prove to be a useful imaging agent, and its merits continue to be explored.

Pyrimidines, and 1-(2'-deoxy-2'-fluoro-β-D-arabinofuranosyluracil) iodouracil (FIAU) in particular, also allow imaging of the success and expression of gene therapy. Such tracers use compounds that are preferentially phosphorylated by thymidine kinase of herpes simplex virus (HSVtk) reporter gene, rather than the mammalian TK. ^{124}I-Labeled FIAU has been used in early studies (91) and ^{18}F-FIAU more recently (88). FMAU has also been shown to be useful in imaging gene expression and compared to guanine derivatives, the other major class of agents used to image transgenes (92). The limitation of FMAU is that proliferating cells that do not contain the transfected genes also retain the tracer. Although further study is needed comparing FIAU and the acycloguanosine agents, one study in cell lines and tumor-bearing mice found that FIAU produced higher accumulation (93).

Summary

Imaging with thymidine and a number of analogues has been demonstrated in a number of laboratory and clinical settings. Although thymidine can be used to accurately quantitate DNA synthesis, the short physical and biologic half-life of the compound limits its routine use for clinical evaluation. Newer analogues, such as FLT, FMAU, and FBAU, offer metabolically simpler approaches. Each radiopharmaceutical has its strengths and weaknesses. FLT is simplest to synthesize but measures the level of cellular TK1, the first step in the DNA synthetic pathway; it is not incorporated into DNA. FMAU and FBAU are incorporated into DNA but are more difficult to label with ^{18}F. Alternatively, they can be labeled with ^{11}C or ^{77}Br, respectively. Clinical studies with thymidine and the analogues are still in the early stages. Use in the detection of brain tumors appears particularly promising given the low background of these tracers in the normal brain. The relative merits of pyrimidines and FDG for use in other organs and in measuring response require further studies.

References

1. Kubota K, Ishiwata K, Kubota R, et al. Tracer feasibility for monitoring tumor radiotherapy: a quadruple tracer study with fluorine-18-fluorodeoxyglucose or fluorine-18-fluorodeoxyuridine, L-[methyl-^{14}C]methionine, [6-^{3}H]thymidine, and gallium-67. J Nucl Med 1991;32:2118–2123.

2. Cleaver JE. Thymidine metabolism and cell kinetics. Front Biol 1967;6:43–100.

3. Cass CE, Young JD, Baldwin SA. Recent advances in the molecular biology of nucleoside transporters of mammalian cells. Biochem Cell Biol 1998;76(5):761–770.

4. Marz R, Wohlhueter RM, Plagemann PG. Growth rate of cultured Novikoff rat hepatoma cells as a function of the rate of thymidine and hypoxanthine transport. J Membr Biol 1977;34(2-3):277–288.

5. Shields AF, Coonrod DV, Quackenbush RC, Crowley JJ. Cellular sources of thymidine nucleotides: studies for PET. J Nucl Med 1987;28:1435–1440.

6. Tannock I. Cell kinetics and chemotherapy: a critical review. Cancer Treat Rep 1978;62(8):1117–1133.

7. Tubiana M. Tumor cell proliferation kinetics and tumor growth rate. Acta Oncol 1989;28(1):113–121.

8. Wilson GD. Assessment of human tumour proliferation using bromodeoxyuridine– current status. Acta Oncol 1991;30(8):903–910.

9. Christman D, Crawford EJ, Friedkin M, Wolf AP. Detection of DNA synthesis in intact organisms with positron-emitting [*methyl*-^{11}C]thymidine. Proc Natl Acad Sci USA 1972;69(4).

10. Sundoro-Wu BM, Schmall B, Conti PS, Dahl JR, Drumm P, Jacobsen JK. Selective alkylation of pyrimidyl-dianions: synthesis and purification of ^{11}C labeled thymidine for tumor visualization using positron emission tomography. Int J Applied Radiat Isot 1984;35:705–708.

11. Vander Borght T, Labar D, Pauwels S, Lambotte L. Production of [2-^{11}C]thymidine for quantification of cellular proliferation with PET. Appl Radiat Isotopes 1991;42:103–104.

12. Wells P, West C, Jones T, Harris A, Price P. Measuring tumor pharmacodynamic response using PET proliferation probes: the case for 2-[(11)C]-thymidine. Biochim Biophys Acta 2004;1705(2):91–102.

13. Larson SM, Weiden PL, Grunbaum Z, et al. Positron imaging feasibility studies: I. Characteristics of ^{3}H-thymidine uptake in rodent and canine neoplasms. J Nucl Med 1981;22:869–874.

14. Shields AF, Lim K, Grierson J, Link J, Krohn KA. Utilization of labeled thymidine in DNA synthesis: studies for PET. J Nucl Med 1990;31(3):337–342.

15. Poupeye EM, Goethals PP, Dams RF, De Leenheer AP, van Eijkeren ME. Evaluation of ^{11}C]thymidine for measurement of cell proliferation in fast dividing tissues. Nucl Med Biol 1993;20(3):359–362.

16. Shields AF, Mankoff D, Graham MM, et al. Analysis of [2-^{11}C]thymidine blood metabolites for imaging with PET. J Nucl Med 1996b;37:290–296.

17. Martiat P, Ferrant A, Labar D, et al. In vivo measurement of carbon-11 thymidine uptake in non-Hodgkin's lymphoma using positron emission tomography. J Nucl Med 1988;29:1633–1637.

18. van Eijkeren ME, De Schryver A, Goethals P, et al. Measurement of short-term ^{11}C-thymidine activity in human head and neck tumours using positron emission tomography (PET). Acta Oncol 1992;31(5):539–543.

19. De Reuck J, Santens P, Goethals P, et al. [Methyl-^{11}C]thymidine positron emission tomography in tumoral and non- tumoral cerebral lesions. Acta Neurol Belg 1999;99(2):118–125.

20. Pomper MG, Constantinides CD, Barker PB, et al. Quantitative MR spectroscopic imaging of brain lesions in patients with AIDS: correlation with ^{11}C-methyl]thymidine PET and thallium-201 SPECT. Acad Radiol 2002;9(4):398–409.

21. Vander Borght T, Lambotte L, Pauwels S, Labar D, Beckers C, Dive C. Noninvasive measurement of liver regeneration with positron emission tomography and [2-^{11}C]thymidine. Gastroenterology 1991;101(3):794–799.

22. Shields AF, Graham MM, Kozawa SM, et al. Contribution of labeled carbon dioxide to PET imaging of carbon-11-labeled compounds. J Nucl Med 1992;33(4):581–584.

23. Vander Borght T, Pauwels S, Lambotte L, et al. Brain tumor imaging with PET and 2-[carbon-11]thymidine. J Nucl Med 1994;35(6):974–982.

24. Eary JF, Mankoff DA, Spence AM, et al. 2-[C-11]thymidine imaging of malignant brain tumors. Cancer Res 1999;59(3):615–621.

25. Shields AF, Mankoff DA, Link JM, et al. [11]C]Thymidine and FDG to measure therapy response. J Nucl Med 1998;39:1757–1762.

26. Mankoff DA, Shields A, Krohn KA. PET imaging of cellular proliferation. Radiol Clin N Am 2005;43:153–167.

27. Kriss JP, Maruyama Y, Tung LA, Bond SB, Revesz L. The fate of 5-bromodeoxycytodine and 5-iododeoxyuridine in man. Cancer Res 1963;23:260–273.

28. Bergstrom M, Lu L, Fasth KJ, et al. In vitro and animal validation of bromine-76-bromodeoxyuridine as a proliferation marker. J Nucl Med 1998;39(7):1273–1279.

29. Lu L, Bergstrom M, Fasth KJ, Wu F, Eriksson B, Langstrom B. Elimination of nonspecific radioactivity from [76]Br]bromide in PET study with [76]Br]bromodeoxyuridine. Nucl Med Biol 1999;26(7):795–802.

30. Gudjonsson O, Bergstrom M, Kristjansson S, et al. Analysis of [76]Br-BrdU in DNA of brain tumors after a PET study does not support its use as a proliferation marker. Nucl Med Biol 2001;28(1):59–65.

31. Gardelle O, Roelcke U, Vontobel P, et al. [76]Br]Bromodeoxyuridine PET in tumor-bearing animals. Nucl Med Biol 2001;28(1):51–57.

32. Carnochan P, Brooks R. Radiolabelled 5'-iodo-2'-deoxyuridine: a promising alternative to ^{18}F]-2-fluoro-2-deoxy-D-glucose for PET studies of early response to anticancer treatment. Nucl Med Biol 1999;26(6):667–672.

33. Quackenbush RC, Shields AF. Local Re-utilization of thymidine in normal mouse tissues as measured with iododeoxyuridine. Cell Tissue Kinet 1988;21:381–387.

34. Tjuvajev JG, Macapinilac HA, Daghighian F, et al. Imaging of brain tumor proliferative activity with iodine-131-iododeoxyuridine. J Nucl Med 1994;35(9):1407–1417.

35. Tjuvajev J, Muraki A, Ginos J, et al. Iododeoxyuridine uptake and retention as a measure of tumor growth. J Nucl Med 1993;34(7):1152–1162.

36. Blasberg RG, Roelcke U, Weinreich R, et al. Imaging brain tumor proliferative activity with [124]I]iododeoxyuridine. Cancer Res 2000;60(3):624–635.

37. Dupertuis YM, Xiao WH, De Tribolet N, et al. Unlabelled iododeoxyuridine increases the rate of uptake of ^{125}I]iododeoxyuridine in human xenografted glioblastomas. Eur J Nucl Med Mol Imaging 2002;29(4):499–505.

38. Dupertuis YM, Vazquez M, Mach JP, et al. Fluorodeoxyuridine improves imaging of human glioblastoma xenografts with radiolabeled iododeoxyuridine. Cancer Res 2001;61(21):7971–7977.

39. Heidelberger C, Chaudhuri NK, Danneberg P, et al. Fluorinated pyrimidines, a new class of tumor-inhibitory compounds. Nature (Lond) 1957;179:663.

40. Fowler JS, Finn RD, Lambrecht RM, Wolf AP. The synthesis of ^{18}F-5-fluorouracil. VII. J Nucl Med 1973;14:63–64.

41. Dimitrakopoulou-Strauss A, Strauss LG, Schlag P, et al. Fluorine-18-fluorouracil to predict therapy response in liver metastases from colorectal carcinoma. J Nucl Med 1998;39(7):1197–1202.

42. Aboagye EO, Saleem A, Cunningham VJ, Osman S, Price PM. Extraction of 5-fluorouracil by tumor and liver: a noninvasive positron emission tomography study of patients with gastrointestinal cancer. Cancer Res 2001;61(13):4937–4941.

43. Ensminger WD, Rosowsky A, Raso V, et al. A clinical-pharmacological evaluation of hepatic arterial infusions of 5-fluoro-2'-deoxyuridine and 5-fluorouracil. Cancer Res 1978;38(11 pt 1):3784–3792.

44. Ishiwata K, Ido T, Kawashima K, Murakami M, Takahasi T. Studies on ^{18}F-labeled pyrimidines. II. Metabolic investigation of ^{18}F-5-fluorouracil, ^{18}F-5-fluoro-2'-deoxyuridine and ^{18}F-5-fluorouridine. Eur J Nucl Med 1984;9:185–189.

45. Abe Y, Fukuda H, Ishiwata K, et al. Studies on ^{18}F-labeled pyrimidines. Tumor uptakes of ^{18}F-5-fluorouracil, ^{18}F-5-fluorouridine, and ^{18}F-5-fluoro-2'-deoxyuridine in animals. Eur J Nucl Med 1983;8:258–261.

46. Seitz U, Wagner M, Vogg AT, et al. In vivo evaluation of 5-[(18)F]fluoro-2'-deoxyuridine as tracer for positron emission tomography in a murine pancreatic cancer model. Cancer Res 2001;61(10):3853–3857.

47. Kameyama M, Ishiwata K, Tsurumi Y, et al. Clinical application of 18F-FUdR in glioma patients: PET study of nucleic acid metabolism. J Neurooncol 1995;23(1):53–61.

48. Ishiwata K, Ido T, Abe Y, Matsuzawa T, Murakami M. Studies of ^{18}F-labeled pyrimidines III. Biochemical investigation of ^{18}F-labeled pyrimidines and comparison with ^{3}H-deoxythymidine in tumor-bearing rats and mice. Eur J Nucl Med 1985;10:39–44.

49. Tsurumi Y, Kameyama M, Ishiwata K, et al. ^{18}F-Fluoro-2'-deoxyuridine as a tracer of nucleic acid metabolism in brain tumors. J Neurosurg 1990;72(1):110–113.

50. Langen P, Etzold G, Hintsche R, G. K. 3'-Deoxy-3'-fluorothymidine, a new selective inhibitor of DNA-synthesis. Acta Biol Med Germ 1969;23(6):759–766.

51. Grierson JR, Shields AF. Radiosynthesis of 3'-deoxy-3'-[(18)F]fluorothymidine: [(18)F]FLT for imaging of cellular proliferation in vivo. Nucl Med Biol 2000;27(2):143–156.

52. Watanabe KA, Su TL, Klein RS, et al. Nucleosides. 123. Synthesis of antiviral nucleosides: 5-substituted 1- (2-deoxy-2-halogeno-beta-D-arabinofuranosyl)cytosines and -uracils. Some structure-activity relationships. J Med Chem 1983;26(2):152–156.

53. Phillips FS, Feinberg A, Chou T-C, et al. Distribution, metabolism, and excretion of 1-(2-fluoro-2-deoxy-b-D-arabinofuranosyl)thymine and 1-(2-fluoro-2-deoxy-b-D-arabinofuranosyl)-5-idocytosine. Cancer Res 1983;43:3619–3627.

54. Fanucchi MP, Leyland-Jones B, Young CW, Burchenal JH, Watanabe KA, Fox JJ. Phase I trial of 1-(2'-deoxy-2'-fluoro-1-beta-D-arabinofuranosyl)-5- methyluracil (FMAU). Cancer Treat Rep 1985;69(1):55–59.

55. Kong XB, Zhu QY, Vidal PM, et al. Comparisons of anti-human immunodeficiency virus activities, cellular transport, and plasma and intracellular pharmacokinetics of 3'-fluoro-3'-deoxythymidine and 3'-azido-3'-deoxythymidine. Antimicrob Agents Chemother 1992;36(4):808–818.

56. Munch-Petersen B, Cloos L, Tyrsted G, Eriksson S. Diverging substrate specificity of pure human thymidine kinases 1 and 2 against antiviral dideoxynucleosides. J Biol Chem 1991;266(14):9032–9038.

57. Sherley JL, Kelly TJ. Regulation of human thymidine kinase during the cell cycle. J Biol Chem 1988;263(17):8350–8358.

58. Langen P, Kowollik G, Etzold G, Venner H, Reinert H. The phosphorylation of 3'-deoxy-3'-fluorothymidine and its incorporation into DNA in a cell free system with tumor cells. Acta Biol Germ 1972;29:483–494.

59. Balzarini J, Matthes E, Meeus P, Johns DG, De Clercq E. The antiretroviral and cytostatic activity, and metabolism of 3'-azido-2',3'-dideoxythymidine, 3'-fluoro-2',3'-dideoxythymidine and 2',3'-dideoxycytidine are highly cell type-dependent. Adv Exp Med Biol 1989;253B:407–413.

60. Shields A, Grierson JR, Muzik O, et al. Kinetics of 3'-deoxy-3'-[F-18]fluorothymidine uptake and retention in dogs. Mol Imaging Biol 2002;4:83–89.

61. Shields A, Grierson J, Dohmen B, et al. Imaging proliferation in vivo with [F-18]FLT and positron emission tomography. Nature Med 1998;4:1334–1336.

62. Barthel H, Perumal M, Latigo J, et al. The uptake of 3'-deoxy-3'-[(18)F]fluorothymidine into L5178Y tumours in vivo is dependent on thymidine kinase 1 protein levels. Eur J Nucl Med Mol Imaging 2004.

63. Toyohara J, Waki A, Takamatsu S, Yonekura Y, Magata Y, Fujibayashi Y. Basis of FLT as a cell proliferation marker: comparative uptake studies with [3H]thymidine and [3H]arabinothymidine, and cell-analysis in 22 asynchronously growing tumor cell lines. Nucl Med Biol 2002;29(3):281–287.

64. Vesselle H, Grierson J, Muzi M, et al. In vivo validation of 3'-deoxy-3'-[(18)F]fluorothymidine ([(18)F]FLT) as a proliferation imaging tracer in humans: correlation of [(18)F]FLT uptake by positron emission tomography with Ki-67 immunohistochemistry and flow cytometry in human lung tumors. Clin Cancer Res 2002;8(11):3315–3323.

65. Rasey JS, Grierson JR, Wiens LW, Kolb PD, Schwartz JL. Validation of FLT uptake as a measure of thymidine kinase-1 activity in A549 carcinoma cells. J Nucl Med 2002;43(9):1210–1217.

66. Francis DL, Freeman A, Visvikis D, et al. In vivo imaging of cellular proliferation in colorectal cancer using positron emission tomography. Gut 2003;52(11):1602–1606.

67. Schwartz JL, Tamura Y, Jordan R, Grierson JR, Krohn KA. Monitoring tumor cell proliferation by targeting DNA synthetic processes with thymidine and thymidine analogs. J Nucl Med 2003;44(12):2027–2032.

68. Boudinot FD, Smith SG, Funderburg ED, Schinazi RF. Pharmacokinetics of 3'-fluoro-3'-deoxythymidine and 3'-deoxy-2',3'-didehydrothymidine in rats. Antimicrob Agents Chemother 1991;35(4):747–749.

69. Resetar A, Spector T. Glucuronidation of 3'-azido-3'-deoxythymidine: Human and rat enzyme specificity. Biochem Pharmacol 1989;38:1389–1393.

70. Schinazi RF, Boudinot FD, Doshi KJ, McClure HM. Pharmacokinetics of 3'-fluoro-3'-deoxythymidine and 3'-deoxy-2',3'-didehydrothymidine in rhesus monkeys. Antimicrob Agents Chemother 1990;34(6):1214–1219.

71. Shields A, Briston D, Chandupatla S, Douglas K, Mangner TJ, Muzik O. A simplified analysis of [F-18] FLT in the blood. J Nucl Med 2004;45(1052):333P.

72. Visvikis D, Francis D, Mulligan R, et al. Comparison of methodologies for the in vivo assessment of [18]FLT utilisation in colorectal cancer. Eur J Nucl Med Mol Imaging 2004;31(2):169–178.

73. Shields A. Monitoring treatment response. In: Wahl R, editor. Principles and Practice of Positron Emission Tomography. Philadelphia: Lippincott, 2002:252–267.

74. Bendaly E, Sloan A, Dohmen B, et al. Use of 18F-FLT-PET to assess the metabolic activity of primary and metastatic brain disease. J Nucl Med 2002;43:111P–112P.

75. Dohmen B, Shields A, Dittmann H, et al. Use of [18F]FLT for breast cancer imaging. J Nucl Med 2001;42:29P.

76. Dittmann H, Dohmen BM, Paulsen F, et al. [18F]FLT PET for diagnosis and staging of thoracic tumours. Eur J Nucl Med Mol Imaging 2003;30(10):1407–1412.

77. Buck AK, Halter G, Schirrmeister H, et al. Imaging proliferation in lung tumors with PET: 18F-FLT versus 18F-FDG. J Nucl Med 2003;44(9):1426–1431.

78. Halter G, Buck AK, Schirrmeister H, et al. [18F] 3-Deoxy-3'-fluorothymidine positron emission tomography: alternative or diagnostic adjunct to 2-[18F]-fluoro-2-deoxy-D-glucose positron emission tomography in the workup of suspicious central focal lesions? J Thorac Cardiovasc Surg 2004;127(4):1093–1099.

79. van Waarde A, Cobben DC, Suurmeijer AJ, et al. Selectivity of 18F-FLT and 18F-FDG for differentiating tumor from inflammation in a rodent model. J Nucl Med 2004;45(4):695–700.

80. Cobben DC, van der Laan BF, Maas B, et al. 18F-FLT PET for visualization of laryngeal cancer: comparison with 18F-FDG PET. J Nucl Med 2004;45(2):226–231.

81. Cobben DC, Jager PL, Elsinga PH, Maas B, Suurmeijer AJ, Hoekstra HJ. 3'-18F-Fluoro-3'-deoxy-L-thymidine: a new tracer for staging metastatic melanoma? J Nucl Med 2003;44(12):1927–1932.

82. Barthel H, Cleij MC, Collingridge DR, et al. 3'-Deoxy-3'-[18F]fluorothymidine as a new marker for monitoring tumor response to antiproliferative therapy in vivo with positron emission tomography. Cancer Res 2003;63(13):3791–3798.

83. Bading JR, Shahinian AH, Vail A, et al. Pharmacokinetics of the thymidine analog 2'-fluoro-5-methyl-1-beta-D-arabinofuranosyluracil (FMAU) in tumor-bearing rats. Nucl Med Biol 2004;31(4):407–418.

84. Sun H, Mangner T, Collins J, Muzik O, Douglas K, Shields A. Imaging DNA Synthesis in vivo with [F-18]FMAU and positron emission tomography. J Nucl Med 2005;46(2):292–296.

85. Bading JR, Shahinian AH, Bathija P, Conti PS. Pharmacokinetics of the thymidine analog 2'-fluoro-5-[(14)C]-methyl-1-beta-D-arabinofuranosyluracil ([(14)C]FMAU) in rat prostate tumor cells. Nucl Med Biol 2000;27(4):361–368.

86. Conti P, Alauddin M, Fissekis J, Schmall B, Watanabe K. Synthesis of 2'-fluoro-5-[11C]-methyl-1-beta-D-arabinofuranosyluracil ([11C]-FMAU): a potential nucleoside analog for in vivo study of cellular proliferation with PET. Nucl Med Biol 1995;22(6):):783–789.

87. Mangner TJ, Klecker R, Anderson L, Shields A. Synthesis of 2'-[F-18]fluoro-2'-deoxy-ββ-D-arabinofuranosyl nucleosides. J Labelled Comp Radiopharm 2001;44:S912–S914.

88. Mangner T, Klecker R, Anderson L, Shields A. Synthesis of 2'-[18F]fluoro-2'-deoxy-β-D-arabinofuranosyl nucleotides, [18F]FAU, [18F]FMAU, [18F]FBAU and [18F]FIAU, as potential pet agents for imaging cellular proliferation. Nucl Med Biol 2003;30:215–224.

89. Lu L, Bergstrom M, Fasth KJ, Langstrom B. Synthesis of [^{76}Br]bromofluorodeoxyuridine and its validation with regard to uptake, DNA incorporation, and excretion modulation in rats. J Nucl Med 2000;41(10):1746–1752.

90. Lu L, Samuelsson L, Bergstrom M, Sato K, Fasth KJ, Langstrom B. Rat studies comparing ^{11}C-FMAU, ^{18}F-FLT, and ^{76}Br-BFU as proliferation markers. J Nucl Med 2002;43(12):1688–1698.

91. Jacobs A, Tjuvajev JG, Dubrovin M, et al. Positron emission tomography-based imaging of transgene expression mediated by replica-tion-conditional, oncolytic herpes simplex virus type 1 mutant vectors in vivo. Cancer Res 2001;61(7):2983–2995.

92. Alauddin MM, Shahinian A, Gordon EM, Conti PS. Direct comparison of radiolabeled probes FMAU, FHBG, and FHPG as PET imaging agents for HSV1-tk expression in a human breast cancer model. Mol Imaging 2004;3(2):76–84.

93. Brust P, Haubner R, Friedrich A, et al. Comparison of [^{18}F]FHPG and [$^{124/125}$I]FIAU for imaging herpes simplex virus type 1 thymidine kinase gene expression. Eur J Nucl Med 2001;28(6):721–729.

25

Assessment of Treatment Response by FDG-PET

Lale Kostakoglu and Peter E. Valk

Objective and accurate evaluation of tumor response to anticancer treatment may have significant therapeutic and management implications in clinical practice. Monitoring tumor response serves as a guide for the clinician in the decision-making process regarding the continuation or cessation of the current therapy protocol. Tumor response is also used as an endpoint in prospective randomized clinical trials that are designed to establish the efficacy of new cytotoxic anticancer agents. The ability of tumor cell to respond to a therapy modality (chemotherapy, radiotherapy, hormonal and immunotherapy) is variable between different individuals even within the same tumor model as a result of inherent biologic variations, including (1) tumor heterogeneity, (2) quantity and accessibility of various receptors and antigenic systems used for targeted therapy, and (3) expression of specific genes that govern tumor resistance, host immunologic function, and tumor oxygenation. Identification of refractory or nonresponding tumors at an early period during or immediately after therapy may lead to timely institution of an alternative therapy protocol. Nevertheless, evaluation of treatment response is clinically consequential only if effective treatment alternatives exist, so that change in treatment may ultimately increase the probability of response and survival. High-dose combination chemotherapy with non-cross-reacting drugs may be successful in the elimination of tumor cells rapidly enough to prevent the selection of resistant cell clones. Emerging novel therapies including immunotherapy, cancer vaccines, and biologic therapy with cytokines are entering into the therapeutic armamentarium as alternative options to patients with refractory or resistant tumors.

Early recognition of therapeutic ineffectiveness is essential to avoid high cumulative toxicity because antineoplastic drugs have a very narrow therapeutic index; for example, there is a small difference between doses yielding antitumor effect and potentially fatal toxicity. For some tumors, the proportion of patients that benefit from chemotherapy may be as low as 20% even under best circumstances (1). Discontinuation of ineffective therapy early on could also lead to improved survival with the initiation of salvage therapy with lesser tumor burden. This important issue, however, is yet to be proven. In general clinical practice, contrast-enhanced computer tomography (CT) is the imaging modality of choice to monitor tumor response over treatment time, but objective demonstration of lack of response may be difficult by CT as tumor size change does not always correlate with tumor viability. Recently, ^{18}F-2-deoxy-D-glucose-positron emission tomography (FDG-PET) imaging has been suggested as a sensitive and relatively more specific means to reflect tumor biologic changes and glucose metabolism after therapy. With increasingly compelling evidence, FDG-PET imaging provides a reliable means to accurately assess tumor response that may lead to a better management with an effective therapeutic approach.

Conventional Criteria for Evaluation of Response

There are universally accepted guidelines for objective evaluation of response to systemic cancer therapy. The criteria established by the World Health Organization (WHO) for evaluation of response to cytotoxic therapy based on tumor size change measurements have recently been modified and simplified by the Response Evaluation Criteria in Solid Tumor (RECIST) guidelines (2, 3). These criteria are summarized in Table 25.1. RECIST criteria adopted a unidimensional measurement (tumor maximum axial diameter) for response assessment by assuming a spherical tumor shape. This criterion replaced two-dimensional measurements (maximum axial diameter and largest perpendicular diameter) previously used for response assessment by the WHO guidelines, which assumed an elliptic tumor shape. Also, the cutoff point for definitions of partial response and progression of disease has changed with the new RECIST guidelines. The major shortcoming associated with these conventional guide-

Table 25.1. WHO and RECIST definition of tumor response.

	WHO[a]	RECIST[b]
CR	Complete disappearance of all lesions[c]	Complete disappearance of all lesions[c]
PR	At least 50% decrease in tumor size[c]	At least 30% decrease in tumor size[c]
SD	Meets neither PR nor PD criteria	Meets neither PR nor PD criteria
PD	Greater than 25% increase of at least one lesion or new lesion	Greater than 20% increase in size

WHO, World Health Organization; RECIST, Response Evaluation Criteria in Solid Tumors; CR, complete response; PR, partial response; SD, stable disease; PD, progressive disease.
[a]Change in sum of products (product of the longest diameter and the greatest perpendicular diameter).
[b]Change in sums of longest diameters.
[c]Confirmed at 4 weeks; no CR, PR, or SD documented before progression of disease.

lines is that morphologic information is dependent on size measurements; thus, it does not reflect metabolic changes. A normal-size lesion can harbor tumor or, conversely, an enlargement of an organ, particularly lymph nodes, does not necessarily represent tumor involvement. This quandary becomes even more pronounced in the posttherapy setting, where tumor morphologic changes often lag behind tumor metabolic response. Consequently, in the posttherapy identification of resid-

ual viable tumor, the specificity and positive predictive value of anatomic imaging modalities is unfavorable. A tumor can respond without any size change for reasons of replacement by fibrotic tissue or even with enlargement of tumor size from intratumoral hemorrhage in some tumor subtypes (4). Additionally, as size measurements are performed on axial CT images, the accuracy of a lesion measurement can be spurious when the length of a lesion is more than twice its width. Following therapy, the decrease or increase in tumor size can occur with varying diameters that might lead to a change in shape. Consequently, these types of unprecedented changes may have important implications in the evaluation of response to therapy that have not been addressed by either the WHO or RECIST guidelines (2, 3). Any erroneous categorization of patients into progressive or stable disease may result in unnecessary therapy and associated morbidity with delay in initiation of alternative therapy (5).In an attempt to streamline the complete response criteria, a new category of response was designated as unconfirmed/uncertain complete responses (Cru) or to further address the inherent uncertainty associated with posttherapy residual CT masses. CRu has been recognized in the response criteria for both Hodgkin's disease (HD) and non-Hodgkin's lymphoma (NHL) (6, 7). The clinical significance of recognition of CRu is that patients with benign posttherapy masses may be subject to unnec-

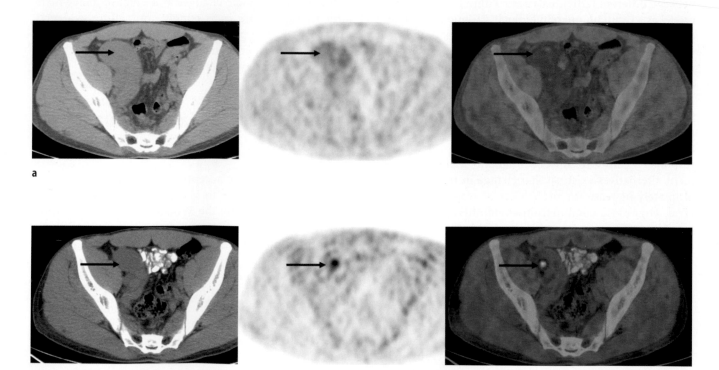

Figure 25.1. Patient with diffuse large cell lymphoma underwent FDG-PET imaging 3 weeks after completion of first-line chemotherapy and subsequently at 4-month follow-up. (**a**) Axial computed tomography (CT; *left*), PET (*middle*), and PET/CT fusion (*right*) images demonstrate that there is minimal FDG uptake [standardized uptake value (SUV), 1.7] in a large pelvic mass (*arrows*) where the patient's original disease was located, which may represent residual lymphoma. (**b**) Follow-up study performed 4 months after completion of first-line therapy reveals a significant reduction in the size of the pelvic mass on the axial CT section (*left*) whereas PET and PET/CT fusion images (*middle* and *right*, respectively) demonstrate a distinct focus of increased FDG uptake within this mass, consistent with recurrent lymphoma (SUV, 5.2) (*arrows*). Following biopsy confirmation of the lesion, patient was started on a second-line therapy.

essary therapy that might only increase morbidity and risk of second malignancies (8).

Evaluation of Treatment Response by FDG-PET Imaging

As alluded to in the previous section, tumor shrinkage assessed by anatomic modalities is not a good indicator of response for detecting resistant clones present even in the significantly shrunken masses after therapy (Figure 25.1). FDG-PET imaging is uniquely suited to evaluate biochemical or metabolic activity of human neoplasms for diagnostic imaging purposes. In the posttherapy setting, FDG-PET imaging may permit differentiation of fibrotic or necrotic changes from viable residual tumor by evaluation of tissue metabolism, thereby permitting more timely institution of therapy targeted toward residual tumor.

Some guidelines have been developed by the European Organization for Research and Treatment of Cancer (EORTC) in an attempt to standardize FDG-PET protocols and response criteria (9). The group made initial recommendations regarding patient preparation and acquisition protocols, methods to measure FDG uptake, and definition of tumor response by FDG-PET criteria.

Patient Preparation

Blood glucose levels have an impact on FDG uptake via the competitive displacement of FDG by plasma glucose. Thus, patients should fast 4 to 6 h to reduce competition with plasma glucose and thus optimize and standardize tumor FDG uptake. There is no consensus as to the optimal procedure for adjusting glucose levels in diabetic patients. Patients with type I diabetes should be scanned in the morning after an overnight fast. In type II diabetes, those with glucose levels exceeding 200 mg/dL should be rescheduled after their blood glucose is better regulated. Generally, administration of insulin is not recommended because it causes preferential muscle and adipose tissue uptake by shunting FDG into the muscles and adipose tissues through the activated insulin-responsive glucose transporters (Glut-4). Because the major glucose transporter for tumors, Glut-1, is not insulin responsive, elevation of insulin does not propagate glucose or FDG uptake by the tumor. The primary route of FDG excretion is renal; thus, good hydration is required before imaging to encourage urinary excretion. It is recommended that patients drink 500 mL water after injection of FDG. Utilization of anxiolytics are strongly recommended in tense patients (diazepam 5–10 mg orally, 30–60 min before to injection), especially patients with head and neck tumors, to avoid FDG uptake caused by muscle tension. Patients should remain still and refrain from speaking during and after injection to reduce uptake by

various muscle groups, particularly by the laryngeal muscles, which may cause a false-positive finding for residual or recurrent head and neck tumor. Patients should be kept warm to avoid brown adipose tissue (BAT) uptake, which may pose diagnostic challenges in the supraclavicular fossa and paravertebral spaces, especially in children and young adults (Figure 25.2). Oral diazepam directly and indirectly acts on abolishing the BAT-FDG uptake, as BAT is rich not only in sympathetic adrenergic receptors but also in peripheral-type benzodiazepine receptors (10, 11).

Timing for FDG-PET Imaging

Monitoring response to therapy requires acquisition of a baseline FDG-PET study before therapy and a repeat study during or after the completion of therapy. Selecting the exact timing of FDG-PET reimaging can be challenging because of the variability in tumor sensitivity to different treatment modalities. In fact, determination of optimal imaging time after therapy is still in evolution for different tumor systems. Imaging 1 to 2 weeks after completion of therapy or individual therapy cycle is recommended to avoid transient fluctuations in FDG metabolism. When FDG-PET imaging is performed during the course of chemotherapy, current data indicate that images can be obtained as early as after one or two cycles of therapy in some tumors (12–17). The timing for FDG-PET imaging after radiotherapy should be longer than after chemotherapy because of differences in cell-killing mechanisms and involvement of normal tissues between these two therapy modalities. This issue is further discussed in the following sections of this chapter.

Response Criteria for FDG-PET Imaging

Another important issue that has to be further investigated is the standardization of the guidelines for evaluation of response to therapy with FDG-PET imaging. Similar to RECIST or WHO, complete response (CR) is usually defined as disappearance of tumor FDG uptake after therapy. Progression of disease (Prog) and no response (NR) can be defined as an increase in FDG uptake and unchanged tumor FDG uptake compared to baseline, respectively. However, definition of partial response (PR) is more challenging and varies widely among studies. The EORTC guidelines recommend using a 25% decrease in uptake as the definition of "partial metabolic response" to therapy (9). However, selection of this cutoff value was not based on substantiated data, and thus there is a strong need for quantitative data validated across various tumor systems. Because metabolic response involves diverse physiologic details that may differ for each tumor system, definition of response criteria for FDG-PET imaging will probably vary among different tumor phenotypes.

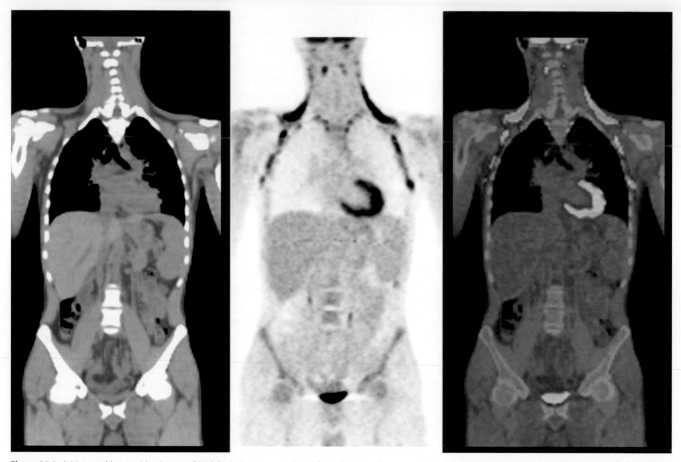

Figure 25.2. A 28-year-old man with a history of Hodgkin's disease was referred for evaluation of response to therapy after completion of chemotherapy. Coronal CT (*left*), PET (*middle*), and PET/CT fusion (*right*) images show symmetrical hypermetabolic foci in the cervical, supraclavicular, axillary, and paravertebral fat planes, bilaterally, consistent with brown adipose tissue (BAT). The BAT uptake renders the interpretation difficult by obscuring the underlying lymph nodes that may harbor viable residual disease. It is essential that this pattern of uptake should be evaluated with a simultaneously acquired CT scan.

Cell Death in Response to Anticancer Therapy

In addition to surgery, there are three main modalities employed in cancer treatment: chemotherapy (including hormonal therapy), radiotherapy, and biologic therapy (cytokines, bacterial adjuvants, vaccines, immunotherapy). The management of most solid tumors involves the use of combined therapy modalities to overcome diverse drug resistance mechanisms and thereby increase the efficacy of treatment. These cytotoxic therapy modalities trigger two independent types of cell death pathways: necrotic cell death and programmed cell death, so-called apoptosis. These two forms of cell death can be differentiated on the basis of their morphologic and biochemical features. In the case of necrotic cell death, any insult resulting in membrane permeability and cytoskeleton derangements can lead to swelling of the cytoplasm, lysosomal rupture, and eventual degradation of chromatin and DNA (18). In this pathway, cytotoxic cellular components induce inflammation in the neighboring cells. In apoptosis, however, the cell undergoes an energy-dependent cellular death initiated by specific molecular

signals in an otherwise normal microenvironment (19) (Table 25.2). Enzymes such as caspases target specific proteins to digest the cytoskeleton with subsequent fragmentation of the nuclear DNA. The final stage is cell shrinkage, chromatin condensation, formation of apoptotic bodies, and phagocytosis by adjoining cells (19). Hence, apoptosis is a dynamic and gradual process with no inflammatory component, in contrast to necrotic death, which is abrupt and always associated with inflammation. Agents that primarily trigger apoptotic pathways would not cause significant inflammatory changes following therapy; the truth, however, is that there is no therapy

Table 25.2. Cell death pathways

Apoptosis	Necrosis
Cells shrink	Cells swell and "explode"
Orderly DNA fragmentation (ladders)	Disorderly DNA fragmentation
Caspase activation (cascade)	No caspase activation
No inflammation	Inflammation
Requires ATP	Caused by lack of ATP
Phagocytosis (no body)	Necrotic "corpse" persists

modality that selectively triggers apoptotic mechanisms without inducing necrotic cell death. The majority of chemotherapic agents exert their cytotoxic effects through perturbation of DNA, RNA, or protein synthesis, which leads to necrotic cell death or discontinuation of cell growth. Additionally, many cytotoxic drugs are able to act through induction of apoptosis: these include alkylating agents (chlorambucil and BCNU), plant alkaloids (Taxol, Taxotere), antimetabolites (fluorouracil and ARA-C) topoisomerase I and II inhibitors (camptothecin and etoposide), and hormonal manipulation (steroids, antiestrogens, and androgen withdrawal) (20).

FDG-PET Imaging in the Evaluation of Cell Death

Accelerated glucose transport resulting from enhanced glucose metabolism is the hallmark of oncogenic transformation in most tumors (21). Many factors influence apparent changes in tumor FDG uptake with therapy, including change in lesion size and perfusion, tumor glucose utilization, tumor heterogeneity, change in cell number and proliferative activity, reversible cell damage, and inflammatory response to therapy. These issues are poorly understood and difficult to evaluate in vivo, and clinical studies of FDG-PET imaging in assessment of treatment effect have taken a pragmatic approach, relating observed changes in tumor FDG concentration to histologic findings and clinical outcome.

The decrease in tumor glucose utilization after therapy is probably a manifestation of impaired cell metabolic functions, including glycolysis resulting from perturbed cell integrity. The absence of any degree of FDG uptake after therapy is a strong evidence of complete response to therapy; however, duration of response may not be predicted by loss of uptake, as relapses are possible from minimal residual disease that is undetectable within the resolution limits of imaging modalities.

Prediction of the ultimate outcome after an initial incomplete response is an elusive quest. Following therapy, cancer cells can be metabolically active with or without the loss of their reproductive or proliferative competence. With the loss of proliferative function, residual cells eventually degenerate and undergo cell death. At this stage, tumor cells can completely escape cytotoxic effects through developing drug resistance mechanisms or continue to undergo necrotic and/or apoptotic cell death (22–24). With the intact proliferative capacity, however, there are two scenarios. If the proliferative capacity is still intact after completion of therapy, progression of disease is inevitable. However, if the proliferative capacity is intact after only a few cycles of therapy, the efficacy of therapy is a function of the dominant component, reproducing cell activity that evades cytotoxic effects versus cytotoxicity which leads to eventual cell death. FDG-PET

imaging can be performed as early as after one cycle of chemotherapy or, more traditionally, after three cycles and at completion of therapy. Considering the foregoing circumstances, residual FDG uptake early during the course of therapy (for example, one to three cycles) may not consistently reflect poor prognosis in all tumor models as the additional cycles of therapy may eventually overcome the cell's initial slow response to therapy. Nonetheless, rapidity of complete response (as early as after one cycle) to therapy is proven to be an independent prognostic factor in some tumor models, particularly in lymphoma, although it is not proven for all tumor models (12–17). Response can also be slow because of inherent cell-killing mechanisms involved with some of the therapy modalities such as immunotherapy and radioimmunotherapy. One caveat to keep in mind is that in patients undergoing hormonal therapy, such as those with breast cancer, hormone-induced changes in tumor metabolism (metabolic flare) may manifest as marked increase in FDG metabolism early after initiation of tamoxifen therapy in responding lesions (25). Accordingly, one should be cautious when extrapolating FDG-PET results obtained for one tumor model or therapy modality to another in the evaluation of response to therapy.

Chemotherapy

Chemotherapy can be given in three essential settings (20). (1) In induction therapy, the aim is to reduce tumor volume to achieve complete response or cure. (2) In adjuvant therapy, the aim is to eradicate any tumor cells that may have escaped after surgery or radiotherapy, as following the removal of the primary tumor (except lymphoma) radiotherapy and/or chemotherapy and/or hormonal therapy (prostate and breast cancer) can be administered to eradicate microscopic disease. (3) In neoadjuvant therapy (also known as preoperative or primary chemotherapy), given preoperatively, the aim for the former is to increase the chance of achieving a better long-term tumor control of a tumor that is operable by effects on microscopic disease and the latter is to increase the chance of tumor control of an inoperable tumor (26). Neoadjuvant chemotherapy is mainly intended to downstage the primary tumor to decrease the morbidity associated with the subsequent treatment (surgery or radiotherapy). Except in a small number of malignancies, such as acute lymphocytic and myelogeneous leukemia, Hodgkin's disease, diffuse large cell lymphoma, testicular cancer, and choriocarcinoma, chemotherapy is primarily used in a neoadjuvant or adjuvant setting because of the tumor biology and advanced stage. Chemotherapy and radiotherapy can also be delivered concurrently primarily to induce spatial cooperation where chemotherapy eradicates subclinical metastases and radiotherapy controls primary disease or to enhance the effects of radiotherapy.

A chemotherapy regimen consists of cycles (often three to six cycles) in which each cycle is 21 or 28 days and drug combinations are administered weekly. The number of cycles can be modified during therapy based on the response and toxicity. Objective evaluation is performed every third cycle, and the therapy is given until a maximum antitumor effect is achieved or until there is significant toxicity (1). The cell-killing effect of cytotoxic tissue insult is most pronounced in rapidly proliferating cells indiscriminately in cancer cells and normal cells. Thus, tissues with rapid turnover, such as bone marrow, and epithelial cells (lung, kidney) can be seriously affected during chemotherapy. During this period, a metabolic imaging tool such as FDG-PET imaging can be of great value to accurately determine tumor response in a timely manner for individually tailored therapy, especially in nonresponding patients for the avoidance of unnecessary morbidity.

Table 25.3. Sources of FDG-PET misinterpretations

	FP disease and conditions	FN disease and conditions
Lymphoma	Sarcoidosis, Tb Benign LPD (e.g., Castleman's) Paget's PTLD Thymic rebound G-CSF (bone marrow uptake)	Extranodal MALT CLL/SLL BM infiltration Lesions less than 5 mm
NSCLC	Sarcoidosis, toxoplasmosis Histoplasmosis, aspergillosis Tb, pneumonia, RT-pneumonitis Esophagitis (post-RT) Pleurodysis	Bronchoalveolar, BALT Primary carcinoid Well-differentiated adenocarcinoma
Breast cancer	Biopsy changes Fibroadenomas (10%)	Mucinous carcinoma Tubular carcinoma Lobular carcinoma, DCIS
H&N cancer	Pneumonia Wartin's tumor Second primary Osteoradionecrosis Stomatitis, esophagitis Recurrent nerve paralysis	Local LN mets close to primary site Necrotic LN metastasis
Colon cancer	IBD Villous adenoma	Mucinous cancer Local LN metastasis close to primary site
Esophagus cancer	Esophagitis (post-RT) GEJ uptake	Local LN metastasis close to primary site Necrotic LN metastasis
Ovarian cancer	Corpus luteum cysts Focal ureteral uptake	Mucinous carcinoma?
Testicular cancer	Thymic rebound Focal ureteral uptake	Mature teratoma

FP, false positive; FN, false negative; LN, lymph nodes; PTLD, post-transplantation, lymphoproliferative disorder; GEJ, gastroesophageal junction; G-CSF, granulocyte colony-stimulating factor; BALT, bronchiole-associated lymphoid tissue; RT, radiotherapy; IBD, inflammatory bowel disease; Tb, tuberculosis; DCIS, ductal carcinoma in situ; BM, bone marrow; H&N, head and neck.

Special Considerations After Chemotherapy

The false-positive and false-negative findings that can be observed after chemotherapy (and radiotherapy) are listed in Table 25.3. The primary side effects of cytotoxic therapy include hematologic toxicity (myelosuppression) and immunosuppression, which may lead to various infectious processes, most notably pneumonia and opportunistic infections. Other therapy-related side effects include pulmonary toxicity (especially with bleomycine and methotrexate), gastrointestinal side effects, follicular lymph node hyperplasia, sarcoid-like granulomas, particularly seen in Hodgkin's disease patients (27), and Epstein–Barr virus-associated posttransplantation lymphoproliferative disorders (PTLD) (28).

FDG uptake can be significantly high in infectious processes, although the pattern of uptake is usually more diffuse than focal (Figure 25.3). The radiographic patterns of PTLD in such cases include mainly pulmonary nodules, which may be misinterpreted as lymphoma progression (29). Although it is yet to be established, inflammatory changes may be prominent in stroma-rich tumors such as breast cancer and lung cancer. FDG-PET findings should be interpreted in light of these therapy-related tissue changes and side effects. Coexisting sarcoidosis is always problematic, especially in patients with lymphoma, because of common presentations on FDG-PET imaging (30). Other sources of false-positive findings include bone marrow (BM) reactive changes in response to hematopoietic colony stimulators [granulocyte colony stimulating factor (G-CSF) and granulocyte-macrophage colony-stimulating factor (GM-CSF)] (Figure 25.4). Cytokines are usually administered a few days after chemotherapy and are given daily for up to 14 days. The clearance half-life of colony stimulators is relatively short (~4 h), and with discontinuation of these agents, leukocyte counts return to pretreatment values over a 1- to 2-week recovery period (31). Therefore, the colony stimulating factor-induced FDG uptake in the BM and in the spleen (in some patients) should return to normal in 2 to 3 weeks after administration of G-CSF or GM-CSF.

Another issue is the thymic rebound seen after cessation of chemotherapy in younger population, especially in patients with lymphoma and testicular cancer (Figure 25.5). The proposed mechanism for thymic hyperplasia after therapy includes initial thymic regression due to apoptotic T cells and thymocyte death via a number of mechanisms caused by the cytotoxic agents or corticosteroids with subsequent hyperplasia and regrowth on reversal of the predisposing cause (32). Thus, thymic rebound usually presents 2 to 6 months after completion of therapy and may persist for 12 to 15 months (our experience).

Radiotherapy

The double-stranded break of nuclear DNA is the most important cellular effect of radiation therapy (RT) (33).

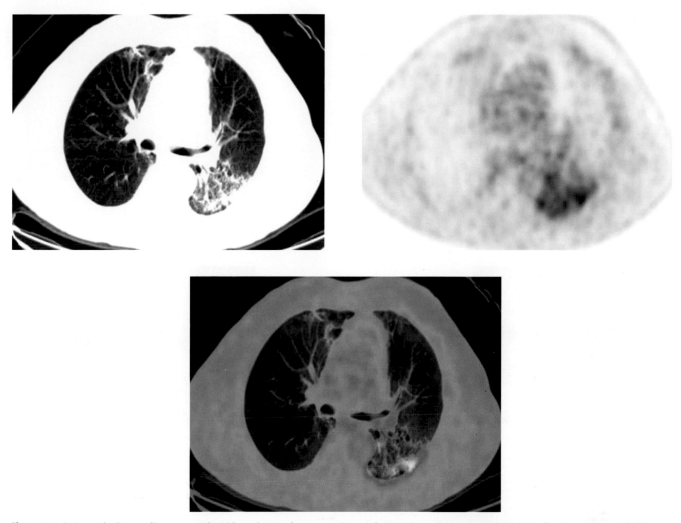

Figure 25.3. Patient with a history of lung cancer referred for evaluation of response to therapy after completion of chemotherapy. Axial CT (*left*), PET (*middle*), and PET/CT fusion (*right*) images demonstrate a diffuse moderately hypermetabolic area in the left posterior lung fields corresponding to the patchy and irregular appearance on the CT slice, consistent with pneumonia.

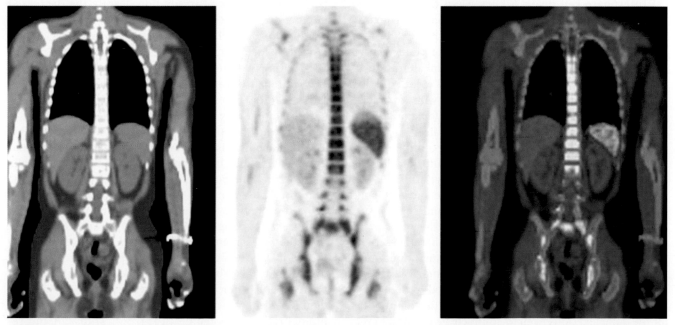

Figure 25.4. A 21-year-old man with a history of non-Hodgkin's lymphoma referred for restaging following granulocyte colony-stimulating factor (G-CSF) administration and chemotherapy. Coronal CT, PET, and PET/CT slices demonstrate diffusely increased metabolic activity within the bone marrow of the vertebral column and pelvis, consistent with colony stimulator-induced bone marrow reactive changes. The increased uptake in the spleen is also related to G-CSF administration rather than involvement by lymphoma.

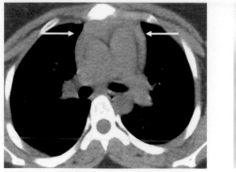
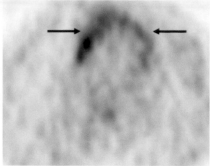
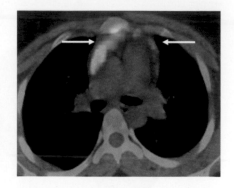

a

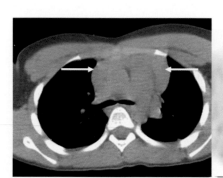
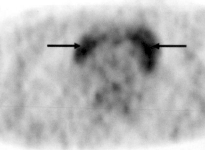
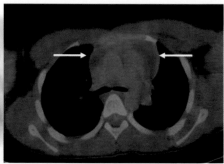

b

Figure 25.5. (**a**) A 20-year-old woman with a history of lymphoma referred for a follow-up PET/CT study 3 months following chemotherapy. Axial CT (*left*), PET (*middle*), and PET/CT (*right*) sections reveal that there is minimally asymmetrical hypermetabolic focus that corresponds to the thymus gland, which appears as a bilobed structure with convex lateral borders in the anterior and superior mediastinum on CT (*white arrows*), consistent with posttherapy thymic rebound (*black arrows*). This pattern of asymmetrical uptake may be confusing in certain clinical presentations, particularly in cases with the original disease located in the superior anterior mediastinum (**b**) A 26-year-old man with a history of testicular cancer referred for posttherapy evaluation. Axial CT (*left*), PET (*middle*), and PET/CT (*right*) sections demonstrate a convex-shaped, symmetrical superior mediastinal hypermetabolic focus that corresponds to the thymus, consistent with thymic rebound (*arrows*).

There are two main types of radiation-induced cell death. (1) Interphase cell death occurs when the cell can no longer maintain metabolic functions and subsequently dies. This type of death is associated with high doses of radiation, generally exceeding 60 Gy. (2) Reproductive cell death occurs when cell injury results in alteration of RNA/DNA that leads to eventual cell death during mitosis. This type of death is associated with low radiation levels, less than 55 Gy. Radiation damage is most commonly indirectly ionizing via free radical intermediaries formed from the radiolysis of cellular water. Radiation can also affect the processes of the cell cycle through apoptosis. As in chemotherapy, after RT, the tumor may still maintain its reproductive and metabolic activity or may be reproductively compromised with sustained metabolic activity. In the latter setting, cells eventually undergo pyknosis or lysis if the proliferative activity is sufficiently damaged, whereas in the former, tumor progression is usually inevitable (22–24). Tumor cells with a high proliferative index would be expected to be lethally damaged after RT; however, it may take up to 3 months for the damaged cancer cell to complete its cell death cycle (4). In vitro assays also demonstrated that irradiated

tumor cells might have a 10-fold increase in FDG uptake in the first 12 days of irradiation despite the decreased number of viable cells (34). This phenomenon may be attributed to giant cell formation after RT, which is characterized by logarithmic increases in the cell's protein and DNA and RNA content (35) but may also be caused by increased repair of cells. Therefore, an FDG-PET study obtained earlier than 3 months after completion of RT or chemoradiotherapy may overestimate FDG uptake values compared to that performed at 3 to 4 months after RT. The dilemma here is the timing of the boost dose of radiation that would be based on the residual tumor metabolic activity. As achieving tumor control is dependent on the timely employment of the fractionated RT protocol, delayed evaluation of the metabolic function may compromise the expected effectiveness of RT.

In the assessment of RT or chemoradiotherapy, one theoretical issue is that posttherapy changes in FDG uptake may also be caused by the radiation-induced altered cellular glucose transport mechanism or vascular damage with resultant poor radiotracer delivery into tumors rather than cell death (34, 36). Transient and reversible cell damage as well as minimal residual disease may obscure assessment of true cell kill in the tumor mass.

After RT, tumor stromal cells and inflammatory cells may also contribute to the overall FDG uptake. In this regard, in vitro or animal studies have contradictory results compared with in vivo human studies. Kubota et al. (37) reported that, in the posttherapy setting, nonneoplastic cellular elements such as macrophages, neutrophils, fibroblasts, and granulation tissue can account for up to 30% of FDG uptake in association with growth and necrosis of the tumor. However, Brown et al. (38) demonstrated that, in human non-small cell lung cancer (NSCLC) tissues, macrophages account for less than 10% (mean, 7.4% ± 6.4%) of all cells in the tumor. Infiltrating inflammatory white blood cells also represent a small fraction of the tissue sample. Therefore, contribution by the inflammatory elements to the overall FDG uptake in vivo is probably not significant. Nonetheless, each tumor may have a different percentage of inflammatory cell population within the tumor mass, and thereby the contribution to the overall FDG uptake may vary greatly among different tumor subtypes. Avril et al. (39) reported that a modest to high degree of infiltration with inflammatory cells was found in approximately 50% of breast cancers. However, no significant relationship was observed between the FDG uptake and the infiltration of inflammatory cells. Haberkorn et al. (40) have suggested that persistence of FDG uptake in patients with colon cancer following RT is the result of inflammatory reactions caused by radiation injury. There is clearly a need for additional in vivo studies in various tumor systems to define the optimal timing of imaging after RT. Considering all these issues, it is recommended to delay FDG-PET studies for at least 3 months after completion of RT to accurately assess therapy outcome (41–44). Generally, FDG uptake observed 6 months after RT is associated with tumor recurrence.

Special Considerations After RT

The false-positive and false-negative findings that can be observed after radiotherapy (and chemotherapy) are listed in Table 25.3. Radiation effects on the normal tissues are divided into acute and chronic effects. Acute radiation effects include direct toxic injury to endothelial and epithelial cells, which starts during therapy, becomes most evident after 1 week of treatment, and lasts 2 to 4 weeks after the completion of RT (45). In this phase, tissues that proliferate rapidly respond acutely to radiation and account for the acute morbidity of the treatment in the form of stomatitis, mucositis, and esophagitis (Figure 25.6). In this early period, the FDG uptake is characteristically diffuse and sometimes patchy in the irradiated mucosal surfaces, particularly in the oral cavity, oropharynx, nasopharynx, and esophagus.

In the lung, acute radiation changes usually appear within 3 to 6 months after completion of RT with a pre-

sentation of ground-glass opacities confined to the mantle radiation ports (45, 46). Acute changes eventually disappear after 6 months, usually replaced by chronic changes in the ensuing 18 months, presenting as retraction or traction bronchiectasis and mediastinal fibrosis, which may resemble granulomatous disease, but these are usually symmetrical and confined to the mantle port (47). In contrast to acute radiation pneumonitis, permanent changes of radiation fibrosis can take months to years to evolve but normally stabilize within 1 to 2 years. Pulmonary fibrosis is the repair process that follows the acute inflammatory response and is characterized by progressive fibrosis of the alveolar septa thickened by bundles of elastic fibers. In the acute period, radiation pneumonitis may reveal significantly high FDG uptake indistinguishable from tumor uptake. In the late phase, however, the FDG uptake is usually low grade, corresponding to fibrotic changes (Figure 25.7). All FDG-PET studies should be correlated with a contemporaneous CT scan in the posttherapy setting to avoid false-positive findings.

The primary osseous complication arising from radiation injury is osteoradionecrosis, which is a major concern, especially for those with head and neck tumors (48). There appears to be a direct relationship between the total dose and the incidence of osteoradionecrosis. When total radiation dose is 70 Gy or more, there is 10 times the relative risk of osteoradionecrosis when compared to the group receiving less than 50 Gy (49). Osseous structures with RT-induced necrosis typically show an intense and focal FDG uptake mimicking a residual or recurrent disease (50).

Hormonal Therapy

Elimination of hormonal stimulation can be an effective treatment for cancers whose growth is under gonadal hormonal control, particularly in breast and prostate cancer (51). For example, in breast cancer, tamoxifen is thought to competitively block estrogen receptors and suppress the genome of the breast cancer cell. Flare response with tamoxifen is a known phenomenon that is characterized by the apparent transient disease progression that occurs within 7 to 10 days of initiation of hormonal therapy (52, 53). Flare response is associated with the initial tumor growth induced by estrogen-like agonist effects of hormonal therapy, which is followed by tumor regression. Hormone-induced flare reaction is considered a reliable predictor of response, as it has been observed that up to 90% of patients who experience a flare reaction demonstrate an objective response (54–56).

Estrogen Receptor Imaging

The responsiveness of a tumor to hormone therapy is an important parameter in breast cancer management. PET

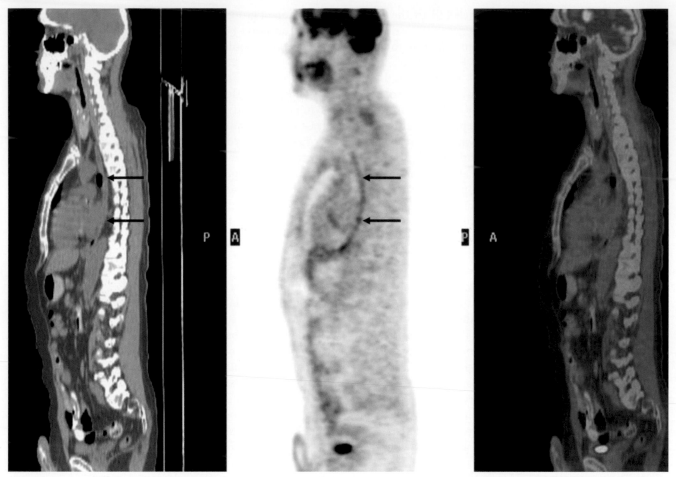

Figure 25.6. Patient with a history of lung cancer referred for evaluation of response to therapy after radiotherapy. Sagittal CT (*left*), PET (*middle*), and PET/CT (*right*) sections demonstrate a tubular uptake pattern in distribution of the esophagus in its entirety immediately anterior to the vertebral column (*arrows*), consistent with postradiotherapy esophagitis.

estrogen receptor (ER) imaging can be used to predict the likelihood of response to hormonal therapy and may guide patient selection process. Several studies reported that a higher level of ^{18}F-fluoroestradiol-17-beta (FES) uptake in advanced tumors predicts a greater chance of response to tamoxifen (25, 57, 58). In a study by Mortimer et al. (25), in the responders, the tumor FDG uptake increased after tamoxifen by 28.4%; only five of these patients had clinical evidence of a flare reaction. In nonresponders, there was no significant change in tumor FDG uptake from baseline. Lesions of responders had higher baseline FES uptake than those of nonresponders.

Serial FES-PET can also assess the functional response to hormonal therapy in the primary tumor or metastasis (59). Inoue et al. (60) performed a pilot study of PET imaging with ^{18}F-fluorotamoxifen (FTX) in 10 patients with ER+ breast tumors for prediction of tumor response to tamoxifen. There was no significant difference of FTX uptake in bone lesions between good and poor responders. However, when bone lesions were excluded, FTX uptakes in tumors with good responses were significantly higher than those with poor responses [mean standardized uptake value (SUV), 2.46 versus

1.37]. These preliminary results show the potential of PET-ER imaging to help guide appropriate, individualized breast cancer treatment and to point the way for future studies and clinical use.

Biologic Therapy

Biologic therapy for cancer involves activation or stimulation of the immune system and its components. Biologic therapy with response modifiers (for example, interferon, IL-2) or monoclonal antibodies (naked) activate host immune effector mechanisms, or, by inducing apoptosis, direct antiproliferative effects also play an important role (61). There are several mechanisms by which this therapy modality can exert its cell-killing effects, including antibody-dependent cellular cytotoxicity (ADCC) and complement-dependent cytotoxicity (CDC), both resulting in lysis of the target cell. Induction of apoptosis is also an important cell death pathway in biologic therapy. This type of therapy is relatively slow acting compared to chemotherapy or radiotherapy because of the differences in cell-killing mechanisms. Rituximab (Rituxan; anti-

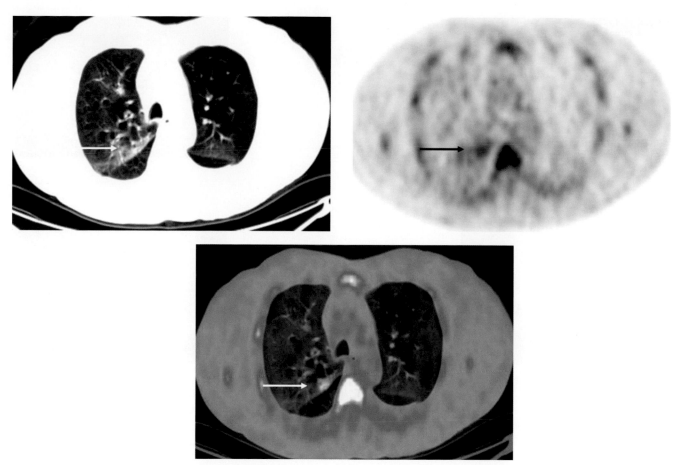

Figure 25.7. Patient with a history of lung cancer referred for a follow-up study 18 months after completion of therapy including radiation and surgery. Axial CT (*left upper*), PET (*middle upper*), and PET/CT (*lower*) sections demonstrate minimally increased FDG uptake corresponding to the pleural thickening seen in the right lung fissure secondary to scarring caused by radiation therapy, consistent with chronic inflammatory changes.

CD20) for lymphoma, alemtuzumab (Campath; anti-CD52) for B-cell chronic lymphocytic leukemia, and trastuzumab (Herceptin; anti-HER2) for metastatic breast cancer are FDA-approved immunotherapy agents.

Radioimmunotherapy (RIT) uses monoclonal antibodies to guide radionuclides for selectively targeting malignant tissues. RIT is fundamentally different from RT. In contrast to RT, RIT delivers radiation continuously with a lower peak rate (62). Radionuclides suited for RIT usually emit β-radiation for irradiation of the tumor cells. The mechanisms by which β-radiation induces cell death are not understood at the molecular level; however, it is deemed that β-irradiation induces apoptosis pathways depending on doses, time points, and dose rates. There are currently two FDA-approved RIT products, ^{131}I-tositumomab (Bexxar; anti-CD20) and ^{90}Y-ibritumomab tiuxetan (Zevalin; anti-CD20), for the treatment of low-grade or transformed low-grade lymphoma. As with immunotherapy, the tumor response to RIT is more gradual as compared to chemotherapy and radiotherapy (Figure 25.8). The prognostic value of FDG- PET imaging was evaluated in 14 patients with NHL treated with ^{131}I-anti-B1 therapy. All patients underwent FDG-PET imaging at baseline and at 5 to 7 days and 1 to 2 months after RIT to estimate the response to RIT. FDG-PET metabolic data obtained 1 to 2 months after RIT correlated well with the ultimate response of NHL to RIT, more significantly than the early FDG-PET data obtained 5 to 7 days after RIT (63). For both immunotherapy and RIT, the earliest response evaluation should be performed at completion of therapy or, ideally, 3 months after completion of therapy for an accurate evaluation. Earlier periods of assessment may give rise to false-positive findings.

One important issue that has not been addressed systematically is the potential value of FDG-PET imaging in the assessment of novel therapeutic regimens in phase II efficacy trials. The patients with a metabolic response can safely continue the new therapeutic regimen, whereas for those without a metabolic response the experimental trial should be discontinued. Using metabolic response as an endpoint may shorten the duration of phase II studies evaluating new cytotoxic drugs and may decrease the morbidity and costs of therapy in nonresponding patients. However, integration of FDG-PET imaging in the evaluation of efficacy of experimental therapy trials will require a very well designed algorithm as well as further establishment of FDG-PET imaging as a reliable surveillance means.

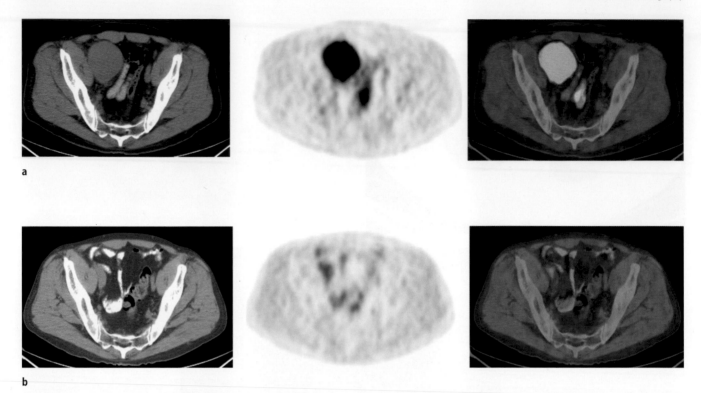

Figure 25.8. Patient with a history of transformed low-grade lymphoma underwent PET/CT imaging before and 3 months after ^{131}I-labeled anti-CD20 antibody. (**a**) Pretherapy axial CT (*left*), PET (*middle*), and PET/CT (*right*) sections demonstrate intense radiotracer uptake in the pelvic mass consistent with patient's known disease. (**b**) Three months following radioimmunotherapy, there is virtually no uptake in the pelvic mass, which has shrunken by more than 80% of its previous volume, consistent with good response to therapy.

Methods of Assessing Tumor FDG Metabolism by PET Imaging

Methods of evaluating changes in tumor FDG metabolism with therapy have been thoroughly discussed by Hoekstra et al. (64), Graham et al. (65), and the PET study group of the European Organization for Research and Treatment of Cancer (EORTC) (9). The following approaches have been compared.

Visual Assessment

Visual assessment based on differential uptake by tumor and adjacent tissues is the most commonly used method in daily clinical practice for the interpretation of FDG-PET images. This approach, however, is subjective, requires substantial experience, and is not sufficient for subtle findings. Alternatively, graded visual assessment results in less interobserver variability provided that there are well-defined criteria for positive and negative findings. Visual assessment alone is valuable for demonstration of response or lack of response to therapy when no visible tumor uptake remains after treatment or when obvious increase in uptake or in the number of lesions occurs during treatment. In cases with subtle changes with treatment, however, a more-objective assessment approach is required. Further, a change in visual contrast can occur

with a change in uptake by adjacent tissue, without change in tumor metabolism. When visual assessment of sequential scans is used, it is important to adjust the image intensities of the studies under comparison in an appropriate fashion, basing intensity on uptake in normal tissues rather than tumor.

Semiquantitative Assessment

Although visual evaluation is valuable and practical, one of the greatest advantages of FDG-PET imaging is the ability to use quantitative data, particularly, in predicting outcome by determining tumor aggressiveness and monitoring response to therapy. Nevertheless, the optimal method of quantitation for prognosis and assessing response to therapy has not yet been defined. There are several ways to approach quantitation, from simple tumor-to-background calculations to intricate kinetic analysis with dynamic PET acquisition and blood sampling. Currently, it is not clear that more-advanced quantitation techniques are superior to more-basic methods in the prediction of prognosis and response to therapy, mainly because of the insufficient FDG-PET data obtained, thus far, on assessment of various tumors. All quantitation methods entail attenuation correction to avoid the variability in FDG uptake resulting from the differences in tumor depth in the body.

Tumor-to-Normal Tissue Ratio (T/N Ratio)

Determination of tumor-to-normal ratio is the simplest means of quantitation. This method can be applied to images even after reconstruction without the requirement of additional procedures or information. The T/N ratio is a semiquantitative index that compares tumor activity to activity in normal tissues. The T/N ratio does not require accurate quantitative determination of tissue activity and is not affected by the injected dose, patient weight, or blood glucose level. It is also not affected by change in distribution of tracer to other tissues, as may occur with extravasation of injected tracer or decrease in renal excretion. Change in T/N ratio is similar to change in visual contrast and could occur with change in normal tissue uptake without change in tumor activity. Also, it may be difficult to select an appropriate adjacent normal reference site, particularly in the abdomen and pelvis where there is marked variation in uptake between studies and between patients. In prediction of treatment response in colorectal cancer metastatic to the liver, Findlay et al. (66) found that change in tumor-to-liver ratio was a better predictor than change in SUV. Although this approach is more objective than visual assessment, it has significant limitations regarding the placement of regions of interest versus background, use of count statistics, and reconstruction algorithms.

Standardized Uptake Value

The SUV, which is less commonly referred to as the differential uptake ratio, is a semiquantitative measure that is derived from determination of tissue activity obtained from a single static image. It is defined as the tissue concentration of injected tracer as determined by PET imaging, divided by the injected dose, and multiplied by a calibration factor. Tracer concentration is usually determined from the image pixel that shows the highest lesion activity (SUV_{max}). SUV has also been determined from mean lesion activity (SUV_{mean}), but this approach has fallen into disuse. For comparison purposes, SUV_{max} should be used as SUV_{mean} measurements may be subject to the size of region of interest.

The most frequently used calibration factor has been body weight. SUV normalized to body weight has been shown to be dependent on patient weight, resulting in overestimation in more-obese individuals, probably due to lower uptake of FDG in fat than in other tissues (67, 68). FDG uptake in the fat is lower than in muscle; therefore, the lean body weight or body surface area have also been used and correlate better with FDG metabolic rate (65). Unlike visual analysis and tumor-to-background ratio determination, the SUV method involves no assumptions regarding normal tissue activity.

The problems involved in evaluation of SUV have been discussed by Hamberg et al. (69) and by Keyes et al. (70). Time interval between FDG injection and imaging (uptake period) is probably the largest single source of variation in determining tumor SUV and affects all methods equally. Hamberg et al. (69) showed that FDG uptake in carcinoma of the lung did not reach equilibrium until more than 3 h after injection, notably greater than the usual 60-min postinjection scanning time. For reproducible and accurate results, posttherapy PET imaging must be acquired at the same time interval (with 5–10 min difference) after injection as the pretherapy study. Blood glucose concentration has also been shown to change tissue uptake of tracer, which decreases with increasing glucose concentration in the blood (71–73). Correction for plasma glucose improves reproducibility, but differences in SUV remain in some patients.

SUV measurements can also be affected by the lesion size and tumor necrosis (Figure 25.9). Underestimation of FDG concentration is inevitable for lesions smaller than two times the scanner resolution at full-width half-maximum (partial-volume effect). Nonetheless, partial-volume effects can be corrected with the use of appropriate recovery coefficients, although these methods do not land themselves as practical application in daily routine.

Quantitative Kinetic Methods

Nonlinear Regression

Kinetic modeling by fitting tissue time–activity curves to a two-compartment model, using an arterial input function and nonlinear regression, may be used to determine the metabolic rate of glucose in moles/min/mL (64, 65). The time course of radioactivity in tissue (uptake curve) is obtained by dynamic imaging, and the arterial blood radioactivity curve (input function) is obtained by arterial blood sampling. By this method, the individual rate constants describing glucose delivery and phosphorylation, as well as the net influx constant, K_i, can be determined. This technique is too invasive and demanding of resources for routine clinical use but provides the most detailed description of glucose uptake. The use of image-derived input functions, obtained immediately following injection from regions of-interest over the left ventricle or the aorta, can simplify the procedure and make it less invasive.

Patlak Graphical Analysis

Patlak graphical analysis (PGA) can be used to provide a linear solution for K_i for tracers that are irreversibly bound after uptake (74); this is approximately true for FDG during the first hour postinjection. PGA permits a simplified dynamic imaging procedure with fewer data frames, and requires only the integral of the blood–activity curve during the early postinjection period, so that arterialized venous blood samples or image data can be used to obtain the input function. This is a linear method

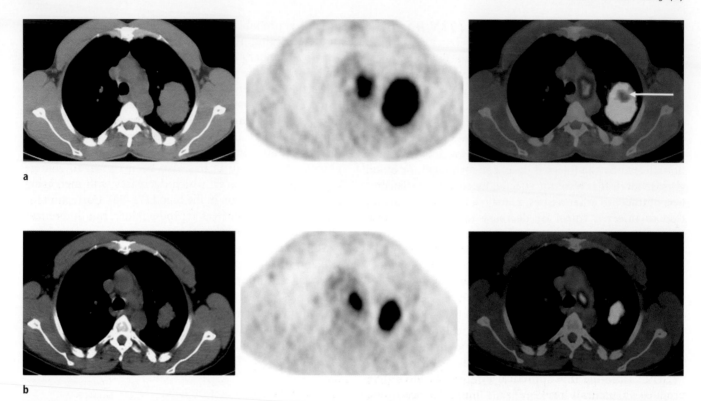

Figure 25.9. A patient with lung carcinoma underwent PET/CT imaging before and one months after completion of neoadjuvant therapy. (**a**) Pretherapy axial CT (*left*), PET (*middle*), and PET/CT (*right*) sections demonstrate intense radiotracer uptake in a large left lung mass with an SUV_{max} of 10, consistent with patient's known disease. Note the relatively decreased FDG uptake within the lesion, which is more evident on the PET/CT fusion image (*right*), consistent with central necrosis (*arrow*). (**b**) Following neoadjuvant therapy, there is significant reduction in the lesion size and apparently in FDG uptake. However, the measured SUV_{max} is 13, which appears to be in contrast to the visual interpretation of partial response to therapy. This discrepancy is probably due to the necrotic component seen on the pretherapy image, which compromises SUV measurements and renders pre- and post-therapy comparison difficult.

with no noise amplification, so that it can be applied on a pixel-by-pixel basis to provide parametric images of glucose consumption.

Simplified Kinetic Method

In the simplified kinetic method (SKM), a single blood sample is used to scale an average blood time–activity curve, and uptake is defined as tissue activity divided by the integral of blood activity up to the midpoint of the PET image. Hunter et al. (75) compared measurements obtained in patients with advanced-stage NSCLC by this method to the metabolic rate of glucose determined by nonlinear regression (NLR) and with the calculated SUV. The SKM was an improvement over the SUV and approached NLR in accuracy. The method is simpler than PGA and is more suitable for routine clinical application than other kinetic methods.

Weber et al. (76) assessed the reproducibility of serial FDG-PET measurements to define objective criteria for the evaluation of treatment-induced changes in 16 patients). Patients underwent PET studies twice within 10 days before therapy. SUVs, FDG net influx constants (K_i), glucose normalized SUVs (SUV(gluc)), and influx constants ($K(i,gluc)$) were determined for various lesions. None of the parameters showed a significant increase or decrease at the two examinations. The differences of re-

peated measurements were approximately normally distributed for all parameters with a standard deviation (SD) of the mean percentage difference of about 10%. This study showed no clear advantage in using tracer kinetics over SUVs. Similarly, in another study by the same investigators, changes in tumor SUVs were compared with changes in FDG net-influx constants (K_i) and tumor-to-muscle ratios (t/m) in patients with advanced-stage NSCLC undergoing response evaluation 3 weeks after initiation of therapy (77). A reduction of SUVs by more than 20% was used as a criterion for a metabolic response in advanced NSCLC. Median time to progression and overall survival were significantly longer for metabolic responders than for metabolic nonresponders determined by SUVs (163 versus 54 days and 252 days versus 151 days, respectively). Similar results were obtained when K_i was used to assess tumor glucose use, whereas changes in t/m showed considerable overlap between responding and nonresponding tumors. Changes of SUVs that are outside the 95% normal range may be used to define a metabolic response to therapy in NSCLC.

Comparison of Methods

The SUV has been found to correlate with the metabolic rate for FDG determined by PET imaging in cancer patients, with correlation coefficients of 0.91 (78) and

0.84 (79). The correlation improved further when the SUV was normalized to body surface area rather than weight. Wahl et al. (80) showed in breast cancer patients that a reduction in SUV in responding patients was accompanied by a comparable or greater decrease in K_i as determined by PGA. Graham et al. (81) compared SUV, normalized to body weight and to body surface area, SKM, and PGA to NLR in 40 patients who had hepatic metastases from colorectal cancer. They concluded that PGA was the most accurate mean for distinguishing tumor from normal tissue, closely followed by SKM. It was best to normalize SUV to body surface area. If one blood sample is obtained during the study, SKM can be used instead of SUV. For prediction of survival, measurement of Patlak slope was best, closely followed by SUV normalized to body surface area. Both SUV measurement and kinetic methods are subject to error, because not all the assumptions made in determination of metabolic rate of FDG are adhered to in the simplified kinetic methods, particularly in assessment of arterial input function. The EORTC PET study group has recommended the use of SUV normalized to body surface area as the basic standard for measurement (9). It is important to use the most practical PET protocol that can produce an accurate result so as to minimize scan time and blood sampling. Patient compliance and ease of performance in active clinical departments are important factors in ensuring success in clinical trials and therapeutic applications. When SUV is being used to evaluate treatment effect, the number of potential variables between studies should be minimized by a strict protocol. Time interval after injection and blood sugar should be maintained within 10% SD limits between the two studies. Reproducibility is likely to be greater in such intrapatient comparisons than in studies that compare results between patients.

The EORTC group also suggested using a 25% decrease in uptake as the definition of "partial metabolic response" to treatment (9). However, selection of this cutoff value was arbitrary, and it seems likely that a meaningful response threshold will need to be defined for each tumor individually by clinical trials.

In semiquantitative evaluation of FDG uptake in patients with hepatic metastases from colorectal cancer, Findlay et al. (66) found that change in T/N ratio was a more accurate predictor of response than change in SUV. In one nonresponding patient who showed a decrease in SUV with treatment, tracer extravasation at the time of injection may have caused underestimation of the SUV without affecting the T/B ratio. Two patients with responding tumors and increasing SUV were more difficult to explain. However, it was noted that in both patients the liver SUV showed a similar increase with treatment, suggesting that there may have been an error in dose calculation in the pretreatment study. Such "miscalculation" could have resulted from higher uptake at a competing site, as the pretreatment scans showed unusually high uptake in the myocardium in one case and the kidneys in the other, which was not seen in the posttreat-

ment scans. These discrepant results raise the possibility that error may result from the introduction of the additional variables in an attempt to devise a semiquantitative method.

Image resolution is limited to 5 to 8 mm full-width at half-maximum, which is well short of the resolution capability of CT or magnetic resonance imaging (MRI). As a result, measured FDG uptake in tumor foci smaller than approximately 1.5 cm will be significantly underestimated because of partial-volume effects (82). On the other hand, high tumor uptake of FDG usually produces higher image contrast between tumor and normal or fibrotic tissue than is seen by CT, offsetting the effect of the resolution difference. Despite resolution limits, the sensitivity and specificity of PET imaging for many malignant tumors is higher than the sensitivity and specificity of CT or MRI, particularly in the posttherapy setting when CT falls short in differentiating benign posttreatment changes from viable tumor. Recently introduced PET/CT systems may allow for more-accurate assessment of tumor response by providing a combination of anatomic CT and metabolic PET information. In addition, most PET/CT systems have 4 to 16 slice multidetector CT capability, which facilitates orthogonal lesion measurements in three planes.

Patient Population to Monitor Response to Therapy

This section is a review of the patient population that should be selected for assessment of the response to therapy using FDG-PET imaging. For further information regarding clinical studies in each tumor category, please refer to the relevant chapters in this volume.

Lymphoma

Improvements in chemotherapy and combination therapy have lead to better overall response and survival rates for patients with HD. Nonetheless, in lymphoma, prognosis and therapy success also depend on histologic subtypes, accurate staging, and follow-up.

Therapy and Patient Population to Monitor Response

At early stages, both HD and nonbulky aggressive NHL are curable in a large proportion of patients (75%–90%) with combination therapy. At advanced stages (stage III or IV), only 30% to 50% of patients achieve complete remission; thus, evaluation of response to therapy is critical in this population as alternative therapy regimens can be started as soon as tumor nonresponse is recognized (83, 84). The patients who are first-line chemotherapy failures may benefit from different chemotherapy combinations

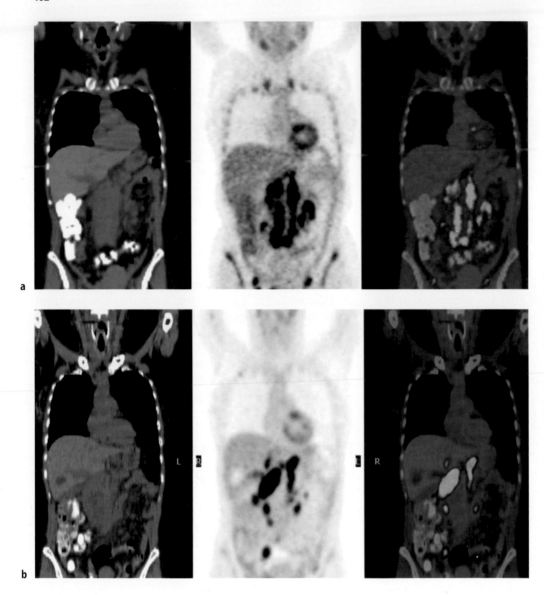

Figure 25.10. Patient with aggressive non-Hodgkin's lymphoma (NHL) in the retroperitoneal, iliac, and inguinal lymph nodes underwent PET/CT imaging before and 3 weeks after completion of first-line therapy. (**a**) Pretherapy axial CT (*left*), PET (*middle*), and PET/CT (*right*) sections demonstrate intense radiotracer uptake in the retroperitoneal, common, internal, and external iliac as well as inguinal lymph nodes, consistent with active lymphoma. (**b**) Two months following first-line therapy, the PET/CT images still demonstrate residual uptake in the retroperitoneal, iliac, and right external iliac lymph node stations with decrease in intensity and extent of FDG uptake and increase in other locations. This pattern is consistent with posttherapy residual disease that requires further therapy. The patient subsequently underwent high-dose therapy and stem cell transplant.

or high-dose chemotherapy with stem cell support (Figure 25.10). Thus, there is an advantage to assess the response to therapy early during chemotherapy (after cycles 1 to 3) because early evidence for persistent disease may suggest intervention with such alternative therapy options (Figure 25.11).

In indolent lymphoma, various combinations of therapy have not convincingly prolonged remission duration or survival (85). Immunotherapy or radioimmunotherapy in relapsing patients has proven to induce longer remissions, but survival benefits are yet to be established. In this patient population, there is no standard therapeutic algorithm; thus, regarding surveillance of response to therapy, avoiding unwarranted toxicity is a more-relevant objective than seeking a more-effective therapy option with better survival benefits in this tumor subtype.

Non-Small Cell Lung Carcinoma

Lung cancer is a major cause of cancer-related death in both men and women. Although FDG-PET response data are available for NSCLC, SCLC has not been studied to a large extent.

Therapy and Patient Population to Monitor Response to Therapy

There are two main limitations in the evaluation of the response to therapy in NSCLC. The first issue relates to the observer variability in defining response using anatomic imaging modalities in the presence of poorly defined boundaries of tumor that may have a significant inflammatory component because of the rich stroma in lung

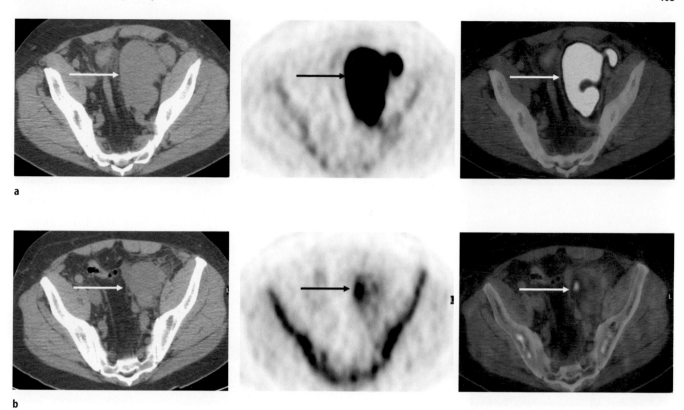

Figure 25.11. Patient with aggressive NHL in the pelvis underwent PET/CT imaging before and after three cycles of chemotherapy. (**a**) Pretherapy axial CT (*left*), PET (*middle*), and PET/CT (*right*) sections demonstrate intense radiotracer uptake in a left pelvic mass, consistent with active disease (*arrow*). (**b**) Following three cycles of chemotherapy, the PET/CT images still demonstrate residual uptake within the pelvic mass (*arrow*). Despite significant decrease in intensity and extent of uptake, this uptake is consistent with residual disease that may require therapy modification or which may change without completion of the entire therapy protocol.

cancer. Consequently, antitumor effect may be underrated by CT evaluation as persistent abnormalities do not necessarily exclude a complete response. The second issue is the fact that, in NSCLC, there is poor correlation between response rate and survival, and therefore determination of complete response may not translate into a long-term benefit for the patient. FDG-PET imaging can be a useful solution for the former problem; however, the latter issue remains as a limitation even with the higher predictive values obtained with PET compared to CT imaging.

In locally advanced disease (stage IIIAN2–IIIB) disease, data from pilot studies demonstrate a survival benefit for patients undergoing induction therapy followed by surgery compared with surgery alone (86). The 5-year survival for patients with no residual disease following induction therapy and surgery is 54% compared with 17% for those treated with induction therapy regardless of the response status. Hence, it would be beneficial to offer surgery only to those patients with objective evidence of response to therapy (87).

In advanced-stage (IIIB and IV) lung cancer, cisplatin- or carboplatin-based therapy yields only a modest prolongation of survival (median, 1.5–3 months). In selected patients with advanced-stage disease, at progression after platinum-based chemotherapy, there may be a survival benefit with docetaxel over supportive care alone (88).

Hence, in locally advanced NSCLC (stage IIIAN2), the role of FDG-PET imaging in the assessment of response to first-line chemotherapy is more useful in locally advanced disease than in advanced disease (Figure 25.12). Nonetheless, response evaluation in NSCLC is primarily geared toward identifying those patients who would not respond to therapy to prevent unnecessary toxicity rather than seeking a more-optimal alternative protocol, particularly in those undergoing second-line therapy with docetaxel.

Ideally, results of postinduction FDG-PET studies should lead to the selection of the proper patient group for surgery with curative intent; however, in clinical practice most patients are operated after induction therapy regardless of the degree of response, unless there is progression of disease, which occurs in one-third of patients (89). This issue should be further addressed with randomized trials to lead to possible management change.

FDG-PET imaging can also be useful in the assessment of the volumetric response to radiation therapy for locally advanced lung cancer. Chemotherapy-assisted novel hyperfractionated accelerated radiotherapy may be used in selected patients. Identification of nonresponders may allow physicians to limit more-aggressive and more-toxic approaches to the subgroups of patients who would benefit from them (90).

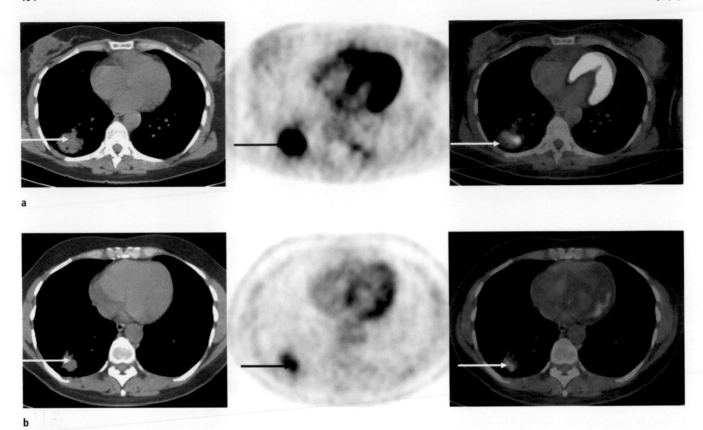

Figure 25.12. Patient with adenocarcinoma of the lung underwent PET/CT imaging before and after neoadjuvant therapy. (**a**) Pretherapy axial CT (*left*), PET (*middle*), and PET/CT (*right*) sections demonstrate intense radiotracer uptake in the right posterior lung corresponding to a large mass, consistent with active disease (*arrows*). (**b**) Following neoadjuvant therapy, the PET/CT images demonstrate minimal residual uptake within the lung mass, consistent with minimal residual disease (*arrows*). The patient subsequently underwent resection of the lower lung. The patient has been disease free for more than 12 months.

Breast Cancer

Breast cancer is the most common cancer in females and also the principal cause of death from cancer among women globally (91). Despite steadily increasing incidence rates, the leveling off in mortality and subsequent decline may be the result of earlier diagnosis and advances in treatment and management during follow-up.

Therapy and Patient Population to Monitor Response

Therapy of operable breast cancer involves surgery with or without radiotherapy or adjuvant chemotherapy. The effects of radiotherapy and surgery in early breast cancer are discussed in an overview of the randomized trials (92).

Tumors larger than 5 cm or smaller tumors with evidence of locally advanced disease (T3–T4, N2, N3) are currently treated with neoadjuvant chemotherapy. There is evidence from controlled studies that such therapy significantly increases the number of patients who can be offered breast-conserving surgery (93). The patient population who would benefit most from monitoring response to therapy includes those patients undergoing preoperative neoadjuvant (primary) therapy for locally advanced

breast cancer (Figure 25.13). Early results of clinical trials have suggested that patients with unresponsive tumors may achieve an improved survival from crossover to a non-cross-resistant regimen (94). It is essential, therefore, to accurately identify patients who would benefit from alternative treatments during the course of chemotherapy. FDG-PET imaging may assist in the selection of individualized therapy by providing the ability to determine tumor sensitivity to therapy.

Limited information from clinical trials suggests that the rapidity of response varies among patients. To induce maximal tumor response to most neoadjuvant chemotherapy, programs recommend four cycles of a single regimen before surgery. Some programs recommend even more cycles of the same treatment, using a "plateau" of response before surgery. Treatment is continued until no further change in the tumor is clinically apparent from one cycle to the next (or the tumor had responded completely). Particularly in this setting, FDG-PET imaging is a valuable tool to evaluate the response to therapy compared to anatomic imaging modalities. Demonstration of nonresponse early in the course of treatment may avoid the morbidity and cost of ineffective chemotherapy but may not change patient outcome. Accordingly, prediction of response has less significance in management of breast cancer than in lymphoma.

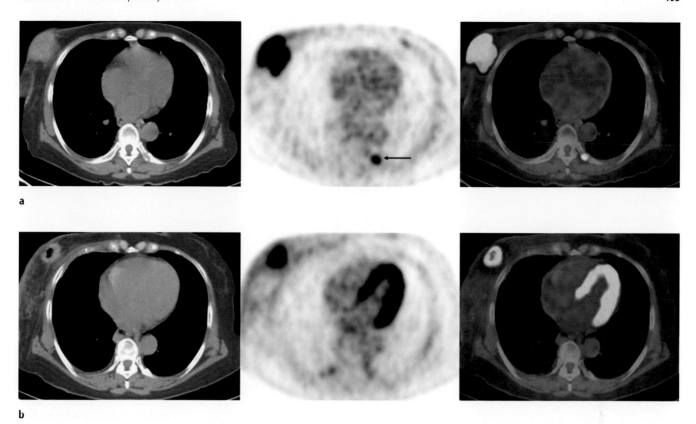

Figure 25.13. Patient with locally advanced infiltrating ductal breast cancer underwent PET/CT imaging before and after neoadjuvant therapy. (**a**) Pretherapy axial CT (*left*), PET (*middle*), and PET/CT (*right*) sections demonstrate intense radiotracer uptake in the right upper quadrant corresponding to a large mass, consistent with patient's known disease. Note the fracture in a posterior left rib (small arrow). (**b**) Following neoadjuvant therapy, the PET/CT images demonstrate significant decrease in intensity and extent of uptake in the corresponding lesion, consistent with good response to therapy. The patient subsequently underwent right modified mastectomy.

Monitoring response to therapy offers minimal benefit to patients who are undergoing induction therapy for metastatic disease because of the short median survival with conventional chemotherapy in this group (12–24 months). Additionally, the alternative therapy options are limited, which renders this group suboptimal for therapy surveillance. Thus, in this group of patients, the evaluation of response to therapy should be geared toward avoidance of toxicity in nonresponding patients. There is no proven method with which to monitor response to postoperative adjuvant chemotherapy because of the removal of reference tumor activity in the primary tumor and lymph nodes. Hence, FDG-PET imaging is not of value to determine therapeutic efficacy in this patient population.

Head and Neck Carcinoma

Head and neck cancer accounts for approximately 5% of all new cancer cases in the Western world each year. Most head and neck malignancies are squamous cell carcinomas (IINSCC). In patients with early-stage disease, both radiation and surgery are often curative whereas the overall survival for patients with locally advanced disease is less than 50%. Mortality is most frequently the result of locoregional recurrence, and death from distant metasta-

sis is less common. In locally advanced disease, potentially curative therapy exists with the combined-modality treatment with chemotherapy, radiotherapy, and surgery (95).

Therapy and Patient Population to Monitor Response

The patient population who would benefit from monitoring response to treatment includes those who are planned to undergo sequential preoperative neoadjuvant chemotherapy and RT or neoadjuvant chemotherapy concurrent with RT. The main benefit of neoadjuvant chemotherapy is to preserve organ structure and function, similar to that in locally advanced breast cancer (Figure 25.14) (96, 97). This sequential therapy setting also allows for modifications in management during the course of treatment in nonresponders to optimally balance benefits with treatment-associated toxicities (98).

The median survival for patients with local or metastatic recurrent HNSCC is less than 1 year (99). FDG-PET imaging may have a role in the evaluation of response to therapy for guiding various alternative therapy strategies, although its impact would probably be not as significant as in those undergoing neoadjuvant therapy before surgery at initial presentation.

Pre

Post

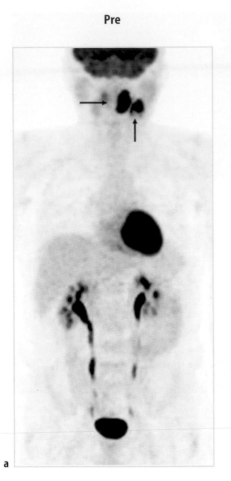

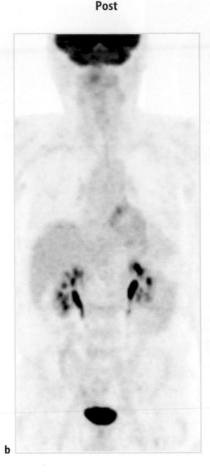

Figure 25.14. Patient with locally advanced head and neck squamous cell carcinoma underwent FDG-PET imaging before and following neoadjuvant therapy. (**a**) Coronal PET image reveals intense FDG uptake in the left tonsil (*horizontal arrow*) as well as in a left jugular lymph node (*vertical arrow*), consistent with primary disease and local lymph node metastasis, respectively. (**b**) One month following completion of neoadjuvant chemoradiotherapy, coronal PET image demonstrates complete resolution of the primary disease in the left tonsil and metastatic disease in the left jugular lymph node, consistent with complete response to therapy.

a   b

Esophageal Carcinoma

Despite significant decrease in surgical mortality rates in patients with esophageal cancer, long-term survival has not changed dramatically. Recent evidence reveals that the addition of neoadjuvant treatment may improve resection rates, reduce the risk of recurrence, and thereby improve survival. Because of the poor prognosis and the risks associated with surgical intervention, accurate assessment of therapy is essential for optimal treatment planning.

Therapy and Patient Population to Monitor Response

For patients with locally advanced esophageal cancer (T3 or T4 status: stage III), neoadjuvant chemoradiotherapy consisting of cisplatin, 5-fluorouracil (5-FU), and radiation followed by esophagectomy has been the mainstay of treatment (100–102). The rationale behind such a treatment is to improve the rate of curative resection by downstaging tumor, early eradication of micrometastases, and increasing radiosensitivity. Improved 5-year survival rates approaching 60% have been reported following pathologic complete response with concomitant RT and chemother-

apy (100). However, local failure, local recurrence, and regional lymph-node metastasis are frequently detected after definitive chemoradiotherapy (103, 104). FDG-PET imaging is particularly useful in patients undergoing neoadjuvant therapy because it allows accurate stratification of patients into surgical and multimodality protocols, an important aspect because approximately 50% of patients do not respond to current chemotherapy regimens (101). In patients failing to respond to neoadjuvant treatment, the risk of treatment-related morbidity and mortality can be avoided if FDG-PET imaging accurately determines therapy failure. Furthermore, these patients can be placed on alternative therapy modalities early during the course of disease.

Currently, the role of postoperative adjuvant chemoradiotherapy is undefined although it may be beneficial in patients with positive resection margins. As with HNSCC, in an adjuvant therapy setting, the role of FDG-PET imaging is limited in monitoring the response to therapy.

Two-thirds of patients have a recurrence within 1 year, and in approximately one-third of patients the recurrence is within the primary surgical field (105). FDG-PET imaging may allow a highly sensitive and specific means to distinguish postoperative scar from tumor recurrence in the follow-up period. FDG-PET imaging is also valuable for further investigation of abnormal masses seen with other imaging modalities, especially in asymptomatic patients. Treatment of recurrent esophageal cancer may include palliative therapy with the use of any standard treatment as well as clinical trials of novel therapy modalities. Although evaluation of response to therapy using FDG-PET imaging is of limited value, it may allow early changes in the treatment of unresponsive tumors or discontinuation of therapy in nonresponding patients.

Colorectal Carcinoma

Adenocarcinoma of the colon affects 5% of the population in the United States and in most of Western countries. Potentially curative resection at disease presentation can be performed only in 70% to 80% of the patients, and overall survival at 5 years is less than 60%. Improvements in surgical and adjuvant therapies, more extensive screening programs, and recent advances in detection techniques including imaging modalities have resulted in a decline in colon cancer mortality in the United States (106).

Therapy and Patient Population to Monitor Response to Therapy

FDG-PET imaging is most useful in monitoring advanced-stage colorectal cancer. Advanced disease is associated with a poor prognosis. Chemotherapy has demonstrated effective palliation, improvement of quality of life, and symptom improvement in such patients. Systemic chemotherapy doubles the survival of these patients compared to untreated controls. For nearly four decades, fluorouracil (5-FU) has been the mainstay of treatment (107). The use of 5-FU in combination with radiation therapy in primary unresectable colorectal cancer is also associated with improved survival.

Chemotherapeutic options in the treatment of advanced colorectal cancer have markedly improved during the last years, partly as a result of the high-dose 5-FU regimen, but also from the development of new cytotoxic agents and drug combinations. Today most patients are treated by a sequential therapeutic concept using the newer drugs mainly for second- or third-line therapy. Combination of oxaliplatin with 5-FU/FA could downstage previously unresectable liver metastases for potentially curative surgery in some patients (108).

FDG-PET imaging is promising in the optimization of therapy, particularly as more targeted therapies become available. FDG-PET imaging enables very early and more specific indication of response to preoperative therapies or of the presence of residual disease after surgical tumor resection or interventional tumor ablation of metastases or recurrences (109). FDG-PET imaging may play an important role in avoiding major surgery in patients for whom curative surgery is intended after chemotherapy or radiotherapy.

Testicular Cancer

The incidence of testicular cancer is increasing in the developed countries. There are two main types of germinal cell testicular cancers: seminomatous germ cell tumors (SGCT) and nonseminomatous germ cell tumors (NSGCT). SGCT are the most common testicular tumors (40%). Metastatic spread is more common with NSGCT (70%) than with SGCT (30%) at initial diagnosis (110).

Therapy and Patient Population to Monitor Response to Therapy

Early-stage (stages I–IIB) SGCTs are treated with RT with negligible relapse rates. Although it is still controversial, early-stage NSGCTs (stages I–IIB) are usually treated surgically, including retroperitoneal lymph node dissection (RPLND), followed by adjuvant chemotherapy or close surveillance (110). Primary chemotherapy is not recommended in patients with high-risk clinical stage I NSGCT because it has not yet been proven to be beneficial and can cause significant long-term sequelae (111). Hence, there is no definite role for FDG-PET imaging in monitoring response to therapy for early-stage disease for both subtypes of testicular cancers.

The treatment of choice in patients with high-burden, advanced GCTs (stages IIC–III; bulky lymph node metastases measuring more than 5 cm in diameter and/or visceral metastases) is cis-platinum-based chemotherapy followed by residual tumor resection or RPLND (112). Following chemotherapy, there are two groups of patients who should be distinguished based on response to therapy. The first group includes the majority of patients (~80%), who will be likely to obtain an initial complete response (low-risk group). The focus of clinical trials in this group is to reduce chemotherapy-induced toxicity while maintaining a high cure rate. The second group comprises those patients not achieving a durable response (~20%; poor-risk group), of whom nearly 25% may be cured with salvage cisplatin plus ifosfamide chemotherapy instead of RPLND (110, 113).

The primary objective of therapy trails in this high-risk group of patients is to improve the proportion of complete response. Not surprisingly, residual radiographic abnormalities are observed in up to 78% of patients who undergo induction chemotherapy for disseminated GCTs (110). In the high-risk group, accurate differentiation of patients with favorable outcome (complete responders) from those with unfavorable outcome (partial or nonresponders) by FDG-PET imaging may select the proper patient population for alternative therapy strategies and potentially avoid unnecessary RPLND.

In the low-risk group, early determination of response to therapy by FDG-PET imaging may also be helpful to determine the optimal number of treatment cycles that would result in low or no morbidity. One exception is teratoma, which is a relatively chemotherapy-resistant histologic subtype; thus, cytotoxic agents may not be appropriate for treatment. Hence, FDG-PET imaging does not seem to offer clinical benefit in the management of patients with teratomas.

As many as 30% of patients with testicular cancer may relapse (110, 112). The durable complete remission rate may exceed 50% with high-dose chemotherapy at first relapse (114). The choice of salvage surgery versus autologous bone marrow transplantation (BMT) for refractory disease may be based on the extent of metastatic disease. FDG-PET imaging can be particularly useful in the accurate assessment of disease burden to categorize patients into different treatment protocols. In the group who are planned to undergo high-dose chemotherapy followed by autologous BMT, prediction of response to therapy is crucial as not all patients would benefit from this treatment. In these patients, FDG-PET imaging may play a role guiding the appropriate therapy approach by accurately identifying the nonresponders for whom the morbidity and the cost of the treatment could be spared. Furthermore, FDG-PET imaging may be indicated to localize recurrence in patients with rising blood levels of tumor markers as well as in patients with multiple residual masses who have marker-negative active disease.

Ovarian Cancer

Ovarian carcinoma is the seventh most common female cancer worldwide and the fifth leading cause of cancer death among women in the developed countries. Approximately 70% of women present with advanced disease at diagnosis. Most patients respond to initial therapy; however, the relapse rate is high, and the 5-year survival with advanced disease is only 25% to 35% (115).

Therapy and Patient Population to Monitor Response to Therapy

The treatment of ovarian cancer is based on the stage of the disease. The mainstay of early-stage ovarian cancer [International Federation of Gynecology and Obstetrics (FIGO) stages IA–IIA] is comprehensive surgical staging followed by appropriate adjuvant chemotherapy (116). In this group of patients, FDG-PET imaging may not offer a benefit in the clinical decision-making process as monitoring response may not be feasible in patients who are technically tumor free following surgery.

The majority of ovarian cancer patients present with advanced disease at diagnosis (FIGO stages IIB–IV). In these patients, aggressive cytoreductive surgery followed by adjuvant chemotherapy and/or external irradiation is the treatment of choice. Because these patients would have detectable tumor mass after surgery (usually =1–2 cm), FDG-PET imaging may have a role in evaluating the response to adjuvant therapy. In those patients who demonstrate no response, FDG-PET imaging may guide clinicians to change management toward areas of research, including non-cross-resistant antineoplastic agents, intraperitoneal therapy, anticancer vaccines, gene therapy, and antiangiogenic therapy (116).

In patients with recurrent disease, although expected survival is not long, depending on the response to therapy, secondary non-cross-resistant chemotherapies can be administered, or resistance-modulating agents, biologic therapies, and high-dose chemotherapy with bone marrow support (ABMT) can be offered. FGD-PET imaging may have a niche in early differentiation of responders from nonresponders in this group of patients, although it may not affect survival to any large extent.

Integrated PET/CT Imaging and Monitoring Response to Therapy

PET/CT fusion imaging, which integrates two disparate imaging modalities, is now gaining momentum toward becoming the diagnostic tool in the staging and restaging algorithm of cancer. Following therapy, subtle metabolic findings on FDG-PET imaging, that would otherwise have been disregarded, may result in the detection of residual disease after correlation with the simultaneously acquired morphologic data. Alternatively, equivocal CT findings, which could represent either recurrent tumor or posttherapy fibrosclerosis, can now be distinguished with the help of the additional information provided by FDG-PET data.

In the posttherapy setting, PET/CT fusion can improve the accuracy of PET imaging in distinguishing recurrent disease from benign posttherapy changes, delineating the anatomic location of metastatic disease, and monitoring the response to therapy by solving a myriad of problems inherent in the posttherapy assessment of cancer. As PET/CT systems are not yet in widespread use, the data regarding the impact of fusion imaging on posttherapy patient management are still in evolution. Nevertheless, the preliminary data evinced that PET/CT findings have resulted in the accurate staging and improved evaluation of response to therapy, in particular, in patients with head and neck cancer or genitourinary and gastrointestinal malignancies (117, 118).

In a recent study, the accurate spatial localization offered by the PET/CT fusion technique provided a better assessment of response to treatment and changed clinical management in up to 30% of cancer patients (117). PET/CT imaging provides accurate information about the anatomic planes and excludes false-positive findings. There are fewer anatomic landmarks in the neck than in other parts of the body; hence, combining PET with CT imaging provides crucial information in differentiating physiologic activity from viable tumor in the cervical muscles, brown fat at the base of the neck (118), vocal cords, lymphoid tissues, mucosal surfaces, and salivary glands, as all these conditions can demonstrate non-specific high FDG uptake. In patients who have paralysis of the recurrent laryngeal nerve, unilaterally increased FDG uptake in the posterior arytenoid muscle in the non-paralyzed vocal cord can also be a potential source of false-positive finding without the guidance of PET/CT fusion (119).

In conclusion, the role of FDG-PET imaging may be significant in differentiating between benign tissue changes and persistent residual tumor; nevertheless further studies are necessary to establish its role in the evaluation of response to therapy.

Summary

The role of FDG-PET imaging in treatment evaluation is most firmly established in posttreatment assessment of Hodgkin's disease and non-Hodgkin's lymphoma. Clinical effectiveness has also been demonstrated for postchemotherapy or postradiation assessment of head and neck cancer and postchemotherapy assessment of testicular tumors. Early results support the use of FDG-PET imaging for preoperative assessment of neoadjuvant therapy in lung cancer to avoid pneumonectomy for tumors that have not responded to treatment. A potentially valuable application of PET imaging lies in the early prediction of treatment response in lymphoma and in neoadjuvant therapy of locally advanced primary breast cancer and locally advanced esophageal cancer.

Treatment evaluation of colorectal cancer metastatic to the liver by PET demonstrates different approaches but has little potential for clinical impact. Semiquantitative assessment of FDG uptake by means of SUV appears to be sufficiently accurate for assessment of uptake in serial studies of the same patient under the same conditions.

Acknowledgments

This work was in part supported by a grant from the Sutter Institute of Medical Research, Sacramento, CA.

References

1. Nygren P, SBU group. Swedish Council on Technology Assessment in Health Care. What is cancer chemotherapy? Acta Oncol 2001;40:166–174.

2. World Health Organization. WHO handbook for reporting results of cancer treatment. Geneva: World Health Organization, 1979:48.

3. Therasse P, Arbuck SC, Eisenhauer EA, Wanders J, Kaplan RS, Rubinstein L, et al. New guidelines to evaluate the response to treatment in solid tumors. J Natl Cancer Inst 2000;92:205–216.

4. Choi H, Charnsangavej C, de Castro Faria S, et al. CT evaluation of the response of gastrointestinal stromal tumors after imatinib mesylate treatment: a quantitative analysis correlated with FDG-PET findings. Am J Radiol 2004;183:1619–1628.

5. Mazumdar M, Smith A, Schwartz LH. A statistical simulation study finds discordance between WHO criteria and RECIST guideline. J Clin Epidemiol 2004;57:358–365.

6. Cheson BD, Horning SJ, Coiffier B, et al. Report of an international workshop to standardize response criteria for non-Hodgkin's lymphomas. NCI Sponsored International Working Group. J Clin Oncol 1999;17:1244–1253.

7. Hasenclever D, Diehl V, for the International Prognostic Factors Project on Advanced Hodgkin's Disease. A prognostic score for advanced Hodgkin's disease. N Engl J Med 1998;339:1506–1514.

8. Ng AK, Bernardo MV, Weller E, et al. Second malignancy after Hodgkin disease treated with radiation therapy with or without chemotherapy: long-term risks and risk factors. Blood 2002;100:1989–1996.

9. Young H, Baum R, Cremerius U, Herholz K, Hoekstra O, Lammertsma AA, et al. Measurement of clinical and subclinical tumour response using [^{18}F]-fluorodeoxyglucose and positron emission tomography: review and 1999 EORTC recommendations. European Organization for Research and Treatment of Cancer (EORTC) PET Study Group. Eur J Cancer 1999;35(13):1773–1782.

10. Girardier L, Seydoux J. Neural control of brown adipose tissue. In: Trayburn P, Nicholls D, editors. Brown Adipose Tissue. London: Arnold, 1986:123–147.

11. Anholt R, De Souza E, Oster-Granite M, Snyder S. Peripheral-type benzodiazepine receptors: autoradiographic localization in whole-body sections of neonatal rats. J Pharmacol Exp Ther 1985;233:517–526.

12. Findlay M, Young H, Cunningham D, et al. Noninvasive monitoring of tumor metabolism using fluorodeoxyglucose and positron emission tomography in colorectal cancer liver metastases: correlation with tumor response to fluorouracil. J Clin Oncol 1996;14:700–708.

13. Jansson T, Westlin JE, Ahlstrom H, et al. Positron emission tomography studies in patients with locally advanced and/or metastatic breast cancer: a method for early therapy evaluation? J Clin Oncol 1995;13:1470–1477.

14. Romer W, Hanauske AR, Ziegler S, et al. Positron emission tomography in non-Hodgkin's lymphoma: assessment of chemotherapy with fluorodeoxyglucose. Blood 1998;91:4464–4471.

15. Kostakoglu L, Coleman M, Leonard JP, Kuji I, Zoe H, Goldsmith SJ. Positron emission tomography predicts prognosis after one cycle of chemotherapy in aggressive lymphoma and Hodgkin's disease. J Nucl Med 2002;43:1018–1027.

16. Schelling M, Avril N, Nahrig J, et al. Positron emission tomography using [(18)F]fluorodeoxyglucose for monitoring primary chemotherapy in breast cancer. J Clin Oncol 2000;18:1689–1695.

17. Bassa P, Kim E, Inoue T, et al. Evaluation of preoperative chemotherapy using PET with fluorine-18-fluorodeoxyglucose in breast cancer. J Nucl Med. 1996;37:931–938.

18. Fukuda K, Kojiro M, Chiu Jen-Fu. Demonstration of extensive chromatin cleavage in transplanted Morris hepatoma tissue: apoptosis or necrosis? Am J Pathol 1993;142:935–946.

19. Kerr JF, Wyllie AH, Currie AR. Apoptosis: a basic biological phenomenon with wide-ranging implications in tissue kinetics. Br J Cancer 1972;26(4):239–257.

20. Wang TH, Wang HS, Soong YK, et al. Paclitaxel-induced cell death: where the cell cycle and apoptosis come together. Cancer (Phila) 2000;88:2619–2628.

21. Warburg O. On respiratory impairment in cancer cells. Science 1956;124:269–270.

22. Suit HD. Clinical radiation biology. In: Choi NC, Grillo HC, editors. Thoracic Oncology. New York: Raven Press, 1983:51–58.

23. Thompson LH, Suit HD. Proliferation kinetics of x-irradiation mouse L cells studies with time-lapse photography. Int J Radiat Biol 1969;15:347–363.

24. Sinclair WK. X-ray induced heritable damage (small colony formation) in cultured mammalian cells. Radiat Res 1964;21:584–611.

25. Mortimer JE, Dehdashti F, Siegel BA, Trinkaus K, Katzenellenbogen JA, Welch MJ. Metabolic flare: indicator of hormone responsiveness in advanced breast cancer. J Clin Oncol 2001;19:2797–2803.

26. Nygren P, Glimelius B. The Swedish Council on Technology Assessment in Health Care. SBU report on cancer chemotherapy. Project objectives, the working process, key definitions and general aspects on cancer trial methodology and interpretation. Acta Oncol 2001;40:155–165.

27. de Hemricourt E, De Boeck K, Hilte F, Abib A, Kockx M, Vandevivere J, De Bock R. Sarcoidosis and sarcoid-like reaction following Hodgkin's disease. Report of two cases. Mol Imaging Biol 2003;5(1):15–19.

28. Koss MN. Pulmonary lymphoid disorders. Semin Diagn Pathol 1995;12:158–171.

29. Marom EM, McAdams HP, Butnor KJ, Coleman RE. Positron emission tomography with fluoro-2-deoxy-D-glucose (FDG-PET) in the staging of post transplant lymphoproliferative disorder in lung transplant recipients. J Thorac Imaging 2004;19:74–78.

30. van der Hoeven JJ, Krak NC, Hoekstra OS, et al. ^{18}F-2-Fluoro-2-deoxy-D-glucose positron emission tomography in staging of locally advanced breast cancer. J Clin Oncol 2004;22:1253–1259.

31. Duhrsen U, Villeval JL, Boyd J, et al. Effects of recombinant human granulocyte colony-stimulating factor on hematopoietic progenitor cells in cancer patients. Blood 1988;72:2074–2081.

32. Wang GJ, Cai L. Relatively low-dose cyclophosphamide is likely to induce apoptotic cell death in rat thymus through Fas/Fas ligand pathway. Mutat Res 1999;427:125–133.

33. Rafla S, Rotman M. Effects of radiation on cells In: Rafla S, Rotman M, editors. Introduction to Radiotherapy. St. Louis: Mosby, 1974:51–55.

34. Higashi K, Clavo AC, Wahl RL et al. In vitro assessment of 2-fluoro-2-deoxy-D-glucose, L-methionine and thymidine as agents to monitor the early response of a human adenocarcinoma cell line to radiotherapy. J Nucl Med 1993;34:773–779.

35. Whitmore GF, Till JE, Gwatkin RB, Siminovitch L, Graham AF. Increase of cellular constituents in X-irradiated mammalian cells. Biochim Biophys Acta 1958;30:583–590.

36. Song CW, Sung JH, Clement JJ, Levitt SH. Vascular changes in neuroblastoma of mice following x-irradiation. Cancer Res 1974;34:2344–2350.

37. Kubota R, Yamada S, Kubota K, et al. Intratumoral distribution of fluorine-18 fluorodeoxyglucose in vivo: high accumulation in macrophages and granulation tissue studied by autoradiography. J Nucl Med 1992;33:1972–1980.

38. Brown RS, Leung JY, Kison PV, et al. Glucose transporters and FDG uptake in untreated primary human non-small cell lung cancer. J Nucl Med 1999;40:556–565.

39. Avril N, Menzel M, Dose J, et al. Glucose metabolism of breast cancer assessed by 18F-FDG-PET: histologic and immunohistochemical tissue analysis. J Nucl Med 2001;42:9–16.

40. Haberkorn U, Strauss LG, Dimitrakopoulou A, Engenhart R,Oberdorfer F, Ostertag H, et al. PET studies of fluorodeoxyglucose metabolism in patients with recurrent colorectal tumors receiving radiotherapy. J Nucl Med 1991;32(8):1485–1490.

41. Collins BT, Gardner LJ, Verma AK, Lowe VJ, Dunphy FR, Boyd JH. Correlation of fine needle aspiration biopsy and fluoride-18 fluorodeoxyglucose positron emission tomography in the assessment of locally recurrent and metastatic head and neck neoplasia. Acta Cytol 1998;42:1325–1329.

42. Haberkorn U, Strauss LG, Dimitrakopoulou A, et al. PET studies for fluorodeoxyglucose metabolism in patients with recurrent colorectal tumors receiving radiotherapy. J Nucl Med 1991;32:1485–1490.

43. Brun E, Kjellen E, Tennvall J, et al. FDG-PET studies during treatment: prediction of therapy outcome in head and neck squamous cell carcinoma. Head Neck 2002;24:127–135.

44. Conessa C, Clement P, Foehrenbach H, et al. Value of positron-emission tomography in the post-treatment follow-up of epidermoid carcinoma of the head and neck. Rev Laryngol Otol Rhinol 2001;122:253–258.

45. Choi YW, Munden RF, Erasmus JJ, Park KJ, Chung WK, Jeon SC, Park CK. Effects of radiation therapy on the lung: radiologic appearances and differential diagnosis. Radiographics 2004;24:985–997.

46. Senan S, De Ruysscher D, Giraud P, Mirimanoff R, Budach V. Literature-based recommendations for treatment planning and execution in high-dose radiotherapy for lung cancer. Radiother Oncol 2004;71:139–146.

47. Theuws JC, Seppenwoolde Y, Kwa SL, et al. Changes in local pulmonary injury up to 48 months after irradiation for lymphoma and breast cancer. Int J Radiat Oncol Biol Phys 2000;47:1201–1208.

48. Marx RE. Osteoradionecrosis: a new concept of its pathophysiology. J Oral Maxillofac Surg 1983;41:283–288.

49. Clayman L. Clinical controversies in oral and maxillofacial surgery: Part two. Management of dental extractions in irradiated jaws: a protocol without hyperbaric oxygen therapy. J Oral Maxillofac Surg 1997;55:275–281.

50. Liu SH, Chang JT, Ng SH, Chan SC, Yen TC. False positive fluorine-18 fluorodeoxy-D-glucose positron emission tomography finding caused by osteoradionecrosis in a nasopharyngeal carcinoma patient. Br J Radiol 2004;77:257–260.

51. Dunn BK, Wickerham DL, Ford LG. Prevention of hormone-related cancers: breast cancer. J Clin Oncol 2005;23(2):357–367.

52. .Reddel RR, Sutherland RL. Tamoxifen stimulation of human breast cancer cell proliferation in vitro: a possible model for tamoxifen tumour flare. Eur J Cancer Clin Oncol 1984;20:1419–1424.

53. Noguchi S, Motomura K, Inaji H, et al. Up-regulation of estrogen receptor by tamoxifen in human breast cancer. Cancer (Phila) 1993;71:1266–1272.

54. Coleman RE, Mashiter G, Whitaker KB, et al. Bone scan flare predicts successful systemic therapy for bone metastases. J Nucl Med 1988;29:1354–1359.

55. Vogel CL, Schoenfelder J, Shemano I, et al. Worsening bone scan in the evaluation of antitumor response during hormonal therapy of breast cancer. J Clin Oncol 1995;13:1123–1128.

56. van Schelven WD, Pauwels EKJ. The flare phenomenon: far from fair and square. Eur J Nucl Med 1994;21:377–380.

57. Campbell FC, Blamey RW, Elston CW. Quantitative oestradiol receptor values in primary breast cancer and response of metastases to endocrine therapy. Lancet 1981;1:1317–1319.

58. Dehdashti F, Flanagan FL, Mortimer JE. Positron emission tomographic assessment of "metabolic flare" to predict response of metastatic breast cancer to antiestrogen therapy. Eur J Nucl Med 1999;26:51–56.

59. McGuire AH, Dehdashti F, Siegel BA. Positron tomographic assessment of 16alpha-[F-18]-fluoro-17beta-estradiol uptake in metastatic breast carcinoma. J Nucl Med 1991;32:1526–1531.

60. Inoue T, Kim EE, Wallace S, Yang DJ, Wong FC, Bassa P, et al. Positron emission tomography using [^{18}F]fluorotamoxifen to evaluate therapeutic responses in patients with breast cancer: preliminary study. Cancer Biother Radiopharm 1996;11(4):235–245

61. Ottaiano A, Mollo E, Di Lorenzo G, Pisano C, et al. Prospective clinical trials of biotherapies in solid tumors: a 5-year survey. Cancer Immunol Immunother 2005;54:44–50.

62. Press OW. Physics for practitioners: the use of radiolabeled monoclonal antibodies in B-cell non-Hodgkin's lymphoma. Semin Hematol 2000;37:2–8.

63. Torizuka T, Zasadny KR, Kison PV, Rommelfanger SG, Kaminski MS, Wahl RL. Metabolic response of non-Hodgkin's lymphoma to ^{131}I-anti-B1 radioimmunotherapy: evaluation with FDG-PET. J Nucl Med 2000;41:999–1005.

64. Hoekstra CJ, Paglianiti I, Hoekstra OS, Smit EF, Postmus PE, Teule GJ, et al. Monitoring response to therapy in cancer using [^{18}F]-2-fluoro-2-deoxy-D-glucose and positron emission tomography: an overview of different analytical methods. Eur J Nucl Med 2000;27(6):731–743.

65. Graham MM, Peterson LM, Hayward RM. Comparison of simplified quantitative analyses of FDG uptake. Nucl Med Biol 2000;27(7):647–655.

66. Findlay M, Young H, Cunningham D, Iveson A, Cronin B, Hickish T, et al. Noninvasive monitoring of tumor metabolism using fluorodeoxyglucose and positron emission tomography in colorectal cancer liver metastases: correlation with tumor response to fluorouracil. J Clin Oncol 1996;14:700–708.

67. Zasadny KR, Wahl RL. Standardized uptake values of normal tissues at PET with 2-[fluorine-18]-fluoro-2-deoxy-D-glucose: variations with body weight and a method for correction. Radiology 1993;189(3):847–850.

68. Kim CK, Gupta NC. Dependency of standardized uptake values of fluorine-18 fluorodeoxyglucose on body size: comparison of body surface area correction and lean body mass correction. Nucl Med Commun 1996;17(10):890–894.

69. Hamberg LM, Hunter GJ, Alpert NM, Choi NC, Babich JW, Fischman AJ. The dose uptake ratio as an index of glucose metabolism: useful parameter or oversimplification? J Nucl Med 1994;35(8):1308–1312.

70. Keyes JW Jr. SUV: standard uptake or silly useless value? J Nucl Med 1995;36(10):1836–1839.

71. Langen KJ, Braun U, Rota KE, et al. The influence of plasma glucose levels on fluorine-18-fluorodeoxyglucose uptake in bronchial carcinomas. J Nucl Med 1993;34:355–359.

72. Lindholm P, Minn H, Leskinen-Kallio S, Bergman J, Ruotsalainen U, Joensuu H. Influence of the blood glucose concentration on FDG uptake in cancer: a PET study. J Nucl Med 1993;34(1):1–6.

73. Wahl RL, Henry CA, Ethier SP. Serum glucose: effects on tumor and normal tissue accumulation of 2-[F-18]-fluoro-2-deoxy-D-glucose in rodents with mammary carcinoma. Radiology 1992;183(3):643–647.

74. Patlak CS, Blasberg RG, Fenstermacher JD. Graphical evaluation of blood-to-brain transfer constants from multiple-time uptake data. J Cereb Blood Flow Metab 1983;3:1–7.

75. Hunter GJ, Hamberg LM, Alpert NM, Choi NC, Fischman AJ. Simplified measurement of deoxyglucose utilization rate. J Nucl Med 1996;37:950–955.

76. Weber WA, Ziegler SI, Thodtmann R, Hanauske AR, Schwaiger M. Reproducibility of metabolic measurements in malignant tumors using FDG-PET. J Nucl Med 1999;40:1771–1777.

77. Weber WA, Petersen V, Schmidt B, Tyndale-Hines L, Link T, Peschel C, Schwaiger M. Positron emission tomography in non-small-cell lung cancer: prediction of response to chemotherapy by quantitative assessment of glucose use. J Clin Oncol 2003;21:2651–2657.

78. Minn H, Leskinen-Kallio S, Lindholm P, Bergman J, Ruotsalainen U, Teras M, et al. [18F]Fluorodeoxyglucose uptake in tumors: kinetic vs. steady-state methods with reference to plasma insulin. J Comput Assist Tomogr 1993;17:115–123.

79. Kole AC, Niewig OE, Pruim J, Paans AM, Plukker JT, Hoekstra O, et al. Standardized uptake value and quantification of metabolism for breast cancer imaging with FDG and [11C]tyrosine PET. J Nucl Med 1997;38:692–696.

80. Wahl RL, Zasadny K, Helvie M, Hutchins GD, Weber B, Cody R. Metabolic monitoring of breast cancer chemohormonotherapy using positron emission tomography: initial evaluation. J Clin Oncol 1993;11(11):2101–2111.

81. Graham MM, Peterson LM, Hayward RM. Comparison of simplified quantitative analyses of FDG uptake. Nucl Med Biol 2000;27(7):647–655.

82. Kessler RM, Ellis JR Jr, Eden M. Analysis of emission tomographic scan data: limitations imposed by resolution and background. J Comput Assist Tomogr 1984;8:514–522.

83. Brandt L, Kimby E, Nygren P, Glimelius B. A systematic overview of chemotherapy effects in Hodgkin's disease. Acta Oncol 2001;40:185–197.

84. Coiffier B, Gisselbrecht C, Vose JM, et al. prognostic factors in aggressive malignant lymphomas. Description and validation of prognostic index that could identify patients requiring a more intensive therapy. J Clin Oncol 1991;9:211–219.

85. Brandt L, Kimby E, Nygren P, Glimelius B, et al. A systematic overview of chemotherapy effects in indolent non-Hodgkin's lymphoma. Acta Oncol 2001;40(2-3):213–223.

86. Sorenson S, Glimelius B, Nygren P, et al. A systematic overview of chemotherapy effects in non-small cell lung cancer. Acta Oncol 2001;40(2-3):327–339.

87. Martini N, Kris MG, Flehinger BJ, et al. Preoperative chemotherapy for stage IIIa (N2) lung cancer: the Sloan-Kettering experience with 136 patients. Ann Thorac Surg 1993;551365–1373.

88. Sorenson S, Glimelius B, Nygren P. A systematic overview of chemotherapy effects in non-small cell lung cancer. Acta Oncol 2001;40:327–339.

89. Sekine I, Tamura T, Kunitoh H, et al. Progressive disease rate as a surrogate endpoint of phase II trials for non-small-cell lung cancer. Ann Oncol 1999;10:731–733.

90. Hicks RJ, Mac Manus MP, Matthews JP, Hogg A, Binns D, Rischin D, et al. Early FDG-PET imaging after radical radiotherapy for non-small-cell lung cancer: inflammatory changes in normal tissues correlate with tumor response and do not confound therapeutic response evaluation. Int J Radiat Oncol Biol Phys 2004;60:412–418.

91. Bray F, McCarron P, Parkin DM. The changing global patterns of female breast cancer incidence and mortality. Breast Cancer Res 2004;6:229–239.

92. Effects of radiotherapy and surgery in early breast cancer. An overview of the randomized trials. Early Breast Cancer Trialists' Collaborative Group. N Engl J Med 1995;333:1444–1455.

93. Schwartz GF, Hortobagyi GN. Proceedings of the consensus conference on neoadjuvant chemotherapy in carcinoma of the breast, April 26–28, 2003, Philadelphia, Pennslyvania. Cancer (Phila) 2004;100:2512–2532.

94. Smith IC, Heys SD, Hutcheon AW. Neoadjuvant chemotherapy in breast cancer: significantly enhanced response with docetaxel. J Clin Oncol 2002;20:1456–1466.

95. Parker SL, Tong T, Bolden S, Wingo PA. Cancer statistics. CA Cancer J Clin 1997;47:5–27.

96. Wanebo HJ, Chougule P, Akerley WL III, et al. Preoperative chemoradiation coupled with aggressive resection as needed ensures near total control in advanced head and neck cancer. Am J Surg 1997;174:518–522.

97. Wendt TG, Grabenbauer GG, Rodel CM, et al. Simultaneous radiochemotherapy versus radiotherapy alone in advanced head and neck cancer: a randomized multicenter study. J Clin Oncol 1998;16:1318–1324.

98. Posner M. Sequential chemotherapy for the curative treatment of squamous cell cancer of the head and neck: a new paradigm. Oncol Spectrums 2001;2:193–202.

99. Scantz SP, Harrison LB, Forastiere AA. Tumors of the nasal cavity and paranasal sinuses, nasopharynx, oral cavity and oropharynx. In: Devita VT, Helmann S, Rosenberg SA, editors. Cancer Principles and Practice of Oncology. Philadelphia: Lippincott-Raven, 1997:741–801.

100. Geh JI, Crellin AM, Glynne-Jones R. Preoperative (neoadjuvant) chemoradiotherapy in oesophageal cancer. Br J Surg 2001;88:338–356.

101. Law S, Fok M, Chow S, Chu KM, Wong J. Preoperative chemotherapy versus surgical therapy alone for squamous cell carcinoma of the esophagus: a prospective randomized trial. J Thorac Cardiovasc Surg 1997;114:210–217.

102. Posner MC, Gooding WE, Lew JL, et al. Complete 5-year follow-up of a prospective phase II trial of preoperative chemoradiotherapy for esophageal cancer. Surgery (St. Louis) 2001;130:620–628.

103. Kavanagh B, Anscher M, Leopold K, et al. Patterns of failure following combined modality therapy for esophageal cancer, 1984–1990. Int J Radiat Oncol Biol Phys 1992;24:633–642.

104. Gill PG, Denham JW, Jamieson GG, et al. Pattern of treatment failure and prognostic factors associated with the treatment of

esophageal carcinoma with chemotherapy and radiotherapy either as sole treatment followed by surgery. J Clin Oncol 1992;10:1037–1043.

105. Flamen P, Lerut A, Van Cutsem E, et al. The utility of positron emission tomography for the diagnosis and staging of recurrent esophageal cancer. J Thorac Cardiovasc Surg 2000;120:1085–1092.

106. Chu KC, Tarone RE, Chow WH, et al. Temporal patterns in colorectal cancer incidence, survival and mortality from 1950 through 1960. J Natl Cancer Inst 1994;86:997–1006.

107. Bertino JR. Biomodulation of 5-fluorouracil with antifolates. Semin Oncol 1997;24(suppl 18):S18-52–S18-56.

108. Reimer P, Ruckle-Lanz H. New therapeutic options in chemotherapy of advanced colorectal cancer (in German). Med Klin 2001;96:593–598.

109. Schlag PM, Amthauer H, Stroszczynski C, Felix R. Influence of positron emission tomography on surgical therapy planning in recurrent colorectal cancer. Chirurg 2001;72:995–1002.

110. Bosl GJ, Sheinfeld J, Bajorin DF, Motzer RJ. Cancer of the testis. In: Devita VTJ, Helmann S, Rosenberg SA, editors. Cancer Principles and Practice of Oncology. Philadelphia: Lippincott-Raven, 1997:1397–1425.

111. Foster RS, Nichols CR. Testicular cancer: what's new in staging, prognosis, and therapy. Oncology 1999;13:1689–1694.

112. Einhorn LH. Curing metastatic testicular cancer. Proc Natl Acad Sci U S A 2002;99:4592–4595.

113. Bajorin DF, Sarosdy MF, Pfister DG, et al. Randomized trial of etoposide and cisplatin versus etoposide and carboplatin in patients with good-risk germ cell tumors: a multiinstitutional study. J Clin Oncol 1993;11:598–606.

114. Bhatia S, Abonour R, Porcu P, et al. High-dose chemotherapy as initial salvage chemotherapy in patients with relapsed testicular cancer. J Clin Oncol 2000;18:3346–3351.

115. McGuire WP, Brady MF, Ozols RF. The Gynecologic Oncology Group experience in ovarian cancer. Ann Oncol 1999;10(suppl 1):29–34.

116. Ozols RF, Schwartz PE, Eifel PJ. Ovarian cancer, fallopian tube carcinoma, and peritoneal carcinoma. In: Devita VT, Helmann S, Rosenberg SA, eds. Cancer Principles and Practice of Oncology. Philadelphia: Lippincott-Raven, 1997:1502–1539.

117. Wahl RL, Towsend DW, Meltzer CC, Von Schulthess GK, Fishman EK. CT/PET fusion imaging. Radiology 2002;225:42 (abstract).

118. Hany TF, Gharehpapagh E, Kamel EM, Buck A, Himms-Hagen J, Schulthess GK. Brown adipose tissue: a factor to consider in symmetrical tracer uptake in the neck and upper chest region. Eur J Nucl Med 2002;29:1393–1398.

119. Kamel EM, Goerres GW, Burger C, von Schulthess GK, Steinert HC. Recurrent laryngeal nerve palsy in patients with lung cancer. Detection with PET/CT image fusion: report of six cases. Radiology 2002;224:153–156.

26

PET in Clinical Cardiology

Frank M. Bengel and Markus Schwaiger

Following its introduction nearly 30 years ago, positron emission tomography (PET) was first established in academic centers for clinical research in neurology, cardiology, and oncology. Based on promising early results, clinical applications for PET in the heart have emerged and were established even before its breakthrough in clinical oncology. Although other cardiac imaging techniques such as single photon emission tomography (SPECT), ultrasound, or magnetic resonance are available on a broader clinical basis, PET is still widely regarded as the noninvasive gold standard for the assessment of myocardial viability and perfusion. Comparison of alternative techniques with this gold standard has refined their application and allowed for improved imaging algorithms. Moreover, PET has substantially contributed to a better understanding of various disease processes. The recent advances in molecular biology and the progress in disease-specific therapy have prompted an increasing need for characterization of physiologic and biologic parameters in the heart. The available techniques, along with new tracer approaches aiming at molecular targets such as specific proteins and genes, are suited to monitor therapeutically induced changes in tissue function. The uniqueness of this biologic information will contribute to define the clinical future of PET and noninvasive cardiovascular imaging in general.

This chapter highlights the potential of PET for clinical imaging in cardiovascular disease. The methodologic background is outlined first, before major practical applications (with a special focus on viability as the predominant clinical use of PET), along with novel biologic imaging targets in the heart, are summarized. Finally, an outlook into the future, comprising molecular tracers and advanced imaging technology, is given.

Cardiac PET Imaging: Methodologic Considerations

Radiopharmaceuticals

PET radiopharmaceuticals that have been applied for cardiac imaging in the clinical setting are summarized in Table 26.1. At present, these tracers target myocardial perfusion, metabolism, innervation, and receptors. Further tracers are under evaluation that aim at imaging of molecular structures such as extracellular matrix, membrane molecules, or intracellular products of gene expression. Those tracers are expected to broaden the spectrum of biologic targets for clinical cardiac PET imaging in the future and are briefly outlined at the end of the chapter.

Kinetic Modeling: Noninvasive Quantification of Biologic Processes

PET is a truly quantitative imaging tool because it allows measuring localized tracer concentration in the body in absolute units of radioactivity concentration. For cardiac imaging, software algorithms are available that are similar to those applied for SPECT image display; these allow for reangulation of tomographic images along cardiac short and long axes and for calculation of polar maps (Figure 26.1).

With the use of tracer kinetic modeling, there is the potential for substantial improvement of the type and quality of information obtained from biologic data. Kinetic modeling of tracers in the heart requires acquisi-

Table 26.1. Positron emission tomography (PET) radiopharmaceuticals for clinical cardiac imaging.

Tracer	Biologic target
Blood flow:	
^{13}N-Ammonia	Perfusion, blood flow
^{15}O-Water	Blood flow, perfusible tissue
^{82}Rb-Chloride	Perfusion, viability
^{62}Cu-PTSM	Perfusion
Metabolism:	
^{18}F-Fluorodeoxyglucose (FDG)	Glucose transport, hexokinase
^{11}C-Glucose	Glucose metabolism
^{11}C-Acetate	Oxidative metabolism (TCA cycle turnover)
^{11}C-Palmitate	Fatty acid metabolism
^{18}F-FTHA	Fatty acid uptake
Innervation:	
^{18}F-Fluorodopamine	Presynaptic sympathetic function
^{11}C-Hydroxyephedrine	Presynaptic catecholamine uptake
^{11}C-Epinephrine	Presynaptic catecholamine uptake and storage
^{11}C-Phenylephrine	Presynaptic catecholamine uptake and metabolism
Receptors:	
^{11}C-CGP12177	Beta-adrenergic receptor (nonselective)
^{11}C-MQNB	Muscarinic receptor

tion of a dynamic imaging sequence, which generates serial images at various time points after radiotracer injection. This method allows for extraction of time–activity curves for myocardial segments. The time course of tracer concentration in arterial blood (arterial input function) is additionally obtained using a region of interest in the left ventricular cavity or left atrium. Time–activity curves are subsequently fitted to suitable compartmental models that characterize the tissue kinetics of the respective tracers (Figure 26.2). For ^{15}O-water as a freely diffusible perfusion tracer, this is a single-compartment model with one extravascular compartment (1). For the perfusion tracer ^{13}N-ammonia, which is taken up by tissue and subsequently used for glutamine synthesis, two- and three-compartment models can be applied for quantification of myocardial blood flow (2, 3). Kinetics of the metabolic tracer ^{18}F-fluorodeoxyglucose (FDG) can be described by a three-compartment model (4). FDG is taken up by myocardium via specific membrane glucose transporters (GLUT). Intracellularly, FDG is then phosphorylated by hexokinase to FDG-6-phosphate. FDG-6-phosphate is not a good substrate for further metabolism and thus is trapped intracellularly. Compartmental modeling of tracer kinetics may be applied to other available tracers and yields rate constants that quantitatively reflect biologic processes. This unique feature of PET is a key element to its acceptance as the noninvasive imaging gold standard in cardiology.

Camera Technology

Because dedicated PET scanners are expensive and not widely available, alternative strategies have been developed for imaging the cardiac distribution of positron-emitting tracers with conventional gamma cameras. One is SPECT imaging of single 511-keV photons with an ultrahigh-energy collimator (5), and the other is coincidence imaging with a dual-headed gamma camera (6). These approaches are characterized by lower spatial resolution and detection sensitivity and do not allow for dynamic imaging and absolute quantification. However, especially for 511-keV SPECT imaging of FDG in the assessment of myocardial viability, considerable evidence has accumulated supporting the clinical usefulness of this simplified strategy (7).

Meanwhile, detector technology for dedicated PET systems has undergone further developments allowing for improvements in spatial resolution, dead time, and sensitivity. Dedicated small animal scanner systems, which allow for imaging of mouse hearts, represent the current epitome of high-resolution PET imaging (8). Developments in these preclinical systems have significantly contributed to a parallel refinement of clinically applicable systems. Most recently, hybrid PET/CT cameras have been introduced that combine high-end PET systems with multislice CT, allowing for application not only in oncology but also in cardiology. The impact of this approach for combined morphologic and biologic imaging is expected to be significant and is outlined elsewhere in this volume.

Assessment of Myocardial Blood Flow and Perfusion

In a clinical setting, PET perfusion imaging is mainly done using static qualitative images. Its purpose is then similar to that of SPECT, namely, to detect myocardial ischemia or infarction in patients with suspected or known coronary artery disease (CAD), to determine extent and severity of the disease, to determine individual risk for cardiac events, and to thereby guide further decision making (9–11). For the standard evaluation of patients with known or suspected coronary artery disease (CAD), the application of PET remains limited to selected cases because of high costs and limited availability. Introduction of hybrid PET/CT systems may increase the clinical use of PET, but alternative techniques more widely available, such as SPECT and echocardiography, are sufficient for detection of ischemia in most situations.

The major impact of PET perfusion studies results from the potential for noninvasive quantification of myocardial blood flow, allowing evaluation of coronary microcirculatory regulation at the resistance level (12). Therefore,

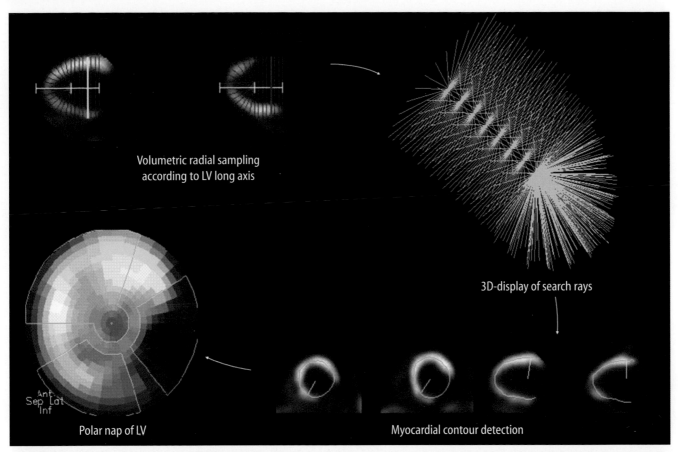

Volumetric radial sampling
according to LV long axis

3D-display of search rays

Polar nap of LV

Myocardial contour detection

Figure 26.1. Semiquantitative analysis of myocardial PET to obtain polar maps of the left ventricular (*LV*) myocardium. Using the displayed three-dimensional volumetric sampling algorithm, the left ventricular distribution of any PET-derived biologic information can be obtained. Polar maps show apex in the center, base in the periphery, anterior wall on the top, inferior wall on the bottom, septum on the left, and lateral wall on the right.

quantitative measurements of myocardial perfusion with PET allows the assessment of earliest alterations in vascular integrity before development of clinically overt CAD and ischemia. Quantitative PET measurements of myocardial blood flow have significantly refined the understanding of risk factors and preventive interventions for coronary atherosclerosis, but a clear-cut clinical application for such measurements of global microvascular reactivity needs to be defined.

Stress Testing

Myocardial perfusion PET imaging is accomplished using flow tracers at rest and during stress (13, 14). In the clinical setting, dipyridamole or adenosine are applied for pharmacologic vasodilation at doses commonly used for SPECT imaging. Physical exercise is not practical for stress flow measurements with PET because of the short half-life of the PET perfusion tracers and the potential for patient motion during imaging.

Adenosine or dipyridamole act directly on vascular smooth muscle cells via specific adenosine receptors, causing relaxation, vasodilation, and flow increase.

Hyperemic flow in response to these agents normally increases 2.5- to 5 fold from baseline and is believed to reflect an integrated response of the coronary circulatory system, partially mediated by direct smooth muscle effects and partially mediated by additional, shear stress-related endothelial activation (15).

More recently, stress tests that are specific for endothelial function have been applied for PET imaging; these are based on sympathetic stimulation, either by cold pressor test or by mental stress (16–18). Release of norepinephrine from stimulated sympathetic neurons activates alpha-adrenoceptors on endothelium that mediate nitric oxide (NO) release. This vasodilatory signal results in a 30% to 50% increase of baseline flow in case of intact endothelium. Alpha-adrenergic stimulation of vascular smooth muscle cells, which causes vasoconstriction, is normally counteracted but may outweigh endothelium-derived vasodilation in case of impaired endothelial integrity. The resulting flow decrease then indicates endothelial dysfunction. PET flow measurements during sympathetic stimulation are therefore thought to provide specific information about coronary endothelial function and have been applied mainly in research studies of vascular reactivity.

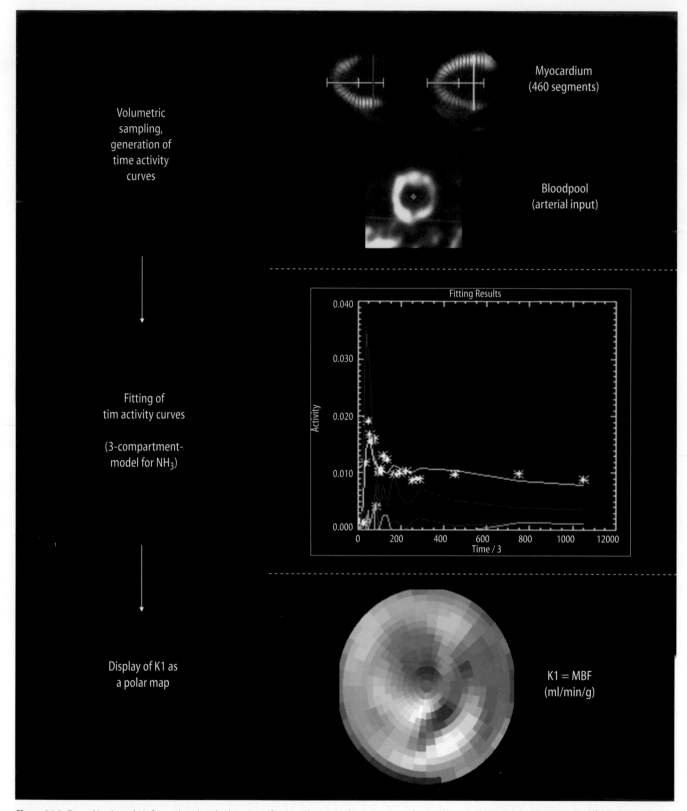

Volumetric
sampling,
generation of
time activity
curves

Myocardium
(460 segments)

Bloodpool
(arterial input)

Fitting of
tim activity curves

(3-compartment-
model for NH$_3$)

Display of K1 as
a polar map

K1 = MBF
(ml/min/g)

Figure 26.2. Tracer kinetic analysis for noninvasive absolute quantification. Shown are the major steps of quantification of myocardial blood flow using ^{13}N-ammonia (NH$_3$), as applied at the TU München. *Top:* Myocardial segments and blood pool are identified using software-based volumetric sampling. Time–activity curves are extracted from the dynamic imaging dataset for these segments. *Middle:* Individual time–activity curves are fitted to a validated three-compartment model for ammonia (*white stars*, measured activity over time in representative myocardial segment; *white line*, fitted myocardial time–activity curve; *red line*, blood pool curve). *Bottom:* After the fitting procedure, the influx constant K1, which reflects absolute myocardial blood flow, is depicted for all myocardial segments as a two-dimensional polar map.

PET Perfusion Imaging in Coronary Artery Disease

For the clinical workup of CAD, relative regional perfusion abnormalities are assessed from static images using ^{13}N-ammonia and [82]Rb (Figure 26.3). The usefulness of ^{15}O-water as a freely diffusible tracer in this regard is less well documented because myocardial images are only obtained after application of factor analysis or blood pool subtraction techniques.

The diagnostic accuracy of PET for detection of CAD is thought to be higher compared to SPECT (19) because of higher spatial resolution, higher extraction fraction of PET flow tracers, and accurate correction of attenuation artefacts. Incremental prognostic value of qualitative perfusion PET over clinical variables for prediction of cardiac events has been demonstrated (20). Additionally, it has been shown that the more-expensive PET perfusion imaging approach can still be cost effective in a specific setting when compared to other strategies, including exercise electrocardiogram (ECG) and SPECT (21).

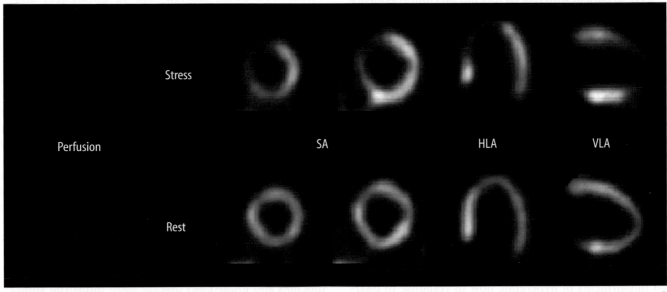

a

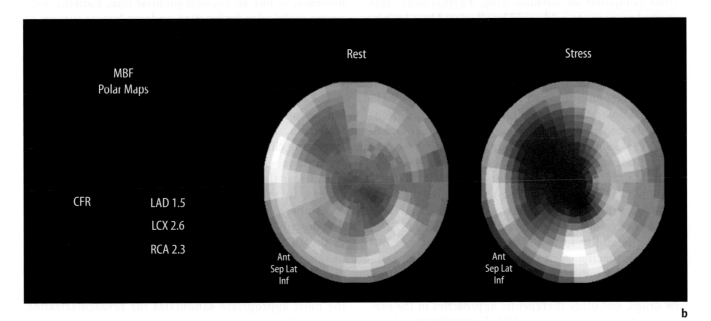

b

Figure 26.3. Myocardial perfusion study using ^{13}N-ammonia PET at *rest* and during adenosine-mediated vasodilation (*stress*). Stress-induced hypoperfusion is observed in the PET images (**a**) in the apex and distal anteroseptal wall. Polar maps (**b**) allow for quantification of defect size. Additional calculation of quantitative myocardial blood flow reveals reduced coronary flow reserve (*CFR*) in the territory of the left anterior descending artery (*LAD*), consistent with the visual impression from static images.

pathophysiologic discussion is less important because many studies have shown that revascularization results in functional improvement of both stunned and hibernating myocardium (66).

PET-Based Diagnostic Approach to Viability Assessment

The level of myocardial blood flow alone does not completely identify reversible contractile dysfunction. At both ends of the spectrum of flow, the situation is clear because normal levels are seen with stunning and severely reduced levels with irreversible injury. An intermediate level of flow reduction, however, can be associated with either reversible or irreversible contractile dysfunction because myocardial hibernation may be present in addition to necrosis or tissue fibrosis (67).

If functional recovery is to occur following revascularization, cellular homeostasis, energy metabolism, and membrane integrity must be preserved. Metabolic activity as the source for energy is a particular aspect of myocardial viability that can be evaluated by PET imaging. Preserved metabolism in a dysfunctional area indicates viability and thus reversibility, whereas reduction of metabolism in parallel with flow reduction reflects scar or fibrosis and thus irreversible injury.

FDG myocardial imaging is the most common clinical application of cardiac PET (13, 14). For assessment of myocardial viability, FDG is combined with resting perfusion measurements using either a second PET tracer in the same session or (if performed at a place remote from a cyclotron and not equipped with a perfusion tracer generator) in combination with perfusion SPECT. Residual metabolic activity is an indicator of myocardial viability and thus of reversibility of contractile dysfunction.

Metabolic Standardization

Diagnostic quality of myocardial FDG images is influenced by the concentration of tracer in myocardium and blood. Myocardial FDG uptake depends on glucose and insulin plasma concentrations as well as the rate of glucose utilization. High glucose plasma concentrations degrade the quality of the myocardial FDG uptake image (68–70). Myocardial glucose uptake also depends on cardiac work, plasma levels of free fatty acids (FFA) and other substrates, insulin, catecholamine, and oxygen supply (71). Standardization of the metabolic environment is necessary for clinical myocardial FDG imaging (72). There are several accepted methods for improvement of myocardial FDG-PET image quality.

- Oral glucose loading is commonly used to stimulate insulin secretion and myocardial glucose utilization. It enhances the image quality with more homogeneously

distributed FDG uptake than observed during the fasting state (73). Despite the gain in image quality after oral glucose loading (50–75 g), diagnostically suboptimal images may still be obtained in 20% to 25% of the patients with CAD (69, 73). Abnormal glucose handling or even type 2 diabetes that remained undetected accounts for the poor image quality in many of these patients.

- Euglycemic hyperinsulinemic clamping is an approach that mimics the postabsorptive steady state (69, 74) and has become an alternative to oral glucose loading for enhancing glucose utilization (69). Insulin clamping stimulates uptake of glucose and FDG in the myocardium and in skeletal muscle and yields images of consistently high diagnostic quality, even in patients with diabetes.

- Several studies (70, 75–77) have demonstrated that oral administration of a nicotinic acid or its derivatives are another effective technique to improve FDG image quality. When combined with small amounts of short-acting insulin, good image quality can also be obtained in diabetic subjects (75). Nicotinic acid inhibits peripheral lipolysis and thus reduces plasma FFA concentrations. Acipimox is a nicotinic acid derivative that is 20 times more potent and has more favorable kinetics than nicotinic acid (78). Importantly, with the exception of flushing, no side effects of Acipimox have been observed.

Clinical Impact of PET Viability Assessment

Differentiation of reversible and irreversible dysfunction in patients with severe impairment of left ventricular (LV) function has become a major clinical question for the decision-making process concerning revascularization. Qualitative evaluation of PET flow studies demonstrated decreased ^{13}N-ammonia and increased FDG uptake (mismatch) in viable myocardium, which has been considered the scintigraphic hallmark of hibernation (79) (Figures 26.5, 26.6). Although such a pattern suggests reduced blood flow, the intensity of tracer uptake depends not only on flow but also on left ventricular wall thickness (partial-volume effect). Therefore, reduction of ^{13}N-ammonia uptake may reflect wall thinning or admixture of viable and scarred myocardium (80). Flow measurements with ^{15}O-water are less sensitive to partial-volume effect but may overestimate transmural flow (81). In contrast, the FDG signal indicates a relatively higher FDG extraction compared to ^{13}N-ammonia. Sensitive and specific identification of viable myocardium can be performed, using information on both blood flow and glucose metabolism, as was first shown by Tillisch et al. (79), who compared relative FDG uptake before and after revascularization in patients with advanced CAD and impaired regional and global function. Maintained FDG uptake in

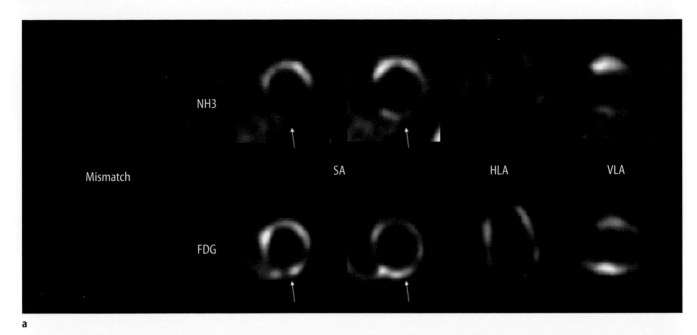

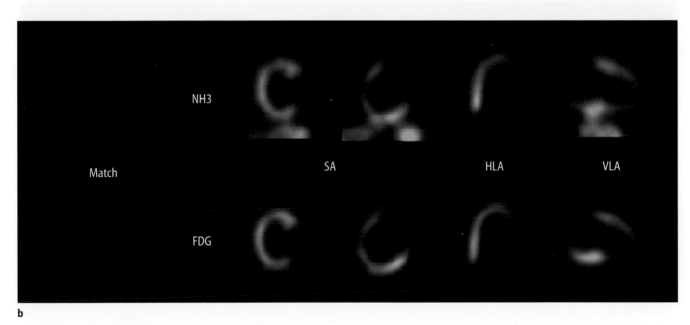

Figure 26.5. Patterns of myocardial blood flow and glucose utilization assessed by PET with ¹³N-ammonia (*NH₃*) and ¹⁸F-deoxyglucose (*FDG*) in patients with advanced coronary artery disease (CAD). Shown on the (**a**) is a mismatch pattern with reduced blood flow but preserved FDG uptake in the inferior wall, indicative of hibernating myocardium. Shown on the (**b**) is a matched perfusion/metabolism defect in the lateral wall, indicating irreversibly damaged scar tissue. *SA*, short-axis slices; *HLA*, horizontal long-axis slices; *VLA*, vertical long-axis slices.

dysfunctional segments with reduced flow was associated with functional recovery after revascularization, whereas segments with concordantly decreased flow and metabolism did not recover. Subsequently, a large number of similar studies confirmed the predictive value of FDG imaging for recovery of contractile function after revascularization (79, 80, 82–87).

The observed increase of global left ventricular ejection fraction in patients with evidence of viability at PET imaging varies among studies and is on average 30% to 40% (88). Functional improvement after revascularization seems to be dependent on several factors. First, the initial depression of global function needs to be taken into account. A higher increase was observed in patients with severe dysfunction compared to those with only mild dysfunction (88). Pagano et al. (89) have demonstrated a significant correlation between extent of PET mismatch pattern and increase of ejection fraction after revascularization. The amount of scar and fibrotic tissue, which may coexist together with viable hibernating myocardium as already mentioned, is another determinant of the magnitude of regional contractile improvement. Finally, the

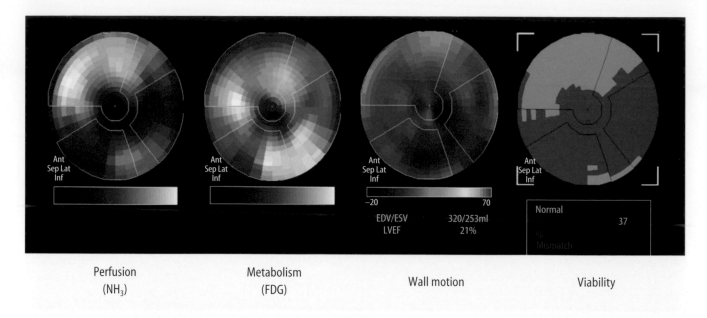

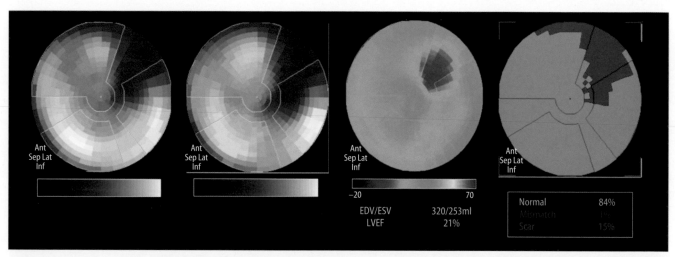

Figure 26.6. Polar maps of perfusion, metabolism, and wall motion (from gated FDG-PET) for the two patients presented in Figure 26.5. Based on information in flow and metabolism maps, a color-coded viability map is additionally calculated (*right*) and displays regional extent of normal, mismatch, and scar areas along with their global extent in percent (%) of the entire map. The polar map and gated analysis provide additional information: The mismatch pattern on the *top* appears to be large and is combined with severe dysfunction and elevated ventricular volumes. The scar area in the *bottom* case, on the other hand, is small; there is only little regional dysfunction and global ejection fraction and volumes are almost normal. 📖

time after revascularization at which recovery is evaluated does also influence the extent of recovery (90). In a recent study, Haas et al. (66) performed repetitive measurements until 1 year after revascularization and demonstrated that functional recovery of hibernating myocardium may require several months.

The importance of PET studies of tissue viability goes beyond the predictive value for functional recovery. Di Carli et al. (91), for example, have demonstrated that the percent of mismatch identified preoperatively predicts recovery of function and improvement in heart failure symptoms and daily life activity. Following revascularization of patients with a mismatch pattern, a significant improvement in New York Heart Association (NYHA) class

was observed (92). When patients are treated medically, NYHA class remains unchanged.

More importantly, the pattern of blood flow and glucose metabolism contains predictive information about long-term survival after surgery. Retrospective data analysis revealed a high incidence of cardiovascular complications in chronic coronary patients who had regionally decreased blood flow with maintained FDG uptake and who did not undergo revascularization (92, 93). When patients with a PET mismatch pattern underwent revascularization, however, a significantly lower mortality was observed. In contrast, the incidence of cardiovascular complications was similar in patients with scintigraphic evidence of scar or normal myocardium, regardless of revascularization (94).

These data indicate that the mismatch pattern identifies a subgroup of patients at increased risk for cardiovascular complications. The prognostic information appears independent of the traditional markers such as left ventricular ejection fraction or NYHA classification, which were not different among the investigated subgroups. Similarly, a study by Beanlands et al. (95) showed higher preoperative mortality when revascularization was delayed for more than 35 days after identification of a PET mismatch pattern. This finding supports the notion that mismatch is an unstable state that requires immediate therapeutic attention. In this study, functional recovery in patients with delayed revascularization was also attenuated.

The degree of functional recovery of patients after revascularization can be predicted on the basis of the scintigraphic pattern, as well as extent of mismatch (80, 91). Di Carli et al. (91) reported an 80% likelihood of functional improvement in the presence of mismatch that exceeded 18% of the left ventricle. Pagano et al. (96) indicated that the extent of viability correlated not only with functional recovery but also with survival. Patients with severe left ventricular dysfunction were at higher risk for complications associated with revascularization. Dreyfus et al.(97) reported that assessment of tissue viability in such patients before surgery improved the selection process for revascularization by lowering perioperative mortality. More recently, Haas et al. (98) confirmed the impact of PET on peri- and postoperative outcome by comparing two groups of patients with three-vessel disease and impaired left ventricular function. Patients selected for revascularization based on PET had significantly lower perioperative complications and better long-term outcome than a group of patients who underwent revascularization without prior viability assessment.

By providing sensitive prediction of functional recovery together with other prognostic information, PET is likely to influence patient management and therapy planning. As reported by Beanlands et al. (99), PET altered clinical decision making in 57% of 87 patients with chronic left ventricular dysfunction. Forty-two percent of patients initially scheduled for bypass surgery were assigned to medical treatment after PET, and approximately half of the patients who were initially assigned for conservative therapy underwent revascularization. Moreover, 63% of patients assigned to cardiac transplantation were submitted to bypass surgery after PET (99). The use of PET to guide clinical decision making and to select appropriate therapy based on individual patient risk and likelihood for recovery may avoid costly procedures such as bypass surgery and cardiac transplantation in patients who are unlikely to benefit. Thereby, the use of viability imaging may contribute to an overall reduction of costs in the management of patients with chronic coronary disease and impaired left ventricular function. Large prospective randomized trials are currently underway to obtain further evidence for the clinical usefulness of viability imaging.

Comparison with Other Imaging Modalities

There have been numerous comparisons of regional FDG uptake with electrocardiographic criteria, standard ^{201}Tl and ^{99m}Tc-flow agent imaging, stress echocardiography, and magnetic resonance imaging (MRI) (54, 100). Early observations of discordant results between the FDG distribution and ^{201}Tl redistribution pattern emphasized the limitations of ^{201}Tl imaging for assessment of tissue viability, which are partially overcome by reinjection techniques (101). The use of ^{99m}Tc-sestamibi also provides clinically useful information regarding tissue viability (102). In combination with nitrate administration, the predictive value of ^{99m}Tc-sestamibi imaging is improved (103). Comparing scintigraphic data with those obtained after positive inotropic interventions (imaged by echocardiography or magnetic resonance), it appeared that tracer retention provides a more-sensitive marker of viable myocardium, whereas assessment of contractile reserve is more specific for reversible myocardial dysfunction (54).

The most attractive test for detection of myocardial scar at present is MRI of the late enhancement of gadolinium. This technique allows for accurate visualization of extent, location, and transmurality of scar (104). It compares well to results of PET for scar detection (105) but seems to detect nontransmural tissue damage with higher sensitivity. Scar detection by late enhancement MRI has a good accuracy for predicting functional recovery following revascularization (106). An important clinical issue is the presence or absence of hibernating myocardium in nontransmural scar because of the balance and transition between programs of cell survival and cell death (107). PET has the advantage to identify hibernating myocardium with the pattern of perfusion–metabolism mismatch whereas MRI with delayed gadolinium enhancement would delineate the scar but would not differentiate adjacent normal from hibernating myocardium.

Targeting of Cardiovascular Molecular Mechanisms

Growing knowledge of the molecular mechanisms of cardiovascular disorders plays a key role in the improvement of prevention, diagnosis, and treatment of heart disease. With novel developments in radiopharmaceuticals and imaging technology, PET is increasingly playing a role as a mediator between molecular cardiac research and clinical management of patients with heart disease. The most promising applications of PET in the near future focus on molecular biologic targets within the myocardium and the coronary vasculature. A variety of molecular-targeted imaging approaches are available and are briefly outlined here to demonstrate that PET is both capable and necessary for the transference of new biologic knowledge to clinical practice. With this new array of molecular imaging approaches, PET constitutes a strong basis for an advanced and individually tailored approach to cardiovascular disease.

Substrate Metabolism

A variety of other tracers exists, besides FDG, to study quantitatively the complex interaction between substrate delivery, uptake, and turnover in myocardial tissue under various physiologic and pathophysiologic conditions (Figure 26.7). The myocardial metabolic rate of glucose can be quantified from dynamic FDG-PET imaging (4) or, as shown more recently, using ^{11}C-glucose (108). Additionally, utilization of fatty acids as the second major substrate can be investigated using ^{11}C-palmitate (109) or ^{18}F-fluorothioheptadecanoic acid (FTHA) (110).

[11]C-Acetate is another metabolic tracer that is extracted proportionally to myocardial blood flow, then rapidly enters the tricarboxylic acid cycle (TCA) and is metabolized to carbon dioxide. Washout of ^{11}C-acetate from the myocardium is used to quantify turnover in TCA as the final common pathway of all substrates and thus measure the overall oxidative metabolism (111, 112). When combined with measures of cardiac contractile performance, PET using ^{11}C-acetate can evaluate noninvasively cardiac efficiency (defined as the relation between work and oxygen consumption) (113). This approach is of special interest in heart failure (114), where it can be used to measure effects of drugs such as beta-blockers or dobutamine (113, 115) and thereby to optimize medical therapy.

Combination of metabolic tracers allows detailed insights into substrate utilization and metabolic regulation of the myocardium. Peterson et al. (116), for example, combined ^{11}C-glucose, ^{11}C-palmitate, and ^{11}C-acetate and found increased myocardial oxygen consumption and decreased efficiency in obese young women. Insulin resistance was also observed, which was associated with increase myocardial fatty acid utilization, and it was concluded that insulin resistance and increased myocardial fatty acid utilization may contribute to the development of impaired cardiac performance in obesity (116). Although this application is limited primarily to research, metabolism can be evaluated quantitatively with PET, not only in the myocardium but also in other organs such as liver, peripheral muscle, or adipose tissue (117).

Autonomic Innervation

The autonomic nervous system plays a central role for regulation of cardiac performance under various physiologic and pathophysiologic conditions. The importance of alterations of autonomic innervation in the pathophysiology of heart disease has been increasingly emphasized. Using radiolabeled catecholamine analogues such as ^{11}C-hydroxyephedrine (HED) or ^{18}F-fluorodopamine, PET provides noninvasive information about global and regional myocardial sympathetic innervation. It has substantially facilitated and refined the study of the heart's

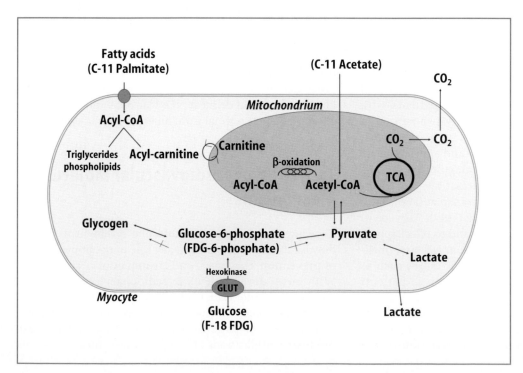

Figure 26.7. Pathways of myocardial metabolism that can be investigated by PET.

nervous system in health and disease and has significantly contributed to a continuous improvement of the understanding of cardiac pathophysiology.

In transplanted hearts, which are completely denervated early after surgery, partial sympathetic reinnervation has been identified (Figure 26.8) (118). The reinnervated heart was used as a model for determination of physiologic effects of innervation by intraindividual comparison of denervated and reinnervated myocardium, substantiating the role of innervation for regulation of myocardial blood flow, metabolism, and contractility (119–121). Furthermore, alterations of cardiac innervations were identified in diabetes mellitus, ischemic syndromes (122), arrhythmogenic diseases (123, 124), and heart failure (125, 126), and their pathophysiologic relevance (127) as well as prognostic implications (128) were identified in subsets of patients.

Knowledge of the physiologic and functional importance of myocardial innervation has been refined in these PET studies, and pathophysiologic correlates and consequences of altered autonomic innervation have been identified. Further studies need to define the prognostic importance of cardiac neuroimaging in various diseases. This achievement will set the basis for future clinical applications by supporting the usefulness in clinical decision making and the utilization as a surrogate endpoint in clinical trials.

Receptors

Various myocardial receptors play a role in cardiovascular pathophysiology and represent attractive targets for imaging. PET imaging with the nonselective beta-

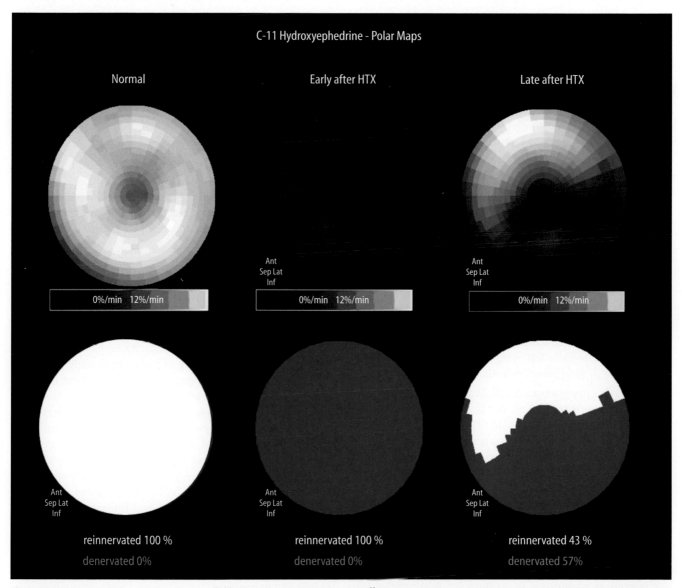

Figure 26.8. Polar maps of left ventricular myocardial distribution of the catecholamine analogue ^{11}C-hydroxyephedrine in a healthy normal individual (*left*), a denervated transplant recipient early after cardiac transplantation (*HTX*) (*middle*), and a reinnervated transplant recipient several years after HTX (*right*). Color-coded maps on the bottom show degree of innervation (*white*) and denervation (*blue*) when compared with a normal database.

adrenergic receptor ligand ^{11}C-CGP12177 allowed visualization of downregulation of beta-receptors in heart failure (129) and identification of the pre-/postsynaptic imbalance in cardiomyopathy and arrhythmogenic diseases (123, 130). Other adrenergic receptor ligands have been introduced but not yet applied in humans (131, 132). PET imaging with the muscarinic receptor agonist ^{11}C-MQNB allowed studying the upregulation of myocardial muscarinic receptors in heart failure as a potential adaptive mechanism (133). Other attractive targets for myocardial receptor imaging include the opiate, endothelin, angiotensin, and adenosine receptor systems. Limitations of these tracers are related to low selectivity and affinity, high nonspecific binding, lack of hydrophilicity (to avoid binding to internalized inactive receptors), and pharmacologic/toxicologic effects. In addition, the complex synthesis of some of these tracers has precluded more widespread investigations. These limitations need to be overcome in the future for potential clinical applications.

Further Molecular Probes

The advances and success of molecular imaging have led to the development of promising novel tracer approaches that go beyond identification of flow, metabolism, and innervation and target further specific biomechanisms involved in disease. Developments in oncologic and cardiologic research are often overlapping and may benefit from each other:

- The regulation of blood vessel growth is one area of interest for both oncologists and cardiologists. $\alpha_v\beta_3$ integrins are expressed on the endothelial surface during tumor angiogenesis and metastatic growth and can be imaged in the experimental setting using ^{18}F-labeled arginine-glycine-aspartate (RGD) peptides (134). These proteins are also a target for cardiac molecular imaging, when expressed during angio-/vasculogenesis, during rapid development of coronary stenosis, for example, restenosis following interventions, or during migration of inflammatory cells.

- Apoptosis (programmed cell death) is a therapeutic target in tumor treatment, but is also known to play a pivotal role in cardiovascular disease, for example, in transplant rejection and plaque destabilization (135). Recently, annexin-V, a molecule that specifically binds to phosphatidylserine, which is expressed on the surface of cells entering the apoptotic cascade, has been successfully labeled with ^{18}F and is now available for PET imaging (136). Previously, this tracer was available only as a ^{99m}Tc-labeled compound for conventional scintigraphy.

- Within recent years, reporter gene imaging has emerged as an innovative strategy for noninvasive visualization of genetic processes (137). The principle of reporter gene imaging is based on vector-mediated overexpression of a transgene in host tissue that encodes for an enzyme, a receptor, or a transport protein which is normally not present in the target area. Suitable reporter probes are labeled substrates or ligands for the reporter gene product. PET can be applied to detect accumulation of the probe, which specifically identifies the reporter gene product (138). The signal imaged reflects indirectly expression and transcriptional regulation of the reporter gene. An example is the use of the herpes viral thymidine kinase reporter gene and radiolabeled nucleosides as reporter probes. This technique allows monitoring the success of gene therapy when reporter genes are combined and coexpressed with therapeutic genes. It may also allow monitoring endogenous gene expression, if reporter genes are expressed under control of specific promoters that are sensitive to endogenous gene products (139). Alternatively, the reporter gene can be expressed in stem cells before their transplantation for subsequent in vivo monitoring of the fate of these therapeutic cells (140).

For all these new molecular PET assays, the amount of signal produced, target-to-background ratios, morphologic localization, and specific activity of the probe are critical issues that need to be overcome for clinical application. The ultimate challenge is to develop molecular imaging probes not only for imaging myocardial biology but also for characterization of atherosclerotic plaques. Molecular targets for identification of vulnerable plaques, which are prone to rupture and thus linked with adverse outcome, include inflammation, apoptosis, proteolytic enzyme activity, adhesion molecule expression, and thrombogenicity.

Summary and Future Perspectives

With the availability of a broad spectrum of tracers for biologic mechanisms, PET as a research tool has substantially improved the understanding of pathophysiologic processes in the heart. Clinical applications for assessment of myocardial perfusion and viability have emerged in the past. The knowledge derived from cardiac PET imaging has been successfully transferred to other imaging techniques, leading to refinement of their application and to increased acceptance of noninvasive cardiac imaging in general.

Clinically, PET is still regarded as the gold standard in cardiac biologic imaging. It continues to serve as an endpoint for evaluation of medical and interventional therapeutic approaches. Costs remain a limitation for the use as a standard procedure, but new technical developments may allow for a more-effective, widespread use of positron-emitting tracers and for integrated biologic–morphologic imaging.

PET will continue to have an impact as a powerful research tool. Availability of micro-PET systems for small

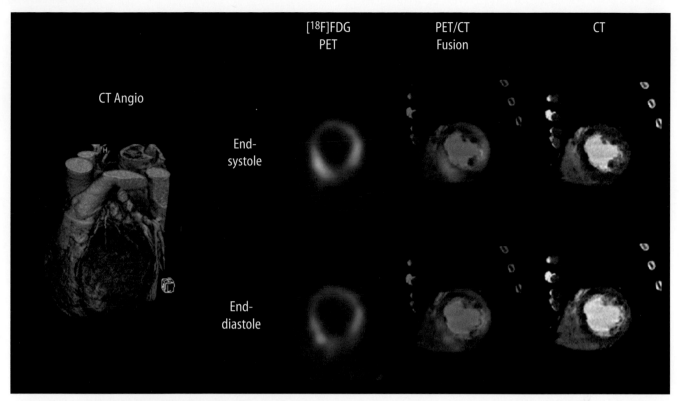

Figure 26.9. Display of an FDG viability study, combined with CT angiography (*left*) and gated analysis (*right*) using a hybrid PET/CT system.

animal imaging facilitates the evaluation of novel molecular probes and their translation to human application (141). Combination of PET with morphologic imaging with integrated PET/CT systems (Figure 26.9) allow in vivo characterization and localization of molecular mechanisms. The vision is an integrated approach for diagnosis of coronary disease, allowing characterization of vascular morphology, plaque morphology and biology, and tissue alterations. With these ongoing developments, PET imaging will remain a forerunner for the future development of the entire field of cardiovascular imaging.

References

1. Bergmann SR, Herrero P, Markham J, Weinheimer CJ, Walsh MN. Non-invasive quantitation of myocardial blood flow in human subjects with oxygen-15 labeled water and positron emission tomography. J Am Coll Cardiol 1989;14:639–652.

2. Hutchins G, Schwaiger M, Rosenspire K, Krivokapich J, Schelbert H, Kuhl D. Noninvasive quantification of regional myocardial blood flow in the human heart using N-13 ammonia and dynamic positron emission tomographic imaging. J Am Coll Cardiol 1990;15:1032.

3. Kuhle WG, Porenta G, Huang SC, et al. Quantification of regional myocardial blood flow using ^{13}N-ammonia and reoriented dynamic positron emission tomographic imaging. Circulation 1992;86:1004–1017.

4. Gambhir SS, Schwaiger M, Huang SC, et al. Simple noninvasive quantification method for measuring myocardial glucose utilization in humans employing positron emission tomography and fluorine-18 deoxyglucose. J Nucl Med 1989;30:359–366.

5. Bax JJ, Valkema R, Visser FC, et al. FDG SPECT in the assessment of myocardial viability. Comparison with dobutamine echo. Eur Heart J 1997;18(suppl D):D124–D129.

6. Nowak B, Zimny M, Schwarz ER, et al. Diagnosis of myocardial viability by dual-head coincidence gamma camera fluorine-18 fluorodeoxyglucose positron emission tomography with and without non-uniform attenuation correction. Eur J Nucl Med. 2000;27:1501–1508

7. Bax J, Patton J, Poldermans D, Elhendy A, Sandler M. 18-Fluorodeoxyglucose imaging with positron emission tomography and single photon emission computed tomography: cardiac applications. Semin Nucl Med 2000;30:281–298.

8. Inubushi M, Jordan MC, Roos KP, et al. Nitrogen-13 ammonia cardiac positron emission tomography in mice: effects of clonidine-induced changes in cardiac work on myocardial perfusion. Eur J Nucl Med Mol Imaging 2004;31:110–116.

9. Gould K, Goldstein R, Mullani N, et al. Clinical feasibility, sensitivity and specifically of positron cardiac imaging without a cyclotron using generator produced Rb-82 for the diagnosis of coronary artery disease. J Nucl Med 1986;27:976.

10. Schelbert HR, Wisenberg G, Phelps ME, et al. Noninvasive assessment of coronary stenoses by myocardial imaging during pharmacologic coronary vasodilation: VI. Detection of coronary artery disease in man with intravenous 13-NH$_3$ and positron computed tomography. Am J Cardiol 1982;49:1197–1207.

11. Tamaki N, Yonekura Y, Senda M, et al. Myocardial positron computed tomography with ^{13}N-ammonia at rest and during exercise. Eur J Nucl Med 1985;11:246–251.

12. Schelbert HR. Positron emission tomography and the changing paradigm in coronary artery disease. Z Kardiol 2000;89(suppl 4):IV55–IV60.

13. Bacharach SL, Bax JJ, Case J, et al. PET myocardial glucose metabolism and perfusion imaging. Part 1: Guidelines for data acquisition and patient preparation. J Nucl Cardiol 2003;10:543–556.

14. Schelbert HR, Beanlands R, Bengel F, et al. PET myocardial perfusion and glucose metabolism imaging. Part 2: Guidelines for interpretation and reporting. J Nucl Cardiol 2003;10:557–571.

15. Buus NH, Bottcher M, Hermansen F, Sander M, Nielsen TT, Mulvany MJ. Influence of nitric oxide synthase and adrenergic inhibition on adenosine-induced myocardial hyperemia. Circulation 2001;104:2305–2310.

16. Campisi R, Czernin J, Schoder H, et al. Effects of long-term smoking on myocardial blood flow, coronary vasomotion, and vasodilator capacity. Circulation 1998;98:119–125.

17. Meeder JG, Peels HO, Blanksma PK, et al. Comparison between positron emission tomography myocardial perfusion imaging and intracoronary Doppler flow velocity measurements at rest and during cold pressor testing in angiographically normal coronary arteries in patients with one-vessel coronary artery disease. Am J Cardiol 1996;78:526–531.

18. Schoder H, Silverman DH, Campisi R, et al. Regulation of myocardial blood flow response to mental stress in healthy individuals. Am J Physiol Heart Circ Physiol 2000;278:H360–H366.

19. Schwaiger M, Melin J. Cardiological applications of nuclear medicine. Lancet 1999;354:661–666.

20. Marwick T, Shan K, Patel S, Go R, Lauer M. Incremental value of rubidium-82 positron emission tomography for prognostic assessment of known or suspected coronary artery disease. Am J Cardiol 1997;80:865–870.

21. Patterson RE, Eisner RL, Horowitz SF. Comparison of cost-effectiveness and utility of exercise ECG, single photon emission computed tomography, positron emission tomography, and coronary angiography for diagnosis of coronary artery disease. Circulation 1995;91:54–65.

22. Muzik O, Duvernoy C, Beanlands R, et al. Assessment of diagnostic performance of quantitative flow measurements in normal subjects and patients with angiographically documented CAD by means of nitrogen-13 ammonia and using PET. J Am Coll Cardiol 1998;31:534–540.

23. Czernin J, Barnard RJ, Sun KT, et al. Effect of short-term cardiovascular conditioning and low-fat diet on myocardial blood flow and flow reserve. Circulation 1995;92:197–204.

24. Dayanikli F, Grambow D, Muzik O, Mosca L, Rubenfire M, Schwaiger M. Early detection of abnormal coronary flow reserve in asymptomatic men at high risk for coronary artery disease using positron emission tomography. Circulation 1994;90:808–817.

25. DiCarli M, Czernin J, Hoh CK, et al. Relation among stenosis severity, myocardial blood flow, and flow reserve in patients with coronary artery disease. Circulation 1995;91:1944–1951.

26. Pitkanen O, Raitakari O, Niinikoski H, et al. Coronary flow reserve is impaired in young men with familial hypercholesterolemia. J Am Coll Cardiol 1996;28:1705–1711.

27. Pitkanen O, Nuutila P, Raitakari O, et al. Coronary flow reserve is reduced in young men with IDDM. Diabetes 1998;47:248–254.

28. Momose M, Abletshauser C, Neverve J, et al. Dysregulation of coronary microvascular reactivity in asymptomatic patients with type 2 diabetes mellitus. Eur J Nucl Med 2002;29:1675–1679.

29. Uren NG, Camici PG, Melin JA, et al. Effect of aging on myocardial perfusion reserve. J Nucl Med 1995;36:2032–2036.

30. Czernin J, Muller P, Chan S, et al. Influence of age and hemodynamics on myocardial blood flow and flow reserve. Circulation 1993;88:62–69.

31. Kaufmann P, Gnecchi-Ruscone T, Schafers K, Luscher T, Camici P. Low density lipoprotein cholesterol and coronary microvascular dysfunction in hypercholesterolemia. J Am Coll Cardiol 2000;36:103–109.

32. Czernin J, Sun K, Brunken R, Bottcher M, Phelps M, Schelbert H. Effect of acute and long-term smoking on myocardial blood flow and flow reserve. Circulation 1995;91:2891–2897.

33. Campisi R, Czernin J, Schoder H, Sayre J, Schelbert H. L-Arginine normalizes coronary vasomotion in long-term smokers. Circulation 1999;99:491–497.

34. Yokoyama I, Momomura S, Ohtake T, et al. Reduced myocardial flow reserve in non-insulin-dependent diabetes mellitus. J Am Coll Cardiol 1997;30:1472–1477.

35. Yokoyama I, Ohtake T, Momomura S, et al. Hyperglycemia rather than insulin resistance is related to reduced coronary flow reserve in NIDDM. Diabetes 1998;47:119–124.

36. Pitkanen OP, Nuutila P, Raitakari OT, et al. Coronary flow reserve is reduced in young men with IDDM. Diabetes 1998;47:248–254.

37. Sundell J, Nuutila P, Laine H, et al. Dose-dependent vasodilating effects of insulin on adenosine-stimulated myocardial blood flow. Diabetes 2002;51:1125–1130.

38. Gould KL, Martucci JP, Goldberg DI, et al. Short-term cholesterol lowering decreases size and severity of perfusion abnormalities by positron emission tomography after dipyridamole in patients with coronary artery disease: a potential noninvasive marker of healing coronary endothelium. Circulation 1994;89:1530–1538.

39. Gould KL, Ornish D, Scherwitz L, et al. Changes in myocardial perfusion abnormalities by positron emission tomography after long-term, intense risk factor modification. JAMA 1995;274:894–901.

40. Kaufmann P, Gnecchi-Ruscone T, di Terlizzi M, Schafers K, Luscher T, Camici P. Coronary heart disease in smokers: vitamin C restores coronary microcirculatory function. Circulation 2000;102:1233–1238.

41. Parodi O, Neglia D, Sambuceti G, Marabotti C, Palombo C, Donato L. Regional myocardial blood flow and coronary reserve in hypertensive patients. The effect of therapy. Drugs 1992;1:48–55.

42. Guethlin M, Kasel AM, Coppenrath K, Ziegler S, Delius W, Schwaiger M. Delayed response of myocardial flow reserve to lipid-lowering therapy with fluvastatin. Circulation 1999;99:475–481.

43. Campisi R, Nathan L, Pampaloni MH, et al. Noninvasive assessment of coronary microcirculatory function in postmenopausal women and effects of short-term and long-term estrogen administration. Circulation 2002;105:425–430.

44. Duvernoy CS, Rattenhuber J, Seifert-Klauss V, Bengel F, Meyer C, Schwaiger M. Myocardial blood flow and flow reserve in response to short-term cyclical hormone replacement therapy in post-menopausal women. J Gend Specif Med 2001;4:21–27, 47.

45. Kaul T, Agnohotri A, Fields B, Riggins L, Wyatt D, Jones C. Coronary artery bypass grafting in patients with an ejection fraction of twenty percent or less. J Thorac Cardiovasc Surg 1996;111:1001–1012.

46. Miller D, Stinson E, Alderman E. Surgical treatment of ischemic cardiomyopathy: is it ever too late? Am J Surg 1981;141:688–693.

47. Mickleborough L, Maruyama H, Takagi Y, Mohammed S, Sun Z, Ebisuzaki L. Results of revascularization in patients with severe left ventricular dysfunction. Circulation 1995;92:73–79.

48. Luciani G, Faggian T, Razzolini R, Livi U, Bortoletti U, Mazzucco A. Severe ischemic left ventricular failure: coronary operation or heart transplantation. Ann Thorac Surg 1993;55:719–723.

49. Passamani E, Davis KB, Gillespie MJ, Killip T, Piata C. A randomized trial of coronary artery bypass surgery: survival of patients with a low ejection fraction. N Engl J Med 1985;312:1665–1671.

50. Wijns W, Vatner S, Camici P. Hibernating myocardium. N Engl J Med 1998;339:173–181.

51. Rahimtoola SH. The hibernating myocardium. Am Heart J 1989;117:211–221.

52. Diamond G, Forrester J, de Luz P, Wyatt H, Swan H. Post extrasystolic potentiation of ischemic myocardium by atrial stimulation. Am Heart 1978;95:204–209.

53. Dyke S, Cohn R, Gorlin R, Sonnenblick E. Detection of residual myocardial function in coronary artery disease using post-extra systolic potentiation. Circulation 1974;50:694–699.

54. Bax JJ, Wijns W, Cornel JH, Visser FC, Boersma E, Fioretti PM. Accuracy of currently available techniques for prediction of functional recovery after revascularization in patients with left ventricular dysfunction due to chronic coronary artery disease: comparison of pooled data. J Am Coll Cardiol 1997;30:1451–1460.

55. Kloner R, Bolli R, Marban E, Reinlib L, Braunwald E. Medical and cellular implications of stunning, hibernation, and preconditioning: an NHLBI Workshop. Circulation 1998;97:1848–1867.

56. Heyndricks GR, Millard RW, McRitchie RJ, Maroko PR, Vatner SF. Regional myocardial function and electrophysiological alterations after brief coronary artery occlusion in conscious dogs. J Clin Invest 1975;56:978–985.

57. Bolli R. Myocardial 'stunning' in man. Circulation 1992;86:1671–1691.

58. Kloner RA, Allen J, Cox TA, Zheng Y, Ruiz CE. Stunned left ventricular myocardium after exercise treadmill testing in coronary artery disease. Am J Cardiol 1991;68:329–334.

59. Heusch G. Hibernating myocardium. Physiol Rev 1998;78:1055–1085.

60. Els%osser A, Schlepper M, Klovekorn WP, et al. Hibernating myocardium: an incomplete adaptation to ischemia. Circulation 1997;96:2920–2931.

61. Schwarz ER, Schoendube FA, Kostin S, et al. Prolonged myocardial hibernation exacerbates cardiomyocyte degeneration and impairs recovery of function after revascularization. J Am Coll Cardiol 1998;31:1018–1026.

62. Depre C, Vanoverschelde JL, Gerber B, Borgers M, Melin JA, Dion R. Correlation of functional recovery with myocardial blood flow, glucose uptake, and morphologic features in patients with chronic left ventricular ischemic dysfunction undergoing coronary artery bypass grafting. J Thorac Cardiovasc Surg 1997;113:371–378.

63. Vanoverschelde JL, Wijns W, Depre C, et al. Mechanisms of chronic regional postischemic dysfunction in humans. New insights from the study of noninfarcted collateral-dependent myocardium. Circulation 1993;87:1513–1523.

64. Iida H, Tamura Y, Kitamura K, Bloomfield P, Eberl S, Ono Y. Histochemical correlates of ^{15}O-water-perfusable tissue fraction in experimental canine studies of old myocardial infarction. J Nucl Med 2000;41:1737–1745.

65. Yamamoto Y, de Silva R, Rhodes C, et al. A new strategy for the assessment of viable myocardium and regional myocardial blood flow using ^{15}O-water and dynamic positron emission tomography. Circulation 1992;86:167–178.

66. Haas F, Augustin N, Holper K, et al. Time course and extent of improvement of dysfunctioning myocardium in patients with coronry artery disease and severely depressed left ventricular function after revascularization: correlation with positron emission tomographic findings. J Am Coll Cardiol 2000;36:1927–1934.

67. Gewirtz H, Fischman AJ, Abraham S, Gilson M, Strauss HW, Alpert NM. Positron emission tomographic measurements of absolute regional myocardial blood flow permits identification of nonviable myocardium in patients with chronic myocardial infarction. J Am Coll Cardiol 1994;23:851–859.

68. Krivokapich J, Huang SC, Selin CE, Phelps ME. Fluorodeoxyglucose rate constants, lumped constant, and glucose metabolic rate in rabbit heart. Am J Physiol 1987;252:H777–H787.

69. Knuuti MJ, Nuutila P, Ruotsalainen U, et al. Euglycemic hyperinsulinemic clamp and oral glucose load in stimulating myocardial glucose utilization during positron emission tomography. J Nucl Med 1992;33:1255–1262.

70. Knuuti MJ, Yki Jarvinen H, Voipio-Pulkki LM, et al. Enhancement of myocardial [fluorine-18]fluorodeoxyglucose uptake by a nicotinic acid derivative. J Nucl Med 1994;35:989–998.

71. Neely JR, Morgan HE. Relationship between carbohydrate and lipid metabolism and the energy balance of the heart muscle. Annu Rev Physiol 1974;36:412–459.

72. Knuuti J, Schelbert HR, Bax JJ. The need for standardisation of cardiac FDG PET imaging in the evaluation of myocardial viability in patients with chronic ischaemic left ventricular dysfunction. Eur J Nucl Med Mol Imaging 2002;29:1257–1266.

73. Berry JJ, Baker JA, Pieper KS, Hanson MW, Hoffman JM, Coleman RE. The effect of metabolic milieu on cardiac PET imaging using fluorine-18-deoxyglucose and nitrogen-13-ammonia in normal volunteers. J Nucl Med 1991;32:1518–1525.

74. DeFronzo RA, Tobin JD, Andres R. Glucose clamp technique: a method for quantifying insulin secretion and resistance. Am J Physiol 1979;237:E214–E223.

75. Schinkel AF, Bax JJ, Valkema R, et al. Effect of diabetes mellitus on myocardial ^{18}F-FDG SPECT using acipimox for the assessment of myocardial viability. J Nucl Med 2003;44:877–883.

76. Schroder O, Hor G, Hertel A, Baum RP. Combined hyperinsulinaemic glucose clamp and oral acipimox for optimizing metabolic conditions during 18F-fluorodeoxyglucose gated PET cardiac imaging: comparative results. Nucl Med Commun 1998;19:867–874.

77. Stone CK, Holden JE, Stanley W, Perlman SB. Effect of nicotinic acid on exogenous myocardial glucose utilization. J Nucl Med 1995;36:996–1002.

78. Musatti L, Maggi E, Moro E, Valzelli G, Tamassia V. Bioavailability and pharmacokinetics in man of acipimox, a new antilipolytic and hypolipemic agent. J Int Med Res 1981;9:381–386.

79. Tillisch J, Brunken R, Marshall R, et al. Reversibility of cardiac wall motion abnormalities predicted by positron tomography. N Engl J Med 1986;314:884–888.

80. vom Dahl J, Eitzman DT, al-Aouar ZR, et al. Relation of regional function, perfusion, and metabolism in patients with advanced coronary artery disease undergoing surgical revascularization. Circulation 1994;90:2356–2366.

81. Gerber B, Melin J, Bol A, et al. Nitrogen-13-ammonia and oxygen-15-water estimates of absolute myocardial perfusion in left ventricular ischemic dysfunction. J Nucl Med 1998;39:1655–1662.

82. Tamaki N, Yonekura Y, Yamashita K, et al. Value of rest-stress myocardial positron tomography using nitrogen-13-ammonia for the preoperative prediction of reversible asynergie. J Nucl Med 1989;30:1302–1310.

83. Tamaki N, Ohtani H, Yamashita K, et al. Metabolic activity in the areas of new fill-in after thallium-201 reinjection: comparison with positron emission tomography using fluorine-18-deoxyglucose. J Nucl Med 1991;32:673–678.

84. Marwick TH, MacIntyre WJ, Lafont A, Nemec JJ, Salcedo EE. Metabolic responses of hibernating and infarcted myocardium to revascularization. A follow-up study of regional perfusion, function, and metabolism. Circulation 1992;85:1347–1353.

85. Lucignani G, Paolini G, Landoni C, et al. Presurgical identification of hibernating myocardium by combined use of technetium-99m hexakis 2-methoxyisobutylisonitrile single photon emission tomography and fluorine-18 fluoro-2-deoxy-D-glucose positron emission tomography in patients with coronary artery disease. Eur J Nucl Med 1992;19:874–881.

86. Gropler RJ, Geltman EM, Sampathkumaran K, et al. Comparison of carbon-11-acetate with fluorine-18-fluorodeoxyglucose for delineating viable myocardium by positron emission tomography. J Am Coll Cardiol 1993;22:1587–1597.

87. Carrel T, Jenni R, Haubold-Reuter S, von Schulthess G, Pasic M, Turina M. Improvement of severely reduced left ventricular function after surgical revascularization in patients with preoperative myocardial infarction. Eur J Cardiothorac Surg 1992;6:479–484.

88. Schelbert HR. Assessment of myocardial viability with positron emission tomography. In: Iskandrian AE, van der Wall EE, editors. Myocardial Viability. Dordrecht: Kluwer, 2000:47–72.

89. Pagano D, Townend JN, Littler WA, Horton R, Camici PG, Bonser RS. Coronary artery bypass surgery as treatment for ischemic heart failure: the predictive value of viability assessment with quantitative positron emission tomography for symptomatic and functional outcome. J Thorac Cardiovasc Surg 1998;115:791–799.

90. Vanoverschelde J, Depre C, Gerber B, et al. Time course of functional recovery after coronary artery bypass graft surgery in patients with chronic left ventricular ischemic dysfunction. Am J Cardiol. 2000;85:1432–1439.

91. DiCarli MF, Asgarzadie F, Schelbert HR, et al. Quantitative relation between myocardial viability and improvement in heart failure symptoms after revascularization in patients with ischemic cardiomyopathy. Circulation 1995;92:3436–3444.

92. DiCarli MF, Davidson M, Little R, et al. Value of metabolic imaging with positron emission tomography for evaluating prognosis in patients with coronary artery disease and left ventricular dysfunction. Am J Cardiol 1994;73:527–533.

93. Eitzman D, Al-Aouar Z, Kanter H, et al. Clinical outcome of patients with advanced coronary artery disease after viability studies with positron emission tomography. J Am Coll Cardiol 1992;20:559–565.

94. Allman KC, Shaw LJ, Hachamovitch R, Udelson JE. Myocardial viability testing and impact of revascularization on prognosis in patients with coronary artery disease and left ventricular dysfunction: a meta-analysis. J Am Coll Cardiol 2002;39:1151–1158.

95. Beanlands R, Hendry P, Masters R, deKemp R, Woodend K, Ruddy T. Delay in revascularization is associated with increased mortality rate in patients with severe left ventricular dysfunction and viable myocardium on fluorine 18-fluorodeoxyglucose positron emission tomography imaging. Circulation 1998;98(19 suppl):II51–II56.

96. Pagano D, Lewis M, Townend J, Davies P, Camici P, Bonser R. Coronary revascularization for postischaemic heart failure: how myocardial viability affects survival. Heart 1999;82:684–688.

97. Dreyfus GD, Duboc D, Blasco A, et al. Myocardial viability assessment in ischemic cardiomyopathy: benefits of coronary revascularization. Ann Thorac Surg 1994;57:1402–1407; discussion 1407–1408.

98. Haas F, Haehnel CJ, Picker W, et al. Preoperative positron emission tomographic viability assessment and perioperative and postoperative risk in patients with advanced ischemic heart disease. J Am Coll Cardiol 1997;30:1693–1700.

99. Beanlands RS, deKemp RA, Smith S, Johansen H, Ruddy TD. F-18-fluorodeoxyglucose PET imaging alters clinical decision making in patients with impaired ventricular function. Am J Cardiol 1997;79:1092–1095.

100. Baer FM, Voth E, Schneider CA, Theissen P, Schicha H, Sechtem U. Comparison of low-dose dobutamine-gradient-echo magnetic resonance imaging and positron emission tomography with [^{18}F]fluorodeoxyglucose in patients with chronic coronary artery disease. A functional and morphological approach to the detection of residual myocardial viability. Circulation 1995;91:1006–1015.

101. Dilsizian V, Rocco TP, Freedman NM, Leon MB, Bonow RO. Enhanced detection of ischemic but viable myocardium by the reinjection of thallium after stress-redistribution imaging. N Engl J Med 1990;323:141–146.

102. Udelson JE, Coleman PS, Metherall J, et al. Predicting recovery of severe regional ventricular dysfunction: comparison of resting scintigraphy with thallium-201 and technetium-99m-sestamibi. Circulation 1994;89:2552–2561.

103. Cornel JH, Arnese M, Forster T, Postma-Tjoa J, Reijs AE, Fioretti PM. Potential and limitations of Tc-99m sestamibi scintigraphy for the diagnosis of myocardial viability. Herz 1994;19:19–27.

104. Kim RJ, Wu E, Rafael A, et al. The use of contrast-enhanced magnetic resonance imaging to identify reversible myocardial dysfunction. N Engl J Med 2000;343:1445–1453.

105. Klein C, Nekolla SG, Bengel FM, et al. Assessment of myocardial viability with contrast-enhanced magnetic resonance imaging: comparison with positron emission tomography. Circulation 2002;105:162–167.

106. Selvanayagam JB, Kardos A, Francis JM, et al. Value of delayed-enhancement cardiovascular magnetic resonance imaging in predicting myocardial viability after surgical revascularization. Circulation 2004;110:1535–1541.

107. Knuesel PR, Nanz D, Wyss C, et al. Characterization of dysfunctional myocardium by positron emission tomography and magnetic resonance: relation to functional outcome after revascularization. Circulation 2003;108:1095–1100.

108. Herrero P, Weinheimer CJ, Dence C, Oellerich WF, Gropler RJ. Quantification of myocardial glucose utilization by PET and 1-carbon-11-glucose. J Nucl Cardiol 2002;9:5–14.

109. Schön HR, Schelbert HR, Robinson G, et al. C-11 labeled palmitic acid for the noninvasive evaluation of regional myocardial fatty acid metabolism with positron emission tomography. I. Kinetics of C-11 palmitic acid in normal myocardium. Am Heart J 1981;103:532–547.

110. Maki MT, Haaparanta M, Nuutila P, et al. Free fatty acid uptake in the myocardium and skeletal muscle using fluorine-18-fluoro-6-thia-heptadecanoic acid. J Nucl Med 1998;39:1320–1327.

111. Brown M, Marshall DR, Sobel BE, Bergmann SR. Delineation of myocardial oxygen utilization with carbon-11-labeled acetate. Circulation 1987;76:687–696.

112. Buxton DB, Schwaiger M, Nguyen A, Phelps ME, Schelbert HR. Radiolabelled acetate as a tracer of myocardial tricarboxylic acid cycle flux. Circ Res 1988;63:628–634.

113. Beanlands RS, Bach DS, Raylman R, et al. Acute effects of dobutamine on myocardial oxygen consumption and cardiac efficiency measured using carbon-11 acetate kinetics in patients with dilated cardiomyopathy. J Am Coll Cardiol 1993;22:1389–1398.

114. Bengel FM, Permanetter B, Ungerer M, Nekolla S, Schwaiger M. Noninvasive estimation of myocardial efficiency using positron emission tomography and C-11 acetate: comparison between the normal and failing human heart. Eur J Nucl Med 2000;27:319––326.

115. Beanlands RS, Nahmias C, Gordon E, et al. The effects of beta(1)-blockade on oxidative metabolism and the metabolic cost of ventricular work in patients with left ventricular dysfunction: a double-blind, placebo-controlled, positron-emission tomography study. Circulation 2000;102:2070–2075.

116. Peterson LR, Herrero P, Schechtman KB, et al. Effect of obesity and insulin resistance on myocardial substrate metabolism and efficiency in young women. Circulation 2004;109:2191–2196.

117. Nuutila P, Knuuti MJ, Raitakari M, et al. Effect of antilipolysis on heart and skeletal muscle glucose uptake in overnight fasted humans. Am J Physiol 1994;267:E941–E946.

118. Bengel FM, Ueberfuhr P, Ziegler SI, Nekolla S, Reichart B, Schwaiger M. Serial assessment of sympathetic reinnervation after orthotopic heart transplantation: a longitudinal study using positron emission tomography and C-11 hydroxyephedrine. Circulation 1999;99:1866–1871.

119. Di Carli MF, Tobes MC, Mangner T, et al. Effects of cardiac sympathetic innervation on coronary blood flow. N Engl J Med 1997;336:1208–1215.

120. Bengel FM, Ueberfuhr P, Schiepel N, Nekolla SG, Reichart B, Schwaiger M. Effect of sympathetic reinnervation on cardiac performance after heart transplantation. N Engl J Med 2001;345:731–738.

121. Bengel FM, Ueberfuhr P, Ziegler SI, et al. Noninvasive assessment of the effect of cardiac sympathetic innervation on metabolism of the human heart. Eur J Nucl Med 2000;27:1650–1657.

122. Allman KC, Wieland DM, Muzik O, Degrado TR, Wolfe ER Jr, Schwaiger M. Carbon-11 hydroxyephedrine with positron emission tomography for serial assessment of cardiac adrenergic neuronal function after acute myocardial infarction in humans. J Am Coll Cardiol 1993;22:368–375.

123. Wichter T, Schafers M, Rhodes CG, et al. Abnormalities of cardiac sympathetic innervation in arrhythmogenic right ventricular cardiomyopathy: quantitative assessment of presynaptic norepinephrine reuptake and postsynaptic beta-adrenergic receptor density with positron emission tomography. Circulation 2000;101:1552–1558.

124. Calkins H, Allman K, Bolling S, et al. Correlation between scintigraphic evidence of regional sympathetic neuronal dysfunction and ventricular refractoriness in the human heart. Circulation 1993;88:172–179.

125. Ungerer M, Hartmann F, Karoglan M, et al. Regional in vivo and in vitro characterization of autonomic innervation in cardiomyopathic human heart. Circulation 1998;97:174–180.

126. Vesalainen RK, Pietila M, Tahvanainen KU, et al. Cardiac positron emission tomography imaging with [^{11}C]hydroxyephedrine, a specific tracer for sympathetic nerve endings, and its functional correlates in congestive heart failure. Am J Cardiol 1999;84:568–574.

127. Di Carli MF, Bianco-Batlles D, Landa ME, et al. Effects of autonomic neuropathy on coronary blood flow in patients with diabetes mellitus. Circulation 1999;100:813–819.

128. Pietila M, Malminiemi K, Ukkonen H, et al. Reduced myocardial carbon-11 hydroxyephedrine retention is associated with poor prognosis in chronic heart failure. Eur J Nucl Med 2001;28:373–376.

129. Merlet P, Delforge J, Syrota A, et al. Positron emission tomography with ^{11}C CGP-12177 to assess beta-adrenergic receptor concentration in idiopathic dilated cardiomyopathy. Circulation 1993;87:1169–1178.

130. Schafers M, Dutka D, Rhodes CG, et al. Myocardial presynaptic and postsynaptic autonomic dysfunction in hypertrophic cardiomyopathy. Circ Res 1998;82:57–62.

131. Law MP, Osman S, Pike VW, et al. Evaluation of [^{11}C]GB67, a novel radioligand for imaging myocardial alpha 1-adrenoceptors with positron emission tomography. Eur J Nucl Med 2000;27:7–17.

132. Momose M, Reder S, Raffel D, et al. Evaluation of cardiac β-adrenoceptors in the isolated perfused rat heart using (s)-[^{11}C] CGP12388. J Nucl Med 2004;45:471–477.

133. Le Guludec D, Cohen-Solal A, Delforge J, Delahaye N, Syrota A, Merlet P. Increased myocardial muscarinic receptor density in idiopathic dilated cardiomyopathy: an in vivo PET study. Circulation 1997;96:3416–3422.

134. Haubner R, Wester HJ, Reuning U, et al. Radiolabeled alpha(v)beta3 integrin antagonists: a new class of tracers for tumor targeting. J Nucl Med 1999;40:1061–1071.

135. Strauss HW, Narula J, Blankenberg FG. Radioimaging to identify myocardial cell death and probably injury. Lancet 2000;356:180–181.

136. Grierson JR, Yagle KJ, Eary JF, et al. Production of [F-18]fluoroannexin for imaging apoptosis with PET. Bioconjug Chem 2004;15:373–379.

137. Avril N, Bengel FM. Defining the success of cardiac gene therapy: how can nuclear imaging contribute? Eur J Nucl Med Mol Imaging 2003;30:757–771.

138. Bengel FM, Anton M, Richter T, et al. Noninvasive imaging of transgene expression using positron emission tomography in a pig model of myocardial gene transfer. Circulation 2003;108:2127–2133.

139. Doubrovin M, Ponomarev V, Beresten T, et al. Imaging transcriptional regulation of p53-dependent genes with positron emission tomography in vivo. Proc Natl Acad Sci U S A 2001;98:9300–9305.

140. Wu JC, Chen IY, Sundaresan G, et al. Molecular imaging of cardiac cell transplantation in living animals using optical bioluminescence and positron emission tomography. Circulation 2003;108:1302–1305.

141. Phelps ME. PET: the merging of biology and imaging into molecular imaging. J Nucl Med 2000;41:661–681.

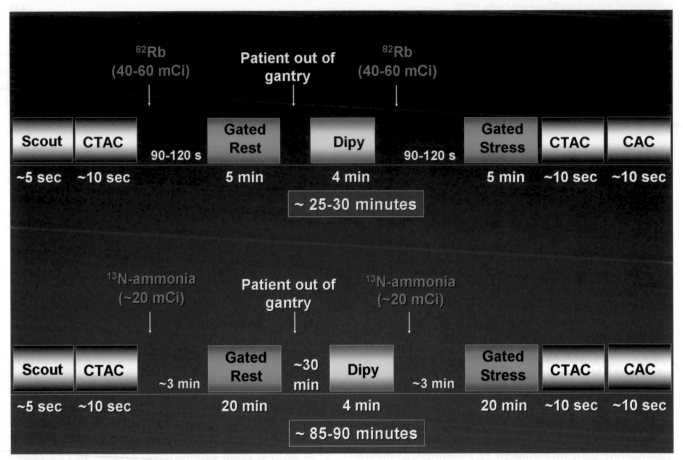

Figure 27.3. Positron emission tomography-computed tomography (PET/CT) imaging protocols for the evaluation of regional myocardial perfusion with dipyridamole and ⁸²Rb (*top panel*) or ¹³N-ammonia (*lower panel*).

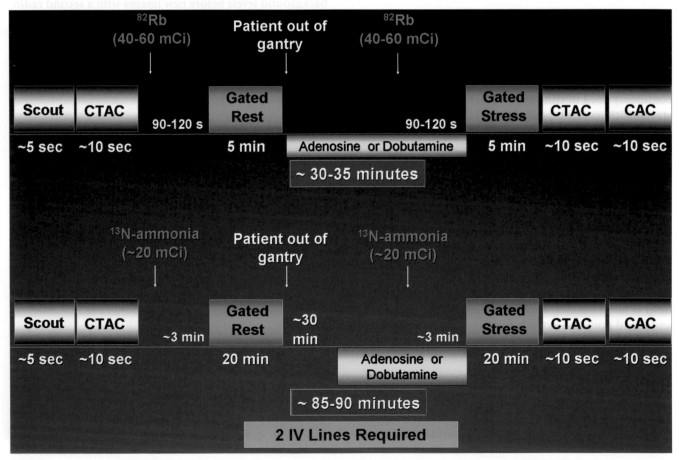

Figure 27.4. PET/CT imaging protocols for the evaluation of regional myocardial perfusion with adenosine or dobutamine and ⁸²Rb (*top panel*) or ¹³N-ammonia (*lower panel*).

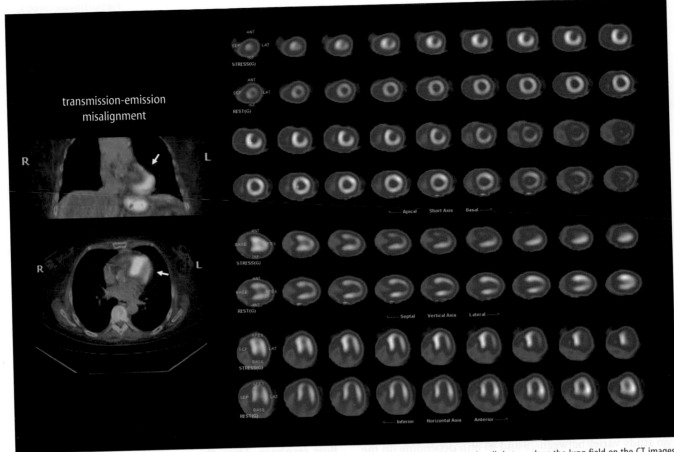

Figure 27.5. Fused CT and 82Rb emission images (*left panel*) demonstrate significant misalignment in the anterolateral wall that overlaps the lung field on the CT images. Reconstructed 82Rb images (*right panel*) demonstrate a small but severe perfusion defect in the anterolateral wall that shows complete reversibility and suggests ischemia in the diagonal coronary territory.

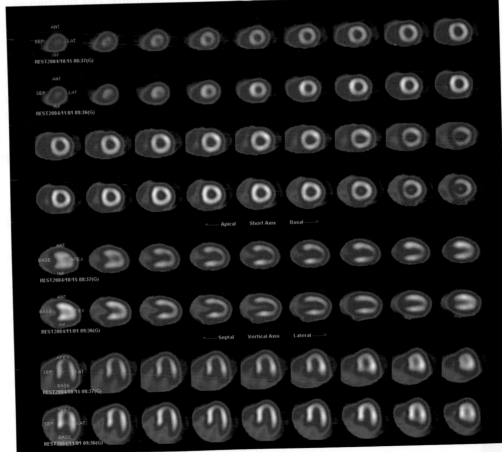

Figure 27.6. Reconstructed 82Rb images of the same patient as in Figure 27.5 after correction of the misalignment between the CT-based transmission and 82Rb demonstrate normal perfusion both at rest and during stress.

Table 27.1. Effects of 3-hydroxy-3-methylgluaryl (HMG)-CoA (HMG-CoA) reductase inhibitor treatment on myocardial perfusion in patients with and without obstructive coronary artery disease (CAD).

Author	Methodology	Endpoint	Magnitude of benefit
Baller (24)	PET	In hyperemic MBF	31%
Janatuinen (25)	PET	In hyperemic MBF	27%
Guethlin (12)	PET	In hyperemic MBF	35%
Huggins (26)	PET	In hyperemic MBF	46%
Yokoyama (13)	PET	In hyperemic MBF	20%
Schwartz (28)	SPECT	PD size and severity	22%

PET, positron emission tomography; SPECT, single photon emission computed tomography; MBF, myocardial blood flow; PD, size:perfusion defect.
Source: Reprinted from J Nucl Cardiol, vol. 11. Campisi R, Di Carli MF. Assessment of coronary flow reserve and microcirculation: a clinical perspective, pages 3–11. Copyright 2004, with permission from The American Society of Nuclear Cardiology.

chemic burden and improve symptoms in patients with obstructive CAD (28). The beneficial effects of statins on vascular function have been linked to improved endothelial function, decreased platelet aggregativity and thrombus deposition, and reduced vascular inflammation.

Clinical CAD

PET has proven to be a powerful and efficient noninvasive imaging modality to evaluate regional myocardial perfusion in patients with known or suspected obstructive CAD. Several technical advantages account for the improved diagnostic power of PET, including (1) routine measured (depth independent) attenuation correction, which decreases false positives and thus increases specificity (Figures 27.9, 27.10); (2) high spatial and contrast resolution (heart-to-background ratio) that allows improved detection of small perfusion defects, thereby decreasing false negatives and increasing sensitivity; and (3) high temporal resolution allowing fast dynamic imaging of tracer kinetics, which makes absolute quantification of myocardial perfusion (in mL/min/g of tissue) possible. In addition, the use of short-lived radiopharmaceuticals allows fast, sequential assessment of regional myocardial perfusion (for example, rest and stress), thereby improving laboratory efficiency and patient throughput (see Figures 27.3, 27.4).

Although these technical advantages have been recognized for a long time, the use of PET for routine detection of CAD has only gained momentum in recent years. Recent FDA approval of PET radiotracers [for example, ^{82}Rb, ^{13}N-ammonia, and ^{18}F-2-deoxy-D-glucose (FDG)]

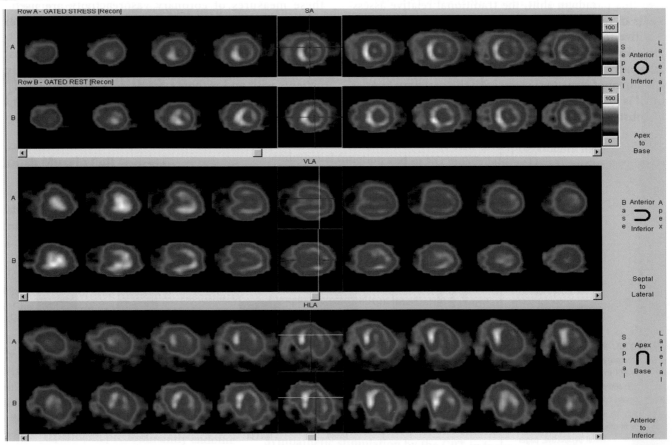

Figure 27.9. Reconstructed ^{99m}Tc-sestamibi single photon emission computed tomography (SPECT) images in a female patient demonstrates a moderately large area of moderate ischemia throughout the anterior and lateral walls.

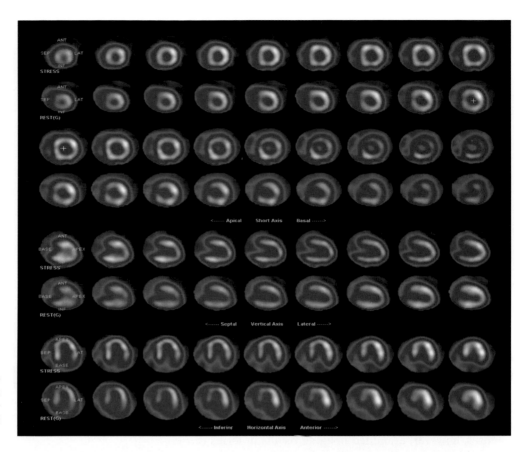

Figure 27.10. Reconstructed [82]Rb PET/CT images in the patient shown in Figure 27.9 demonstrate normal myocardial perfusion both at rest and during stress.

and the subsequent changes in reimbursement are responsible for much of the recent growth in clinical cardiac PET.

Diagnostic Accuracy of PET for Detection of Obstructive CAD

The experience with PET for detecting obstructive CAD has been extensively documented in seven studies including 663 patients (Table 27.2) (1). In these studies, regional myocardial perfusion was assessed with ^{13}N-ammonia or [82]Rb. The average sensitivity for detecting more than 50% angiographic stenosis was 89% (range, 83%–100%), whereas the average specificity was 86% (range, 73%–100%).

Comparative Studies of PET Versus SPECT

Only two studies have performed a head-to-head comparison of the diagnostic accuracy of [82]Rb-PET and ^{201}Tl-single photon emission computed tomography (-SPECT) in the same patient population (1). Go and colleagues (29) compared PET and SPECT in 202 patients. Their results showed a higher sensitivity with PET (76% versus 93%), and no significant changes for specificity (80% versus 78% for SPECT and PET, respectively). In another study, Stewart et al. (30) compared PET and SPECT in 81

patients. They observed a higher specificity for PET (53% versus 83% for SPECT and PET, respectively) and no significant differences in sensitivity (84% versus 86% for SPECT and PET, respectively). Diagnostic accuracy was higher with PET (89% versus 78%).

Table 27.2. Sensitivity and specificity of PET for detecting obstructive CAD.

Year	Author	Radiotracer	Prior MI (%)	Sensitivity (%)	Specificity (%)
1992	Marwick (56)	[82]Rb	49	90 (63/70)	100 (4/4)
1992	Grover-McKay (57)	[82]Rb	13	100 (16/16)	73 (11/15)
1991	Stewart (30)	[82]Rb	42	83 (50/60)	86 (18/21)
1990	Go (29)	[82]Rb	47	93 (142/152)	78 (39/50)
1989	Demer (58)	[82]Rb/ ^{13}N-ammonia	34	83 (126/152)	95 (39/41)
1988	Tamaki (59)	^{13}N-Ammonia	75	98 (47/48)	100 (3/3)
1986	Gould (60)	[82]Rb/ ^{13}N-ammonia	NR	95 (21/22)	100 (9/9)
	Total			89	86

Source: Reprinted from J Nucl Cardiol, vol. 11. Di Carli MF. Advances in positron emission tomography, pages 719–732. Copyright 2004, with permission from The American Society of Nuclear Cardiology.

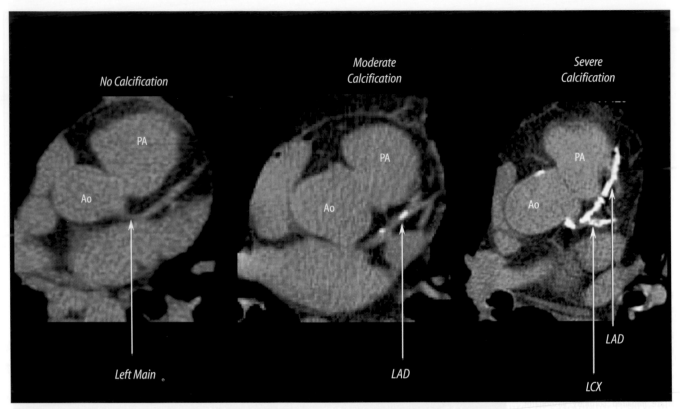

Figure 27.13. Cross-sectional images of the heart obtained with gated multidetector CT (MDCT) demonstrate different degrees of coronary artery calcification. (Reprinted from J Nucl Cardiol, vol. 11. Di Carli MF. Advances in positron emission tomography, pages 719–732. Copyright 2004, with permission from The American Society of Nuclear Cardiology.)

frequently found in advanced atherosclerotic lesions. Conventional plain chest X-ray, cine fluoroscopy, coronary angiography, ultrasound, and magnetic resonance imaging (MRI) can identify calcium in blood vessels; however, only electron beam CT (EBCT) and MDCT are able to accurately quantify the coronary calcium plaque burden (Figure 27.13). Because arterial calcification almost always represents atherosclerosis, detection of coronary artery calcium by means of CT is a sensitive, although not specific, marker for obstructive CAD (43). Indeed, a recent study by Berman and colleagues (44) reported that only 13% of patients with coronary artery calcium scores of 400 or more (Agatston method) showed mild to moderate ischemic defects by stress SPECT. This number was slightly higher (20%) among those with calcium scores of 1,000 or more. Together, these data suggest that coronary calcium may provide an assessment of preclinical CAD. Importantly, recent data from Shaw et al. (45) suggest that coronary calcium provides independent incremental information over that provided by traditional risk factors in the prediction of all-cause mortality.

MDCT can also delineate the burden of noncalcified coronary plaques, even in arterial segments without significant luminal narrowing (Figure 27.14). Achenbach et al. (46) showed a sensitivity of 78% and a specificity of 87% for MDCT compared to intravascular ultrasound (IVUS). These numbers were higher for plaques localized

in proximal coronary segments (91% and 89%, respectively). However, MDCT systematically underestimates plaque volume per segment as compared with IVUS (46).

Integrating Structure and Biology to Identify Vulnerable Plaques

Because there is marked heterogeneity in the composition of human atherosclerotic plaques, it would be clinically desirable to have reliable noninvasive imaging tools that can characterize the composition of such plaques, thereby allowing determining their risk for complications (for example, erosion and rupture). Such imaging tools would provide mechanistic insights into atherothrombotic processes, better risk stratification, optimal selection of therapeutic targets, and the means for monitoring therapeutic responses. Several imaging modalities have been employed to study atherosclerotic plaques and their composition. PET/CT appears attractive because it allows image fusion of structure and biology, thereby allowing characterization of plaques. Rudd and colleagues (47) have recently demonstrated a relationship between anatomic plaque, FDG uptake as a marker of inflammation, and patients' symptoms. Eight patients with symptomatic carotid atherosclerosis were imaged using FDG-PET and coregistered CT (Figure 27.15). The net FDG accumu-

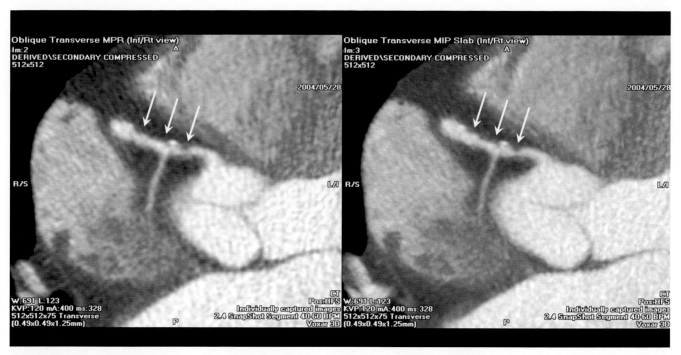

Figure 27.14. Contrast-enhanced 16-detector-row CT coronary angiographic images in oblique maximum intensity projection in a patient with chest pain demonstrate a long eccentric segment of noncalcified plaque (*arrows*) in the proximal right coronary artery.

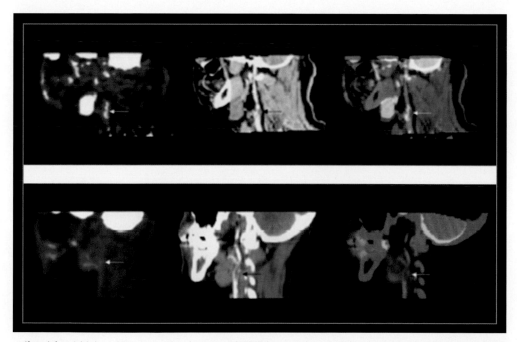

Figure 27.15. *Upper row* (from *left* to *right*) shows PET, contrast CT, and coregistered PET/CT images in the sagittal plane from a 63-year-old man who had experienced two episodes of left-sided hemiparesis. Angiography demonstrated stenosis of the proximal right internal carotid artery, which was confirmed on the CT image (*black arrow*). The *white arrows* show FDG uptake at the level of the plaque in the carotid artery. As expected, there was high FDG uptake in the brain, jaw muscles, and facial soft tissues. *Lower row* (from *left* to *right*) demonstrates a low level of FDG uptake in an asymptomatic carotid stenosis. The *black arrow* highlights the stenosis on the CT angiogram, and the *white arrows* demonstrate minimal [18F]-fluorodeoxyglucose accumulation at this site on the FDG-PET and coregistered PET/CT images. (Reproduced from Rudd JH, Warburton EA, Fryer TD, et al. Imaging atherosclerotic plaque inflammation with [18F]-fluorodeoxyglucose positron emission tomography. Circulation 2002;105:2708–2727, with permission of the American Heart Association.)

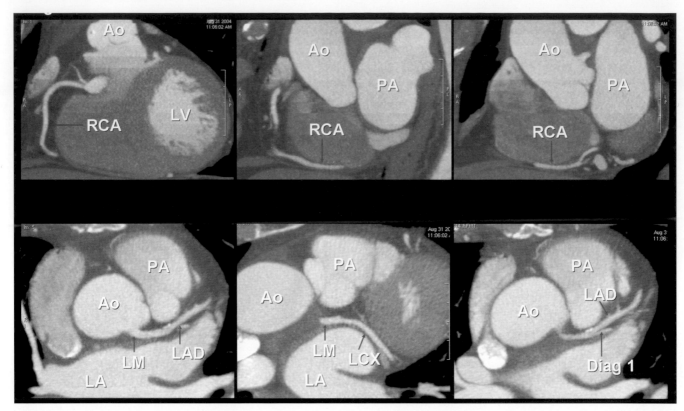

Figure 27.16. Contrast-enhanced 16-detector-row CT coronary angiographic images in transverse and oblique maximum intensity projection in a patient without obstructive CAD. The cross-sectional, multiplane, high spatial resolution MDCT images enable detailed noninvasive evaluation of the coronary artery tree. (Reprinted from J Nucl Cardiol, vol. 11. Di Carli MF. Advances in positron emission tomography, pages 719–732. Copyright 2004, with permission from The American Society of Nuclear Cardiology.)

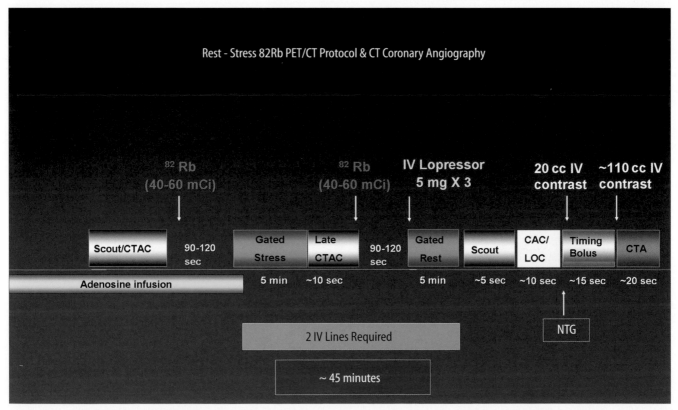

Figure 27.17. PET/CT imaging protocol for the combined evaluation of regional myocardial perfusion with [82]Rb and coronary angiography.

lation rate in symptomatic lesions was 27% higher than in contralateral asymptomatic lesions. There was no measurable FDG uptake into normal carotid arteries. Autoradiography of excised plaques confirmed accumulation of FDG in macrophage-rich areas of the plaque. This study provides proof that it is possible to measure the inflammatory burden of atherosclerotic plaques in vivo with PET/CT, thereby allowing definition of the underlying biology and possibly the risk of complication of such plaques.

CT Coronary Angiography

Breath-hold cardiac CT with retrospective ECG gating can provide detailed information regarding angiographic stenoses of the coronary artery tree, especially its proximal and mid portions. The overall diagnostic quality of noninvasive CT coronary angiography is largely dependent on spatial resolution, patient heart rate during examination and temporal resolution of the scanner, choice of appropriate reconstruction time point within the cardiac cycle, and quality of contrast enhancement (48). The coronary arteries and disease manifestation within these vessels are minute and difficult targets for imaging. Recent multidetector-row (16 and now 64) CT scanners provide high spatial and temporal resolution cardiac scan protocols with submillimeter (0.5–0.75 mm) section collimation and an in-plane spatial resolution of up to 0.5 x 0.5 mm (Figure 27.16).

The accuracy of CT coronary angiography for noninvasive detection of coronary artery stenosis is an area of active research. Published data using 4-slice CT scanners have reported sensitivities between 80% and 90%, especially for the proximal coronary segments (49–53). However, the number of uninterpretable vessels due to image degradation caused by motion (particularly in the RCA territory), increased calcium deposition, small vessel diameter, or breathing artifacts in the published studies remains high (range, 6%–32%). Newer CT technology with increased temporal and spatial resolution appears promising to overcome some of limitations observed using 4-slice CT scanners (46). The advent of faster CT scanners with added detector elements (for example, 16-slice CT and higher) increases the number of assessable coronary arteries and improves the overall accuracy of noninvasive CT coronary angiography for stenosis detection (54, 55). Of note, the majority of published studies agree with regard to the high negative predictive value of a negative CT coronary angiogram (as high as 97% with 16-detector row CT) (49–53).

Assessment of Extent and Severity of Myocardial Ischemia and Scar

Because not all coronary stenoses are flow limiting, however, the myocardial perfusion PET data complement the CT anatomic information by providing instant assessment about the clinical significance (i.e., ischemic burden) of such stenoses (Figures 27.17, 27.18). Image fusion of the functional PET data with the coronary CT information can also help identify the culprit stenosis in a patient presenting with chest pain. Presence of severe calcification is a limitation of contrast-enhanced CT coronary angiography because beam-hardening artifacts and partial-volume effects can completely obscure the cross section of the vessel and prevent assessment of the degree of luminal narrowing. Owing to similar effects, metal objects such as stents, surgical clips, and sternal wires can also interfere with the evaluation of underlying coronary stenoses. Thus, the functional myocardial perfusion PET data are also very useful for sorting out the presence of flow-limiting stenoses within areas of heavy calcification or prior stenting (Figure 27.19).

Conclusions

Positron emission tomography provides accurate diagnosis of the extent, severity, and anatomic location of coronary artery disease. A review of the current literature indicates that the sensitivity and specificity of myocardial perfusion with pharmacologic stress vary from 90% to 95% in both men and women. An additional advantage of PET is the possibility of quantifying regional perfusion and coronary flow reserve. Experimental and clinical evidence indicates that these measurements of coronary flow reserve have a nonlinear inverse correlation with the anatomic severity of stenosis. These measurements are useful for assessing the functional implications of coronary stenoses of intermediate severity (50%–80%), especially in patients with extensive CAD.

The high relative cost of PET requires a careful selection of patients. The great sensitivity and, above all, the high specificity of PET for diagnosing CAD make it a particularly useful tool for the assessment of obese patients and women with a low to intermediate probability of having CAD. This important clinical role is expected to grow with the availability of PET/CT scanners that allow a true integration (fusion) of structure and function (42), which will allow a comprehensive examination of the heart's anatomy and function in ways never before possible (1).

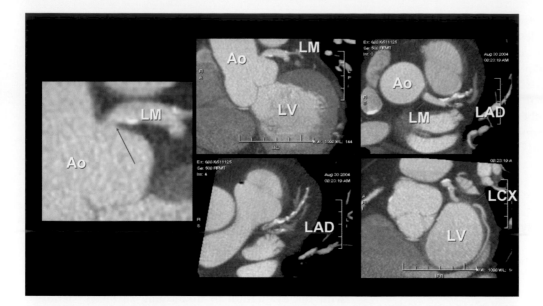

a

b

Figure 27.18. (**a**) Contrast-enhanced 16-detector-row CT coronary angiographic images in transverse and oblique maximum intensity projection in a patient with ostial obstruction of the left main (*LM*) and moderate calcification of the distal LM and proximal left anterior descending artery (*LAD*). (**b**) Stress and rest myocardial perfusion images obtained with [82]Rb demonstrate the anatomic stenoses seen on MDCT are not flow limiting. The functional PET data gave instant access to the functional significance of the anatomic MDCT data. This patient was counseled about the presence of coronary atherosclerosis and the benefits of risk factor modification and was treated with aggressive medical therapy. (Reprinted from J Nucl Cardiol, vol. 11. Di Carli MF. Advances in positron emission tomography, pages 719–732. Copyright 2004, with permission from The American Society of Nuclear Cardiology.)

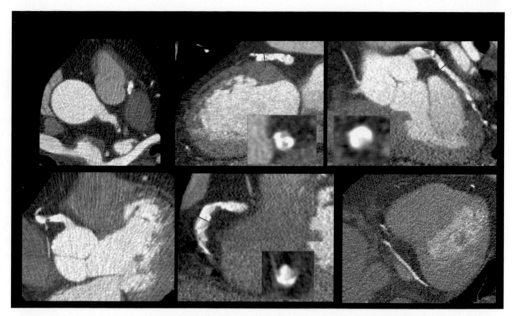

a

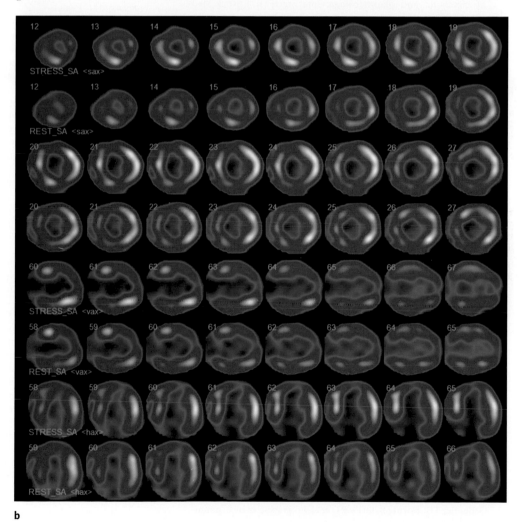

Figure 27.19. (**a**) Contrast-enhanced 16-detector row CT coronary angiographic images in transverse and oblique maximum intensity projection in a patient with suspected CAD. The images demonstrate extensive calcified (proximal LAD) and complex plaque (calcified and noncalcified) (proximal right coronary artery, *RCA*), as well as a total occlusion of the distal RCA. The presence of coronary calcium limits the ability of contrast-enhanced CT coronary angiography to determine the degree of luminal narrowing. (**b**) Stress and rest myocardial perfusion images obtained with ^{82}Rb demonstrate a dilated left ventricle (*LV*), a large area of scar in the LAD territory, and a moderate amount of ischemia in the posterior descending artery (PDA) and obtuse marginal (OM) territory. The myocardial perfusion PET information added diagnostic sensitivity to the MDCT data and provided the additional information needed for an appropriate management decision. The patient underwent double-vessel percutaneous coronary intervention (PCI) of the left circumflex and right coronary arteries.

b

References

1. Di Carli MF. Advances in positron emission tomography. J Nucl Cardiol 2004;11:719–732.
2. Schelbert HR, Phelps ME, Huang SC, et al. N-13 ammonia as an indicator of myocardial blood flow. Circulation 1981;63:1259–1272.
3. Hutchins GD, Schwaiger M, Rosenspire KC, Krivokapich J, Schelbert H, Kuhl DE. Noninvasive quantification of regional blood flow in the human heart using N-13 ammonia and dynamic positron emission tomographic imaging. J Am Coll Cardiol 1990;15:1032–1042.
4. Schelbert HR. Evaluation of myocardial blood flow in cardiac disease. In: Skorton DJ, Schelbert HR, Wolf GL, Brundage BH, editors. Cardiac Imaging. A Companion to Braunwald's Heart Disease. Philadelphia: Saunders, 1991:1093–1112.
5. Selwyn AP, Allan RM, L'Abbate A, et al. Relation between regional myocardial uptake of rubidium-82 and perfusion: absolute reduction of cation uptake in ischemia. Am J Cardiol 1982;50:112–121.
6. Mullani NA, Goldstein RA, Gould KL, et al. Myocardial perfusion with rubidium-82. I. Measurement of extraction fraction and flow with external detectors. J Nucl Med 1983;24:898–906.
7. Goldstein RA, Mullani NA, Marani SK, Fisher DJ, Gould KL, O'Brien HA Jr. Myocardial perfusion with rubidium-82. II. Effects of metabolic and pharmacologic interventions. J Nucl Med 1983;24:907–915.
8. Yoshida K, Mullani N, Gould KL. Coronary flow and flow reserve by PET simplified for clinical applications using rubidium-82 or nitrogen-13-ammonia. J Nucl Med 1996;37:1701–1712.
9. Smits P, Thien T. Cardiovascular effects of sulphonylurea derivatives. Implications for the treatment of NIDDM? Diabetologia 1995;38:116–121.
10. Buus NH, Bottcher M, Hermansen F, Sander M, Nielsen TT, Mulvany MJ. Influence of nitric oxide synthase and adrenergic inhibition on adenosine-induced myocardial hyperemia. Circulation 2001;104:2305–2310.
11. Czernin J, Barnard RJ, Sun KT, et al. Effect of short-term cardiovascular conditioning and low-fat diet on myocardial blood flow and flow reserve. Circulation 1995;92:197–204.
12. Guethlin M, Kasel AM, Coppenrath K, Ziegler S, Delius W, Schwaiger M. Delayed response of myocardial flow reserve to lipid-lowering therapy with fluvastatin. Circulation 1999;99:475–481.
13. Yokoyama I, Momomura S, Ohtake T, et al. Improvement of impaired myocardial vasodilatation due to diffuse coronary atherosclerosis in hypercholesterolemics after lipid-lowering therapy. Circulation 1999;100:117–122.
14. Nabel EG, Ganz P, Gordon JB, Alexander RW, Selwyn AP. Dilation of normal and constriction of atherosclerotic coronary arteries caused by the cold pressor test. Circulation 1988;77:43–52.
15. Zeiher AM, Drexler H, Wollschlaeger H, Saurbier B, Just H. Coronary vasomotion in response to sympathetic stimulation in humans: importance of the functional integrity of the endothelium [see comments]. J Am Coll Cardiol 1989;14:1181–1190.
16. Zeiher AM, Drexler H, Wollschlager H, Just H. Endothelial dysfunction of the coronary microvasculature is associated with coronary blood flow regulation in patients with early atherosclerosis. Circulation 1991;84:1984–1992.
17. Kichuk MR, Seyedi N, Zhang X, et al. Regulation of nitric oxide production in human coronary microvessels and the contribution of local kinin formation. Circulation 1996;94:44–51.
18. Campisi R, Di Carli MF. Assessment of coronary flow reserve and microcirculation: a clinical perspective. J Nucl Cardiol 2004;11:3–11.
19. Schindler TH, Hornig B, Buser PT, et al. Prognostic value of abnormal vasoreactivity of epicardial coronary arteries to sympathetic stimulation in patients with normal coronary angiograms. Arterioscler Thromb Vasc Biol 2003;23:495–501.
20. Schindler TH, Nitzsche EU, Munzel T, et al. Coronary vasoregulation in patients with various risk factors in response to cold pressor testing: contrasting myocardial blood flow responses to short- and long-term vitamin C administration. J Am Coll Cardiol 2003;42:814–822.
21. Halcox JP, Schenke WH, Zalos G, et al. Prognostic value of coronary vascular endothelial dysfunction. Circulation 2002;106:653–658.
22. Schachinger V, Zeiher AM. Prognostic implications of endothelial dysfunction: does it mean anything? Coron Artery Dis 2001;12:435–443.
23. Al Suwaidi J, Reddan DN, Williams K, et al. Prognostic implications of abnormalities in renal function in patients with acute coronary syndromes. Circulation 2002;106:974–980.
24. Baller D, Notohamiprodjo G, Gleichmann U, Holzinger J, Weise R, Lehmann J. Improvement in coronary flow reserve determined by positron emission tomography after 6 months of cholesterol-lowering therapy in patients with early stages of coronary atherosclerosis. Circulation 1999;99:2871–2875.
25. Janatuinen T, Laaksonen R, Vesalainen R, et al. Effect of lipid-lowering therapy with pravastatin on myocardial blood flow in young mildly hypercholesterolemic adults. J Cardiovasc Pharmacol 2001;38:561–568.
26. Huggins GS, Pasternak RC, Alpert NM, Fischman AJ, Gewirtz H. Effects of short-term treatment of hyperlipidemia on coronary vasodilator function and myocardial perfusion in regions having substantial impairment of baseline dilator reverse. Circulation 1998;98:1291–1296.
27. Yokoyama I, Yonekura K, Inoue Y, Ohtomo K, Nagai R. Long-term effect of simvastatin on the improvement of impaired myocardial flow reserve in patients with familial hypercholesterolemia without gender variance. J Nucl Cardiol 2001;8:445–451.
28. Schwartz RG, Pearson TA, Kalaria VG, et al. Prospective serial evaluation of myocardial perfusion and lipids during the first six months of pravastatin therapy: coronary artery disease regression single photon emission computed tomography monitoring trial. J Am Coll Cardiol 2003;42:600–610.
29. Go RT, Marwick TH, MacIntyre WJ, et al. A prospective comparison of rubidium-82 PET and thallium-201 SPECT myocardial perfusion imaging utilizing a single dipyridamole stress in the diagnosis of coronary artery disease [see comments]. J Nucl Med 1990;31:1899–1905.
30. Stewart RE, Schwaiger M, Molina E, et al. Comparison of rubidium-82 positron emission tomography and thallium-201 SPECT imaging for detection of coronary artery disease. Am J Cardiol 1991;67:1303–1310.
31. Van Train KF, Garcia EV, Maddahi J, et al. Multicenter trial validation for quantitative analysis of same-day rest-stress technetium-99m-sestamibi myocardial tomograms. J Nucl Med 1994;35:609–618.
32. Di Carli M, Czernin J, Hoh CK, et al. Relation among stenosis severity, myocardial blood flow, and flow reserve in patients with coronary artery disease. Circulation 1995;91:1944–1951.
33. Uren NG, Melin JA, De Bruyne B, Wijns W, Baudhuin T, Camici PG. Relation between myocardial blood flow and the severity of coronary-artery stenosis. N Engl J Med 1994;330:1782–1788.
34. Beanlands RS, Muzik O, Melon P, et al. Noninvasive quantification of regional myocardial flow reserve in patients with coronary atherosclerosis using nitrogen-13 ammonia positron emission tomography. Determination of extent of altered vascular reactivity. J Am Coll Cardiol 1995;26:1465–1475.
35. Yoshinaga K, Katoh C, Noriyasu K, et al. Reduction of coronary flow reserve in areas with and without ischemia on stress perfusion imaging in patients with coronary artery disease: A study using oxygen 15-labeled water PET. J Nucl cardiol 2003;10:275–283.
36. Parkash R, deKemp RA, Ruddy TDT, et al. Potential utility of rubidium 82 PET quantification in patients with 3-vessel coronary artery disease. J Nucl Cardiol 2004;11:440–449.
37. Ornish D, Brown SE, Scherwitz LW, et al. Can lifestyle changes reverse coronary heart disease? The Lifestyle Heart Trial [see comments]. Lancet 1990;336:129–133.

38. Brown G, Albers JJ, Fisher LD, et al. Regression of coronary artery disease as a result of intensive lipid- lowering therapy in men with high levels of apolipoprotein B [see comments]. N Engl J Med 1990;323:1289–1298.

39. Kane JP, Malloy MJ, Ports TA, Phillips NR, Diehl JC, Havel RJ. Regression of coronary atherosclerosis during treatment of familial hypercholesterolemia with combined drug regimens. Jama 1990;264:3007–3012.

40. Gould KL, Ornish D, Scherwitz L, et al. Changes in myocardial perfusion abnormalities by positron emission tomography after long-term, intense risk factor modification [see comments]. JAMA 1995;274:894–901.

41. Watts GF, Lewis B, Brunt JN, et al. Effects on coronary artery disease of lipid-lowering diet, or diet plus cholestyramine, in the St Thomas' Atherosclerosis Regression Study (STARS). Lancet 1992;339:563–569.

42. Townsend DW, Beyer T. A combined PET/CT scanner: the path to true image fusion. Br J Radiol 2002;75(special):S24–S30.

43. O'Rourke RA, Brundage BH, Froelicher VF, et al. American College of Cardiology/American Heart Association Expert Consensus Document on electron-beam computed tomography for the diagnosis and prognosis of coronary artery disease. J Am Coll Cardiol 2000;36:326–340.

44. Berman DS, Wong ND, Gransar H, et al. Relationship between stress-induced myocardial ischemia and atherosclerosis measured by coronary calcium tomography. J Am Coll Cardiol 2004;44:923–930.

45. Shaw LJ, Raggi P, Schisterman E, Berman DS, Callister TQ. Prognostic value of cardiac risk factors and coronary artery calcium screening for all-cause mortality. Radiology 2003;228:826–833.

46. Achenbach S, Moselewski F, Ropers D, et al. Detection of calcified and noncalcified coronary atherosclerotic plaque by contrast-enhanced, submillimeter multidetector spiral computed tomography: a segment-based comparison with intravascular ultrasound. Circulation 2004;109:14–17.

47. Rudd JH, Warburton EA, Fryer TD, et al. Imaging atherosclerotic plaque inflammation with [18F]-fluorodeoxyglucose positron emission tomography. Circulation 2002;105:2708–2711.

48. Schoepf UJ, Becker CR, Ohnesorge BM, Yucel EK. CT of coronary artery disease. Radiology 2004;232:18–37.

49. Achenbach S, Moshage W, Ropers D, Nossen J, Daniel WG. Value of electron-beam computed tomography for the noninvasive detection of high-grade coronary-artery stenoses and occlusions. N Engl J Med 1998;339:1964–1971.

50. Achenbach S, Giesler T, Ropers D, et al. Detection of coronary artery stenoses by contrast-enhanced, retrospectively electrocardiographically-gated, multislice spiral computed tomography. Circulation 2001;103:2535–2538.

51. Nieman K, Oudkerk M, Rensing BJ, et al. Coronary angiography with multi-slice computed tomography. Lancet 2001;357:599–603.

52. Becker CR, Knez A, Leber A, et al. Detection of coronary artery stenoses with multislice helical CT angiography. J Comput Assist Tomogr 2002;26:750–755.

53. Kopp AF, Schroeder S, Kuettner A, et al. Non-invasive coronary angiography with high resolution multidetector-row computed tomography. Results in 102 patients. Eur Heart J 2002;23:1714–1725.

54. Nieman K, Cademartiri F, Lemos PA, Raaijmakers R, Pattynama PM, de Feyter PJ. Reliable noninvasive coronary angiography with fast submillimeter multislice spiral computed tomography. Circulation 2002;106:2051–2054.

55. Ropers D, Baum U, Pohle K, et al. Detection of coronary artery stenoses with thin-slice multi-detector row spiral computed tomography and multiplanar reconstruction. Circulation 2003;107:664–666.

56. Marwick TH, Nemec JJ, Stewart WJ, Salcedo EE. Diagnosis of coronary artery disease using exercise echocardiography and positron emission tomography: comparison and analysis of discrepant results. J Am Soc Echocardiogr 1992;5:231–238.

57. Grover-McKay M, Ratib O, Schwaiger M, et al. Detection of coronary artery disease with positron emission tomography and rubidium-82. Am Heart J 1992;123:646–652.

58. Demer LL, Gould KL, Goldstein RA, et al. Assessment of coronary artery disease severity by positron emission tomography. Comparison with quantitative arteriography in 193 patients. Circulation 1989;79:825–835.

59. Tamaki N, Yonekura Y, Senda M, et al. Value and limitation of stress thallium-201 single photon emission computed tomography: comparison with nitrogen-13 ammonia positron tomography. J Nucl Med 1988;29:1181–1188.

60. Gould KL, Goldstein RA, Mullani NA, et al. Noninvasive assessment of coronary stenoses by myocardial perfusion imaging during pharmacologic coronary vasodilation. VIII. Clinical feasibility of positron cardiac imaging without a cyclotron using generator-produced rubidium-82. J Am Coll Cardiol 1986;7:775–789.

28

PET in Clinical Neurology

Yen F. Tai and Paola Piccini

Positron emission tomography (PET) offers a noninvasive method to investigate in vivo brain functions. It allows us to quantify neuroreceptor binding, cerebral metabolism, and blood flow. A vast array of tracers has enabled investigators to study various brain functions. Table 28.1 lists some of the commonly used tracers and their specific applications. PET studies may be carried out at rest, while performing certain tasks, or after administration of challenge compounds.

PET is increasingly used in clinical neurology to aid with disease diagnosis in early or equivocal cases and to monitor disease progression and treatment response. This chapter discusses the main uses of PET in clinical neurology.

PET in Clinical Neurology

Dementia

Alzheimer's disease is the most common form of dementia. It is characterized pathologically by deposition of amyloid plaques and neurofibrillary tangles. ^{18}F-2-Fluoro-2-deoxyglucose (FDG)-PET has been widely used to study regional cerebral glucose metabolism in dementia. Early Alzheimer's disease patients typically exhibit posterior cingulate and temporoparietal hypometabolism of glucose (1). With progression of disease, there may also be frontal involvement (2). The degree of hypometabolism correlates with the severity of dementia (3). The glucose hypometabolism in Alzheimer's disease results from a combination of neuronal cell loss and decreased synaptic activity, although the latter appears to be of greater importance (4). In one study that included 138 patients with symptoms of dementia and postmortem histopathologic examination, visual analysis of FDG PET was able to identify Alzheimer's disease with a sensitivity of 94% and specificity of 73% (5). Patients with mild cognitive impairment (MCI) represent a large group of subjects whose cognitive deficits are not severe enough to cause

significant functional impairment or fulfill the diagnosis of dementia. MCI is heterogeneous in aetiologies and clinical presentations. It is estimated that the rate of progression from MCI to dementia is between 6% and 25% per year (6). MCI can present with PET findings similar to that found in Alzheimer's (7). De Santi et al. (8) attempted to find out the best discriminant of various subtypes of cognitive impairment using atrophy-corrected FDG-PET. They found that entorhinal cortex glucose metabolism is the most accurate variable to differentiate MCI from controls, whereas temporal cortex glucose metabolism is best at separating MCI from Alzheimer's disease.

FDG-PET has been used to detect subjects at risk for Alzheimer's disease even before onset of symptoms. Cognitively normal carriers of the apolipoprotein E type 4 allele, a common susceptibility gene for Alzheimer's disease, showed similar pattern of glucose hypometabolism as Alzheimer's disease patients (9, 10), and such abnormalities can be seen several decades before possible onset of dementia (11). After a mean follow-up of 2 years, the cortical metabolic abnormality continued to decline despite relative preservation of cognitive function (12, 13). PET may be able to help predict which of these at-risk subjects will progress to Alzheimer's disease. Entorhinal cortex hypometabolism on FDG-PET in cognitively normal elderly can predict the progression to mild cognitive impairment (14). Arnaiz et al. showed that combined glucose hypometabolism in the left temporoparietal area and performance on the block design, a measure of visuospatial function, correctly predicted future development of Alzheimer's disease from MCI in 90% of cases (15).

FDG-PET in dementia with Lewy bodies (DLB) reveals changes similar to those seen in Alzheimer's disease, plus additional hypometabolism in primary and associative visual cortices (16). In a PET study with postmortem confirmatory diagnosis, the antemortem occipital glucose hypometabolism can help distinguish DLB from Alzheimer's disease with 90% sensitivity and 80% specificity (17). In addition, patients with DLB can present with parkinsonism and are more likely than Alzheimer's

Table 28.1. Common positron emission tomography (PET) tracers used to study neurologic disorders.

Application	Tracer
Cerebral blood flow	[^{15}O]-H$_2$O
Oxygen metabolism	[^{15}O]-O$_2$
Glucose metabolism	[^{18}F]-2-Fluoro-2-deoxyglucose ([^{18}F]-FDG)
Presynaptic dopamine transporter	[^{11}C]-Methylphenidate, ^{11}C-RTI
Dopamine storage	[^{18}F]-6-Fluoro-L-dopa (^{18}F-DOPA)
Dopamine D1 receptors	[^{11}C]-SCH23390
Dopamine D2 receptors	[^{11}C]-Raclopride
Central benzodiazepine binding	[^{11}C]-Flumazenil
Opioid binding	[^{11}C]-Diprenorphine

disease patients to have nigrostriatal dopaminergic dysfunction. Reduced putamenal uptake of [^{18}F]-6-fluoro-L-DOPA ([^{18}F]-DOPA) is able to differentiate DLB from Alzheimer's disease with a sensitivity of 86% and specificity of 100% (18). FDG-PET in multiinfarct dementia shows multiple focal areas of hypometabolism, the extent of which is greater than actual pathology seen on postmortem examination, probably caused by degeneration of axons following infarct with disconnection of remote structures (19, 20). Frontotemporal dementia is associated with hypometabolism in frontal and temporal lobes (20).

Alzheimer's disease is also characterised by degeneration of the nucleus basalis of Meynert and its ascending cholinergic innervation (21). These cholinergic neurons express the enzyme acetylcholine esterase (AChE), which degrades acetylcholine. The uptake of ^{11}C-methyl-piperidin-4-yl propionate ([^{11}C]-PMP), an AChE analogue, is markedly reduced in the neocortex, hippocampus, and amygdala in early-onset Alzheimer's disease, whereas the reduction is found only in temporoparietal cortex and amygdala in late-onset Alzheimer's disease (22). There was also significant correlation between cortical AChE activity and Mini-Mental State Examination scores. This, together with known beneficial effects of AChE inhibitors in Alzheimer's disease (23), suggests cholinergic deficiency plays an important role in the pathogenesis of Alzheimer's

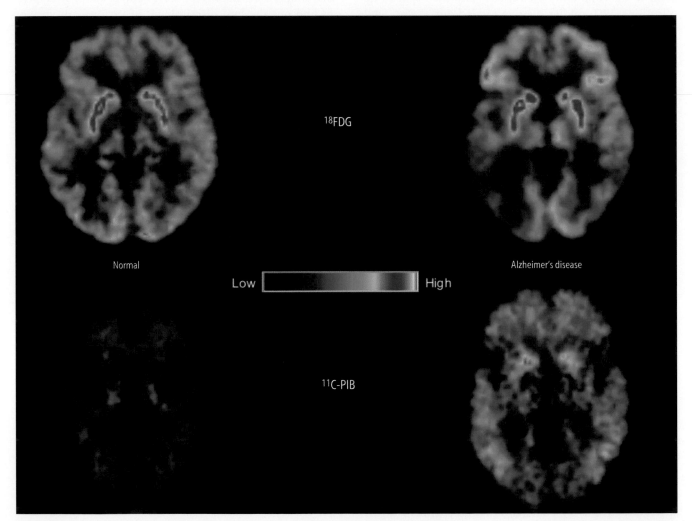

Figure 28.1. FDG cerebral glucose metabolism parametric maps (*upper panel*) of a normal individual and an Alzheimer's disease patient. The latter displays characteristic temporoparietal hypometabolism. The *lower panel* shows minimal [^{11}C]-6-OH-BTA-1 ([^{11}C]-PIB) uptake in the normal individual, but marked widespread uptake in the cortex and striatum of the same Alzheimer's disease patient, indicating deposition of amyloid plaques. (Images courtesy of Paul Edison.)

disease. ^{11}C-PMP PET may help distinguish patients with Alzheimer's disease from Parkinson's disease dementia. The latter group showed even greater and more widespread reduction in cortical AChE activity than Alzheimer's disease patients matched for severity of dementia (24).

Despite the characteristic patterns seen in many of the dementia syndromes, there can be considerable overlap in their PET findings. Recent efforts in developing a specific ligand for beta-amyloid plaques, such as ^{11}C-6-OH-BTA-1 (^{11}C-PIB) (Figure 28.1), may further enhance the sensitivity of PET for early diagnosis of Alzheimer's disease and provide a biologic marker of disease progression. A recent PET study by Klunk et al. (25) showed significant ^{11}C-PIB uptake particularly in frontal, temporal, parietal, and occipital cortices and the striatum in mild Alzheimer's disease patients. ^{11}C-PIB binding correlated inversely with glucose metabolism. However, no correlation between ^{11}C-PIB uptake and cognitive performance was observed. Further studies are needed to determine the sensitivity and specificity of this tracer in diagnosing Alzheimer's disease and in predicting conversion of MCI to Alzheimer's disease. PET can provide important information that will help identify at-risk individuals, such as those carrying the apolipoprotein E type 4 allele or with MCI who will eventually develop Alzheimer's disease, because timely administration of putative neuroprotective agents may prevent or delay phenoconversion.

Movement Disorders

The majority of movement disorders are characterized by dopaminergic dysfunction in the basal ganglia neuronal circuit. The tracer [18F]-DOPA is frequently used to study presynaptic dopaminergic function. Following intravenous administration, [18F]-DOPA is taken up by dopaminergic neurons and converted to [18F]-dopamine by aromatic amino acid decarboxylase (AADC). The influx rate constant (K_i) of [18F]-DOPA reflects dopa uptake into the neurons, AADC activity, and dopamine storage capacity. [18F]-DOPA-PET therefore provides an in vivo indicator of the function and integrity of presynaptic dopaminergic terminals. Tracers that bind to presynaptic dopamine transporters, such as ^{11}C-methylphenidate, and dopamine terminal vesicular monoamine transporters, such as ^{11}C-dihydrotetra-

benazine, have also been utilised as markers of presynaptic dopaminergic function.

Differentiating various types of Parkinsonian syndromes clinically, especially in the early stages of the disease, can be difficult. Conventional imaging methods such as magnetic resonance imaging (MRI) may not reveal any abnormality. PET can be employed as an adjunct to clinical diagnosis in equivocal cases. Parkinson's disease is characterized by loss of dopaminergic neurons in pars compacta of substantia nigra, particularly in the ventrolateral tier, with lesser involvement in the dorsomedial tier (26). The loss of their corresponding striatal projection terminals is detected by [18F]-DOPA-PET as progressive decline in [18F]-DOPA K_i in the putamen in a caudal-rostral direction. The greatest decrease is seen in the putamen contralateral to the side with most severe symptoms. The caudate nucleus also becomes affected later on (27). By contrast, the more-diffuse loss of nigrostriatal dopaminergic terminals in multiple system atrophy (MSA) and progressive supranuclear palsy is displayed by [18F]-DOPA-PET as bilateral, symmetrical reduction in K_i in the entire striatum. Corticobasal degeneration (CBD) shows equivalent reduction in [18F]-DOPA K_i of caudate and putamen, but the striatum opposite the most affected limbs has the greatest reduction. ^{18}F-DOPA-PET can distinguish Parkinson's disease from the striatonigral degeneration form of MSA in 70% of cases, and Parkinson's disease from progressive supranuclear palsy in 90% of cases (28); it is, however, less effective in discriminating between the atypical Parkinsonian syndromes.

FDG-PET is also helpful in the differential diagnosis of various Parkinsonian disorders. FDG-PET in Parkinson's disease reveals normal or raised glucose metabolism in striatum but decreased metabolism in temporoparietal areas (29). Progressive supranuclear palsy shows bilateral striatal and frontal hypometabolism, whereas decreases in striatal, brainstem and cerebellar metabolism are found in MSA. In CBD, there is asymmetrical hypometabolism of striatum, thalamus, frontal, and temporoparietal cortex, with the hemisphere contralateral to the most affected limb displaying the greatest reduction. Principal-component analysis can reveal an abnormal pattern of relative hypermetabolism in lentiform nucleus and hypometabolism in premotor cortex in Parkinson's disease (30). Table 28.2 summarizes major PET findings in Parkinsonian syndromes.

Table 28.2. Summary of major PET findings in Parkinsonian syndromes.

PET tracer	Parkinson's disease	Progressive supranuclear palsy	Multiple system atrophy (MSA)	Corticobasal degeneration (CBD)
^{18}F-DOPA	Asymmetrical reduction (caudal putamen > rostral putamen > caudate)	Symmetrical reduction (caudate = putamen)	Symmetrical reduction (caudate = putamen)	Asymmetrical reduction (caudate = putamen)
[18F]-FDG	Normal/raised in striatum Asymmetrically reduced in temporoparietal cortex	Reduced in striatum and frontal cortex bilaterally	Reduced in striatum, brainstem, and cerebellum	Asymmetrical reduction in striatum, thalamus, and frontal and temporoparietal cortex

PET has been developed as a marker of disease severity and progression in Parkinson's disease. Striatal [^{18}F]-DOPA K_i is shown to correlate with postmortem dopaminergic cell density in the substantia nigra (31). Cross-sectional putamenal K_i in Parkinson's disease also correlates with motor disability (32). Longitudinal progression study of Parkinson's disease found a 9% to 12% annual decline in striatal [^{18}F]-DOPA K_i (33, 34), although it follows a negative exponential decline course in which the rate of decline is fastest earlier in the course of the disease (35). Several neuroprotective/restorative trials involving dopamine agonists (36), neural transplantation (37), and infusion of neurotrophic factors (38) have used ^{18}F-DOPA-PET as a marker of response to treatment, although the outcome may sometimes be confounded by potential interactions between the PET ligands and medical interventions (39).

[^{18}F]-DOPA-PET has also identified subclinical dopaminergic dysfunction in subjects at risk for familial parkinsonism, such as those carrying mutations in *parkin* (40) and *PINK1* (41) genes. Parkinson's disease patients with *parkin* mutations have more extensive pre- and post-synaptic dopaminergic dysfunction than other Parkinson's disease patients matched for age and disease severity (Figure 28.2) (42), but their presynaptic striatal dopaminergic function progressed at a slower rate than idiopathic Parkinson's disease patients (40). This finding suggests that dopamine cell loss in these patients occurs early but progresses very slowly, allowing compensatory mechanisms to take place.

PET has also been widely used to study hyperkinetic movement disorders. Huntington's disease is an autosomal dominant disorder arising from expanded CAG repeats in the IT15 gene on chromosome 4. The GABAergic medium spiny projection neurons in the striatum, which express dopamine D1 and D2 receptors, are preferentially lost in Huntington's disease. Using [^{11}C]-SCH23390 and [^{11}C]-raclopride PET, Turjanski et al. (43)

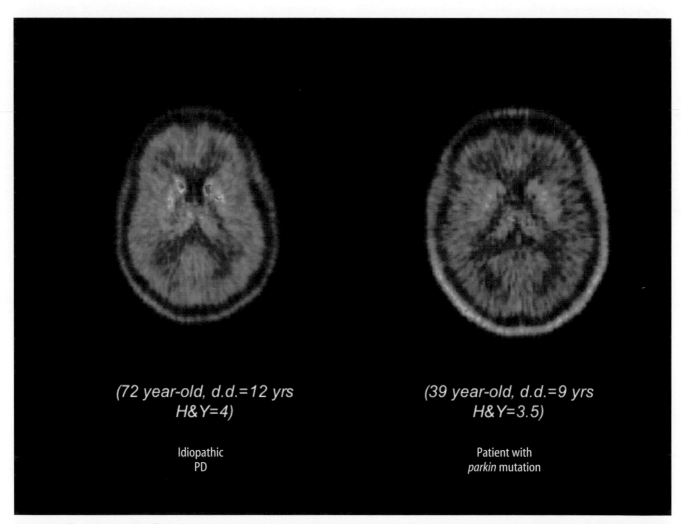

(72 year-old, d.d.=12 yrs H&Y=4)

Idiopathic
PD

(39 year-old, d.d.=9 yrs H&Y=3.5)

Patient with
parkin mutation

Figure 28.2. [^{18}F]-6-Fluoro-L-DOPA ([^{18}F]-DOPA) integrated image of a subject with idiopathic Parkinson's disease (*left*) compared with that of a subject with young-onset familial Parkinson's disease linked to *parkin* mutation. The latter showed a greater reduction in striatal ^{18}F-DOPA uptake despite having similar disease duration and severity. *H&Y*, Hoehn & Yahr stage; *d.d.*, disease duration.

found a parallel reduction in striatal D1 and D2 receptor binding in Huntington's disease. Striatal [^{11}C]-raclopride binding decreases by approximately 5% per year in Huntington's disease and correlates cross-sectionally with the duration and clinical severity of the disease (44). Huntington's disease patients also showed striatal glucose hypometabolism, with the cortex becomes progressively involved with increasing severity of disease, reflecting the widespread nature of Huntington's disease pathology (45). Huntington's disease gene carriers can be identified using genetic tests but existing methods, including CAG repeat length, do not accurately predict disease onset (46). PET studies of presymptomatic Huntington's disease carriers have found reduced striatal D2 binding and glucose metabolism in only a proportion of them (47, 48). It is likely that those nearer the onset of disease are more likely to display PET abnormalities. Larger trials are ongoing to ascertain the accuracy of PET in identifying such subjects, as intervention at this early stage using putative neuroprotective agents may yield most benefits.

Striatal D2 receptor binding and glucose metabolism are also reduced in chorea because of other degenerative conditions [e.g., neuroacanthocytosis (49)] but are preserved or even increased in nondegenerative chorea [e.g., systemic lupus erythematosus (49) and Sydenham's chorea (50)].

Epilepsy

Antiepileptic drugs fail to adequately control seizures in 25% of patients who have epilepsy (51). Epilepsy surgery can play a significant role in the treatment of intractable focal epilepsy, as demonstrated in temporal lobe epilepsy where surgery results in a significant improvement in the control of the seizures and quality of life (52). Modern MRI is able to identify the source of seizure in the majority of patients with focal epilepsy. However, 20% to 30% of potential surgical candidates with focal epilepsy have normal MRI (53). These patients are also less likely to become seizure free if they do undergo epilepsy surgery (54). Subtle structural abnormalities, which may only be evident through histologic examination, are often not detected by MRI (55). The main clinical use of PET in epilepsy is to localize epileptogenic foci in potential surgical candidates with focal epilepsy and to corroborate findings from other investigational modalities such as electroencephalography (EEG).

In focal epilepsy, the glucose metabolism and cerebral blood flow in the region of the epileptogenic focus are increased during the ictal period (56). Immediately after seizures, the hyperperfusion gradually returns to baseline, but the glucose hypermetabolism persists for another 24 to 48 h (57). Interictal PET shows decreased glucose metabolism and blood flow in the epileptogenic focus. It is important to perform PET with concomitant scalp EEG recording to correlate PET findings with ictal status of the patients. Interictal studies in patients with temporal lobe epilepsy using FDG-PET have found a 60% to 90% incidence of temporal lobe hypometabolism (53). However, the area with abnormal cerebral blood flow and metabolism is considerably larger than the actual structural abnormality, possibly reflecting reduced synaptic inhibition or deafferentiation of neighboring neurons in areas of epileptic propagation (53). Therefore, false localizations may occur, but this can be minimized by using quantitative rather than qualitative assessment of regional cerebral glucose metabolism. Nevertheless, this problem has limited the usefulness of FDG-PET as a localizing tool for epileptogenic foci.

Gamma-aminobutyric acid (GABA) is the principal inhibitory neurotransmitter in brain. The binding of ^{11}C-flumazenil ([^{11}C]-FMZ), a specific ligand for the benzodiazepine binding site of the GABA$_A$–central benzodiazepine receptor complex (58), is reduced by 30% in epileptogenic foci (59). An autoradiographic and histopathologic study of sclerotic hippocampi revealed that this is the result of a combination of neuronal loss and decreased density of GABA$_A$ receptors (60). There is also a good correlation between in vivo hippocampal [^{11}C]-FMZ binding and ex vivo [^{3}H]-FMZ autoradiographic measurements in individual patients who have undergone surgery for hippocampal sclerosis (61). One study examined 100 patients with focal epilepsy who had undergone presurgical evaluation using FDG and [^{11}C]-FMZ PET. The latter demonstrated abnormality in 94% of patients with temporal lobe epilepsy. The [^{11}C]-FMZ abnormality coincided with MRI abnormality in 81% of the cases. The area with abnormal [^{11}C]-FMZ binding is usually smaller than that seen on FDG-PET but larger than the abnormality detected on MRI (62). Another study that examined the focus-localizing ability of [^{11}C]-FMZ and FDG-PET, using extra- and intracranial EEG recordings as reference, found the former to be more sensitive and accurate (63). Correction for partial-volume effect caused by atrophy increased the sensitivity of [^{11}C]-FMZ-PET in detecting unilateral hippocampal sclerosis from 65% to 100% in one study (64).

[^{11}C]-FMZ PET can detect abnormalities in patients with normal MRI. In a study of patients with refractory temporal lobe epilepsy and normal MRI, 16 of 18 patients showed abnormal [^{11}C]-FMZ binding in the temporal lobe, and in 7 of these the findings were concordant with clinical and EEG data (65). Three patients subsequently underwent anterior temporal lobe resection with significant clinical improvement. The authors have also identified increased white matter [^{11}C]-FMZ binding in patients with temporal lobe epilepsy and neocortical epilepsy but normal MRI, which they attributed to possible presence of heterotopic white matter neurons or